Mathematics for the Trades
A GUIDED APPROACH

athematics r the Trades

UIDED APPROACH

ert A. Carman
a City College

Hal M. Saunder
Santa Barbara School District

QA 39.2 C3 1981
Carman, Robert A.
Mathematics for the trades

JOHN WILEY & SONS

New York • Chichester • Brisbane • Toronto • Singapore

For Chrissy with love and for Clarence
Fielstra with admiration and affection

This book was printed and bound by Halliday Lithograph Corp. It was set in 10/12 Century Schoolbook by York Graphic Services. The designer was Judith Fletcher Getman. The drawings were designed and executed by John Balbalis with the assistance of the Wiley Illustration Department. Linda Sadovnick supervised production. Eugene Patti was the manuscript editor.

Library of Congress Cataloging in Publication Data

Carman, Robert A
Mathematics for the trades.

Includes index.
1. Mathematics—1961- I. Saunders, Hal M., joint author. II. Title.
QA39.2.C35 512'.142 79-11491
ISBN 0-471-13481-3

Printed in the United States of America.

10 9 8

PREFACE

This book provides the practical mathematics skills needed in a wide variety of trade and technical areas, including electronics, auto mechanics, construction trades, air conditioning, machine technology, welding, drafting, and many other occupations. It is especially intended for students who have a poor math background and for adults who have been out of school for a time. Most of these students have had little success in mathematics, some openly fear it, and all need a direct, practical approach that emphasizes careful, complete explanations and actual on-the-job applications. This book is intended to provide practical help with real math, beginning at each student's own individual level of ability.

Those who have difficulty with mathematics will find in this book several special features designed to make it most effective for them. These include:

- Careful attention to readability. Reading specialists have helped plan both the written text and the visual organization.
- A diagnostic pretest and performance objectives keyed to the text at the beginning of each unit. These clearly indicate the content of each unit and provide the student with a sense of direction.
- Each unit ends with a problem set covering the work of the unit. The format is clear and easy to follow. It respects the individual needs of each reader, providing immediate feedback at each step to assure understanding and continued attention. The emphasis is on *explaining* concepts rather than simply *presenting* them. This is a practical presentation rather than a theoretical one.
- Special attention has been given to on-the-job math skills, using a wide variety of real problems and situations. Many problems parallel those that appear on professional and apprenticeship exams. The answers to all problems are given in the back of the book.
- Supplementary unit exams are available in an accompanying teacher's manual.
- A light, lively conversational style of writing and a pleasant, easy-to-understand visual approach are used. The use of humor is designed to appeal to students who have in the past found mathematics to be dry and uninteresting.

Field-testing with a wide variety of students indicates that this approach is successful—the book works and students learn, many of them experiencing success in mathematics for the first time.

Flexibility of use was a major criterion in the design of the book. Field-testing indicates that the book can be used successfully in a variety of course formats. It can be used as a textbook in traditional lecture-oriented courses. It is very effective in situations where an instructor wishes to modify a traditional course by devoting a portion of class time to independent study. The book is especially useful in programs of individualized or self-paced instruction, whether in a learning lab situation, with tutors, with audio tapes, or in totally independent study.

A careful introduction to the metric system is included in the text, and both metric and English units are used throughout.

Electronic pocket calculators are a valuable tool for workers in trade and technical areas, and we have recognized their extensive use by including special calculator problem sets wherever these are appropriate. The realistic problems included in these sets involve large numbers, repeated calculations, and large quantities of information. They are representative of actual trades situations where a calculator is needed. Students are taught the skill of estimating and checking answers. Detailed instruction on the use of calculators is included in a special appendix.

An accompanying teacher's resource book provides

- Information on a variety of self-paced and individualized course formats.
- Additional unit exams.
- Answers to all exams that appear in the teacher's resource book.

It is a pleasure to acknowledge the help of many people who have contributed to the development of this book. 'Lyn Carman spent countless hours interviewing trades workers, union leaders, apprentices, teachers, and training program directors so that their experience could be used in developing realistic problems, an effective format, and appropriate content. Her contributions were invaluable to us. We are indebted to the following teachers who read preliminary versions of the text and offered many helpful suggestions:

Robert Ahntholz, Coordinator—Learning Center, Central City Occupational Center, Los Angeles, California

B. H. Dwiggins, Systems Development Engineer, Technovate, Inc., Tacoma, Washington

Donald Fama, Chairperson, Mathematics-Engineering Science Department, Cayuga Community College, Auburn, New York

Ronald J. Gryglas, Tool and Die Institute, Chicago, Illinois

Vincent J. Hawkins, Chairperson, Department of Mathematics, Warwick Public Schools, Warwick, Rhode Island

Bernard Jenkins, Lansing Community College, Lansing, Michigan

David C. Mitchell, Seattle Central Community College, Seattle, Washington

Emma M. Owens, Tri-County Technical College, Pendleton, South Carolina

Martin Prolo, San Jose City College, San Jose, California

Richard C. Spangler, Developmental Instruction Coordinator, Tacoma Community College, Tacoma, Washington

Arthur Theobald, Bergen County Vocational School, Wayne, New Jersey

Joseph Weaver, Associate Professor, State University of New York, Delhi, New York

This book has benefited greatly from their excellence as teachers.

Finally, through every step of the seemingly endless sequence of researching, interviewing, testing, writing, and rewriting that makes a textbook, we have benefited from the patience, understanding, and concern of our wives, 'Lyn and Chris. They have made it a better book and a more pleasant experience than we could have otherwise had.

Santa Barbara, California

Robert A. Carman
Hal M. Saunders

CONTENTS

HOW TO USE THIS BOOK

In this book you will find many questions, not only at the end of each chapter or section, but on every page. This textbook is designed for those who need to learn the practical math used in the trades, and who want it explained carefully and completely at each step. The questions and explanations are designed so that you can:

- ★ Start at the beginning or where you need to start.
- ★ Work on only what you need to know.
- ★ Move as fast or as slow as you wish.
- ★ Skip material you already understand.
- ★ Do as many practice problems as you need.
- ★ Test yourself often to measure your progress.

In other words, if you find mathematics difficult and you want to be guided carefully through it, this book is designed for you.

This is no ordinary book. You cannot browse in it; you don't read it. You *work* your way through it. The ideas are arranged step by step in short portions or *frames*. Each frame contains information, careful explanations, examples, and questions to test your understanding. Read the material in each frame carefully, follow the examples, and answer the questions that lead to the next frame. Correct answers move you quickly through the book. Incorrect answers lead you to frames that provide further explanation. You move through this book frame by frame. Because we know that every person is different and has different needs, each major section of the book starts with a preview that will help you to determine the parts on which you need to work.

As you move through the book you will notice that material not directly connected to the frames appears in special boxes. Read these at your leisure. They contain information that you may find interesting, helpful in your work, or even fun.

Most students hesitate to ask the questions that nag at their understanding. They are fearful that "dumb questions" will humiliate them and reveal their lack of understanding. To relieve you of worry over dumb questions (or DAQs), we will ask and answer them for you. Thousands of students have taught us that "dumb questions" can produce smart students. Watch for DAQ.

According to an old Spanish proverb, the world is an ocean and he who cannot swim will sink to the bottom. A recent study published by the U.S. Office of Education revealed that two-thirds of the skilled and semiskilled job opportunities on today's labor market are available only to those who have an understanding of the basic principles of arithmetic, algebra, and geometry. If the modern world of work is an ocean, the skill needed to keep afloat or even swim to the top is clearly mathematics. It is the purpose of this book to help you learn these basic skills.

Now, turn to frame 1 on page 3 and let's begin.

R. A. C.
H. M. S.

1 Arithmetic of Whole Numbers

PREVIEW 1

Objective	Sample Problems		Where To Go for Help Page	Frame
Upon successful completion of this unit you will be able to:				
1. Add and subtract whole numbers.	(a) $67 + 58$	= ________	6	**5**
	(b) $7009 + 1598$	= ________		
	(c) $82 - 45$	= ________	14	**11**
	(d) $4035 - 1967$	= ________		
	(e) $14 + 31 - 67 + 59 + 22 + 37 - 19$	= ________		
2. Multiply and divide whole numbers.	(a) 64×37	= ________	21	**18**
	(b) 305×243	= ________		
	(c) 908×705	= ________		
	(d) $2006 \div 6$	= ________	30	**27**
	(e) $7511 \div 37$	= ________		
3. Do word problems involving whole numbers.	A metal casting weighs 680 lb; 235 lb of metal are removed during shaping. What is its finished weight?	= ________		

(Answers to these preview problems are on page 2. Don't peek.)

If you are certain you can work *all* of these problems correctly, turn to page 39 for a set of practice problems. If you cannot work one or more of the preview problems, turn to the page indicated. Superstudents—those who want to be tops in their work—will turn to frame **1** and begin work there.

____
Name

____
Date

____
Course/Section

Answers to Preview 1

1. (a) 125 (b) 8607 (c) 37 (d) 2068 (e) 77
2. (a) 2368 (b) 74,115 (c) 640,140 (d) 334 remainder 2
   (e) 203
3. 445 lb

Arithmetic of Whole Numbers

"WHEN DO WE LEARN HOW TO MAKE A FAST BUCK?"

1 The average grade school student uses numbers to tell time, count marbles, and keep track of lunch money. He might be surprised to learn that to earn a living, fast buck or slow, most adults need to develop technical skills based on their ability to read, write, and work with numbers. We may be living in an age of electronic computers and calculators, but most of the simple arithmetic used in industry, business, and the skilled trades is still done by hand. In fact, most trade and technical areas require you to *prove* that you can do the calculations by hand before you can get a job.

In this book we will take a practical, how-to-do-it look at the basic operations of arithmetic: addition, subtraction, multiplication, and division, including fractions, decimal numbers, negative numbers, powers, and roots. There are no fast-buck formulas here, but a lot of help for people who need to use mathematics in their daily work.

The simplest kind of number is the *counting number*—the number we use for counting the number of objects in a group. For example, how many letters are in this collection?

Count them. Write your answer here ______, then turn to frame **3**.

---

2 Hi.

What are you doing here? Lost? Window shopping? Just passing through? Nowhere in this book are you directed to frame **2**. (Notice that little **2** to the left above? That's a frame number.) Remember, in this book you move from frame to frame as directed, but not necessarily in 1–2–3 order. Go slow, follow directions, and you'll never get lost.

Now, return to **1** for a fresh start.

---

3 We counted 23. Notice that we have counted the letters by grouping them into sets of ten:

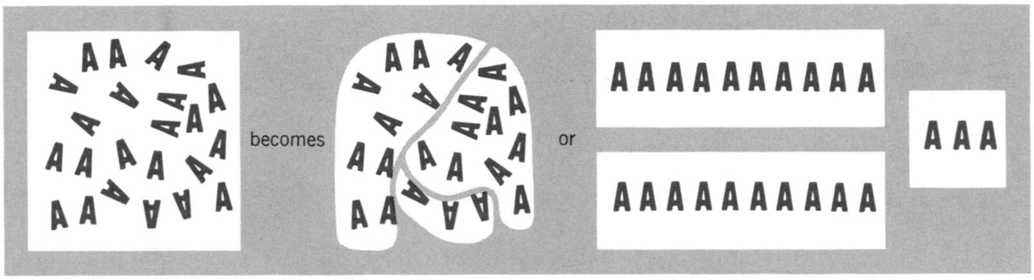

2 tens + 3 ones
20 + 3 or 23

Mathematicians call this the *expanded form* of a number. For example,

46	=	40 + 6	=	4 tens + 6 ones
274	=	200 + 70 + 4	=	2 hundreds + 7 tens + 4 ones
305	=	300 + 5	=	3 hundreds + 0 tens + 5 ones

Notice that only ten numerals or number symbols—0, 1, 2, 3, 4, 5, 6, 7, 8, and 9—are needed to write any number. These ten basic numerals are called the *digits* of the number. The digits 4 and 6 are used to write 46, the number 274 is a three-digit number, and so on.

Write out the following three-digit numbers in expanded form:

(a) 362 = ______ + ______ + ______ = ____ hundreds + ____ tens + ____ ones

(b) 425 = ______ + ______ + ______ = ____ hundreds + ____ tens + ____ ones

(c) 208 = ______ + ______ + ______ = ____ hundreds + ____ tens + ____ ones

Check your work in **4**.

---

4 (a) 362 = *300* + *60* + *2* = *3* hundreds + *6* tens + *2* ones
(b) 425 = *400* + *20* + *5* = *4* hundreds + *2* tens + *5* ones
(c) 208 = *200* + *0* + *8* = *2* hundreds + *0* tens + *8* ones

Notice that the 2 in 362 means something very different from the 2 in 425 or 208. In 362 the 2 signifies two ones. In 425 the 2 signifies two tens. In 208 the 2 signifies two hundreds. Ours is a *place-value* system of naming numbers: the value of any digit depends on the place where it is located.

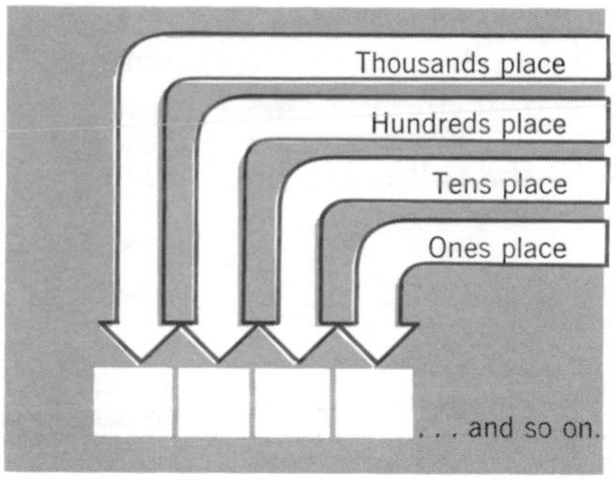

Being able to write a number in expanded form will help you to understand and remember the basic operations of arithmetic—even though you'll never find it on a blueprint or in a technical handbook.

This expanded-form idea is used in naming numbers, especially very large numbers.

Any large number given in numerical form may be translated to words by using the following diagram.

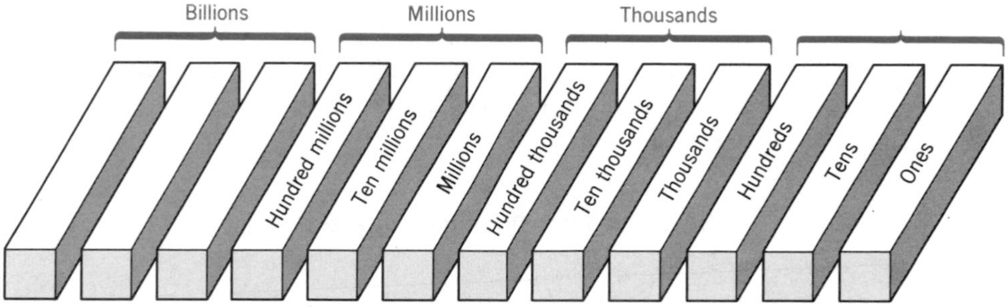

The number 14,237 can be placed in the diagram like this

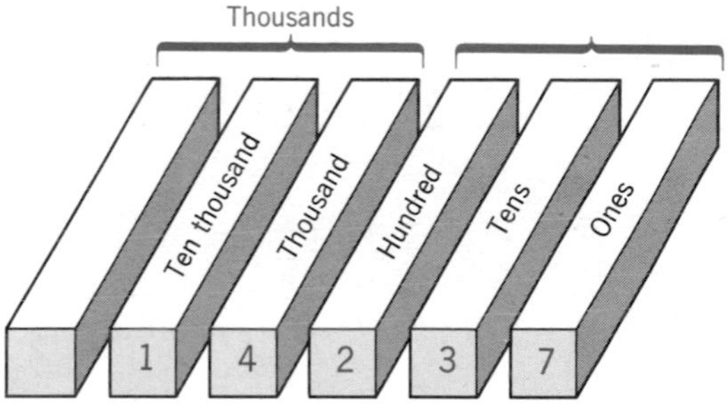

and is read "fourteen thousand, two hundred thirty-seven."

The number 47,653,290,866 becomes

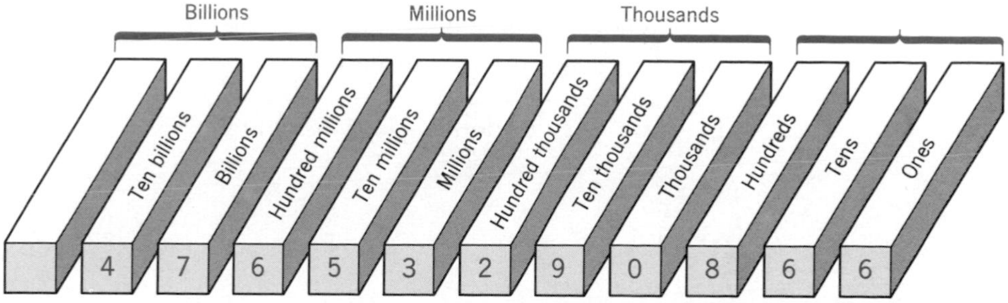

and is read "forty-seven billion, six hundred fifty-three million, two hundred ninety thousand, eight hundred sixty-six."

In each block of three digits read the digits in the normal way ("forty-seven," "six hundred fifty-three") and add the name of the block ("billion," "million"). Notice that the word "and" is not used in naming these numbers.

Use the diagram to name the following numbers:

(a) 4072 (b) 1,360,105
(c) 3,000,210 (d) 21,010,031,001

Check your answers in frame **5**.

---

5 (a) four thousand, seventy-two
(b) one million, three hundred sixty thousand, one hundred five
(c) three million, two hundred ten
(d) twenty-one billion, ten million, thirty-one thousand, one

1-1 ADDITION OF WHOLE NUMBERS

Adding whole numbers is fairly easy provided you have stored in your memory a few simple addition facts. It is most important that you be able to add simple one-digit numbers.

The following sets of problems in one-digit addition are designed to give you some practice. Work quickly. You should be able to answer all problems in a set in the time shown.

Exercises 1-1 One-Digit Addition

A. Add:

7	5	2	5	8	2	3	8	9	7
3	6	9	7	8	5	6	7	3	6
6	8	9	3	7	2	9	9	7	4
4	5	6	5	7	7	4	9	2	7
9	2	5	8	4	9	6	4	8	8
7	6	5	9	5	5	6	3	2	3
5	6	7	7	5	6	2	3	6	9
8	7	5	9	4	5	8	7	8	8
7	5	9	4	3	8	4	8	5	7
4	9	2	6	8	6	9	4	8	8

Average time = 90 seconds
Record = 35 seconds

B. Add. Try to do all addition mentally:

2	7	3	4	2	6	3	5	9	5
5	3	6	5	7	7	4	7	6	2
4	2	5	8	9	8	4	8	3	8

6	5	4	8	6	9	7	4	8	1
2	4	2	1	8	3	1	9	4	8
7	5	9	9	8	5	6	1	6	7

1	9	3	1	7	2	9	9	8	5
9	9	1	6	9	9	8	5	3	4
2	1	4	3	6	1	2	1	3	7

Average time = 90 seconds
Record = 41 seconds

The answers are on page 541.
When you have had the practice you need, turn to **6** and continue.

---

6 The ability to do arithmetic with one-digit numbers is very important. It is the key to performing any mathematical computation—even if you do the work on an electronic calculator. Suppose you need to find the total time spent on a job by two workers. You need to find the sum

31 hours + 48 hours = ________

What is the first step? Start adding digits? Punch in some numbers on your trusty electronic calculator? Rattle your abacus? None of these. The first step is to *estimate* your answer. The most important rule in any mathematical calculation is:

Know the answer to any calculation before you calculate it.

Never do an arithmetic calculation until you know roughly what the answer is going to be. Always know where you are going.

Make an estimate of the answer to the problem above. Write your estimate here:

____

Then turn to **7** to continue.

---

7 31 hours + 48 hours is approximately 30 hours + 50 hours or 80 hours, not 8 or 8000 or 800 hours. Having the estimate will keep you from making any major (and embarrassing) mistakes. Once you have a reasonable estimate of the answer, you are ready to do the arithmetic work.

Calculate 31 + 48 = ________

You don't really need an air-conditioned $50 solid-state electronic calculator for that, do you? Work it out and check your answer in **8**.

---

8 You should have set it up like this:

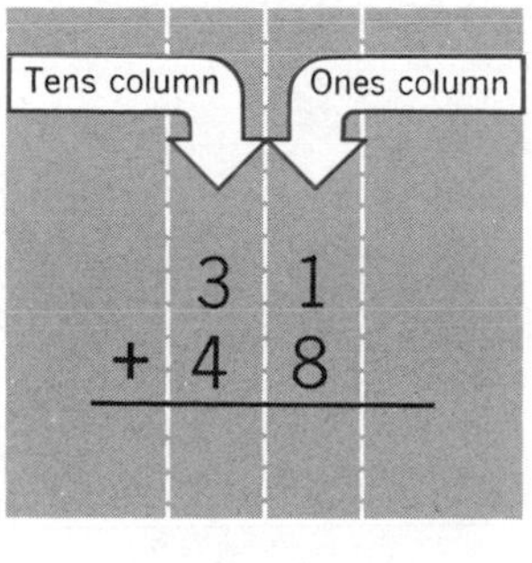

1. The numbers to be added are arranged vertically (up and down) in columns.
2. The right end or ones digits are placed in the ones column, the tens digits are placed in the tens column, and so on.

Avoid the confusion of

```
  31            31
+ 48     or   +48
```

The most frequent cause of errors in arithmetic is carelessness, especially in simple procedures such as lining up the digits correctly.

Once the digits are lined up the problem is easy.

$$\begin{array}{r} 31 \\ +48 \\ \hline 79 \end{array}$$

Does the answer agree with your original estimate? Yes. The estimate, 80, is roughly equal to the actual sum, 79.

What we have just shown you is often called the *Guess n' check* method of doing mathematics.

Step 1 **Estimate** the answer.

Step 2 **Work** the problem carefully.

Step 3 **Check** your answer against the estimate. If they disagree, repeat both steps 1 and 2.

Most students hesitate at estimating, either because they are unwilling to take the time to do it or because they are afraid they might do it incorrectly. Relax. You are the only one who will know your estimate. Do it in your head, do it quickly, and make it reasonably accurate. Step 3 helps you to detect incorrect answers before you finish the problem. The *Guess n' check* method means you never work in the dark; you always know where you are going.

Estimating is especially important in practical math, where a wrong answer is not just a mark on a piece of paper. An error may mean time and money lost.

Here is a slightly more difficult problem:

27 lb + 58 lb = ________

Try it, then turn to **9**.

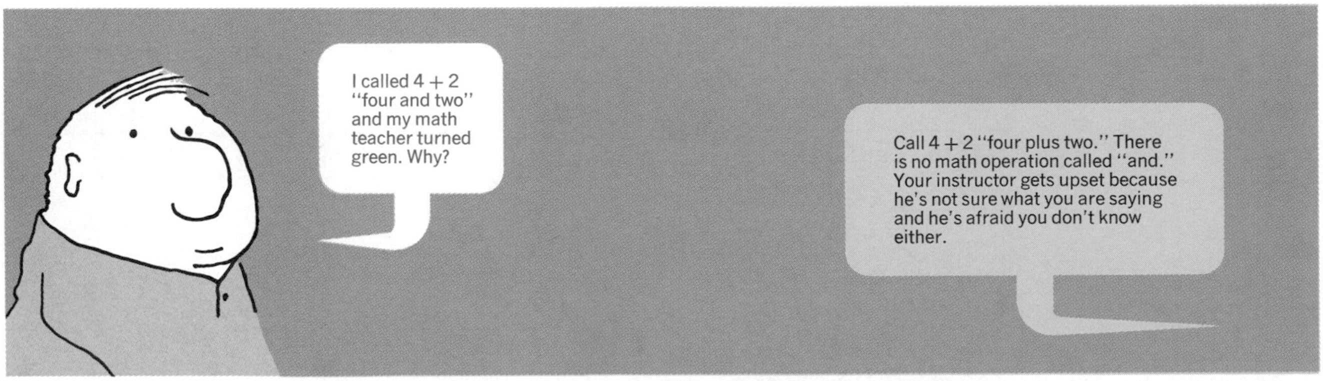

9 **First,** estimate the answer. 27 + 58 is roughly 30 + 60 or about 90. The answer is about 90 lb.

Second, line up the digits in columns.

$$\begin{array}{r} 27 \\ +58 \\ \hline \end{array}$$

The numbers to be added, 27 and 58 in this case, are usually called *addends.*

Third, add carefully.

$$\begin{array}{r} \overset{1}{2}7 \\ +58 \\ \hline 85 \end{array}$$

Finally, check your answer by comparing it with the estimate. The estimate 90 lb is roughly equal to the answer 85 lb—at least you have an answer in the right ballpark.

What does that little 1 above the tens digit mean? What really happens when you "carry" a digit? Let's look at it in detail. In expanded notation,

27 → 2 tens + 7 ones
+58 → 5 tens + 8 ones
= 7 tens + 15 ones
= 7 tens + 1 ten + 5 ones
= 8 tens + 5 ones
= 85

The 1 that is carried over to the tens column is really a ten.

Here is a short set of problems. Add, and be sure to estimate your answers first.

(a) 429 + 738 = __________ (b) 446 + 867 = __________

(c) 2368 + 744 = __________ (d) 409 + 72 = __________

Compare your work with ours in **10**.

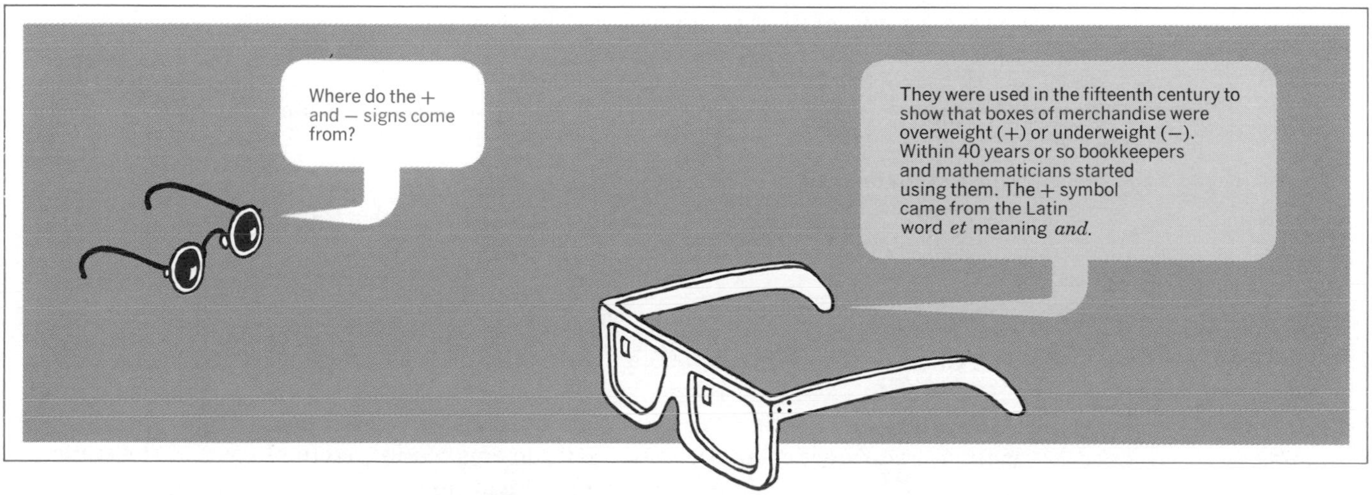

10 (a) **Estimate:** 400 + 700 = 1100

Line up the digits:

$$\begin{array}{r} 429 \\ +738 \\ \hline \end{array}$$

Calculate:

Step 1

$$\begin{array}{r} \scriptstyle 1 \\ 429 \\ +738 \\ \hline 7 \end{array} \qquad 9 + 8 = 17 \qquad \text{Write 7; carry 1 ten}$$

Step 2

$$\begin{array}{r} \scriptstyle 1 \\ 429 \\ +738 \\ \hline 67 \end{array} \qquad 1 + 2 + 3 = 6 \qquad \text{Write 6}$$

Step 3

$$\begin{array}{r} \scriptstyle 1 \\ 429 \\ +738 \\ \hline 1167 \end{array} \qquad 4 + 7 = 11 \qquad \text{Write 11}$$

Check: The estimate 1100 and the answer 1167 are roughly equal.

(b) **Estimate:** $400 + 900 = 1300$

Calculate:

Step 1

$$\begin{array}{r} \scriptstyle 1 \\ 446 \\ +867 \\ \hline 3 \end{array} \qquad 6 + 7 = 13 \qquad \text{Write 3; carry 1 ten}$$

Step 2

$$\begin{array}{r} \scriptstyle 11 \\ 446 \\ +867 \\ \hline 13 \end{array} \qquad 1 + 4 + 6 = 11 \qquad \text{Write 1; carry 1 hundred}$$

Step 3

$$\begin{array}{r} \scriptstyle 11 \\ 446 \\ +867 \\ \hline 1313 \end{array} \qquad 1 + 4 + 8 = 13 \qquad \text{Write 13}$$

Check: The estimate 1300 and the answer 1313 are roughly equal.

(c) **Estimate:** $2400 + 700 = 3100$

Calculate:

$$\begin{array}{r} \scriptstyle 111 \\ 2368 \\ +\ 744 \\ \hline 3112 \end{array}$$

Check: The estimate 3100 and the answer 3112 are roughly equal.

(d) **Estimate:** $400 + 100 = 500$

Calculate:

$$\begin{array}{r} \scriptstyle 1 \\ 409 \\ +\ 72 \\ \hline 481 \end{array}$$

Check: The estimate 500 and the answer 481 are roughly equal.

Estimating answers is a very important part of any mathematics calculation, especially for the practical mathematics used in engineering, technology, and the trades. A successful builder, painter, or repairman must make accurate estimates of job costs—business success depends on it. If you work in a technical trade, getting and keeping your job may depend on your ability to get the correct answer *every* time.

If you use an electronic calculator to do the actual arithmetic, it is even more important to get a careful estimate of the answer first. If you plug in a wrong number to the calculator, accidentally hit a wrong key, or unknowingly use a failing battery, the calculator may give you a wrong answer—lightning fast, but wrong. The estimate is your best insurance that a wrong answer will be caught immediately. Convinced?

Now, work the following problems for practice in adding whole numbers.

Exercises 1-2 Addition of Whole Numbers

A. Add:

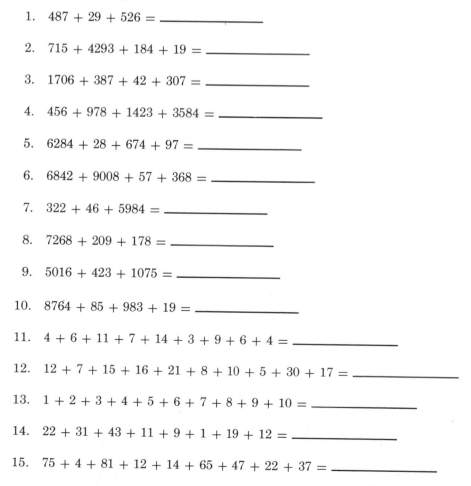

1. 47 23	2. 27 38	3. 45 35	4. 38 65	5. 75 48
6. 26 98	7. 48 84	8. 67 69	9. 189 204	10. 508 495
11. 684 706	12. 432 399	13. 621 388	14. 747 59	15. 375 486
16. 4237 1288	17. 5076 4385	18. 7907 1395	19. 3785 7643	20. 6709 9006
21. 18745 6972	22. 40026 7085	23. 10674 397	24. 9876 4835	25. 78044 97684

B. Arrange vertically and add:

1. 487 + 29 + 526 = ____________
2. 715 + 4293 + 184 + 19 = ____________
3. 1706 + 387 + 42 + 307 = ____________
4. 456 + 978 + 1423 + 3584 = ____________
5. 6284 + 28 + 674 + 97 = ____________
6. 6842 + 9008 + 57 + 368 = ____________
7. 322 + 46 + 5984 = ____________
8. 7268 + 209 + 178 = ____________
9. 5016 + 423 + 1075 = ____________
10. 8764 + 85 + 983 + 19 = ____________
11. 4 + 6 + 11 + 7 + 14 + 3 + 9 + 6 + 4 = ____________
12. 12 + 7 + 15 + 16 + 21 + 8 + 10 + 5 + 30 + 17 = ____________
13. 1 + 2 + 3 + 4 + 5 + 6 + 7 + 8 + 9 + 10 = ____________
14. 22 + 31 + 43 + 11 + 9 + 1 + 19 + 12 = ____________
15. 75 + 4 + 81 + 12 + 14 + 65 + 47 + 22 + 37 = ____________

C. Applied Problems

1. In setting up his latest wiring job, an electrician cut the following lengths of wire: 387 ft, 913 ft, 76 ft, 2640 ft, and 845 ft. Find the total length of wire used.

2. The Acme Lumber Co. made four deliveries of 1″ × 6″ flooring: 3280 fbm, 2650 fbm, 2465 fbm, and 2970 fbm. What was the total number of board feet of flooring delivered? (The abbreviation for "board feet" is fbm, which is short for "feet board measure.")

3. The stockroom has eight boxes of No. 10 hexhead cap screws. How many screws of this type are in stock if the boxes contain 346, 275, 84, 128, 325, 98, 260, and 120 screws respectively?

4. In calculating her weekly expenses, a contractor found that she had spent the following amounts: materials, $386; labor, $537; salaried help, $93; overhead expense, $132. What was her total expense for the week?

5. The head machinist at Tiger Tool Co. is responsible for totaling time cards to determine job costs. He found that five different jobs this week took 78, 428, 143, 96, and 384 minutes each. What was the total time in minutes for the five jobs?

6. On a home construction job, a carpenter laid 1480 wood shingles the first day, 1240 the second, 1560 the third, 1320 the fourth, and 1070 the fifth day. How many shingles did he lay in the five days?

7. Eight individually powered machines in a small production shop have motors using 420, 260, 875, 340, 558, 564, 280, and 310 watts each. What is the total wattage used when (a) the total shop is in operation? (b) the three largest motors are running? (c) the three smallest motors are running?

D. Calculator Problems

You probably own an electronic calculator and, of course, you are eager to put it to work doing practical mathematics calculations. In this textbook we will include problem sets for calculator users. These problems will be taken from real-life situations and, unlike most textbook problems, will involve big numbers and lots of calculations. If you think that having an electronic brain-in-a-box means you do not need to know the basic operations of arithmetic, you will be disappointed. The calculator helps you to work faster, but it will not tell you *what* to do or *how* to do it.

Detailed instruction on using an electronic calculator appears in Appendix A on page 527.

Here are a few helpful hints for calculator users:

1. Always *estimate* your answer before doing a calculation.
2. *Check* your answer by comparing it with the estimate or by the other methods shown in this text. Be certain your answer makes sense.
3. If you doubt the reliability of the calculator (they do break down, you know), put a problem in it whose answer you know, preferably a problem like the one you are solving.

1. The following table lists the number of widget fasteners made by each of the five machines at the Ace Widget Co. during the last ten working days.

Day	Machine A	B	C	D	E	Daily Totals
1	347	402	406	527	237	
2	451	483	312	563	316	
3	406	511	171	581	289	
4	378	413	0	512	291	
5	399	395	452	604	342	
6	421	367	322	535	308	
7	467	409	256	578	264	
8	512	514	117	588	257	
9	302	478	37	581	269	
10	391	490	112	596	310	
Machine Totals						

(a) Complete the table by finding the number of fasteners produced each day. Enter these totals under the column "Daily Totals" on the right.

(b) Find the number of fasteners produced by each machine during the ten-day period and enter these totals along the bottom row marked "Machine Totals."

(c) Does the sum of the daily totals equal the sum of the machine totals?

2. Add the following as shown:

(a)	(b)	(c)
\$ 67429	\$216847	\$693884
6070	9757	675489
4894	86492	47039
137427	4875	276921
91006	386738	44682
399	28104	560487

(d) \$4299 + \$137 + \$20 + \$177 + \$63 + \$781 + \$1008 + \$671 = ________

3. Joe's Air Conditioning Installation Co. has not been successful, and he is wondering if he should sell it and move to a better location. During the first six months of the year his expenses were:

Rent \$1620
Supplies \$2540
Part-time helper \$2100
Transportation \$948
Taxes \$143
Advertising \$250
Miscellaneous \$187

His monthly income was:

January \$609
February \$1151
March \$1269
April \$1381
May \$1687
June \$1638

(a) What was his total expense for the six-month period?

(b) What was his total income for the six-month period?

(c) Rotate your calculator to learn what Joe should do about this unhappy situation.

4. A mapper is a person employed by an electrical utility company who has the job of reading diagrams of utility installations and listing the materials to be installed or removed by engineers. Part of a typical job list might look like this:

INSTALLATION (in Feet of Conductor)

Location Code	#12 BHD (Bare, hard-drawn copper wire)	#TX (Triplex)	410 AAC (All-aluminum conductor)	110 ACSR (Aluminum-core steel-reinforced conductor)	6B (No. 6, bare conductor)
A3	1740	40	1400		350
A4	1132		5090		2190
B1	500			3794	
B5		87	3995		1400
B6	4132	96	845		
C4		35		3258	2780
C5	3949		1385	1740	705

(a) How many total feet of each kind of conductor must the installer have to complete the job?

(b) How many feet of conductor are to be installed at each of the seven locations?

When you have completed these exercises, check your answers on page 541 and then continue in frame **11**.

1-2 SUBTRACTION OF WHOLE NUMBERS

11 Subtraction is the reverse of the process of addition.

Addition: $3 + 4 = \square$

Subtraction: $3 + \square = 7$

Written this way, a subtraction problem asks the question, "How much must be added to a given number to produce a required amount?"

Most often, however, the numbers in a subtraction problem are written using a minus sign (−):

$17 - 8 = \square$ means that there is a number $\square$ such that $8 + \square = 17$

Write in the answer to this subtraction problem, then turn to **12**.

---

12 $8 + 9 = 17$ or $17 - 8 = 9$ ⇐ Difference

When you set up the subtraction vertically, the subtrahend is the bottom number and the minuend is the top number.

The *difference* is the name given to the answer in a subtraction problem.

The ability to solve simple subtraction problems depends on your knowledge of the addition of one-digit numbers.

For example, to solve the problem

9 − 4 = ________ you probably go through a chain of thoughts something like this:

> "Nine minus four. Four added to what number gives nine? Five? Try it: four plus five equals nine. Right."

Subtraction problems involving small whole numbers will be easy for you if you know your addition tables.

Here is a more difficult subtraction problem:

47 − 23 = ________

What is the first step?
Work the problem and continue in **13**.

---

13 The **first** step is to estimate the answer—remember?

47 − 23 is roughly 40 − 20 or 20.

The difference, your answer, will be about 20—not 2 or 10 or 200.

The **second** step is to write the numbers in a vertical format as you did with addition. Be careful to keep the ones digits in line in one column, the tens digits in a second column, and so on.

$$\begin{array}{r} 47 \\ -\,23 \\ \hline \end{array}$$

Notice that the minuend is written above the subtrahend—larger number on top.

Once the numbers have been arranged in this way the difference may be written immediately.

Step 1

$$\begin{array}{r} 47 \\ -23 \\ \hline 4 \end{array}$$

ones digits: 7 − 3 = 4

Step 2

$$\begin{array}{r} 47 \\ -\,23 \\ \hline 24 \end{array}$$

tens digits: 4 − 2 = 2

The difference is 24 and this agrees with our estimate.

With some problems it is necessary to rewrite the larger numbers before the problem can be solved. For example, try this one:

64 − 37 = ________

Check your work in **14**.

---

14 **First,** estimate the answer. 64 − 37 is roughly 60 − 40 or 20.

Second, arrange the numbers vertically in columns.

$$\begin{array}{r} 64 \\ -37 \\ \hline \end{array}$$

Because 7 is larger than 4 we must "borrow" one ten from the 6 tens in 64. We are actually rewriting 64 (6 tens + 4 ones) as 5 tens + 14 ones. In actual practice our work would look like this:

Step 1

$$\begin{array}{r} 64 \\ -37 \\ \hline \end{array}$$

Step 2

$$\begin{array}{r} {\scriptstyle 5\,14} \\ \not{6}\not{4} \\ -37 \\ \hline 7 \end{array}$$

Borrow one ten, change the 6 in the tens place to 5, change 4 to 14, subtract 14 − 7 = 7.

Step 3

$$\begin{array}{r} {\scriptstyle 5\,14} \\ \not{6}\not{4} \\ -37 \\ \hline 27 \end{array}$$

14 − 7 = 7

5 − 3 = 2

Double-check subtraction problems by adding the answer and the smaller number; their sum should equal the larger number.

Step 4 Check:

$$\begin{array}{r} 37 \\ +27 \\ \hline 64 \end{array}$$

Try these problems for practice.

(a) $\begin{array}{r} 71 \\ -39 \\ \hline \end{array}$ (b) $\begin{array}{r} 263 \\ -127 \\ \hline \end{array}$ (c) $\begin{array}{r} 426 \\ -128 \\ \hline \end{array}$ (d) $\begin{array}{r} 902 \\ -465 \\ \hline \end{array}$

Solutions are in **15**.

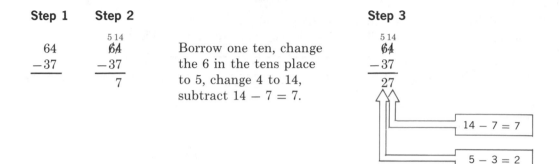

15 (a) **Estimate:** $70 - 40 = 30$

Step 1	Step 2	
71	$\overset{6\,11}{\not{7}\not{1}}$	Borrow one ten from 70,
-39	-39	change the 7 in the tens place to 6,
	32	change the 1 in the ones place to 11.

$11 - 9 = 2$ Write 2

$6 - 3 = 3$ Write 3

Check: The answer 32 is approximately equal to the estimate 30.

(b) **Estimate:** $200 - 100 = 100$

Step 1	Step 2	
263	$\overset{5\,13}{2\not{6}\not{3}}$	Borrow one ten from 60,
-127	-127	change the 6 in the tens place to 5,
	136	change the 3 in the ones place to 13.

$13 - 7 = 6$ Write 6

$5 - 2 = 3$ Write 3

$2 - 1 = 1$ Write 1.

Check: The answer is approximately equal to the estimate.

(c) **Estimate:** $400 - 100 = 300$

Step 1	Step 2	Step 3
426	$\overset{1\,16}{4\not{2}\not{6}}$	$\overset{3\,11\,16}{\not{4}\not{2}\not{6}}$
-128	-128	-128
	8	298

Notice that in this case we must borrow twice. Borrow one ten from the 20 in 426 to make 16. Then borrow one hundred from the 400 in 426 to make 110.

$16 - 8 = 8$ Write 8

$11 - 2 = 9$ Write 9

$3 - 1 = 2$ Write 2.

Check: The answer 298 is approximately equal to the estimate 300.

(d) **Estimate:** $900 - 500 = 400$

Step 1	Step 2	Step 3
902	$\overset{8\,10}{\not{9}\not{0}2}$	$\overset{8\,9\,12}{\not{9}\not{0}\not{2}}$
-465	-465	-465
		437

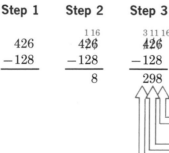

$12 - 5 = 7$ Write 7

$9 - 6 = 3$ Write 3

$8 - 4 = 4$ Write 4

Check: The answer 437 is roughly equal to the estimate 400.

In problem (d) we first borrow one hundred from 900 to get a 10 in the tens place. Then we borrow one 10 from the tens place to get a 12 in the ones place.

Now, do you want more worked examples of problems containing zeros, similar to this last one? If so, go to **17**.

Otherwise, go to **16** for a set of practice problems.

---

16 **Exercises 1-3 Subtraction of Whole Numbers**

A. Subtract:

1.	2.	3.	4.	5.	6.
13	12	8	8	11	16
7	5	6	0	7	7

7.	8.	9.	10.	11.	12.
10	5	12	11	10	14
7	5	9	8	2	6

13.	14.	15.	16.	17.	18.
12	15	9	14	9	11
3	6	0	5	6	5

19.	20.	21.	22.	23.	24.
15	12	13	18	16	12
7	7	6	0	9	4

25.
0
0

B. Subtract:

1.	2.	3.	4.	5.	6.
40	78	51	36	42	52
27	49	39	17	27	16

7.	8.	9.	10.	11.	12.
70	34	56	546	409	476
48	9	18	357	324	195

13.	14.	15.	16.	17.	18.
747	400	803	632	438	6218
593	127	88	58	409	3409

19.	20.	21.	22.	23.	24.
6084	13042	57022	5007	10000	48093
386	524	980	266	386	500

25.
27004
4582

C. Applied Problems

1. In planning for a particular job, a painter buys $264 worth of materials. When the job is completed, he returns five empty paint drums for a credit of $17. What was the net amount of his bill?

2. How many sq ft of plywood remain from an original supply of 8000 sq ft after 5647 sq ft are used?

3. A storage rack at the Tiger Tool Company contains 3540 ft of 1″ stock. On a certain job 1782 ft are used. How much is left?

4. Five pieces measuring 26 cm, 47 cm, 38 cm, 27 cm, and 32 cm are cut from a steel bar that was 200 cm long. Allowing for a total of 1 cm for waste in cutting, what is the length of the piece remaining?

5. Taxes on a group of factory buildings owned by the Ace Manufacturing Company amounted to $875,977 eight years ago. Taxes on the same buildings last year amounted to $1,206,512. Find the increase in taxes.

6. In order to pay their bills, the Edwards Plumbing Company made the following withdrawals from their bank account: $42, $175, $24, $217, and $8. If the original balance was $3610, what was the amount of the new balance?

7. Which total volume is greater, four drums containing 72, 45, 39, and 86 liters, or three drums containing 97, 115, and 74 liters? By how much is it greater?

8. Determine the missing dimension (*L*) in the drawings below.

(a)

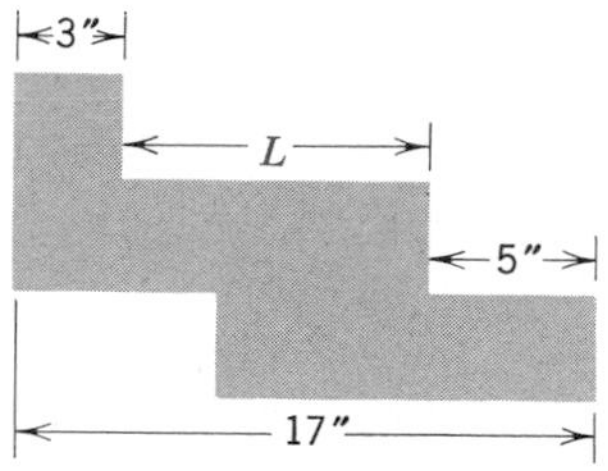

(b)

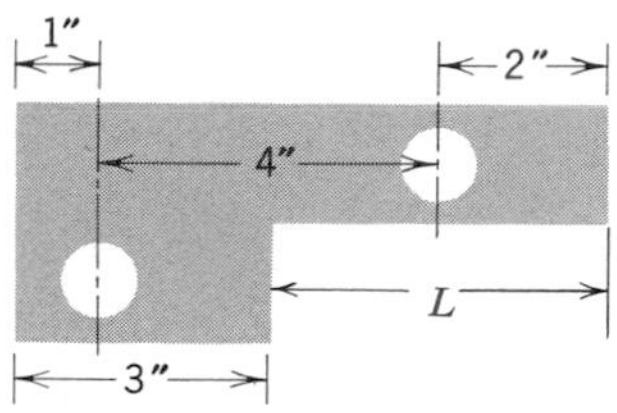

(c)

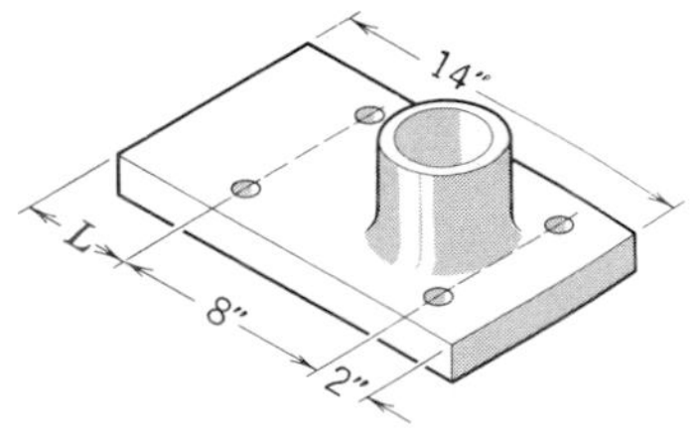

(d)

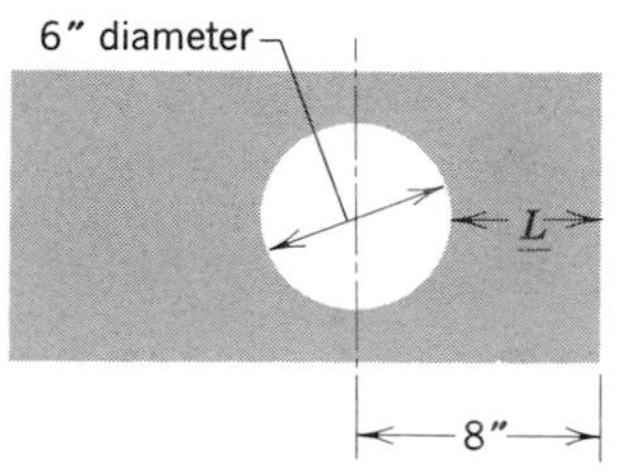

D. Calculator Problems

1. The Karroll Plumbing Co. has 10 trucks and, for the month of April, the following mileage was recorded on each.

Truck No.	Mileage at Start	Mileage at End
1	58352	60027
2	42135	43302
3	76270	78007
4	40006	41322
5	08642	10002
6	35401	35700
7	79002	80101
8	39987	40122
9	10210	11671
10	71040	73121

Find the mileage traveled by each truck during the month of April and the total mileage of all vehicles.

2. Which sum is greater?

987654321		123456789
87654321		123456780
7654321		123456700
654321		123456000
54321	or	123450000
4321		123400000
321		123000000
21		120000000
1		100000000

3. If your income is \$8245 per year and you pay \$959 in taxes, what is your take-home pay?

4. The income of the Smith Construction Company for the year is \$37,672 and the total expenses are \$20,867. Find the difference, Smith's profit, for that year.

5. Balance the following checking account record.

Date	Deposits	Withdrawals	Balance
7/1			\$6375
7/3		\$379	
7/4	\$1683		
7/7	\$ 474		
7/10	\$ 487		
7/11		\$2373	
7/15		\$1990	
7/18		\$ 308	
7/22		\$1090	
7/26		\$ 814	
8/1			*A*

(a) Find the new balance, *A*.

(b) Keep a running balance by filling in each blank in the balance column.

When you have completed these exercises, check your answers on page 541, then turn to **18** to study the multiplication of whole numbers.

---

17 Let's work through a few examples of subtraction problems involving zero digits.

Example 1: 400 − 167 = ?

(a) **Step 1** **Step 2** **Step 3**

$$\begin{array}{r} 400 \\ -167 \\ \hline \end{array} \qquad \begin{array}{r} {}^{3\,10} \\ \not{4}00 \\ -167 \\ \hline \end{array} \qquad \begin{array}{r} {}^{3\,9\,10} \\ \not{4}\not{0}\not{0} \\ -167 \\ \hline 233 \end{array}$$

Do you see that we have rewritten 400 as 300 + 90 + 10?

Check:

$$\begin{array}{r} 167 \\ +233 \\ \hline 400 \end{array}$$

Example 2: 5006 − 2487 = ?

(b) Step 1	Step 2	Step 3	Step 4
	4 10	4 9 10	4 9 9 16
5006	5006	5006	5006
−2487	−2487	−2487	−2487
			2519

Check:

$$\begin{array}{r} 2487 \\ +2519 \\ \hline 5006 \end{array}$$

Here is an example involving repeated borrowing.

Example 3: 24632 − 5718 = ?

(c) Step 1	Step 2	Step 3	Step 4
	2 12	3 16 2 12	1 13 16 2 12
24632	24632	24632	24632
− 5718	− 5718	− 5718	− 5718
	14	914	18914

Check:

$$\begin{array}{r} 5718 \\ +18914 \\ \hline 24632 \end{array}$$

Any subtraction problem that involves borrowing should always be checked in this way. It is very easy to make a mistake in this process.

Now turn back to **16** for some practice problems on subtraction.

1-3 MULTIPLICATION OF WHOLE NUMBERS

18 In a certain football game, the West Newton Waterbugs scored five touchdowns at six points each. How many total points did they score through touchdowns? We can answer the question several ways:

1. Count points, ...
2. Add touchdowns, 6 + 6 + 6 + 6 + 6 = ?

   or

3. Multiply 5 × 6 = ?

We're not sure about the mathematical ability of the West Newton scorekeeper, but most people would multiply. Multiplication is a short-cut method of counting or repeated addition.

How many points did they score?

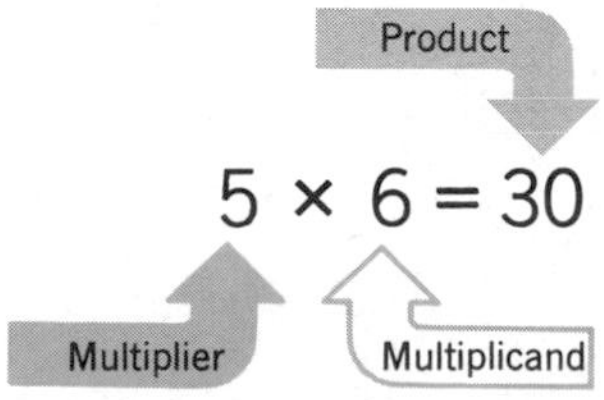

In a multiplication problem the *product* is the name given to the result of the multiplication. The numbers being multiplied are called the *factors* of the product.

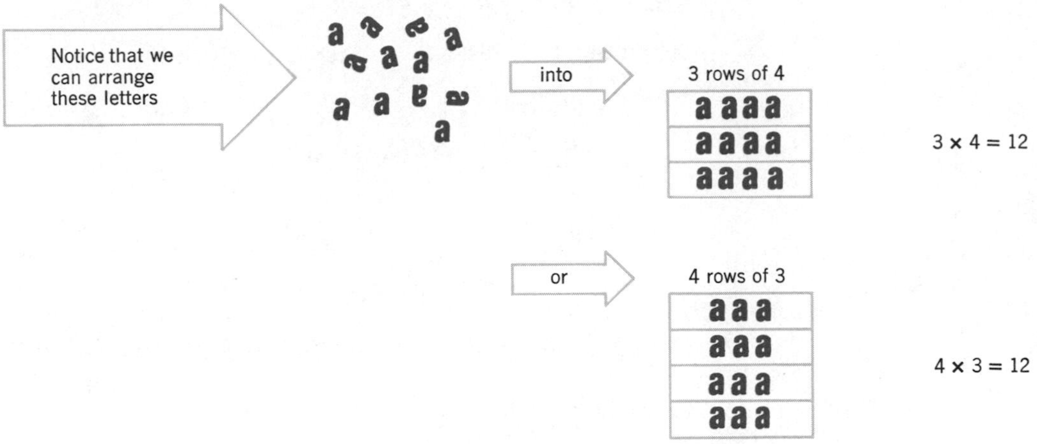
Notice that we can arrange these letters
into
3 rows of 4
3 × 4 = 12
or
4 rows of 3
4 × 3 = 12

Changing the order of the factors does not change their product. This is the *commutative* property of multiplication.

In order to become skillful at multiplication, you must know the one-digit multiplication table from memory. Even if you use an electronic calculator for your work, you need to know the one-digit multiplication table to make estimates and to check your work. Complete the table below by multiplying the number at the top by the number at the side and placing their product in the proper square. We have multiplied $3 \times 4 = 12$ and $2 \times 5 = 10$ for you.

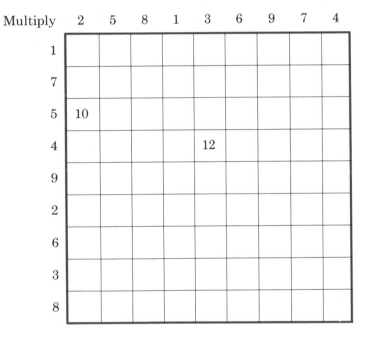

Multiply	2	5	8	1	3	6	9	7	4
1									
7									
5	10								
4					12				
9									
2									
6									
3									
8									

Check your work in **20**.

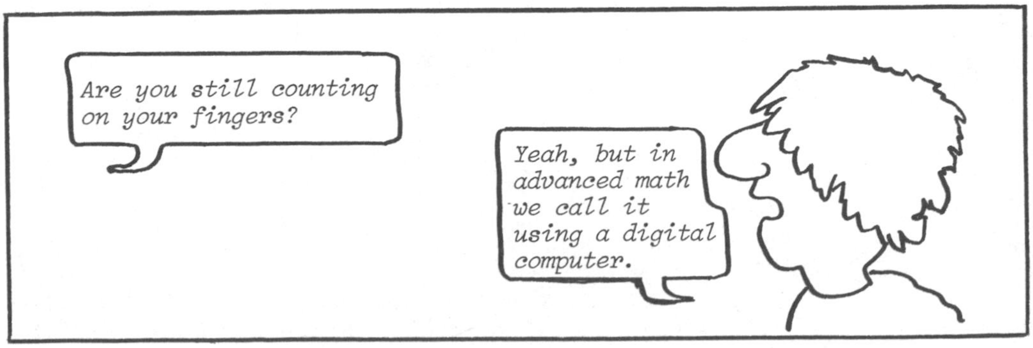
Are you still counting on your fingers?
Yeah, but in advanced math we call it using a digital computer.

19 **Exercises 1-4 One-Digit Multiplication**

Multiply as shown. Work quickly; you should be able to answer all problems in a set correctly in the time indicated. (These times are for community college students enrolled in a developmental math course.)

A. Multiply:

6	4	9	6	3	9	7	8	2	8
$\underline{2}$	$\underline{8}$	$\underline{7}$	$\underline{6}$	$\underline{4}$	$\underline{2}$	$\underline{0}$	$\underline{3}$	$\underline{7}$	$\underline{1}$
6	8	5	5	2	3	9	7	3	1
$\underline{8}$	$\underline{2}$	$\underline{9}$	$\underline{6}$	$\underline{5}$	$\underline{3}$	$\underline{8}$	$\underline{5}$	$\underline{6}$	$\underline{4}$
7	5	4	7	4	8	6	9	8	6
$\underline{4}$	$\underline{3}$	$\underline{9}$	$\underline{7}$	$\underline{2}$	$\underline{5}$	$\underline{7}$	$\underline{6}$	$\underline{8}$	$\underline{4}$
5	3	5	9	9	6	1	8	4	7
$\underline{4}$	$\underline{0}$	$\underline{5}$	$\underline{3}$	$\underline{9}$	$\underline{1}$	$\underline{1}$	$\underline{6}$	$\underline{4}$	$\underline{9}$

Average time: 100 sec Record: 37 sec

B. Multiply:

2	6	3	5	6	4	4	8	2	7
$\underline{8}$	$\underline{5}$	$\underline{3}$	$\underline{7}$	$\underline{3}$	$\underline{5}$	$\underline{7}$	$\underline{6}$	$\underline{6}$	$\underline{9}$
8	0	2	3	1	5	6	9	5	8
$\underline{4}$	$\underline{6}$	$\underline{9}$	$\underline{8}$	$\underline{9}$	$\underline{5}$	$\underline{4}$	$\underline{5}$	$\underline{2}$	$\underline{9}$
3	7	5	6	9	2	7	8	9	2
$\underline{5}$	$\underline{7}$	$\underline{8}$	$\underline{9}$	$\underline{4}$	$\underline{4}$	$\underline{6}$	$\underline{8}$	$\underline{0}$	$\underline{2}$
5	9	1	8	6	4	9	0	2	7
$\underline{5}$	$\underline{3}$	$\underline{7}$	$\underline{7}$	$\underline{6}$	$\underline{3}$	$\underline{9}$	$\underline{4}$	$\underline{1}$	$\underline{8}$

Average time: 100 sec Record: 36 sec

Check your answers on page 542.
Hop ahead to **21** to continue.

20 Here is the completed multiplication table:

MULTIPLICATION TABLE

×	2	5	8	1	3	6	9	7	4
1	2	5	8	1	3	6	9	7	4
7	14	35	56	7	21	42	63	49	28
5	10	25	40	5	15	30	45	35	20
4	8	20	32	4	12	24	36	28	16
9	18	45	72	9	27	54	81	63	36
2	4	10	16	2	6	12	18	14	8
6	12	30	48	6	18	36	54	42	24
3	6	15	24	3	9	18	27	21	12
8	16	40	64	8	24	48	72	56	32

If you are not able to perform these one-digit multiplications quickly from memory, you should practice until you can do so. A multiplication table is given on page 575. Use it if you need it.

Notice that the product of any number and 1 is that same number. For example,

$1 \times 2 = 2$
$1 \times 6 = 6$

or even

$1 \times 753 = 753$

Zero has been omitted from the table because the product of any number and zero is zero. For example,

$$\begin{aligned} 0 \times 2 &= 0 \\ 0 \times 7 &= 0 \\ 395 \times 0 &= 0 \end{aligned}$$

If you want some more practice in one-digit multiplication, turn back to **19**. Otherwise go to **21**.

---

21 The multiplication of larger numbers is based on the one-digit number multiplication table. Find the product

$34 \times 2 =$ ______

Remember the procedure you followed for addition. What are the first few steps in this multiplication?

Try it, then go to **22**.

---

22 **First,** estimate the answer: $30 \times 2 = 60$. The actual product of the multiplication will be about 60.

Second, arrange the factors to be multiplied vertically, with ones digits in a single column, tens digits in a second column, and so on.

$$\begin{array}{rcc} & 3 & 4 \\ \times & & 2 \\ \hline \end{array}$$

Finally, to make the process clear, let's write it in expanded form.

$$\begin{array}{r} 34 \\ \times\ 2 \\ \hline \end{array} \Rightarrow \begin{array}{r} 3 \text{ tens} + 4 \text{ ones} \\ \times\ 2 \\ \hline 6 \text{ tens} + 8 \text{ ones} \end{array} = 60 + 8 = 68$$

Check: The guess 60 is roughly equal to the answer 68.

Now write the following multiplication in expanded form.

$$\begin{array}{r} 28 \\ \times\ 3 \\ \hline \end{array}$$

Check your work in **23**.

---

23 **Estimate:** $30 \times 3 = 90$ The answer is about 90.

$$\begin{array}{r} 28 \\ \times\ 3 \\ \hline \end{array} \Rightarrow \begin{array}{r} 2 \text{ tens} + 8 \text{ ones} \\ \times\ 3 \\ \hline \end{array}$$

$$\begin{aligned} & 6 \text{ tens} + 24 \text{ ones} \\ &= 6 \text{ tens} + 2 \text{ tens} + 4 \text{ ones} \\ &= 8 \text{ tens} + 4 \text{ ones} \\ &= 80 + 4 \\ &= 84 \end{aligned}$$

Check: 90 is roughly equal to 84.

Of course we do not normally use the expanded form; instead we simplify the work like this:

$$\begin{array}{r} {}^{2} \\ 28 \\ \times\ 3 \\ \hline 84 \end{array}$$

$3 \times 8 = 24$ Write 4 and carry 2 tens
3×2 tens $= 6$ tens 6 tens + 2 tens = 8 tens
Write 8

Now try these problems to be certain you understand the process. Multiply as shown.

(a) $\begin{array}{r} 43 \\ \times\ 5 \\ \hline \end{array}$ (b) $\begin{array}{r} 73 \\ \times\ 4 \\ \hline \end{array}$ (c) $\begin{array}{r} 29 \\ \times\ 6 \\ \hline \end{array}$ (d) $\begin{array}{r} 258 \\ \times\ 7 \\ \hline \end{array}$

Check your answers in **24**.

---

24 (a) **Estimate:** $40 \times 5 = 200$ The answer is roughly 200.

$$\begin{array}{r} {}^{1} \\ 43 \\ \times\ 5 \\ \hline 215 \end{array}$$

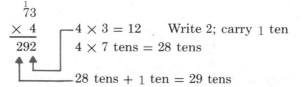

$5 \times 3 = 15$ Write 5; carry 1 ten
5×4 tens $= 20$ tens
20 tens + 1 ten = 21 tens

Check: The answer 215 is roughly equal to the estimate 200.

(b) **Estimate:** $70 \times 4 = 280$

$$\begin{array}{r} {}^{1} \\ 73 \\ \times\ 4 \\ \hline 292 \end{array}$$

$4 \times 3 = 12$ Write 2; carry 1 ten
4×7 tens $= 28$ tens
28 tens + 1 ten = 29 tens

Check: The answer 292 is roughly equal to the estimate 280.

(c) **Estimate:** $30 \times 6 = 180$

$$\begin{array}{r} {}^{5}\\ 29 \\ \times\ 6 \\ \hline 174 \end{array}$$

$6 \times 9 = 54$ Write 4; carry 5 tens

6×2 tens $= 12$ tens

12 tens + 5 tens = 17 tens

Check: The answer 174 is roughly equal to the estimate 180.

(d) **Estimate:** $300 \times 7 = 2100$

$$\begin{array}{r} {}^{4\,5}\\ 258 \\ \times\ \ 7 \\ \hline 1806 \end{array}$$

$7 \times 8 = 56$ Write 6; carry 5 tens

7×5 tens $= 35$ tens

35 tens + 5 tens = 40 tens Write 0; carry 4 hundreds

7×2 hundreds $= 14$ hundreds

14 hundreds + 4 hundreds = 18 hundreds

Check: The answer and estimate are roughly equal.

Two-Digit Multiplications

Calculations involving two-digit multipliers are done in exactly the same way. For example, to multiply

$$\begin{array}{r} 89 \\ \times 24 \\ \hline \end{array}$$

First, estimate the answer. $90 \times 20 = 1800$

Second, multiply by the ones digit 4.

$$\begin{array}{r} {}^{3}\\ 89 \\ \times\ 24 \\ \hline 356 \end{array}$$

$4 \times 9 = 36$ Write 6; carry 3 tens

4×8 tens $= 32$ tens

32 tens + 3 tens = 35 tens

Third, multiply by the tens digit 2.

$$\begin{array}{r} {}^{1}\\ {}^{3}\\ 89 \\ \times\ 24 \\ \hline 356 \\ 178\ \\ \hline 2136 \end{array}$$

$2 \times 9 = 18$ Write 8; carry 1

$2 \times 8 = 16$

$16 + 1 = 17$

Fourth, add the products obtained.

Finally, check it. The estimate and the answer are roughly the same, at least in the same ballpark.

Notice that the product in the third step, 178, is written one digit space over from the product from the second step. When we multiplied 2×9 to get 18 in step 3, we were actually multiplying $20 \times 9 = 180$, but the zero in 180 is usually omitted to save time.

Try these:

(a) $\begin{array}{r} 64 \\ \times 37 \\ \hline \end{array}$ (b) $\begin{array}{r} 327 \\ \times 145 \\ \hline \end{array}$ (c) $\begin{array}{r} 342 \\ \times 102 \\ \hline \end{array}$

Multiply as shown and check your work in **25**.

THE SEXY SIX

Here are the six most often missed one-digit multiplications:

Inside digits

$9 \times 8 = 72$
$9 \times 7 = 63$
$9 \times 6 = 54$
$8 \times 7 = 56$
$8 \times 6 = 48$
$7 \times 6 = 42$

It may help you to notice that in these multiplications the "inside" digits, such as 8 and 7 are consecutive, and the digits of the answer add to nine: $7 + 2 = 9$. This is true for *all* one-digit numbers multiplied by 9.

Be certain you have these memorized.

(There is nothing very sexy about them, but we did get your attention, didn't we?)

---

25 (a) **Estimate:** $60 \times 40 = 2400$

```
   64
  ×37
  448 ←  7 × 4 = 28   Write 8; carry 2
         7 × 6 = 42   Add carry 2 to get 44; write 44
 192  ←  3 × 4 = 12   Write 2; carry 1
         3 × 6 = 18   Add carry 1 to get 19; write 19
 2368 ← Answer
```

(b) **Estimate:** $300 \times 150 = 45{,}000$

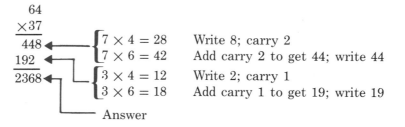

```
   327
  ×145
  1635 ←  5 × 7 = 35   Write 5; carry 3
          5 × 2 = 10   Add carry 3 to get 13; write 3, carry 1
          5 × 3 = 15   Add carry 1 to get 16; write 16
 1308  ←  4 × 7 = 28   Write 8, carry 2
          4 × 2 = 8    Add carry 2 to get 10; write 0, carry 1
          4 × 3 = 12   Add carry 1 to get 13; write 13
 327   ←  1 × 327 = 327
 47415
```

(c) **Estimate:** $300 \times 100 = 30{,}000$

```
   342
  ×102
   684 ← 2 × 342 = 684
  000  ← 0 × 342 = 000
 342   ← 1 × 342 = 342
 34884
```

Be very careful when there are zeros in the multiplier; it is very easy to misplace one of those zeros. Do not skip any steps, and be sure to estimate your answer first.

Go to **26** for a set of practice problems on the multiplication of whole numbers.

A. Multiply:

1.	7	2.	7	3.	6	4.	8	5.	9	6.	29
	6		8		8		9		7		3
7.	72	8.	47	9.	64	10.	39	11.	58	12.	94
	8		9		5		4		5		6
13.	17	14.	47	15.	77	16.	48	17.	64	18.	90
	9		6		4		15		27		56
19.	86	20.	34	21.	66	22.	59	23.	29	24.	78
	83		57		25		76		32		49
25.	94										
	95										

B. Multiply:

1.	305	2.	3006	3.	8043	4.	809	5.	3706
	123		125		37		47		102
6.	708	7.	684	8.	2043	9.	2008	10.	563
	58		45		670		198		107
11.	809	12.	609	13.	500	14.	542	15.	7009
	9		7		50		600		504

C. Practical Problems

1. A plumber receives $19 per hour. How much is he paid for 40 hours of work?
2. What is the total length of wire on 14 spools if each spool contains 150 ft?
3. How many total lineal feet of redwood are there in 65 2″ by 4″ boards each 20 ft long?
4. An auto body shop does 17 paint jobs at $159 each and 43 paint jobs at $267 each. How much money does the shop receive from these jobs?
5. Three different-sized boxes of envelopes contain 50, 100, and 500 envelopes, respectively. How many envelopes total are there in 18 boxes of the first size, 16 of the second size, and 11 of the third size?
6. A machinist needs 25 lengths of steel each 9″ long. What is the total length of steel that he needs? No allowance is required for cutting.
7. The Ace Machine Company advertises that one of its machinists can produce 2 parts per hour. How many such parts can 27 machinists produce if they work 45 hours each?
8. What is the horizontal distance in inches covered by 12 stair steps if each is 11″ wide?

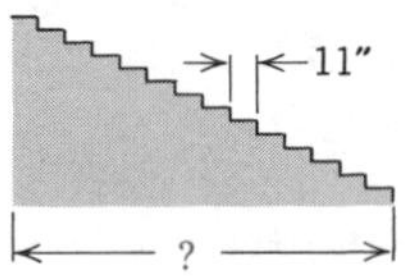

D. Calculator Problems

1. If an investment company started in business with \$100,000 of capitol on January 1 and, because of mismanagement, lost an average of \$273.47 every day for a year, what would be its financial situation on the last day of the year?

2. Which of the following pay schemes gives you the most money over a one-year period?

   (a) \$100 per day
   (b) \$700 per week
   (c) \$400 for the first month and a \$400 raise each month
   (d) 1¢ for the first two-week pay period, 2¢ for the second two-week period, 4¢ for the third two-week period, and so on, the pay doubling each two weeks.

3. Multiply:

   (a) $12{,}345{,}679 \times 9 =$
   $12{,}345{,}679 \times 18 =$
   $12{,}345{,}679 \times 27 =$

   (b) $15{,}873 \times 7 =$
   $15{,}873 \times 14 =$
   $15{,}873 \times 21 =$

   (c) $1 \times 1 =$
   $11 \times 11 =$
   $111 \times 111 =$
   $1111 \times 1111 =$
   $11111 \times 11111 =$

   (d) $6 \times 7 =$
   $66 \times 67 =$
   $666 \times 667 =$
   $6666 \times 6667 =$
   $66666 \times 66667 =$

   Can you see the pattern in each of these?

4. In the Aztec Machine Shop there are 9 lathes each weighing 2285 lb, 5 milling machines each weighing 2570 lb, and 3 drill presses each weighing 395 lb. What is the total weight of these machines?

5. The Omega Calculator Company makes five models of electronic calculator. The following table gives the weekly (5-day) production output.

Model	Alpha	Beta	Gamma	Delta	Tau
Cost of production of each model	\$6	\$17	\$32	\$49	\$178
Number produced during typical day	117	67	29	37	18

   Find the weekly (five days) production costs for each model.

Check your answers on page 542.
Turn to 27 to continue.

MULTIPLICATION SHORT CUTS

There are hundreds of quick ways to multiply various numbers. Most of them are only quick if you are already a math whiz. If not, they will confuse you more than help you. Here are a few that are easy to do and easy to remember.

1. To multiply by 10, annex a zero on the right end of the multiplicand. For example,

   $34 \times 10 = 340$
   $256 \times 10 = 2560$

   Multiplying by 100 or 1000 is similar.

   $34 \times 100 = 3400$
   $256 \times 1000 = 256000$

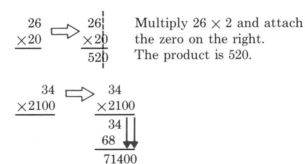

2. To multiply by a number ending in zeros, carry the zeros forward to the answer. For example,

   $$\begin{array}{r} 26 \\ \times 20 \\ \hline \end{array} \Rightarrow \begin{array}{r} 26 \\ \times 20 \\ \hline 520 \end{array}$$

   Multiply 26×2 and attach the zero on the right. The product is 520.

   $$\begin{array}{r} 34 \\ \times 2100 \\ \hline \end{array} \Rightarrow \begin{array}{r} 34 \\ \times 2100 \\ \hline 34 \\ 68 \\ \hline 71400 \end{array}$$

3. If both multiplier and multiplicand end in zeros, bring all zeros forward to the answer.

   $$\begin{array}{r} 230 \\ \times 200 \\ \hline \end{array} \Rightarrow \begin{array}{r} 230 \\ 200 \\ \hline 46000 \end{array}$$

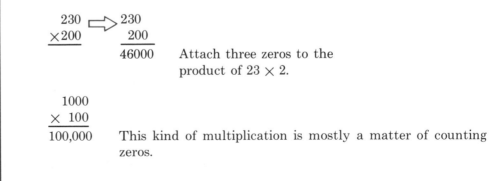

   Attach three zeros to the product of 23×2.

   $$\begin{array}{r} 1000 \\ \times\ 100 \\ \hline 100{,}000 \end{array}$$

   This kind of multiplication is mostly a matter of counting zeros.

1-4 DIVISION OF WHOLE NUMBERS

27 Division is the reverse process of multiplication. It enables us to separate a given quantity into equal parts. The mathematical phrase $12 \div 3$ is read "twelve divided by three," and it asks us to separate a collection of 12 objects into 3 equal parts. The mathematical phrases

$12 \div 3 \qquad 3\overline{)12} \qquad \frac{12}{3} \qquad \text{and} \qquad 12/3$

all represent division and they are all read "twelve divided by three."

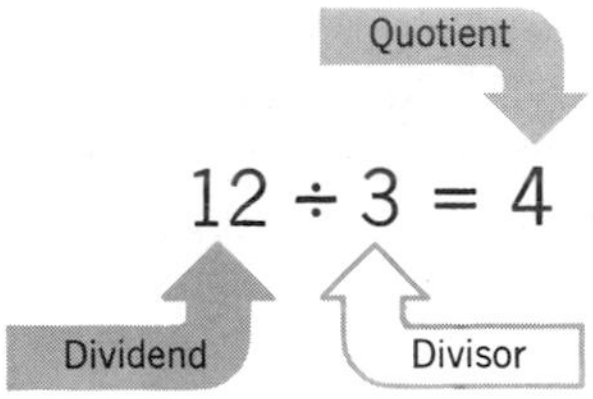

In this division problem, 12, the number being divided, is called the *dividend;* 3, the number used to divide, is called the *divisor;* and 4, the result of the division, is called the *quotient,* from a Latin word meaning "how many times."

One way to perform division is to reverse the multiplication process.

$24 \div 4 = \square$ means that $4 \times \square = 24$

If the one-digit multiplication tables are firmly in your memory, you will recognize immediately that $\square = 6$.

Try these.

$35 \div 7 =$ ______ $42 \div 6 =$ ______

$28 \div 4 =$ ______ $56 \div 7 =$ ______

$45 \div 5 =$ ______ $18 \div 3 =$ ______

$70 \div 10 =$ ______ $63 \div 9 =$ ______

$30 \div 5 =$ ______ $72 \div 8 =$ ______

Check your answers in **28**.

---

28 $35 \div 7 = 5$ $42 \div 6 = 7$
$28 \div 4 = 7$ $56 \div 7 = 8$
$45 \div 5 = 9$ $18 \div 3 = 6$
$70 \div 10 = 7$ $63 \div 9 = 7$
$30 \div 5 = 6$ $72 \div 8 = 9$

You should be able to do all of these quickly by working backward from the one-digit multiplication tables.

How do we divide dividends that are larger than 9×9 and therefore not in the multiplication table? Obviously, we need a better procedure.

Here is a step-by-step explanation of the division of whole numbers:

Divide: $96 \div 8 =$ ______.

First, *estimate* the answer: $90 \div 8$ is about 10 since $8 \times 10 = 80$. The quotient or answer will be about 10.

Second, arrange the numbers in this way:

$8\overline{)96}$

Third, divide using the following step-by-step procedure.

Step 1 $\begin{array}{r} 1 \\ 8\overline{)\,96} \end{array}$ 8 into 9? Once. Write 1 in the answer space above the 9.

Step 2 $\begin{array}{r} 1 \\ 8\overline{)\ 96} \\ \underline{8} \end{array}$ Multiply $8 \times 1 = 8$ and write the product 8 under the 9.

Step 3 $\begin{array}{r} 1 \\ 8\overline{)\ 96} \\ \underline{-8\downarrow} \\ 16 \end{array}$ Subtract $9 - 8 = 1$ and write 1. Bring down the next digit 6.

Step 4 $\begin{array}{r} 12 \\ 8\overline{)\ 96} \\ \underline{-8} \\ 16 \end{array}$ 8 into 16? Twice. Write 2 in the answer space above the 6.

Step 5 $\begin{array}{r} 12 \\ 8\overline{)\ 96} \\ \underline{-8} \\ 16 \\ \underline{-16} \end{array}$ Multiply $8 \times 2 = 16$ and write the product 16 under the 16.

Step 6 $\begin{array}{r} 12 \\ 8\overline{)\ 96} \\ \underline{-8} \\ 16 \\ \underline{-16} \\ 0 \end{array}$ Subtract $16 - 16 = 0$. Write 0.

0 ← The remainder is zero.

Finally, *check* your answer. The answer 12 is approximately equal to the original estimate 10. As a second check, multiply the divisor 8 and the quotient 12. Their product should be the original dividend number, $8 \times 12 = 96$, which is correct.

Practice this step-by-step division process by finding $112 \div 7$.

Our work is shown in **29**.

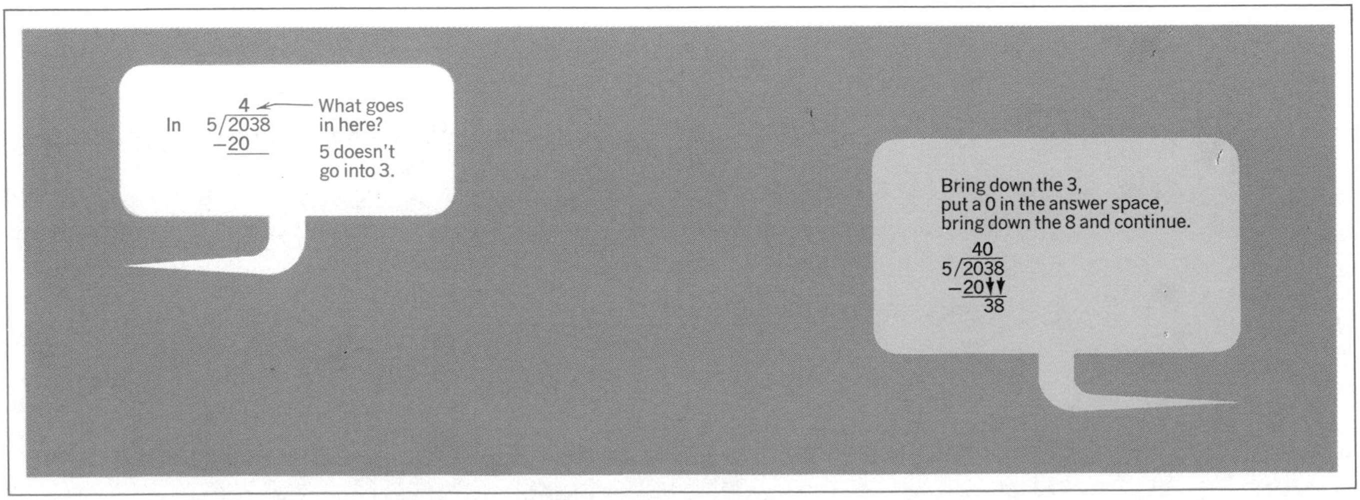

29 **Estimate:** 112 ÷ 7 is roughly 140 ÷ 7 or 20. The answer will be roughly 20.

```
    016
7) 112
  - 7↓
    42
  - 42
     0
```

Step 1 7 into 1. Won't go, so put a zero in the answer space.

Step 2 7 into 11. Once. Write 1 in the answer space as shown.

Step 3 $7 \times 1 = 7$. Subtract 7 from 11.

Step 4 $11 - 7 = 4$. Write 4. Bring down the 2.

Step 5 7 into 42. Six. Write 6 in the answer space.

Step 6 $7 \times 6 = 42$. Subtract 42. The remainder is zero.

Check: The answer 16 is roughly equal to our original estimate of 20. Check again by multiplying. $7 \times 16 = 112$. Finding a quick estimate of the answer will help you avoid making mistakes.

⇨ Notice that once the first digit of the answer has been obtained, there will be an answer digit for every digit of the dividend.

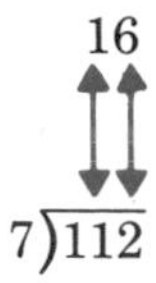

Another example: 203 ÷ 7 = ?

Estimate: 203 ÷ 7 is roughly 210 ÷ 7 or 30. The answer will be roughly 30.

```
    029
7) 203
  -14↓
    63
  - 63
     0
```

Step 1 7 into 2. Won't go, so put a zero in the answer space.

Step 2 7 into 20. Twice. Write 2 in the answer space as shown.

Step 3 $7 \times 2 = 14$. Subtract 14 from 20.

Step 4 $20 - 14 = 6$. Write 6. Bring down the 3.

Step 5 7 into 63. Nine times. Subtract 63. The remainder is zero.

Check: The answer is 29, roughly equal to the original estimate of 30. Double-check by multiplying: $7 \times 29 = 203$.

The remainder is not always zero of course. For example,

Write the answer as 38*r*1 to indicate that the quotient is 38 and the remainder is 1.

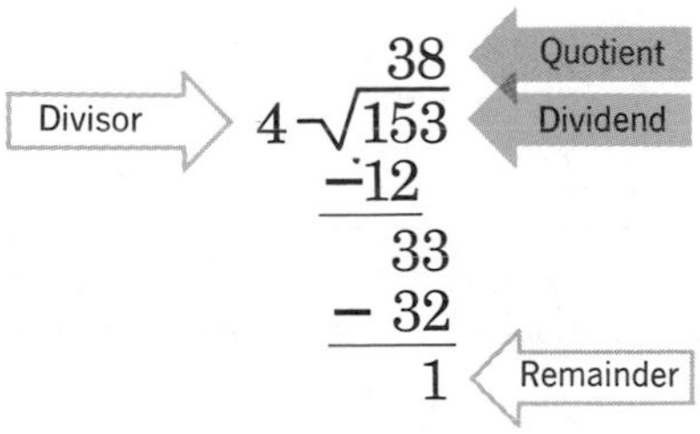

Ready for some guided practice? Try these problems:

(a) 3174 ÷ 6	(b) 206 ÷ 6	(c) 5084 ÷ 31
(d) 59 ÷ 8	(e) 341 ÷ 43	(f) 7528 ÷ 37

Check your work in **30**.

30 (a) $3174 \div 6 =$ ______

Estimate: $6 \times 500 = 3000$ The answer will be roughly 500.

```
     0529
 6) 3174
   -30
    17
   - 12
     54
   - 54
      0
```

Step 1 6 into 3? No. Write a zero. 6 into 31? 5 times. Write 5 above the 1.

Step 2 $6 \times 5 = 30$. Subtract $31 - 30 = 1$.

Step 3 Bring down 7. 6 into 17? Twice. Write 2 above the 7.

Step 4 $6 \times 2 = 12$. Subtract $17 - 12 = 5$.

Step 5 Bring down 4. 6 into 54? 9 times. Write 9 above the 4.

Step 6 $6 \times 9 = 54$. Subtract $54 - 54 = 0$.

Quotient = 529
The answer is 529.

Check: The estimate, 500, and the answer, 529, are roughly equal.

Double-check: $6 \times 529 = 3174$

(b) $206 \div 6 =$ ______

Estimate: $6 \times 30 = 180$ The answer is about 30.

```
     034
 6) 206
   -18
    26
  - 24
     2
```

Step 1 6 into 2? No. Write a zero. 6 into 20? Three times. Write 3 above the zero.

Step 2 $6 \times 3 = 18$. Subtract $20 - 18 = 2$.

Step 3 Bring down 6. 6 into 26? 4 times. Write 4 above the 6.

Step 4 $6 \times 4 = 24$. Subtract $26 - 24 = 2$. The remainder is 2.

Quotient = 34
The answer is 34 with a remainder of 2.

Check: The estimate 30 and the answer are roughly equal.

Double-check: $6 \times 34 = 204$ $204 + 2 = 206$

(c) $5084 \div 31 =$ ______

Estimate: This is roughly the same as $5000 \div 30$ or $500 \div 3$ or about 200. The quotient will be about 200.

```
       164
 31) 5084
    -31
     198
    -186
     124
   - 124
       0
```

Step 1 31 into 5? No. 31 into 50? Yes, once. Write 1 above the 0.

Step 2 $31 \times 1 = 31$. Subtract $50 - 31 = 19$.

Step 3 Bring down 8. 31 into 198? (That is, about the same as 3 into 19.) Yes, 6 times; write 6 above the 8.

Step 4 $31 \times 6 = 186$. Subtract $198 - 186 = 12$.

Step 5 Bring down 4. 31 into 124? (That is, about the same as 3 into 12.) Yes, 4 times. Write 4 above the 4.

Step 6 $31 \times 4 = 124$. Subtract $124 - 124 = 0$.

The quotient is 164.

Check: The estimate is reasonably close to the answer.

Double-check: $31 \times 164 = 5084$.

Notice that in Step 3 it is not at all obvious how many times 31 will go into 198. Again, you must make an educated guess and check your guess as you go along.

(d) $59 \div 8 =$ ______

Estimate: $7 \times 8 = 56$. The answer will be about 7.

$$\begin{array}{r} 7 \\ 8\overline{)\;59} \\ -56 \\ \hline 3 \end{array}$$

The quotient is 7 with a remainder of 3.

Check it, and double-check by multiplying and adding the remainder.

(e) $341 \div 43 =$ ______

Estimate: This is roughly $300 \div 40$ or $30 \div 4$ or about 7. The answer will be about 7 or 8.

$$\begin{array}{r} 7 \\ 43\overline{)\;341} \\ -301 \\ \hline 40 \end{array}$$

Your first guess would probably be that 43 goes into 341 8 times (try 4 into 34), but $43 \times 8 = 344$, which is larger than 341. The quotient is 7 with a remainder of 40.

Check it.

(f) $7528 \div 37 =$ ______

Estimate: This is roughly $7000 \div 30$ or $700 \div 3$ or about 200. The answer is roughly 200, or at least you know it is closer to 200 than to 20 or 2000.

$$\begin{array}{r} 203 \\ 37\overline{)\;7528} \\ -74\downarrow\downarrow \\ \hline 128 \\ -\;111 \\ \hline 17 \end{array}$$

On the second step notice that 37 does not go into 12. Write a zero in the answer space above the 2 and bring down the 8.

The quotient is 203 with a remainder of 17.

Check: The answer 203 and the estimate 200 are roughly equal. A careless person might get an answer of 23 and, if he had no estimate to compare it with, would not know it was incorrect.

Double-check: $37 \times 203 = 7511$ $\quad 7511 + 17 = 7528$

Now hop to **31** for a set of problems on the division of whole numbers.

---

31 **Exercises 1-6 Division of Whole Numbers**

A. Divide:

1. $63 \div 7 =$ ______
2. $92 \div 8 =$ ______
3. $72 \div 0 =$ ______
4. $37 \div 5 =$ ______
5. $71 \div 7 =$ ______
6. $6 \div 6 =$ ______
7. $\frac{32}{4} =$ ______
8. $\frac{28}{7} =$ ______
9. $\frac{54}{9} =$ ______
10. $245 \div 7 =$ ______

11. $167 \div 7 =$ ______

12. $228 \div 4 =$ ______

13. $310 \div 6 =$ ______

14. $3310 \div 3 =$ ______

15. $\frac{147}{7} =$ ______

16. $7\overline{)364}$ ______

17. $6\overline{)222}$ ______

18. $4\overline{)201}$ ______

19. $322 \div 14 =$ ______

20. $382 \div 19 =$ ______

21. $936 \div 24 =$ ______

22. $700 \div 28 =$ ______

23. $730 \div 81 =$ ______

24. $\frac{901}{17} =$ ______

25. $31\overline{)682}$ ______

B. Divide:

1. $61\overline{)7320}$ ______
2. $33\overline{)303}$ ______
3. $16\overline{)904}$ ______
4. $2001 \div 21 =$ ______
5. $2016 \div 21 =$ ______
6. $1000 \div 7 =$ ______
7. $2000 \div 9 =$ ______
8. $2400 \div 75 =$ ______
9. $14\overline{)4275}$ ______
10. $71\overline{)6005}$
11. $53\overline{)6307}$
12. $3\overline{)9003}$
13. $7\overline{)3507}$
14. $6\overline{)48009}$
15. $6\overline{)3624}$
16. $3\overline{)62160}$
17. $15\overline{)3000}$
18. $67\overline{)3354}$
19. $24\overline{)2596}$
20. $47\overline{)94425}$
21. $38\overline{)22800}$
22. $231\overline{)14091}$
23. $411\overline{)42020}$
24. $603\overline{)48843}$
25. $111\overline{)11111}$

C. Applied Problems

1. A machinist has a piece of bar stock 243″ long. If he must cut 9 equal pieces, how long will each piece be? (Assume no waste to get a first approximation.)
2. How many rafters 32 inches long can be cut from a piece of lumber 192 inches long?
3. If subflooring is laid at the rate of 85 sq ft per hour, how many hours will be needed to lay 1105 sq ft?
4. The illustration below shows a stringer for a short flight of stairs. Determine the missing dimensions H and W.

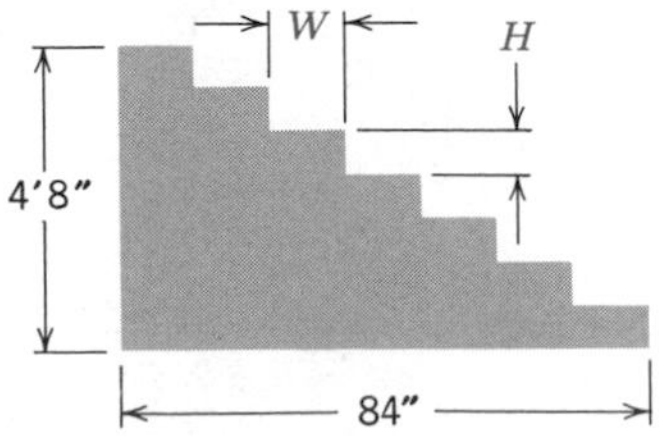

5. How many joists spaced 16″ o.c. (on center) are required for a floor 36′ long? (Add one joist for a starter.)

6. If rafters are placed 24″ o.c., how many are required for both sides of a gable roof that is 48′ long? Add one rafter on each side for a starter.

7. A stairway in a house is to be built with 18 risers. If the distance from the top of the first floor to the top of the second floor is 10′6″, what is the height of each step?

8. What is the average horsepower of six engines having the following HP ratings: 385, 426, 278, 434, 323, and 392? (**Hint:** To find the average, add the six numbers and divide their sum by six.)

9. Mr. Martinez, owner of the Maya Restaurant, hired a plumber who worked for 18 hours installing some new equipment. The total bill for wages and supplies used was \$274. If the cost for materials was \$76, calculate the plumber's hourly wage.

10. A truck traveling at the average rate of 54 miles per hour covered a distance of 486 miles in one run. Allowing 2 hours for refueling and meals, how long did it take to complete the trip?

D. Calculator Problems

1. Calculate the following:

(a) $\frac{1347 \times 46819}{3}$ (b) $\frac{76459 + 93008 + 255}{378}$

(c) $\frac{4008 + 408 + 48}{48}$ (d) $\frac{9909 \times 9090}{3303}$

2. In Problem 1 of the calculator problems on page 19, (a) calculate the miles per day for each truck if there were 22 working days in April. (Round to the nearest mile.) (b) Calculate the total cost (to the nearest dollar) of fuel used by the fleet of trucks if it costs 91¢ per gallon and the trucks averaged 12 miles to the gallon.

3. The specified shear strength of a $\frac{1}{8}$ inch 2117-T3 (AD) rivet is 344 lb. If it is necessary to assure a strength of 6587 lb to a riveted joint in an aircraft structure, how many rivets must be used?

4. A train of 74 railway cars weighed a total of 9,108,734 lb. What was the average weight of a car?

5. A high-speed stamping machine can produce small flat parts at the rate of 96 per minute. If an order calls for 297,600 parts to be stamped, how many hours will it take to complete the job? (60 minutes = 1 hour.)

6. Here are four problems from the national apprentice examination for painting and decorating contractors:

(a) 11,877,372 ÷ 738 (b) 87,445,005 ÷ 435
(c) 1,735,080 ÷ 760 (d) 206,703 ÷ 579

Check your answers on page 543.
Turn to **32** for a set of problems on the arithmetic of whole numbers, with many practical applications of this mathematics.

Where did the ÷ sign come from?
How do you know all this garbage?
Easy one. The fraction $\frac{a}{b}$ means $a \div b$ and ÷ looks like a fraction.
I read old Sears' catalogs.

1 Arithmetic of Whole Numbers

PROBLEM SET 1

32 Answers are on page 543.

Arithmetic of Whole Numbers

A. Perform the arithmetic as shown:

1. $24 + 69$	2. $38 + 45$	3. $456 + 72$
4. $43 + 817$	5. $396 + 538$	6. $2074 + 906$
7. $43 - 28$	8. $93 - 67$	9. $734 - 85$
10. $315 - 119$	11. $543 - 348$	12. $3401 - 786$
13. 67×21	14. 45×82	15. 207×63
16. 314×926	17. 5236×44	18. 4018×392
19. $42\overline{)2394}$	20. $34\overline{)2108}$	21. $1293 \div 160$
22. $11309 \div 263$	23. $\dfrac{1314}{73}$	24. $\dfrac{23 \times 51}{17}$
25. $\dfrac{29 \times 74}{13 \times 42}$		

B. Practical Problems

1. From a roll of number 12 wire, an electrician cut the following lengths: 6 ft, 8 ft, 20 ft, and 9 ft. How many feet of wire did he use?
2. A machine shop bought 14 steel rods of $\frac{7}{8}$-in. diameter steel, 23 rods of $\frac{1}{2}$-in. diameter, 8 rods of $\frac{1}{4}$-in. diameter, and 19 rods of 1 in. diameter. How many rods were purchased?
3. In estimating the floor area of a house, John listed the rooms as: living room 346 sq ft, dining room 210 sq ft, four bedrooms each 164 sq ft, two bathrooms each 96 sq ft, kitchen 208 sq ft, and the family room 280 sq ft. What is the total floor area of the house?
4. A metal casting weighs 680 lb, and 235 lb of metal are removed during shaping. What is its finished weight?
5. Roberto has 210 ft of #14 wire left on a roll. If he cuts it into pieces 35 ft long, how many pieces will he get?
6. How many hours will be required to install the flooring of a 2160 sq ft house if 90 sq ft can be installed in one hour?
7. The weights of seven cement platforms are: 210, 215, 245, 217, 220, 227, and 115 lb. What is the average weight per platform? (Add up and divide by 7.)
8. A room has 26 sq yd of floor space. If carpeting costs $13 per sq yd, what would it cost to carpet the room?
9. A portable TV can be bought for $25 down and 12 payments of $15 each. What is its total cost?
10. A 500 ft spool of #14-2TF wire is being used on an electrical wiring job. If 248 ft of wire are used on the job, how much wire is left on the spool?
11. Texas has more miles of highway than any other state. In a recent survey, Texas had a paved road mileage of 194,292; California was next with 126,327 miles, Illinois third with 125,559 miles, and Minnesota fourth with 117,810. What was the total paved road mileage for these four states?

Name ______________

Date ______________

Course/Section ______________

12. During the first three months of the year, the Write Rite Typewriter Company reported the following sales:
January $3572
February $2716
March $4247
What was their sales total for this quarter of the year?
13. A fire control pumping truck can move 156,000 gallons of water in 4 hours of continuous pumping. What is the flow rate (gallons per minute) for this truck?
14. A pile of lumber contains 170 boards 10 ft long, 118 boards 12 ft long, 206 boards 8 ft long, and 19 boards 16 ft long.
(a) How many boards are in the pile?
(b) How many linear feet of lumber are in the pile?
15. If the liquid pressure on a surface is 17 psi (pounds per square inch), what is the total force on a surface area of 167 sq in.?
16. A rule of thumb useful to auto mechanics is that a car with a pressure radiator cap will boil at a temperature of three times the cap rating plus 212. What temperature would a Ford LTD reach with a 17 psi cap?
17. If it takes 45 minutes to set up a job and cut the teeth on a gear blank, how many hours will be needed for a job that requires cutting 32 such gear blanks?
18. To construct a certain pipe system, the following material is needed:
6 elbows at 39¢ each
5 tees at 43¢ each
3 couplings at 27¢ each
4 pieces of pipe each 28 inches long
6 pieces of pipe each 24 inches long
4 pieces of pipe each 17 inches long
The pipe costs 18¢ per lineal foot. Find the total cost of the material for the system.
19. A gallon is a volume of 231 cubic inches. If a storage tank holds 380 gallons of oil, what is the volume of the tank in cubic inches?
20. The Ace Machine Tool Co. received an order for 15,500 flanges. If 2 dozen flanges are packed in a box, how many boxes are needed to ship the order?
21. At 8 a.m. the revolution counter on a diesel engine reads 460089. At noon it reads 506409. What is the average rate, in revolutions per minute, for the machine?

2 Fractions

PREVIEW 2

Objective	Sample Problems		Where To Go for Help	
Upon successful completion of this unit you will be able to:			Page	Frame
1. Work with fractions.	(a) Write as a mixed number $\frac{31}{4}$	= __________	43	1
	(b) Write as an improper fraction $3\frac{7}{8}$	= __________	45	6
	Write as an equivalent fraction (c) $\frac{5}{16} = \frac{?}{64}$	= __________	48	10
	(d) $1\frac{3}{4} = \frac{?}{32}$	= __________		
	(e) Reduce to lowest terms $\frac{10}{64}$	= __________	50	12
	(f) Which is larger, $1\frac{7}{8}$ or $\frac{5}{3}$?	= __________	52	15
2. Multiply and divide fractions.	(a) $\frac{7}{8} \times \frac{5}{32}$	= __________	54	18
	(b) $4\frac{1}{2} \times \frac{2}{3}$	= __________		
	(c) $\frac{3}{5}$ of $1\frac{1}{2}$	= __________	59	24
	(d) $\frac{3}{4} \div \frac{1}{2}$	= __________		
	(e) $2\frac{7}{8} \div 1\frac{1}{4}$	= __________		
	(f) $4 \div \frac{1}{2}$	= __________		
3. Add and subtract fractions.	(a) $1\frac{3}{16} + \frac{3}{4}$	= __________	64	29
	(b) $\frac{3}{4} - \frac{1}{5}$	= __________		

____
Name

____
Date

____
Course/Section

(c) $4 - 1\frac{5}{16}$ = ______________

(Answers to these problems are at the bottom of this page.)

If you are certain you can work *all* of these problems correctly, turn to page 75 for a set of practice problems. If you cannot work one or more of the preview problems, turn to the page indicated after the problem. Superstudents—those who want to be tops in their work—will turn to frame **1** and begin work there.

Answers to Preview 2

1. (a) $7\frac{3}{4}$ (b) $\frac{31}{8}$ (c) $\frac{20}{64}$ (d) $\frac{56}{32}$ (e) $\frac{5}{32}$ (f) $1\frac{7}{8}$
2. (a) $\frac{35}{256}$ (b) 3 (c) $\frac{9}{10}$ (d) $1\frac{1}{2}$ (e) $2\frac{3}{10}$ (f) 8
3. (a) $1\frac{15}{16}$ (b) $\frac{11}{20}$ (c) $2\frac{11}{16}$

2 Fractions

B.C. by permission of Johnny Hart and Field Enterprises, Inc.

2-1 WORKING WITH FRACTIONS

1 The word *fraction* comes from a Latin word meaning *to break,* and fraction numbers are used when we need to break down standard measuring units into smaller parts. Anyone doing practical mathematics problems will find that fractions are used in many different technical areas. And unfortunately fraction arithmetic *must* be done with pencil and paper, since most electronic calculators cannot handle fractions directly.

Look at the piece of lumber below. What happens when we break it into equal parts?

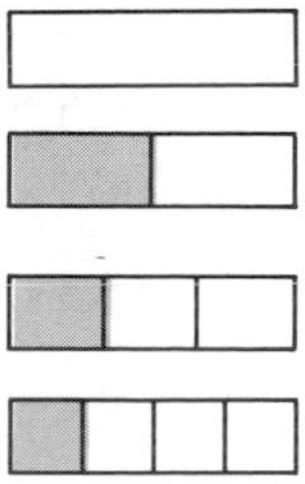

2 parts, each one-half of the whole

3 parts, each one-third of the whole

4 parts, each one-fourth of the whole

Divide this area ☐ into fifths by drawing vertical lines.

Try it, then go to **2**.

2 Notice that the five parts or "fifths" are equal in area. 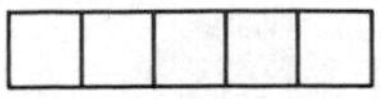

A fraction is normally written as the division of two whole numbers: $\frac{2}{3}$, $\frac{3}{4}$, or $\frac{9}{16}$. One of the five equal areas above would be "one-fifth" or $\frac{1}{5}$ of the entire area.

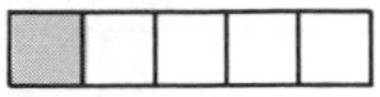

How would you label this portion of the area? 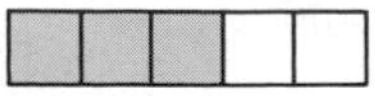 = ?

Continue in **3**.

---

3 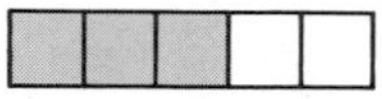 $\frac{3}{5} = \frac{3 \text{ shaded parts}}{5 \text{ total parts}}$

The fraction $\frac{3}{5}$ implies an area equal to three of the original portions.

$$\frac{3}{5} = 3 \times \left(\frac{1}{5}\right)$$

There are three equal parts and the name of each part is $\frac{1}{5}$ or one-fifth. In this collection of letters **HHHHPPT** what fraction are **H** s?

Count the total number of letters and decide what portion are **H** s.

Check your answer in **4**.

---

4 Fraction of **H** s $= \frac{\text{number of H s}}{\text{total number of letters}} = \frac{4}{7}$ (read it "four sevenths")

The fraction of **P**s is $\frac{2}{7}$, and the fraction of **T**s is $\frac{1}{7}$.

The two numbers that form a fraction are given special names to simplify talking about them. In the fraction $\frac{3}{5}$ the upper number 3 is called the *numerator* from the Latin *numero* meaning *number*. It is a count of the number of parts. The lower number 5 is called the *denominator* from the Latin *nomen* or *name*. It tells us the name of the part being counted. The numerator and denominator are called the *terms* of the fraction.

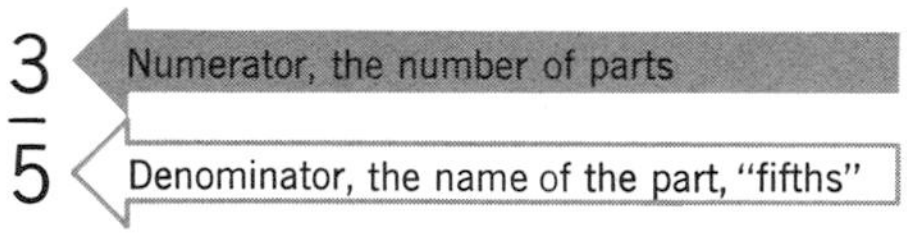

A textbook costs \$6 and I have \$5. What fraction of its cost do I have? Write the answer as a fraction.

________ numerator = ________, denominator = ________

Check your answer in **5**.

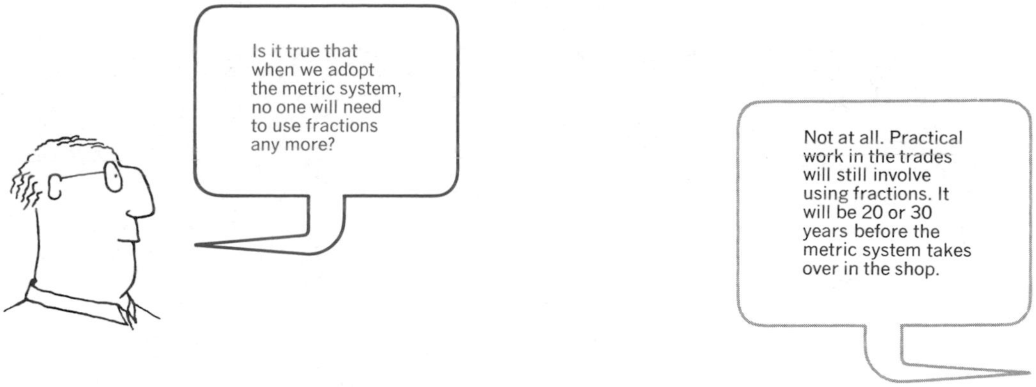

---

5 $\underbrace{\$\,\$\,\$\,\$\,\$}_{5}\,\$$ $\$5$ is $\frac{5}{6}$ of the total cost.

numerator = 5, denominator = 6

Complete these sentences by writing in the correct fraction.

(a) If we divide a length into eight equal parts, each part will be ________ of the total length.

(b) Then three of these parts will represent ________ of the total length.

(c) Eight of these parts will be ________ of the total length.

(d) Ten of these parts will be ________ of the total length.

Check your answers in **6**.

---

6 (a) $\frac{1}{8}$ (b) $\frac{3}{8}$ (c) $\frac{8}{8}$ (d) $\frac{10}{8}$

The original length is used as a standard, and any other length—smaller or larger—can be expressed as a fraction of it. A *proper fraction* is a number less than 1, as you would suppose a fraction should be. It represents a quantity less than the standard. For example, $\frac{1}{2}$, $\frac{2}{3}$, and $\frac{17}{20}$ are all proper fractions. Notice that for a proper fraction, the numerator is less than the denominator—the top number is less than the bottom number in the fraction.

An *improper fraction* is a number greater than 1 and represents a quantity greater than the standard. If a standard length is 8 inches, a length of 11 inches would be $\frac{11}{8}$ of the standard. Notice that for an improper fraction the numerator is greater than the denominator—top number greater than the bottom number in the fraction.

Circle the proper fractions in the following list.

$\frac{3}{2}$ $\frac{3}{4}$ $\frac{7}{8}$ $\frac{5}{4}$ $\frac{15}{12}$ $\frac{1}{16}$ $\frac{35}{32}$ $\frac{7}{50}$ $\frac{65}{64}$ $\frac{105}{100}$

Go to **7** when you have finished.

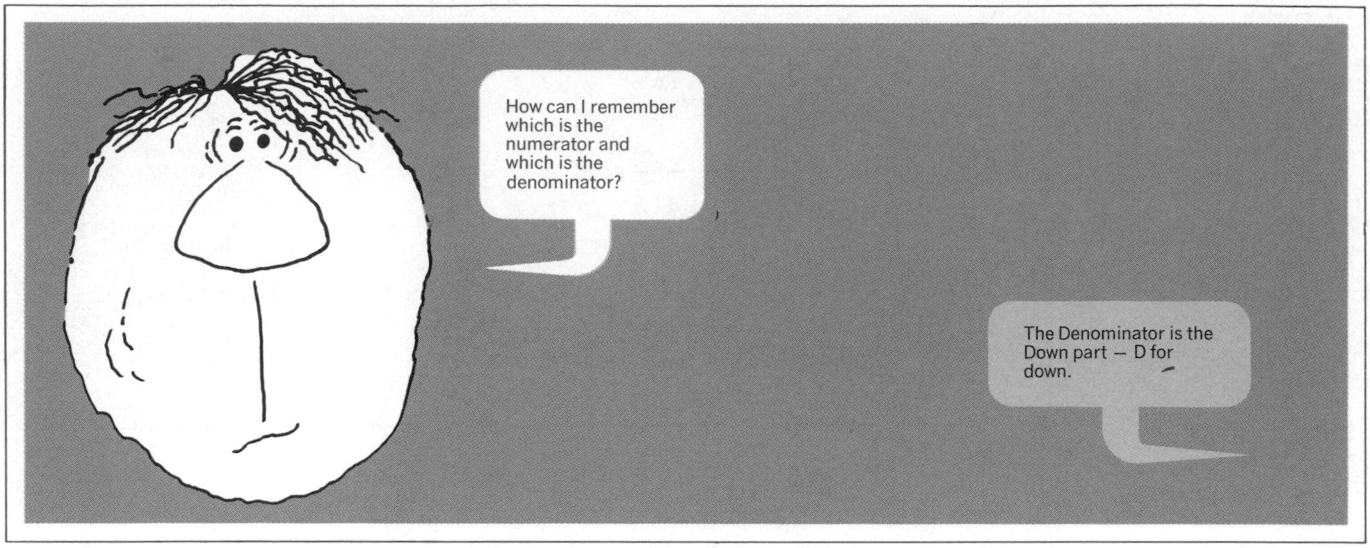

7 You should have circled the following proper fractions: $\frac{3}{4}, \frac{7}{8}, \frac{1}{16}, \frac{7}{50}$. In each fraction the numerator (top number) is less than the denominator (bottom number). Each of these fractions represents a number less than 1.

The improper fraction $\frac{7}{3}$ can be shown graphically as follows:

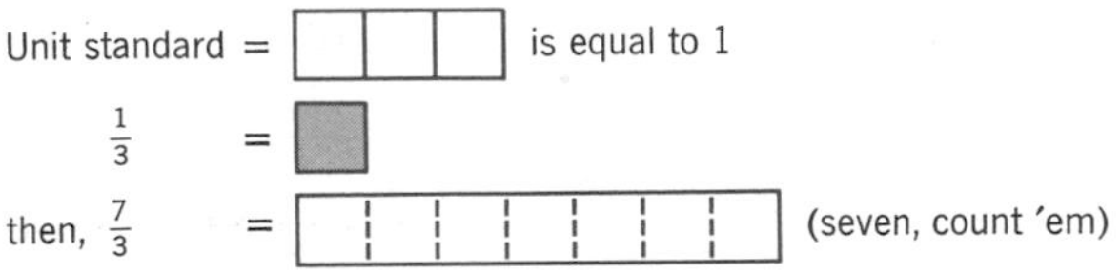

We can rename this number by regrouping.

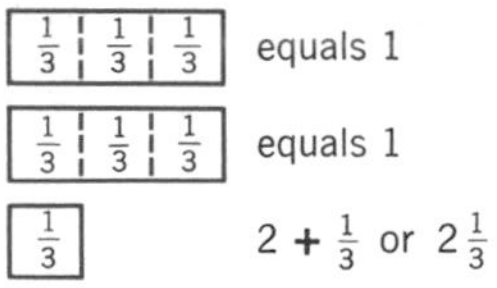

A *mixed number* is an improper fraction written as the sum of a whole number and a proper fraction.

$\frac{7}{3} = 2 + \frac{1}{3}$ or $2\frac{1}{3}$ We usually omit the + sign and write $2 + \frac{1}{3}$ as $2\frac{1}{3}$, and read it as "two and one-third." The numbers $1\frac{1}{2}$, $2\frac{3}{5}$, and $16\frac{2}{3}$ are all written as mixed numbers.

To write an improper fraction as a mixed number, divide numerator by denominator and form a new fraction as shown:

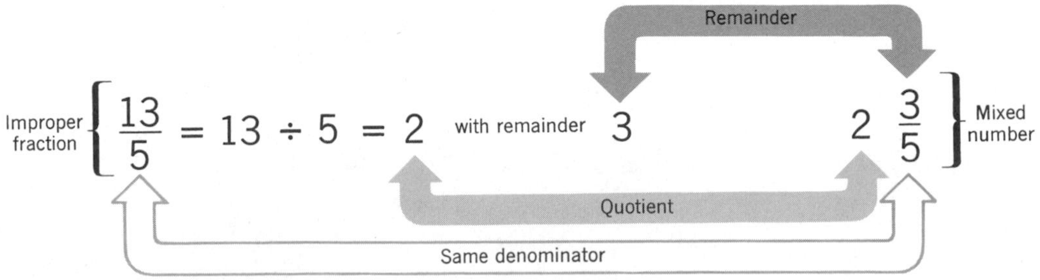

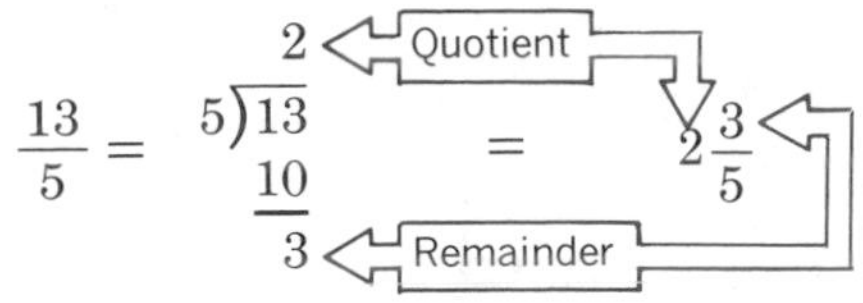

Now you try it. Rename $\frac{23}{4}$ as a mixed number. $\frac{23}{4}$ = ________

Follow the procedure shown above, then hop to **8**.

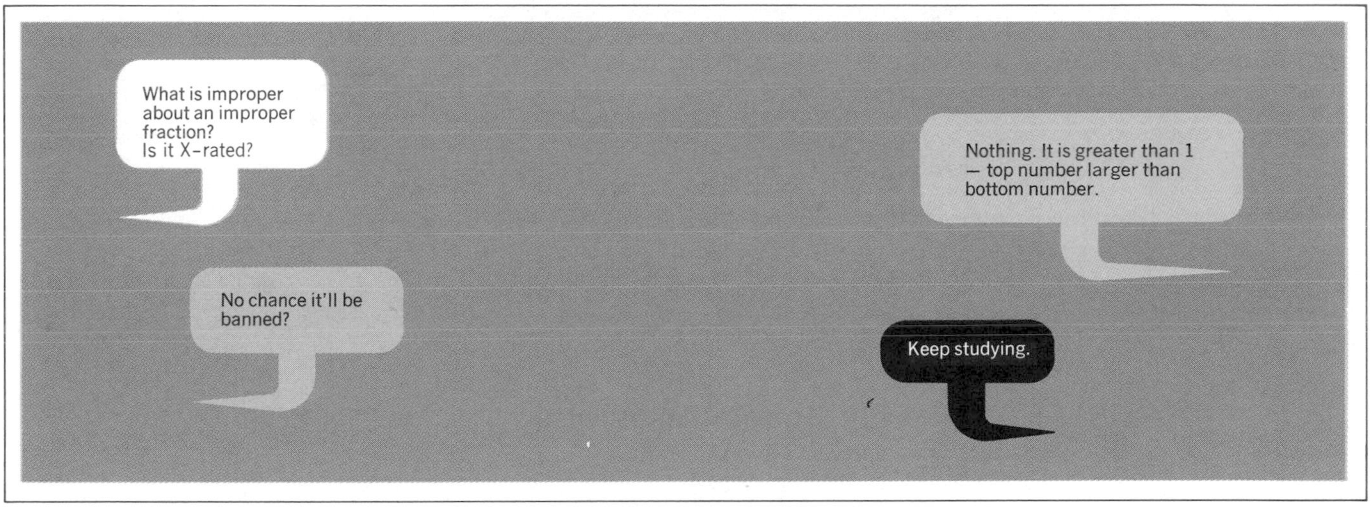

---

8 $\frac{23}{4} = 23 \div 4 = 5$ with remainder $3 \longrightarrow 5\frac{3}{4}$

If in doubt, check your work with a diagram like this:

Now try these for practice. Write each improper fraction as a mixed number.

(a) $\frac{9}{5}$ (b) $\frac{13}{4}$ (c) $\frac{27}{8}$ (d) $\frac{31}{4}$ (e) $\frac{41}{12}$ (f) $\frac{17}{2}$

The answers are in **9**.

---

9 (a) $\frac{9}{5} = 1\frac{4}{5}$ (b) $\frac{13}{4} = 3\frac{1}{4}$ (c) $\frac{27}{8} = 3\frac{3}{8}$

(d) $\frac{31}{4} = 7\frac{3}{4}$ (e) $\frac{41}{12} = 3\frac{5}{12}$ (f) $\frac{17}{2} = 8\frac{1}{2}$

The reverse process, rewriting a mixed number as an improper fraction, is equally simple.

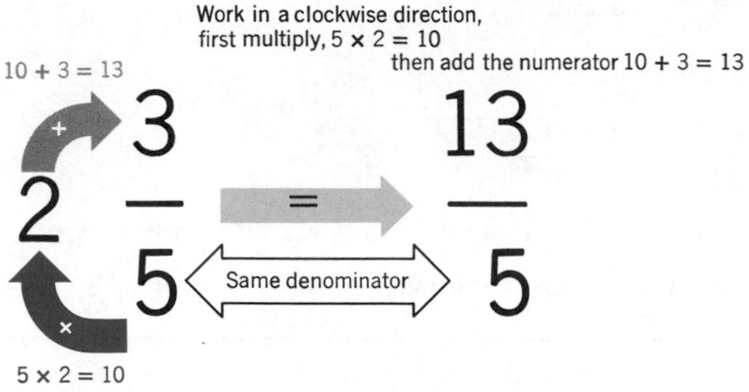

Graphically, $2\frac{3}{5} = 1 + 1 + \frac{3}{5}$

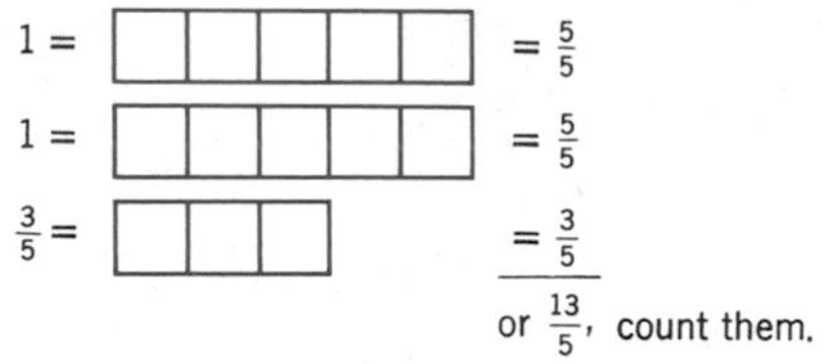

Now you try it. Rewrite these mixed numbers as improper fractions.

(a) $3\frac{1}{5}$ (b) $4\frac{3}{8}$ (c) $1\frac{1}{16}$ (d) $5\frac{1}{2}$ (e) $15\frac{3}{8}$ (f) $9\frac{3}{4}$

Check your answers in **10** .

---

10 (a) $3\frac{1}{5} = \frac{16}{5}$ (b) $4\frac{3}{8} = \frac{35}{8}$ (c) $1\frac{1}{16} = \frac{17}{16}$

(d) $5\frac{1}{2} = \frac{11}{2}$ (e) $15\frac{3}{8} = \frac{123}{8}$ (f) $9\frac{3}{4} = \frac{39}{4}$

Equivalent Fractions

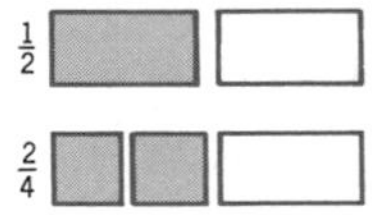

Two fractions are said to be *equivalent* if they represent the same number. For example, $\frac{1}{2} = \frac{2}{4}$ since both fractions represent the same portion of some standard amount.

There is a very large set of fractions equivalent to $\frac{1}{2}$.

$$\frac{1}{2} = \frac{2}{4} = \frac{3}{6} = \frac{4}{8} = \frac{5}{10} = \cdots = \frac{46}{92} = \frac{61}{122} = \frac{1437}{2874} \text{ and so on.}$$

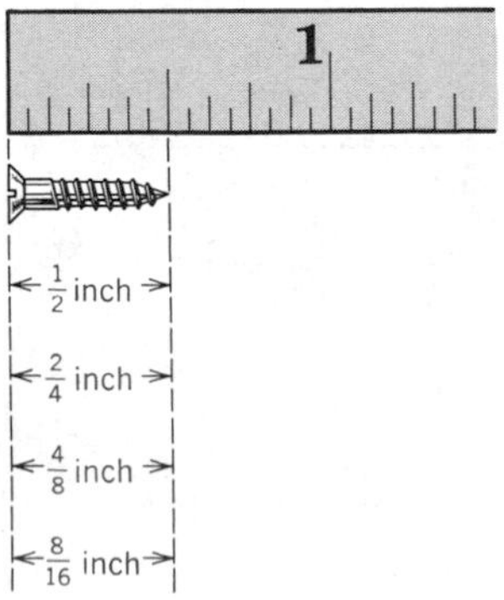

Each fraction is the same number, and we can use these fractions interchangeably in any mathematics problem.

To obtain a fraction equivalent to any given fraction, multiply the original numerator and denominator by the same nonzero number. For example,

Multiply top and bottom by 3

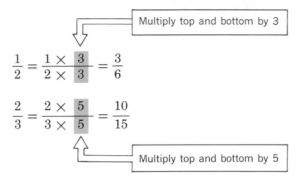

$$\frac{1}{2} = \frac{1 \times 3}{2 \times 3} = \frac{3}{6}$$

$$\frac{2}{3} = \frac{2 \times 5}{3 \times 5} = \frac{10}{15}$$

Multiply top and bottom by 5

Rewrite the fraction $\frac{3}{4}$ as an equivalent fraction with denominator equal to 20.

$$\frac{3}{4} = \frac{?}{20}$$

Check your work in **11**.

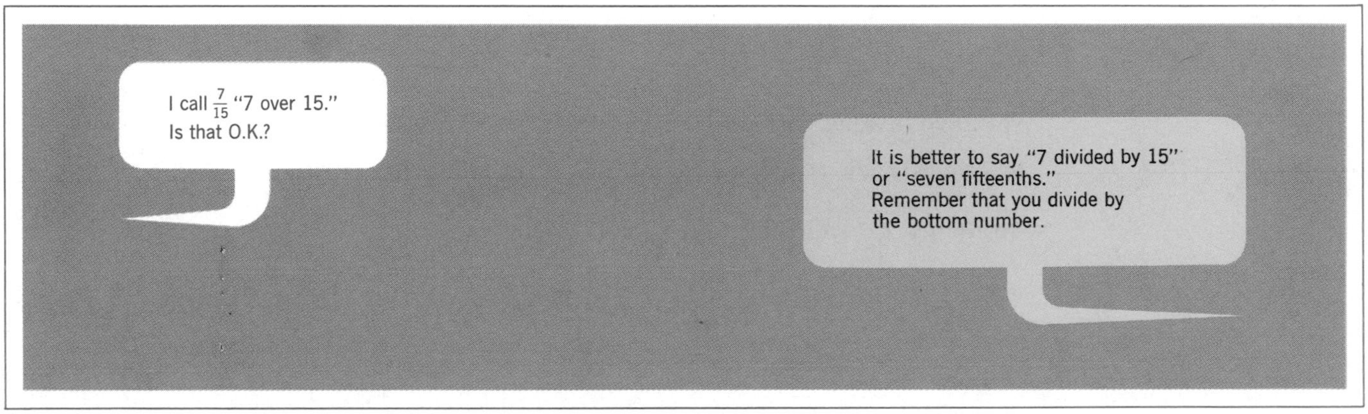
I call $\frac{7}{15}$ "7 over 15." Is that O.K.?
It is better to say "7 divided by 15" or "seven fifteenths." Remember that you divide by the bottom number.

---

11 $\frac{3}{4} = \frac{3 \times ?}{4 \times ?} = \frac{3 \times 5}{4 \times 5} = \frac{15}{20}$

$4 \times ? = 20$ so ? must be 5

The number value of the fraction has not changed; we have simply renamed it.

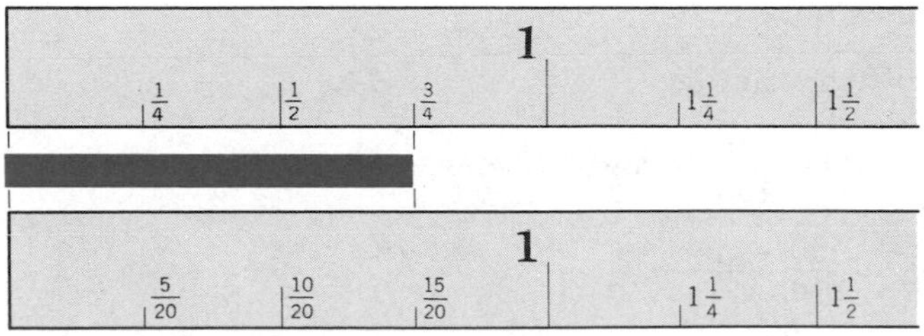
1
$\frac{1}{4}$
$\frac{1}{2}$
$\frac{3}{4}$
$1\frac{1}{4}$
$1\frac{1}{2}$
1
$\frac{5}{20}$
$\frac{10}{20}$
$\frac{15}{20}$
$1\frac{1}{4}$
$1\frac{1}{2}$

Practice with these.

(a) $\frac{5}{8} = \frac{?}{32}$ (b) $\frac{7}{16} = \frac{?}{48}$ (c) $1\frac{2}{3} = \frac{?}{12}$

Look in **12** for the answers.

12 (a) $\frac{5}{8} = \frac{5 \times 4}{8 \times 4} = \frac{20}{32}$ (b) $\frac{7}{16} = \frac{7 \times 3}{16 \times 3} = \frac{21}{48}$

(c) $1\frac{2}{3} = \frac{5}{3} = \frac{5 \times 4}{3 \times 4} = \frac{20}{12}$

Reducing to Lowest Terms

Very often in working with fractions you will be asked to *reduce a fraction to lowest terms.* This means to replace it with the most simple fraction in its set of equivalent fractions. To reduce $\frac{15}{30}$ to its lowest terms means to replace it by $\frac{1}{2}$. They are equivalent.

$$\frac{15}{30} = \frac{15 \div 15}{30 \div 15} = \frac{1}{2}$$

We have divided both the top and the bottom of the fraction by 15.

In general, you would reduce a fraction to lowest terms this way:

To reduce $\frac{30}{48}$ to lowest terms,

First, find the largest number that divides both top and bottom of the fraction evenly.

$30 = 5 \cdot 6$
$48 = 8 \cdot 6$

In this case, 6 is the largest number that divides both parts of the fraction evenly. The number 6 is called a *factor* of the number 30.

Second, eliminate this common factor by dividing.

$$\frac{30}{48} = \frac{30 \div 6}{48 \div 6} = \frac{5}{8}$$

The fraction $\frac{5}{8}$ is the simplest fraction equivalent to $\frac{30}{48}$. No number greater than 1 divides both 5 and 8 evenly.

Another example:

To reduce $\frac{90}{105}$ to lowest terms,

$$\frac{90}{105} = \frac{90 \div 15}{105 \div 15} = \frac{6}{7}$$

This process of eliminating a common factor is usually called *cancelling.* When you cancel a factor, you divide both top and bottom of the fraction by that number. We would write $\frac{90}{115}$ as $\frac{\overset{6}{\cancel{90}}}{\underset{7}{\cancel{115}}} = \frac{6}{7}$

Careful When you cancel, cancel only the same factor.

Reduce the following fractions to lowest terms.

(a) $\frac{6}{8}$ (b) $\frac{12}{16}$ (c) $\frac{2}{4}$ (d) $\frac{4}{12}$

(e) $\frac{15}{84}$ (f) $\frac{21}{35}$ (g) $\frac{12}{32}$ (h) $\frac{6}{64}$

The answers are in **13**.

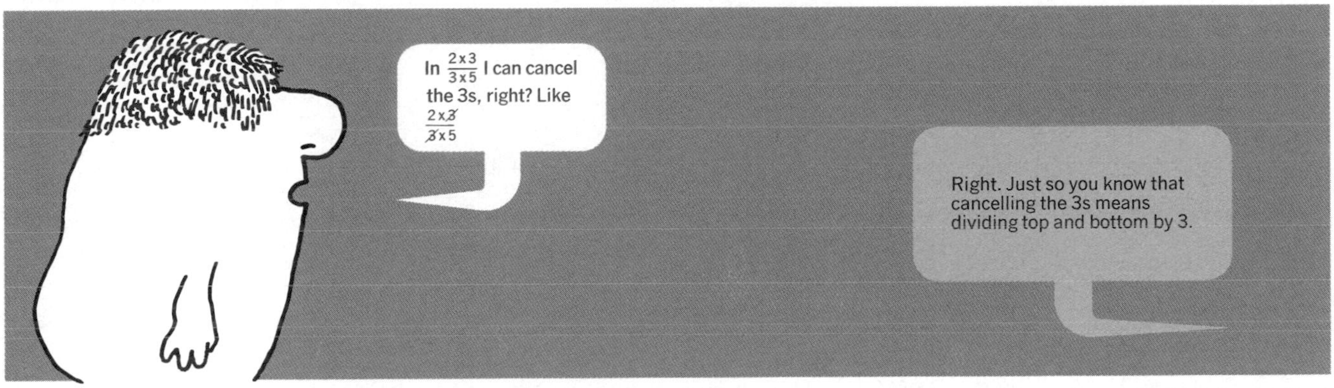

---

13 (a) $\frac{6}{8}=\frac{6\div 2}{8\div 2}=\frac{3}{4}$ (b) $\frac{12}{16}=\frac{12\div 4}{16\div 4}=\frac{3}{4}$

(c) $\frac{2}{4}=\frac{2\div 2}{4\div 2}=\frac{1}{2}$ (d) $\frac{4}{12}=\frac{4\div 4}{12\div 4}=\frac{1}{3}$

(e) $\frac{15}{84}=\frac{15\div 3}{84\div 3}=\frac{5}{28}$ (f) $\frac{21}{35}=\frac{21\div 7}{35\div 7}=\frac{3}{5}$

(g) $\frac{12}{32}=\frac{12\div 4}{32\div 4}=\frac{3}{8}$ (h) $\frac{6}{64}=\frac{6\div 2}{64\div 2}=\frac{3}{32}$

This one is a little tricky. Reduce $\frac{6}{3}$ to lowest terms. Check your answer in **14**.

---

14 $\frac{6}{3}=\frac{6\div 3}{3\div 3}=\frac{2}{1}$ or simply 2

Any whole number may be written as a fraction by using a denominator equal to 1.

$3=\frac{3}{1}$

$4=\frac{4}{1}$

and so on. Writing numbers in this way will be helpful when you learn to do arithmetic with fractions.

Comparing Fractions

If you were offered your choice between $\frac{2}{3}$ of a certain amount of money and $\frac{5}{8}$ of it, which would you choose? Which is the larger fraction, $\frac{2}{3}$ or $\frac{5}{8}$?

Can you decide? Try. Rewriting the fractions as equivalent fractions would help. The answer is in **15**.

---

15 To compare two fractions, rename each by changing them to equivalent fractions with the same denominator.

$$\frac{2}{3} = \frac{2 \times 8}{3 \times 8} = \frac{16}{24} \quad \text{and} \quad \frac{5}{8} = \frac{5 \times 3}{8 \times 3} = \frac{15}{24}$$

Now compare the new fractions: $\frac{16}{24}$ is greater than $\frac{15}{24}$.

Notice: (1) The new denominator is the product of the original ones ($24 = 8 \times 3$).

(2) Once both fractions are written with the same denominator, the one with the larger numerator is the larger fraction. (16 of the fractional parts is more than 15 of them.)

Which of the following quantities is the larger?

(a) $\frac{3}{4}$ in. and $\frac{5}{7}$ in.

(b) $\frac{7}{8}$ and $\frac{19}{21}$

(c) 3 and $\frac{40}{13}$

(d) $1\frac{7}{8}$ lb and $\frac{5}{3}$ lb

(e) $2\frac{1}{4}$ ft and $\frac{11}{5}$ ft

(f) $\frac{5}{16}$ and $\frac{11}{35}$

Check your answer in **16**.

---

16 (a) $\frac{3}{4} = \frac{21}{28}, \frac{5}{7} = \frac{20}{28}$; $\frac{21}{28}$ is larger than $\frac{20}{28}$ so $\frac{3}{4}$ in. is larger than $\frac{5}{7}$ in.

(b) $\frac{7}{8} = \frac{147}{168}, \frac{19}{21} = \frac{152}{168}$; $\frac{152}{168}$ is larger than $\frac{147}{168}$ so $\frac{19}{21}$ is larger than $\frac{7}{8}$.

(c) $3 = \frac{39}{13}$; $\frac{40}{13}$ is larger than $\frac{39}{13}$ so $\frac{40}{13}$ is larger than 3.

(d) $1\frac{7}{8} = \frac{15}{8} = \frac{45}{24}, \frac{5}{3} = \frac{40}{24}$; $\frac{45}{24}$ is larger than $\frac{40}{24}$ so $1\frac{7}{8}$ lb is larger than $\frac{5}{3}$ lb.

(e) $2\frac{1}{4} = \frac{9}{4} = \frac{45}{20}, \frac{11}{5} = \frac{44}{20}$; $\frac{45}{20}$ is larger than $\frac{44}{20}$ so $2\frac{1}{4}$ ft is larger than $\frac{11}{5}$ ft.

(f) $\frac{5}{16} = \frac{175}{560}, \frac{11}{35} = \frac{176}{560}$; $\frac{176}{560}$ is larger than $\frac{175}{560}$ so $\frac{11}{35}$ is larger than $\frac{5}{6}$.

Now turn to **17** for some practice in working with fractions.

---

17 **Exercises 2-1 Fractions**

A. Write as an improper fraction:

1. $2\frac{1}{3}$	2. $7\frac{1}{2}$	3. $8\frac{3}{8}$	4. $1\frac{1}{16}$	5. $2\frac{7}{8}$
6. 2	7. $2\frac{2}{3}$	8. $4\frac{3}{64}$	9. $4\frac{5}{6}$	10. $1\frac{13}{16}$

B. Write as a mixed number:

1. $\frac{17}{2}$	2. $\frac{8}{5}$	3. $\frac{11}{8}$	4. $\frac{40}{16}$	5. $\frac{3}{2}$

6. $\frac{11}{3}$ 7. $\frac{100}{6}$ 8. $\frac{4}{3}$ 9. $\frac{80}{32}$ 10. $\frac{5}{2}$

C. Reduce to lowest terms:

1. $\frac{12}{16}$ 2. $\frac{4}{6}$ 3. $\frac{6}{16}$ 4. $\frac{18}{4}$ 5. $\frac{4}{10}$

6. $\frac{35}{30}$ 7. $\frac{24}{30}$ 8. $\frac{10}{4}$ 9. $4\frac{3}{12}$ 10. $\frac{34}{32}$

D. Complete these:

1. $\frac{7}{8} = \frac{?}{16}$ 2. $\frac{3}{4} = \frac{?}{16}$ 3. $\frac{1}{8} = \frac{?}{64}$

4. $\frac{3}{8} = \frac{?}{64}$ 5. $1\frac{1}{4} = \frac{?}{16}$ 6. $2\frac{7}{8} = \frac{?}{32}$

7. $3\frac{3}{5} = \frac{?}{10}$ 8. $1\frac{1}{16} = \frac{?}{32}$ 9. $1\frac{40}{60} = \frac{?}{3}$

10. $4 = \frac{?}{6}$

E. Which is larger?

1. $\frac{3}{5}$ or $\frac{4}{7}$ 2. $\frac{3}{2}$ or $\frac{13}{8}$ 3. $1\frac{1}{2}$ or $1\frac{3}{7}$

4. $\frac{3}{4}$ or $\frac{13}{16}$ 5. $\frac{7}{8}$ or $\frac{5}{6}$ 6. $2\frac{1}{2}$ or $1\frac{11}{8}$

7. $1\frac{2}{5}$ or $\frac{6}{4}$ 8. $\frac{3}{16}$ or $\frac{25}{60}$ 9. $\frac{13}{5}$ or $\frac{5}{2}$

10. $3\frac{1}{2}$ or $2\frac{7}{4}$

F. Practical Problems

1. Tom, the apprentice carpenter, measured the length of a 2 by 4 as $15\frac{6}{8}$ in. Express this measurement in lowest terms.
2. An electrical light circuit in John's welding shop had a load of 2800 watts. He changed the circuit to ten 150-watt bulbs and six 100-watt bulbs. What fraction represents the comparison of the new load to the old load?
3. The numbers $\frac{22}{7}$, $\frac{19}{6}$, $\frac{47}{15}$, $\frac{25}{8}$, and $\frac{41}{13}$ are all reasonable approximations to the number π. Which is the largest approximation? Which is the smallest approximation?
4. Which is thicker, a $\frac{3}{16}''$ sheet of metal or a $\frac{13}{64}''$ fastener?
5. Is it possible to have a $\frac{7}{8}''$ pipe with an inside diameter of $\frac{29}{32}''$?
6. In the sheet metal part shown below the seven rivets are equally spaced. Write the distance between rivets as a mixed number.

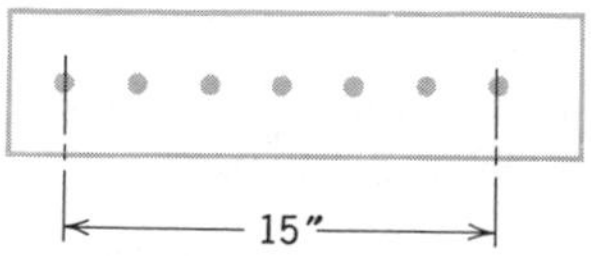

7. A printer has 15 rolls of newsprint in the warehouse. What fraction of this total will remain if 6 rolls are used?

8. A machinist who had been producing 40 parts per day increased the output to 60 parts per day by going to a faster machine. How much faster is the new machine? Express your answer as a mixed number.

When you have had the practice you need, turn to **18**.

2-2 MULTIPLICATION OF FRACTIONS

18 The simplest arithmetic operation with fractions is multiplication and, happily, it is easy to show graphically. The multiplication of a whole number and a fraction may be illustrated this way.

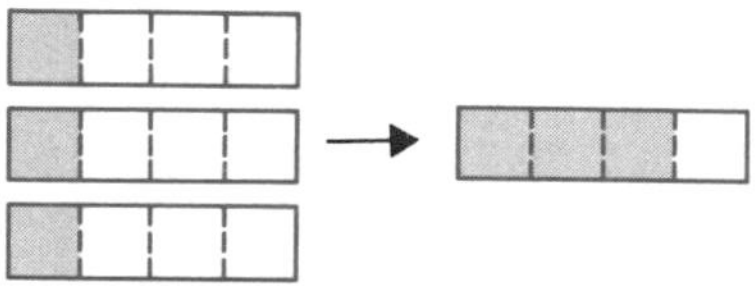

$$3 \times \frac{1}{4} = \frac{1}{4} + \frac{1}{4} + \frac{1}{4} = \frac{3}{4}$$

three segments each $\frac{1}{4}$ unit long.

Any fraction such as $\frac{3}{4}$ can be thought of as a product:

$$3 \times \frac{1}{4}$$

The product of two fractions can also be shown graphically. To find the product $\frac{1}{2} \times \frac{1}{3}$, first find $\frac{1}{2}$ of the area of a square.

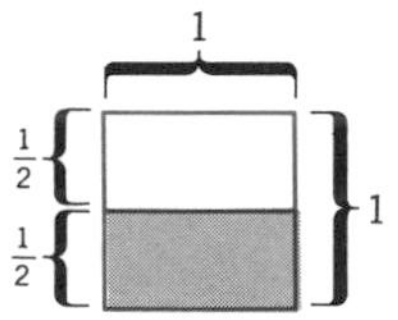

Then find $\frac{1}{3}$ of the area of the square.

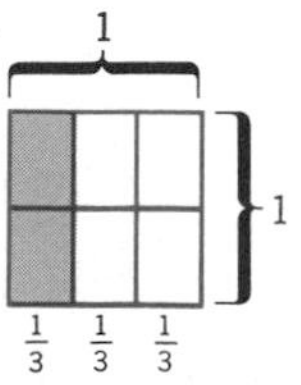

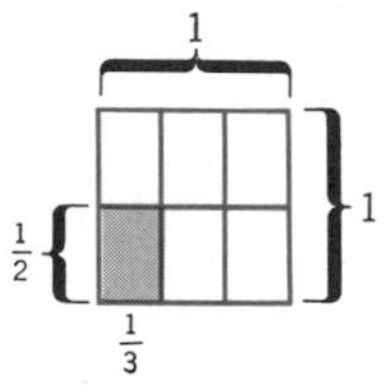

The product $\frac{1}{2} \times \frac{1}{3}$ is

$$\frac{1}{2} \times \frac{1}{3} = \frac{1}{6} = \frac{1 \text{ shaded area}}{6 \text{ equal areas in the square}}$$

$\frac{1}{2} \times \frac{1}{3}$ means $\frac{1}{2}$ of $\frac{1}{3}$

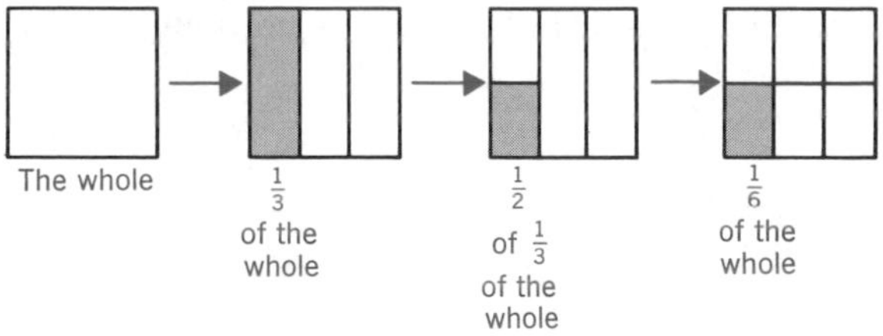

In general, we calculate this product as

$$\frac{1}{2}\times\frac{1}{3}=\frac{1\times 1}{2\times 3}$$

Multiply the numerators (top)

Multiply the denominators (bottom)

$$\frac{1}{2}\times\frac{1}{3}=\frac{1}{6}$$

The product of two fractions is a fraction whose numerator is the product of their numerators and whose denominator is the product of their denominators.

Multiply $\frac{5}{6}\times\frac{2}{3}$

(a) $\frac{5}{6}\times\frac{2}{3}=\frac{10}{18}$ — If you think this is correct go to frame **19**.

(b) $\frac{5}{6}\times\frac{2}{3}=\frac{5}{9}$ — If you think this is correct go to frame **20**.

(c) I don't know how to do it and I can't figure out how to draw the little boxes. — Go to **21**.

---

19 Good.

$$\frac{5}{6}\times\frac{2}{3}=\frac{5\times 2}{6\times 3}=\frac{10}{18}$$

Now reduce this answer to lowest terms and turn to **20**.

---

20 Excellent.

$$\frac{5}{6}\times\frac{2}{3}=\frac{5\times 2}{6\times 3}=\frac{10}{18}=\frac{5\times \cancel{2}}{9\times \cancel{2}}=\frac{5}{9}$$

Always reduce your answer to lowest terms. In this problem you probably recognized that both 10 and 18 were evenly divisible by 2, so you cancelled out that common factor. It will save you time and effort if you cancel common factors, such as the 2 above, *before* you multiply this way:

$$\frac{5\times \overset{1}{\cancel{2}}}{\underset{3}{\cancel{6}}\times 3}=\frac{5}{9}$$

Another kind of fraction multiplication problem:

$$3 \times \frac{1}{4} = \frac{3}{1} \times \frac{1}{4}$$
$$= \frac{3 \times 1}{1 \times 4}$$
$$= \frac{3}{4}$$

That was not really very different, was it?

Now test your understanding with these problems. Multiply as shown. Change any mixed numbers to improper fractions *before* you multiply.

(a) $\frac{7}{8} \times \frac{2}{3} =$ ______ (b) $\frac{8}{12} \times \frac{3}{16} =$ ______ (c) $\frac{3}{32} \times \frac{4}{15} =$ ______

(d) $\frac{15}{4} \times \frac{9}{10} =$ ______ (e) $\frac{3}{2} \times \frac{2}{3} =$ ______ (f) $1\frac{1}{2} \times \frac{2}{5} =$ ______

(g) $4 \times \frac{7}{8} =$ ______ (h) $3\frac{5}{6} \times \frac{3}{10} =$ ______ (i) $\frac{4}{3} \times \frac{3}{4} =$ ______

Remember to reduce your answer to lowest terms. The step-by-step answers are in **22**.

---

21 Now, now, don't panic. You needn't draw the little boxes to do the calculation. Try it this way:

$$\frac{5}{6} \times \frac{2}{3} = \frac{5 \times 2}{6 \times 3}$$

Multiply the numerators

Multiply the denominators

Finish the calculation and then return to **18** and choose an answer.

---

22 (a) $\frac{7}{8} \times \frac{2}{3} = \frac{7 \times \cancel{2}}{(4 \times \cancel{2}) \times 3} = \frac{7}{4 \times 3} = \frac{7}{12}$

Eliminate common factors before you multiply.

Your work will look like this when you learn to do these operations mentally:

$$\frac{7}{\underset{4}{\cancel{8}}} \times \frac{\overset{1}{\cancel{2}}}{3} = \frac{7}{12}$$

(b) $\frac{8}{12} \times \frac{3}{16} = \frac{\overset{1}{\cancel{8}} \times \overset{1}{\cancel{3}}}{(4 \times \underset{1}{\cancel{3}}) \times (\underset{1}{\cancel{8}} \times 2)} = \frac{1}{4 \times 2} = \frac{1}{8}$

or $\frac{\overset{1}{\cancel{8}}}{\underset{4}{\cancel{12}}} \times \frac{\overset{1}{\cancel{3}}}{\underset{2}{\cancel{16}}} = \frac{1}{8}$

(c) $\frac{\overset{1}{\cancel{3}}}{\underset{8}{\cancel{32}}} \times \frac{\overset{1}{\cancel{4}}}{\underset{5}{\cancel{15}}} = \frac{1}{40}$ (d) $\frac{\overset{3}{\cancel{15}}}{4} \times \frac{9}{\underset{2}{\cancel{10}}} = \frac{27}{8} = 3\frac{3}{8}$

(e) $\frac{\overset{1}{\cancel{3}}}{\underset{1}{\cancel{2}}} \times \frac{\overset{1}{\cancel{2}}}{\underset{1}{\cancel{3}}} = 1$

(f) $1\frac{1}{2} \times \frac{2}{5} = \frac{3}{\underset{1}{\cancel{2}}} \times \frac{\overset{1}{\cancel{2}}}{5} = \frac{3}{5}$

If you don't remember how to change a mixed number to an improper fraction, see frame **9**.

(g) $4 \times \frac{7}{8} = \frac{\overset{1}{\cancel{4}}}{1} \times \frac{7}{\underset{2}{\cancel{8}}} = \frac{7}{2} = 3\frac{1}{2}$

(h) $3\frac{5}{6} \times \frac{3}{10} = \frac{23}{\underset{2}{\cancel{6}}} \times \frac{\overset{1}{\cancel{3}}}{10} = \frac{23}{20} = 1\frac{3}{20}$

(i) $\frac{\overset{1}{\cancel{4}}}{\underset{1}{\cancel{3}}} \times \frac{\overset{1}{\cancel{3}}}{\underset{1}{\cancel{4}}} = 1$

Now turn to **23** for a set of practice problems.

---

23 **Exercises 2-2 Multiplication of Fractions**

A. Multiply and reduce the answer to lowest terms:

1. $\frac{1}{2} \times \frac{1}{4}$ 2. $\frac{2}{5} \times \frac{2}{3}$ 3. $\frac{4}{5} \times \frac{1}{6}$ 4. $6 \times \frac{1}{2}$

5. $\frac{8}{9} \times 3$ 6. $\frac{11}{12} \times \frac{4}{15}$ 7. $\frac{8}{3} \times \frac{5}{12}$ 8. $\frac{7}{8} \times \frac{13}{14}$

9. $\frac{12}{8} \times \frac{15}{9}$ 10. $\frac{4}{7} \times \frac{49}{2}$ 11. $4\frac{1}{2} \times \frac{2}{3}$ 12. $6 \times 1\frac{1}{3}$

13. $2\frac{1}{6} \times 1\frac{1}{2}$ 14. $\frac{5}{7} \times 1\frac{7}{15}$ 15. $4\frac{3}{5} \times 15$ 16. $10\frac{5}{6} \times 3\frac{3}{10}$

17. $34 \times 2\frac{3}{17}$ 18. $7\frac{9}{10} \times 1\frac{1}{4}$ 19. $11\frac{6}{7} \times \frac{7}{8}$

20. $18 \times 1\frac{5}{27}$ 21. $\frac{1}{2} \times \frac{1}{2} \times \frac{1}{2}$ 22. $1\frac{4}{5} \times \frac{2}{3} \times \frac{1}{4}$

23. $\frac{1}{4} \times \frac{2}{3} \times \frac{2}{5}$ 24. $2\frac{1}{2} \times \frac{3}{5} \times \frac{8}{9}$ 25. $\frac{2}{3} \times \frac{3}{2} \times 2$

B. Multiply (*Hint:* "Of" means "multiply.")

1. $\frac{1}{2}$ of $\frac{1}{3}$ = ________ 2. $\frac{1}{4}$ of $\frac{3}{8}$ = ________

3. $\frac{2}{3}$ of $\frac{3}{4}$ = ________ 4. $\frac{7}{8}$ of $\frac{1}{2}$ = ________

5. $\frac{1}{2}$ of $1\frac{1}{2}$ = ________

6. $\frac{3}{4}$ of $1\frac{1}{4}$ = ________

7. $\frac{5}{8}$ of $2\frac{1}{10}$ = ________

8. $\frac{5}{3}$ of $1\frac{2}{3}$ = ________

9. $\frac{4}{3}$ of $\frac{3}{4}$ = ________

10. $\frac{2}{5}$ of $1\frac{1}{4}$ = ________

C. Practical Problems

1. Find the width of floor space covered by 38 boards with $3\frac{5}{8}$-inch exposed surface each.

2. There are 14 risers in the stairs from the basement to the first floor of a house. Find the distance if the risers are $7\frac{1}{8}$ inches high.

3. Shingles are laid so that 5 inches are exposed in each layer. How many feet of roof will be covered by 28 courses?

4. A board $5\frac{3}{4}$ inches wide is cut to three-quarters of its original width. Find the new width.

5. **What length of 2″ x 4″ material will be required to make 6 bench legs each 2′$4\frac{1}{4}$″ long?**

6. Find the total length of 12 pieces of wire each $9\frac{3}{16}$ in. long.

7. If a car averages $22\frac{4}{5}$ miles to a gallon of gas, how many miles can it travel on 14 gallons of gas?

8. What is the shortest bar that can be used for making six chisels each $6\frac{1}{8}$ in. in length?

9. How many pounds of grease are contained in a barrel if a barrel holds $46\frac{1}{2}$ gallons, and a gallon of grease weighs $7\frac{2}{3}$ lb?

10. An electrician cut eight pieces of copper tubing from a coil. Each piece is $14\frac{3}{4}$ ft long. What is the total length used?

11. How far will a nut advance if it is given 18 turns on a $\frac{1}{4}$-in. 24-NF (National Fine thread) bolt? (*Hint:* The first number, $\frac{1}{4}$ in., gives the diameter of the bolt. The second number gives the width of the groove, or the pitch, of the bolt. On a 24-NF thread, the nut advances $\frac{1}{24}$ in. for each complete turn. On a 16-NF thread, the nut advances $\frac{1}{16}$ in. for each turn.)

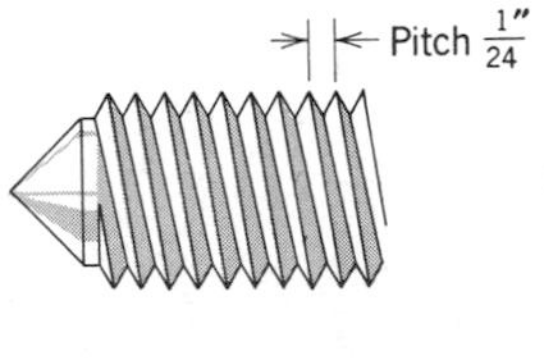

12. What width of floor space can be covered by 48 boards each with $4\frac{3}{8}$ in. of exposed surface?

13. If $\frac{3}{8}$ in. on a drawing represents 1 ft, how many inches on the drawing will represent 26 ft?

14. What is the volume of a rectangular box with interior dimensions $12\frac{1}{2}$ in. long, $8\frac{1}{8}$ in. wide, and $4\frac{1}{4}$ in. deep? (*Hint:* Multiply these dimensions.)

15. How long will it take to machine 45 pins if each pin requires $6\frac{3}{4}$ minutes? Allow one minute per pin for placing stock in the lathe.

16. There are 231 cu in. in a gallon. How many cu in. are needed to fill a container with a rated capacity of $4\frac{1}{3}$ gallons?

17. In a print shop, one unit of labor is equal to $\frac{1}{6}$ hr. How many hours are involved in 64 units of work?

18. Find the total width of 36 two by fours if the finished width of each board is actually $3\frac{1}{2}$ in.

19. A photograph must be reduced to $\frac{4}{5}$ of its original size to fit the space available in a newspaper. Find the length of the reduced photograph if the original was $6\frac{3}{4}''$ long.

Check your answers on page 544 then continue in **24**.

2-3 DIVISION OF FRACTIONS

24 Addition and multiplication are both reversible arithmetic operations. For example,

2×3 and 3×2 both equal 6
$4 + 5$ and $5 + 4$ both equal 9

The order in which you add or multiply is not important. This reversibility is called the commutative property of addition and multiplication.

$8 \div 4$ or $\frac{8}{4}$

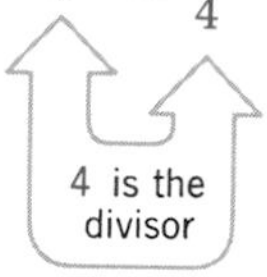

In division this kind of exchange is not allowed, and because it is not allowed many people find division very troublesome. In the division of fractions it is particularly important that you set up the process correctly.

The phrase "8 divided by 4" can be written as $8 \div 4$ or $\frac{8}{4}$.

In this problem you are being asked to divide a set of 8 objects into sets of 4 objects.

In the division $5 \div \frac{1}{2}$, which number is the divisor?

Check your answer in **25**.

---

25 The divisor is $\frac{1}{2}$.

The division $5 \div \frac{1}{2}$ is read "5 divided by $\frac{1}{2}$," and it asks how many $\frac{1}{2}$-unit lengths are included in a length of 5 units.

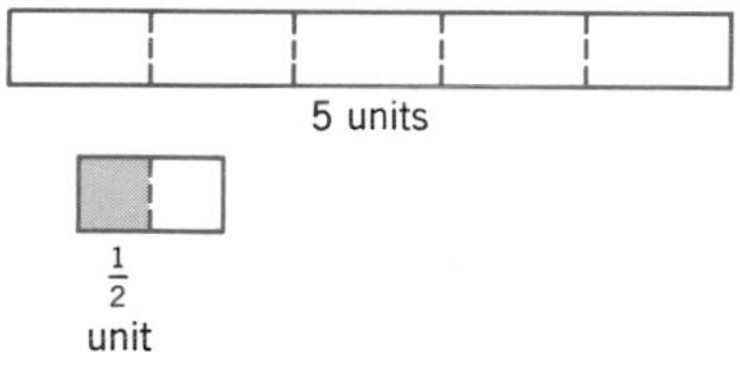

Division answers the question, "How many of the divisor are in the dividend?"

$8 \div 4 = \square$ asks you to find how many fours are in 8. It is easy to see that $\square = 2$.

$5 \div \frac{1}{2} = \square$ asks you to find how many $\frac{1}{2}$s are in 5. Do you see that $\square = 10$?

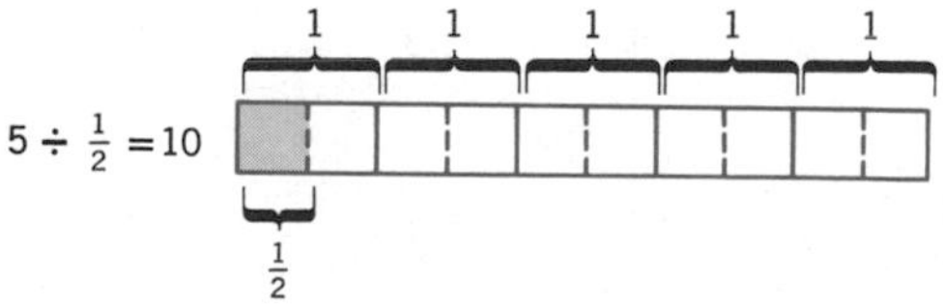

There are ten $\frac{1}{2}$-unit lengths contained in the 5-unit length.

Using a drawing of this sort to solve a division problem is difficult and clumsy. We need a simple rule. Here it is:

> **To divide by a fraction, invert the divisor and multiply.**

For example,

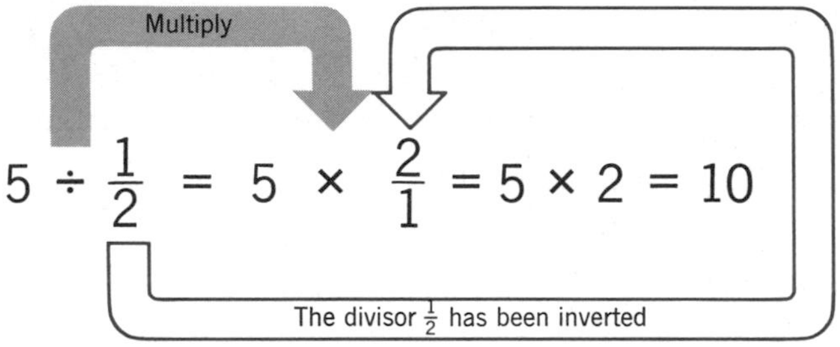

Another example is,

$$\frac{3}{5} \div \frac{2}{3} = \frac{3}{5} \times \frac{3}{2} = \frac{3 \times 3}{5 \times 2} = \frac{9}{10}$$

We have converted a division problem that is difficult to picture into a simple multiplication.

The final, and very important, step in every division is checking the answer. To check, multiply the divisor and the quotient and compare this answer with the original fraction.

If $\frac{3}{5} \div \frac{2}{3} = \frac{9}{10}$ Then $\frac{2}{3} \times \frac{9}{10}$ should equal $\frac{3}{5}$

$$\frac{3}{5} \div \frac{2}{3} = \frac{9}{10}$$

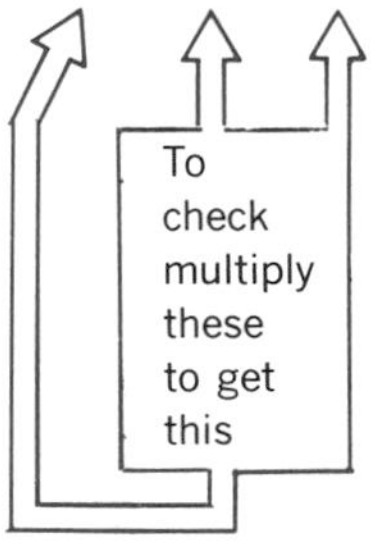

Check: $\frac{\overset{1}{\cancel{2}}}{\underset{1}{\cancel{3}}} \times \frac{\overset{3}{\cancel{9}}}{\underset{5}{\cancel{10}}} = \frac{1 \times 3}{1 \times 5} = \frac{3}{5}$

You try it:

$\frac{7}{8} \div \frac{3}{2} =$ ________

Solve by inverting the divisor and multiplying.
Check your work in **26** .

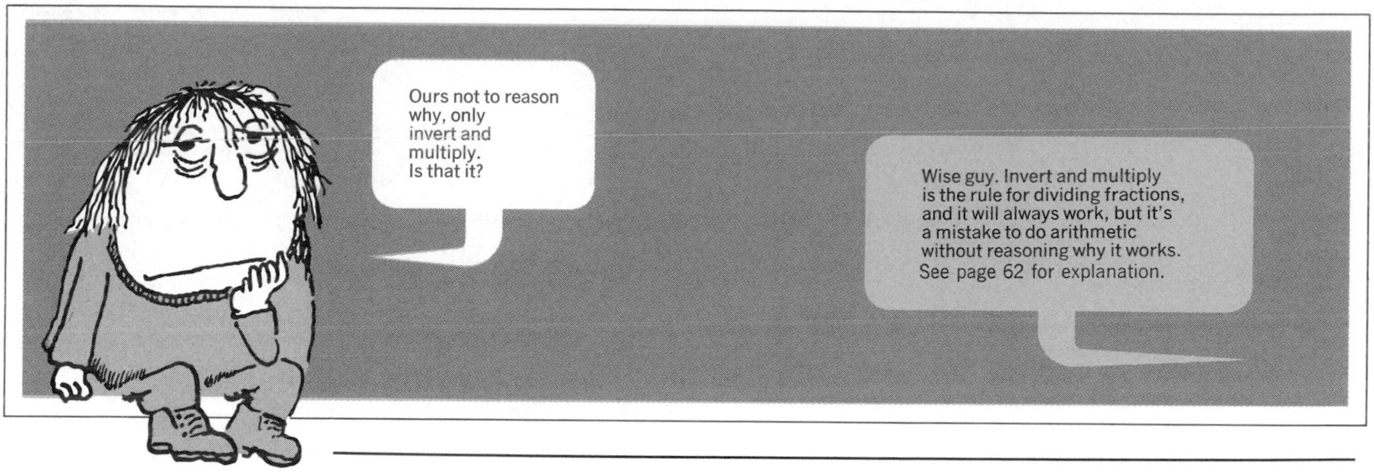

26 Invert

$$\frac{7}{8} \div \frac{3}{2} = \frac{7}{8} \times \frac{2}{3} = \frac{7 \times \overset{1}{\cancel{2}}}{\underset{4}{\cancel{8}} \times 3} = \frac{7}{12}$$

Multiply

Check: $\frac{\overset{1}{\cancel{3}}}{2} \times \frac{7}{\underset{4}{\cancel{12}}} = \frac{7}{8}$

The chief source of confusion for many people in dividing fractions is deciding which fraction to invert. It will help if you

(1) put every division problem in the form

(dividend) ÷ (divisor)

first, then invert the divisor, and finally multiply to obtain the quotient, and

(2) check your answer by multiplying. The product

(divisor) × (quotient or answer)

should equal the dividend.

Here are a few problems to test your understanding.

(a) $\frac{2}{5} \div \frac{3}{8} =$ ________ (b) $\frac{7}{40} \div \frac{21}{25} =$ ________

(c) $3\frac{3}{4} \div \frac{5}{2} =$ ________ (d) $4\frac{1}{5} \div 1\frac{4}{10} =$ ________

(e) $3\frac{2}{3} \div 3 =$ ________ (f) Divide $\frac{3}{4}$ by $2\frac{5}{8}$ ________

(g) Divide 8 by $\frac{1}{2}$ ________ (h) Divide $1\frac{1}{4}$ by $1\frac{7}{8}$ ________

Work carefully, check each answer, then turn to **27** for our worked solutions.

WHY DO WE INVERT AND MULTIPLY WHEN WE DIVIDE FRACTIONS?

The division $8 \div 4$ can be written $\frac{8}{4}$. Similarly, $\frac{1}{2} \div \frac{2}{3}$ can be written $\dfrac{\frac{1}{2}}{\frac{2}{3}}$

To simplify this fraction, multiply by $\dfrac{\frac{3}{2}}{\frac{3}{2}}$ (which is equal to 1).

$$\frac{\frac{1}{2}}{\frac{2}{3}} = \frac{\frac{1}{2} \times \frac{3}{2}}{\frac{2}{3} \times \frac{3}{2}} = \frac{\frac{1}{2} \times \frac{3}{2}}{1} = \frac{1}{2} \times \frac{3}{2}$$

$$\frac{2}{3} \times \frac{3}{2} = \frac{2 \times 3}{3 \times 2} = \frac{6}{6} = 1$$

Therefore $\frac{1}{2} \div \frac{2}{3} = \frac{1}{2} \times \frac{3}{2}$. We have inverted the fraction $\frac{2}{3}$ and multiplied by it.

---

27 (a) $\frac{2}{5} \div \frac{3}{8} = \frac{2}{5} \times \frac{8}{3} = \frac{16}{15} = 1\frac{1}{15}$

The answer is $1\frac{1}{15}$.

Check: $\frac{\cancel{3}^{1}}{\cancel{8}_{1}} \times \frac{\cancel{16}^{2}}{\cancel{15}_{5}} = \frac{2}{5}$

(b) $\frac{7}{40} \div \frac{21}{25} = \frac{\cancel{7}^{1}}{\cancel{40}_{8}} \times \frac{\cancel{25}^{5}}{\cancel{21}_{3}} = \frac{5}{24}$

Check: $\frac{\cancel{21}^{7}}{\cancel{25}_{5}} \times \frac{\cancel{5}^{1}}{\cancel{24}_{8}} = \frac{7}{40}$

(c) $3\frac{3}{4} \div \frac{5}{2} = \frac{15}{4} \div \frac{5}{2} = \frac{\cancel{15}^{3}}{\cancel{4}_{2}} \times \frac{\cancel{2}^{1}}{\cancel{5}_{1}} = \frac{3}{2} = 1\frac{1}{2}$

Check: $\frac{5}{2} \times \frac{3}{2} = \frac{15}{4} = 3\frac{3}{4}$

(d) $4\frac{1}{5} \div 1\frac{4}{10} = \frac{21}{5} \div \frac{14}{10} = \frac{\cancel{21}^{3}}{\cancel{5}_{1}} \times \frac{\cancel{10}^{2}}{\cancel{14}_{2}} = \frac{6}{2} = 3$

Check: $1\frac{4}{10} \times 3 = \frac{\cancel{14}^{7}}{\cancel{10}_{5}} \times \frac{3}{1} = \frac{21}{5} = 4\frac{1}{5}$

(e) $3\frac{2}{3} \div 3 = \frac{11}{3} \div \frac{3}{1} = \frac{11}{3} \times \frac{1}{3} = \frac{11}{9} = 1\frac{2}{9}$

Check: $3 \times \frac{11}{9} = \frac{\overset{1}{\cancel{3}}}{1} \times \frac{11}{\underset{3}{\cancel{9}}} = \frac{11}{3} = 3\frac{2}{3}$

(f) $\frac{3}{4} \div 2\frac{5}{8} = \frac{3}{4} \div \frac{21}{8} = \frac{\overset{1}{\cancel{3}}}{\underset{1}{\cancel{4}}} \times \frac{\overset{2}{\cancel{8}}}{\underset{7}{\cancel{21}}} = \frac{2}{7}$

Check: $2\frac{5}{8} \times \frac{2}{7} = \frac{\overset{3}{\cancel{21}}}{\underset{4}{\cancel{8}}} \times \frac{\overset{1}{\cancel{2}}}{\underset{1}{\cancel{7}}} = \frac{3}{4}$

(g) $8 \div \frac{1}{2} = \frac{8}{1} \times \frac{2}{1} = 16$

Check: $\frac{1}{\underset{1}{\cancel{2}}} \times \overset{8}{\cancel{16}} = 8$

(h) $1\frac{1}{4} \div 1\frac{7}{8} = \frac{5}{4} \div \frac{15}{8} = \frac{\overset{1}{\cancel{5}}}{\underset{1}{\cancel{4}}} \times \frac{\overset{2}{\cancel{8}}}{\underset{3}{\cancel{15}}} = \frac{2}{3}$

Check: $1\frac{7}{8} \times \frac{2}{3} = \frac{\overset{5}{\cancel{15}}}{\underset{4}{\cancel{8}}} \times \frac{\overset{1}{\cancel{2}}}{\underset{1}{\cancel{3}}} = \frac{5}{4} = 1\frac{1}{4}$

Turn to **28** for a set of practice problems on dividing fractions.

Exercises 2-3 Dividing Fractions

28 A. Divide and reduce the answer to lowest terms:

1. $\frac{5}{6} \div \frac{1}{2}$
2. $6 \div \frac{2}{3}$
3. $\frac{5}{12} \div \frac{4}{3}$
4. $8 \div \frac{1}{4}$
5. $\frac{6}{16} \div \frac{3}{4}$
6. $\frac{1}{2} \div \frac{1}{2}$
7. $\frac{3}{16} \div \frac{6}{8}$
8. $\frac{3}{4} \div \frac{5}{16}$
9. $1\frac{1}{2} \div \frac{1}{6}$
10. $6 \div 1\frac{1}{2}$
11. $3\frac{1}{7} \div 2\frac{5}{14}$
12. $3\frac{1}{2} \div 2$
13. $6\frac{2}{5} \div 5\frac{1}{3}$
14. $10 \div 1\frac{1}{5}$
15. $8 \div \frac{1}{2}$
16. $\frac{2}{3} \div 6$
17. $12 \div \frac{2}{3}$
18. $\frac{3}{4} \div \frac{7}{8}$
19. $\dfrac{5}{\frac{2}{3}}$
20. $\dfrac{1\frac{1}{2}}{2\frac{1}{2}}$

B. Practical Problems

1. How many feet are represented by a 4-in. line, if it is drawn to a scale of $\frac{1}{2}$ in. = 1 foot?
2. If $\frac{1}{4}''$ on a drawing represents 1′0″, how many inches on the drawing will be required to represent 18′0″?
3. How many boards $4\frac{5}{8}$ in. wide will it take to cover a floor $18\frac{1}{2}$ ft wide?
4. How many supporting columns 7 ft $4\frac{1}{2}$ in. long can be cut from 6 pieces, each 22 ft long?
5. How many pieces of $\frac{1}{2}$-in. plywood are there in a stack 3 ft 6 in. high?
6. If $\frac{1}{4}$ in. represents 1′0″ on a drawing how many feet will be represented by a line $10\frac{1}{8}$ in. long?

7. If we allow $2\frac{5}{8}''$ for the thickness of a course of brick, including mortar joints, how many courses of brick will there be in a wall $3'11\frac{1}{4}''$ high?

8. How many lengths of pipe $2\frac{5}{8}$ ft long can be cut from a pipe 21 ft long?

9. How many pieces $6\frac{1}{4}$ in. long can be cut from 35 metal rods each 40 in. long? Disregard waste.

10. The architectural drawing for a room measures $3\frac{5}{8}$ in. by $4\frac{1}{4}$ in. If $\frac{1}{8}$ in. is equal to 1 ft on the drawing, what are the actual dimensions of the room?

Check your answers on page 544 then continue in frame **29**.

2-4 ADDITION AND SUBTRACTION OF FRACTIONS

Addition

29 At heart, adding fractions is a matter of counting:

$$\frac{1}{5} + \frac{3}{5} = \frac{1+3}{5} = \frac{4}{5}$$

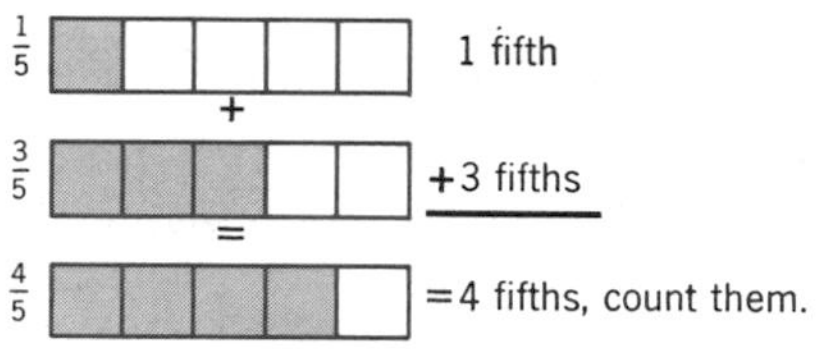

Add $\frac{1}{8} + \frac{3}{8} =$ ______

This is easy to see with measurements:

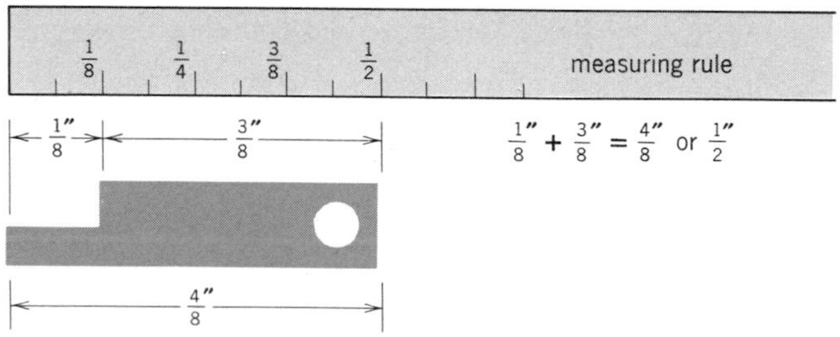

$\frac{1''}{8} + \frac{3''}{8} = \frac{4''}{8}$ or $\frac{1''}{2}$

Add $\frac{2}{7} + \frac{3}{7} =$ ______

Check your answer in **30**.

30

$$\frac{2}{7} + \frac{3}{7} = \frac{2+3}{7} = \frac{5}{7}$$

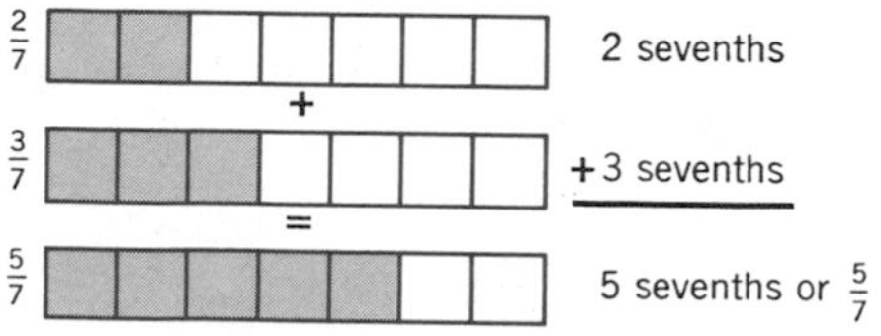

Fractions having the same denominator are called *like* fractions. In the problem above, $\frac{2}{7}$ and $\frac{3}{7}$ both have denominator 7 and are like fractions. Adding like fractions is easy: *first,* add the numerators to find the numerator of the sum and *second,* use the denominator the fractions have in common as the denominator of the sum.

$$\frac{2}{9} + \frac{5}{9} = \frac{2+5}{9}$$

Add numerators

Same denominator

Adding three or more like fractions is easy:

$\frac{3}{12} + \frac{1}{12} + \frac{5}{12} =$ ________

Add these fractions, then turn to **31**

---

31 $\frac{3}{12} + \frac{1}{12} + \frac{5}{12} = \frac{3+1+5}{12} = \frac{9}{12} = \frac{3}{4}$

Notice that we reduce the sum to lowest terms.

Try these problems for exercise:

(a) $\frac{1}{8} + \frac{3}{8} =$ ________

(b) $\frac{7}{9} + \frac{5}{9} =$ ________

(c) $2\frac{1}{2} + 3\frac{3}{2} =$ ________

(d) $\frac{1}{7} + \frac{4}{7} + \frac{5}{7} + 1\frac{2}{7} + \frac{8}{7} =$ ________

(e) $2 + 3\frac{1}{2} =$ ________

(f) $\frac{3}{5} + 1\frac{1}{5} + 3 =$ ________

Go to **32** to check your work.

---

32 (a) $\frac{1}{8} + \frac{3}{8} = \frac{1+3}{8} = \frac{4}{8} = \frac{1}{2}$, reduced to lowest terms

(b) $\frac{7}{9} + \frac{5}{9} = \frac{7+5}{9} = \frac{12}{9} = \frac{4}{3} = 1\frac{1}{3}$

(c) $2\frac{1}{2} + 3\frac{3}{2} = \frac{5}{2} + \frac{9}{2} = \frac{5+9}{2} = \frac{14}{2} = 7$

Write mixed numbers as fractions and then add. (Of course you might have realized that $3\frac{3}{2} = 4\frac{1}{2}$. The addition $2\frac{1}{2} + 4\frac{1}{2}$ is easy.)

(d) $\frac{1}{7} + \frac{4}{7} + \frac{5}{7} + 1\frac{2}{7} + \frac{8}{7} = \frac{1}{7} + \frac{4}{7} + \frac{5}{7} + \frac{9}{7} + \frac{8}{7}$

$= \frac{1+4+5+9+8}{7} = \frac{27}{7} = 3\frac{6}{7}$

(e) $2 + 3\frac{1}{2} = 2 + 3 + \frac{1}{2} = 5\frac{1}{2}$. Remember, $3\frac{1}{2}$ means $3 + \frac{1}{2}$

(f) $\frac{3}{5} + 1\frac{1}{5} + 3 = \frac{3}{5} + \frac{6}{5} + \frac{15}{5} = \frac{3+6+15}{5} = \frac{24}{5} = 4\frac{4}{5}$

or $\frac{3}{5} + 1 + \frac{1}{5} + 3 = 1 + 3 + \frac{3}{5} + \frac{1}{5} = 4 + \frac{4}{5} = 4\frac{4}{5}$

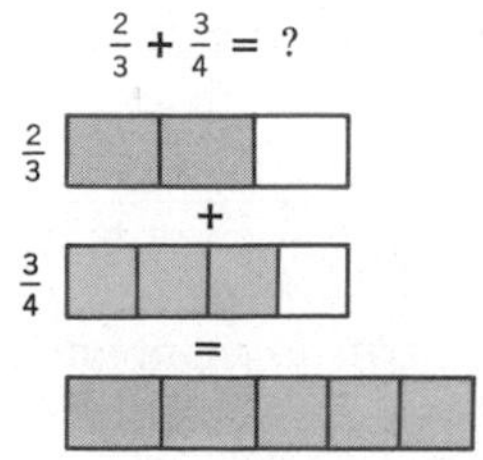

How do we add fractions whose denominators are not the same?

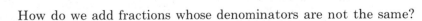

The problem is to find a name for this new number. One way to find it is to change these fractions to equivalent fractions with the same denominator.

$$\frac{3}{4} = \frac{3 \times 3}{4 \times 3} = \frac{9}{12}$$

$$\frac{2}{3} = \frac{2 \times 4}{3 \times 4} = \frac{8}{12}$$

(You should remember that we discussed equivalent fractions back in frame **10**. Return for a quick review if you need it.)

Now add $\frac{2}{3} + \frac{3}{4}$ using the equivalent fractions, then continue in **33**.

---

33 $\frac{2}{3} + \frac{3}{4} = \frac{8}{12} + \frac{9}{12} = \frac{17}{12} = 1\frac{5}{12}$

We change the original fractions to equivalent fractions with the same denominator and then add as before.

How do you know what number to use as the new denominator? In general, you cannot simply guess at the best new denominator. We need a method for finding it from the denominators of the fractions to be added. The new denominator is called the *Least Common Denominator,* abbreviated *LCD.*

Suppose we want to add the fractions $\frac{1}{8} + \frac{5}{12}$.

The first step is to find the LCD of the denominators 8 and 12. To find the LCD, follow this procedure.

Step 1 Write the denominators.

8 12

Step 2 Divide each denominator by any number that will divide evenly one or more of the denominators. In this example, the number 2 will divide both denominators.

$$\begin{array}{r|ll} 2 & 8 & 12 \\ \hline & 4 & 6 \end{array}$$

Step 3 Repeat the process, dividing by any number that will divide evenly at least one of the new numbers. Any number that is not divisible should be brought down to the next row.

Step 4 When no more divisions are possible, multiply all of the divisors. This product is the LCD.

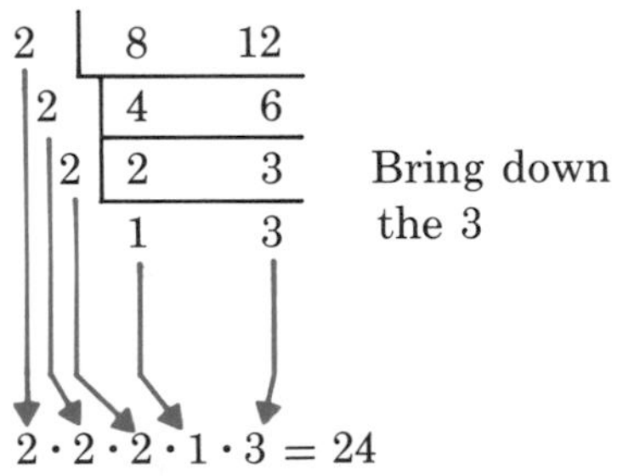

LCD = 24

The Least Common Denominator of 8 and 12 is 24. It is the smallest number that both 8 and 12 divide evenly.

Once you have found the LCD, add the original fractions by rewriting them with this new denominator.

$$\frac{1}{8} = \frac{?}{24} \qquad \frac{1}{8} = \frac{1 \times 3}{8 \times 3} = \frac{3}{24}$$

$$\frac{5}{12} = \frac{?}{24} \qquad \frac{5}{12} = \frac{5 \times 2}{12 \times 2} = \frac{10}{24}$$

Now we can add the fractions:

$$\frac{1}{8} + \frac{5}{12} = \frac{3}{24} + \frac{10}{24}$$
$$= \frac{13}{24}$$

Use the method described above to find the LCD of the numbers 12 and 15. Check your work in **34**.

ANOTHER METHOD FOR FINDING THE LCD

Here is an alternate method for finding the LCD that you might find easier than the method described in frame **33**. Follow these two steps:

Step 1 Choose the larger denominator and write down a few multiples of it.

Example: To find the LCD of 12 and 15, first write down a few of the multiples of the larger number, 15. The multiples are 15, 30, 45, 60, 75, and so on.

Step 2 Test each multiple until you find one that is evenly divisible by the smaller denominator.

Example: 15 is not evenly divisible by 12. 30 is not evenly divisible by 12. 45 is not evenly divisible by 12. 60 *is* evenly divisible by 12. The LCD is 60.

This method of finding the LCD has the advantage that you can use it with an electronic calculator. For the example above the calculator steps would look like this:

[15] [÷] [12] [=]	**1.25**	*Not* an integer, therefore, *not* evenly divisible by 12.
[15] [×] [2] [=] [÷] [12]	**2.5**	Second multiple: answer is *not* an integer.
[15] [×] [3] [=] [÷] [12]	**3.75**	Third multiple: answer is *not* an integer.
[15] [×] [4] [=] [÷] [12]	**5**	Fourth multiple: answer *is* an integer. therefore the LCD is 4×15 or 60.

34 **Step 1** Write the denominators: 12 15

Step 2 Divide by 2. Repeat until no number is divisible by 2.

$$\begin{array}{r|rr} 2 & 12 & 15 \\ \hline 2 & 6 & 15 \\ \hline 2 & 3 & 15 \\ \hline \end{array}$$

Divide 12 by 2. 15 is not evenly divisible by 2, so bring it down unchanged.

Step 3 Divide by 3.

Step 4 Multiply.

LCD = 60

$$\begin{array}{r|rr} 2 & 12 & 15 \\ \hline 2 & 6 & 15 \\ \hline 3 & 3 & 15 \\ \hline & 1 & 5 \end{array}$$

$$2 \cdot 2 \cdot 3 \cdot 1 \cdot 5 = 60$$

Of course the process is exactly the same no matter how many numbers you begin with. To find the LCD of 4, 6, and 8:

$$\begin{array}{r|rrr} 2 & 4 & 6 & 8 \\ \hline 2 & 2 & 3 & 4 \\ \hline 2 & 1 & 3 & 2 \\ \hline & 1 & 3 & 1 \end{array}$$

Bring down the 3.

LCD $= 2 \cdot 2 \cdot 2 \cdot 1 \cdot 3 \cdot 1 = 24$

Ready for a bit of practice in finding LCDs?

Find the LCD of 12 and 18. Check your work in **35**

---

35 **Step 1** 12 18

Step 2 Divide by 2.

$$\begin{array}{r|rr} 2 & 12 & 18 \\ \hline 2 & 6 & 9 \\ \hline & 3 & 9 \end{array}$$

9 is not divisible by 2. Bring down the 9.

Step 3 Divide by 3.

$$\begin{array}{r|rr} 2 & 12 & 18 \\ \hline 2 & 6 & 9 \\ \hline 3 & 3 & 9 \\ \hline & 1 & 3 \end{array}$$

Step 4 Multiply

LCD $= 2 \cdot 2 \cdot 3 \cdot 1 \cdot 3 = 36$

Notice that we start the dividing with 2 and continue until the remaining numbers cannot be divided further.

Practice by finding the LCD for each of the following sets of numbers.

(a) 2 and 4	(b) 8 and 4	(c) 6 and 3
(d) 5 and 4	(e) 9 and 15	(f) 15 and 24
(g) 4, 5, and 6	(h) 4, 8, and 12	(i) 12, 15, 21

Check your answers in **36**.

36 (a) 4 (b) 8 (c) 6 (d) 20 (e) 45
(f) 120 (g) 60 (h) 24 (i) 420

In order to use the LCD to add fractions, rewrite the fractions with the LCD as the new denominator, then add the new equivalent fractions.

For example, add $\frac{1}{6} + \frac{5}{8}$.

First, find the LCD. The LCD of 6 and 8 is 24.

Next, rewrite the two fractions with denominator 24.

$$\frac{1}{6} = \frac{1 \times 4}{6 \times 4} = \frac{4}{24}$$

$$\frac{5}{8} = \frac{5 \times 3}{8 \times 3} = \frac{15}{24}$$

Finally, add the new equivalent fractions.

$$\frac{1}{6} + \frac{5}{8} = \frac{4}{24} + \frac{15}{24}$$
$$= \frac{19}{24}$$

Try it. Add $\frac{3}{8} + \frac{1}{12}$.

Check your work in **37**.

---

37 The LCD of 8 and 12 is 24.

$$\frac{3}{8} = \frac{3 \times 3}{8 \times 3} = \frac{9}{24}$$

$$\frac{1}{12} = \frac{1 \times 2}{12 \times 2} = \frac{2}{24}$$

$$\frac{3}{8} + \frac{1}{12} = \frac{9}{24} + \frac{2}{24}$$
$$= \frac{11}{24}$$

Ready for some practice at this?
Find the LCD, rewrite the fractions, and add.

(a) $\frac{1}{2} + \frac{1}{4}$ (b) $\frac{3}{8} + \frac{1}{4}$ (c) $\frac{1}{6} + \frac{2}{3}$

(d) $\frac{3}{4} + \frac{1}{6}$ (e) $\frac{2}{5} + \frac{1}{4}$ (f) $\frac{5}{6} + \frac{3}{8}$

Check your work in **38**.

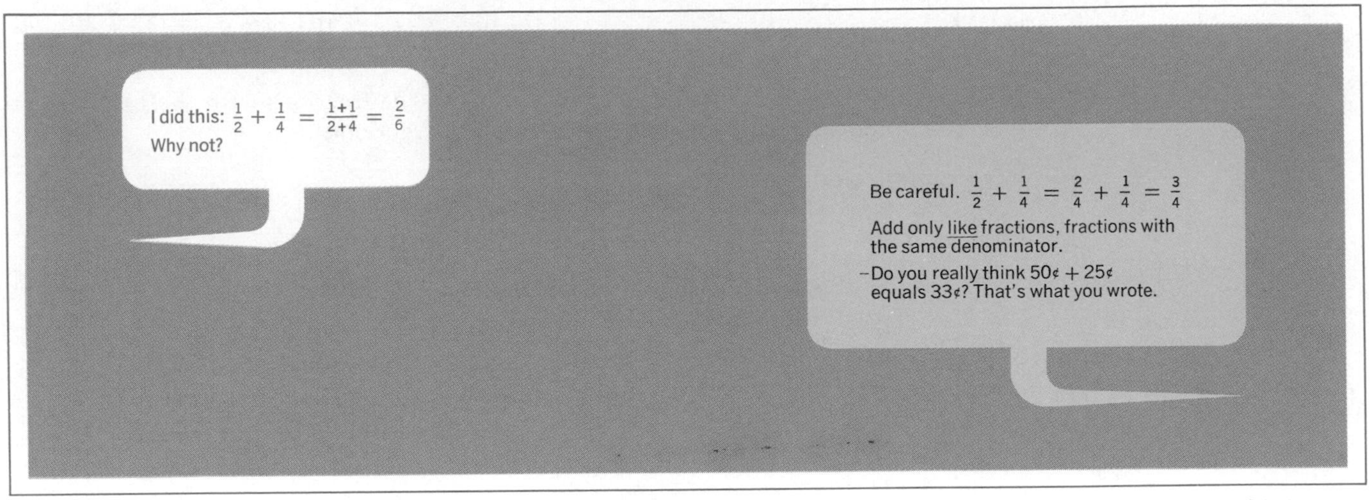

---

38 (a) The LCD of 2 and 4 is 4.

$$\frac{1}{2} = \frac{?}{4} = \frac{1 \times 2}{2 \times 2} = \frac{2}{4}$$

$$\frac{1}{2} + \frac{1}{4} = \frac{2}{4} + \frac{1}{4} = \frac{3}{4}$$

(b) The LCD of 8 and 4 is 8.

$$\frac{1}{4} = \frac{?}{8} = \frac{1 \times 2}{4 \times 2} = \frac{2}{8}$$

$$\frac{3}{8} + \frac{1}{4} = \frac{3}{8} + \frac{2}{8} = \frac{5}{8}$$

(c) The LCD of 6 and 3 is 6.

$$\frac{2}{3} = \frac{?}{6} = \frac{2 \times 2}{3 \times 2} = \frac{4}{6}$$

$$\frac{1}{6} + \frac{2}{3} = \frac{1}{6} + \frac{4}{6} = \frac{5}{6}$$

(d) The LCD of 4 and 6 is 12.

$$\frac{3}{4} = \frac{?}{12} = \frac{3 \times 3}{4 \times 3} = \frac{9}{12} \qquad \frac{1}{6} = \frac{?}{12} = \frac{1 \times 2}{6 \times 2} = \frac{2}{12}$$

$$\frac{3}{4} + \frac{1}{6} = \frac{9}{12} + \frac{2}{12} = \frac{11}{12}$$

(e) The LCD of 5 and 4 is 20.

$$\frac{2}{5} = \frac{?}{20} = \frac{2 \times 4}{5 \times 4} = \frac{8}{20} \qquad \frac{1}{4} = \frac{?}{20} = \frac{1 \times 5}{4 \times 5} = \frac{5}{20}$$

$$\frac{2}{5} + \frac{1}{4} = \frac{8}{20} + \frac{5}{20} = \frac{13}{20}$$

(f) The LCD of 6 and 8 is 24.

$$\frac{5}{6} = \frac{?}{24} = \frac{5 \times 4}{6 \times 4} = \frac{20}{24} \qquad \frac{3}{8} = \frac{?}{24} = \frac{3 \times 3}{8 \times 3} = \frac{9}{24}$$

$$\frac{5}{6} + \frac{3}{8} = \frac{20}{24} + \frac{9}{24} = \frac{29}{24} = 1\frac{5}{24}$$

Subtraction

Once you have mastered the process of adding fractions, subtraction is very simple indeed. To find $\frac{3}{8} - \frac{1}{8}$, notice that the denominators are the same. We can subtract the numerator and write this difference over the common denominator

$$\frac{3}{8} - \frac{1}{8} = \frac{3-1}{8}$$

(Subtract numerators; Same denominator)

$$= \frac{2}{8} \quad \text{or} \quad \frac{1}{4}$$

To find $\frac{3}{4} - \frac{1}{5}$, first find the LCD of 4 and 5.

The LCD of 4 and 5 is 20.

Find equivalent fractions with a denominator of 20:

$$\frac{3}{4} = \frac{?}{20} = \frac{3 \times 5}{4 \times 5} = \frac{15}{20}$$

$$\frac{1}{5} = \frac{?}{20} = \frac{1 \times 4}{5 \times 4} = \frac{4}{20}$$

Subtract the equivalent fractions:

$$\frac{3}{4} - \frac{1}{5} = \frac{15}{20} - \frac{4}{20} = \frac{15-4}{20} = \frac{11}{20}$$

The procedure is exactly the same as for addition.

If two fractions to be added are given as mixed numbers, add the whole number parts first, then add the fraction parts as shown above. For example,

$$2\frac{1}{2} + 1\frac{1}{3} = (2 + 1) + \left(\frac{1}{2} + \frac{1}{3}\right)$$

$$= 3 + \left(\frac{3}{6} + \frac{2}{6}\right)$$

$$= 3 + \frac{5}{6}$$

$$= 3\frac{5}{6}$$

Another example:

$$2\frac{4}{5} + 3\frac{5}{6} = (2 + 3) + \left(\frac{24}{30} + \frac{25}{30}\right)$$

$$= 5 + \frac{49}{30}$$

$$= 5 + 1 + \frac{19}{30}$$

$$= 6\frac{19}{30}$$

If two fractions to be subtracted are given as mixed numbers, it is usually easiest to rewrite them as improper fractions before subtraction. For example,

$$3\frac{1}{4} - 1\frac{3}{8} = \frac{13}{4} - \frac{11}{8}$$

$$= \frac{26}{8} - \frac{11}{8}$$

$$= \frac{15}{8} \text{ or } 1\frac{7}{8}$$

If one of the fractions is a whole number, write it as a fraction first, then add or subtract as with any other fraction. For example,

$$3 - \frac{2}{5} = \frac{3}{1} - \frac{2}{5}$$
$$= \frac{15}{5} - \frac{2}{5}$$
$$= \frac{13}{5} \text{ or } 2\frac{3}{5}$$

Now turn to **39** for a set of problems on adding and subtracting fractions.

39

Exercises 2-4 Adding and Subtracting Fractions

A. Add or subtract as shown:

1. $\frac{1}{16} + \frac{3}{16}$
2. $\frac{5}{12} + \frac{11}{12}$
3. $\frac{5}{16} + \frac{7}{16}$
4. $\frac{2}{6} + \frac{3}{6}$
5. $\frac{3}{4} - \frac{1}{4}$
6. $\frac{13}{16} - \frac{3}{16}$
7. $\frac{3}{5} - \frac{1}{5}$
8. $\frac{5}{12} - \frac{2}{12}$
9. $\frac{5}{16} + \frac{3}{16} + \frac{7}{16}$
10. $\frac{1}{8} + \frac{3}{8} + \frac{7}{8}$
11. $\frac{1}{4} + \frac{1}{2}$
12. $\frac{7}{16} + \frac{3}{8}$
13. $\frac{3}{8} + \frac{1}{12}$
14. $\frac{5}{12} + \frac{3}{16}$
15. $\frac{1}{2} - \frac{3}{8}$
16. $\frac{5}{8} - \frac{1}{16}$
17. $\frac{15}{16} - \frac{1}{2}$
18. $\frac{7}{16} - \frac{1}{32}$
19. $\frac{1}{2} + \frac{1}{4} - \frac{1}{8}$
20. $\frac{11}{16} - \frac{1}{8} - \frac{1}{3}$
21. $1\frac{1}{2} + \frac{1}{4}$
22. $2\frac{7}{16} + \frac{1}{4}$
23. $2\frac{1}{2} + 1\frac{5}{8}$
24. $2\frac{8}{32} + 1\frac{1}{16}$
25. $2\frac{1}{2} - 1\frac{3}{4}$

B. Add or subtract as shown:

1. $8 - 2\frac{7}{8}$
2. $3 - 1\frac{3}{16}$
3. $3\frac{5}{8} - \frac{13}{16}$
4. $\frac{1}{2} + \frac{1}{3} + \frac{1}{4} + \frac{1}{5}$
5. $\frac{1}{2} + \frac{1}{4} + \frac{1}{8}$
6. $6\frac{1}{2} + 5\frac{3}{4} + 8\frac{1}{8}$
7. $\frac{7}{8} - 1\frac{1}{4} + 2\frac{1}{2}$
8. $1\frac{3}{8}$ subtracted from $4\frac{3}{4}$
9. $2\frac{3}{16}$ less than $4\frac{7}{8}$
10. $6\frac{2}{3}$ reduced by $1\frac{1}{4}$
11. $2\frac{3}{5}$ less than $6\frac{1}{2}$
12. By how much is $1\frac{8}{7}$ larger than $1\frac{7}{8}$?

C. Practical Problems

1. The time sheet for operations on a machine part listed the following: chucking, $\frac{3}{4}$ min; spotting and drilling, $3\frac{1}{3}$ min; facing, $1\frac{2}{3}$ min; grinding, $4\frac{1}{2}$ min; reaming, $\frac{2}{5}$ min. What was the total time for the operations?

2. A counter top is made of $\frac{5}{8}$-in. particle board and is covered with $\frac{3}{16}$-in. laminated plastic. What width of metal edging is needed to finish off the edge?

3. What is the total length of a certain machine part that is made by joining four pieces that measure $3\frac{1}{8}$ in., $1\frac{5}{32}$ in., $2\frac{7}{16}$ in., and $1\frac{1}{4}$ in.?

4. A blueprint requires four separate pieces of wood measuring $5\frac{3}{8}$ in., $8\frac{1}{4}$ in., $6\frac{9}{16}$ in., and $2\frac{5}{8}$ in. How long a piece of wood is needed to cut these pieces if we allow $\frac{1}{2}$ in. for waste?

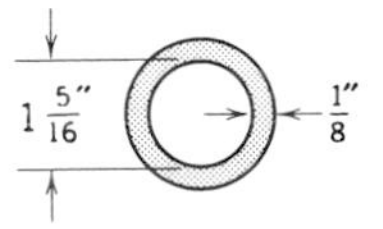

5. What is the outside diameter of tubing whose inside diameter is $1\frac{5}{16}$ in. and whose wall thickness is $\frac{1}{8}$ in.?

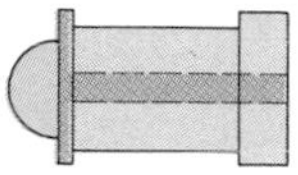

6. How long a bolt is needed to go through a piece of tubing $\frac{5}{8}$ in. long, a washer $\frac{1}{16}$ in. thick, and a nut $\frac{1}{4}$ in. thick?

7. The front and rear axles of a car need to be aligned. If the length measurement on one side of the car is 8 ft $5\frac{1}{4}$ in. and 8 ft $4\frac{7}{8}$ in. on the other side, how much shifting must be done to bring the axles into alignment?

8. While installing water pipes, a plumber used pieces of pipe measuring $2\frac{3}{4}$ ft, $4\frac{1}{3}$ ft, $3\frac{1}{2}$ ft, and $1\frac{1}{4}$ ft. How much pipe would remain if these pieces were cut from a 14 ft length of pipe? (Ignore waste in cutting.)

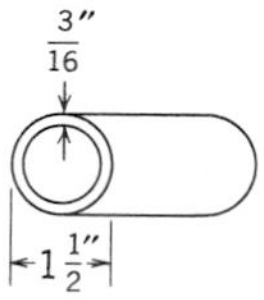

9. A piece of electrical iron pipe conduit has a diameter of $1\frac{1}{2}$ in. and a wall thickness of $\frac{3}{16}$ in. What is its inside diameter?

10. The diameter of a steel shaft is reduced $\frac{7}{1000}$ in. The original diameter of the shaft was $\frac{850}{1000}$ in. What is the new diameter of the shaft?

11. If a piece of $\frac{3}{8}$ in. I.D. (inside diameter) copper tubing measures $\frac{9}{16}$ in. O.D. (outside diameter), what is the wall thickness?

12. Find the missing dimension in the drawing below.

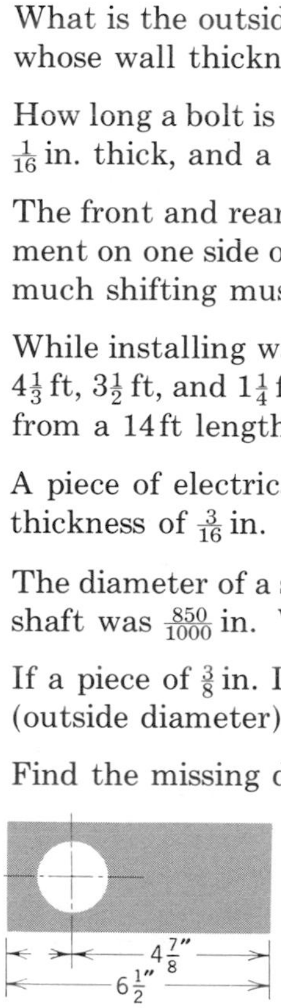

13. Two splice plates are cut from a piece of sheet steel that has an overall length of $18\frac{5}{8}$ in. The plates are $9\frac{1}{4}$ in. and $6\frac{7}{16}$ in. long. How much material remains from the original piece if each saw cut removes $\frac{1}{16}$ in.?

14. A printer has $2\frac{3}{4}$ rolls of a certain kind of paper in stock. He must do three jobs that require $\frac{5}{8}$ roll, $1\frac{1}{2}$ rolls, and $\frac{3}{4}$ roll respectively. Does he have enough?

15. Newspaper ads are sold by the column inch (c.i.). What is the total number of c.i. for a month in which a plumbing contractor has had ads of $6\frac{1}{2}$ c.i., $5\frac{3}{4}$ c.i., $3\frac{1}{4}$ c.i., $4\frac{3}{4}$ c.i., and 5 c.i.?

16. A cabinet 30″ high must have a $4\frac{1}{2}$″ base and a $1\frac{3}{4}$″ top. How much space is left for drawers?

Check your answers on page 545.

Turn to **40** for a set of problems on the arithmetic of fractions, with many practical applications.

2 Fractions

PROBLEM SET 2

40 Answers are on page 545.

Fractions

A. Write as an improper fraction:

1. $1\frac{1}{8}$ 2. $4\frac{1}{5}$ 3. $1\frac{2}{3}$ 4. $2\frac{3}{16}$

5. $3\frac{3}{32}$ 6. $2\frac{1}{16}$ 7. $1\frac{5}{8}$ 8. $3\frac{7}{16}$

Write as a mixed number:

9. $\frac{40}{16}$ 10. $\frac{19}{2}$ 11. $\frac{25}{3}$ 12. $\frac{9}{8}$

13. $\frac{50}{32}$ 14. $\frac{21}{16}$ 15. $\frac{100}{8}$ 16. $\frac{35}{15}$

Reduce to lowest terms:

17. $\frac{6}{32}$ 18. $\frac{8}{32}$ 19. $\frac{12}{32}$ 20. $\frac{18}{24}$

21. $\frac{5}{30}$ 22. $1\frac{12}{21}$ 23. $1\frac{16}{20}$ 24. $3\frac{10}{25}$

Complete these:

25. $\frac{3}{4} = \frac{?}{12}$ 26. $\frac{7}{16} = \frac{?}{64}$ 27. $2\frac{3}{4} = \frac{?}{16}$ 28. $1\frac{3}{8} = \frac{?}{32}$

29. $5\frac{2}{3} = \frac{?}{12}$ 30. $1\frac{4}{5} = \frac{?}{10}$ 31. $1\frac{1}{4} = \frac{?}{12}$ 32. $2\frac{3}{5} = \frac{?}{10}$

Circle the large number:

33. $\frac{7}{16}$ or $\frac{2}{15}$ 34. $\frac{2}{3}$ or $\frac{4}{7}$ 35. $\frac{13}{16}$ or $\frac{7}{8}$ 36. $1\frac{1}{4}$ or $\frac{7}{6}$

37. $\frac{13}{32}$ or $\frac{3}{5}$ 38. $\frac{2}{10}$ or $\frac{3}{16}$ 39. $1\frac{7}{16}$ or $\frac{7}{4}$ 40. $\frac{3}{32}$ or $\frac{1}{9}$

B. Multiply or divide as shown:

1. $\frac{1}{2} \times \frac{3}{16}$ 2. $\frac{3}{4} \times \frac{2}{3}$ 3. $\frac{7}{16} \times \frac{4}{3}$ 4. $\frac{3}{64} \times \frac{1}{12}$

5. $1\frac{1}{2} \times \frac{5}{6}$ 6. $3\frac{1}{16} \times \frac{1}{5}$ 7. $\frac{3}{16} \times \frac{5}{12}$ 8. $4 \times \frac{3}{8}$

9. $\frac{3}{4} \times 10$ 10. $\frac{1}{2} \times 1\frac{1}{3}$ 11. $18 \times 1\frac{1}{2}$ 12. $16 \times 2\frac{1}{8}$

13. $\frac{1}{2} \div \frac{1}{4}$ 14. $\frac{2}{5} \div \frac{1}{2}$ 15. $4 \div \frac{1}{8}$ 16. $8 \div \frac{3}{4}$

Name

Date

Course/Section

17. $\frac{2}{3} \div 4$ 18. $1\frac{1}{2} \div 2$ 19. $3\frac{1}{2} \div 5$ 20. $1\frac{1}{4} \div 1\frac{1}{2}$

C. Add or subtract as shown:

1. $\frac{3}{8} + \frac{7}{8}$ 2. $\frac{1}{2} + \frac{3}{4}$ 3. $\frac{3}{32} + \frac{1}{8}$ 4. $\frac{3}{8} + 1\frac{1}{4}$

5. $\frac{9}{16} - \frac{3}{16}$ 6. $\frac{7}{8} - \frac{1}{2}$ 7. $\frac{11}{16} - \frac{1}{4}$ 8. $1\frac{1}{2} - \frac{3}{32}$

9. $2\frac{1}{8} + 1\frac{1}{4}$ 10. $1\frac{5}{8} + \frac{13}{16}$ 11. $6 - 1\frac{1}{2}$ 12. $3 - 1\frac{7}{8}$

13. $3\frac{2}{3} - 1\frac{7}{8}$ 14. $2\frac{1}{4} - \frac{5}{6}$

15. $\frac{1}{2} + \frac{1}{3} + \frac{1}{5}$ 16. $1\frac{1}{2} + 1\frac{1}{4} + 1\frac{1}{5}$ 17. $3\frac{1}{2} - 2\frac{1}{3}$

18. $2\frac{3}{5} - 1\frac{4}{15}$ 19. $2 - 1\frac{3}{5}$ 20. $4\frac{5}{6} - 1\frac{1}{2}$

D. Practical Problems

1. In a welding job three pieces of 2″ I-beam with lengths $5\frac{7}{8}$″, $8\frac{1}{2}$″, and $22\frac{3}{4}$″ are needed. What is the total length of I-beam needed? (Do not worry about the waste in cutting.)
2. How many pieces of $10\frac{5}{16}$″ bar can be cut from a stock 20′ bar? The metal is torch cut and an allowance of $\frac{3}{16}$″ kerf (waste) should be made for each piece.
3. If an I-beam is to be $24\frac{3}{8}$″ long with a tolerance of $\pm 1\frac{1}{4}$″, find the longest and shortest acceptable lengths.
4. In squaring a damaged auto frame, the mechanic measured the diagonals as $77\frac{3}{16}$″ and $69\frac{5}{8}$″. How much is the difference to be equalized?
5. A shaft $1\frac{7}{8}$″ in diameter is turned down on a lathe to a diameter of $1\frac{3}{32}$″. What is the difference in diameters?
6. A bar $14\frac{5}{16}$″ long is cut from a piece $25\frac{1}{4}$″ long. If $\frac{3}{32}$″ is wasted in cutting, will there be enough left to make another bar $10\frac{3}{8}$″ long?
7. A cubic foot contains roughly $7\frac{1}{2}$ gallons. How many cubic feet are there in a tank containing $34\frac{1}{2}$ gallons?
8. Find the total width of the three pieces of steel plate shown below.

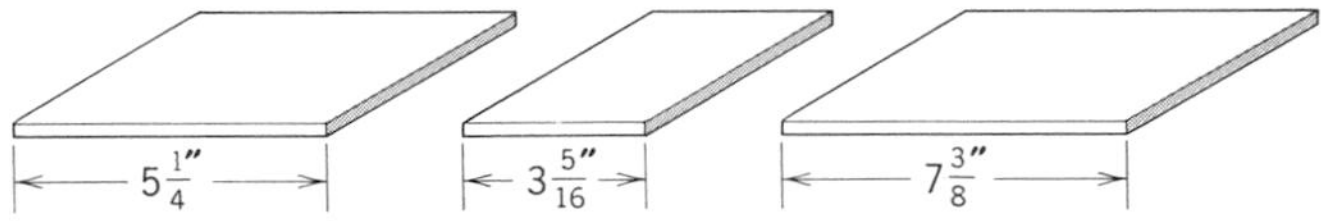

9. What would be the total length of the bar formed by welding together the five pieces of bar stock shown below?

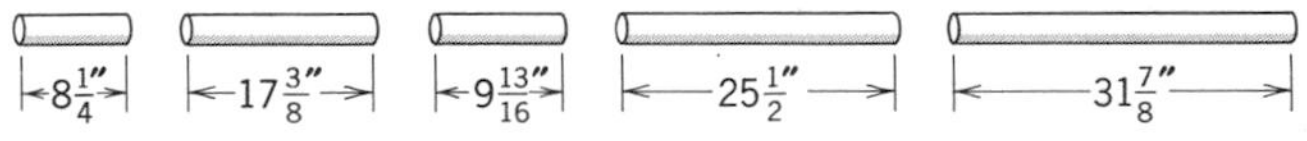

10. The Ace Machine Shop has the job of producing 32 zinger bars. Each zinger bar must be turned on a lathe from a piece of stock $4\frac{7}{8}$″ long. How many feet of stock will they need?
11. What is the thickness of a table top made of $\frac{3}{4}$″ plywood and covered with a $\frac{3}{16}$″ sheet of glass?

12. For the wooden form shown below find the lengths A, B, C, and D.

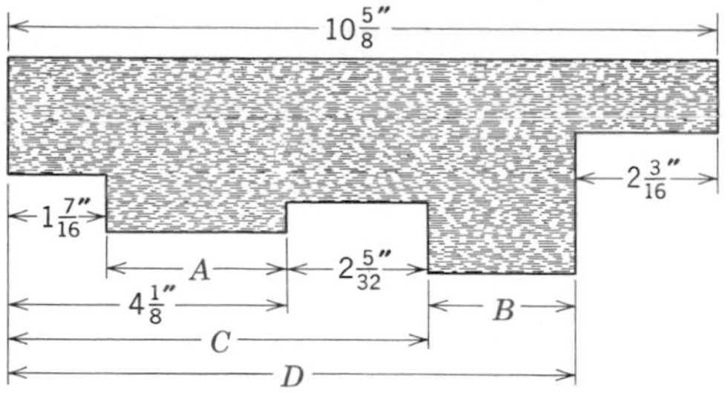

13. Find the spacing x between the holes shown below.

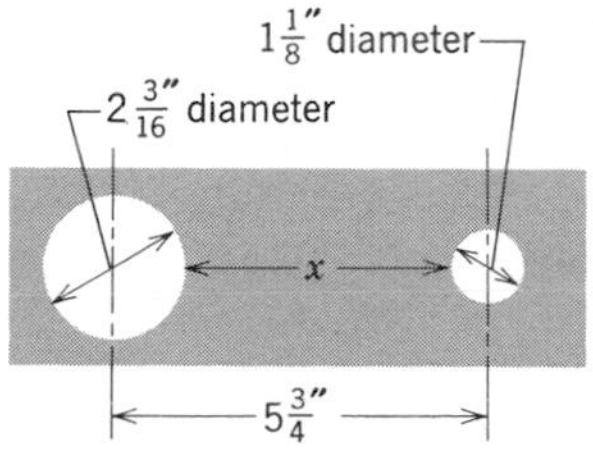

14. **Find the height of the five-course (five bricks high) brick wall shown if each brick is $2\frac{1}{2}''$ by $3\frac{7}{8}''$ by $8\frac{1}{4}''$ and all mortar joints are $\frac{1}{2}''$.**

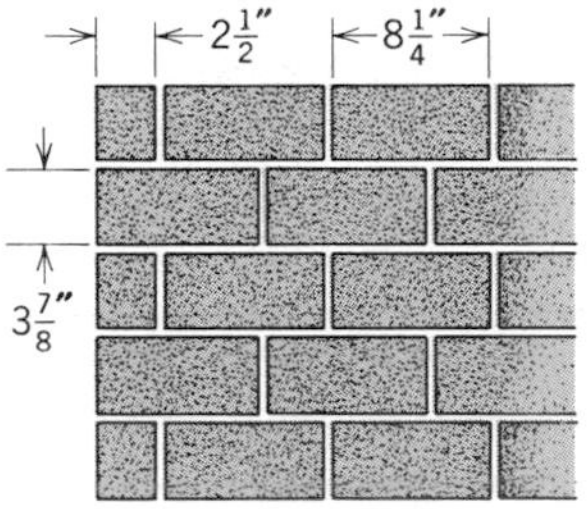

15. If the wall in Problem 14 has 28 stretchers (bricks laid lengthwise), what is its length?
16. An electrical wiring job requires the following lengths of 14/2 BX cable: seven pieces each $6\frac{1}{2}$ ft long, four pieces each $34\frac{3}{4}''$ long, and nine pieces $19\frac{3}{8}''$ long. What is the total length of cable needed?

3 Decimal Numbers

PREVIEW 3

Objective	Sample Problems		Where To Go for Help: Page	Frame
Upon successful completion of this unit you will be able to:				
1. Add, subtract, multiply, and divide decimal numbers.	(a) $5.82 + 0.096$	= ______	84	4
	(b) $3.78 - 0.989$	= ______	85	5
	(c) $27 - 4.03$	= ______		
	(d) 7.25×0.301	= ______	89	8
	(e) $104.2 \div 0.032$	= ______	92	11
	(f) $0.09 \div 0.0004$	= ______		
	(g) $20.4 \div 6.7$ (round to three decimal places)	= ______		
2. Find averages.	Find the average of 4.2, 4.8, 5.7, 2.5, 3.6, 5.0	= ______	98	19
3. Work with decimal fractions.	(a) Write as a decimal number $\frac{3}{16}$	= ______	101	22
	(b) $1\frac{2}{3} + 1.785$	= ______		
	(c) $4.1 \times 2\frac{1}{4}$	= ______		
	(d) $1\frac{5}{16} \div 4.3$ (round to three decimal places)	= ______		
4. Work with exponents.	(a) 4^3	= ______	108	28
	(b) $2^5 \times 3^2$	= ______		
	(c) 2.05^2	= ______		
5. Find square roots.	(a) $\sqrt{169}$	= ______	112	34
	(b) $\sqrt{14.5}$ (to two decimal places)	= ______		
6. Work with negative numbers.	(a) $(-4) - (-7)$	= ______	116	41
	(b) $8 - 15$	= ______		
	(c) $(-1.5) \times (-3.1)$	= ______		
	(d) $(5.25) \div (-2.1)$	= ______		

Name ______

Date ______

Course/Section ______

(Answers to these preview problems are on page 80.)

If you are certain you can work *all* of these problems correctly, turn to page 123 for a set of practice problems. If you cannot work one or more of the preview problems, turn to the page indicated after the problem. Superstudents—those who want to be winners in their work—will turn to frame **1** and begin work there.

Answers to Preview 3

1. (a) 5.916 (b) 2.791 (c) 22.97 (d) 2.18225
   (e) 3256.25 (f) 225 (g) 3.045
2. 4.3
3. (a) 0.1875 (b) 3.452 (c) 9.225 (d) 0.305
4. (a) 64 (b) 288 (c) 4.2025
5. (a) 13 (b) 3.81 rounded
6. (a) 3 (b) −7 (c) 4.65 (d) −2.5

3 Decimal Numbers

"Because batteries go dead in pocket calculators, that's why."

3-1 ADDITION AND SUBTRACTION OF DECIMALS

1 By now you know that whole numbers are written in a form based on powers of ten. A number such as

237

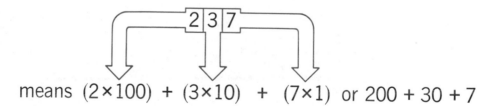

means $(2 \times 100) + (3 \times 10) + (7 \times 1)$ or $200 + 30 + 7$

This way of writing numbers can be extended to fractions. A *decimal* number is a fraction whose denominator is 10 or some multiple of 10.

For example,

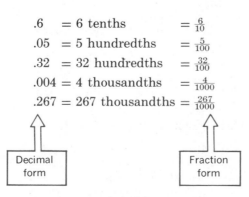

.6	= 6 tenths	$= \frac{6}{10}$
.05	= 5 hundredths	$= \frac{5}{100}$
.32	= 32 hundredths	$= \frac{32}{100}$
.004	= 4 thousandths	$= \frac{4}{1000}$
.267	= 267 thousandths	$= \frac{267}{1000}$
↑ Decimal form		↑ Fraction form

The decimal number .267 can be written in expanded form as

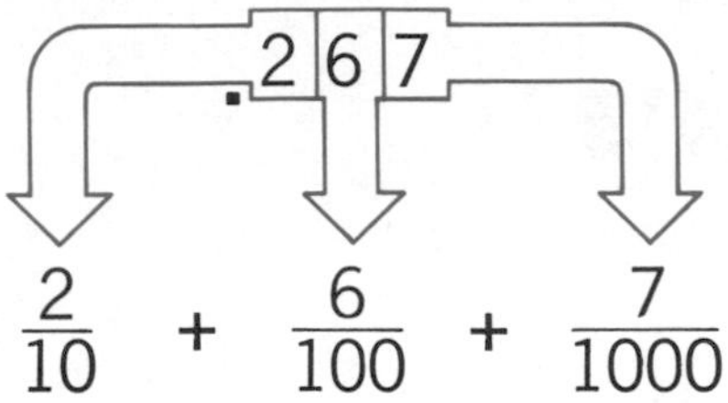

Write the decimal number .526 in expanded form. Check your answer in 2.

---

2 $.526 = \frac{5}{10} + \frac{2}{100} + \frac{6}{1000}$

We would usually write a decimal number less than one with a zero to the left of the decimal point.

.526 would be written 0.526
.4 would be written 0.4
.001 would be written 0.001

It is easy to mistake .4 for 4, but the decimal point in 0.4 cannot be overlooked.

That zero out front will help you remember where the decimal point is located.

A decimal number may have both a whole-number part and a fraction part. For example, the number 324.576 means

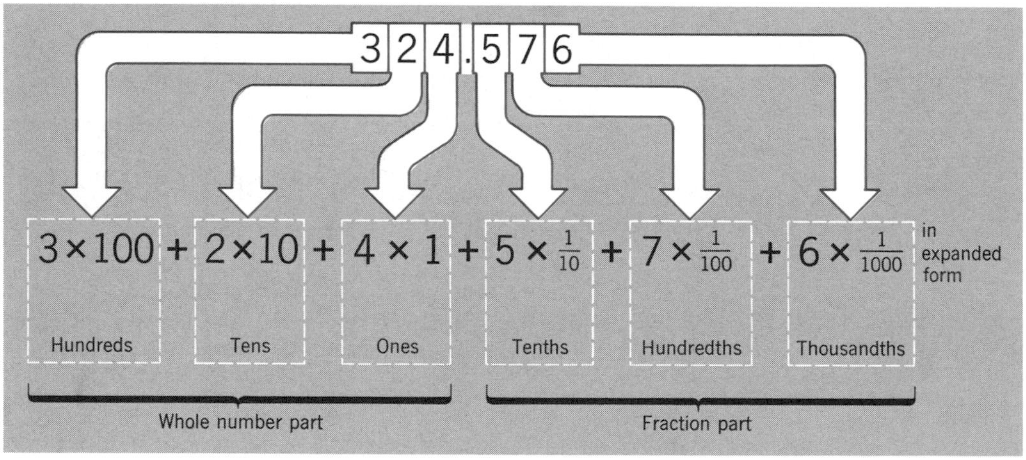

You are already familiar with this way of interpreting decimal numbers from working with money.

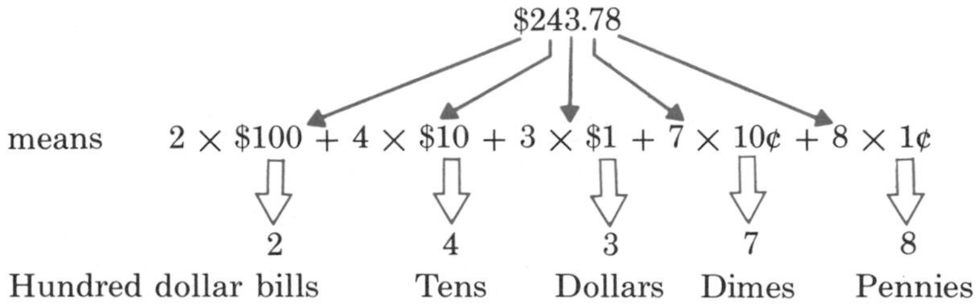

Try to get the idea clear in your mind. Write the following in expanded form.

(a) 86.42 (b) 43.607 (c) 14.5060 (d) 235.22267

Compare your answers with ours in 3.

3 (a) $86.42 = 8 \times 10 + 6 \times 1 + 4 \times \frac{1}{10} + 2 \times \frac{1}{100}$
$= 80 + 6 + \frac{4}{10} + \frac{2}{100}$

(b) $43.607 = 4 \times 10 + 3 \times 1 + 6 \times \frac{1}{10} + 0 \times \frac{1}{100} + 7 \times \frac{1}{1000}$
$= 40 + 3 + \frac{6}{10} + \frac{0}{100} + \frac{7}{1000}$

(c) $14.5060 = 1 \times 10 + 4 \times 1 + \frac{5}{10} + \frac{0}{100} + \frac{6}{1000} + \frac{0}{10000}$

(d) $235.22267 = 2 \times 100 + 3 \times 10 + 5 \times 1 + \frac{2}{10} + \frac{2}{100} + \frac{2}{1000} + \frac{6}{10000} + \frac{7}{100000}$

Notice that the denominators in the decimal fractions change by a factor of 10. For example,

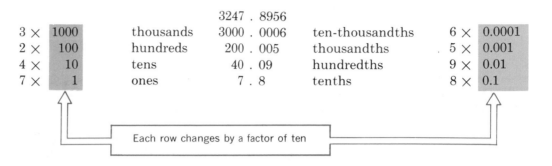

		3247 . 8956			
3 ×	1000	thousands	3000 . 0006	ten-thousandths	6 × 0.0001
2 ×	100	hundreds	200 . 005	thousandths	5 × 0.001
4 ×	10	tens	40 . 09	hundredths	9 × 0.01
7 ×	1	ones	7 . 8	tenths	8 × 0.1

Each row changes by a factor of ten

$1 \times 10 = 10$ $\qquad 0.01 \times 10 = 0.1$
$10 \times 10 = 100$ $\qquad 0.001 \times 10 = 0.01$
$100 \times 10 = 1000$ $\qquad 0.0001 \times 10 = 0.001$

In the decimal number 86.423 the digits 4, 2, and 3 are called *decimal digits.*

The number 43.6708 has four decimal digits: 6, 7, 0, and 8.
The number 5376.2 has one decimal digit: 2.
All digits to the right of the decimal point, those that name the fractional part of the number, are decimal digits.

How many decimal digits are included in each of these numbers?
(a) 1.4 (b) 315.7 (c) 0.425 (d) 324.0075

Count them, then turn to **4**.

4 (a) one (b) one (c) three (d) four

We will use the idea of decimal digits often in doing arithmetic with decimal numbers.

The decimal point is simply a way of separating the whole-number part from the fraction part. It is a place marker. In whole numbers the decimal point usually is not written, but it is understood to be there.

The whole number 2 is written 2. as a decimal.

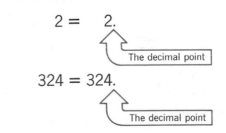

2 = 2.

The decimal point

or 324 = 324.

The decimal point

This is very important. Many people make big mistakes in arithmetic because they do not know where that decimal point should go.

Very often additional zeros are attached to the decimal number without changing its value. For example,

$8.5 = 8.50 = 8.5000$ and so on
$6 = 6. = 6.0 = 6.000$ and so on

The value of the number is not changed but the additional zeros may be useful, as we shall see.

Addition

Because decimal numbers represent fractions with denominators equal to multiples of ten, addition is very simple.

$$\begin{aligned} 2.34 &= 2 + \tfrac{3}{10} + \tfrac{4}{100} \\ +5.23 &= 5 + \tfrac{2}{10} + \tfrac{3}{100} \\ & \overline{7 + \tfrac{5}{10} + \tfrac{7}{100}} = 7.57 \end{aligned}$$

Adding like fractions

Of course we do not need this clumsy business in order to add decimal numbers. As with whole numbers, we may arrange the digits in vertical columns and add directly.

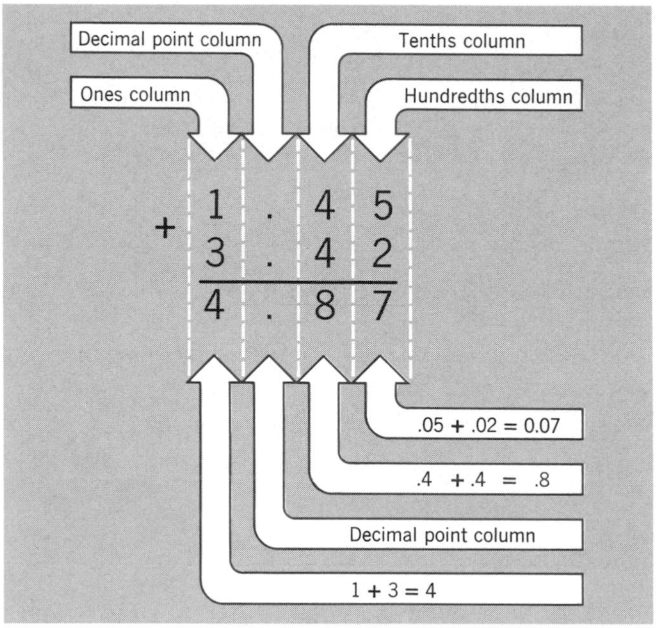

Digits of the same power of ten are placed in the same vertical column. Decimal points are always lined up vertically.

If one of the numbers is written with fewer decimal digits than the other, attach as many zeros as needed so that both have the same number of decimal digits.

2.345		2.345
+1.5	becomes	+1.500

Except for the preliminary step of lining up decimal points, addition of decimal numbers is exactly the same process as addition of whole numbers.

Add the following decimal numbers.

(a) \$4.02 + \$3.67 = ________ (b) 13.2 + 1.57 = ________

(c) 23.007 + 1.12 = ________ (d) 14.6 + 1.2 + 3.15 = ________

(e) 5.7 + 3.4 = ______ (f) \$9 + \$0.72 + \$6.09 = ______

(g) 0.07 + 6.79 + 0.3 + 3 = ______ (h) 0.36 + 17 + 3.9 + 0.6 = ______

Arrange each sum vertically, placing the decimal points in the same column, then add as with whole numbers.

Check your work in **5**.

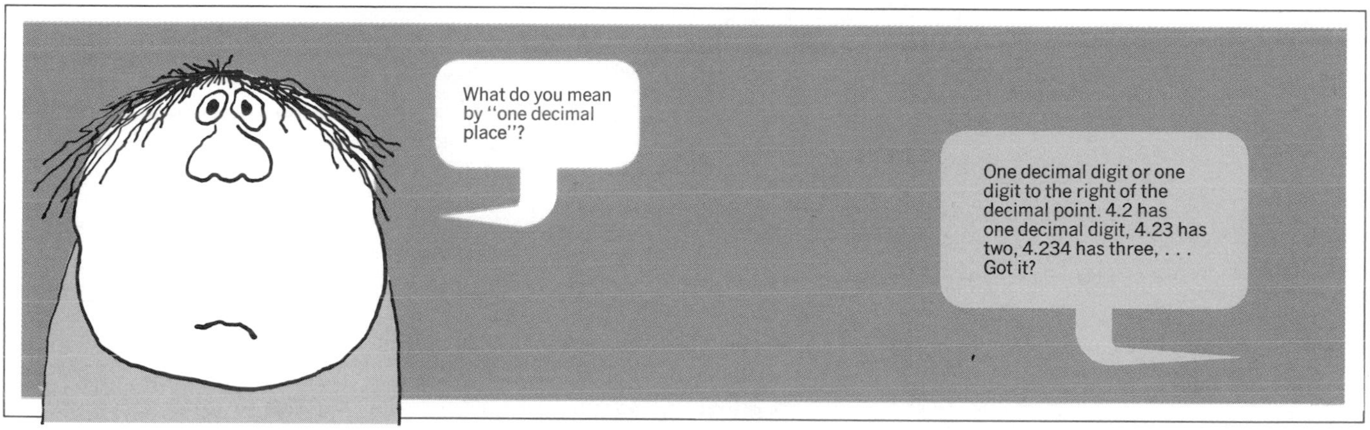

---

5 (a)

```
 $4.02      ← Decimal points in line vertically
 $3.67
 -----
 $7.69     0.02 + 0.07 = 0.09    Add cents
           0.0  + 0.6  = 0.6     Add 10¢ units
           4    + 3    = 7       Add dollars
```

As a check, notice that the sum is roughly \$4 + \$3 or \$7, which agrees with the actual answer. Always check your answer by first estimating it, then comparing your estimate or rough guess with the final answer.

(b)

```
 13.20  ← Decimal points in line
  1.57    Annex a zero to provide the same number of
 -----    decimal digits as in the other addend
 14.77  ← Place answer decimal point in the same vertical line
```

Check: 13 + 1 = 14, which agrees roughly with the answer.

(c)

```
   23.007
 +  1.120  ← Attach extra zero
 --------
   24.127
```

(d)

```
   14.60  ←
    1.20  ← Attach extra zeros
 +  3.15
 -------
   18.95
```

(e)

```
   1
   5.7
 + 3.4
 -----
   9.1    0.7 + 0.4 = 1.1     Write 0.1
                              Carry 1
          Carry 1 + 5 + 3 = 9
```

(f)

```
     1
 $ 9.00
   0.72
   6.09
 ------
 $15.81
```

(g)

```
  1 1
  0.07
  6.79
  0.30  ←
  3.00  ← Attach extra zeros
 -----
 10.16
```

(h)

```
  11
  0.36
 17.00  ←
  3.90  ←
  0.60  ← Attach extra zeros
 -----
 21.86
```

Careful You must line up the decimal points carefully to be certain of getting a correct answer.

Subtraction

Subtraction is equally simple if you are careful to line up decimal points carefully and attach any needed zeros before you begin work.

For example, $437.56 − $41 = ______

is
```
   3 13
  $437.56   ← Decimal points in a vertical line
 −$ 41.00   ← Attach zeros (remember that $41 is $41. or $41.00)
  $396.56
```

or again 19.452 − 7.3617 = ______

```
   3 15110   ← Decimal points in a vertical line
  19.4520    ← Attach zero
 − 7.3617
  12.0903
    ↑ Answer decimal point in same vertical line
```

Try these problems to test yourself on the subtraction of decimal numbers.

(a) $37.66 − $14.57 = ______ (b) 248.3 − 135.921 = ______

(c) 6.4701 − 3.2 = ______ (d) 7.304 − 2.59 = ______

(e) $20 − $7.74 = ______ (f) 36 − 11.132 = ______

(g) 10 − 0.037 = ______

Work carefully. The answers are in **6**.

---

6 (a)
```
     5 16   ← Line up decimal points
  $37.66
 −$14.57
  $23.09
```
Check: $14.57 + $23.09 = $37.66

(b)
```
  248.300   ← Line up decimal points; Attach zeros
 −135.921
  112.379
     ↑ Answer decimal point in the same vertical line
```
Check: 135.921 + 112.379 = 248.300

(c)
```
  6.4701
 −3.2000   ← Attach zeros
  3.2701
```
Check: 3.2000 + 3.2701 = 6.4701

(d)
```
  6 1210
  7.304
 −2.590   ← Attach zero
  4.714
```
Check: 2.590 + 4.714 = 7.304

(e)
```
   19 910
  $20.00   ← Attach zeros
 −$ 7.74
  $12.26
```
Check: $7.74 + $12.26 = $20

(f)
```
  5 9910
  36.000   ← Attach zeros
 −11.132
  24.868
```
Check: 11.132 + 24.868 = 36.000

```
         9 9 9 10
(g)      1 0.0 0 0  ←── Attach zeros
      −    0.037
      ───────────
           9.963       Check:  0.037 + 9.963 = 10.000
```

Notice that each problem is checked by comparing the sum of the answer and the number subtracted with the first number. You should also start by estimating the answer. Whether you use a calculator or work it out with pencil and paper, checking your answer is important if you are to avoid careless mistakes.

Now, for a set of practice problems on addition and subtraction of decimal numbers, turn to **7**.

---

7 **Exercises 3-1 Addition and Subtraction of Decimals**

A. Add or Subtract as shown:

1. 14.21 + 6.8	2. 75.6 + 2.57
3. \$2.83 + \$12.19	4. \$52.37 + \$98.74
5. 0.687 + 0.93	6. 0.096 + 5.82
7. 507.18 + 321.42	8. 212.7 + 25.46
9. 45.6725 + 18.058	10. 390 + 72.04
11. 19 − 12.03	12. 7.83 − 6.79
13. \$33.40 − \$18.04	14. \$20.00 − \$13.48
15. 75.08 − 32.75	16. 40 − 3.82
17. \$30 − \$7.98	18. \$25 − \$0.61
19. 130 − 16.04	20. 19 − 5.78
21. 37 + 0.09 + 3.5 + 4.605	22. 183 + 3.91 + 45 + 13.2
23. \$14.75 + \$9 + \$3.76	24. 148.002 + 3.4
25. 68.708 + 27.18	

B. Practical Problems

1. What is the combined thickness of these 5 shims: 0.008 in., 0.125 in., 0.150 in., 0.185 in., and 0.005 in.?

2. The combined weight of a spool and the wire it carries is 13.6 lb. If the weight of the spool is 1.75 lb, what is the weight of the wire?

3. The following are diameters of some common household wires: #10 is 0.102 in., #11 is 0.090 in., #12 is 0.081 in., #14 is 0.064 in., and #16 is 0.051 in.

   a. The diameter of #16 wire is how much smaller than #14 wire's?
   b. Is #12 wire larger or smaller than #10 wire? What is the difference in their diameters?
   c. John measured the thickness of a wire with a micrometer as 0.066 in. Assuming the manufacturer was slightly off, what wire size did John have?

4. In estimating a finishing job, a painter included the following items:

Material	\$377.85
Trucking	\$ 15.65
Profit	\$250
Labor	\$745.50
Overhead	\$ 37.75

   What was his total estimate for the job?

5. Find A, B, and C.

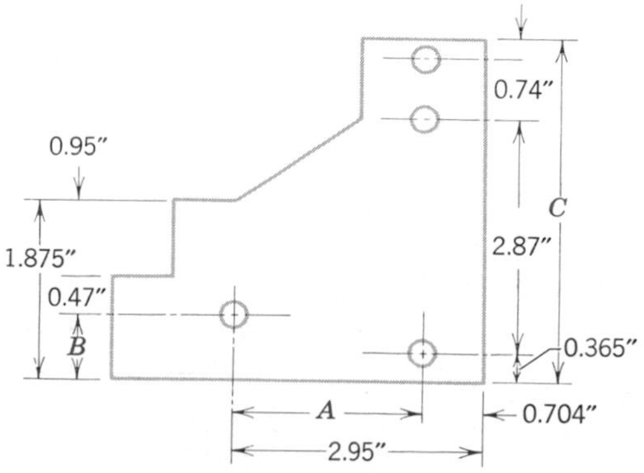

6. A piece of pipe 3.4′ long is shortened by 0.7′. How long are the two pieces remaining after it is cut if the width of the saw cut is 0.125″?

7. A certain machine part is 2.345 in. thick. What is its thickness after 0.078 in. is ground off?

C. Calculator Problems

1. Balance this checkbook by finding the closing balance as of November 4.

Date	Balance	Withdrawals	Deposits
Oct. 1	\$367.21		
Oct. 3		\$167.05	
Oct. 4		104.97	
Oct. 8			\$357.41
Oct. 16		87.50	
Oct. 18		9.43	
Oct. 20		30.09	
Oct. 22			364.85
Oct. 27		259.47	
Oct. 30		100.84	
Nov. 2		21.88	
Nov. 4	?		

2. What is the actual cost of the following car?

"Sticker Price"	\$5745.00
Dealer preparation	106.50
Shipping/handling	249.90
Air conditioning	475.40
Power steering	139.00
AM-FM radio	119.95
CB radio	105.98
White sidewall tires	35.50
Tax and license	518.62
Less trade-in	\$1780.00

3. Add as shown:

(a)	(b)	(c)
0.0067	1379.4	14.07
0.032	204.5	67.81
0.0012	16.75	132.99
0.0179	300.04	225.04
0.045	2070.08	38.02
0.5	167.99	4
0.05	43.255	16.899
0.0831	38.81	7.007
0.004	19.95	4.6

(d) 0.002 + 17.1 + 4.806 + 9.9981 − 3.1 + 0.701 − 1.001 − 14 − 8.09 + 1.0101

Check your answers on page 545, then continue in frame 8.

3-2 MULTIPLICATION OF DECIMAL NUMBERS

A decimal number is really a fraction with a power of 10 as denominator. For example,

8 $0.5 = \frac{5}{10}$, $0.3 = \frac{3}{10}$, and $0.85 = \frac{85}{100}$

Multiplication of decimals is easy to understand if we think of it in this way:

$0.5 \times 0.3 = \frac{5}{10} \times \frac{3}{10} = \frac{15}{100} = 0.15$

To estimate the answer remember that since both 0.5 and 0.3 are less than 1, their product must be less than one.

Of course it would be very, very clumsy and time-consuming to calculate every decimal multiplication this way. We need a simpler method. Here is the procedure most often used:

Step 1 Multiply the two decimal numbers as if they were whole numbers. Pay no attention to the decimal points.

Step 2 The sum of the decimal digits in the two numbers being multiplied will give you the number of decimal digits in the answer.

For example, to multiply 3.2 by 0.41:

Step 1 Multiply, ignoring the decimal points.

$$\begin{array}{r} 32 \\ \times 41 \\ \hline 1312 \end{array}$$

Step 2 Count decimal digits in each number: 3.2 has *one* decimal digit (the 2), and 0.41 has *two* decimal digits (the 4 and the 1). The total number of decimal digits in the two factors is three. The answer will have *three* decimal digits. Count over *three* digits from right to left in the answer.

1.312 three decimal digits

Check: 3.2×0.41 is roughly $3 \times \frac{1}{2}$ or about $1\frac{1}{2}$. The answer 1.312 agrees with our rough guess. Remember, even if you use a calculator to do the actual work of arithmetic, you must *always* estimate your answer first and check it afterward.

Try these simple decimal multiplications:

(a) $2.5 \times 0.5 =$ ______ (b) $0.1 \times 0.1 =$ ______

(c) $10 \times 0.6 =$ ______ (d) $2 \times 0.4 =$ ______

(e) $1 \times 0.1 =$ ______ (f) $2 \times 0.003 =$ ______

(g) $0.01 \times 0.02 =$ ______ (h) $0.04 \times 0.005 =$ ______

Check your answers in **9**

"Wilkens, you've just been replaced by a pocket calculator"
BURBANK

"From Cartoon Features Syndicate"

---

9 (a) $2.5 \times 0.5 =$ ________

First, multiply $25 \times 5 = 125$. Second, count decimal digits.

2.5×0.5

One decimal digit	One decimal digit

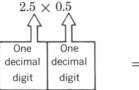

= a total of *two* decimal digits.

Count over *two* decimal digits from the right:

1.25

two decimal digits. The product is 1.25.

Check: $2 \times \frac{1}{2}$ is about 1, so the answer seems reasonable.

(b) $0.1 \times 0.1 \qquad 1 \times 1 = 1$

Count over *two* decimal digits from the right. Since there are not two decimal digits in the product, attach a few zeros on the left.

0.01

two decimal digits

So $0.1 \times 0.1 = 0.01$.

Check: $\frac{1}{10} \times \frac{1}{10} = \frac{1}{100}$.

(c) $10 \times 0.6 \qquad 10 \times 6 = 60$

Count over *one* decimal digit from the right: 6.0 so that $10 \times 0.6 = 6.0$.

Notice that multiplication by 10 simply shifts the decimal place one digit to the right.

$10 \times 6.2 = 62$
$10 \times 0.075 = 0.75$
$10 \times 8.123 = 81.23$ and so on.

(d) $2 \times 0.4 \qquad 2 \times 4 = 8$

Count over *one* decimal digit.

.8 $\quad 2 \times 0.4 = 0.8$

(e) 1×0.1 $1 \times 1 = 1$

One decimal digit

Count over 1 decimal digit.

.1 $1 \times 0.1 = 0.1$

(f) 2×0.003 $2 \times 3 = 6$

Three decimal digits

Count over 3 decimal digits.

0.006 $2 \times 0.003 = 0.006$

(g) 0.01×0.02 $1 \times 2 = 2$

Two decimal digits | Two decimal digits

Count over 4 decimal digits.

0.0002 $0.01 \times 0.02 = 0.0002$

Total of 4 decimal digits

Four decimal digits

(h) 0.04×0.005 $4 \times 5 = 20$

Two decimal digits | Three decimal digits

Count over 5 decimal digits.

0.00020

Total of 5 decimal digits

Five decimal digits

Remember:

Do not try to do this entire process mentally until you are certain you will not misplace zeros.

Always estimate before you begin the arithmetic, and check your answer against your estimate.

Multiplication of larger decimal numbers is performed in exactly the same manner. Try these:

(a) $4.302 \times 12.05 =$ ________

(b) $6.715 \times 2.002 =$ ________

(c) $3.144 \times 0.00125 =$ ________

Look in **10** for the answers.

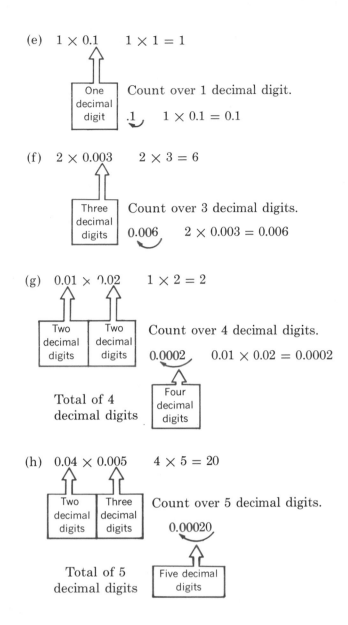

10 (a) **Estimate:** $4 \times 12 = 48$ The answer will be about 48.

Multiply

$$\begin{array}{r} 4302 \\ \times 1205 \\ \hline 5183910 \end{array}$$

(If you cannot do this multiplication correctly, turn to page 21 in Section 1-1 for help with the multiplication of whole numbers.)

The two factors being multiplied have a total of five decimal digits (three in 4.302 and two in 12.05). Count over five decimal digits from the right in the answer.

51.83910

So that $4.302 \times 12.05 = 51.8391$

Check: The answer 51.8391 is approximately equal to the estimate of 48.

(b) **Estimate:** 6.7×2 is about 7×2 or 14.

Multiply		
	6.715	6.715 has *three* decimal digits
	×2.002	2.002 has *three* decimal digits
	13.443430	a total of *six* decimal digits

six decimal digits

$6.715 \times 2.002 = 13.44343$ which agrees with our estimate.

Check: The answer and the estimate are roughly equal.

(c) **Estimate:** 3×0.001 is about 0.003

Multiply		
	3.144	3.144 has *three* decimal digits
	×0.00125	0.00125 has *five* decimal digits
	.00363000	a total of *eight* decimal digits

eight decimal digits.

$3.144 \times 0.00125 = 0.00363$

Check: The answer and estimate are roughly equal.

One major convenience of a pocket calculator is that it counts decimal digits and automatically adjusts the answer in any calculation. But to get an estimate of the answer, you must still understand the process.

Now turn to **11** for a look at the division of decimal numbers.

3-3 DIVISION OF DECIMAL NUMBERS

11 Division of decimal numbers is very similar to the division of whole numbers. For example,

$6.8 \div 1.7$ can be written $\frac{6.8}{1.7}$

and if we multiply both top and bottom of the fraction by 10,

$$\frac{6.8}{1.7} = \frac{6.8 \times 10}{1.7 \times 10} = \frac{68}{17}$$

But you should know how to divide these whole numbers.

$68 \div 17 = 4$

Therefore $6.8 \div 1.7 = 4$

To divide decimal numbers, use the following procedure:

		Example:
Step 1	Write the divisor and dividend in standard long division form.	$6.8 \div 1.7$ $1.7\overline{)6.8}$
Step 2	Shift the decimal point in the divisor to the right so as to make the divisor a whole number.	$1.7.\overline{)\quad}$
Step 3	Shift the decimal point in the dividend the same amount. (Add zeros if necessary.)	$1.7.\overline{)6.8.}$
Step 4	Place the decimal point in the answer space directly above the new decimal position in the dividend.	$17.\overline{)68.}$
Step 5	Complete the division exactly as you would with whole numbers. The decimal points in divisor and dividend may now be ignored.	4. $17.\overline{)68.}$ 68

$6.8 \div 1.7 = 4$

Notice in steps 2 and 3 that we have simply multiplied both divisor and dividend by 10.

Repeat the process above with this division:

$1.38 \div 2.3$

Work carefully, then compare your work with ours in **12**.

---

12 Let's do it step-by-step.

$2.3\overline{)1.38}$

$2.3\overline{)1.38}$ Shift both decimal points one digit to the right to make the divisor (2.3) a whole number (23).

$23.\overline{)13.8}$ Place the decimal point in the answer space.

```
     .6      Divide as usual.     23 goes into 138 6 times.
23.)13.8
    138      6 × 23 = 138
```

$1.38 \div 2.3 = 0.6$

Check: $1.38 \div 2.3$ is roughly $1 \div 2$ or 0.5

Double-check: $2.3 \times 0.6 = 1.38$. Always multiply the answer by the divisor to double-check the work. Do this even if you are using an electronic calculator.

How would you do this one?

$2.6 \div 0.052 = ?$

Look in **13** for the solution after you have tried it.

---

13 $0.052.\overline{)2.6}$

To shift the decimal place three digits in the dividend, we must attach two zeros to its right.

$0.052.\overline{)2.600.}$ Now place the decimal point in the answer space above that in the dividend.

```
       50.
52.)2600.
    260 ←——— 5 × 52 = 260
      0
      0
```

2.6 ÷ 0.052 = 50 **Check:** 0.052 × 50 = 2.6

Shifting the decimal point three digits and attaching zeros to the right of the decimal point in this way is equivalent to multiplying both divisor and dividend by 1000.

Try these problems:

(a) 3.5 ÷ 0.001 = ______ (b) 9 ÷ 0.02 = ______

(c) .365 ÷ 18.25 = ______ (d) 8.8 ÷ 3.2 = ______

(e) 7.230 ÷ 6 = ______ (f) 3 ÷ 4 = ______

The answers are in **14**.

---

14 (a)

```
0.001.)3.500.

       3500.
    1.)3500.
```

3.5 ÷ 0.001 = 3500

Check: 0.001 × 3500 = 3.5

(b)

```
0.02)9.00.

   450.
2.)900.
```

Shift the decimal point two places to the right. Divide 900 by 2.

9 ÷ 0.02 = 450

Check: 0.02 × 450 = 9

This is a problem that is very troublesome for most people.

(c)

```
18.25.).36.5

           .02
1825.)36.50
      36.50
```

1825 does not go into 365, so place a zero above the 5.

2 × 1825 = 3650

0.365 ÷ 18.25 = 0.02

Check: 18.25 × 0.02 = 0.365

(d)

```
3.2.)8.8.

     2.75
32.)88.00
    64        2 × 32 = 64
    240
    224       7 × 32 = 224
     160
     160      5 × 32 = 160
```

8.8 ÷ 3.2 = 2.75

Check: The estimated answer is 9 ÷ 3 or 3. **Double-check:** 3.2 × 2.75 = 8.8

```
      1.205
(e) 6)7.230     
      6
      1 2
      1 2
        30
        30
```

The divisor 6 is a whole number, so we can bring the decimal point in 7.23 up to the answer space.

Check: $1.205 \times 6 = 7.230$

```
       .75
(f) 4)3.00
      2 8
       20
       20
```

Check: $0.75 \times 4 = 3.00$

If the dividend is not exactly divisible by the divisor, we must either stop the process after some pre-set number of decimal places in the answer, or we must round the answer. We do not generally indicate a remainder in decimal division.

Turn to **15** for some rules for rounding.

Rounding

15 *Rounding* is a process of approximating a number. To round a number means to find another number roughly equal to the given number but expressed less precisely. For example,

$432.57 = $400 rounded to the nearest hundred dollars,
= $430 rounded to the nearest ten dollars,
= $433 rounded to the nearest dollar, and so on.

1.376521 seconds is equal to 1.377 seconds rounded to three decimal digits,

or 1.4 seconds rounded to the nearest tenth of a second,

or 1 second rounded to the nearest whole number second.

There are exactly 5280 feet in 1 mile. To the nearest thousand feet how many feet are in one mile? To the nearest hundred feet?

Check your answers in **16** .

HOW TO NAME DECIMAL NUMBERS

The decimal number 3,254,935.4728 should be interpreted as

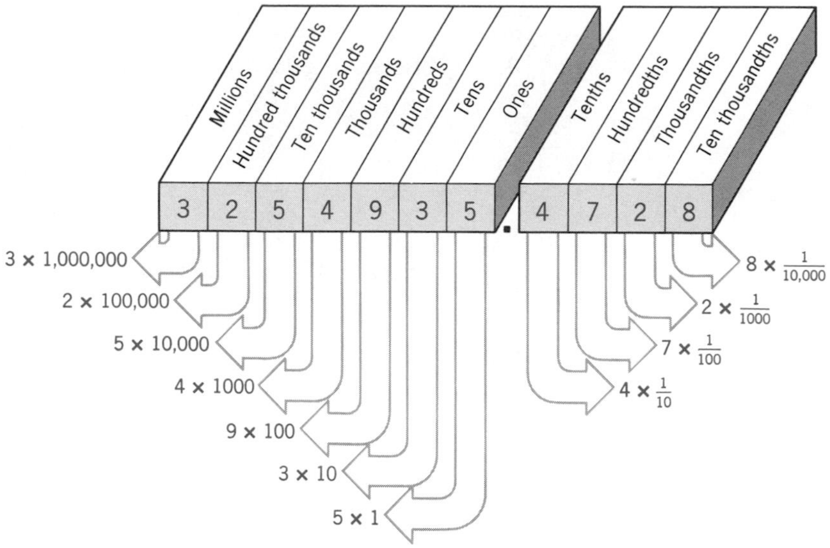

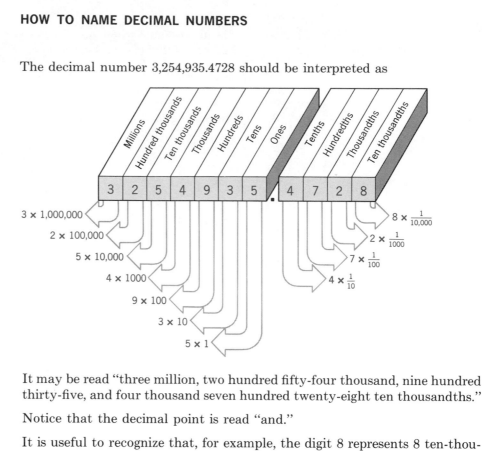

It may be read "three million, two hundred fifty-four thousand, nine hundred thirty-five, and four thousand seven hundred twenty-eight ten thousandths."

Notice that the decimal point is read "and."

It is useful to recognize that, for example, the digit 8 represents 8 ten-thousandths or $\frac{8}{10,000}$ and the digit 7 represents 7 hundredths or $\frac{7}{100}$. Most often, however, this number is read more simply as "three million, two hundred fifty-four thousand, nine hundred thirty-five, *point* four, seven, two, eight." This way of reading the number is easiest to write, to say, and to understand.

---

16 5280 ft = 5000 ft rounded to the nearest thousand feet. In other words, 5280 is closer to 5000 than to 4000 or 6000.

5280 ft = 5300 ft rounded to the nearest hundred feet. In other words, 5280 is closer to 5300 than to 5200 or 5400.

For most rounding, follow this simple rule:

Example:

Step 1	Determine the number of digits or the place to which the number is to be rounded. Mark it with a $_\wedge$.	Round 3.462 to one decimal place 3.4$_\wedge$62
Step 2	If the digit to the right of the mark is less than 5, replace all digits to the right of the mark by zeros. If the zeros are decimal digits, you may discard them.	2.8$_\wedge$32 becomes 2.8$_\wedge$00 or 2.8
Step 3	If the digit to the right of the mark is equal to or larger than 5, increase the digit to the left by 1.	3.4$_\wedge$62 becomes 3.5

Try it with these.

(a) Round 74.238 to two decimal places.
(b) Round 8.043 to two decimal places.
(c) Round 156 to the nearest hundred.
(d) Round 6.07 to the nearest tenth.

Follow the rules, round them, then check your work in **17**.

---

17 (a) 74.238 = 74.24 to two decimal places
(Write 74.23‸8 and note that 8 is larger than 5, so increase the 3 to 4.)

(b) 8.043 = 8.04 to two decimal places
(Write 8.04‸3 and note that 3 is less than 5 so drop it.)

(c) 156 = 200 to the nearest hundred
(Write 1‸56 and note that the digit to the right of the mark is 5 so increase the 1 to 2 and attach two zeros.)

(d) 6.07 = 6.1 to the nearest tenth
(Write 6.0‸7 and note that 7 is greater than 5 so increase the 0 to 1.)

There are a few very specialized situations where this rounding rule is not used:

1. Some engineers use a more complex rule when rounding a number that ends in 5.
2. In business, fractions of a cent are usually rounded up to determine selling price. 3 items for 25¢ or $8\frac{1}{3}$¢ each is rounded to 9¢ each.

Our rule will be quite satisfactory for most of your work in arithmetic.

To round answers in a division problem, first continue the division so that your answer has one place more than the rounded answer will have, then round it. For example, in the division problem

4.7 ÷ 1.8 = ?

to get an answer rounded to one decimal place, first, divide to two decimal places

4.7 ÷ 1.8 = 2.61

then round back to one decimal place

4.7 ÷ 1.8 = 2.6, rounded.

For the following problem, divide and round your answer to two decimal places.

6.84 ÷ 32.7 = ________

Careful now.

Check your work in **18**.

---

18 32.7.)6.8.4

```
         .209     Carry the answer to three decimal places.
327.)68.400
     65 4↓↓       2 × 327 = 654
      3 000
      2 943       9 × 327 = 2943
```

0.209 rounded to two decimal places is 0.21.

6.84 ÷ 32.7 = 0.21, rounded to two decimal places.

Check: $32.7 \times 0.21 = 6.867$, which is approximately equal to 6.84. (The check will not be exact because we have rounded.)

The ability to round numbers is especially important for people who work in the practical, trade, or technical areas. You will need to round answers to practical problems if they are obtained "by hand" or with a calculator. We will discuss rounding further in the section of this book on measurement.

Now turn to **19** to continue.

3-4 AVERAGES

19 Suppose you needed to know the diameter of a steel connecting pin. As a careful and conscientious worker, you would probably measure its diameter several times with a micrometer, and you might come up with a sequence of numbers like this:

1.3731″, 1.3728″, 1.3736″, 1.3749″, 1.3724″, 1.3750″

What is the actual diameter of the pin? The best answer is to find the *average* value or arithmetic mean of these measurements.

$$\text{Average} = \frac{\text{sum of the measurements}}{\text{number of measurements}}$$

For the problem above,

$$\begin{aligned}\text{Average} &= \frac{1.3731 + 1.3728 + 1.3736 + 1.3749 + 1.3724 + 1.3750}{6} \\ &= \frac{8.2418}{6} \\ &= 1.3736333\ldots \\ &= 1.3736 \text{ rounded to four decimal places}\end{aligned}$$

When you calculate the average of a set of numbers, the usual rule is to round the answer to the same number of decimal places as the least accurate number in the set. If the numbers to be averaged are all whole numbers, the average will be a whole number.

Example: The average of 4, 6, 4, and 5 is $\frac{19}{4} = 4.75$ or 5 when rounded.

If one number of the set has fewer decimal places than the rest, round off to agree with the least accurate number.

Example: The average of 2.41, 3.32, 5.23, 3.51, 4.1, and 4.12 is $\frac{22.69}{6} = 3.78166\ldots$

and this answer should be rounded to 3.8 to agree in accuracy with the least accurate number 4.1.

To avoid confusion, we will usually give directions for rounding.

Try it. Find the average for each of these sets of numbers.

(a) 8, 9, 11, 7, 5

(b) 0.4, 0.5, 0.63, 0.2

(c) 2.35, 2.26, 2.74, 2.55, 2.6, 2.31

Check your work in **20**.

---

20 (a) Average $= \dfrac{8+9+11+7+5}{5} = \dfrac{40}{5} = 8$

(b) Average $= \dfrac{0.4+0.5+0.63+0.2}{4} = \dfrac{1.73}{4} = 0.4325 = 0.4$ rounded

(c) Average $= \dfrac{2.35+2.26+2.74+2.55+2.6+2.31}{6}$

$= \dfrac{14.81}{6}$

$= 2.468333\ldots$

$= 2.5$ rounded to agree with 2.6, the least accurate number in the set.

Now turn to **21** for a set of exercises on the multiplication and division of decimal numbers.

---

21 **Exercises 3-2 Multiplication and Division of Decimals**

A. Multiply or divide as shown:

1. 0.01×0.001
2. 10×2.15
3. 0.04×100
4. 0.3×0.3
5. 0.7×1.2
6. 0.005×0.012
7. 0.003×0.01
8. 7.25×0.301
9. 2×0.035
10. $0.2 \times 0.3 \times 0.5$
11. $0.6 \times 0.6 \times 6.0$
12. $2.3 \times 1.5 \times 1.05$
13. $3.618 \div 0.6$
14. $3.60 \div 0.03$
15. $4.40 \div 0.22$
16. $6.5 \div 0.05$
17. $0.0405 \div 0.9$
18. $0.378 \div 0.003$
19. $3 \div 0.05$
20. $10 \div 0.001$
21. $4 \div 0.01$
22. $2.59 \div 70$
23. $44.22 \div 6.7$
24. $104.2 \div 0.0320$
25. $484 \div 0.8$

B. Divide and round as indicated.
Round to two decimal digits:

1. $10 \div 3$
2. $5 \div 6$
3. $2.0 \div 0.19$
4. $3 \div 0.081$
5. $0.023 \div 0.19$
6. $12.3 \div 4.7$
7. $2.37 \div 0.07$
8. $6.5 \div 1.31$

Round to the nearest tenth:

9. $100 \div 3$
10. $21.23 \div 98.7$
11. $1 \div 4$
12. $100 \div 9$
13. $0.006 \div 0.04$
14. $1.008 \div 3$

Round to three decimal places:

15. $10 \div 70$
16. $0.09 \div 0.402$
17. $0.091 \div 0.0014$
18. $3.41 \div 0.257$
19. $6.001 \div 2.001$
20. $123.21 \div 0.1111$

C. Word Problems

1. A color television set is advertised for \$420. It can also be bought "on time" for 24 payments of \$22.75 each. How much extra do you pay by purchasing it on the installment plan?

2. The telephone rates between Santa Barbara and Zanzibar are $10.75 for the first 3 minutes and $1.95 for each additional minute. What would be the cost of an 11 minute telephone call?

3. Find the average weight of five castings that weigh 17 lb, 21 lb, 12 lb, 20.6 lb, and 23.4 lb.

4. If you work 8.2 hours on Monday, 10.1 hours on Tuesday, 8.5 hours on Wednesday, 9.4 hours on Thursday, 6.5 hours on Friday, and 4.2 hours on Saturday, what is the average number of hours worked per day?

5. For the following four machine parts, find *W*, the number of pounds per part; *C*, the cost of the metal per part; and *T*, the total cost.

Metal Parts	Number of Feet Needed	Number of Pounds per Foot	Cost per Pound	Pounds (*W*)	Cost per Part (*C*)
A	3.7	4.5	$ .98		
B	10.2	2.3	$ .89		
C	9	0.95	$1.05		
D	0.75	3.7	$2.15		

6. How much do 15.7 sq ft of No. 16 gauge steel weigh if 1 sq ft weighs 2.55 lb?

7. A 4 ft-by-8 ft sheet of $\frac{1}{4}$-in. plywood has an area of 32 sq ft. If the weight of $\frac{1}{4}$-in. plywood is 1.5 lb/sq ft, what is the weight of the sheet?

8. Each inch of 1 in. diameter cold-rolled steel weighs 0.22 lb. How much would a piece weigh that was 38 in. long?

D. Calculator Problems

1. Divide: 9.87654321 $\div$ 1.23456789.

   Notice anything interesting? (Divide it out to 7 decimal digits.) You should be able to get the correct answer even if your calculator will not accept a nine-digit number.

2. Divide: (a) $\frac{1}{81}$ (b) $\frac{1}{891}$ (c) $\frac{1}{8991}$

   Notice a pattern? (Divide them out to about 8 decimal places.)

3. The outside diameter of a steel casting is measured six times at different positions with a vernier caliper. The measurements are 4.2435″, 4.2426″, 4.2441″, 4.2436″, 4.2438″, and 4.2432″. Find the average diameter of the casting.

4. To determine the thickness of a sheet of paper, five batches of 12 sheets each are measured with a micrometer. The thickness of each of the five batches of 12 is 0.7907 mm, 0.7914 mm, 0.7919 mm, 0.7912 mm, and 0.7917 mm. Find the average thickness of a single sheet of paper.

5. The John Hancock Towers office building in Boston had a serious problem when it was built: When the wind blew hard the pressure caused its windows to pop out! The only reasonable solution was to replace all 10,344 windows in the building at a cost of $6,000,000. What was the replacement cost per window? Round to the nearest dollar.

When you have finished these exercises, check your answers on page 546, then continue in **22**.

22 Since decimal numbers are fractions, they may be used, as fractions are used, to represent a part of some quantity. For example, recall that

"$\frac{1}{2}$ of 8 equals 4"   means $\frac{1}{2} \times 8 = 4$

and, therefore,

"0.5 of 8 equals 4"   means $0.5 \times 8 = 4$

The word *of* used in this way indicates multiplication, and decimal numbers are often called "decimal fractions."

It is very useful to be able to convert any number from fraction form to decimal form. Simply divide the top by the bottom of the fraction. If the division has no remainder, the decimal number is called a *terminating* decimal. For example,

$\frac{5}{8} = \underline{\quad ? \quad}$

```
   .625
8)5.000 ←—Attach as many zeros as needed
  4 8
  ---
   20
   16
   --
    40
    40
    --
     0 ←—Zero remainder; therefore the decimal terminates or ends
```

$\frac{5}{8} = 0.625$

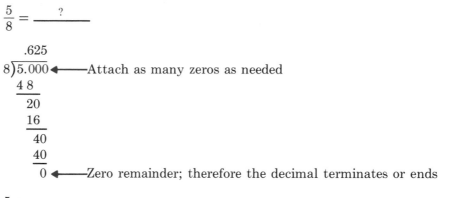

If the decimal does not terminate, you may round it to any desired number of decimal digits. For example,

$\frac{2}{13} = \underline{\quad ? \quad}$

```
    .1538
13)2.0000 ←—Attach zeros
   1 3
   ---
    70
    65
    --
     50
     39
     --
     110
     104
     ---
       6 ←—Remainder not equal to zero
```

$\frac{2}{13} = 0.154$ rounded to three decimal places.

Convert the following fractions to decimal form and round to two decimal places if they are not terminating decimals.

(a) $\frac{4}{5}$ (b) $\frac{2}{3}$ (c) $\frac{17}{7}$ (d) $\frac{5}{6}$ (e) $\frac{7}{16}$ (f) $\frac{5}{9}$

Our work is in 23.

23 (a) $\frac{4}{5} = \underline{\quad ? \quad}$

$$\begin{array}{r} .8 \\ 5\overline{)4.0} \end{array}$$

$\frac{4}{5} = 0.8$

(b) $\frac{2}{3} = \underline{\quad ? \quad}$

$$\begin{array}{r} .666\ldots \\ 3\overline{)2.000} \\ \underline{1\,8} \\ 20 \\ \underline{18} \\ 20 \\ \underline{18} \\ 2 \end{array}$$

$\frac{2}{3} = 0.67$, rounded to two decimal places.

Notice that, in order to round to *two* decimal places, we must carry the division out to at least *three* decimal digits.

(c) $\frac{17}{7} = \underline{\quad ? \quad}$

$$\begin{array}{r} 2.428 \\ 7\overline{)17.000} \\ \underline{14} \\ 3\,0 \\ \underline{2\,8} \\ 20 \\ \underline{14} \\ 60 \\ \underline{56} \\ 4 \end{array}$$

$\frac{17}{7} = 2.43$, rounded to two decimal digits.

(d) $\frac{5}{6} = \underline{\quad ? \quad}$

$$\begin{array}{r} .833\ldots \\ 6\overline{)5.000} \\ \underline{4\,8} \\ 20 \\ \underline{18} \\ 20 \\ \underline{18} \\ 2 \end{array}$$

$\frac{5}{6} = 0.83$, rounded to two decimal places.

(e) $\frac{7}{16} = \underline{\quad ? \quad}$

$$\begin{array}{r} .4375 \\ 16\overline{)7.0000} \\ \underline{6\,4} \\ 60 \\ \underline{48} \\ 120 \\ \underline{112} \\ 80 \\ \underline{80} \\ 0 \end{array}$$

$\frac{7}{16} = 0.4375$, or 0.44 rounded to two decimal digits.

(f) $\frac{5}{9} = \underline{\quad ? \quad}$

$$\begin{array}{r} .55\ldots \\ 9\overline{)5.00} \\ \underline{4\,5} \\ 50 \\ \underline{45} \\ 5 \end{array}$$

$\frac{5}{9} = 0.555\ldots$ or 0.56, rounded.

Decimal numbers that do not terminate will repeat a sequence of digits. This kind of decimal number is called a *repeating* decimal. For example,

$\frac{1}{3} = 0.3333\ldots$

where the three dots are read "and so on" and they tell us that the digit 3 continues without end.

Similarly, $\frac{2}{3} = 0.6666\ldots$ and $\frac{3}{11}$ is

$$\begin{array}{r} .2727 \\ 11\overline{)3.0000} \\ \underline{2\,2} \\ 80 \\ \underline{77} \\ 30 \\ \underline{22} \\ 80 \\ \underline{77} \\ 3 \end{array}$$

3 ←The remainder 3 is equal to the original dividend. This tells us that the decimal quotient repeats itself.

$\frac{3}{11} = 0.272727\ldots$

In Problem (c) above you may have noticed that

$\frac{17}{7} = 2.\overbrace{428571}\overbrace{428571}\ldots$ if you did this division with a calculator. The digits 428571 repeat endlessly.

Mathematicians often use a shorthand notation to show that a decimal repeats.

Write $\frac{1}{3} = 0.\overline{3}$ or $\frac{2}{3} = 0.\overline{6}$

where the bar means that the digits under the bar repeat endlessly.

$\frac{3}{11} = 0.\overline{27}$ means $0.272727\ldots$

and $\frac{17}{7} = 2.\overline{428571}$

Write $\frac{41}{33}$ as a repeating decimal using the "bar" notation.

Check your answer in **24**.

24

```
     1.24
33)41.00
   33
   80
   66
   140
   132
     8
```

These remainders are the same so we know that further division will produce a repeat of the digits 24 in the answer.

$$\frac{41}{33} = 1.242424\ldots = 1.\overline{24}$$

Some fractions are used so often in practical work that it is important for you to know their decimal equivalents.

DECIMAL-FRACTION EQUIVALENTS

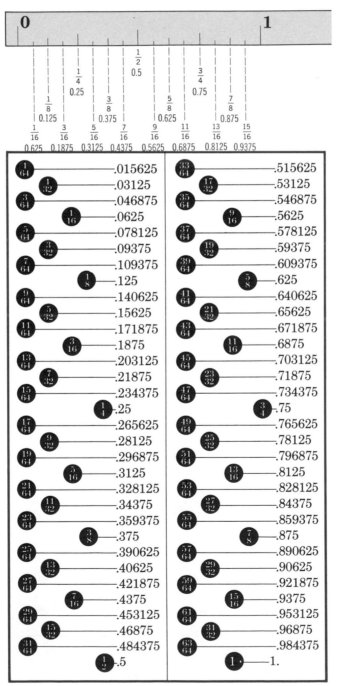

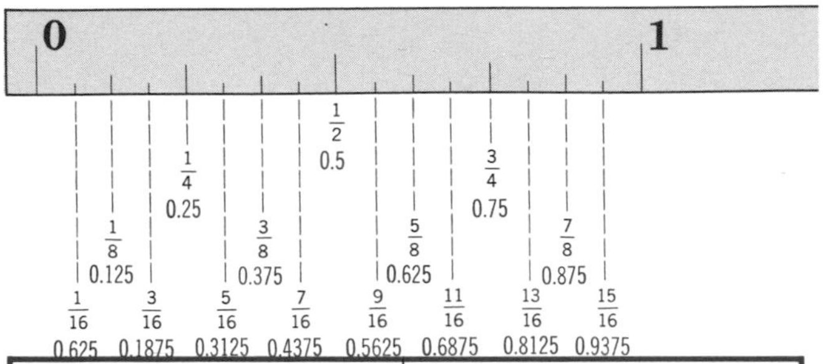

1/64	.015625	33/64	.515625
1/32	.03125	17/32	.53125
3/64	.046875	35/64	.546875
1/16	.0625	9/16	.5625
5/64	.078125	37/64	.578125
3/32	.09375	19/32	.59375
7/64	.109375	39/64	.609375
1/8	.125	5/8	.625
9/64	.140625	41/64	.640625
5/32	.15625	21/32	.65625
11/64	.171875	43/64	.671875
3/16	.1875	11/16	.6875
13/64	.203125	45/64	.703125
7/32	.21875	23/32	.71875
15/64	.234375	47/64	.734375
1/4	.25	3/4	.75
17/64	.265625	49/64	.765625
9/32	.28125	25/32	.78125
19/64	.296875	51/64	.796875
5/16	.3125	13/16	.8125
21/64	.328125	53/64	.828125
11/32	.34375	27/32	.84375
23/64	.359375	55/64	.859375
3/8	.375	7/8	.875
25/64	.390625	57/64	.890625
13/32	.40625	29/32	.90625
27/64	.421875	59/64	.921875
7/16	.4375	15/16	.9375
29/64	.453125	61/64	.953125
15/32	.46875	31/32	.96875
31/64	.484375	63/64	.984375
1/2	.5	1	1.

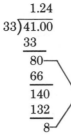

When you work in many trade or technical jobs, you may find that you are expected to have these decimal-fraction equivalents memorized.

Quick now, without looking at the table, fill in the blanks in the problems below:

$\frac{1}{2}$ = ______

$\frac{1}{4}$ = ______ $\frac{3}{4}$ = ______

$\frac{1}{8}$ = ______ $\frac{2}{8}$ = ______ $\frac{3}{8}$ = ______ $\frac{4}{8}$ = ______

$\frac{5}{8}$ = ______ $\frac{6}{8}$ = ______ $\frac{7}{8}$ = ______

$\frac{1}{16}$ = ______ $\frac{2}{16}$ = ______ $\frac{3}{16}$ = ______ $\frac{4}{16}$ = ______

$\frac{5}{16}$ = ______ $\frac{6}{16}$ = ______ $\frac{7}{16}$ = ______

$\frac{8}{16}$ = ______ $\frac{9}{16}$ = ______ $\frac{10}{16}$ = ______

$\frac{11}{16}$ = ______ $\frac{12}{16}$ = ______ $\frac{13}{16}$ = ______

$\frac{14}{16}$ = ______ $\frac{15}{16}$ = ______

Check your work against the table when you are finished, then turn to **25** and continue.

---

25 Convert each of the following fractions into decimal form.

(a) $\frac{95}{100}$ (b) $\frac{1}{20}$ (c) $\frac{7}{10}$

(d) $\frac{4}{1000}$ (e) $\frac{11}{1000}$ (f) $\frac{327}{10000}$

(g) $\frac{473}{1000}$ (h) $\frac{3}{50}$ (i) $\frac{1}{25}$

(j) $\frac{27}{64}''$ (to the nearest thousandth of an inch)

Divide them out, then check your work in **26**.

HOW TO WRITE A REPEATING DECIMAL AS A FRACTION

A repeating decimal is one in which some sequence of digits is endlessly repeated. For example, $0.333\ldots = 0.\overline{3}$ and $0.272727\ldots = 0.\overline{27}$ are repeating decimals. The bar over the number is a shorthand way of showing that those digits are repeated.

What fraction is equal to $0.\overline{3}$? To answer this form a fraction with numerator equal to the repeating digits and denominator equal to a number formed with the same number of 9s.

$0.\overline{3} = \frac{3}{9} = \frac{1}{3}$

$0.\overline{27} = \frac{27}{99} = \frac{3}{11}$ Two digits in $0.\overline{27}$; therefore use 99 as the denominator.

$0.\overline{123} = \frac{123}{999} = \frac{41}{333}$ Three digits in $0.\overline{123}$; therefore use 999 as the denominator.

26 (a) 0.95 (b) 0.05 (c) 0.7
(d) 0.004 (e) 0.011 (f) 0.0327
(g) 0.473 (h) 0.06 (i) 0.04
(j) 0.422″

You will do many of these calculations quicker if you remember that dividing by a multiple of ten is equivalent to shifting the decimal point to the left. To divide by a multiple of ten, move the decimal point to the left as many digits as there are zeros in the multiple of ten. For example,

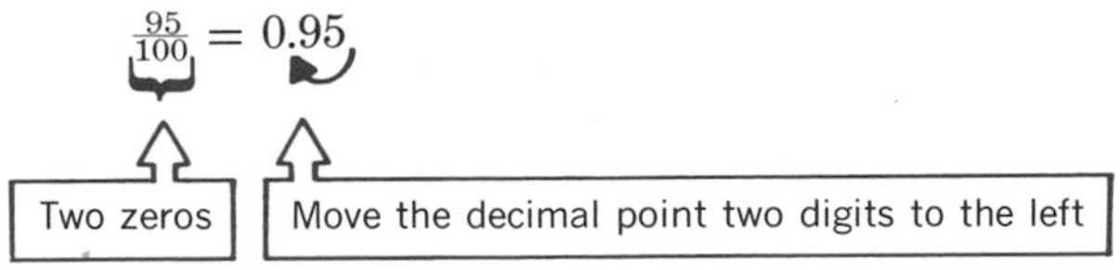

and

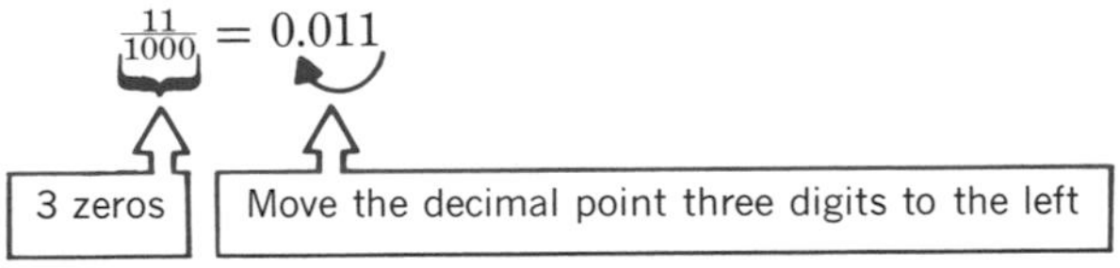

Now turn to **27** for a set of exercises on decimal fractions.

---

27 **Exercises 3-3 DECIMAL FRACTIONS**

A. Write as decimal numbers. (Round to two decimal digits.)

1. $\frac{1}{4}$	2. $\frac{2}{3}$	3. $\frac{3}{4}$	4. $\frac{2}{5}$
5. $\frac{4}{5}$	6. $\frac{5}{6}$	7. $\frac{2}{7}$	8. $\frac{4}{7}$
9. $\frac{6}{7}$	10. $\frac{3}{8}$	11. $\frac{6}{8}$	12. $\frac{1}{10}$
13. $\frac{3}{10}$	14. $\frac{2}{12}$	15. $\frac{5}{12}$	16. $\frac{3}{16}$
17. $\frac{6}{16}$	18. $\frac{9}{16}$	19. $\frac{13}{16}$	20. $\frac{3}{20}$

B. Calculate in decimal form. (Round to agree with the number of decimal digits in the decimal number.)

1. $2\frac{3}{5} + 1.785$	2. $\frac{1}{5} + 1.57$
3. $3\frac{7}{8} - 2.4$	4. $1\frac{3}{16} - 0.4194$
5. $2\frac{1}{2} \times 3.15$	6. $1\frac{3}{25} \times 2.08$
7. $3\frac{4}{5} \div 2.65$	8. $3.72 \div 1\frac{1}{4}$

C. Practice Problems

1. If one tablet of calcium pantothenate contains 0.5 gram, how much is contained in $2\frac{3}{4}$ tablets? How many tablets are needed to make up 2.6 grams?

2. Estimates of matched or tongue-and-groove (T & G) flooring—stock that has a tongue on one edge, must allow for the waste from milling. To allow for this waste, $\frac{1}{4}$ of the area to be covered must be added to the estimate when 1″ × 4″ flooring is used. If 1″ × 6″ flooring is used, $\frac{1}{6}$ of the area must be added to the estimate.

   (a) A house has a floor size 28′ long by 12′ wide. How many sq ft 1″ × 4″ T & G flooring will be required to lay the floor?

   (b) A motel contains 12 units or rooms, and each room is 22′ × 27′ or 594 sq ft. On 5 rooms the builder is going to use 4″ stock and on the other 7 rooms 6″ stock—both being T & G. How many square feet of each size will be used?

(c) A contractor is building a motel with 24 rooms. Eight rooms will be 16′ × 23′, nine rooms 18′ × 26′, and the rest of the rooms 14′ × 20′. How much did she pay for flooring, using 1″ × 4″ T & G at $157 per M. bd. ft?

3. If a carpenter lays $10\frac{1}{2}$ squares of shingles in $4\frac{1}{2}$ days, how many squares does he do in one day?

4. A bearing journal measures 1.996 in. If the standard readings are in $\frac{1}{32}$-in. units, what is its probable standard size as a fraction?

5. Complete the following invoice for upholstery fabric:

(a) $5\frac{1}{4}$ yards @ 87¢ = ______

(b) $23\frac{3}{4}$ yds @ $1.12 = ______

(c) $31\frac{5}{6}$ yds @ 73¢ = ______

(d) $16\frac{2}{3}$ yds @ $1.37 = ______

Total ______

6. A land developer purchased a piece of land containing 437.49 acres. He plans to divide it into a 45 acre recreational area and lots of $\frac{3}{4}$ acre each. How many lots will he be able to form from this piece of land?

7. Find the corner measurement *A* needed to make an octagonal end on a square bar as shown in the diagram.

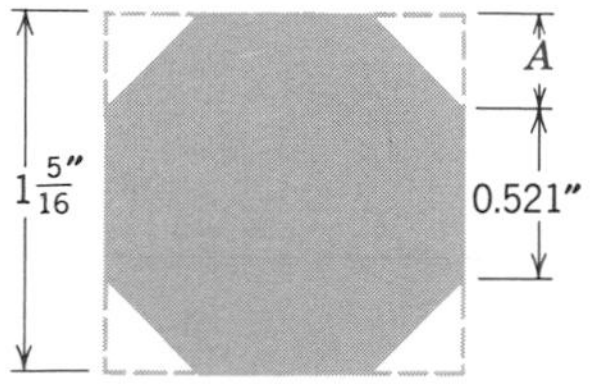

D. Calculator Problems

1. In the table on page 108 the first column lists the U.S. Standard sheet metal gauge number. The second column gives the equivalent thickness as a fraction in inches. Complete the third column, giving the thickness as a decimal number rounded to the nearest thousandth of an inch.

2. Calculate the total cost of the following building materials:

8,024 ft of flooring	@ $ 95.75 per 1000 ft	______
10,227 ft of siding	@ $ 18.30 per 100 ft	______
24,390 bricks	@ $112.00 per 1000	______
19,200 sq ft of shingles	@ $ 9.72 per 100 ft	______
	Total	______

3. Machine helpers at the CALMAC Tool Co. earn $4.72 per hour. The company pays time and one-half for hours over 8 per day. Sunday hours are paid at double time. How much would you earn if you worked the following hours: Monday, 9; Tuesday, 8; Wednesday, $9\frac{3}{4}$; Thursday 8; Friday, $10\frac{1}{2}$; Saturday, $5\frac{1}{4}$; Sunday, $4\frac{3}{4}$?

When you have finished these exercises, check your answers on page 546, and then continue in **28**.

Gauge No.	Fraction Thickness, inches	Decimal Thickness, inches	Gauge No.	Fraction Thickness, inches	Decimal Thickness, inches
7-0	$\frac{1}{2}$		14	$\frac{5}{64}$	
6-0	$\frac{15}{32}$		15	$\frac{9}{128}$	
5-0	$\frac{7}{16}$		16	$\frac{1}{16}$	
4-0	$\frac{13}{32}$		17	$\frac{9}{160}$	
3-0	$\frac{3}{8}$		18	$\frac{1}{20}$	
2-0	$\frac{11}{32}$		19	$\frac{7}{160}$	
0	$\frac{5}{16}$		20	$\frac{3}{80}$	
1	$\frac{9}{32}$		21	$\frac{11}{320}$	
2	$\frac{17}{64}$		22	$\frac{1}{32}$	
3	$\frac{1}{4}$		23	$\frac{9}{320}$	
4	$\frac{15}{64}$		24	$\frac{1}{40}$	
5	$\frac{7}{32}$		25	$\frac{7}{320}$	
6	$\frac{13}{64}$		26	$\frac{3}{160}$	
7	$\frac{3}{16}$		27	$\frac{11}{640}$	
8	$\frac{11}{64}$		28	$\frac{1}{64}$	
9	$\frac{5}{32}$		29	$\frac{9}{640}$	
10	$\frac{9}{64}$		30	$\frac{1}{80}$	
11	$\frac{1}{8}$		31	$\frac{7}{640}$	
12	$\frac{7}{64}$		32	$\frac{13}{1280}$	
13	$\frac{3}{32}$		—	—	—

3-6 EXPONENTS AND SQUARE ROOTS

28 When the same number appears many times in a multiplication, writing the product may become monotonous, tiring, and even inaccurate. It is easy, for example, to miscount the twos in

$$131{,}072 = 2 \times 2 \times 2 \times 2 \times 2 \times 2 \times 2 \times 2 \times 2 \times 2 \times 2 \times 2 \times 2 \times 2 \times 2 \times 2 \times 2$$

or the tens in

$$100{,}000{,}000{,}000 = 10 \times 10 \times 10 \times 10 \times 10 \times 10 \times 10 \times 10 \times 10 \times 10 \times 10$$

Products of this sort are usually written in a shorthand form as 2^{17} and 10^{11}. In this *exponential* form the raised number 17 shows the number of times 2 is to be used as a factor in the multiplication. For example,

$2 \times 2 = 2^2$ — Product of two factors of 2

$2 \times 2 \times 2 = 2^3$ — Product of three factors of 2

$2 \times 2 \times 2 \times 2 = 2^4$ — Product of four factors of 2

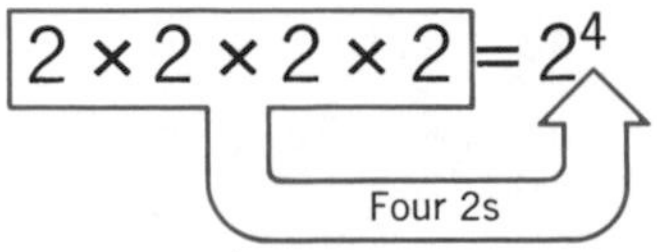

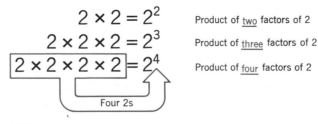

Write $3 \times 3 \times 3 \times 3 \times 3$ in exponential form.

$3 \times 3 \times 3 \times 3 \times 3 =$ ______

Check your answer in **29**.

---

29 $\underbrace{3 \times 3 \times 3 \times 3 \times 3} = 3^5$

Five factors of 3 (points to the exponent 5)

In this expression, 3 is called the *base,* and 5 is called the *exponent.* The exponent 5 tells you how many times the base 3 must be used as a factor in the multiplication.

Find the value of $4^3 =$ ___?___

Pick an answer and turn to the frame shown after the answer:

(a) 12 Go to **30**.
(b) 64 Go to **31**.
(c) 81 Go to **32**.

---

30 Your answer is incorrect; 4^3 is *not* equal to 12.

The raised 3 in 4^3 tells you to multiply 4 by itself. Use the number 4 *three* times as a factor in a multiplication.

$4^3 = \underbrace{4 \times 4 \times 4}$

means use three factors of 4

Once you have set up the multiplication in this way it is easy to do it.

$4 \times 4 \times 4 = (4 \times 4) \times 4 = 16 \times 4 = ?$

Now, hop back to **29** and choose the correct answer.

---

31 Excellent!

$4^3 = 4 \times 4 \times 4 = (4 \times 4) \times 4 = 16 \times 4 = 64$

It is important that you be able to read exponential forms correctly.
2^2 is read "two to the second power" or "two squared"
2^3 is read "two to the third power" or "two cubed"
2^4 is read "two to the fourth power"
2^5 is read "two to the fifth power"
and so on.

Students studying basic electronics or those going on to another mathematics or science class will find it is important to understand exponential notation.

Do these following problems for practice in using exponents.

(a) Write in exponential form.

$5 \times 5 \times 5 \times 5$	= ____	base = ____	exponent = ____
7×7	= ____	base = ____	exponent = ____
$10 \times 10 \times 10 \times 10 \times 10$	= ____	base = ____	exponent = ____
$3 \times 3 \times 3 \times 3 \times 3 \times 3 \times 3$	= ____	base = ____	exponent = ____
$9 \times 9 \times 9$	= ____	base = ____	exponent = ____
$1 \times 1 \times 1 \times 1$	= ____	base = ____	exponent = ____

(b) Write as a product of factors and multiply out.

2^6 = ______________ = ____ base = ____ exponent = ____

10^7 = ______________ = ____ base = ____ exponent = ____

3^4 = ______________ = ____ base = ____ exponent = ____

5^2 = ______________ = ____ base = ____ exponent = ____

6^3 = ______________ = ____ base = ____ exponent = ____

4^5 = ______________ = ____ base = ____ exponent = ____

12^3 = ______________ = ____ base = ____ exponent = ____

1^5 = ______________ = ____ base = ____ exponent = ____

The correct answers are in **33**. Go there when you finish these problems.

---

32 This answer is not correct. Apparently you found the product $3 \times 3 \times 3 \times 3$.

Do it this way:

$4^3 = \underbrace{4 \times 4 \times 4}$

3 factors of 4

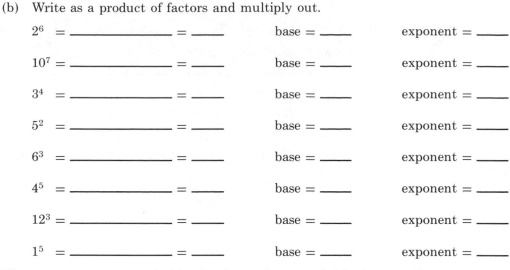

The raised 3 tells you how many factors of 4 are to be multiplied together.

$4 \times 4 \times 4 = (4 \times 4) \times 4 = 16 \times 4 = ?$

Complete this and return to **29** to continue.

---

33 (a) $5 \times 5 \times 5 \times 5$ = 5^4 base = 5 exponent = 4

7×7 = 7^2 base = 7 exponent = 2

$10 \times 10 \times 10 \times 10 \times 10$ = 10^5 base = 10 exponent = 5

$3 \times 3 \times 3 \times 3 \times 3 \times 3 \times 3$ = 3^7 base = 3 exponent = 7

$9 \times 9 \times 9$ = 9^3 base = 9 exponent = 3

$1 \times 1 \times 1 \times 1$ = 1^4 base = 1 exponent = 4

(b) $2^6 = 2 \times 2 \times 2 \times 2 \times 2 \times 2$ = 64 base = 2 exponent = 6

$10^7 = 10 \times 10 \times 10 \times 10 \times 10 \times 10 \times 10$ = 10,000,000 base = 10 exponent = 7

$3^4 = 3 \times 3 \times 3 \times 3$ = 81 base = 3 exponent = 4

$5^2 = 5 \times 5$ = 25 base = 5 exponent = 2

$6^3 = 6 \times 6 \times 6$ = 216 base = 6 exponent = 3

$4^5 = 4 \times 4 \times 4 \times 4 \times 4$ = 1024 base = 4 exponent = 5

$12^3 = 12 \times 12 \times 12 \quad = \underline{1728} \qquad \text{base} = \underline{12} \quad \text{exponent} = \underline{3}$

$1^5 = 1 \times 1 \times 1 \times 1 \times 1 \quad = \underline{1} \qquad \text{base} = \underline{1} \quad \text{exponent} = \underline{5}$

Any power of 1 is equal to 1, of course.

$1^2 = 1 \times 1 = 1$
$1^3 = 1 \times 1 \times 1 = 1$
$1^4 = 1 \times 1 \times 1 \times 1 = 1$ and so on

Notice that when the base is ten, the product is easy to find.

$10^2 = 10 \times 10 = 100$
$10^3 = 10 \times 10 \times 10 = 1000$
$10^4 = 10000$
$10^5 = 100000$

The exponent number is always exactly equal to the number of zeros in the final product.

Continue the pattern to find the value of 10^0 and 10^1.

$10^5 = 100000$	$10^2 = 100$
$10^4 = 10000$	$10^1 = 10$
$10^3 = 1000$	$10^0 = 1$

Of course, this is true for any base:

$2^1 = 2$	$2^0 = 1$
$3^1 = 3$	$3^0 = 1$
$4^1 = 4$	$4^0 = 1$ and so on

To evaluate an arithmetic expression containing exponential numbers, first evaluate the exponential number, then do the other arithmetic. For example,

$$2^2 \times 3^3 = 4 \times 27$$
$$= 108$$

since $2^2 = 2 \times 2 = 4$ and $3^3 = 3 \times 3 \times 3 = 27$.

Calculate: (First evaluate the exponential number, then multiply.)

(a) $2^4 \times 5^3$ (b) $3^4 \times 2^7 \times 1^6$
(c) $2^3 \times 3^2 \times 4^0$ (d) $2^2 \times 3^2 \times 4^3$

Check your answers in **34**

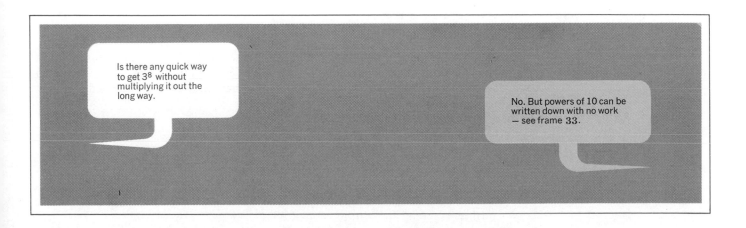

34 (a) 2000 (b) 10,368 (c) 72 (d) 2304

What is interesting about the numbers

1, 4, 9, 16, 25, 36, 49, 64, 81, 100, . . . ?

Do you recognize them?

These numbers are the squares or second powers of the counting numbers,

$1^2 = 1$
$2^2 = 4$
$3^2 = 9$
$4^2 = 16$, and so on. 1, 4, 9, 16, 25, . . . are called *perfect squares.*

If you have memorized the multiplication table for one-digit numbers, you will recognize them immediately. The number 3^2 is read "three squared." What is "square" about $3^2 = 9$? The name comes from an old Greek idea about the nature of numbers. Ancient Greek mathematicians called certain numbers "square numbers" or "perfect squares" because they could be represented by a square array of dots.

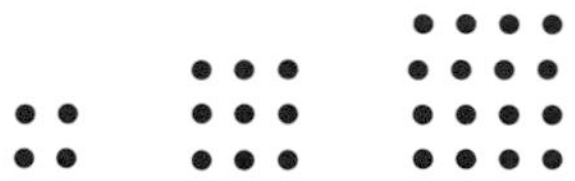

4 9 16 . . . and so on

The number of dots along the side of the square was called the "root" or origin of the square number. We call it the *square root.* For example, the square root of 16 is 4, since $4 \times 4 = 16$.

What is the square root of 64?

(a) 32 Go to **35**.
(b) 8 Go to **36**.

PERFECT SQUARES

$1^2 = 1$	$6^2 = 36$	$11^2 = 121$	$16^2 = 256$
$2^2 = 4$	$7^2 = 49$	$12^2 = 144$	$17^2 = 289$
$3^2 = 9$	$8^2 = 64$	$13^2 = 169$	$18^2 = 324$
$4^2 = 16$	$9^2 = 81$	$14^2 = 196$	$19^2 = 361$
$5^2 = 25$	$10^2 = 100$	$15^2 = 225$	$20^2 = 400$

35 Sorry, you are not correct.
You simply cannot divide 64 in half to find its square root!

The square root of 64 is some number □ such that □ × □ = 64. For example, the square root of 25 is equal to 5 because $5 \times 5 = 25$. To "square" means to multiply by itself, and to find a square root means to find a number that, when multiplied by itself, gives the original number.
Go back to **34** and try again.

---

36 Right! The square root of 64 is equal to 8 because $8 \times 8 = 64$.

The sign $\sqrt{}$ is used to indicate the square root.

$\sqrt{16} = 4$ read it "square root of 16"

$\sqrt{9} = 3$ read it "square root of 9"

$\sqrt{169} = 13$ read it "square root of 169"

Find $\sqrt{81} =$ ______ $\sqrt{361} =$ ______ $\sqrt{289} =$ ______

Try using the table of perfect squares on page 112 if you do not recognize these. Check your answers in **37**.

---

37 $\sqrt{81} = 9$ **Check:** $9 \times 9 = 81$

$\sqrt{361} = 19$ **Check:** $19 \times 19 = 361$

$\sqrt{289} = 17$ **Check:** $17 \times 17 = 289$

Always check your answer as shown.

How do you find the square root of any number? The surest and simplest way is to use an electronic calculator or to consult a table of square roots. The table of square roots given on page 539 in this book lists the square roots of all whole numbers from 1 to 200, rounded to four decimal places. Here is a section of the table:

Number	Square Root
1	1.0000
2	1.4142
3	1.7321
4	2.0000
5	2.2361
6	2.4495
7	2.6458
⋮	⋮

Notice that $\sqrt{2} = 1.4142$ approximately, because $1.4142 \times 1.4142 = 1.99996164$ or about 2.

Use this table of square roots to find the following.

(a) $\sqrt{10}$ (b) $\sqrt{50}$ (c) $\sqrt{140}$ (d) $\sqrt{175}$

Check your answers in **38**.

---

38 (a) 3.1623 (b) 7.0711 (c) 11.8322 (d) 13.2288

To find the square root of a decimal number or a number that does not appear on the table, either use a calculator or break the problem into parts like this:

$\sqrt{300} = \sqrt{3 \times 100} = \sqrt{3} \times \sqrt{100} = 1.7321 \times 10 = 17.321$

Check: $17.321 \times 17.321 = 300.017041$

$\cong 300$

or

$$\sqrt{2.5} = \sqrt{\frac{25}{10}} = \frac{\sqrt{25}}{\sqrt{10}} = \frac{5.0000}{3.1623} = 1.5811, \text{ rounded}$$

Check: $1.5811 \times 1.5811 = 2.49987721$
$\qquad\qquad\qquad\qquad \cong 2.5$

Always check your answer.

If you do not have a calculator or a table of square roots, you can find the square root of any number using an arithmetic method. If you need to learn to find square roots this "long way," turn to frame **39**. Otherwise, turn to frame **40** for a set of problems on exponents and square roots.

39

THE ARITHMETIC METHOD FOR FINDING SQUARE ROOTS

Here is a very commonly used method of finding the square root of any decimal number.

Example:

Step 1 Mark off the number into two-digit periods, working from the decimal point outward. If necessary add zeros to the right of the decimal point to complete a two-digit period.

Find $\sqrt{1234.5}$

$\sqrt{12\,34.50}$

Step 2 Bring up the decimal point.

$\sqrt{12\,34.50}$ (with the decimal point above)

Step 3 Starting with the leftmost pair of digits, write the largest square less than or equal to the end pair below that pair. Write the square root of that square above the leftmost pair of digits.

```
     3 .
  √1234.50
   9
```

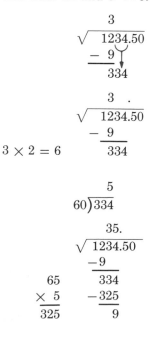

9 is the largest square less than 12 and $3^2 = 9$.

Step 4 Subtract as in a division problem.

Step 5 Bring down the next pair of digits.

```
      3
  √ 1234.50
   − 9
     334
```

Step 6 Double the partial answer.

```
              3  .
           √ 1234.50
            − 9
3 × 2 = 6     334
```

Step 7 Mentally attach a 0 to this number (6) and divide it into the new dividend (334).

```
    5
60)334
```

Step 8 Place the quotient (5) in the answer and in the divisor (65). Multiply the new divisor by the quotient and subtract this from the previous dividend (334).

```
          35.
        √ 1234.50
         −9
  65      334
×  5     −325
 325        9
```

Step 9 Return to Step 5 and repeat the process until the answer is as accurate as desired.

Be careful in Step 9 to add only *pairs* of zeros, so that the number can be divided into two-digit periods.

	35 .1 3
	$\sqrt{1234.5000}$
	− 9
65	334
× 5	−325
	950
701	−701
× 1	24900
7023	−21069
3	

$\sqrt{1234.5} \cong 35.13$

Exercises:

1. $\sqrt{41.4}$
2. $\sqrt{1.45}$
3. $\sqrt{113.6}$
4. $\sqrt{0.075}$
5. $\sqrt{425}$
6. $\sqrt{20.1}$
7. $\sqrt{8.04}$
8. $\sqrt{0.006}$
9. $\sqrt{651}$
10. $\sqrt{210}$

The answers are on page 547.

40

Exercises 3-4 Exponents and Square Roots

A. Find the value of these:

1. 2^4	2. 3^2	3. 4^3	4. 5^3
5. 10^3	6. 7^2	7. 2^8	8. 6^2
9. 8^3	10. 3^4	11. 5^4	12. 10^5
13. 2^3	14. 3^5	15. 9^3	16. 6^0
17. 5^1	18. 1^4	19. 2^5	20. 8^2
21. 14^2	22. 21^2	23. 15^3	24. 16^2
25. $2^2 \times 3^3$	26. $2^6 \times 3^2$	27. $3^2 \times 5^3$	28. $2^3 \times 7^3$
29. $6^2 \times 5^2 \times 3^1$	30. $2^{10} \times 3^2$		

B. Calculate. (Round to two decimal places.)

1. $\sqrt{81}$	2. $\sqrt{144}$	3. $\sqrt{36}$	4. $\sqrt{16}$
5. $\sqrt{25}$	6. $\sqrt{9}$	7. $\sqrt{256}$	8. $\sqrt{400}$
9. $\sqrt{225}$	10. $\sqrt{49}$	11. $\sqrt{324}$	12. $\sqrt{121}$
13. $\sqrt{4.5}$	14. $\sqrt{500}$	15. $\sqrt{12.4}$	16. $\sqrt{700}$
17. $\sqrt{210}$	18. $\sqrt{321}$	19. $\sqrt{810}$	20. $\sqrt{92.5}$
21. $\sqrt{1000}$	22. $\sqrt{2000}$	23. $\sqrt{25000}$	24. $\sqrt{2500}$
25. $\sqrt{150}$	26. $\sqrt{300}$	27. $\sqrt{3000}$	28. $\sqrt{30000}$
29. $\sqrt{1.25}$	30. $\sqrt{1.008}$		

C. Practical Problems. (Round to the nearest tenth if necessary.)

1. Find the length of side of a square whose area is 184.96 sq ft. (*Hint:* The area of a square is equal to the square of one of its sides. The side of a square equals the square root of its area.)
2. A square building covers an area of 1269 sq ft. What is the length of each side of the building?
3. A square room takes exactly 169 1 square-foot tiles to cover the floor. What are the dimensions of the room?
4. Find the length of side of a square tool crib having a floor area of 252 sq ft.

5. A square piece of sheet metal has an area of 12,368 sq in. How long is each side?

6. The length of a rafter is 15 ft and the run is 12 ft. What is the rise? (*Hint:* For a right triangle the sides are related as shown below.)

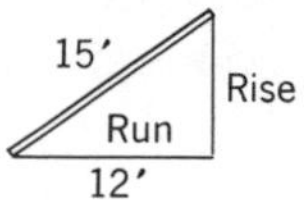

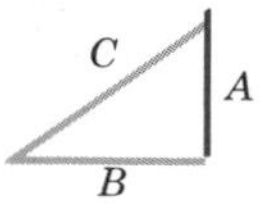

$C = \sqrt{A^2 + B^2}$

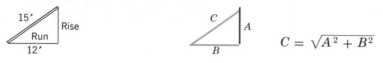

7. Find the diagonal of the square shown if the sides are 1.8 inches long.

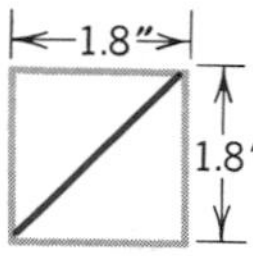

diagonal $= \sqrt{1.8^2 + 1.8^2}$

8. A painter must reach a window that is 12 ft from the ground. However a bush 5 ft wide prevents her from putting the ladder close to the house. How long must her ladder be to reach the window?

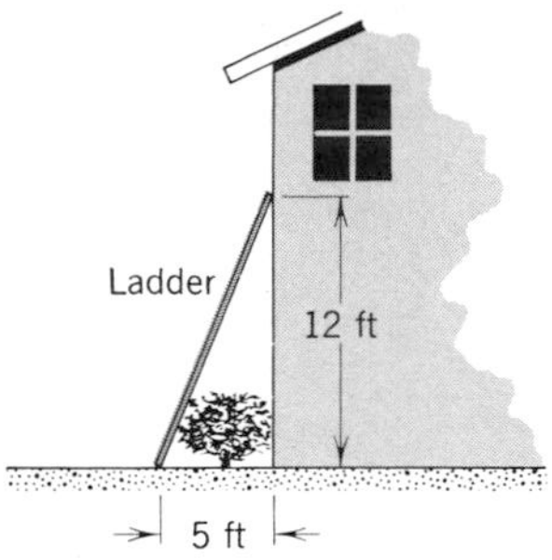

9. A contractor is building houses around a small lake and needs to know the exact length of the lake. He does not want to wade and swim the length of the lake carrying a tape measure so he sets up a right triangle as shown. How long is the lake?

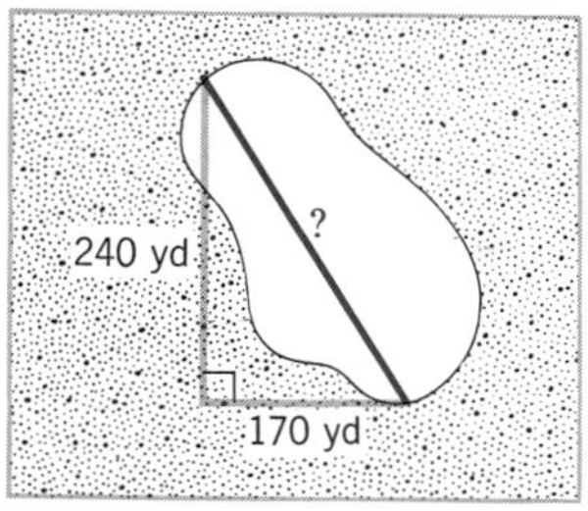

When you have finished these exercises, check your answers on page 547, and then continue in **41**.

3-7 NEGATIVE NUMBERS

41 What kind of number would you use to name each of the following: a golf score two strokes below par, a $2 loss in a poker game, a debt of $2, a loss of two yards on a football play, a temperature two degrees below zero, the year 2 B.C.?

The answer is that they are all *negative numbers,* all equal to -2.

If we display the whole numbers as points along a line we have:

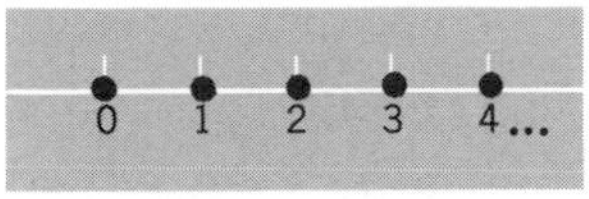

It is reasonable that we should be able to extend the line to the left and talk about numbers less than zero. Notice that every positive number is paired with a negative number.

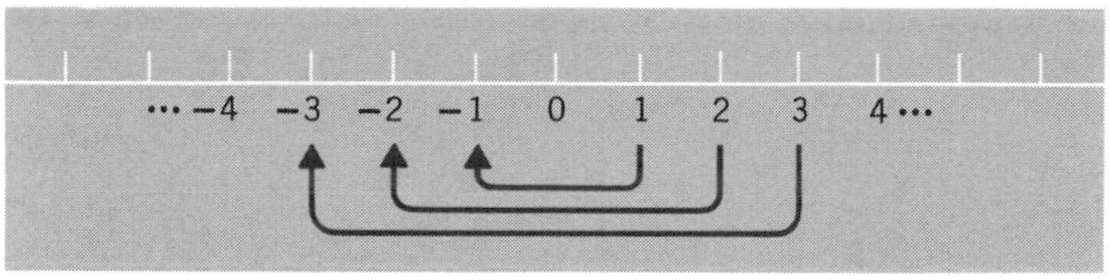

This number line can be drawn any length (it should be infinitely long but we can draw only part of it), and we may choose any point on the line as a zero position. Using any convenient length unit, we mark off equal intervals to the right and left of the zero point. We call all numbers to the right of the zero *positive* numbers and all numbers to the left of the zero *negative* numbers. On the number line above we have labeled only the positive and negative integers . . . -3, -2, -1, 0, 1, 2, 3, . . . , but of course other, non-integer numbers such as $\frac{1}{2}$, -0.2, 4.2, or -3.17 may be located there as well.

Mark the following points on the number line above:

(a) -4 (b) $-2\frac{1}{2}$ (c) -0.4 (d) -1.2

Compare your answer with ours in **42**.

42

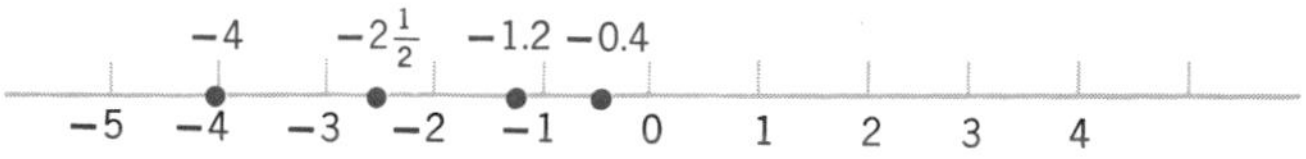

Signed numbers are a bookkeeping concept. They were used in bookkeeping by the ancient Greeks, Chinese, and Hindus more than 2000 years ago. Chinese merchants wrote positive numbers in black and negative numbers in red in their account books. (Being "in the red" still means losing money or having a negative income.) The Hindus used a dot or circle to show that a number was negative. We use a minus sign ($-$) to show that a number is negative, and because this symbol is used also for subtraction, some confusion results.

The + and − signs have a dual role in mathematics. With positive numbers + means "add" and − means "subtract." But these signs also are used to indicate position on a number line. The number $+2$ names a position on the line two units to the *right* of the zero. The number -2 names a position on the line two units to the *left* of the zero. A golf score -2 is two strokes below par; the year 50 B.C. might be written as -50; nitrogen freezes at a temperature of $-210°C$. If you receive a check for \$8.00, your net gain is $+8$. If you send a check for \$8.00, your net gain is -8.

Read -8 as "negative 8" rather than "minus 8" and it will be less confusing for you.

Addition and subtraction of signed numbers may be pictured using the number line. A positive number may be represented on the line by an arrow to the right or positive direction. Think of a positive arrow as a "gain" of some quantity such as money. The sum of two numbers is the net "gain."

For example, the sum 4 + 3 or (+4) + (+3) is

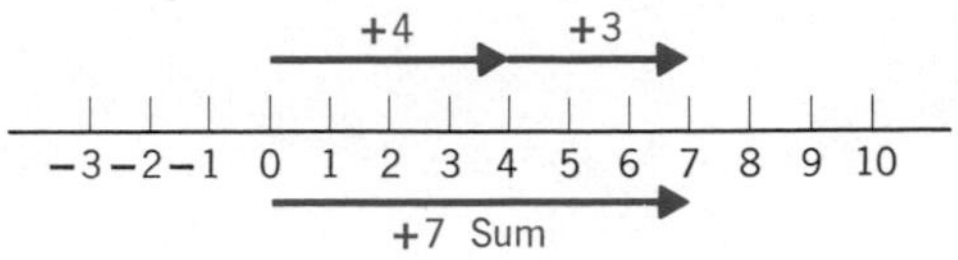

The numbers +4 and +3 are represented by arrows directed to the right. The +3 arrow begins where the +4 arrow ends. The sum 4 + 3 is the arrow that begins at the start of the +4 arrow and ends at the tip of the +3 arrow. In terms of money, a "gain" of $4 followed by a "gain" of $3 produces a net gain of $7.

A negative number may be represented on the number line by an arrow to the left or negative direction. Think of a negative arrow as a loss of some quantity.

For example the sum $(-4) + (-3)$ is -7.

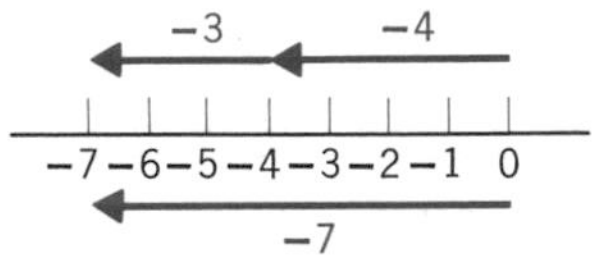

The numbers −4 and −3 are represented by arrows directed to the left. The −3 arrow begins at the tip of the −4 arrow. Their sum, −7, is the arrow that begins at the start of the −4 arrow and ends at the tip of the −3 arrow. A loss of $4 followed by another loss of $3 makes a total loss of $7.

Try setting up number line diagrams for the following two sums.

(a) $4 + (-3)$ (b) $(-4) + 3$

Check your work in **43**.

---

43 (a) $4 + (-3) = 1$

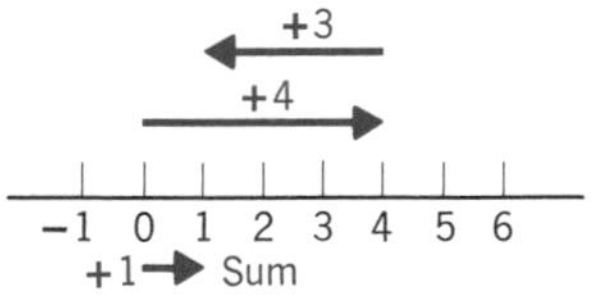

The number 4 is represented by an arrow directed to the right. The number −3 is represented by an arrow starting at the tip of +4 and directed to the left. A gain of $4 followed by a loss of $3 produces a net gain of $1.

(b) $(-4) + 3 = -1$

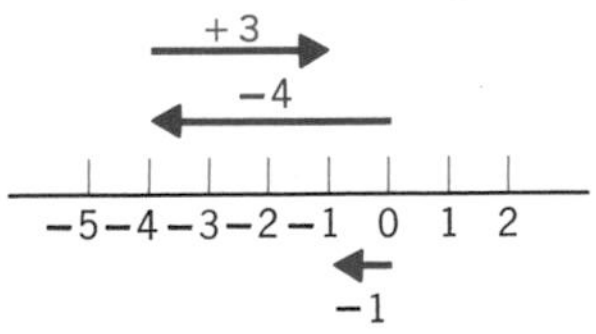

The number −4 is represented by an arrow directed to the left. The number 3 is represented by an arrow starting at the tip of −4 and directed to the right. A loss of $4 followed by a gain of $3 produces a net loss of $1.

Number lines are a good way of seeing what is happening when we add signed numbers, but we need a nice easy rule that will allow us to do the addition with no pictures or drawings.

Here it is:

Rules for Adding Signed Numbers:

1. If both numbers have the *same* sign, add them and give the sum the sign of the original numbers.
2. If the two numbers have opposite signs, subtract them and give the difference the sign of the larger number.

For example,

$(-7) + (-3) = -10$ — -7 and -3 have the same sign. Add the numbers and give the sum (10) the sign of the original numbers ($-$). The answer is -10.

$5 + (-11) = -6$ — $+5$ and -11 have opposite signs. Take the difference (6) and give it the sign ($-$) of the larger number (11). The answer is -6.

Use these rules to add the following numbers.

(a) $6 + 4$	(b) $(-6) + (-4)$	(c) $(-5) + 2$
(d) $(-1) + 5$	(e) $4 + (-16)$	(f) $(-11) + 11$
(g) $(-7) + (-9)$		

Check your work in **44**.

---

44 (a) $6 + 4 = 10$ — The signs are the same—both positive. Add ($6 + 4 = 10$) and attach the common sign ($+$).

(b) $(-6) + (-4) = -10$ — The signs are the same—both negative. Add ($6 + 4 = 10$) and attach the common sign ($-$).

(c) $(-5) + 2 = -3$ — The signs are different. Subtract ($5 - 2 = 3$) and attach the sign ($-$) of the larger number (5).

(d) $(-1) + 5 = 4$ — The signs are different. Subtract ($5 - 1 = 4$) and attach the sign ($+$) of the larger number (5).

(e) $4 + (-16) = -12$ — The signs are different. Subtract ($16 - 4 = 12$) and attach the sign ($-$) of the larger number (16).

(f) $(-11) + 11 = 0$ — The signs are different. Subtract ($11 - 11 = 0$). Zero has no sign.

(g) $(-7) + (-9) = -16$ — The signs are the same. Add ($7 + 9 = 16$) and attach the common sign ($-$).

Subtraction of Signed Numbers

Every beginning arithmetic student knows that $4 - 3 = 1$. But what do you do with $3 - 4$? How can we subtract a larger from a smaller number? Using signed numbers it is not only possible, it is easy. First, rewrite the subtraction:

$$3 - 4 = 3 + (-4)$$

and now add the two signed numbers as we showed in the last frame.

$$3 + (-4) = -1$$

Another example: $7 - 12 = 7 + (-12) = -5$

Solve the following problems in this same way.

(a) $11 - 6$	(b) $4 - 13$	(c) $(-2) - 9$
(d) $0 - 3$	(e) $6 - 7$	(f) $(-8) - 6$

Check your answers in **45**.

---

45 (a) $11 - 6 = 11 + (-6) = 5$
(b) $4 - 13 = 4 + (-13) = -9$
(c) $(-2) - 9 = (-2) + (-9) = -11$
(d) $0 - 3 = 0 + (-3) = -3$
(e) $6 - 7 = 6 + (-7) = -1$
(f) $(-8) - 6 = (-8) + (-6) = -14$

Subtracting a positive number is the same as adding its negative counterpart.

Subtracting a negative number is the same as adding its positive counterpart.

For example, $6 - (-4) = 6 + (+4) = 10$

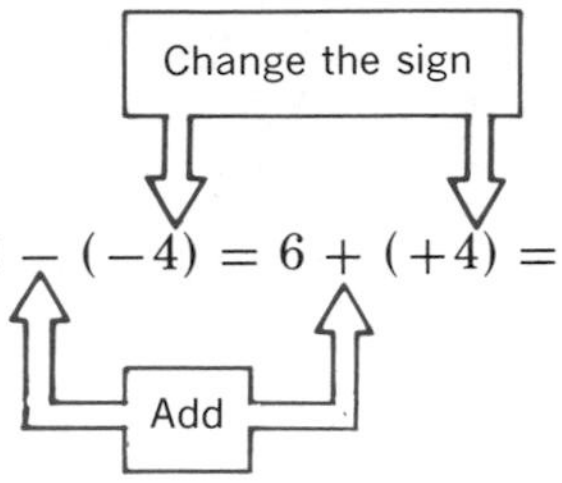

and $(-5) - (-3) = (-5) + (+3) = -2$

Try these problems for practice.

(a) $7 - (-5)$ (b) $(-3) - (-2)$ (c) $8 - (-10)$
(d) $(-6) - (-2)$ (e) $0 - (-9)$ (f) $(-4) - (-9)$

Check your work in **46**.

---

46 (a) $7 - (-5) = 7 + (+5) = 7 + 5 = 12$
(b) $(-3) - (-2) = (-3) + (+2) = -3 + 2 = -1$
(c) $8 - (-10) = 8 + (+10) = 8 + 10 = 18$
(d) $(-6) - (-2) = (-6) + (+2) = -6 + 2 = -4$
(e) $0 - (-9) = 0 + (+9) = 9$
(f) $(-4) - (-9) = (-4) + (+9) = -4 + 9 = 5$

To summarize the rule for subtraction, change all subtraction problems to addition problems and use the rules for adding signed numbers. When you make the change remember that subtracting a positive number is the same as adding its negative counterpart and subtracting a negative number is the same as adding its positive counterpart.

Never use two signs together without parentheses. The phrase

$+4$ added to -2

is written $+4 + (-2)$

but *never* $+4 + -2$.

Multiplication of Signed Numbers

The product of two positive numbers is easy to find. For example,

$(+3) \times (+4) = +12$

The product of numbers with opposite signs can be understood in terms of repeated addition. For example,

$$(+3) \times (-4) = -12$$

because $3 \times (-4) = (-4) + (-4) + (-4)$

In the same way $(-3) \times (+4) = -12$

because $4 \times (-3) = (-3) + (-3) + (-3) + (-3)$

Finally, the product of two negative numbers is always positive. For example,

$(-3) \times (-4) = +12$

Use the same procedure to find

(a) $(+5) \times (-7) =$ ______

(b) $(-2) \times (-8) =$ ______

(c) $(-6) \times (+5) =$ ______

Look in **47** for the answers.

---

47 (a) -35 (b) $+16$ (c) -30

From these examples we can develop a general rule for the multiplication and division of signed numbers. Here it is:

If both numbers have the *same* sign, the result of the operation will be a *positive* number; if both numbers have the *opposite* sign, the result of the operation will be a negative number.

We can summarize it this way:

Factors or Divisors	Product or Quotient
Same sign	+
Opposite sign	−

For example,

$$\text{Same sign}\begin{cases}(4) \times (3) = +12\\(-4) \times (-3) = +12\end{cases}$$

$$\text{Opposite sign}\begin{cases}(-4) \times (3) = -12\\(4) \times (-3) = -12\end{cases}$$

For division:

$$\text{Same sign}\begin{cases}(8) \div (2) = +4\\(-8) \div (-2) = +4\end{cases}$$

$$\text{Opposite sign}\begin{cases}(-8) \div (2) = -4\\(8) \div (-2) = -4\end{cases}$$

Work the following problems for practice in this very important rule.

(a) $(-1.2) \times (-3)$	(b) $(0.02) \times (-1.5)$	(c) $(-2.4) \times (3.5)$
(d) $(-1.5) \times (-2)$	(e) $(-3)^2$	(f) $(-2)^3$
(g) $(-4.5) \div (-5)$	(h) $(-20) \div (4)$	(i) $(3.6) \div (-2)$
(j) $(-7.2) \div (6)$	(k) $(-27) \div (-3)$	(l) $(4.2) \div (-0.07)$

Check your answers in **48**.

48

(a) $+3.6$	(b) -0.03	(c) -8.4
(d) $+3$	(e) $+9$	(f) -8
(g) $+0.9$	(h) -5	(i) -1.8
(j) -1.2	(k) $+9$	(l) -60

For more practice in the arithmetic of signed numbers turn to **49** for problem exercises.

49 **Exercises 3-5 Negative Numbers**

A. Find the value of the following:

1. $3 - 5$	2. $9 - 13$	3. $-11 - 8$
4. $-17 - 3$	5. $-23 - (-7)$	6. $6.4 - 8.5$
7. $5.7 + (-3.1)$	8. $-1.1 - (-3.2)$	9. $-0.05 + 0.24$
10. $0.5 - 2$	11. $3 \times (-7)$	12. $(-5) \times (-9)$
13. $(-2) \times 3$	14. $7 \times (-8)$	15. $1 \times (-1)$
16. $(0.5) \times (-3.1)$	17. $(-3.5) \times (2.5)$	18. $(-1.2) \times (-3)$
19. $(-1.5) \times (-1.5)$	20. $(-72) \div 12$	21. $36 \div (-9)$
22. $(-1) \div (-1)$	23. $42 \div (-6)$	24. $(-63) \div (-0.7)$
25. $(-144) \div (-1.2)$		

B. Word Problems

1. The height of Mt. Whitney is +14,410 ft. The lowest point in Death Valley is −288 ft. What is the difference in altitude between these two points?

2. The highest temperature ever recorded on earth was +136°F at Libya in North Africa. The coldest was −127°F at Vostok in Antarctica. What is the difference in temperature between these two extremes?

3. During a series of plays in a football game, a running back carried the ball for the following yardages: 7, −3, 2, −1, 8, −5, and −2. Find his total yards and his average yards per carry.

4. At 8 p.m. the temperature was 2°F in Anchorage, Alaska. At 4 a.m. it was −37°F. By how much did the temperature drop in that time?

5. The Jiffy Print Shop has a daily work quota that it tries to meet. During the first two weeks of last month its daily comparisons to this quota showed the following: +\$30, −\$27, −\$15, +\$8, −\$6, +\$12, −\$22, +\$56, −\$61, +\$47. (In each case "+" means they were above the quota and "−" means they were below.) How far below or above their quota was the two-week total?

6. A vacation trip involved a drive from a spot in Death Valley 173 ft below sea level (−173 ft) to the San Bernardino Mountains 4250 ft above sea level (+4250 ft). What was the vertical ascent for this trip?

7. Dave's Auto Shop was \$2465 in the "red" (−\$2465) at the end of April—they owed this much to their creditors. At the end of December they were \$648 in the black. How much money did they make in profit between April and December to achieve this turnaround?

Now turn to frame **50** for a set of practice problems on decimal numbers.

3 Decimal Numbers

PROBLEM SET 3

50 Answers are on page 547.

Decimal Numbers

A. Add or subtract as shown:

1. $4.39 + 18.8$
2. $18.8 + 156.16$
3. $\$7.52 + \$11.77$
4. $26 + 0.06$
5. $3.68 - 1.74$
6. $\$12.46 - \$8.51$
7. $104.06 - 15.80$
8. $16 - 3.45$
9. $264.3 + 12.804$
10. $0.232 + 5.079$
11. $165.4 + 73.61$
12. $245.94 + 7.07$
13. $116.7 - 32.82$
14. $4.07 - 0.085$
15. $0.42 + 1.452 + 31.8$
16. $\frac{1}{2} + 4.21$
17. $3\frac{1}{5} + 1.08$
18. $1\frac{1}{4} - 0.91$
19. $3.045 - 1\frac{1}{8}$
20. $8.1 + 0.47 - 1\frac{4}{5}$

B. Multiply or divide as shown:

1. 0.004×0.02
2. 0.06×0.05
3. 1.4×0.6
4. 3.14×12
5. $0.2 \times 0.6 \times 0.9$
6. 5.3×0.4
7. 6.02×3.3
8. $3.224 \div 2.6$
9. $187.568 \div 3.04$
10. $0.078 \div 0.3$
11. $0.6 \times 3.15 \times 2.04$

Round to two decimal digits:

12. $0.007 \div 0.03$
13. $3.005 \div 2.01$
14. $17.8 \div 6.4$
15. $0.0041 \div 0.019$

Round to three decimal places:

16. $0.04 \div 0.076$
17. $234.1 \div 465.8$
18. $17.6 \div 0.082$

Round to the nearest tenth:

19. $0.08 \div 0.053$
20. $3.05 \div 0.13$

C. Write as a Decimal Number:

1. $\frac{1}{16}$
2. $\frac{7}{8}$
3. $\frac{5}{32}$
4. $\frac{7}{16}$
5. $\frac{11}{8}$
6. $1\frac{3}{16}$
7. $2\frac{11}{32}$
8. $1\frac{1}{6}$
9. $2\frac{2}{3}$
10. $1\frac{1}{32}$
11. $2\frac{5}{16}$

Write as a fraction in lowest terms:

12. 0.06
13. 0.08
14. 0.35
15. 0.64
16. 1.25
17. 2.75
18. 3.0625
19. 2.01
20. 14.44

Write in decimal form, calculate, round to two decimal digits:

21. $4.82 \div \frac{1}{4}$
22. $11.5 \div \frac{3}{8}$
23. $1\frac{3}{16} \div 0.62$
24. $2\frac{3}{4} \div 0.035$
25. $0.45 \times 2\frac{1}{8}$
26. $0.068 \times 1\frac{7}{8}$

Name ______________

Date ______________

Course/Section ______________

D. Find the numerical value of each expression:

1. 2^6	2. 3^5	3. 17^2	4. 23^3
5. 0.5^2	6. 1.2^3	7. 0.02^2	8. 0.03^3
9. 0.001^3	10. 2.01^2	11. 4.02^2	12. $\sqrt{1369}$
13. $\sqrt{784}$	14. $\sqrt{4.41}$	15. $\sqrt{0.16}$	16. $\sqrt{5.29}$

Round to two decimal digits:

17. $\sqrt{80}$	18. $\sqrt{106}$	19. $\sqrt{310}$	20. $\sqrt{1.8}$
21. $\sqrt{4.20}$	22. $\sqrt{3.02}$	23. $\sqrt{1.09}$	

E. Calculate:

1. $-7 - 6$	2. $-3 - 18$
3. $-14 - 21$	4. $-6 - 30$
5. $2.4 - (-7.1)$	6. $3.5 - (-6.8)$
7. $-17.4 + 4.9$	8. $-3.5 + 12.81$
9. $(-1.2) \times (-0.4)$	10. $(-3.1) \times (0.09)$
11. $(4.5) \times (-0.06)$	12. $(-7.2) \times (0.05)$
13. $(91.02) \div (-11.1)$	14. $(-8.015) \div (-0.07)$
15. $(-3.424) \div (0.016)$	16. $(-0.007) - (-0.091)$

F. Practical Problems

1. A $\frac{5}{16}''$ bolt weighs 0.43 lb. How many bolts are there in a 125 lb keg?
2. Twelve equally spaced holes are to be drilled in a plywood strip $34\frac{3}{4}$ in. long, with 2 in. remaining on each end. What is the distance, to the nearest hundredth of an inch, from center to center of two consecutive holes?
3. A plumber finds the length of pipe needed to complete a bend by performing the following multiplication:

   Length needed = 0.01745 × (radius of the bend) × (angle of bend)

   What length of pipe is needed to complete a 35° bend with a radius of 16 in.?
4. During the first six months of the year the Busy Bee Printing Co. spent the following amounts on paper:

January	\$470.16	February	\$390.12	March	\$176.77
April	\$200.09	May	\$506.45	June	\$189.21

   What is the average monthly expenditure for paper?
5. The *feed* of a drill is the distance it advances with each revolution. If a drill makes 310 rpm and drills a hole 2.125″ deep in 0.800 minutes, what is the feed?

   $$\left(\textit{Hint:}\quad \text{Feed} = \frac{\text{total depth}}{(\text{rpm}) \times (\text{time of drilling in minutes})}\right)$$

6. The resistance of an armature while it is cold is 0.208 ohm. After running for several minutes, the resistance increases to 1.340 ohms. Find the increase in resistance of the armature.
7. The Ace Place Paint Co. sells paint in steel drums each weighing 36.4 lb empty. If one gallon of paint weighs 9.06 lb, what is the total weight of a 50 gal drum of paint?
8. The diameter of No. 12 bare copper wire is 0.08081″ and the diameter of No. 15 bare copper wire is 0.05707″. How much larger is No. 12 wire than No. 15 wire?
9. What is the decimal equivalent of each of the following drill bit diameters?
   (a) $\frac{3}{16}$ in. (b) $\frac{5}{32}$ in. (c) $\frac{3}{8}$ in. (d) $\frac{13}{64}$ in.
10. What is the fractional equivalent in 64ths of an inch of the following dimensions?

    (*Hint:* Use the table on page 104.)

    (a) 0.135 in. (b) 0.091 in. (c) 0.295 in. (d) 0.185 in.

11. The following table lists the thickness in inches of several sizes of sheet steel:

U.S. Gage	Thickness(in.)
35	0.0075
30	0.0120
25	0.0209
20	0.0359
15	0.0673
10	0.1345
5	0.2092

(a) What is the difference in thickness between 30 gage and 25 gage sheet?
(b) What is the difference in thickness between 7 sheets of 25 gage and 4 sheets of 20 gage sheet?
(c) What length of $\frac{3}{16}''$ diameter rivet is needed to join one thickness of 25 gage sheet to a strip of $\frac{1}{4}''$ stock? Add $1\frac{1}{2}$ times the diameter of the rivet to the length of the rivet in order to assure that the rivet is long enough that a proper rivet head can be formed.
(d) What length of $\frac{5}{32}''$ rivet is needed to join two sheets of 20 gage sheet steel? (Don't forget to add the $1\frac{1}{2}$ times rivet diameter.)

12. When the Better Builder Co. completed a small construction job, they found that the following expenses had been incurred: labor, \$672.25; gravel, \$86.77; sand, \$39.41; cement, \$180.96; and bricks, \$204.35. What total bill should they give the customer if they want to make \$225 profit on the job?
13. One lineal foot of 12 in. I-beam weighs 25.4 lb. What is the length of a beam that weighs 444.5 lb?
14. In order to determine the average thickness of a metal sheet, a sheet-metal worker measures it at five different locations. His measurements are: 0.0401 in., 0.0417 in., 0.0462 in., 0.0407 in., and 0.0428 in. What is the average thickness of the sheet?
15. A dimension in a technical drawing is given as $2\frac{1}{2}$ inches with a tolerance of $\pm\frac{1}{32}$ inch. What are the maximum and minimum permissible dimensions written in decimal form?
16. The four employees of the Busted Body Shop earned the following amounts last week: \$211.76, \$196.21, \$208.18, and \$276.35. What is the average weekly pay for the employees of the shop?
17. If Romex cable sells for \$172.05 per 1850 ft, what is the cost of 1210 ft of this cable?
18. In Lucy's first week as an electrician, she earned \$273.75 for 37.5 hours of work. What is her hourly rate of pay?
19. A welder finds that 2.083 cu ft of acetylene gas are needed to make one bracket. How much gas will be needed to make 27 brackets?
20. What is the outside diameter of a 6.50×14 tire?

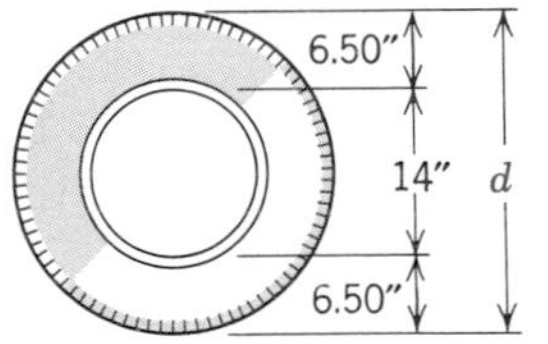

21. How many 2.34″ spacer blocks can be cut from a 2″ by 2″ square bar 48″ long? Allow $\frac{1}{8}''$ waste for each saw cut.
22. The Fixum Auto Shop charges customers a flat hourly rate of \$11.50 for labor on all jobs. They pay their motor overhaul mechanic \$6.25 per hour, brake mechanic \$4.20 per hour, and general mechanic helpers \$2.65 per hour. A motor overhaul takes 23.5 hours of labor, and a brake relining takes 6.25 hours of labor.
(a) What is the total cost to the customer for labor for these two jobs?
(b) What does the shop pay the brake mechanic?
(c) What does the shop pay the motor overhaul mechanic?
(d) What does the shop pay the helper for the two jobs?
(e) What amount of money does the shop keep to cover overhead and profit?

23. In Hibbing, Minnesota, the temperature dropped 24°C overnight. If the thermometer read $-14°C$ before the plunge, what did it read afterward?

24. Morgan Plumbing had earnings of \$3478 in January and expenses of \$4326. What was their net profit or loss? (Indicate profit with a + sign and loss with a − sign.)

4 Percent

PREVIEW 4

Objective	Sample Problems		Where To Go for Help: Page	Frame
1. Write fractions and decimal numbers as percents.	(a) Write $\frac{1}{4}$ as a percent.	______	129	**1**
	(b) Write 0.46 as a percent.	______		
	(c) Write 5 as a percent.	______		
	(d) Write $1\frac{7}{8}$ as a percent.	______		
2. Convert percents to decimal numbers.	(a) Write 35% as a decimal number.	______	133	**5**
	(b) Write 0.25% as a decimal number.	______		
	(c) Write 112% as a decimal number.	______		
3. Solve problems involving percent.	(a) Find $37\frac{1}{2}$% of 600.	______	136	**9**
	(b) Find 120% of 45.	______		
	(c) What percent of 80 is 5?	______	140	**12**
	(d) 12 is 16% of what number?	______	143	**13**
	(e) If a measurement is given as 2.778 ± 0.025 in., state the tolerance as a percent.	______	147	**16**
	(f) If a certain kind of solder is 52% tin, how many pounds of tin are needed to make 20 lb of solder?	______		

(Answers to these sample problems are at the bottom of this page.)

If you are certain you can work *all* of the problems correctly, turn to page 161 for a set of practice problems. If you cannot work one or more of the sample problems, turn to the page indicated for some help. If you want to be tops in your work, turn to frame **1** and begin work there.

Name ______

Date ______

Course/Section ______

Answers to Preview 4

1. (a) 25% (b) 46% (c) 500% (d) 187.5%
2. (a) 0.35 (b) 0.0025 (c) 1.12
3. (a) 225 (b) 54 (c) 6.25% or $6\frac{1}{4}$% (d) 75
   (e) 0.90% (f) 10.4 lb

4 Percent

4-1 INTRODUCTION TO PERCENT

1 The cartoon character may not be a great musician, but at least his arithmetic is correct. In this kind of practical calculation it is usually most convenient to write the comparison number, $\frac{1}{100}$th, as a percent, 1%. Percent calculations are a very important part of any work in business, technical skills, or the trades.

The word *percent* comes from the Latin phrase *per centum* meaning "by the hundred" or "for every hundred." A number written as a percent is being compared with a second number called the standard or *base*. For example, what part of the base is length A?

We could answer the question with a fraction, a decimal, or a percent. First, divide the base with 100 equal parts

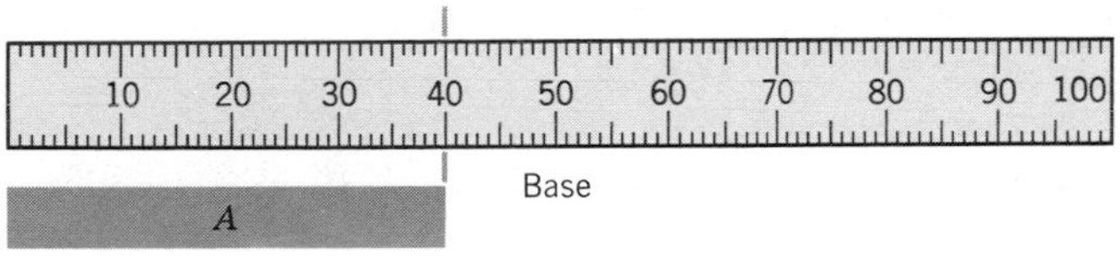

then compare length A with it.

The length of A is 40 parts out of the 100 parts that make up the base.

A is $\frac{40}{100}$ or 0.40 or 40% of A.

$\frac{40}{100} = 40\,\%$ $\qquad$ $40\,\%$ means 40 parts in 100 or $\frac{40}{100}$.

For the following diagram, what part of the base is length B?

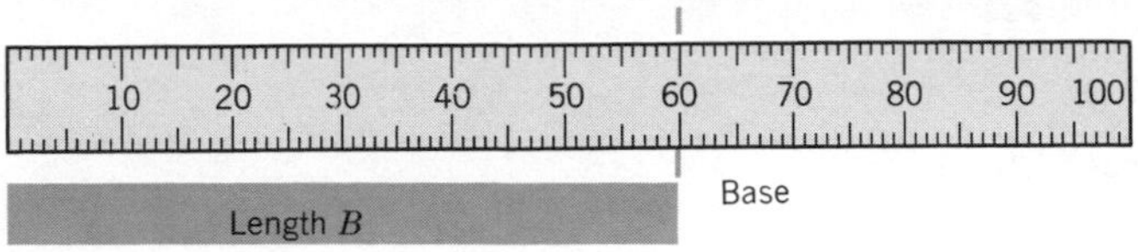

Answer with a percent.

Turn to **2** to check your answer.

---

2 B is $\frac{60}{100}$ or 60%

The compared number may be larger than the base. For example,

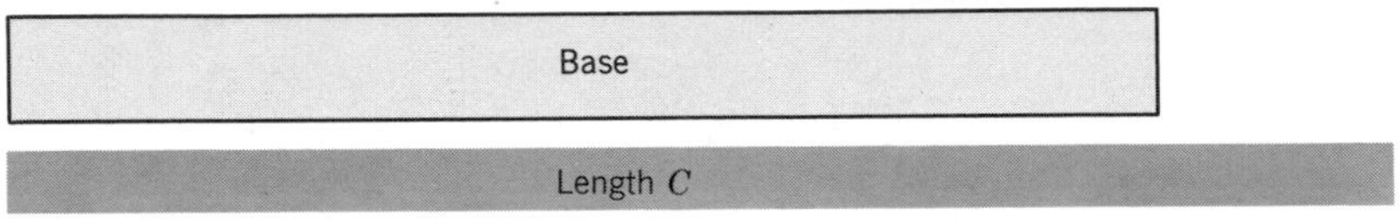

In this case, divide the base into 100 parts and extend it in length.

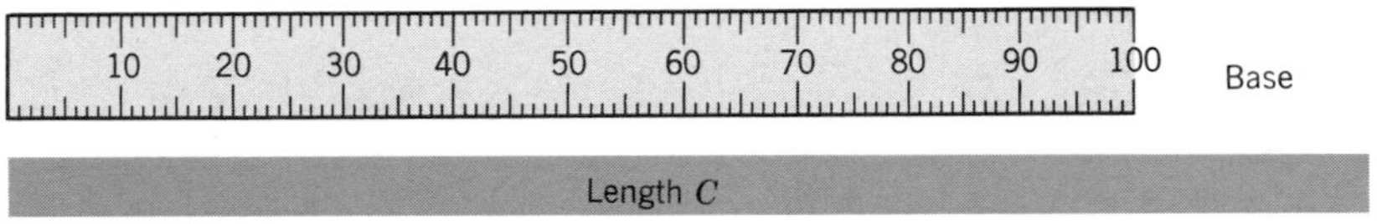

The length of C is 120 of the 100 parts that make up the base.

C is $\frac{120}{100}$ or 120 % of the base.

Fractions, decimals, and percents are all alternative ways to compare two numbers. For example, in the following drawing what fraction of the rectangle is shaded? What part of 12 is 3?

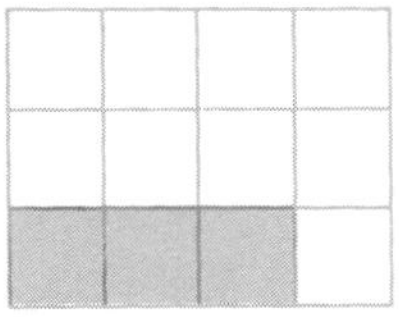

First, we can write the answer as a fraction:

$$\frac{\text{3 shaded squares}}{\text{12 squares}} = \frac{3}{12} = \frac{1}{4}$$

Second, by dividing 4 into 1 we can write this fraction as a decimal:

$$\frac{1}{4} = 0.25 \qquad 4\overline{)1.00}\ \ (0.25)$$

Finally, we can rewrite the fraction with a denominator of 100:

$$\frac{1}{4} = \frac{1 \times 25}{4 \times 25}$$

$$\frac{1}{4} = \frac{25}{100} \quad \text{or} \quad 25\%$$

Changing Decimal Numbers to Percent

Converting a number from fraction or decimal form to a percent is a very useful skill.

> To write a decimal number as a percent, multiply the decimal number by 100%.

For example,

$0.60 = 0.60 \times 100\% = 60\%$

Multiply by 100%

$$\begin{array}{r} 0.60 \\ \times\ 100 \\ \hline 60.00 \end{array}$$

$0.375 = 0.375 \times 100\% = 37.5\%$
$3.4 = 3.4 \times 100\% = 340\%$
$0.02 = 0.02 \times 100\% = 2\%$

Notice that multiplying by 100 is equivalent to shifting the decimal point two places to the right. For example,

$0.60 = 0.60 = 60.\%$ or 60%

Shift decimal point two places right

$0.056 = 0.056 = 5.6\%$

Rewrite the following decimal numbers as percents.

(a) 0.75	(b) 1.25	(c) 0.064
(d) 3	(e) 0.05	(f) 0.004

Check your answers in **3**.

---

3 (a) $0.75 = 0.75 \times 100\% = 75\%$ — 0.75 becomes 75%

(b) $1.25 = 1.25 \times 100\% = 125\%$ — 1.25 becomes 125%

(c) $0.064 = 0.064 \times 100\% = 6.4\%$ — 0.064 becomes 6.4%

(d) $3 = 3 \times 100\% = 300\%$ — $3 = 3.00$ or 300%

(e) $0.05 = 0.05 \times 100\% = 5\%$ — 0.05 becomes 5%

(f) $0.004 = 0.004 \times 100\% = 0.4\%$ — 0.004 becomes 0.4%

Changing Fractions to Percent

> To rewrite a fraction as a percent, first change to decimal form by dividing, then multiply by 100%.

For example,

$\frac{1}{2}$ is 1 divided by 2, or $2\overline{)1.00}$ giving 0.50, so that

$\frac{1}{2} = 0.50 = 0.50 \times 100\% = 50\%$

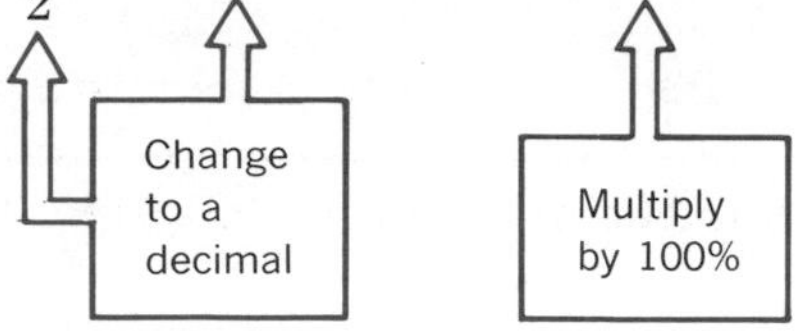
Change to a decimal
Multiply by 100%

More examples:

$\frac{3}{4} = 0.75 = 0.75 \times 100\% = 75\%$

$\frac{3}{20} = 0.15 = 0.15 \times 100\% = 15\%$ since $\frac{3}{20} = 20\overline{)3.00}$ giving 0.15

$1\frac{7}{20} = \frac{27}{20} = 1.35 = 1.35 \times 100\% = 135\%$ since $\frac{27}{20} = 20\overline{)27.00}$ giving 1.35

Your turn. Rewrite $\frac{5}{16}$ as a percent.

Check your work in **4**.

---

4 $\frac{5}{16}$ means $16\overline{)5.0000}$ giving 0.3125 so that

$\frac{5}{16} = 0.3125 = 31.25\%$

This is often written as $31\frac{1}{4}\%$.

Some fractions cannot be converted to an exact decimal. For example,

$\frac{1}{3} = 0.3333\ldots$ where the 3s continue endlessly. We can round to get an approximate percent.

$\frac{1}{3} \cong 0.3333 \cong 33.33\%$ or $33\frac{1}{3}\%$

The fraction $\frac{1}{3}$ is roughly equal to 33.33% and exactly equal to $33\frac{1}{3}\%$.

Rewrite the following fractions as percents.

(a) $\frac{4}{5}$ (b) $\frac{2}{3}$ (c) $3\frac{1}{8}$

(d) $\frac{5}{12}$ (e) $\frac{5}{6}$ (f) $1\frac{1}{3}$

Check your work in **5**.

---

5 (a) $\frac{4}{5} = 0.80 = 80\%$

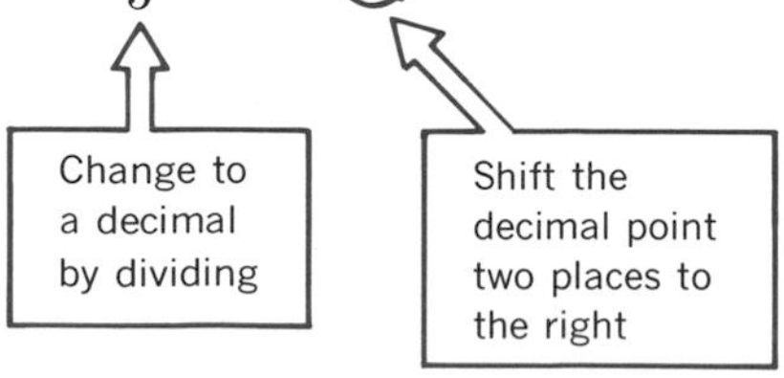
Change to a decimal by dividing
Shift the decimal point two places to the right

(b) $\frac{2}{3} \cong 0.6666 \cong 66.66\%$ or exactly $66\frac{2}{3}\%$

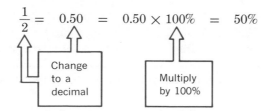

(c) $\frac{1}{8}$ is $8\overline{)1.000}$ with quotient 0.125 so $3\frac{1}{8} = 3.125$

$3\frac{1}{8} = 3.125 = 312.5\%$

(d) $\frac{5}{12}$ is $12\overline{)5.000}$ with quotient 0.4166 . . . so

$\frac{5}{12} = 0.4166\ldots \cong 41.66\%$ or exactly $41\frac{2}{3}\%$

(e) $\frac{5}{6}$ is $6\overline{)5.000}$ with quotient 0.833 . . . so

$\frac{5}{6} \cong 0.833 \cong 83.3\%$ or exactly $83\frac{1}{3}\%$

(f) $1\frac{1}{3} = 1.333\ldots \cong 133.3\%$ or exactly $133\frac{1}{3}\%$

Changing Percents to Decimal Numbers

In order to use percent numbers when you solve practical problems, it is often necessary to change a percent to a decimal number.

PERCENT EQUIVALENTS

Percent	Decimal	Fraction
5%	.05	$\frac{1}{20}$
$6\frac{1}{4}\%$	.0625	$\frac{1}{16}$
$8\frac{1}{3}\%$	$.08\overline{3}$	$\frac{1}{12}$
10%	.10	$\frac{1}{10}$
$12\frac{1}{2}\%$	.125	$\frac{1}{8}$
$16\frac{2}{3}\%$	$.1\overline{6}$	$\frac{1}{6}$
20%	.20	$\frac{1}{5}$
25%	.25	$\frac{1}{4}$
30%	.30	$\frac{3}{10}$
$33\frac{1}{3}\%$	$.3\overline{3}$	$\frac{1}{3}$
$37\frac{1}{2}\%$	.375	$\frac{3}{8}$
40%	.40	$\frac{2}{5}$
50%	.50	$\frac{1}{2}$
60%	.60	$\frac{3}{5}$
$62\frac{1}{2}\%$	.625	$\frac{5}{8}$
$66\frac{2}{3}\%$	$.6\overline{6}$	$\frac{2}{3}$
70%	.70	$\frac{7}{10}$
75%	.75	$\frac{3}{4}$
80%	.80	$\frac{4}{5}$
$83\frac{1}{3}\%$	$.8\overline{3}$	$\frac{5}{6}$
$87\frac{1}{2}\%$	.875	$\frac{7}{8}$
90%	.90	$\frac{9}{10}$
100%	1.00	$\frac{10}{10}$

> To change a percent to a decimal number divide by 100%.

For example,

$$50\% = \frac{50\%}{100\%} = \frac{50}{100} = 0.50 \text{ since } \begin{array}{r}0.50\\100\overline{)50.00}\end{array}$$

Divide by 100%

$$6\% = \frac{6\%}{100\%} = \frac{6}{100} = 0.06 \text{ since } \begin{array}{r}0.06\\100\overline{)6.00}\end{array}$$

$$0.2\% = \frac{0.2\%}{100\%} = \frac{0.2}{100} = 0.002 \text{ since } \begin{array}{r}0.002\\100\overline{)0.200}\end{array}$$

Notice that in each of these examples division by 100% is the same as moving the decimal point two digits to the left. For example,

$50\% = 50.\% = 0.50$

Shift the decimal point two places left

$5\% = 05.\% = 0.05$

$0.2\% = 00.2\% = 0.002$

If a fraction is part of the percent number, write it as a decimal number before dividing by 100%. For example,

$$8\frac{1}{2}\% = 8.5\% = \frac{8.5\%}{100\%} = 0.085$$

Now try these. Write each percent as a decimal number.

(a) 4%	(b) 112%	(c) 0.5%
(d) $9\frac{1}{4}\%$	(e) 45%	(f) $12\frac{3}{4}\%$

Our answers are in **6**.

---

6 (a) $4\% = \frac{4\%}{100\%} = \frac{4}{100} = 0.04 \text{ since } \begin{array}{r}0.04\\100\overline{)4.00}\end{array}$

This may also be done by shifting the decimal point: $4\% = 04.\% = 0.04$.

(b) $112\% = 112.\% = 1.12$

(c) $0.5\% = 00.5\% = 0.005$

(d) $9\frac{1}{4}\% = 9.25\% = 09.25\% = 0.0925$

(e) $45\% = 45.\% = 0.45$

(f) $12\frac{3}{4}\% = 12.75\% = 0.1275$

Did you notice that in (b) a percent greater than 100% gives a decimal number greater than 1?

100% = 1
200% = 2
300% = 3 and so on.

If you have difficulty getting a feel for percents greater than 100%, think of it this way.

10% of 4 is $\frac{1}{10}$ of 4 or 0.4

50% of 4 is $\frac{1}{2}$ of 4 or 2

100% of 4 is all of 4 or 4

200% of 4 is twice 4 or 8

300% of 4 is three times 4 or 12

. . . and so on

A table of the most often used fractions with their percent equivalents is given on page 133.

Work the following problems for practice in using percent.

A. Write each number as a percent.

1. 0.40	2. 0.10	3. 0.95	4. 0.03
5. 0.3	6. 0.015	7. 0.60	8. 7.75
9. 1.2	10. 4	11. 6.04	12. 9
13. $\frac{1}{4}$	14. $\frac{1}{5}$	15. $\frac{7}{20}$	16. $\frac{3}{8}$
17. $\frac{5}{6}$	18. $2\frac{3}{8}$	19. $3\frac{7}{10}$	20. $1\frac{4}{5}$

B. Write each percent as a decimal number.

1. 7%	2. 3%	3. 56%	4. 15%
5. 1%	6. $7\frac{1}{2}$%	7. 90%	8. 0.3%
9. 150%	10. $1\frac{1}{2}$%	11. $6\frac{1}{3}$%	12. $\frac{1}{2}$%
13. $12\frac{1}{4}$%	14. $125\frac{1}{5}$%	15. 1.2%	16. 240%

Check your answers in **7**.

---

7 A.

1. 40%	2. 10%	3. 95%	4. 3%
5. 30%	6. 1.5%	7. 60%	8. 775%
9. 120%	10. 400%	11. 604%	12. 900%
13. 25%	14. 20%	15. 35%	16. 37.5%
17. $83\frac{1}{3}$%	18. 237.5%	19. 370%	20. 180%

B.

1. 0.07	2. 0.03	3. 0.56	4. 0.15
5. 0.01	6. 0.075	7. 0.90	8. 0.003
9. 1.5	10. 0.015	11. 0.063	12. 0.005
13. 0.1225	14. 1.252	15. 0.012	16. 2.4

Now turn to frame **8** for a set of problems involving percent calculations.

8 **Exercises 4-1 Percent Calculations**

A. Convert to a percent:

1. 0.32	2. 1	3. 0.5	4. 2.1
5. $\frac{1}{4}$	6. 3.75	7. 40	8. 0.675
9. 2	10. 0.075	11. $\frac{1}{2}$	12. $\frac{1}{6}$
13. 0.335	14. 0.001	15. 0.005	16. $\frac{3}{10}$
17. $\frac{3}{2}$	18. $\frac{3}{40}$	19. $3\frac{3}{10}$	20. $\frac{1}{5}$

B. Convert to a decimal number:

1. 6%	2. 45%	3. 1%	4. 33%
5. 71%	6. 456%	7. $\frac{1}{2}$%	8. 0.05%
9. $6\frac{1}{4}$%	10. $8\frac{3}{4}$%	11. 30%	12. 2.1%
13. 800%	14. 8%	15. 0.25%	16. $16\frac{1}{3}$%

When you have had the practice you need, turn to frame **9**.

4-2 PERCENT PROBLEMS

9 In all of your work with percent you will find that there are three basic types of problems. These three form the basis for all percent problems that arise in business, technology, or the trades. In this section we will show you how to solve any percent problem, and we will examine each of the three types of problems.

All percent problems involve three quantities:

B, the *base* or total amount, a standard used for comparison,
P, the *percentage* or part being compared with the base,
R, the *rate* or *percent,* a percent number.

For any percent problem these three quantities are related by the equation

$P = R \times B$	The percent problem formula

The three types of percent problems involve finding one of these three quantities when the other two are known.

In every problem follow the following five steps.

Step 1 **Translate** the problem sentence into a math statement. For example, the question

30% of what number is 16? should be translated

30% × □ = 16

or 30% × □ = 16

In this case, 30% is the percent or rate; □, the unknown quantity, is the total or base; and 16 is the percentage or part of the total.

Notice that the words and phrases in the problem question become math symbols. The word "of" is translated *multiply.* The word "is" (and similar verbs such as "will be" and "becomes") are translated *equals.* Use a □ or a letter of the alphabet or a question mark for the unknown quantity you are asked to find.

Step 2 **Rearrange** the equation so that the unknown quantity is alone on the left of the equals sign and the other quantities are on its right.

The equation $30\% \times \square = 16$

becomes $\square = 16 \div 30\% = \dfrac{16}{30\%} = \dfrac{16}{0.30}$

The following *Equation Finder* may help you to do this rearranging.

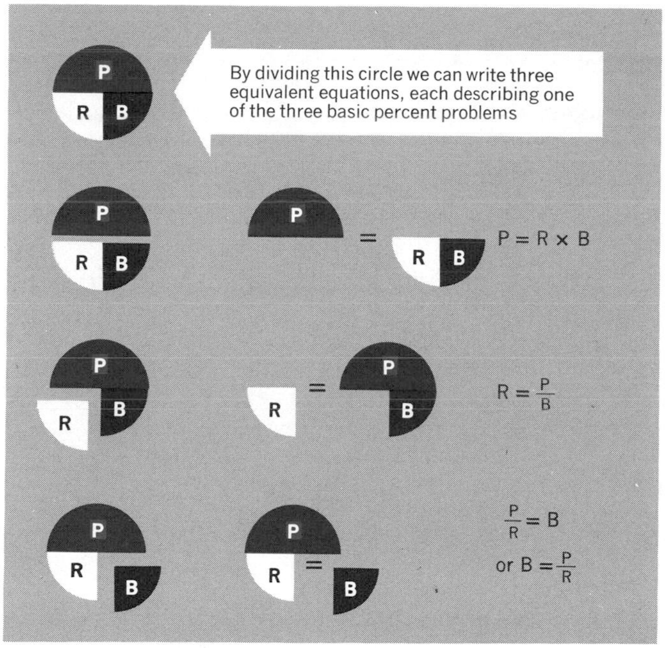

By dividing this circle we can write three equivalent equations, each describing one of the three basic percent problems
P
R
B
P = R × B
R = P/B
P/R = B
or B = P/R

The three equations $P = R \times B$, $R = \dfrac{P}{B}$, and $B = \dfrac{P}{R}$ are all equivalent.

Step 3 **Estimate.** Make a reasonable estimate of the answer. This is a guess, but it is a careful, educated guess. A good estimate of the answer at the start of the problem will help you catch any careless mistakes. Estimating is especially important when you are doing your calculations with an electronic calculator.

Step 4 **Solve** the problem by doing the arithmetic.

Careful Never do arithmetic, never multiply or divide, with percent numbers. All percents must be rewritten as fractions or decimals before you can use them in a multiplication or division.

Step 5 **Check** and **double-check** your answer.

Check your answer against the original estimate. Are they the same, or at least close? If they do not agree, at least roughly, you have probably made a mistake and should repeat your work.

Double-check your answer by putting it back into the original problem to see if it makes sense. If possible, use the answer to calculate one of the other numbers in the equation as a check.

Now let's look very carefully at each type of percent problem. We'll explain each, give examples, show you how to solve them, and work through a few problems together.

Turn to **10**.

10 The first of the three types of percent problems involves finding the percentage P when the base B and the percent R are given. P-type problems are usually stated in the form

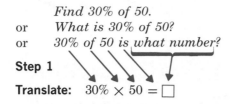

Find 30% of 50.
or *What is 30% of 50?*
or *30% of 50 is what number?*

Step 1

Translate: $30\% \times 50 = \square$

The problem sentence is translated into an arithmetic equation. Notice that the word "of" means *multiply* and the word "is" becomes *equals*. Use any letter or symbol for the unknown number.

Step 2

Rearrange: Rewrite the percent equation so that you can calculate the unknown number.

$P = 0.30 \times 50$

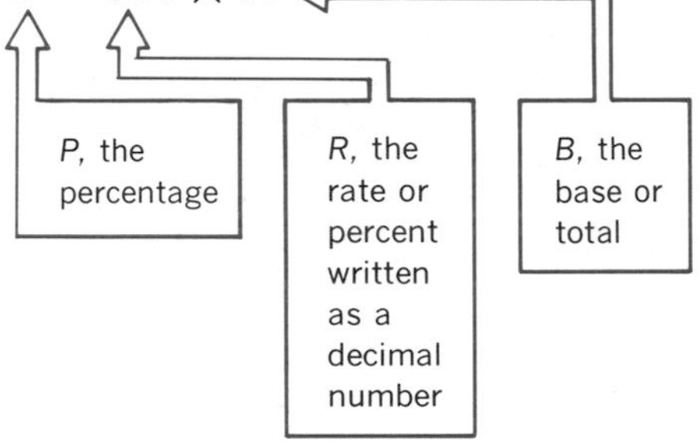

Here we have rewritten 30% as a decimal number so that we can do arithmetic with it.

Divide by 100% $\quad 30\% = \dfrac{30\%}{100\%} = 0.30$

or shift the decimal point $\quad 30\% = 30.0\% = 0.30$

Never do arithmetic, never multiply or divide, with percent numbers. All percents must be rewritten as fractions or decimals before they can be used in a multiplication or division.

Step 3

Estimate: To get a rough estimate of the answer, you might recall that 30% is about one-third, and one-third of 50 is roughly 17. Your answer will be about 17—at least it will be closer to 17 than to 1.7 or 170.

Step 4

Solve: For this equation no fancy solving is needed. Simply multiply.

$P = 0.30 \times 50$
$P = 15$

Step 5

Check: The answer 15 agrees reasonably well with the estimate 17. The answer and the estimate are not exactly the same, but they are close enough.

Don't be intimidated by numbers. If the problem involves very large or very difficult-looking numbers, reduce it to a simpler problem. For example, the problem

Find $14\frac{17}{25}\%$ of 6.41

looks difficult until you realize that it is essentially the same problem as

Find 10% of 6

which is fairly easy.

Before you begin any actual arithmetic problem involving percent you should

(a) Know roughly the size of the answer.
(b) Have a plan for solving the problem based on a simpler problem.
(c) Always change percents to decimals or fractions before multiplying or dividing with them. Never use percent numbers in arithmetic calculations.

Now, use the five-step method to solve this *P*-type problem:

$8\frac{1}{2}\%$ of 160 is how much?

Check your answer in **11**.

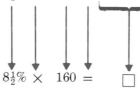

---

11 **Step 1 Translate:** $8\frac{1}{2}\%$ of \$160 is how much?

$$8\frac{1}{2}\% \times 160 = \square$$

Watch for the "is" and "of" words. They are the key to solving percent problems.

Step 2 Rearrange: $8\frac{1}{2}\% = 8.5\%$

so $08.5\% = 0.085$

The equation becomes $0.085 \times 160 = P$ or $P = 0.085 \times 160$.

Step 3 Estimate: $8\frac{1}{2}\%$ is close to 10%, so $8\frac{1}{2}\%$ of 160 is roughly equal to 10% of 160 or one-tenth of 160 or 16. The answer should be about 16.

Step 4 Solve: $P = 0.085 \times 160$
$P = 13.6$

$$\begin{array}{r} 0.085 \\ \times\ 160 \\ \hline 5100 \\ 85 \\ \hline 13.600 \end{array}$$

Step 5 Check: The answer 13.6 is approximately equal to the estimate 16—at least it is in the correct ballpark.

Solve the following percent problems.

(a) Find 2% of 140 lb.
(b) 35% of \$20 is equal to what amount?
(c) What is $7\frac{1}{4}\%$ of \$1000?

(d) Calculate $5\frac{1}{3}\%$ of 3.3.
(e) Find the number that is 120% of 15.

The step-by-step answers are in **12**

---

12 (a) 2% of 140 = ?

$2\% \times 140 = \square$

$P = 2\% \times 140$

$P = 0.02 \times 140$

$P = 2.8$ lb

Estimate: 10% or $\frac{1}{10}$ of 140 = 14

2% of 140 would be about $\frac{1}{5}$th of this or 3.

$$2\% = \frac{2\%}{100\%} = \frac{2}{100} = 0.02$$

Check: 2.8 is roughly equal to 3, our original estimate.

(b) 35% of 20 = ?

$35\% \times 20 = \square$

$P = 0.35 \times 20$

$P = \$7$

Estimate: 35% is roughly $\frac{1}{3}$, and $\frac{1}{3}$ of 20 is about 6 or 7.

Check: Our estimate (6 or 7) is very close to the answer 7.

(c) $7\frac{1}{4}\%$ of \$1000 = ?

$7\frac{1}{4}\% \times 1000 = \square$

$P = 7\frac{1}{4}\% \times 1000$

$P = 0.0725 \times 1000$

$P = \$72.50$

Estimate: 10%, or $\frac{1}{10}$, of \$1000 is \$100.

The answer should be less than \$100.

$$7\tfrac{1}{4}\% = 7.25\% = \frac{7.25}{100} = 0.0725$$

Check: The answer \$72.50 is a bit less than the estimate of about \$100.

(d) $5\frac{1}{3}\%$ of 3.3 = ?

$P = 5\frac{1}{3}\% \times 3.3$

$$P = \frac{16}{300} \times 3.3$$

$P = 0.176$

Estimate: 10% of 3 is 0.3
5% of 3 would be half of this or 0.15. A good guess at the answer would be 0.15.

Check: The answer 0.176 is reasonably close to the estimate 0.15.

Changing $5\frac{1}{3}\%$ to a decimal or fraction form is a bit tricky.

Do it this way:

$$5\tfrac{1}{3}\% = \frac{5\frac{1}{3}}{100} = \frac{\frac{16}{3}}{100} = \frac{16}{300}$$

If you are using a calculator, change $5\frac{1}{3}\%$ to 5.3333333%.

(e) 120% of 15 = ?

$120\% \times 15 = \square$

$P = 120\% \times 15$

$P = 1.2 \times 15$

$P = 18$

Estimate: 100% of 15 is 15, so the answer is certainly more than 15. 200% of 15 is twice 15 or 30. A good estimate would be that the answer is between 15 and 20.

$120\% = \frac{120}{100} = 1.2$

Check: The answer 18 is between 15 and 20 just as we expected our answer to be.

R-Type Problems

The second type of percent problem, the *R*-type problem, requires that you find the *rate* or percent. Problems of this kind are usually stated like this:

5 is what percent of 8?
or *Find what percent 5 is of 8.*
or *What percent of 8 is 5?*

Again, any such problem can be solved using the five-step method.

Step 1 Translate: What percent of 8 is 5?

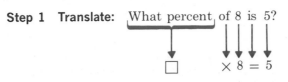

$\square \times 8 = 5$

Step 2 Rearrange: $R \times 8 = 5$

To rearrange the equation and solve for R we can use the equation finder.

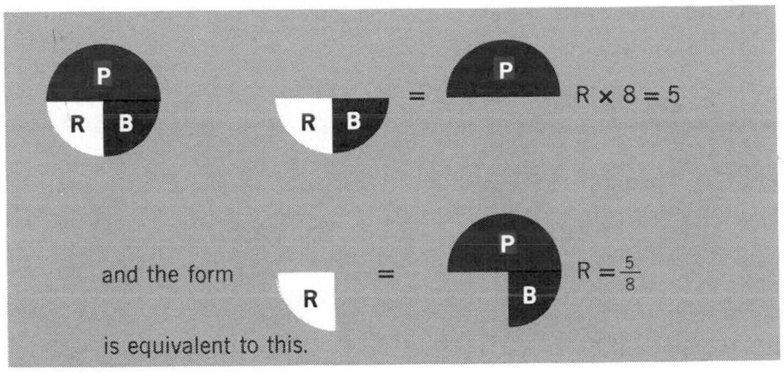

Therefore $R = \dfrac{5}{8}$.

Step 3 Estimate: 4 is exactly one-half or 50% of 8, so 5 is a bit more than 50% of 8. We expect the answer to be somewhat greater than 50%.

Step 4 Solve: The arithmetic is easy. We divide to find R as a decimal.

$R = \dfrac{5}{8}$ $\qquad 8\overline{)5.000} = 0.625$

$R = 0.625$

Now write this decimal number as a percent.

$R = 62.5\%$ $\qquad 0.625 = 62.5\%$

Step 5 Check: The answer 62.5% is reasonably close to our estimate of 50%.

Double-check: Substitute the answer back into the original problem.

"What percent of 8 is 5?"

$(62.5\%) \times 8 = 5$

$0.625 \times 8 = 5$ If you do the arithmetic you will see that this is true.

If you had difficulty converting $\frac{5}{8}$ to a percent you may want to review by returning to frame **3**. Remember, the solution to an R-type problem will always be a fraction or decimal number that must be converted to a percent.

Some people find it helpful to solve percent problems of this type using the "is-of" method.

$$\text{Percent} = \frac{\text{"is" number}}{\text{"of" number}}$$

The "is" number is the quantity closest to the word "is" and the "of" number is the quantity closest to the word "of." For example, to solve the problem

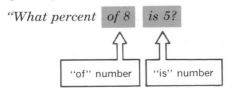

"What percent of 8 is 5?

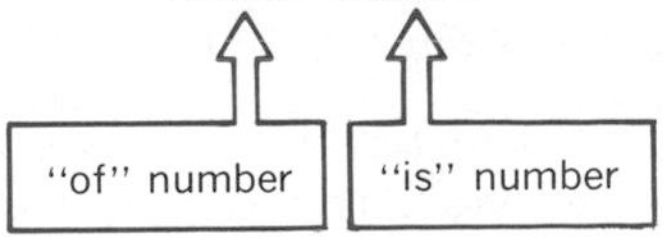

$\text{Percent} = \frac{5}{8}$ or 62.5%

Be careful with this method. Although it usually gives the correct answer for this kind of problem, it can lead to the wrong answer with some problems. Use the Equation Finder to be certain you are solving the problem correctly.

Now try these problems for practice.

(a) What percent of 40 lb is 16 lb?
(b) Find what percent 65 is of 25.
(c) \$9.90 is what percent of \$18.00?
(d) What percent of 2 is $3\frac{1}{2}$?

Check your work in **13**.

---

13 (a) **Step 1** $R \times 40 \text{ lb} = 16 \text{ lb}$

Step 2 $R = \frac{16}{40}$

Step 3 **Estimate:** $\frac{16}{40}$ is about $\frac{1}{3}$ or roughly 33%. The answer will be closer to 30% than to 3% or 300%.

Step 4 $R = 0.4$ $\qquad 40\overline{)16.0}$ gives 0.4

or $R = 40\%$ $\qquad 0.4 = 0.40 = 40\%$

Step 5 **Check:** The answer is reasonably close to the estimate.

Double-check: $40\% \times 40 \text{ lb} = 16 \text{ lb}$

$0.40 \times 40 = 16$ which is correct.

(b) **Step 1** $R \times 25 = 65$

Step 2 $R = \frac{65}{25}$

Step 3 $\frac{65}{25}$ is more than 2, and 2 is 200%. The answer will be over 200%.

Step 4 $R = \frac{65}{25}$

$R = 2.6$

or $R = 260\%$ $\qquad 2.6 = 2.60 = 260\%$

Step 5 Check and double-check it.

The most difficult part of this problem is in deciding whether the percent needed is found from $\frac{65}{25}$ or $\frac{25}{65}$. There is no magic to it. If you read the problem very carefully you will see that it speaks of 65 as a part "of 25." The base or total is 25. The percentage or part is 65.

(c) **Step 1** $\$9.90 = R \times \$18$

Step 2 $R \times 18 = 9.9$

$$R = \frac{9.9}{18}$$

Step 3 \$9.90 is about half of \$18, and one-half is 50%. The answer will be about 50%.

Step 4 $R = \dfrac{9.9}{18}$

$R = 0.55$

or $R = 55\%$ $\quad 0.55 = 55\%$

Step 5 Check and double-check it.

(d) **Step 1** $R \times 2 = 3\frac{1}{2}$

Step 2 $R \times 2 = 3.5$

$$R = \frac{3.5}{2}$$

Step 3 $\frac{3}{2}$ is $1\frac{1}{2}$ or 1.5, and 1.5 is 150%. The answer will be something more than 150%.

Step 4 $R = \dfrac{3.5}{2}$

$R = 1.75$

or $R = 175\%$ $\quad 1.75 = 175\%$

Step 5 Check and double-check it.

B-Type Problems

This third type of percent problem requires that you find the total or base when the percent and the percentage are given. Typically, *B*-type problems are stated like this:

8.7 is 30% of what number?
or *Find a number such that 30% of it is 8.7.*
or *8.7 is 30% of a number. Find the number.*
or *30% of what number is equal to 8.7?*

Again we can use the five-step method to solve these problems.

Step 1 Translate: 30% of what number is equal to 8.7?

$$30\% \times \square = 8.7$$

Step 2 Rearrange: Use the Equation Finder to rewrite the equation.

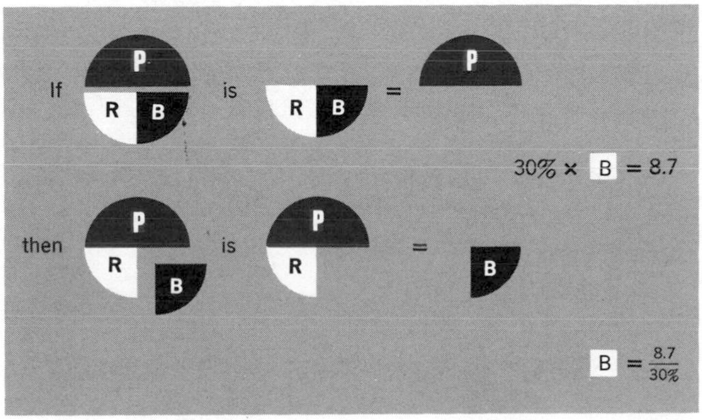

The rearranged equation is

$$B = \frac{8.7}{30\%}$$

Step 3 **Estimate:** 30% is roughly equal to one-third. 8.7 is roughly equal to 9. One-third of what number is equal to 9? 27. Our estimate of the answer is 27.

Step 4 **Solve:** $B = \frac{8.7}{30\%}$

$B = \frac{8.7}{0.30}$ Rewrite 30% as a decimal: 30% = 0.30.

$B = 29$

Step 5 **Check:** The answer 29 and the estimate 27 are close.

Double-check: Substitute the answer back into the original problem.
30% of (29) is equal to 8.7.
$0.30 \times 29 = 8.7$ which is correct.

Careful In Step 4 we cannot divide by 30%. We must change the percent to a decimal number before we do the division.

The "is-of" method can again be used as a short-cut.

$$\text{Base} = \frac{\text{"is" number}}{\text{"of" number}}$$

For example, to solve the problem

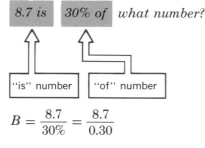

8.7 is 30% of what number?

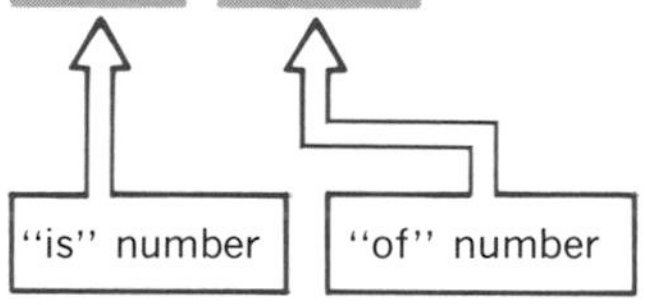

$$B = \frac{8.7}{30\%} = \frac{8.7}{0.30}$$

$B = 29$

If you use this short-cut be very careful to check and double-check your answer in order to catch any errors.

Ready for a few practice problems? Try these.

(a) 16% of what amount is equal to \$5.76?
(b) 41 days is 5% of what time?
(c) \$2 is 8% of the cost. Find the cost.
(d) Find a distance such that $12\frac{1}{2}\%$ of it is $26\frac{1}{4}$ ft.

The correct solutions are in **14**.

---

14 (a) $16\% \times \square = \$5.76$

$B = \frac{5.76}{16\%}$ **Estimate:** $\frac{5.76}{0.16}$ is about $\frac{500}{10}$ or about 50.

$16\% = 0.16$

$B = \frac{5.76}{0.16}$

$B = \$36$ **Check:** The estimate and the answer are reasonably close.
Double-check: $0.16 \times 36 = 5.76$

(b) $41 = 5\% \times \square$

$B = \dfrac{41}{0.05}$ **Estimate:** $\dfrac{40}{0.05}$ is about $\dfrac{4000}{5}$ or about 800.

$B = 820$ days **Check:** The estimate and answer are quite close.

Double-check: $0.05 \times 820 = 41$

(c) $\$2 = 8\% \times \square$

$B = \dfrac{2}{8\%}$ **Estimate:** 8% is close to 10% or one-tenth. One-tenth of what number is 2? 20. The estimate is 20.

$B = \dfrac{2}{0.08}$

$B = \$25$ **Check:** The estimate and answer are close.

Double-check: $2 = 0.08 \times 25$

(d) $12\frac{1}{2}\% \times \square = 26\frac{1}{4}$ ft

$B = \dfrac{26\frac{1}{4}}{12\frac{1}{2}\%}$ **Estimate:** 26 is one-tenth of 260, so 260 is a reasonable estimate of the answer.

$12\frac{1}{2}\% = 12.5\% = 0.125$

$B = \dfrac{26.25}{0.125}$

$B = 210$ **Check:** The estimate and answer are close.

Double-check: $0.125 \times 210 = 26.25$

Notice that we translate any fractions to decimal numbers before doing the arithmetic. Decimal numbers are easier to work with, especially if you are using a calculator.

A Review

Let's review those five steps for solving percent problems:

Step 1 **Translate** the problem sentence into a math statement—an equation.

Step 2 **Rearrange** the arithmetic equation so that the unknown quantity is alone on the left. Use the Equation Finder if necessary.

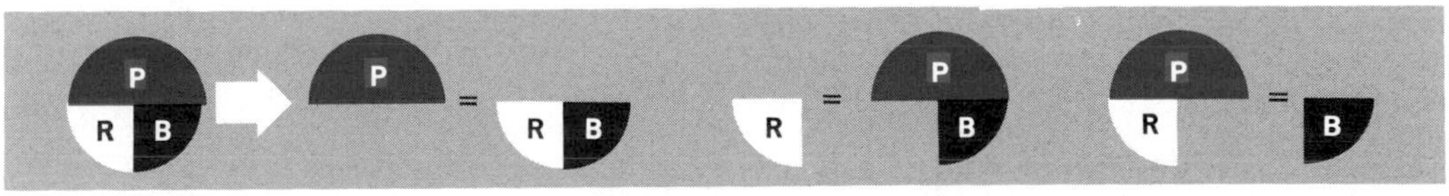

Step 3 **Estimate.** Get a reasonable estimate of the answer.

Step 4 **Solve** the problem by doing the arithmetic. *Always* change percent numbers to decimal numbers first.

Step 5 **Check** your answer by comparing it with the estimate from Step 3.

Double-check the answer if you can by putting it back into the original problem to see if it is correct.

Are you ready for a bit of practice on the three basic kinds of percent problems? Turn to **15** for a problem set.

Exercises 4-2 Percent Problems

Solve

A.

1. 4 is ______% of 5.
2. What percent of 25 is 16? ______
3. 20% of what number is 3? ______
4. 8 is what percent of 8? ______
5. 120% of 45 is ______.
6. 8 is ______ % of 64.
7. 3% of 5000 = ______.
8. 2.5% of what number is 2? ______
9. What percent of 54 is 36? ______
10. 60 is ______% of 12.
11. 17 is 17% of ______.
12. 13 is what percent of 25? ______
13. $8\frac{1}{2}$% of \$250 is ______.
14. 12 is ______% of 2.
15. 6% of 25 is ______.
16. 60% of what number is 14? ______
17. What percent of 16 is 7?______
18. 140 is ______% of 105?

Solve

B.

1. 75 is $33\frac{1}{3}$% of ______.
2. What percent of 10 is 2.5? ______
3. 6% of \$3.29 is ______.
4. 63 is ______% of 35.
5. 12.5% of what number is 20? ______
6. $33\frac{1}{3}$% of \$8.16 = ______.
7. 9.6 is what percent of 6.4? ______
8. \$0.75 is ______% of \$37.50.
9. $6\frac{1}{4}$% of 280 is ______.
10. 1.28 is ______% of 0.32
11. 42.7 is 10% of ______.
12. 260% of 8.5 is ______.
13. $\frac{1}{2}$ is ______% of 25.
14. $7\frac{1}{4}$% of 50 is ______.
15. 287.5% of 160 is ______.
16. 0.5% of ______ is 7.
17. 112% of ______ is 56.
18. $2\frac{1}{4}$% of 110 is ______.

C. Word Problems

1. If you answered 37 problems correctly on a 42-question test, what percent score would you have?
2. The profits from Ed's Plumbing Co. increased by \$14,460, or 30%, this year. What profit did it earn last year?
3. A casting weighed 146 lb out of the mold. It weighed 138 lb after finishing. What percent of the weight was lost in finishing?
4. The Smiths bought a house for \$34,500 and made a down payment of 20%. What was the actual amount of the down payment?
5. If your salary is \$760 per month and you get a 6% raise, what is your new salary?

6. If state sales tax is 6%, what tax would you pay on a purchase of \$7.80?
7. A carpenter earns \$4.95 per hour. If he receives a $7\frac{1}{2}$% pay raise, what is his new pay rate?
8. If 5.45% of your salary is withheld for Social Security, what amount is withheld from monthly earnings of \$645.00?

When you have had the practice you need, continue in frame **16** with the study of some applications of percent to practical problems.

4-3 APPLICATIONS OF PERCENT CALCULATIONS

16 Now that you have seen and solved the three basic percent problems, you need some help in applying what you have learned to practical situations.

Discount

An important kind of percent problem, especially in business, involves the idea of *discount.* In order to stimulate sales, a merchant may offer to sell some item at less than its normal price. A discount is the amount of money by which the normal price is reduced. It is a percentage or part of the normal price. The list price is the normal or original price before the discount is subtracted. The sales price is the new, reduced price. It is always less than the list price.

Sales price = list price − discount

The discount rate is a percent number that enables you to calculate the discount as a part of the list price.

Discount = rate × list price

For example, the list price of a tool is \$18.50. On a special sale it is offered at 20% off. What is the sales price?

Try working this problem, then turn to **17** to see our solution.

---

17 Discount = rate × list price
= 20% of \$18.50
= 0.20 × \$18.50
= \$3.70 **Check:** 20% is $\frac{1}{5}$ and $\frac{1}{5}$ of \$18 is about \$3.

Sales price = list price − discount
= \$18.50 − \$3.70
= \$14.80

Think of it this way:

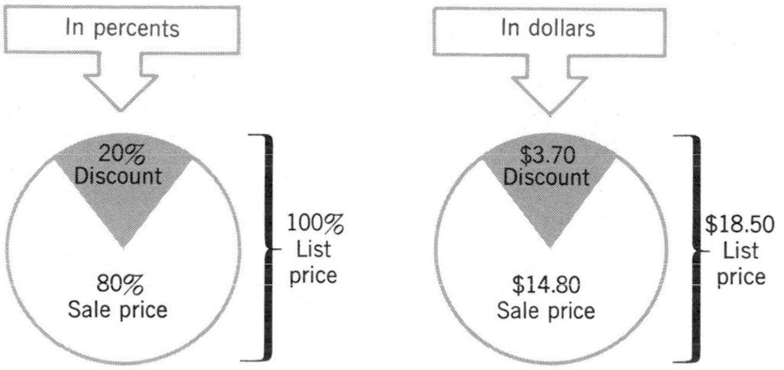

Ready for another problem?

After a 25% discount, the sales price of a power sander is \$144. What was its list price?

Check your work in **18**.

---

18 Sales price = list price − discount

$S = L - D$

Discount = 25% of list price

$$D = 25\% \times L$$
$$D = 0.25L$$

Therefore,

$$S = L - 0.25L$$
$$S = 0.75L$$

But sales price S = \$144 so

$$\$144 = 0.75L$$

$L = \dfrac{\$144}{0.75}$ Divide both sides of equation by 0.75.

$L = \$192$ The list price is \$192.

Check: Discount = 25% of \$192
= 0.25 × \$192
= \$48

Sales price = list price − discount
= \$192 − \$48
= \$144 which is the correct list price as given in the problem.

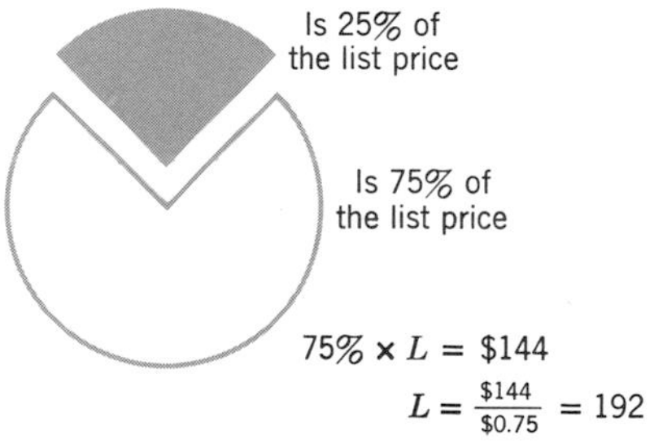

Here are a few problems to test your understanding of the idea of discount.

(a) A pre-inventory sale advertises all tools 20% off. What would be the sale price of a motor drill that cost \$19.95 before the sale?
(b) A compressor is on sale for \$376 and is advertised as "12% off regular price." What was its regular price?
(c) A set of four 740–15 automobile tires is on sale for 15% off list price. What would be the sale price if their list price is \$30.80 each?

Check your answers in **19**.

---

19 (a) Discount = 20% of \$19.95
= 0.20 × \$19.95
= \$3.99

Sales price = list price − discount
= \$19.95 − 3.99
= \$15.96

(b) Discount = 12% of list price

Therefore,

Sales price = 88% of list price

or \$376 = 88% of L
\$376 = 0.88L

$$L = \frac{\$376}{0.88}$$

List price = \$427.27 rounded to the nearest cent

(c) Discount = 15% of list price
= 15% × (4 × \$30.80)
= 0.15 × \$123.20
= \$18.48

Sales price = list price − discount
= \$123.20 − \$18.48
= \$104.72

Sales Tax

Taxes are almost always calculated as a percent of some total amount. *Property taxes* are written as some fraction of the value of the property involved. *Income taxes* are most often calculated from complex formulas that depend on many factors. We cannot consider either income or property taxes here.

A *sales tax* is an amount calculated from the actual price of a purchase and added to the buyer's cost. Retail sales tax rates are set by the individual states in the United States and vary from 0 to 6 or 7% of the sales price. A sales tax of 6% is often stated as "6¢ on the dollar," since 6% of \$1.00 equals 6¢.

If the retail sales tax rate is 6% in California, how much would you pay in Los Angeles for a pair of shoes costing \$26.50?

Check your answer in **20**.

---

20 Sales tax = 6% of \$26.50
= 0.06 × \$26.50
= \$1.59

Actual cost = list price + sales tax
= \$26.50 + \$1.59
= \$28.09

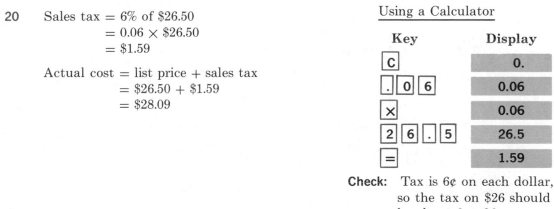

Using a Calculator

Key	Display
C	0.
. 0 6	0.06
×	0.06
2 6 . 5	26.5
=	1.59

Check: Tax is 6¢ on each dollar, so the tax on \$26 should be about 6 × 26 or 156¢ or about \$1.56.

Most stores and sales clerks use tables to look up the sales tax. They need to do the arithmetic only for large purchases beyond the range of the tables. It will pay you to be able to check their work.

Here are a few problems in calculating sales tax.

Find the tax and total cost for each of the following if the sales tax is as shown.

(a) A roll of plastic electrical tape at 59¢ (5%).
(b) A wood worker's vise priced at \$27.50 (6%).
(c) A feeler gauge priced at \$2.95 ($3\frac{1}{2}$%).
(d) A new car priced at \$8785 (4%).
(e) A portable paint compressor priced at \$146.50 ($6\frac{1}{2}$%).
(f) A toggle switch priced at 79¢ (6%).

Check your answers in **21**.

21

	Tax	*Total Cost*	*Calculation*
(a)	3¢	62¢	0.05 × 59¢ = 2.95¢ ≅ 3¢
(b)	\$1.65	\$29.15	0.06 × \$27.50 = \$1.65
(c)	\$0.10	\$3.05	0.035 × \$2.95 = \$0.10325 ≅ \$0.10
(d)	\$351.40	\$9136.40	0.04 × \$8785 = \$351.40
(e)	\$9.52	\$156.02	0.065 × \$146.50 = \$9.5225 ≅ \$9.52
(f)	5¢	84¢	0.06 × 79¢ = 4.74¢ ≅ 5¢

Interest

An important characteristic of modern society is that we have set up complex ways to enable you to use someone else's money. A *lender,* with money beyond his or her needs, supplies cash to a *borrower* whose needs exceed his money. The money is called a loan.

Interest is the amount the lender is paid for the use of his money. Interest is the money you pay to use someone else's money. The more you use and the longer you use it, the more interest you must pay.

When you purchase a house with a bank loan, a car, or refrigerator on an installment loan, or gasoline on a credit card, you are using someone else's money and you pay interest for that use. If you are on the other end of the money game, you may earn interest for money you invest in a savings account or in shares of a business.

Interest can be calculated from the following formula:

> *Interest* = principal · rate · time

For example, suppose you have \$500 in savings earning $5\frac{1}{4}$% annually in your local bank: How much interest do you receive in a year?

Do it this way:

Interest = \$500 × $5\frac{1}{4}$% × 1

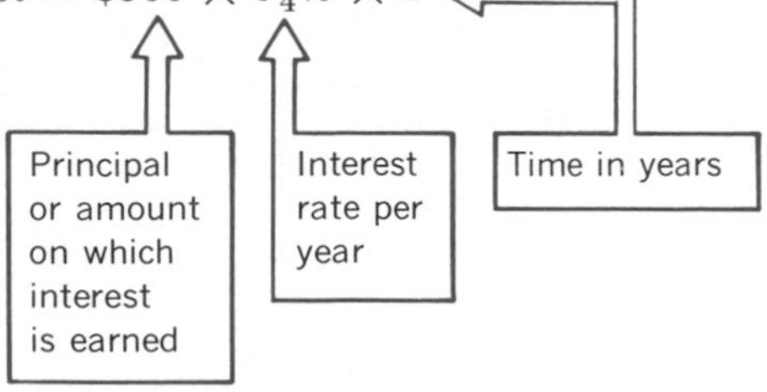

Interest = \$500 × 0.0525
= \$26.25

By placing your \$500 in a bank savings account, you allow the bank to use it in various profitable ways, and you are paid \$26.25 per year for its use.

Most of us play the money game from the other side of the counter. Suppose you find yourself in need of cash and arrange to obtain a loan from a bank. You borrow \$600 at 8% per year for 3 months. How much interest must you pay?

Try to set up and solve this problem exactly as we did in the problem above.

Check your work in **22** .

---

22 $\frac{3 \text{ months}}{12 \text{ months}} = \frac{3}{12}$ year

Using a Calculator

Key	Display
C	0.
6 0 0	600.
×	600.
. 0 8	0.08
×	48.
3	3.
÷	144.
1 2	12.
=	12.

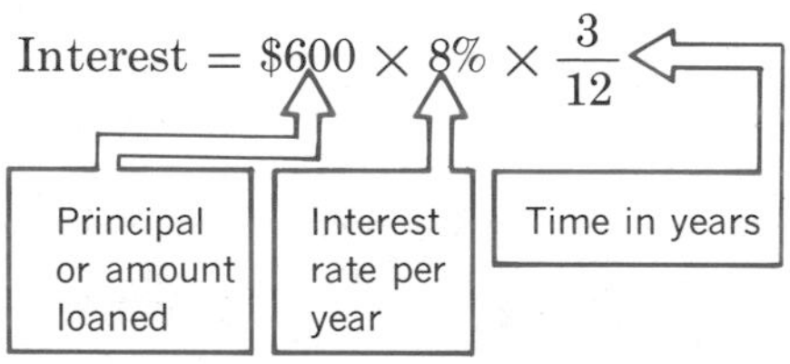

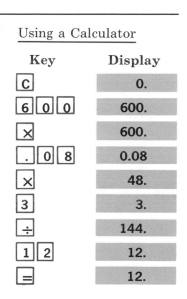

$$\text{Interest} = \$600 \times 0.08 \times \frac{3}{12}$$

$$= \$12$$

Depending on how you and the bank decide to arrange it, you may be required to pay the total principal (\$600) plus interest (\$12) all at once, at the end of three months, or according to some sort of regular payment plan—for example, pay \$204 each month for three months.

Commission

The simplest practical use of percent is in the calculation of a part or percentage of some total. For example, sales persons are often paid on the basis of their success at selling and receive a *commission* or share of the sales receipts. Commission is usually described as a percent of sales income.

Commission = rate or percent × sales income

Suppose your job as a lumber salesman pays 12% commission on all sales. How much do you earn from a sale of \$400?

$$
\begin{aligned}
\text{Commission} &= 12\% \times \$400 \\
&= 0.12 \times \$400 \\
&= \$48
\end{aligned}
$$

Try this one yourself.

At the Happy Bandit Used Car Company each salesman receives a 6% commission on his sales. What would a salesman earn if he sold a 1960 Airedale for $1299.95?

Check your work in **23** .

HOW DOES A CREDIT CARD WORK?

Many forms of small loans, such as credit card loans, charge interest by the month. These are known as "revolving credit" plans. (If you buy very much money this way, *you* do the revolving and may run in circles for years trying to pay it back!) Essentially, you buy now and pay later. If you repay the full amount borrowed within 25 or 30 days there is no charge for the loan.

After the first pay period of 25 or 30 days, you pay a percent of the unpaid balance each month, usually between $\frac{1}{2}$ and 2%. In addition you must pay some minimum amount each month, usually $10 or 10% of the unpaid balance, whichever is larger. Generally you are also charged a small monthly amount for insurance premiums. (The credit card company insure themselves against your defaulting on the loan or disappearing, and you pay for their insurance.)

Let's see how it works. Suppose you go on a short vacation and pay for gasoline, lodging, and meals with your Handy Dandy credit card. A few weeks later you receive a bill for $100. You can pay it within 30 days and owe no interest or you can pay over several months as follows:

Month 1 $\$100 \times 1\frac{1}{2}\% = \$100 \times 0.015 = \$1.50$ owe $101.50

pay $10 and carry $91.50 over to next month

Month 2 $\$91.50 \times 0.015 = \$1.38$ owe $92.88

pay $10 and carry $82.88 over to next month

. . . and so on.

A year later you will have repaid the $100 loan and all interest.

The $1\frac{1}{2}\%$ per month interest rate seems small, but it is equivalent to between 15% and 18%. You pay at a high rate for the convenience of using the credit card and the no-questions-asked ease of getting the loan.

---

23 $$
\begin{aligned}
\text{Commission} &= 6\% \times \$1299.95 \\
&= 0.06 \times \$1299.95 \\
&= \$77.9970 \quad \text{or} \quad \$78.00 \text{ rounded}
\end{aligned}
$$

The salesman earns a commission of $78 on the sale.

Efficiency

When energy is converted from one form to another in any machine or conversion process, it is useful to talk about the *efficiency* of the machine or the process. The efficiency of a process is the ratio of the energy or power output to the energy or power input, expressed as a percent.

$$\text{Efficiency} = \frac{\text{output}}{\text{input}} \times 100\%$$

For example, an auto engine is rated at 175 hp and is found to deliver only 140 hp to the transmission. What is the efficiency of the process?

$$\text{Efficiency} = \frac{\text{output}}{\text{input}} \times 100\%$$
$$= \frac{140\text{ hp}}{175\text{ hp}} \times 100\%$$
$$\text{Efficiency} = 0.8 \times 100\%$$
$$= 80\%$$

Now you try one. A gasoline shop engine rated at 65 hp is found to deliver 56 hp through a belt drive to a pump. What is the efficiency of the drive system?

Check your work in **24**

---

24 $$\text{Efficiency} = \frac{56\text{ hp}}{65\text{ hp}} \times 100\%$$
$$\cong 0.86 \times 100\%$$
$$= 86\%$$

The efficiency of any practical system will always be less than 100%—you can't produce an output greater than the input!

Tolerances

On technical drawings or other specifications, measurements are usually given with a *tolerance* showing the allowed error. For example, the fitting shown in this drawing has a length of

1.370 ± 0.015″

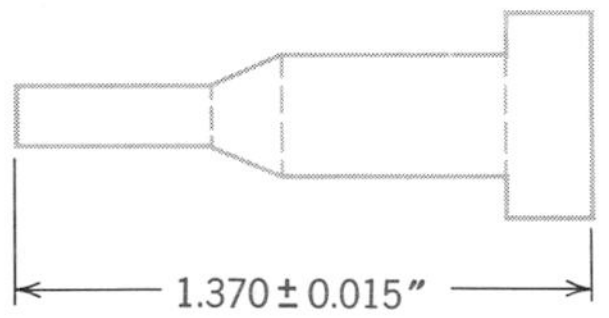

This means that the dimension shown must be between

1.370 + 0.015″ or 1.385″

and

1.370 − 0.015″ or 1.355″

The *tolerance limits* are 1.385″ and 1.355″.

Very often the tolerance is written as a percent.

$$\text{Percent tolerance} = \frac{\text{tolerance}}{\text{measurement}} \times 100\%$$

In this case,

$$\text{Percent tolerance} = \frac{0.015}{1.370} \times 100\%$$
$$\cong 0.0109 \times 100\%$$
$$\cong 1.1\%$$

The dimension would be written 1.370″ ± 1.1%.

Rewrite the following dimension with a percent tolerance:

$1.426 \pm 0.010'' =$ ________

Check your work in **25**

SMALL LOANS

Sooner or later everyone finds it necessary to borrow money. When you do you will want to know beforehand how it works. Suppose you borrow \$200 and the loan company specifies that you repay it at \$25 per month plus interest at 3% per month on the unpaid balance. What interest do you actually pay?

Month 1	$\$200 \times 0.03 = \$\ 6.00$	you pay	$\$25 + \$6.00 = \$\ 31.00$
Month 2	$\$175 \times 0.03 = \$\ 5.25$	you pay	$\$25 + \$5.25 = \$\ 30.25$
Month 3	$\$150 \times 0.03 = \$\ 4.50$	you pay	$\$25 + \$4.50 = \$\ 29.50$
Month 4	$\$125 \times 0.03 = \$\ 3.75$	you pay	$\$25 + \$3.75 = \$\ 28.75$
Month 5	$\$100 \times 0.03 = \$\ 3.00$	you pay	$\$25 + \$3.00 = \$\ 28.00$
Month 6	$\$\ 75 \times 0.03 = \$\ 2.25$	you pay	$\$25 + \$2.25 = \$\ 27.25$
Month 7	$\$\ 50 \times 0.03 = \$\ 1.50$	you pay	$\$25 + \$1.50 = \$\ 26.50$
Month 8	$\$\ 25 \times 0.03 = \$\ 0.75$	you pay	$\$25 + \$0.75 = \$\ 25.75$
	\$27.00		\$227.00

Total interest is \$27.00 — Total of 8 loan payments

They might also set it up as 8 equal payments of $\$227.00 \div 8 = \$28.375$ or \$28.38 per month.

The 3% monthly interest rate seems small but it amounts to about 20% per year.

A bank loan for \$200 at 7% for 8 months would cost you

$$\$200 \times 7\% \times \frac{8}{12}$$

or $$\$200 \times 0.07 \times \frac{8}{12}$$

or \$9.34 Quite a difference.

The loan company demands that you pay more and in return they are less worried about your ability to meet the payments. For a bigger risk, they want a higher interest.

---

25 Percent tolerance $= \dfrac{0.010''}{1.426''} \times 100\%$

$$= 0.0070 \times 100\%$$
$$= 0.70\%$$
$$1.426 \pm 0.010'' = 1.426'' \pm 0.70\%$$

If the dimension is given in metric units, the procedure is exactly the same. For example, the dimension

315 ± 0.25 mm becomes

$$\text{Percent tolerance} = \frac{0.25}{315} \times 100\%$$
$$\cong 0.00079 \times 100\%$$
$$\cong 0.08\%$$

or 315 mm ± 0.08%

Using a Calculator

Key	Display
C	0.
. 2 5	0.25
÷	0.25
3 1 5	315.
×	0.0007936
1 0 0	100.
=	0.07936

If the tolerance is given as a percent, it is easy to work backward to find the tolerance as a dimension value.

$$\text{Tolerance} = \frac{\text{percent tolerance}}{100\%} \times \text{measurement}$$

For example,

2.450″ ± 0.2%

becomes

$$\text{Tolerance} = \frac{0.2\%}{100\%} \times 2.450$$
$$= 0.0049''$$
$$\cong 0.005''$$

or

2.450 ± 0.005″

Using a Calculator

Key	Display
C	0.
. 2	0.2
÷	0.2
1 0 0	100.
×	0.002
2 . 4 5	2.45
=	0.0049

Now for some practice in working with tolerances, convert the following measurement tolerances to percents, and vice versa. Round your answers to two decimal places.

Measurement	Tolerance	Percent tolerance
2.345″	± 0.001″	
	± 0.002″	
	± 0.005″	
	± 0.010″	
	± 0.125″	
274 mm	± 0.10 mm	
	± 0.20 mm	
	± 0.50 mm	
3.475″		± 0.10%
		± 0.20%
123 mm		± 0.05%
		± 0.15%

Check your answers in **26**.

26

Measurement	Tolerance	Percent Tolerance
2.345″	± 0.001″	± 0.04%
	± 0.002″	± 0.09%
	± 0.005″	± 0.21%
	± 0.010″	± 0.43%
	± 0.125″	± 5.33%
274 mm	± 0.10 mm	± 0.04%
	± 0.20 mm	± 0.07%
	± 0.50 mm	± 0.18%
3.475″	± 0.003″	± 0.10%
	± 0.007″	± 0.20%
123 mm	± 0.06 mm	± 0.05%
	± 0.18 mm	± 0.15%

Percent Change

In many situations in technical work and the trades you may need to find a *percent increase* or *percent decrease* in a given quantity. For example, suppose the output of a certain electrical circuit is 20 amps, and the output increases by 10%. What is the new value of the output? Solve it this way:

Increase $= 20 \times 10\%$

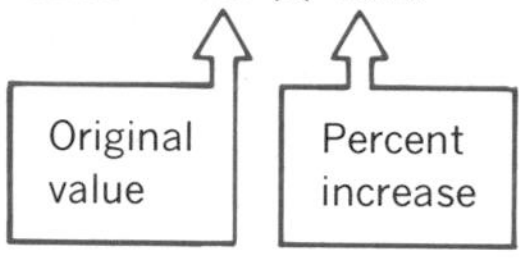

Increase $= 20 \times 0.10$
$= 2$ amps   The new value is 20 amps + 2 amps or 22 amps.

It will save time and help prevent mistakes if you calculate the new value directly. Try it this way:

New value $= 20 \times 110\%$

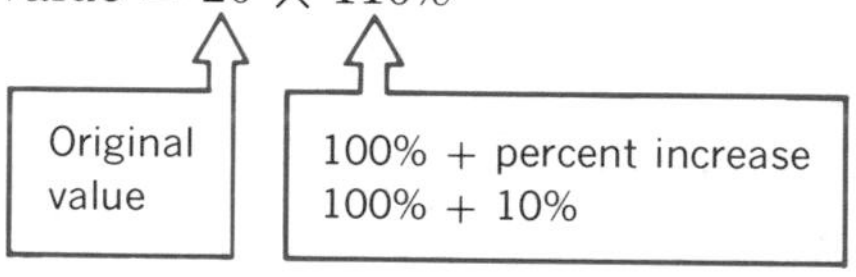

New value $= 20 \times 1.1$
$= 22$ amps

Percent decrease can be calculated in a very similar way. For example, suppose the output of a certain machine at the Acme Gidget Co. is 600 gidgets per day. If the output decreases by 8%, what is the new output?

Decrease $= 600 \times 8\%$

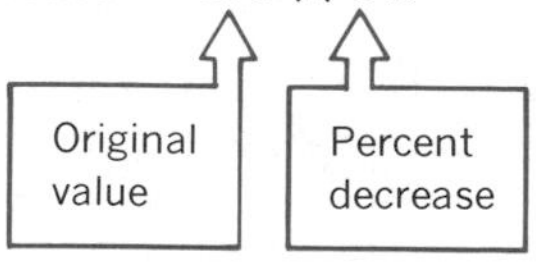

Decrease $= 600 \times 0.08$
$= 48$

New value $= 600 - 48$
$= 552$ gidgets/day

Again, the new value can be calculated directly.

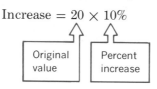

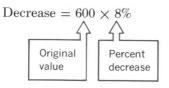

New value = 600 × 92%

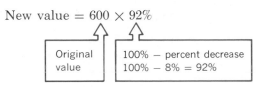

(600: Original value; 92%: 100% − percent decrease, 100% − 8% = 92%)

Try these problems involving percent change.

(a) What is your new pay rate if your old rate of \$6.72 per hour is increased 15%?
(b) Normal line voltage, 115 volts, drops 3.5% during a system malfunction. What is the reduced voltage?
(c) Because of changes in job specifications, the cost of a small construction job is increased 25% from the original cost of \$275. What is the cost of the job now?
(d) By using automatic welding equipment, the time for a given job can be reduced by 40% from its original 30 hours. How long will the job take using the automatic equipment?

Check your answers in **27**

---

27 (a) New pay rate = \$6.72 × 115%
= \$6.72 × 1.15
= \$7.728
≅ \$7.73

(b) Reduced voltage = 115 volts × (100% − 3.5%)
= 115 × 96.5%
= 115 × 0.965
= 110.975
≅ 111 volts

(c) New cost = \$275 × 125%
= 275 × 1.25
= \$343.75

(d) New time = 30 hr × (100% − 40%)
= 30 × 60%
= 30 × 0.60
= 18 hr

In trade and technical work changes in a measured quantity are often specified as a percent increase or decrease. For example, if the reading on a pressure valve increases from 30 psi to 36 psi, the percent increase is

$$\text{Percent increase} = \frac{36\text{ psi} - 30\text{ psi}}{30\text{ psi}}$$

(36 psi − 30 psi: The change; 30 psi: The original value)

$$= \frac{6}{30}$$

$$= 0.2$$

Percent increase = 20%, that is, a 20% increase

Another example: The length of a heating duct is reduced from 110 in. to 96 in. because of design changes. The percent decrease is

$$\text{Percent decrease} = \frac{110\text{ in.} - 96\text{ in.}}{110\text{ in.}}$$

(110 in. − 96 in.: The change; 110 in.: The original value)

$$= \frac{14}{110}$$

$$\cong 0.127$$

$$\cong 12.7\%$$ . . . about a 13% decrease

Now try these problems to sharpen your ability to work with percent changes.

(a) If 8 inches are cut from a 12 foot board, what is the percent decrease in length?
(b) What is the percent increase in voltage when the voltage increases from 65 volts to 70 volts?
(c) The measured value of the power output of a motor is 2.7 hp. If the motor is rated at 3 hp, what is the percent difference between the measured value and the expected value?

Check your answers in **28**.

---

28 (a) Use 12 ft = 144 in.

$$\text{Percent decrease} = \frac{8\text{ in.}}{144\text{ in.}} \begin{array}{l} \leftarrow \text{Change} \\ \leftarrow \text{Original value} \end{array}$$

$$\simeq 0.0555$$
$$\simeq 5.6\%$$

(b)
$$\text{Percent increase} = \frac{70\text{ volts} - 65\text{ volts}}{65\text{ volts}}$$
$$= \frac{5}{65}$$
$$\simeq 0.0769$$
$$\simeq 7.7\%$$

(c)
$$\text{Percent decrease} = \frac{3\text{ hp} - 2.7\text{ hp}}{3\text{ hp}}$$
$$= \frac{0.3}{3}$$
$$= 0.1$$
$$= 10\%$$

Now turn to frame **29** for a set of problems involving percent calculations.

---

29 **Exercises 4-3 Practical Percent Problems**

1. A CB radio is rated at 7.5 watts, and actual measurements show that it delivers 4.8 watts to its antenna. What is its efficiency?

2. An electric motor uses 6 kilowatts at an efficiency of 63%. How much power does it deliver?

3. An engine supplies 110 hp to an electric generator and the generator delivers 70 hp of electrical power. What is the efficiency of the generator?

4. Electrical resistors are rated in ohms and color coded to show both their resistance and percent tolerance. For each of the following resistors find its tolerance limits and actual tolerance.

Resistance, Ohms	Limits, Ohms	Tolerance, Ohms
5300 ± 5%	______ to ______	
2750 ± 2%	______ to ______	
6800 ± 10%	______ to ______	
5670 ± 20%	______ to ______	

5. A 120 hp automobile engine delivers only 81 hp to the driving wheels of the car. What is the efficiency of the transmission and drive mechanism?

6. An electrical resistor is rated at 500 ohms ± 10%. What is the highest value its resistance could have within this tolerance range?

7. On a roofing job 21 of 416 shingles had to be rejected for minor defects. What percent is this?

8. If you earn 12% commission on sales of $4200, what actual amount do you earn?

9. On a cutting operation 2 sq ft of sheet steel is wasted for every 16 sq ft used. What is the percent waste?

10. Four pounds of a certain bronze alloy is one-sixth tin, 0.02 zinc, and the rest copper. Express the portion of each metal in (a) percents and (b) pounds.

11. An iron casting is made in a mold with a hot length of 16.40 in. After cooling, the casting is 16.25 in. long. What is the shrinkage in percent?

12. Because of friction, a pulley block system is found to be only 83% efficient. What actual load can be raised if the theoretical load is one ton?

13. A small gasoline shop engine develops 65 hp at 2000 rpm. At 2400 rpm its power output is increased by 25%. What actual horsepower does it produce at 2400 rpm?

14. On the basis of past experience, a contractor expects to find 4.5% broken bricks in every truck load. If he orders 2000 bricks, will he have enough to complete a job requiring 1900 bricks?

15. Specifications call for a hole in a machined part to be 2.315″ in diameter. If the hole is measured to be 2.318″, what is the machinist's percent error?

16. A welding shop charges 15% over the cost of labor and materials. If a bill totals $36.75, what is the cost of labor and materials?

17. Complete the following table.

Measurement	Tolerance	Percent Tolerance
3.425″	± 0.001″	(a)
3.425″	± 0.015″	(b)
3.425″	(c)	± 0.20%

18. If you receive a pay increase of 12%, what is your new pay rate, assuming that your old rate was $8.65?

19. The pressure in a hydraulic line increases from 40 psi to 55 psi. What is the percent increase in pressure?

20. The cost of the paint used in a redecorating job is $65.70. This is a reduction of 20% from its initial cost. What was the original cost?

When you have completed these exercises, check your answers on page 549, then turn to **30** for a set of practice problems on percent.

4 Percent

PROBLEM SET 4

Percent

30

Answers are on page 549.

A. Write each number as a percent:

1. 0.72	2. 0.86	3. 0.6	4. 0.35
5. 1.3	6. 3.03	7. $\frac{1}{10}$	8. $\frac{7}{10}$
9. $\frac{1}{6}$	10. $2\frac{3}{5}$		

B. Write each percent as a decimal number:

1. 4%	2. 37%	3. 11%	4. 94%
5. $1\frac{1}{4}\%$	6. 0.09%	7. $\frac{1}{5}\%$	8. 1.7%
9. $3\frac{7}{8}\%$	10. 8.02%	11. 115%	12. 210%

C. Solve:

1. 3 is ________% of 5.
2. 5% of $120 is ________.
3. 25% of what number is 1.4? ________.
4. 16 is what percent of 8? ________.
5. 105% of 40 is ________.
6. 6% of $2.57 is ________.
7. $7\frac{1}{4}\%$ of ________ is $2.10.
8. 250% of 50 is ________.
9. 0.05% of ________ is 4.
10. $8\frac{1}{4}\%$ of 1.2 is ________.

D. Solve:

1. Extruded steel rods shrink 12% in cooling from furnace temperature to room temperature. If a standard tie rod is 34″ exactly, how long was it when it was formed?
2. Cast iron contains up to 4.5% carbon, and wrought iron contains up to 0.08% carbon. How much carbon is in a 20 lb bar of each metal?
3. (a) A real estate saleswoman sells a house for $54,500. Her usual commission is 7%. How much does she earn on the sale?
   (b) All salespeople in the Ace Junk Store receive $60 per week plus a 2% commission. If you sold $975 worth of junk in a week what would be your income?
   (c) A salesman at the Wasteland TV Company sold 5 color television sets last week and earned $128.70 in commissions. If his commission is 6%, what does a color TV set cost?
4. What is the selling price of a radial arm saw with a list price of $165.50 if it is on sale at a 35% discount?
5. If the retail sales tax in your state is 6%, what would be the total cost of each of the following items?
   (a) A $1.98 pair of pliers.
   (b) A $3.10 adjustable wrench.
   (c) 69¢ worth of washers.
   (d) A $13.60 textbook.
   (e) A $245.95 shop table.
6. A typewriter sells for $176 after a 12% discount. What was its original or list price?

Name ________

Date ________

Course/Section ________

7. How many board feet of matched 1″ × 6″ boards will be required to lay a subfloor in a house that is 28′ by 26′? Add 20% to the area to allow for waste and matching.
8. In the Easy Does It Metal Shop, one sheet of metal is wasted for every 25 used. What percent of the sheets are wasted?
9. Complete the following table.

Measurement	Tolerance	Percent Tolerance
1.775″	± 0.001″	(a)
1.775″	± 0.05″	(b)
1.775″	(c)	± 0.50%
310 mm	± 0.1 mm	(d)
310 mm	(e)	± 0.20%

10. If 16″ are cut from a 6′ cable, what is the percent decrease in length?
11. A production job is bid at $1275 but cost overruns amount to 15%. What is the actual job cost?
12. After heating, a metal rod has expanded 3.5%, to 15.23 cm. What was its original length?
13. An electrical resistor is rated at 4500 ohms ± 3%. Express this tolerance in ohms and state the actual range of resistance.
14. A 140-hp automobile engine delivers only 96 hp to the driving wheels of the car. What is the efficiency of the transmission and drive mechanism?
15. A steel casting has a hot length of 26.500 in. After cooling the length is 26.255 in. What is the shrinkage expressed as a percent? Round to one decimal place.
16. The parts manager for an automobile dealership can buy a part at a 25% discount off the retail price. If the retail price is $6.75, how much does he pay?

5 Measurement

PREVIEW 5

Objective	Sample Problems			Where To Go For Help	
Upon successful completion of this unit you will be able to:				Page	Frame
1. Do arithmetic with measurement numbers.	(a) $4 \text{ hr} \times 35 \text{ mph}$	= ______		167	**3**
	(b) $1\frac{1}{4} \text{ hr} + 40 \text{ min} + 2 \text{ hr} - 5 \text{ min}$	= ______		167	**3**
	(c) $50 \text{ mi} \div 20 \text{ mph}$	= ______		167	**3**
	(d) $\sqrt{400 \text{ sq ft}}$	= ______		167	**3**
	(e) $\pi(4'')^2$	= ______		167	**3**
2. Convert from one measurement unit to another.	(a) $2\frac{1}{2}$ weeks	= ______	min	177	**12**
	(b) 46 hp	= ______	watts	177	**12**
	(c) 120 cu ft	= ______	cu yd	177	**12**
	(d) 60 lb	= ______	oz	177	**12**
3. Work with metric units.	(a) Estimate the weight in pounds of a casting weighing 45 kg.	= ______	lb	186	**22**
	(b) 20 m	$\cong$ ______	ft	186	**22**
	(c) 3 in.	$\cong$ ______	cm	186	**22**
	(d) $40° \text{ F}$	$\cong$ ______	°C	200	**31**
	(e) 1500 mm = ______ cm	= ______	m	186	**22**
4. Use the common technical measuring instruments.	Read a length rule, micrometer, vernier caliper, and electrical meter.			206	**35**

(Answers to all preview problems are on page 164.)

If you are certain you can work *all* of these problems correctly, turn to page 231 for a set of practice problems. If you cannot work one or more of the preview problems, turn to the page indicated after the problem. If you want to be certain you are successful at this, turn to frame **1** and begin work there.

Name ______

Date ______

Course/Section ______

Answers to Preview 5

1. (a) 140 mi (b) 230 min or 3 hr 50 min
   (c) 2.5 hr (d) 20 ft (e) 50 sq in., rounded
2. (a) 25,200 min (b) 34,000 watts (c) 4.4 cu yd
   (d) 960 oz
3. (a) 99 lb (b) 66 ft (c) 7.6 cm
   (d) 4.4 °C (e) 150 cm, 1.5 m

5 Measurement

5-1 WORKING WITH MEASUREMENT NUMBERS

1 Most of the numbers used in practical or technical work either come from measurements or they lead to measurements. A carpenter must measure the dimensions of a wall to be framed; a machinist may need to weigh materials; an electronic repair technician may measure circuit resistance or voltage output. At one time or another each of these technicians must make measurements, do calculations based on measurements, perhaps round measurement numbers, and worry about the units of measurement. Of course they must also interpret measurement numbers from drawings, blueprints, or other sources.

Every technical worker knows that measurement numbers are never exact. There is a built-in limit of accuracy to every measurement, either because of the person doing the measuring or because of the limitations of the measuring device. For example, look at this line:

If we measure the length of this line using a rough ruler, we would write that its length is

Length = 2 inches

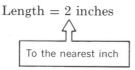

To the nearest inch

But with a better measuring device, we might be able to say with confidence that its length is

Length = 2.1 inches

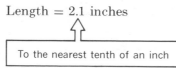

To the nearest tenth of an inch

Using a vernier caliper, we would find that the length is

Length = 2.13 inch

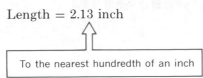

To the nearest hundredth of an inch

We could go on using more and more accurate measuring devices until we find ourselves looking through a microscope trying to decide where the ink line begins. If someone told us that the length of that line was

Length = 2.1304756 inch

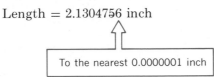

To the nearest 0.0000001 inch

we would know immediately that the length had been measured with incredible accuracy—they couldn't get a number like that by using a folding carpenter's rule!

Which of the following measurements has the greatest accuracy?

(a) 4.15 sec (b) 4.1 sec (c) 4.1452 sec

Choose the correct answer, then turn to **2** and continue.

---

2 Measurement (c), 4.1452 sec, is the most accurate. Measurement (b) is given to the nearest tenth of a second and could have been obtained with a hand timer. Measurement (a) is given to the nearest hundredth of a second and could have been obtained with an electronic stopwatch. But (c) is accurate to the nearest ten-thousandth of a second and this would require some fancy electronic timing.

Very often the technical measurements and numbers shown in drawings and specifications have a *tolerance* attached to them so that you will know the accuracy needed. For example, if a dimension on a machine tool is given as 2.835″ ± 0.004″, this means that the dimension should be 2.835″ to within 4 thousandths of an inch. In other words, the dimension may be no more than 2.839″ and no less than 2.831″. These are the limits of size that are allowed or tolerated.

Suppose the length of a piece of pipe is measured to be 8 ft. Two bits of information are given in this measurement: the size of the measurement (8) and the units of the measurement (ft).

The unit name for any measurement number *must* be included when you use that number, write it, or talk about it. A number usually has no practical physical meaning for a technician or scientist unless there are units attached to it. For example, we know immediately that 20 is a number, 20 ft is a length, 20 lb is a weight or force, and 20 sec is a time interval.

The measurement unit compares the size of the quantity being measured to some standard. For example, if the piece of pipe is measured to be 8 ft in length, it is 8 times the standard 1-ft length.

Length = 8 ft = 8 × 1 ft

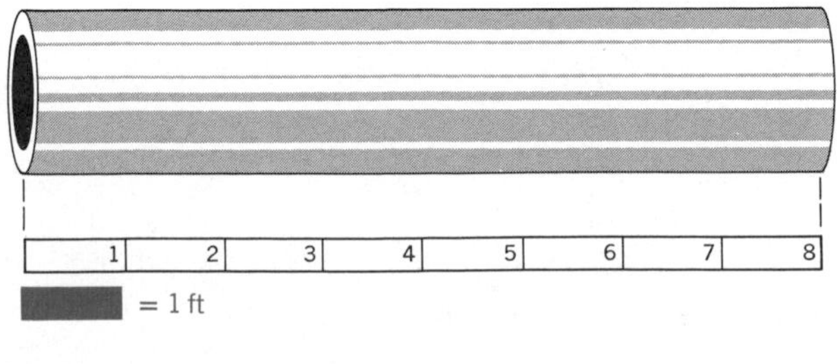

Complete the following equations.

(a) 6 in. = 6 × ______ (b) 10 lb = ______ × ______

(c) $5 = 5 × ______ (d) 14.7 psi = ______ × ______

Check your answers in 3.

(a) 6 in. = 6 × 1 in. (b) 10 lb = 10 × 1 lb

(c) $5 = 5 × $1 (d) 14.7 psi = 14.7 × 1 psi

Addition and Subtraction of Measurement Numbers

3 When measurement numbers are added, subtracted, multiplied, or divided, we must be especially careful to keep track of the units as well as the numbers. For example, to do the addition

4.1 ft + 3.5 ft + 1.27 ft = ___?___

First, check to be certain that all numbers have the same units. In the problem above all numbers have ft units. If one or more of the numbers are given in other units, you will need to convert so that all have the same units.

Second, add the numerical parts.

<u>By Hand</u>

```
  4.1
  3.5
+1.27
-----
 8.87
```

Notice that we line up the decimal points and add as we would with any decimal numbers.

<u>Using a Calculator</u>

Key	Display
C	0.
4 . 1	4.1
+	4.1
3 . 5	3.5
+	7.6
1 . 2 7	1.27
=	8.87

Third, attach the common units to the sum.

8.87 ft

If some of the numbers to be added are in fraction form, you should convert them to decimal form before adding. For example, the sum 3.2 lb + $1\frac{3}{4}$ lb + 2.4 lb becomes

```
  3.2  lb
  1.75 lb
+2.4  lb
--------
  7.35 lb
```

You may need to round the answer if any numbers were converted from fraction to decimal form. We will look at rounding later in this chapter, but first you need to work some practice problems in adding measurement numbers.

Add:

1. 4″ + 5″ + 17″ = ______
2. 2 ft + 3 in. + 7 in. = ______
3. 1.8 sec + 3.5 sec = ______
4. $3\frac{1}{2}$ in. + 1.7 in. + 2.75 in. = ______
5. 4 lb + $2\frac{1}{2}$ lb + $4\frac{1}{4}$ lb = ______
6. 1.375″ + 3.50″ + 2.833″ = ______

7. $0.05'' + 0.15'' + 0.0075'' =$ ________

8. 1 ft 5 in. + 17 in. + 4 in. = ________

9. $4\frac{1}{8}$ in. + 1 ft + $11\frac{1}{4}$ in. = ________

10. $2\frac{1}{5}$ lb + $1\frac{1}{4}$ lb + 4 oz = ________

Check your answers in **4**.

4

1. 26″
2. 34 in. or 2 ft 10 in.
3. 5.3 sec
4. 7.95 in.
5. 10.75 lb
6. 7.708″
7. 0.2075″
8. 38″ or 3 ft 2 in.
9. 27.375 in. or 2 ft 3.375 in.
10. 59.2 oz or 3 lb 11.2 oz (Remember, 16 oz = 1 lb, so $2\frac{1}{5}$ lb = $\frac{11}{5} \cdot 16$ oz or 35.2 oz)

The subtraction of measurement numbers is very similar to their addition. For example, in the subtraction

$7.425'' - 3.5'' =$ ___?___

First, check to be certain that both numbers have the same units. In this problem both numbers have inch units. If one of the numbers was given in other units, we would convert so that they have the same units. (You'll learn more about converting measurement units later in this chapter.)

Second, subtract the numerical parts.

By Hand

$$\begin{array}{r} 7.425 \\ -3.500 \\ \hline 3.925 \end{array}$$

← Attach zeros if necessary

Using a Calculator

Key	Display
C	0.
7 . 4 2 5	7.425
−	7.425
3 . 5	3.5
=	3.925

Third, attach the common units to the difference.

3.925″

If one of the numbers in the subtraction is in fraction form, you may need to convert it to a decimal before subtracting.

Now for a few practice subtraction problems, try these.

1. 5.7 sec − 1.9 sec = ________

2. $11.75'' - 3.825'' =$ ________

3. 4.2 lb − 1.5 lb = ________

4. $2.15'' - 1.005'' =$ ________

5. 1 ft 2 in. − 8 in. = ________

6. $4\frac{1}{2}$ lb − 1.7 lb = ________

7. 1 ft − 3.7 in. = ________

8. $1\frac{1}{5}$ lb − 11.5 oz = ________

9. 30.4 mph − 8.5 mph = ________

10. 30 psi − 14.7 psi = ________

Check your work in **5**.

5

1.	3.8 sec	2.	7.925″
3.	2.7 lb	4.	1.145″
5.	6 in.	6.	2.8 lb
7.	8.3″	8.	7.7 oz
9.	21.9 mph	10.	15.3 psi

Multiplication and Division of Measurement Numbers

With both addition and subtraction of measurement numbers, the answer to the arithmetic calculation has the same units as the numbers being added or subtracted. This is not true when we multiply or divide measurement numbers. For example, to multiply

4.3 ft × 3.6 ft = ___?___

First, multiply the numerical parts

$$\begin{array}{r} 4.3 \\ \times\ 3.6 \\ \hline 15.48 \end{array}$$

Second, multiply the units

1 ft × 1 ft = 1 ft^2 or 1 sq ft

It may help if you remember that this multiplication is actually the following:

$$\begin{aligned} 4.3 \text{ ft} \times 3.6 \text{ ft} &= (4.3 \times 1 \text{ ft}) \times (3.6 \times 1 \text{ ft}) \\ &= (4.3 \times 3.6) \times (1 \text{ ft} \times 1 \text{ ft}) \\ &= 15.48 \quad \times 1 \text{ sq ft} \\ &= 15.48 \text{ sq ft} \end{aligned}$$

The rules of arithmetic say that we can do the multiplications in any order we wish, so we multiply the units part separate from the number part.

Notice that the product 1 ft × 1 ft can be written either as 1 ft^2 or as 1 sq ft.

If both numbers are lengths, but are given in different units, it is best to convert to the same units. For example, to multiply 2 ft × 5 in. convert it to 24 in. × 5 in.

Third, round the answer if necessary. You may want to review the explanation of rounding in Chapter 3, page 95. In the problem above, the answer should be rounded to agree in accuracy with the numbers being multiplied. The product is 15.5 sq ft.

In most calculations in the trades, the accuracy of rounding is obvious from the numbers given. To simplify the process here, we will include rounding instructions with most problems.

Another example:

8.1 in. × 5.3 in. = ___?___ (Round to the nearest tenth.)

First, multiply the numbers:

$$\begin{array}{r} 8.1 \\ \times\ 5.3 \\ \hline 42.93 \end{array}$$

Second, multiply the units 1 in. × 1 in. = 1 $in.^2$ or 1 sq in:

42.93 sq in.

Third, round to one decimal place:

42.9 sq in.

Try these practice problems. (Round to the nearest tenth if necessary.)

1. 3 ft × 4 ft = ________ 2. 1.4″ × 2.5″ = ________

3. 1 ft × 5 in. = ________ 4. $2\frac{1}{4}$ in. × 4 in. = ________

5. 9 ft 4 in. × 12 ft = ________ 6. $6\frac{3}{4}$ in. × 8.4 in. = ________

7. 30 mph × 2 hr = ________ 8. 17 ft/sec × 2.1 sec = ________

Now check your work in **6**.

---

6 1. 12 sq ft 2. 3.5 sq in.
3. 60 sq in. 4. 9 sq in.
5. 112 sq ft or 16,128 sq in. 6. 56.7 sq in.
7. 60 miles 8. 35.7 ft

Did Problems 7 and 8 give you any special difficulty?

Do them this way:

7. $$30 \text{ mph} \times 2 \text{ hr} = 30 \frac{\text{miles}}{\text{hr}} \times 2 \text{ hr}$$
$$= \left(\frac{30 \times 1 \text{ mile}}{1 \cancel{\text{hr}}}\right) \times (2 \times 1 \cancel{\text{hr}})$$
$$= 60 \times 1 \text{ mile}$$
$$= 60 \text{ mi}$$

8. $$17 \text{ ft/sec} \times 2.1 \text{ sec} = \left(\frac{17 \times 1 \text{ ft}}{1 \cancel{\text{sec}}}\right) \times (2.1 \times 1 \cancel{\text{sec}})$$
$$= 17 \times 1 \text{ ft} \times 2.1$$
$$= 35.7 \times 1 \text{ ft}$$
$$= 35.7 \text{ ft}$$

Division with measurement numbers requires that you take the same care with the units that is needed in multiplication. For example, to divide

24 mi ÷ 1.5 hr = ___?___

First, divide the numerical parts as usual.

24 ÷ 1.5 = 16

```
       16.
  1.5)24.0
      15
      --
       90
       90
       --
        0
```

Second, divide the units.

$$1 \text{ mi} \div 1 \text{ hr} = \frac{1 \text{ mi}}{1 \text{ hr}} \quad \text{or} \quad 1 \text{ mph}.$$

In this case the unit $\frac{1 \text{ mi}}{1 \text{ hr}}$ is usually written mi/hr or mph—miles per hour—and the answer to the problem is 16 mph.

Units such as sq ft or mph, which are made up of a combination of simpler units, are called *compound* units.

Remember that

$$24 \text{ mi} \div 1.5 \text{ hr} = \frac{24 \times 1 \text{ mi}}{1.5 \times 1 \text{ hr}}$$
$$= \left(\frac{24}{1.5}\right) \times \left(\frac{1 \text{ mi}}{1 \text{ hr}}\right)$$
$$= 16 \times 1 \text{ mph}$$
$$= 16 \text{ mph}$$

Third, round the answer so it has the same accuracy as the numbers used to obtain it, or as directed in the problem.

Another example:

48 lb ÷ 1.8 sq in. = ___?___ (Round to the nearest whole number of units.)

First, divide the numbers 48 ÷ 1.8 = 26.667 ≅ 27 rounded. (The sign ≅ means "approximately equal to.")

Second, divide the units $1 \text{ lb} \div 1 \text{ sq in.} = \dfrac{1 \text{ lb}}{1 \text{ sq in.}}$

$$= 1 \text{ lb/sq in.}$$
$$= 1 \text{ psi}$$

The answer is 27 psi.

Now try a few practice problems.

1. 4 ft ÷ 2 sec = ________
2. 20 mi ÷ 1.25 hr = ________
3. 24 sq ft ÷ 3 ft = ________
4. 98¢ ÷ 1.50 gal = ________ (Round to the nearest cent.)
5. 40 lb ÷ 2 cu ft = ________
6. $2\frac{1}{4}$ sq ft ÷ 3 in. = ________
7. 70 mi ÷ 35 mph = ________
8. $3\frac{1}{2}$ volts ÷ 0.5 ohm = ________

Check your answers in **7**.

7

1. 2 ft/sec
2. 16 mph
3. 8 ft
4. 65¢/gal (rounded)
5. 20 lb/cu ft
6. 9 ft
7. 2 hr
8. 7 volts/ohm = 7 amp

Did Problems 6, 7, or 8 seem difficult? Do them this way:

6. $2\frac{1}{4} \text{ sq ft} \div 3 \text{ in.} = 2\frac{1}{4} \text{ sq ft} \div \frac{1}{4} \text{ ft}$

$$= 2.25 \text{ sq ft} \div 0.25 \text{ ft}$$
$$= \left(\frac{2.25}{0.25}\right) \times \left(\frac{\text{sq ft}}{\text{ft}}\right)$$
$$= 9 \text{ ft}$$

$$\frac{\text{sq ft}}{\text{ft}} = \frac{\text{ft}^2}{\text{ft}} = \frac{\text{ft} \times \cancel{\text{ft}}}{\cancel{\text{ft}}} = \text{ft}$$

7. $70 \text{ mi} \div 35 \text{ mph} = 70 \text{ mi} \div 35 \dfrac{\text{mi}}{\text{hr}}$

$$= \left(\frac{70}{35}\right) \times \left(\frac{\text{mi}}{\text{mi/hr}}\right)$$
$$= 2 \text{ hr}$$

To divide by the fraction mi/hr, invert and multiply

$$\frac{\text{mi}}{\text{mi/hr}} = \cancel{\text{mi}} \times \frac{\text{hr}}{\cancel{\text{mi}}} = \text{hr}$$

8. Electronics scientists use the unit *amp* for volts/ohm.

$$\frac{1 \text{ volt}}{1 \text{ ohm}} = 1 \text{ amp}$$

Rounding Measurement Numbers

You have already seen many problems where a division does not "come out even" or where a multiplication of measurement numbers produces an answer that has more decimal digits than its accuracy would allow. In such cases we must approximate or *round-off* the answer. For example, suppose you need to find the decimal equivalent of the fraction $\frac{12}{7}$. Punch this division into your trusty calculator and you'll get

$$\frac{12}{7} = 1.7142857$$

Using a Calculator

Key	Display
C	0.
1 2	12.
÷	12.
7	7.
=	1.7142857

which is correct to ten digits. But in any practical situation you will not need all of those digits. You need to round this number to one or two decimal places in order to use it in any practical work.

To *one* decimal digit $\frac{12}{7} \cong 1.7$ (The sign $\cong$ means "approximately equal to.")

To *two* decimal digits $\frac{12}{7} \cong 1.71$

To *three* decimal digits $\frac{12}{7} \cong 1.714$

To *four* decimal digits $\frac{12}{7} \cong 1.7143$

To *five* decimal digits $\frac{12}{7} \cong 1.71429$

If you need to review the rules for rounding, return to Chapter 3, starting on page 95.

When we do calculations with measurement numbers, our answers cannot be *more* accurate than the numbers we used to calculate them. For example, calculate the speed of a flywheel if it turns 45 rotations in 1.3 min. Run these numbers through your calculator and you will find the following:

$$45 \text{ rotations} \div 1.3 \text{ min.} = \frac{45}{1.3} \frac{\text{rotations}}{\text{min}}$$

$$= 34.61538462\ldots \text{ rpm}$$

Using a Calculator

Key	Display
C	0.
4 5	45.
÷	45.
1 . 3	1.3
=	34.615384

Of course the calculator can give the answer to eight or ten digits, but a ten-digit answer is actually *not* correct. We cannot have an answer accurate to ten digits when the time, 1.3 min, is accurate to only two digits. The answer must be rounded to agree in accuracy with the measurement number. The rotation speed is 35 rpm, rounded to two digits.

In this problem the time, 1.3 minutes, was probably measured, perhaps with a stopwatch. It is a measurement number and is therefore only approximate. The number of rotations was probably counted directly, perhaps with a mechanical counter, and it is an *exact* number. Numbers that result from simple counting or that are defined are considered to be exact. Calculations are rounded to agree with the approximate or measurement numbers. For example, there are exactly 12 inches in a foot by definition. The 12 is exact. The number π in geometry is an exact number. Many numbers that appear in scientific and technical formulas are exact.

Round the result of a calculation so that it agrees in accuracy with the least accurate approximate number. The usual rule is to take the arithmetic calculation to one digit more than the final answer will have and then round it. If you are using a calculator, it will produce an answer for the arithmetic calculation that is 6, 8, or even 10 digits long—many more digits than needed for any practical work. Rounding is an important skill for calculator users.

Try these problems. Round carefully.

1. 4.2 ft × 8.7 ft = ________
   (Round to the nearest unit.)
2. 0.23″ × 1.4″ = ________
   (Round to two decimal places.)
3. $(2.7\text{ ft})^2$ = ________
   (Round to one decimal place.)
4. $\frac{1}{2} \times 32\text{ ft/sec}^2 \times (1.7\text{ sec})^2$ = ________
   (Round to the nearest unit.)
5. 82 ft ÷ 3.492 sec = ________
   (Round to the nearest unit.)
6. \$1.79 ÷ 8 oz = ________
   (Round to the nearest cent.)

Check your work in **8**.

---

8 1. 4.2 ft × 8.7 ft = 36.54 sq ft
= 37 sq ft

2. 0.23″ × 1.4″ = 0.322 sq in.
= 0.32 sq in., rounded

3. $(2.7\text{ ft})^2$ = 2.7 ft × 2.7 ft
= 7.29 sq ft
= 7.3 sq ft, rounded

4. $\frac{1}{2} \times 32\text{ ft/sec}^2 \times (1.7\text{ sec})^2 = \frac{1}{2} \times 32\text{ ft/sec}^2 \times 2.89\text{ sec}^2$
= 46.24 ft
= 46 ft, rounded

5. 82 ft ÷ 3.492 sec = 23.48224513 ft/sec, using a calculator
= 23.4 ft/sec, dividing by hand
= 23 ft/sec, rounded

Notice that when dividing by hand we carry the arithmetic to one digit more than we will keep after rounding.

By Hand

6. \$1.79 ÷ 8 oz = 0.223 \$/oz
= 0.22 \$/oz

or 22¢/oz, to the nearest cent

Using a Calculator

Key	Display
C	0.
1 . 7 9	1.79
÷	1.79
8	8.
=	0.22375

Most of the measurements made by technical workers in a shop are made in fractions. On many shop drawings and specifications, the dimensions may be given in decimal form.

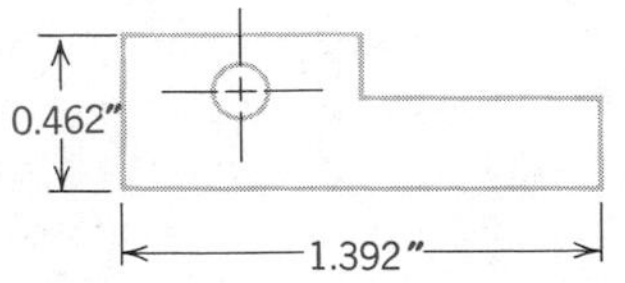

The steel rule usually used in the shop or on the job may be marked in 8ths, 16ths, 32nds, or even 64ths of an inch. A common problem is to rewrite the decimal number to the nearest 32nd or 64th of an inch and to determine how much error is involved in using the fraction number rather than the decimal number. For example, in the drawing above, what is the fraction, to the nearest 32nd of an inch, equivalent to 0.462 inch? The rule is to multiply the decimal number by the fraction $\frac{32}{32}$.

$$0.462 \text{ in.} = 0.462 \text{ in.} \times \frac{32}{32}$$

$$= \frac{0.462 \text{ in.} \times 32}{32}$$

$$= \frac{14.784}{32} \text{ in.}$$ Round the top number to the nearest unit.

$$\cong \frac{15}{32} \text{ in.,}$$ rounded to the nearest 32nd of an inch.

The error involved is the difference between

$$\frac{15}{32} \text{ in. and } \frac{14.784}{32} \text{ in., or } \frac{15 - 14.784}{32} \text{ in.} = \frac{0.216}{32} \text{ in.}$$

$$= 0.00675 \text{ in.}$$

$$\cong 0.0068 \text{ in., rounded}$$

The error involved in using $\frac{15}{32}$ in. instead of 0.462 in. is about 68 ten-thousandths of an inch. In problems of this kind we would usually round the error to the nearest ten-thousandth.

Now you try it. Change 1.392 in. to a fraction expressed in 64ths of an inch, and find the error.

Check your work in **9**.

---

9 $1.392 \text{ in.} = 1.392 \text{ in.} \times \frac{64}{64}$

$$= \frac{1.392 \times 64}{64} \text{ in.}$$

$$= \frac{89.088}{64} \text{ in.}$$

$$\cong \frac{89}{64} \text{ in.,}$$ rounded to the nearest unit

$$\cong 1\frac{25}{64} \text{ in.}$$

The error in using the fraction rather than the decimal is

$$\frac{89.088}{64} \text{ in.} - \frac{89}{64} \text{ in.} = \frac{0.088}{64} \text{ in.}$$

$$\cong 0.0014 \text{ in.}$$ or about fourteen ten-thousandths of an inch.

Try these problems for practice.

A. Express to the nearest 16th of an inch.

1. 0.438 in.	2. 0.30 in.
3. 0.18 in.	4. 2.70 in.

B. Express to the nearest 32nd of an inch.

1. 1.650 in.	2. 0.400 in.
3. 0.720 in.	4. 2.285 in.

C. Express to the nearest 64th of an inch and find the error, to the nearest ten thousandth, involved in using the fraction in place of the decimal number.

1. 0.047 in.	2. 2.106 in.
3. 1.640 in.	4. 0.517 in.

Check your answers in **10**.

A DIFFERENT WAY TO ROUND NUMBERS

Some engineers and technicians use the following rounding rules rather than the ones we presented in Chapter 3, page 95.

	Example
1. Determine the place to which the number is to be rounded. Mark it with a $_\wedge$.	Round 4.786 to three digits. $4.78_\wedge 6$
2. When the digit to the right of the mark is greater than 5, increase the digit to the left by 1.	$4.78_\wedge 6$ becomes 4.79 $1.701_\wedge 72$ becomes 1.702
3. When the digit to the right of the mark is less than 5, drop the digits to the right or replace them by zeros.	$3.81_\wedge 2$ becomes 3.81 $14_\wedge 39$ becomes 1400
4. When the digit to the right of the mark is equal to 5, round the left digit to the nearest *even* value.	$1.41_\wedge 5$ becomes 1.42 $37.00_\wedge 5$ becomes 37.00 $53_\wedge 50$ becomes 5400

10 A. 1. $\frac{7}{16}$ in. 2. $\frac{5}{16}$ in. 3. $\frac{3}{16}$ in. 4. $2\frac{11}{16}$ in.

B. 1. $1\frac{21}{32}$ in. 2. $\frac{13}{32}$ in. 3. $\frac{23}{32}$ in. 4. $2\frac{9}{32}$ in.

C. 1. $\frac{3}{64}$ in. The error is about 0.0001 in.

2. $2\frac{7}{64}$ in. The error is about 0.0034 in.

3. $1\frac{41}{64}$ in. The error is about 0.0006 in.

4. $\frac{33}{64}$ in. The error is about 0.0014 in.

For a set of practice problems over this section of Unit 5, turn to **11**.

---

11 **Exercises 5-1 Working with Measurement Numbers**

A. Add or subtract as shown:

1. 4″ + 7″	2. 14 lb + 36 lb
3. $3\frac{1}{4}$ in. + 1.7 in. + 4.05 in.	4. 16.0 psi + 15.6 psi + 7.0 psi
5. 0.008 in. + 0.016 in. + 0.310 in.	6. 0.64 ft + $1\frac{1}{4}$ in.

7. 64.5 mph − 17.0 mph

8. 4.7 gal − 1.9 gal

9. $\frac{1}{4}$ in. − 0.17 in.

10. 37.0 psi − 11.6 psi

B. Multiply or divide as shown, and round if necessary:

1. 4 ft × 7 ft
2. 3.1 in. × 1.7 in.
3. 3.0 ft × 2.407 ft
4. 21 mph × 1.2 hr
5. $1\frac{1}{4}$ ft × 1.5 ft
6. 4 ft × 8 in.
7. 2.1 ft × 1.7 ft × 3.4 ft
8. 18 sq ft ÷ 2.1 ft
9. 45 mi ÷ 2.3 hr
10. $1.37 ÷ 3.2 lb
11. $40\frac{1}{2}$ mi ÷ 25 mph
12. 2.0 sq ft ÷ 1.073 ft

C. Convert as shown

1. Write each of the following fractions as a decimal number rounded to two decimal places.

(a) $1\frac{7}{8}$ in. (b) $4\frac{3}{64}''$ (c) $3\frac{1}{8}$ sec (d) $\frac{1}{16}$ in.
(e) $\frac{3}{32}$ in. (f) $1\frac{3}{13}$ lb (g) $2\frac{17}{64}$ in. (h) $\frac{19}{32}$ in.
(i) $\frac{1}{7}$ lb

2. Write each of the following decimal numbers as a fraction rounded to the nearest 16th of an inch.

(a) 0.921 in. (b) 2.55 in. (c) 1.80 in. (d) 3.69 in.
(e) 0.802 in. (f) 0.306 in. (g) 1.95 in. (h) 1.571 in.
(i) 0.825 in.

3. Write each of the following decimal numbers as a fraction rounded to the nearest 32nd of an inch.

(a) 1.90 in. (b) 0.85 in. (c) 2.350 in. (d) 0.666 in.
(e) 2.091 in. (f) 0.285 in. (g) 0.600 in. (h) 0.685 in.
(i) 1.525 in.

4. Write each of the following decimal numbers as a fraction rounded to the nearest 64th of an inch and find the error, to the nearest ten-thousandth, involved in using the fraction in place of the decimal number.
(a) 0.235 in. (b) 0.515 in. (c) 1.80 in.
(d) 2.420 in. (e) 3.175 in. (f) 2.860 in.
(g) 1.935 in. (h) 0.645 in. (i) 0.480 in.

D. Practical Problems

1. The specifications for a bracket call for a hole 0.637″ ± 0.005″ in diameter. Will a $\frac{41}{64}''$ hole be within the required tolerance?

2. According to standard American wire size tables, 3/0 wire is 0.4096 inch in diameter. Write this size to the nearest 32nd of an inch. What error is involved in using the fraction rather than the decimal?

3. An auto mechanic converts a metric part to 0.473 inch. The part only comes in fractional sizes given to the nearest 64th of an inch. What would be the closest size?

4. What size wrench, to the nearest 32nd of an inch, will fit a bolt 0.748 in. in diameter?

5. Find the missing distance x in the drawing.

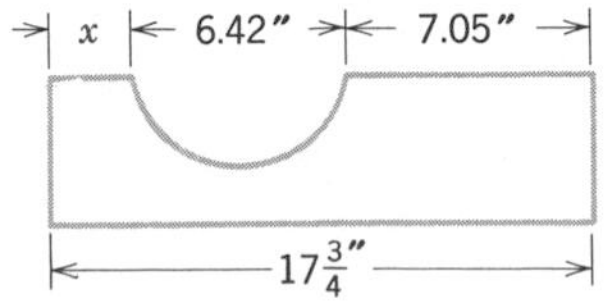

6. An auto mechanic uses shims to adjust the clearance in various parts of a car. If the intake valve is supposed to have an 0.008″ clearance, and the mechanic measures it to be exactly $\frac{1}{32}$″, what size shim does he need?

When you have completed these problems, check your answers on p. 549, then turn to **12** for some information on units and unit conversion.

5-2 UNITS AND UNIT CONVERSION

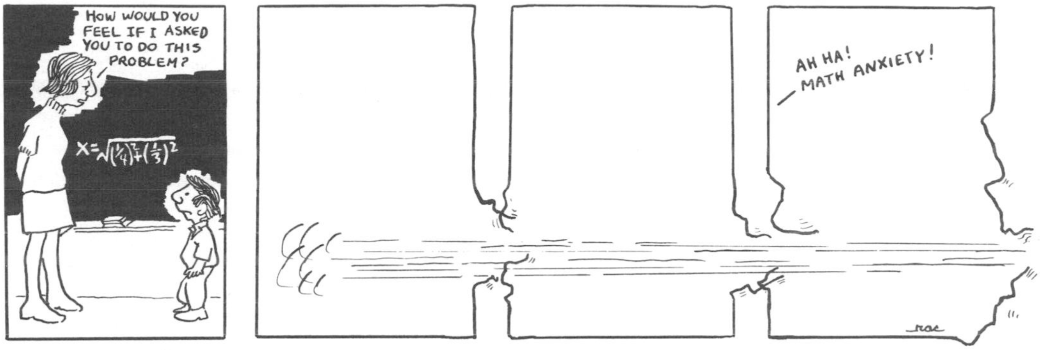

12 When you talk about a 4 ft by 8 ft wall panel, a $1\frac{1}{4}$ in. bolt, 3 lb of solder, or a gallon of paint, you are comparing an object with a *standard*. The units or standards used in science, technology, or the trades have two important characteristics: they are convenient in size and they are standardized. Common units such as the foot, pound, gallon, or hour were originally chosen because they were related to either natural quantities or to body measurements.

A pinch of salt, a drop of water, a handful of sand, or a tank of gas, all are given in *natural* units. The width of a finger, the length of a foot, or a stride length are all distances given in *body-related* units. In either case, there is a need to define and standardize these units before they are useful for business, science, or technology.

In biblical times, the *digit* was a length unit defined as the width of a person's finger. Four digits was called a *hand,* a unit still used to measure horses. Four hands was a *cubit,* the distance from fingertip to elbow, about 18 inches. One *foot* was defined as the length of a man's foot. A *yard* was the distance from nose to tip of outstretched arm. All of these lengths depend on whose finger, hand, arm, or foot is used, of course. In the fifteenth century standardization began, and one yard was defined as the length of a standard iron bar kept in a London vault. One foot was defined as exactly one-third the length of the bar, and one inch was one-twelfth of a foot. (The word *inch* comes from Latin and Anglo-Saxon words meaning "one-twelfth.") Today the inch is legally defined worldwide in terms of metric units.

Because there are a great many units available for writing the same quantity, it is necessary that you be able to convert a given measurement from one unit to another. For example, a length of exactly one mile can be written as

1 mile = 1760 yards, for the traffic engineer
= 5280 ft, for a landscape architect
= 63,360 in., in a science problem
= 320 rods, for a surveyor
= 8 furlongs, for a horse racing fan
= 1609.344 meters, in metric units.

In this section of Chapter 5 we will show you a quick and mistake-proof way to convert measurements from one unit to another. Whatever the units given, if a new unit is defined, you should be able to convert the measurement to the new units.

Try the following simple unit conversion. Convert 4 yards to feet.

4 yd = ________ ft

Check your answer in **13** .

---

13 You should know that 1 yard is defined as exactly 3 feet, therefore

$4 \text{ yd} = 4 \times 3 \text{ ft} = 12 \text{ ft}$

Easy.

Try another, more difficult conversion. Surveyors use a unit of length called a *rod,* defined as exactly $16\frac{1}{2}$ ft.

Convert 11 yd = ________ rods.

Check your answer in **14** .

---

14 The correct answer is 2 rods.

If you were very clever, you first converted 11 yd to 33 ft, and then noticed that 33 ft divided by $16\frac{1}{2}$ equals 2. There are 2 rods in a 33 ft length.

Most people find it difficult to reason through a problem in this way. To help you, we have devised a method of solving *any* unit conversion problem quickly with no chance of error. This is the *unity fraction* method.

The first step in converting a number from one unit to another is to set up a unity fraction from the definition linking the two units. For example, to convert the distance 4 yd to ft units, take the equation relating yd to ft

$1 \text{ yd} = 3 \text{ ft}$

and form the fractions $\dfrac{1 \text{ yd}}{3 \text{ ft}}$ and $\dfrac{3 \text{ ft}}{1 \text{ yd}}$.

These fractions are called unity fractions because they are both equal to one. Any fraction whose top and bottom are equal has the value 1.

Practice this step by forming pairs of unity fractions from each of the following definitions.

(a) 12 in. = 1 ft
(b) 1 lb = 16 oz
(c) 3.8 liters = 1 gallon

Check your answers in **15** .

---

15 (a) 12 in. = 1 ft. The unity fractions are $\dfrac{12 \text{ in.}}{1 \text{ ft}}$ and $\dfrac{1 \text{ ft}}{12 \text{ in.}}$.

(b) 1 lb = 16 oz. The unity fractions are $\dfrac{1 \text{ lb}}{16 \text{ oz}}$ and $\dfrac{16 \text{ oz}}{1 \text{ lb}}$.

(c) 3.8 liters = 1 gallon. The unity fractions are $\dfrac{3.8 \text{ liters}}{1 \text{ gallon}}$ and $\dfrac{1 \text{ gallon}}{3.8 \text{ liters}}$.

The second step in converting units is to multiply the original number by one of the unity fractions. For example, to convert the distance 4 yd to ft units, multiply this way:

$$\begin{aligned} 4 \text{ yd} &= (4 \times 1 \cancel{\text{yd}}) \times \left(\frac{3 \text{ ft}}{1 \cancel{\text{yd}}}\right) \\ &= 4 \times 3 \text{ ft} \\ &= 12 \text{ ft} \end{aligned}$$

You must choose one of the two fractions. Multiply by the fraction that allows you to cancel out the *yd* units you do not want and to keep the *ft* units you do want.

Did multiplying by a fraction give you any trouble? Remember that any number A can be written as $\frac{A}{1}$ so that a multiplication such as

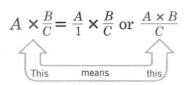

$$A \times \frac{B}{C} = \frac{A}{1} \times \frac{B}{C} \text{ or } \frac{A \times B}{C}$$

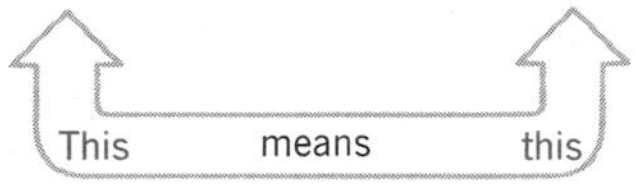

For example,

$$3 \times \frac{2}{5} = \frac{3}{1} \times \frac{2}{5} = \frac{3 \times 2}{5} = \frac{6}{5} \quad \text{or} \quad 1\frac{1}{5}$$

and

$$6 \times \frac{3}{4} = \frac{\overset{3}{\cancel{6}}}{1} \times \frac{3}{\underset{2}{\cancel{4}}} = \frac{9}{2} \quad \text{or} \quad 4\frac{1}{2}$$

(If you need to review the multiplication of fractions turn back to page 54.)

Now try this problem. Use unity fractions to convert 44 oz to lb. Remember that 1 lb = 16 oz.

Check your work in **16**

---

16 Use the equation 1 lb = 16 oz to set up the unity fractions $\frac{1 \text{ lb}}{16 \text{ oz}}$ and $\frac{16 \text{ oz}}{1 \text{ lb}}$. Multiply by the first fraction to convert oz to lb.

$$44 \text{ oz} = (44 \cancel{\text{oz}}) \times \left(\frac{1 \text{ lb}}{16 \cancel{\text{oz}}}\right) = \frac{44 \times 1 \text{ lb}}{16} = \frac{44}{16} \text{ lb} \quad \text{or} \quad 2\frac{3}{4} \text{ lb}$$

It may happen that several unity fractions are needed in the same problem. For example, if you know that

1 rod = $16\frac{1}{2}$ ft and 1 yd = 3 ft,

then to convert 11 yd to rods multiply as follows:

$$11 \text{ yd} = (11 \cancel{\text{yd}}) \times \left(\frac{3 \cancel{\text{ft}}}{1 \cancel{\text{yd}}}\right) \times \left(\frac{1 \text{ rod}}{16\frac{1}{2} \cancel{\text{ft}}}\right) = \frac{11 \times 3 \times 1 \text{ rod}}{16\frac{1}{2}} = \frac{33}{16\frac{1}{2}} \text{ rod}$$

11 yd = 2 rods

Using a Calculator

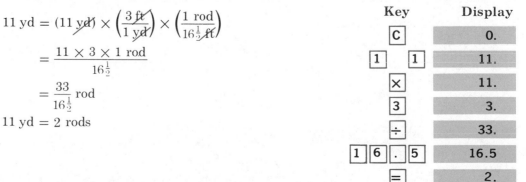

Key	Display
C	0.
1 1	11.
×	11.
3	3.
÷	33.
1 6 . 5	16.5
=	2.

Remember, to divide by a fraction, invert and multiply.

$$\frac{33}{16\frac{1}{2}} = 33 \div 16\tfrac{1}{2} = 33 \div \tfrac{33}{2} = 33 \times \tfrac{2}{33} = 2$$

Invert and multiply

Be careful to set up the unity fraction so that the unwanted units cancel. At first it helps to write both fractions and then choose the one that fits. As you become an expert at unit conversion, you will be able to write down the correct fraction at first try.

Try these practice problems. Use unity fractions to convert.

(a) 6.25 ft = ________ in.

(b) $5\frac{1}{4}$ yd = ________ ft

(c) 2.1 mi = ________ yd (1760 yd = 1 mi)

(d) 13 gallons = ________ barrel (31.5 gal = 1 bbl)

(e) 12.5 bbl = ________ cu ft (1 gal = 231 cu in.
1 cu ft = 1728 cu in.)

(f) 32 psi = ________ atm (1 atm = 14.7 psi)

The solutions are in **17**.

---

17 (a) $6.25 \text{ ft} = (6.25 \cancel{\text{ft}}) \times \left(\frac{12 \text{ in.}}{1 \cancel{\text{ft}}}\right)$

$= 6.25 \times 12$ in.
$= 75$ in.

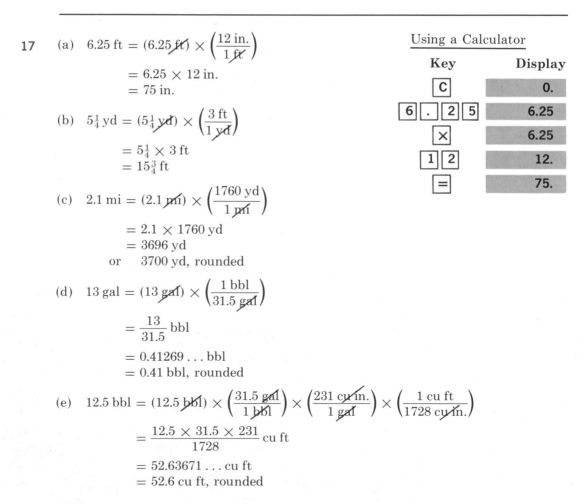

Using a Calculator

Key	Display
C	0.
6 . 2 5	6.25
×	6.25
1 2	12.
=	75.

(b) $5\frac{1}{4} \text{ yd} = (5\frac{1}{4} \cancel{\text{yd}}) \times \left(\frac{3 \text{ ft}}{1 \cancel{\text{yd}}}\right)$

$= 5\frac{1}{4} \times 3$ ft
$= 15\frac{3}{4}$ ft

(c) $2.1 \text{ mi} = (2.1 \cancel{\text{mi}}) \times \left(\frac{1760 \text{ yd}}{1 \cancel{\text{mi}}}\right)$

$= 2.1 \times 1760$ yd
$= 3696$ yd
or 3700 yd, rounded

(d) $13 \text{ gal} = (13 \cancel{\text{gal}}) \times \left(\frac{1 \text{ bbl}}{31.5 \cancel{\text{gal}}}\right)$

$= \frac{13}{31.5}$ bbl

$= 0.41269\ldots$ bbl
$= 0.41$ bbl, rounded

(e) $12.5 \text{ bbl} = (12.5 \cancel{\text{bbl}}) \times \left(\frac{31.5 \cancel{\text{gal}}}{1 \cancel{\text{bbl}}}\right) \times \left(\frac{231 \cancel{\text{cu in.}}}{1 \cancel{\text{gal}}}\right) \times \left(\frac{1 \text{ cu ft}}{1728 \cancel{\text{cu in.}}}\right)$

$= \frac{12.5 \times 31.5 \times 231}{1728}$ cu ft

$= 52.63671\ldots$ cu ft
$= 52.6$ cu ft, rounded

(f) $32 \text{ psi} = (32 \cancel{\text{psi}}) \times \left(\dfrac{1 \text{ atm}}{14.7 \cancel{\text{psi}}}\right)$

$$= \frac{32}{14.7} \text{ atm}$$

$$= 2.17687\ldots \text{ atm}$$

$$= 2.2 \text{ atm, rounded}$$

Using a Calculator

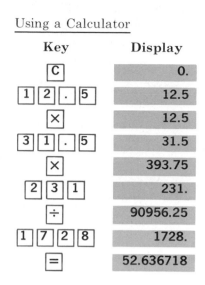

Key	Display
C	0.
1 2 . 5	12.5
×	12.5
3 1 . 5	31.5
×	393.75
2 3 1	231.
÷	90956.25
1 7 2 8	1728.
=	52.636718

Notice in Problems (c), (d), (e), and (f) that we rounded after doing the arithmetic. The final answer should have the same number of significant digits as the original number.

In the English system of units commonly used in the United States in technical work, the basic units for length, weight, and time are named with a single word or abbreviation: ft, lb, and sec. We often define units for other similar quantities in terms of these. For example, 1 mile = 5280 ft, 1 ton = 2000 lb, 1 hr = 3600 sec, and so on. We may also define units for more complex quantities using these basic units. For example, the common unit of speed is miles per hour or mph, and it involves both distance (miles) and time (hour) units. Units named using the product or quotient of two or more simpler units are called *compound* units.

If a car travels 75 miles at a constant rate in 1.5 hours, it is moving at a speed of

$$\frac{75 \text{ miles}}{1.5 \text{ hours}} = \frac{75 \times 1 \text{ mi}}{1.5 \times 1 \text{ hr}}$$

$$= \frac{75}{1.5} \times \frac{1 \text{ mi}}{1 \text{ hr}}$$

$$= 50 \text{ mi/hr}$$

We write this speed as 50 mi/hr or 50 mph, and read it as "50 miles per hour." Ratios and rates are often written with compound units.

Which of the following are expressed in compound units?

(a) diameter, 4.65″

(b) density, 12 lb/cu ft

(c) gas usage, mi/gal or mpg

(d) current, 4.6 amp

(e) rotation rate, revolution/min or rpm

(f) pressure, lb/sq in. or psi

(g) cost ratio, ¢/lb

(h) length, 5 yd 2 ft

(i) area, sq ft

(j) volume, cu in.

Check your answers in **18**.

---

18 Compound units are used in quantities (b), (c), (e), (f), (g), (i), and (j). Two different units are used in (h), but they have not been combined by multiplying or dividing.

Unity fractions can also be used to convert compound units. For example, to convert 50 mph to ft per sec units,

$$50 \text{ mph} = \left(\frac{50 \times 1 \cancel{\text{mi}}}{1 \cancel{\text{hr}}}\right) \times \left(\frac{5280 \text{ ft}}{1 \cancel{\text{mi}}}\right) \times \left(\frac{1 \cancel{\text{hr}}}{3600 \text{ sec}}\right)$$

$$= \frac{50 \times 5280}{3600} \text{ ft/sec}$$

$$= 73.333\ldots \text{ ft/sec}$$

$$= 73 \text{ ft/sec, rounded}$$

Use this procedure to convert a rotation rate of 160 revolutions per minute (rpm) to revolutions per second (rps).

160 rpm = ________ rps

Check your work in **19**.

---

19 $$160 \text{ rpm} = \left(\frac{160 \times 1 \text{ rev}}{1 \cancel{\text{min}}}\right) \times \left(\frac{1 \cancel{\text{min}}}{60 \text{ sec}}\right)$$

$$= \frac{160}{60} \frac{\text{rev}}{\text{sec}}$$

$$= 2.6666\ldots \text{ rps}$$

$$= 2.7 \text{ rps, rounded}$$

Remember to round your answer to agree in precision with the original number.

In Chapter 7 we will study the geometry needed to calculate the area and volume of the various plane and solid figures used in technical work. Here we want to look at the many different units used to measure area and volume.

When the area of a surface is found, the measurement or calculation is given in "square units." If the lengths are measured in inches, the area is given in square inches or sq in.; if the lengths are measured in feet, the area is given in square feet or sq ft.

area = 1 ft × 1 ft
= 1 sq ft

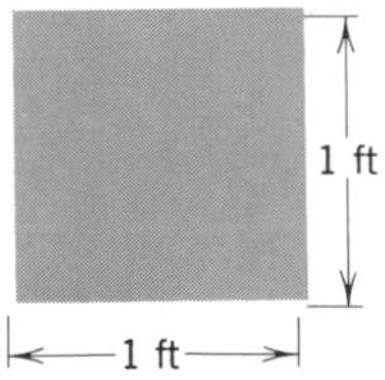

To see the relation between sq ft area units and sq in. area units, rewrite the above area in inch units.

area = 12 in. × 12 in.
= 144 × 1 in. × 1 in.
= 144 sq in.

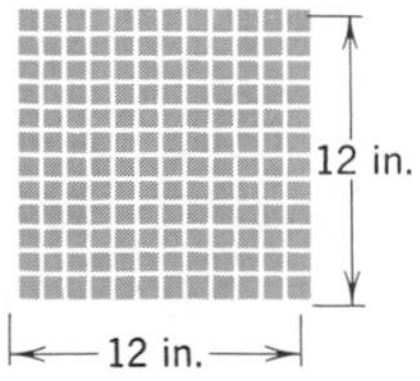

Therefore,

1 sq ft = 144 sq in.

Use this information to convert

15 sq ft = ________ sq in.

Check your work in **20**.

LUMBER MEASURE

Carpenters and workers in the construction trades use a special unit to measure the amount of lumber. The volume of lumber is measured in *board feet.* One board food (bf) of lumber is a piece having an area of 1 sq ft and a thickness of 1 in. or less.

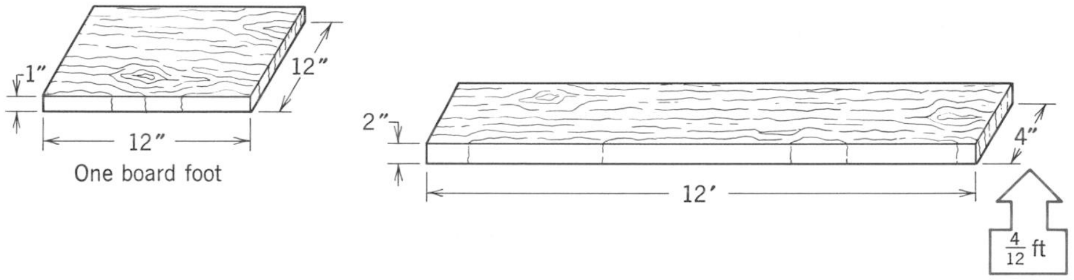

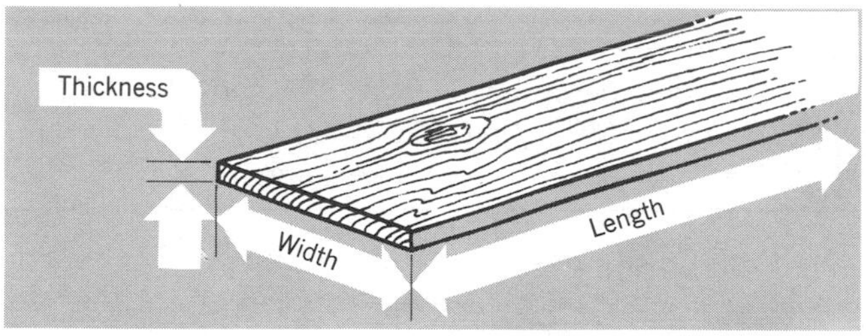

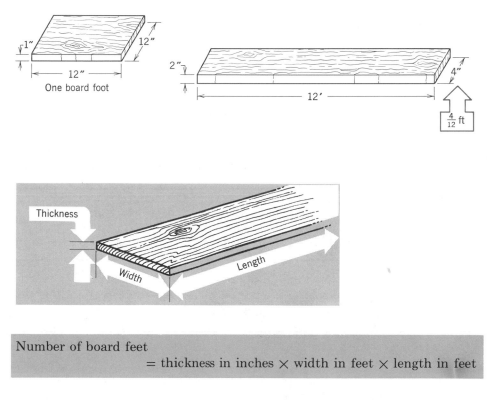

Number of board feet
= thickness in inches × width in feet × length in feet

To find the number of board feet in a piece of lumber, multiply the length in feet by the width in feet by the thickness in inches. A thickness of less than 1 inch should be counted as 1 inch. If the lumber is dressed or finished, use the full size or rough stock dimension to calculate board feet.

For example, a 2″ by 4″ used in framing a house would actually measure $1\frac{1}{2}$″ by $3\frac{1}{2}$″, but we would use the dimensions 2″ by 4″ in calculating board feet. A 12 ft length of 2″ by 4″ would have a volume of 12 ft × $\frac{4}{12}$ ft × 2 in. = 8 board ft or 8 bf. In lumber measure the phrase "per ft" means "per board foot." The phrase "per running foot" means "per foot of length."

Problems:

1. Find the number of board feet in each of the following pieces of lumber.

   (a) $\frac{3}{4}$″ × 6″ × 4′ (b) 2″ × 6″ × 16′

   (c) $\frac{1}{2}$″ × 8″ × 12′ (d) 4″ × 4″ × 8′

   (e) 1″ × 6″ × 14′ (f) $\frac{7}{8}$″ × 12″ × 3′

2. How many board feet are there in a shipment containing 80 boards, 2 in. by 6 in., each 16 ft long?
3. To floor a small building requires 245 boards each $1\frac{1}{2}$ in. by 12 in. by 12 ft long. How many board feet should be ordered?

The answers are on page 550.

20 $15 \text{ sq ft} = (15 \times 1 \cancel{\text{sq ft}}) \times \left(\dfrac{144 \text{ sq in.}}{1 \cancel{\text{sq ft}}}\right)$

$= 15 \times 144$ sq in.

$= 2160$ sq in. or 2200 sq in., rounded.

Using a Calculator

Key	Display
C	0.
1 5	15.
×	15.
1 4 4	144.
=	2160.

If you do not remember the conversion number, 144, try it this way:

$$15 \text{ sq ft} = (15 \times 1 \cancel{\text{ft}} \times 1 \cancel{\text{ft}}) \times \left(\frac{12 \text{ in.}}{1 \cancel{\text{ft}}}\right) \times \left(\frac{12 \text{ in.}}{1 \cancel{\text{ft}}}\right)$$

$= 15 \times 12 \times 12 \times 1$ in. $\times$ 1 in.

$= 2160$ sq in.

$= 2200$ sq in., rounded

A number of other convenient area units have been defined.

1 square yard (sq yd)	= 9 sq ft
1 square rod (sq rod)	= 30.25 sq yd
1 acre = 160 sq rod	= 4840 sq yd
1 sq mile (sq mi)	= 640 acres

Volume units are given in "cubic units." If the lengths are measured in inches, the volume is given in cubic inches or cu in.; if the lengths are measured in feet, the volume is given in cubic feet or cu ft.

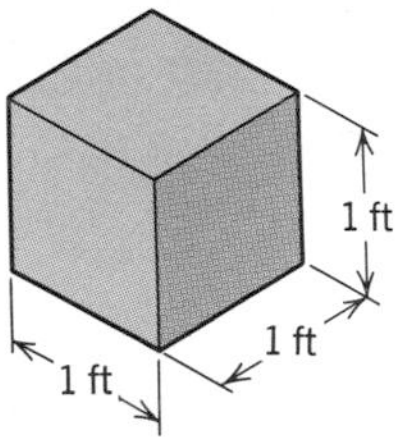

volume $= 1 \text{ ft} \times 1 \text{ ft} \times 1 \text{ ft}$

$= 1$ cu ft

In inch units this same volume is 12 in. $\times$ 12 in. $\times$ 12 in. = 1728 cu in., so that 1 cu ft = 1728 cu in.

A very large number of special volume units have been developed for various uses.

1 cubic yard (cu yd)	= 27 cu ft
1 gallon (gal)	= 231 cu in.
1 barrel (bbl)	= 31.5 gal
1 bushel (bu)	= 2150.42 cu in.
1 fluid ounce (fl oz)	= 1.805 cu in.
1 pint (pt)	= 28.875 cu in. (liquid measure)

Dozens of other volume units are in common use: cup, quart, peck, teaspoonful, etc. The cubic in., cubic ft, and gallon are most used in technical work.

In the next section of this chapter we will study metric units and you will find only a few basic units connected by simple conversion numbers rather than the confusion of having many different length, area, and volume units.

Now turn to frame **21** for a set of problems involving unit conversion.

21 Exercises 5-2 Units and Unit Conversion

A. Convert the units as shown:

1. $4\frac{1}{4}$ ft = ______ in.	2. 33 yd = ______ ft
3. 3.4 mi = ______ ft	4. 17 lb = ______ oz
5. 126 gal = ______ bbl	6. $8\frac{1}{2}$ mi = ______ yd
7. 46 psi = ______ atm	8. 7500 lb = ______ ton
9. 18.5 bbl = ______ gal	10. 32 in. = ______ ft
11. 2640 ft = ______ mi	12. 19 yd = ______ rods

B. Convert the units as shown:

1. 6 sq ft = ______ sq in.	2. $2\frac{1}{2}$ acre = ______ sq ft
3. 17.25 bbl = ______ cu ft	4. 385 sq ft = ______ sq yd
5. 972 cu in. = ______ cu ft	6. 325 rpm = ______ rps
7. $60\,\dfrac{\text{lb}}{\text{cu ft}}$ = ______ $\dfrac{\text{lb}}{\text{cu in.}}$	8. 30 cu ft = ______ bbl
9. 14,520 sq ft = ______ acre	10. $2500\,\dfrac{\text{ft}}{\text{min}}$ = ______ $\dfrac{\text{in.}}{\text{sec}}$
11. 20 sq mi = ______ acre	12. 65 mph = ______ ft/sec

C. Solve:

1. Hydrological engineers often measure very large amounts of water in acre-ft. One acre-ft of water fills a rectangular volume of base one acre and height 1 ft. Convert 1 acre-ft to

   (a) ______ cu ft (b) ______ gallons (c) ______ barrels

   (d) ______ cu yd

2. The board ft defined on page 183 is a volume unit. Use the definition given to complete the following:

   (a) 1 bd ft = ______ cu ft (b) 1 bd ft = ______ cu in.

3. Martian technologists use the gronk, smersh, and pflug as length units. There are 3 gronks in a smersh, and 2 pflugs equal one smersh. Translate a length of 6 gronks to pflugs.

4. The area covered by a given volume of paint will be determined by the consistency of the paint and the nature of the surface. The measure of covering power is called the *spreading rate* of the paint and is expressed in units of sq ft per gallon. The following table gives some typical spreading rates.

Paint	Surface	Spreading Rate	
		One Coat	Two Coats
Oil-base-paint	Wood, smooth	580	320
	Wood, rough	360	200
	Plaster, smooth	300	200
	Plaster, rough	250	160
Latex paint	Wood	500	350
Varnish	Wood	400	260

Use this table to answer the following questions.

(a) How many gallons of oil base paint are needed to cover 2500 sq ft of rough wood with two coats?
(b) At \$7.98 per gallon, how much would it cost to paint a room of 850 sq ft area with latex base paint? The surfaces are finished wood.
(c) How many square feet of rough plaster will 8 gallons of oil base paint cover with one coat?
(d) How many gallons of varnish are needed to cover 600 sq ft of cabinets with two coats?
(e) At \$10.98 per gallon, how much will it cost to paint 750 sq ft of smooth plaster with two coats of oil base paint?

5. The earliest known unit of length to be used in a major construction project is the "megalithic yard" used by the builders of Stonehenge in southwestern Britain about 2600 B.C. If 1 megalithic yard = 2.72 ± 0.05 ft, convert this length to inches. (Round to the nearest tenth of an inch.)

6. In the Bible (Genesis, Chapter 7) Noah built an ark 300 cubits long, 50 cubits wide, and 30 cubits high. In I Samuel, Chapter 17, it is reported that the giant Goliath was "six cubits and a span" in height. If 1 cubit = 18 in. and 1 span = 9 in.,

   (a) What were the dimensions of the ark in English units, and
   (b) How tall was Goliath?

7. A homeowner needing carpeting has calculated that 1265 sq ft of carpet must be ordered. However, carpet is sold in square yards. Make the necessary conversion for her. (Round up to the nearest whole number.)

8. Cast aluminum has a density of 160.0 lb/cu ft. Convert this density to units of lb/cu in. and round to one decimal place.

9. A $\frac{5}{16}''$ twist drill with a periphery speed of 50 ft/min has a cutting speed of 611 rpm. Convert this speed to rps and round to one decimal place.

Check your answers on page 550, then continue in **22** with the study of metric units.

5-3 METRIC UNITS

B.C. by permission of Johnny Hart and Field Enterprises, Inc.

22 Steadily, without fuss and almost without anyone noticing it, the United States is adopting the metric system—the units used by most of the world. Leading the way are American business and industry. General Motors, Ford, Caterpillar Tractor, International Harvester, Honeywell, General Electric, Kodak, IBM, and many other corporations are busily converting their products and activities to metric units. Drug, photographic, and chemical companies have used the metric system for years. We are rapidly joining the 95% of the world's people who already think metric.

For the technical or trades worker, the ability to use metric units is important. In many ways, working with meters, kilograms, and liters is easier and more logical than using our traditional units of feet, pounds, or quarts. This section of Chapter 5 will define and explain the most useful metric units, show how they are related to the corresponding English units, explain how to convert from one unit to another, and provide practice problems designed to help you use the metric system and to "think metric."

The most important common units in the metric system are those for length or distance, speed, weight, volume, area, and temperature. Time units—year, day, hour, minute, second—are the same in the metric as in the English system.

Length

The basic unit of length in the metric system is the *meter,* pronounced *meet-ur* and abbreviated *m.* (This word is sometimes spelled *metre,* but pronounced exactly the same.) One meter is roughly equal to one yard. Originally, in the eighteenth century, the meter was defined as one ten-millionth of the distance from the North Pole to the equator on a line through Paris. Today it is defined as exactly 1,650,763.73 times the wavelength of the orange light emitted by krypton-86 gas.

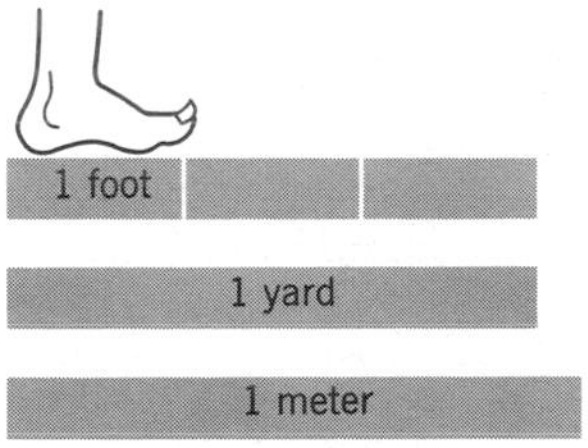

The meter is the appropriate unit to use in measuring your height, the width of a room, length of lumber, or the height of a building.

Estimate the following lengths in meters:

(a) The length of the room in which you are now sitting = ________ m.

(b) Your height = ________ m.

(c) The length of a ping-pong table = ________ m.

(d) The length of a football field = ________ m.

Guess as closely as you can, then turn to **23** and continue.

---

23 (a) A small room might be 3 or 4 meters long, and a large one might be 6 or 8 meters long.
(b) Your height is probably between $1\frac{1}{2}$ and 2 meters. Two meters is roughly 6 ft 7 in.
(c) A ping-pong table is a little less than 3 meters in length.
(d) A football field is about 100 m long.

All other length units used in the metric system are defined in terms of the meter, and these units differ from one another by multiples of ten. For example, the *centimeter,* pronounced *cent-a-meter* and abbreviated *cm,* is defined as exactly one-hundredth of a meter.

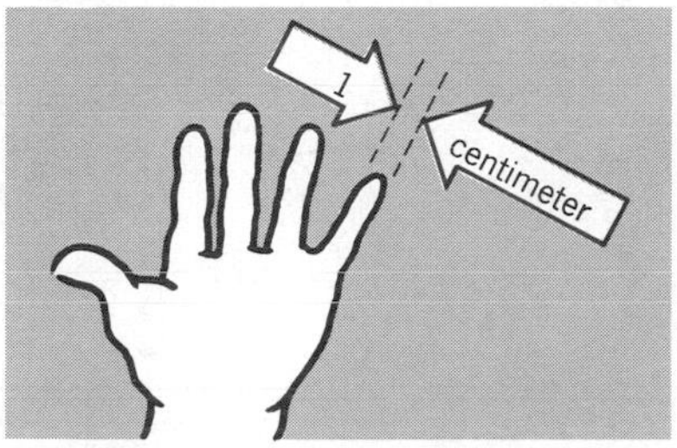

1 centimeter, cm $= \frac{1}{100}$ of a meter $= 0.01$ meter

The *kilometer,* pronounced *kill-o-meter* and abbreviated *km,* is defined as exactly 1000 meters.

1 km = 1000 m

Because metric units increase or decrease in multiples of ten, they may be named using prefixes attached to a basic unit. For length units we have:

Metric Length Unit	Prefix	Multiplier	Money Analogy	Memory Aid
kilometer	kilo-	1000 × 1 meter	$1000	*kilo*watt
hectometer	hecto-	100 × 1 meter	$ 100	
decameter	deca-	10 × 1 meter	$ 10	*deca*de = 10 years
meter	—	1 meter	$ 1	
decimeter	deci-	0.1 × 1 meter	10¢	*deci*mal = $\frac{1}{10}$
centimeter	centi-	0.01 × 1 meter	1¢	*cent*ury = 100 years
millimeter	milli-	0.001 × 1 meter	$\frac{1}{10}$¢	*mill*ennium = 1000 years

Now, for some practice in working with metric length units, turn to page 575 of this book. There you will find a metric measuring rule. Cut it out and use it to measure the following:

(a) Length of this page = ________ cm

(b) Length of a dollar bill = ________ cm

(c) Length of a pencil = ________ cm

(d) Length of your shoe = ________ cm

(e) Diameter of a 25¢ coin = ________ cm

(f) Distance from seat of your chair to the floor = ________ cm

Measure carefully, record your answers above, then turn to **24** to check your work.

The metric system is very interesting. Are you going to use it in your work?
No. I wouldn't touch it with a 3.05 meter pole.

---

24 (a) This page is 27.94 cm long.
(b) The length of a dollar bill is about 15.6 cm.
(c) A new pencil is about 21 cm long.
(d) Your shoe is probably 20 to 30 cm long.
(e) A 25¢ coin is about 2.5 cm in diameter.
(f) The seat of a chair is usually about 42 cm from the floor.

The most often used metric length units are the kilometer, meter, centimeter, and millimeter.

1 **kilo**meter = **1000** meters

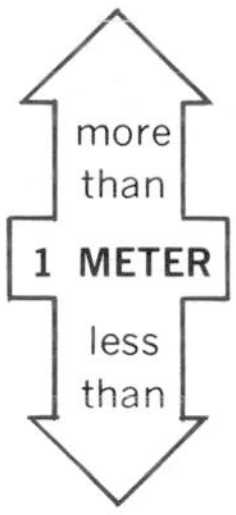

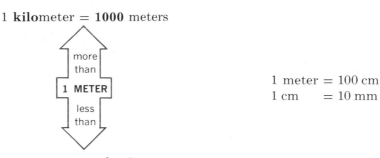

1 meter = 100 cm
1 cm = 10 mm

1 **centi**meter = $\frac{1}{100}$ of a meter
1 **milli**meter = $\frac{1}{1000}$ of a meter

The millimeter (mm) unit of length is very often used on technical drawings, and the centimeter (cm) is handy for shop measurements.

1 cm is roughly the width of a paper clip.
1 mm is roughly the thickness of the wire in a paper clip.
1 meter is roughly 10% more than a yard.

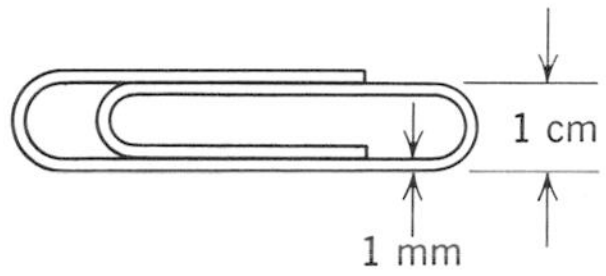

One very great advantage of the metric system is that we can easily convert from one metric unit to another. There are no hard-to-remember conversion factors: 36 in. in a yard, 5280 ft in a mile, 220 yards in a furlong, or whatever.

For example, with English units, if we convert a length of 137 in. to feet or yards, we find:

137 in. = $11\frac{5}{12}$ ft = $3\frac{29}{36}$ yd

Here we must divide by 12 and then by 3 to convert the units. But in the metric system we simply shift the decimal point. The same length in the metric system is

100 cm = 1 m

348 cm = 3.48 m

To divide by 100, shift the decimal point two digits to the left

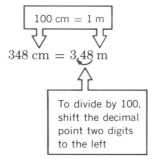

Of course we may also use unity fractions:

$$348 \text{ cm} = 348 \cancel{\text{cm}} \times \left(\frac{1 \text{ m}}{100 \cancel{\text{cm}}}\right)$$

$$= \frac{348}{100} \text{ m}$$

$$= 3.48 \text{ m}$$

Try it. Complete the following unit conversions.

(a) 147 cm = ________ m (b) 3.1 m = ________ cm

(c) 21 cm = ________ m (d) 11.65 m = ________ cm

(e) 7 cm = ________ mm (f) 20.5 mm = ________ cm

Check your answers in **25**.

Why should I study this metric stuff? — I don't use it in my job.

One good reason to study it is that people who can handle metrics will have an advantage in the job market when the U.S. goes metric.

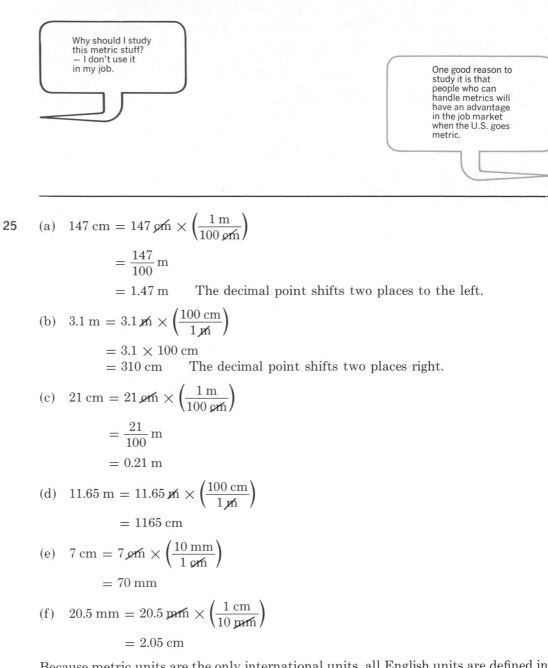

25 (a) $147 \text{ cm} = 147 \cancel{\text{cm}} \times \left(\frac{1 \text{ m}}{100 \cancel{\text{cm}}}\right)$

$= \frac{147}{100} \text{ m}$

$= 1.47 \text{ m}$ The decimal point shifts two places to the left.

(b) $3.1 \text{ m} = 3.1 \cancel{\text{m}} \times \left(\frac{100 \text{ cm}}{1 \cancel{\text{m}}}\right)$

$= 3.1 \times 100 \text{ cm}$

$= 310 \text{ cm}$ The decimal point shifts two places right.

(c) $21 \text{ cm} = 21 \cancel{\text{cm}} \times \left(\frac{1 \text{ m}}{100 \cancel{\text{cm}}}\right)$

$= \frac{21}{100} \text{ m}$

$= 0.21 \text{ m}$

(d) $11.65 \text{ m} = 11.65 \cancel{\text{m}} \times \left(\frac{100 \text{ cm}}{1 \cancel{\text{m}}}\right)$

$= 1165 \text{ cm}$

(e) $7 \text{ cm} = 7 \cancel{\text{cm}} \times \left(\frac{10 \text{ mm}}{1 \cancel{\text{cm}}}\right)$

$= 70 \text{ mm}$

(f) $20.5 \text{ mm} = 20.5 \cancel{\text{mm}} \times \left(\frac{1 \text{ cm}}{10 \cancel{\text{mm}}}\right)$

$= 2.05 \text{ cm}$

Because metric units are the only international units, all English units are defined in terms of the metric system. For length measurements,

1 in.	= 2.54 cm
1 ft	= 30.48 cm
1 yd	= 91.44 cm
1 cm	= 0.3937 in.

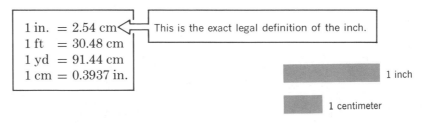

This is the exact legal definition of the inch.

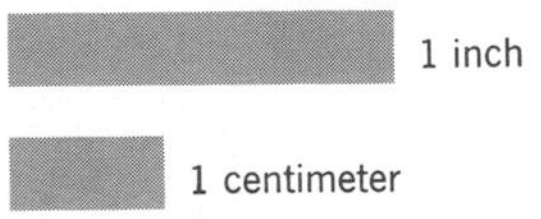

To shift from English to metric units or from metric to English units either use unity fractions or the conversion factors given on page 191.

APPROXIMATE CONVERSION FACTORS: LENGTH

When you know	you can find	if you multiply by
inches, in.	millimeters, mm	25.40
inches, in.	centimeters, cm	2.54
feet, ft	centimeters, cm	30.48
feet, ft	meters, m	0.3048
yards, yd	meters, m	0.9144
miles, mi	kilometers, km	1.6093
millimeters, mm	inches, in.	0.03937
centimeters, cm	inches, in.	0.3937
centimeters, cm	feet, ft	0.0328
meters, m	feet, ft	3.2808
meters, m	yards, yd	1.0936
kilometers, km	miles, mi	0.6214

Convert the following measurements as shown.

1. 4.3 yd = ________ m
2. 16 in. = ________ cm
3. $2\frac{1}{2}$ in. = ________ mm
4. 6 ft = ________ m
5. 1.27 m = ________ ft
6. 37 cm = ________ in.
7. 4.2 m = ________ yd
8. 136 cm = ________ ft
9. 18 ft 6 in. = ________ m
10. $2\frac{1}{4}$ in. = ________ mm

Check your anwers in **26**.

DUAL DIMENSIONING

Some companies involved in international trade use "Dual Dimensioning" on their technical drawings and specifications. With dual dimensioning both inch and metric dimensions are given. For example, a part might be labeled like this:

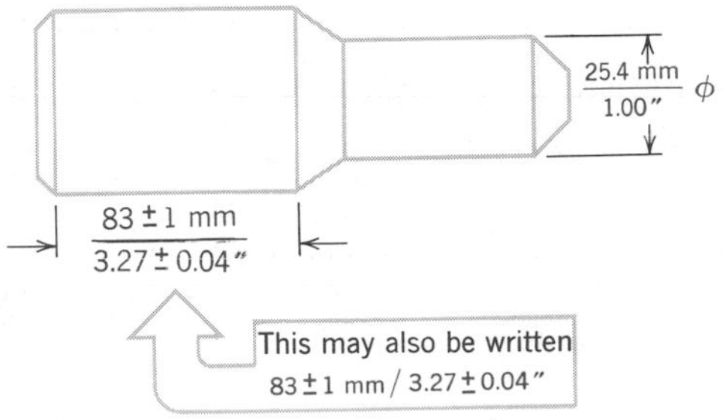

Notice that the metric measurement is written first or on top of the fraction bar. Diameter dimensions are marked with the symbol ∅.

26 1.	3.9 m	2.	40.6 cm
3.	63.5 mm	4.	1.83 m
5.	4.17 ft	6.	14.6 in.
7.	4.6 yd	8.	4.46 ft
9.	5.64 m	10.	57.2 mm

All answers have been rounded, of course.

Area

When the area of a surface is calculated, the units will be given in "square length units." If the lengths are measured in feet, the area will be given in square feet; if the lengths are measured in meters, the area will be given in square meters.

For example, the area of a carpet 3 meters wide and 4 meters long would be

3 m × 4 m = 12 square meters
= 12 sq m

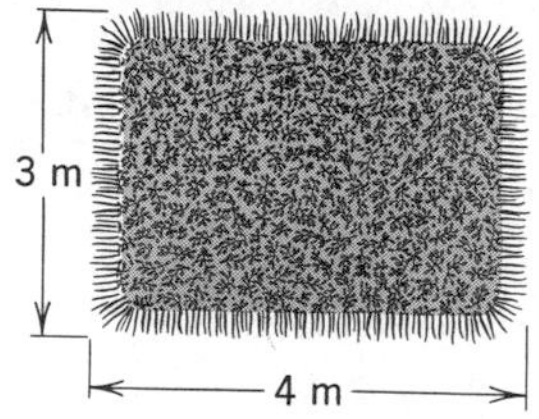

Areas roughly the size of carpets and room flooring are usually measured in square meters. Larger areas are usually measured in *hectares,* a metric surveyor's unit. One hectare is the area of a square 100 meters on each side. The hectare is roughly $2\frac{1}{2}$ acres, or about double the area of a football field.

1 hectare = 100 m × 100 m
= 10,000 square meters
= 2.471 acres

To convert use the approximate conversion factors given below.

APPROXIMATE CONVERSION FACTORS: AREA

When you know	you can find	if you multiply by
square inches, $in.^2$	square centimeters, cm^2	6.452
square feet, ft^2	square meters, m^2	0.093
square yards, yd^2	square meters, m^2	0.836
acres, A	hectares, ha	0.4047
acres, A	square meters, m^2	4047
square centimeters, cm^2	square inches, $in.^2$	0.155
square meters, m^2	square feet, ft^2	10.764
square meters, m^2	square yards, yd^2	1.196
square meters, m^2	hectares, ha	0.0001
hectares, ha	acres, A	2.471

Use this information to solve the following problems.

1. Convert (a) 15 sq ft = ________ sq m

   (b) 21 sq m = ________ sq ft

   (c) $8\frac{1}{2}$ acres = ________ hectares

   (d) 10 sq cm = ________ sq in.

2. Find the area of the following rectangles:

   (a) a floor 5 m by 6.2 m = ________ sq m

   (b) a field 36 m by 123 m = ________ hectares

(c) a window 40 cm by 110 cm = ________ sq cm

(d) a wall 4.5 m by 6.2 m = ________ sq m

Check your answers in **27**.

---

27 1. (a) 1.4 sq m (b) 226 sq ft
(c) 3.4 hectares (d) 1.55 sq in.

2. (a) 31 sq m (b) 0.443 hectares
(c) 4400 sq cm (d) 28 sq m

Speed

In the metric system, ordinary highway speeds are measured in kilometers per hour, abbreviated km/hr or kmh. To "think metric" while you drive, remember that

> 100 kmh is approximately equal to 62 mph.

Machinists measure pulley, drill, and lathe speeds in centimeters per second, abbreviated cm/sec. The following table gives useful conversion factors:

APPROXIMATE CONVERSION FACTORS: SPEED

When you know	you can find	if you multiply by
inches per second, ips	centimeters per second, cm/sec	2.54
feet per second, fps	centimeters per second, cm/sec	30.48
feet per second, fps	meters per second, m/sec	0.3048
feet per minute, fpm	centimeters per second, cm/sec	0.5080
miles per hour, mph	kilometers per hour, kmh	1.6093
centimeters per second, cm/sec	inches per second, ips	0.3937
centimeters per second, cm/sec	feet per second, fps	0.0328
meters per second, m/sec	feet per second, fps	3.2808
kilometers per hour, kmh	miles per hour, mph	0.6214

Use these conversion factors to find the following:

1. (a) 4800 ft/min = ________ cm/sec

   (b) 35 mph = ________ kmh

   (c) 30 m/sec = ________ ft/sec

   (d) 120 cm/sec = ________ in./sec

2. The surface speed of a grinding wheel is 2400 ft/min. Express this in cm/sec.

3. The cutting speed for a soft steel part in a lathe is 165 ft/min. Express this in cm/sec.

   Round your answers to the nearest whole number.

Check your answers in **28**.

---

28 1. (a) 2438 cm/sec (b) 56 km/hr
(c) 98 ft/sec (d) 47 in./sec

2. 1219 cm/sec 3. 84 cm/sec

Volume

When the volume of an object is calculated or measured, the units will normally be given in "cubic length units." If the lengths are measured in feet, the volume will be given in cubic feet; if the lengths are measured in meters, the volume will be given in cubic meters.

For example, the volume of a box 2 m high by 3 m wide by 4 m long would be

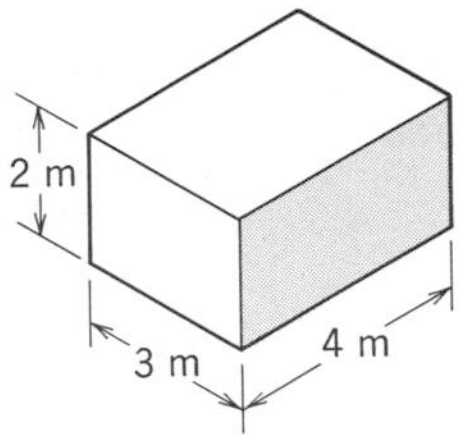

2 m × 3 m × 4 m = 24 cubic meters

and this is often written as 24 m^3.

The cubic meter is an appropriate unit for measuring the volume of large amounts of water, sand, or gravel, or the volume of a room. However, a more useful metric volume unit for most practical purposes is the *liter,* pronounced *leet-ur* and abbreviated ℓ. You may find this volume unit spelled either *liter* in the United States or *litre* elsewhere in the world.

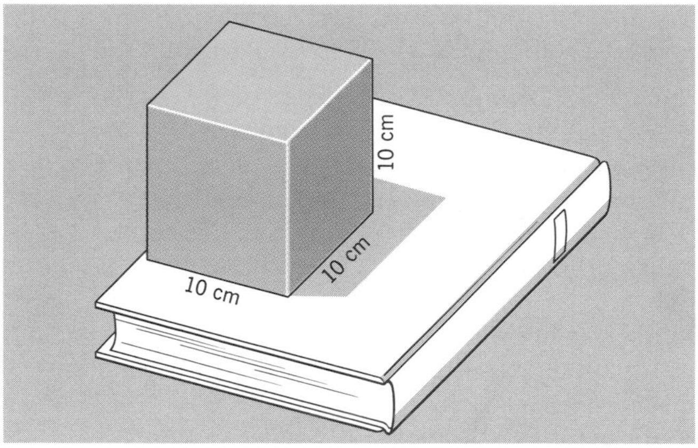

One liter is defined as the volume of a cube 10 cm on each edge.

A volume of one liter is slightly larger than one quart. In fact, in the United States, one quart is legally defined as exactly 0.94635295 liter.

For practical purposes remember that

1 gallon = 4 qt
1 qt ≅ 0.946 ℓ
or

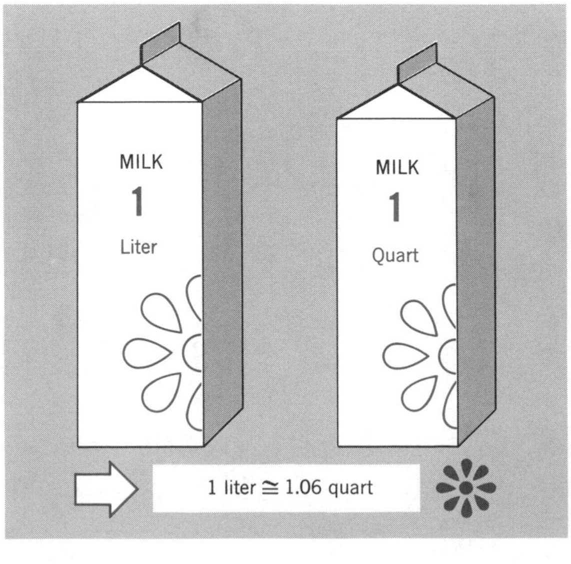

$1\ \ell \cong 1.0567\ \text{qt}$

One liter is roughly 6% more than one quart.

Calculate each of the following volumes in liters. (Round to one decimal place when necessary.)

(a) A rough iron casting 20 cm by 15 cm by 32 cm
(b) A rectangular tank 15 in. by 18 in. by 16 in.
(c) A 20-gallon container of fuel oil

Check your work in **29**.

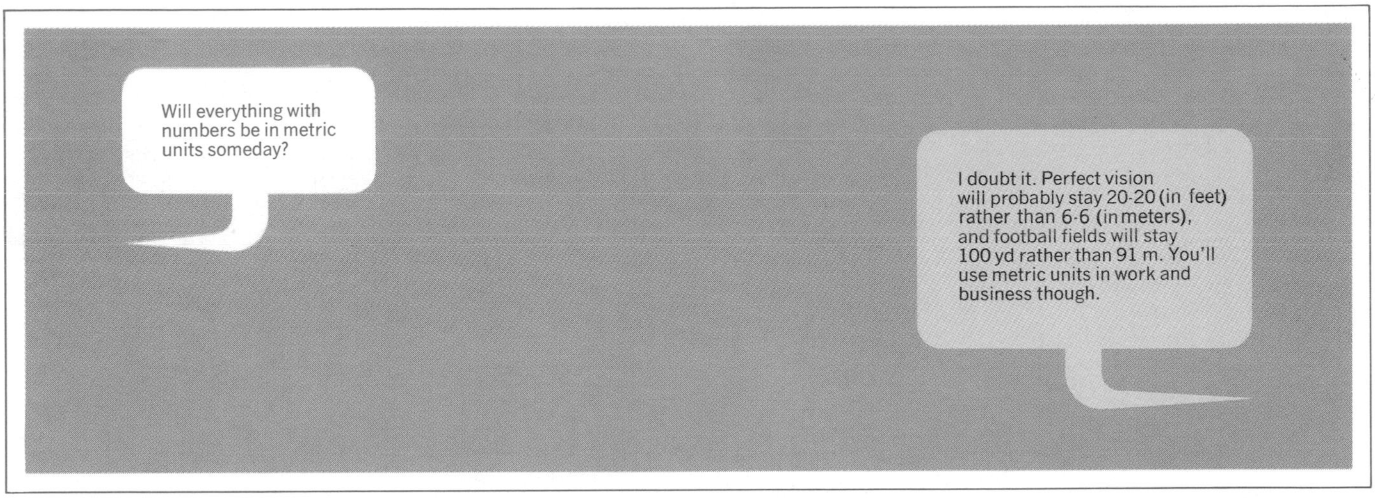

---

29 (a) Volume = 20 cm × 15 cm × 32 cm
= 9600 cubic cm
= 9.6 liters, since 1 liter = 1000 cubic cm

Using a Calculator

Key	Display
[C]	0.
[2] [0]	20.
[×]	20.
[1] [5]	15.
[×]	300.
[3] [2]	32.
[=]	9600.

(b) Convert all dimensions to cm using the conversion factor 1 in. = 2.54 cm or the table on page 192.

Volume = 15 in. × 18 in. × 16 in.
= 38.1 cm × 45.72 cm × 40.64 cm
≅ 70792 cubic cm
≅ 70.8 liters, rounded.

Using a Calculator

Key	Display
C	0.
1 5	15.
×	15.
2 . 5 4	2.54
×	38.1
1 8	18.
×	685.8
2 . 5 4	2.54
×	1741.932
1 6	16.
×	27870.912
2 . 5 4	2.54
=	70792.116

(c) 20 gal = 80 quarts

$$\cong 80\,\text{qt} \times \frac{1\,\ell}{1.06\,\text{qt}}$$

$\cong 75.5\ \ell$

Volumes smaller than the liter are usually measured in units of cubic centimeters, abbreviated cc or cm^3.

1 cubic centimeter = 1 cm × 1 cm × 1 cm
= 1 cm^3 or 1 cc

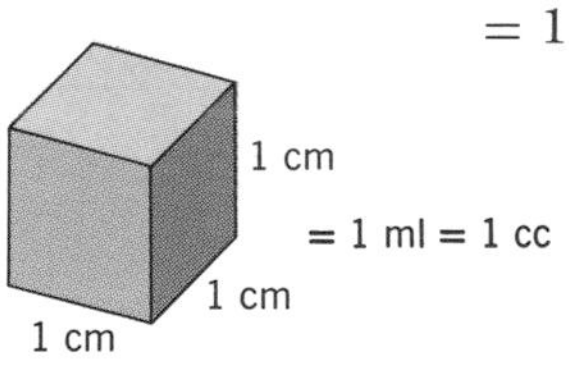

Notice that 1 liter is a volume of

1 liter = 10 cm × 10 cm × 10 cm
= 1000 cm^3

Because 1000 cubic centimeters are needed to make up one liter, a cubic centimeter is $\frac{1}{1000}$ of a liter or a *milli-liter.*

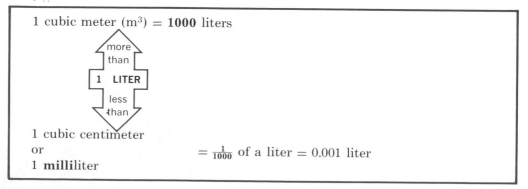

1 cubic meter (m^3) = **1000** liters

1 cubic centimeter
or
1 **milli**liter
= $\frac{1}{1000}$ of a liter = 0.001 liter

To shift from English to metric units or from metric to English units either use unity fractions or the conversion factors given below.

APPROXIMATE CONVERSION FACTORS: VOLUME

When you know	you can find	if you multiply by
cubic inches, in.^{3}	cubic centimeters, cm^3	16.387
cubic feet, ft^3	liters, ℓ	28.317
quarts (liquid), qt	liters, ℓ	0.946
gallons, gal	liters, ℓ	3.785
cubic yards, yd^3	cubic meters, m^3	0.765
fluid ounce, fl oz	cubic centimeters, cm^3	29.574
cubic centimeters, cm^3	cubic inches, in.^{3}	0.061
liters, ℓ	cubic feet, ft^3	0.035
liters, ℓ	quarts, qt	1.057
liters, ℓ	gallons, gal	0.264
cubic meters, m^3	cubic yards, yd^3	1.307

Convert the following measurements as shown.

1. 13.4 gal = ________ liter
2. 18.2 cu ft = ________ liter
3. 9.1 fl oz = ________ cm^3
4. 16.0 cu yd = ________ m^3
5. 5.1 ℓ = ________ qt
6. 135.0 cc = ________ in.^{3}
7. 24.0 m^3 = ________ yd^3
8. 64.0 ℓ = ________ gal
9. 12.4 qt = ________ ℓ
10. 4.2 in.^{3} = ________ cc

(Round to one decimal place when necessary.)

Check your answers in **30**.

30

1. 50.7 ℓ
2. 515.4 ℓ
3. 269.1 cm^3
4. 12.2 m^3
5. 5.4 qt
6. 8.2 in.^{3}
7. 31.4 yd^3
8. 16.9 gal
9. 11.7 ℓ
10. 68.8 cm^3

All answers were rounded to one decimal place.

The weight of an object is the gravitational pull of the earth exerted on that object. The mass of an object is related to how it behaves when pushed or pulled. For scientists the difference between mass and weight can be important. For all practical purposes they are the same.

The basic unit of mass or metric weight in the International Metric System is the *kilogram,* pronounced *kill-o-gram* and abbreviated *kg.* The kilogram is defined as the mass of a standard platinum-iridium metal cylinder kept at the International Bureau of Weights and Measures in Paris. By law in the United States, the pound is defined as exactly equal to the weight of 0.45359237 kg, so that

1 kg weighs about 2.2046 lb

One kilogram weighs about 10% more than 2 lb.

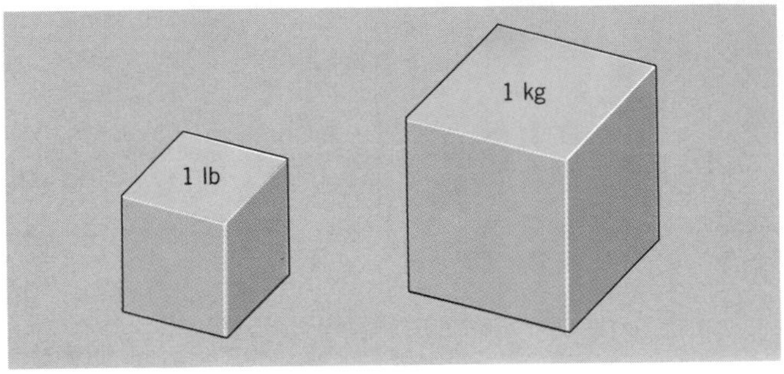

The kilogram was originally designed so that it was almost exactly equal to the weight of one liter of water.

Two smaller metric units of mass are also defined. The *gram,* abbreviated *g,* is defined as

1 gram = $\frac{1}{1000}$ of a kilogram = 0.001 kg

and the *milligram,* abbreviated *mg,* is defined as

1 milligram = $\frac{1}{1000}$ of a gram = 0.001 gram

The gram is a very small unit of weight, equal to roughly $\frac{1}{500}$ of a lb, or the weight of a paper clip. The milligram is often used by druggists and nurses to measure very small amounts of medication or other chemicals, and it is also used in scientific measurements.

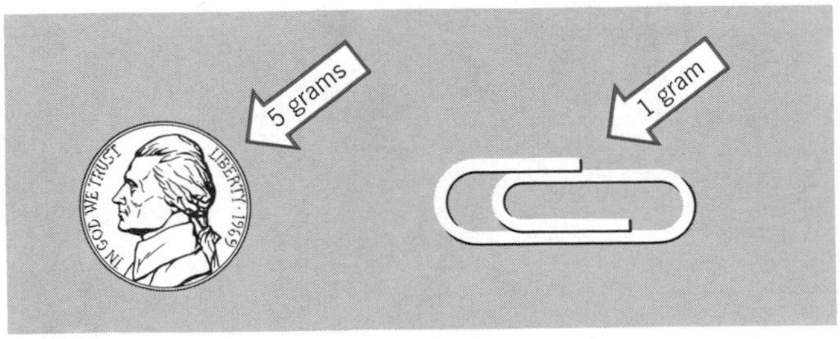

For industrial work the *metric ton* is often used.

1 metric ton = 1000 kg or 2204.62 lb

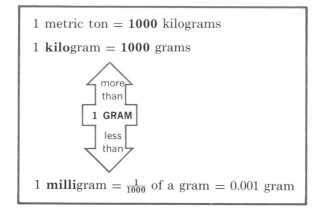

1 metric ton = **1000** kilograms

1 **kilo**gram = **1000** grams

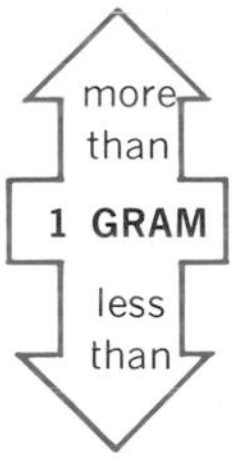
more than
1 GRAM
less than

1 **milli**gram = $\frac{1}{1000}$ of a gram = 0.001 gram

To shift from English to metric units or from metric to English units, either use unity fractions or the conversion factors given below.

APPROXIMATE CONVERSION FACTORS: WEIGHT

When you have	**you can find**	**if you multiply by**
pounds, lb	kilograms, kg	0.454
ounces, oz	grams, g	28.350
tons, T	kilograms, kg	907.19
tons, T	metric tons, t	0.907
kilograms, kg	pounds, lb	2.205
grams, g	ounces, oz	0.0353
kilograms, kg	tons, T	0.0011
metric tons, t	tons, T	1.102

Convert the following measurements as shown.

1. 150 lb = ________ kg
2. 6.0 oz = ________ g
3. 2.50 ton = ________ kg
4. 100.0 kg = ________ lb
5. 1.5 kg = ________ oz
6. 3.5 metric ton = ________ ton
7. 25.0 lb = ________ kg
8. 150 g = ________ lb
9.* 4 lb 5 oz = ________ kg
10. 65 lb = ________ kg

Round to one decimal place.

Check your work in **31**.

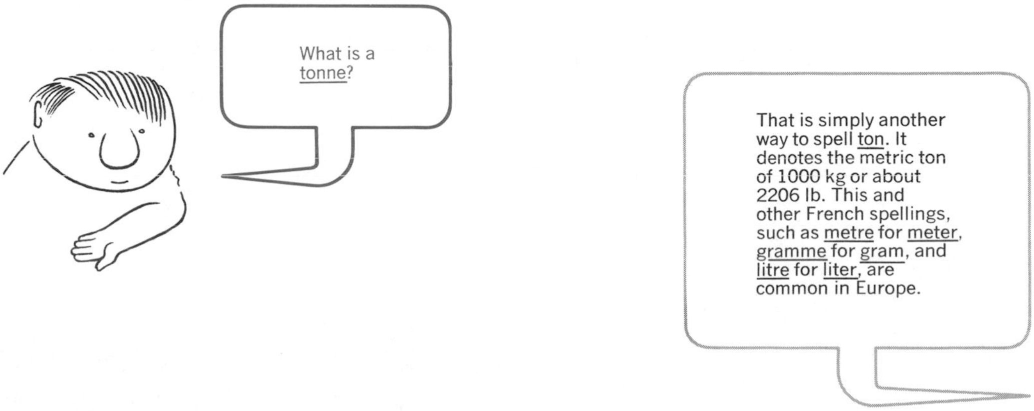
What is a tonne?
That is simply another way to spell ton. It denotes the metric ton of 1000 kg or about 2206 lb. This and other French spellings, such as metre for meter, gramme for gram, and litre for liter, are common in Europe.

31 1. 68.1 kg 2. 170.1 g
3. 2268.0 kg 4. 220.5 lb
5. 52.9 oz 6. 3.9 ton
7. 11.4 kg 8. 0.3 lb
9. 2.0 kg 10. 29.5 kg

All answers are rounded of course. In Problem 9 the answer 1.956 is 2.0 when rounded to one decimal place.

Temperature

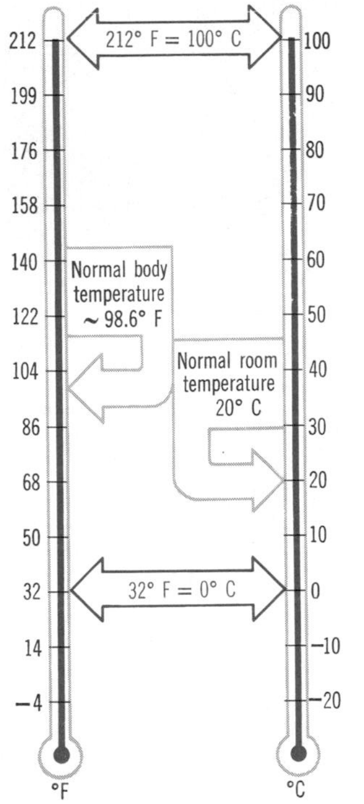

The Fahrenheit temperature scale is commonly used in the United States for weather reports, cooking, and other practical work. On this scale water boils at 212° F and freezes at 32° F. The metric or *Celsius* temperature scale is a simpler scale originally designed for scientific work, but now used worldwide for all temperature measurements. On the Celsius scale water boils at 100° C and freezes at 0° C.

Notice that the range from freezing to boiling is covered by 180 degrees on the Fahrenheit scale and 100 degrees on the Celsius scale. The size and meaning of a temperature degree is different for the two scales. A temperature of zero does not mean "no temperature"—it is simply another point on the scale. Temperatures less than zero are possible, and they are labeled with negative numbers.

Because zero does not correspond to the same temperature on both scales, converting Celsius to Fahrenheit or Fahrenheit to Celsius is not as easy as converting length or weight units. The simplest way to convert from one scale to the other is to use the following chart.

Temperature conversion chart
F = Fahrenheit C = Celsius

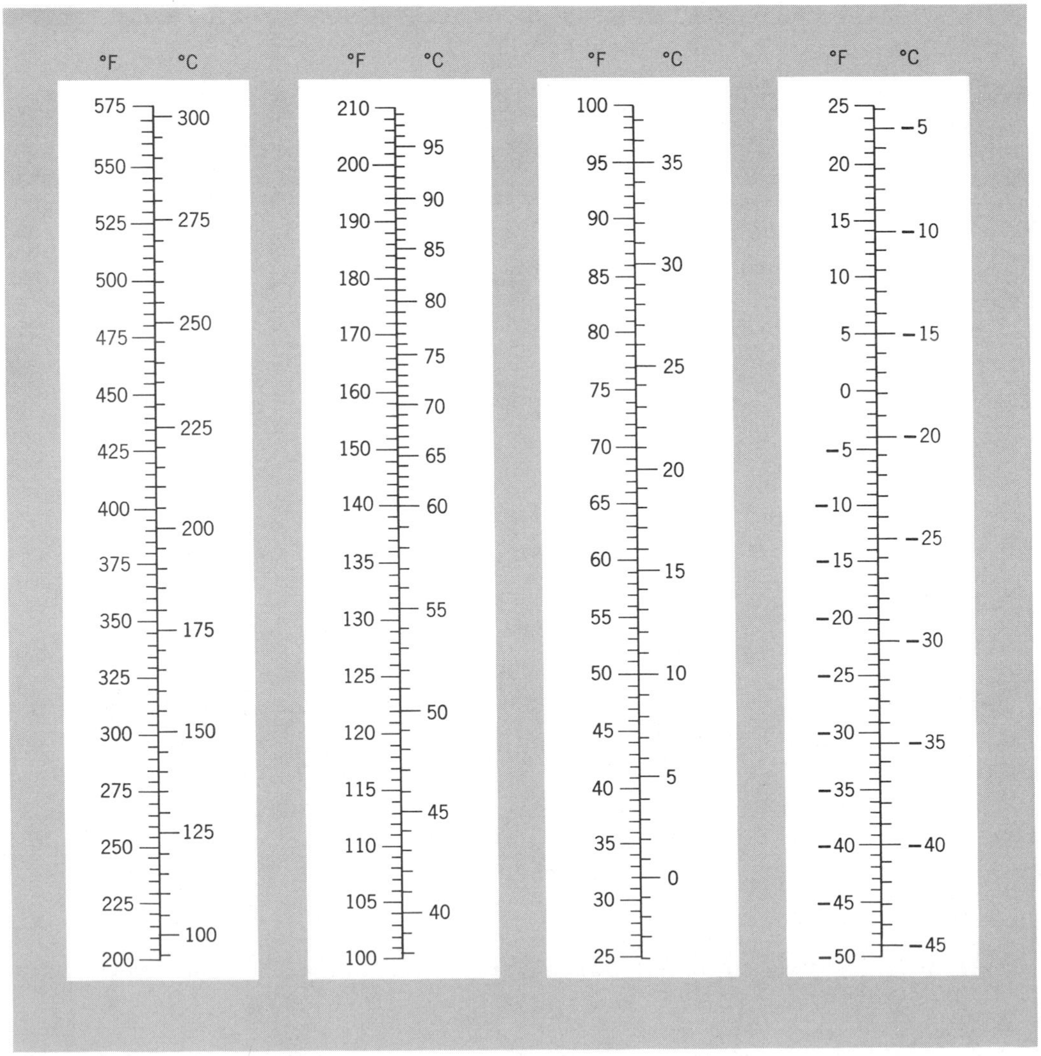

On this chart equal temperatures are placed side by side. For example, to convert 50° F to Celsius.

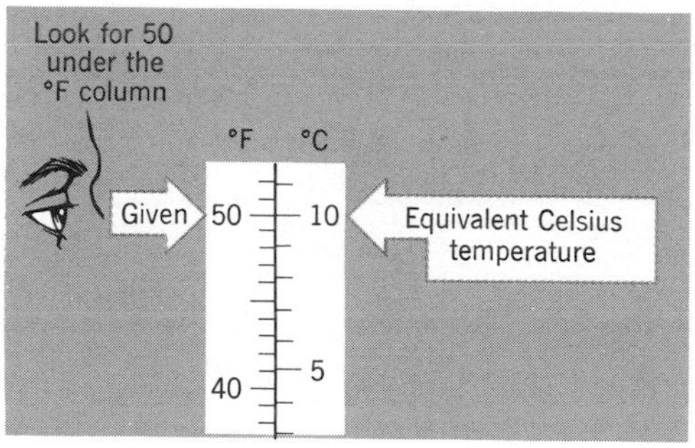

Use this temperature conversion chart to find the following temperatures.

1. 375° F = ________ ° C
2. −20° F = ________ ° C
3. 14° C = ________ ° F
4. 80° C = ________ ° F
5. 37° C = ________ ° F
6. 68° F = ________ ° C
7. −40° C = ________ ° F
8. 525° F = ________ ° C
9. 80° F = ________ ° C
10. −14° C = ________ ° F

Check your answers in **32**.

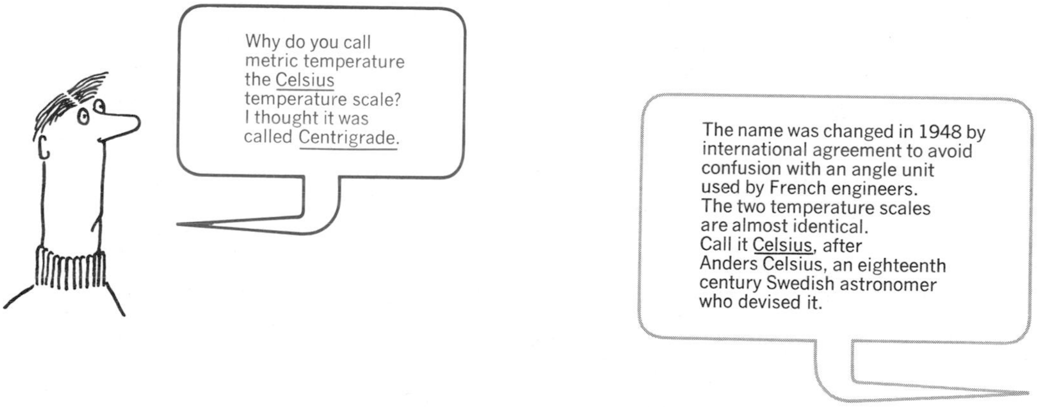

32

1. 191° C
2. −29°C
3. 57.5° F
4. 176° F
5. 98.6° F
6. 20° C
7. −40° F
8. 274° C
9. 27° C
10. 7° F

If you must convert a temperature with more accuracy than this chart allows, or if you are converting a temperature not on the chart, the following temperature conversion formulas should be used.

From Celsius to Fahrenheit:

$$F = \frac{9\,C}{5} + 32$$

By Hand

Example:

8° C gives $F = \dfrac{9 \times 8}{5} + 32$ ← Multiply first

$F = \dfrac{72}{5} + 32$ ← Then divide

$F = 14.4 + 32$ ← Add last

$F = 46.4°$ C

Using a Calculator

Key	Display
C	0.
9	9.
×	9.
8	72.
÷	72.
5	5.
+	14.4
3 2	32.
=	46.4

From Fahrenheit to Celsius:

$$C = \frac{5(F - 32)}{9}$$

By Hand

Example:

136° F gives $C = \dfrac{5(136 - 32)}{9}$ ← Subtract first

$C = \dfrac{5 \times 104}{9}$ ← Then multiply

$C = \dfrac{520}{9}$ ← Divide last

$C = 57.8°$ F

Using a Calculator

Key	Display
C	0.
1 3 6	136.
−	136.
3 2	32.
×	104.
5	5.
÷	520.
9	9.
=	57.777777

Use these conversion formulas to find the following temperatures.

1. 650° F = ________ ° C
2. 160° C = ________ ° F
3. 400° C = ________ ° F
4. 2200° F = ________ ° C
5. −80° F = ________ ° C
6. −50° C = ________ ° F

Round each answer to one decimal place.

Check your work in **33**.

---

33 1. $C = \dfrac{5(650 - 32)}{9}$

$= \dfrac{5(618)}{9}$

$= \dfrac{3090}{9}$

$= 343.3°$ C

2. $F = \dfrac{9 \times 160}{5} + 32$

$= \dfrac{1440}{5} + 32$

$= 288 + 32$

$= 320°$ F

3. $F = \dfrac{9 \times 400}{5} + 32$

$= \dfrac{3600}{5} + 32$

$= 720 + 32$

$= 752°$ F

4. $C = \dfrac{5(2200 - 32)}{9}$

$= \dfrac{5 \times 2168}{9}$

$= \dfrac{10840}{9}$

$= 1204.4°$ C

5. $C = \dfrac{5(-80 - 32)}{9}$

$= \dfrac{5 \times (-112)}{9}$

$= -\dfrac{560}{9}$

$= -62.2°$ C

6. $F = \dfrac{9 \times (-50)}{5} + 32$

$= \dfrac{-450}{5} + 32$

$= -90 + 32$

$= -58°$ F

Now turn to **34** for a set of practice problems designed to help you use the metric system and to "think metric."

Changing from the Fahrenheit to the Celsius temperature scale is a plot by scientists to take over the world!
Sure . . . by degrees.

---

34 Exercises 5-3 Metric Units

A. **Think Metric.** For each problem, circle the measurement closest to the first one given. No calculations are needed.

Remember:
(1) A meter is a little more (about 10%) than a yard.
(2) A kilogram is a little more (about 10%) than two pounds.
(3) A liter is a little more (about 6%) than a quart.

1.	30 cm	(a) 30 in.	(b) 75 in.	(c) 1 ft
2.	5 ft	(a) 1500 cm	(b) 1.5 m	(c) 2 m
3.	1 yd	(a) 90 cm	(b) 110 cm	(c) 100 cm
4.	2 m	(a) 6 ft 6 in.	(b) 6 ft	(c) 2 yd
5.	3 km	(a) 3 mi	(b) 2 mi	(c) 1 mi
6.	200 km	(a) 20 mi	(b) 100 mi	(c) 120 mi
7.	50 kmh	(a) 30 mph	(b) 50 mph	(c) 60 mph
8.	55 mph	(a) 30 kmh	(b) 60 kmh	(c) 90 kmh
9.	100 m	(a) 100 yd	(b) 100 ft	(c) 1000 in.
10.	400 lb	(a) 800 kg	(b) 180 kg	(c) 250 kg
11.	6 oz	(a) 1.7 g	(b) 17 g	(c) 170 g
12.	5 kg	(a) 2 lb	(b) 5 lb	(c) 10 lb
13.	50 ℓ	(a) 12 gal	(b) 120 gal	(c) 50 gal
14.	6 qt	(a) 2 ℓ	(b) 3 ℓ	(c) 6 ℓ

15.	14 sq ft	(a) 1.3 m^2	(b) 13 m^2	(c) 130 m^2
16.	8 sq in.	(a) 50 cm^2	(b) 5 cm^2	(c) 500 cm^2
17.	100° F	(a) 38°C	(b) 212° C	(c) 32° C
18.	60° C	(a) 20° F	(b) 140° F	(c) 100° F
19.	212° F	(a) 100° C	(b) 400° C	(c) 50° C
20.	0° C	(a) 100° F	(b) 32° F	(c) −30° F

B. **Think Metric.** Choose the closest estimate.

1.	Diameter of a penny	(a) 3 cm	(b) 1.5 cm	(c) 5 cm
2.	Length of a man's foot	(a) 3 m	(b) 3 cm	(c) 30 cm
3.	Tank of gasoline	(a) 50 ℓ	(b) 5 ℓ	(c) 500 ℓ
4.	Volume of a wastebasket	(a) 10 ℓ	(b) 100 ℓ	(c) 10 cc
5.	One-half gallon of milk	(a) 1 ℓ	(b) 2 ℓ	(c) $\frac{1}{2}$ ℓ
6.	Hot day in Phoenix, Arizona	(a) 100° C	(b) 30° C	(c) 45° C
7.	Cold day in Minnesota	(a) −80° C	(b) −10° C	(c) 20° C
8.	200 lb barbell	(a) 400 kg	(b) 100 kg	(c) 40 kg
9.	Length of paper clip	(a) 10 cm	(b) 35 mm	(c) 3 mm
10.	Your height	(a) 17 m	(b) 17 cm	(c) 1.7 m
11.	Your weight	(a) 80 kg	(b) 800 kg	(c) 8 kg
12.	Boiling water	(a) 100° C	(b) 212° C	(c) 32° C

C. Convert to the units shown. (Round to one decimal place.)

1. 3.0 in. = ________ cm
2. 4.0 ft = ________ cm
3. 2.0 mi = ________ km
4. 20 ft 6 in. = ________ m
5. $9\frac{1}{4}$ in. = ________ cm
6. 30 kg = ________ lb
7. 152 lb = ________ kg
8. 3 lb 4 oz = ________ kg
9. 8.5 kg = ________ lb
10. 10.4 oz = ________ g
11. 3.0 gal = ________ ℓ
12. 2.5 *l* = ________ qt
13. 6.5 qt = ________ ℓ
14. 62 mph = ________ kmh
15. 8.0 kmh = ________ mph
16. 208° F = ________ ° C
17. 10° C = ________ ° F
18. 35 in./sec = ________ cm/sec
19. 64 ft/sec = ________ m/sec
20. 2.0 m = ________ ft

D. Practical Problems. (Round as indicated.)

1. Welding electrode sizes are presently given in inches. Convert the following set of electrode sizes to millimeters. (1 in. = 25.4 mm)

in.	mm
0.030	
0.035	
0.040	
0.045	

in.	mm
$\frac{1}{16}$	
$\frac{5}{64}$	
$\frac{3}{32}$	
$\frac{1}{8}$	

in.	mm
$\frac{5}{32}$	
$\frac{3}{16}$	
$\frac{3}{8}$	
$\frac{11}{64}$	

(Round to the nearest one-thousandth of a mm.)

2. One cubic foot of water weighs 62.4 lb. (a) What is the weight in pounds of 1 liter of water? (b) What is the weight in kilograms of one cubic foot of water? (c) What is the weight in kilo·rams of one liter of water? (Round to one decimal place.)

3. At the Munich Olympic Games in 1972 Frank Shorter of the United States won the 26.2 mile marathon race in 2 hr 12 min 19.8 sec. What was his average speed in mph and km/hr? (Round to nearest tenth.)

4. The metric unit of pressure is the *pascal,* where $1 \frac{\text{lb}}{\text{in.}^2}$ (psi) $\cong$ 6894 pascal, and $1 \frac{\text{lb}}{\text{ft}^2}$ (psf) $\cong$ 47.88 pascal.

   Convert the following pressures to metric units:

   (a) 15 psi = ________ p (b) 100 psi = ________ p

   (c) 40 psf = ________ p (d) 2500 psf = ________ p

   (Round to the nearest thousand pascals.)

5. Find the difference between

   (a) 1 mile and 1 kilometer ________ km

   (b) 3 miles and 5000 meters ________ ft

   (c) 120 yds and 110 meters ________ ft

   (d) 1 quart and 1 liter ________ qt

   (e) 10 lb and 5 kg ________ lb

6. The Mamiya RB 67 camera uses 120 film and produces $2\frac{1}{4}$-inch by $2\frac{3}{4}$-inch negatives. How would this negative size be given in millimeters in Japan?

7. Which is performing more efficiently, a car getting 12 km per liter of gas or one getting 25 miles per gallon of gas?

8. Industrial paint sells for $5.25 per gallon. What would 1 liter cost?

9. In flooring and roofing, the unit "one square" is sometimes used to mean 10 sq ft. Convert this unit to the metric system. (Round to 1 decimal place.)

   1 square = ________ sq m

10. A *cord* of wood is a volume of wood 8 ft long, 4 ft wide, and 4 ft high. (a) What are the equivalent metric dimensions for a "metric cord" of wood? (b) What is the volume of a cord in cubic meters? (Round to the nearest tenth.)

11. Translate these well-known phrases into metric units.

(a) A miss is as good as ______ km. (1 mile = ______ km)

(b) A ______ cm TV screen. (24 in. = ______ cm)

(c) An ______ cm by ______ cm sheet of paper. ($8\frac{1}{2}$ in. = ______ cm 11 in. = ______ cm)

(d) He was beaten within ______ cm of his life. (1 in. = ______ cm)

(e) Race in the Indy ______ km. (500 miles = ______ km)

(f) Take it with ______ grams of salt. (*Hint:* 437.5 grain = 1 oz.)

(g) ______ grams of prevention is worth ______ grams of cure. (1 oz = ______ g; 1 lb = ______ g)

(h) See the world in ______ grams of sand. (1 grain = ______ gram)

(i) Peter Piper picked ______ liters of pickled peppers. (*Hint:* 1 peck = 8 quarts.)

(j) A cowboy in a ______ liter hat. (10 gal = ______ liter)

(k) He's in ______ ml of trouble. (1 peck = ______ ml)

12. In an automobile, cylinder displacement is measured in either cubic inches or liters. What is the cylinder displacement in liters of a 400 cubic inch engine? (Round to the nearest liter.)

When you have completed these exercises, check your answers on page 550, then turn to **35** to learn about direct measurement.

5-4 DIRECT MEASUREMENTS

35 The simplest kind of measuring device to use is one in which the output is a digital display. For example, in a digital clock or digital volt-meter, the measured quantity is translated into electrical signals, and each digit of the measurement number is displayed.

A digital display is easy to read, but most technical work involves direct measurement where the numerical reading is displayed on some sort of scale. Reading the scale correctly requires skill and the ability to estimate or *interpolate* accurately between scale divisions.

The easiest scales to read are those used in home or industry to measure amounts of electricity, gas, or water. For example, the typical electrical meter shown measures the amount of electrical energy used in kilowatt-hr (kwh).

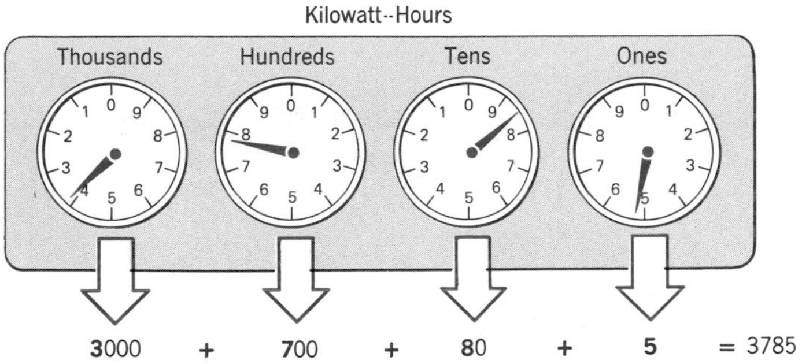

Because of the gear arrangement, adjacent meter pointers rotate in opposite directions.

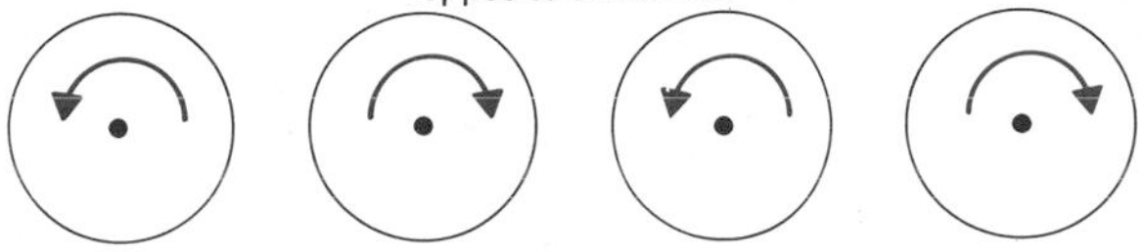

Don't let it confuse you.

On this meter the dial on the right records from 0 to 10 kilowatt-hours. The other dials, moving from right to left, record tens of kwh, hundreds of kwh, and thousands of kwh. To read the meter, record the dial numbers from left to right, always selecting the smaller number when the pointer is between two numbers. The meter above reads 3785 kwh.

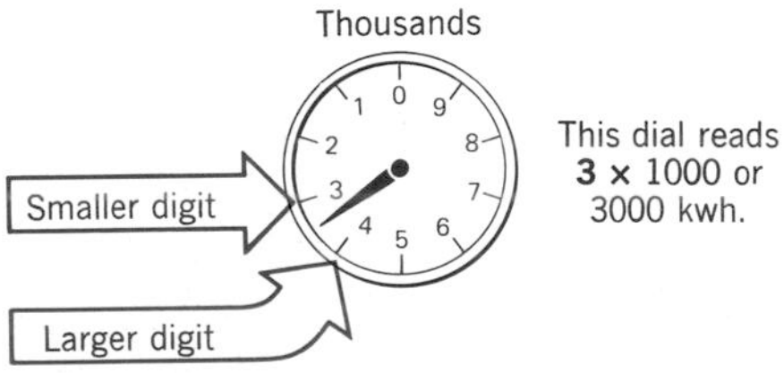

For practice, read the following kwh meters.

1.

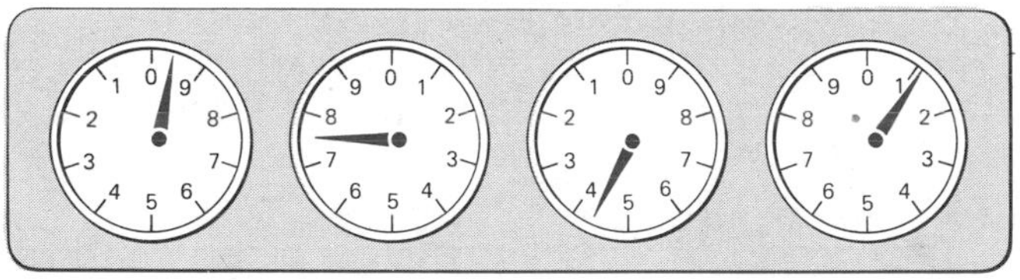

2.

3.

4.

5.

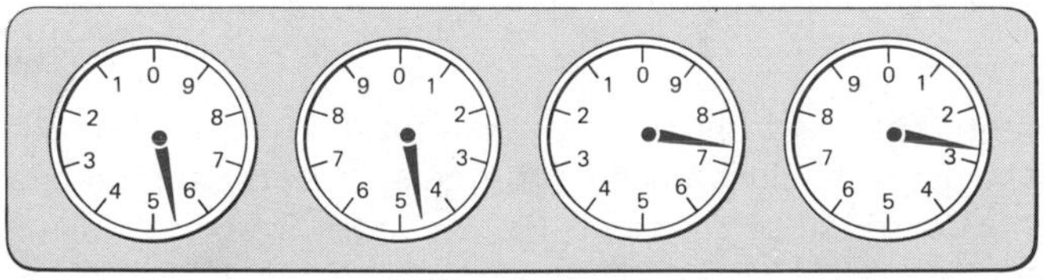

6.

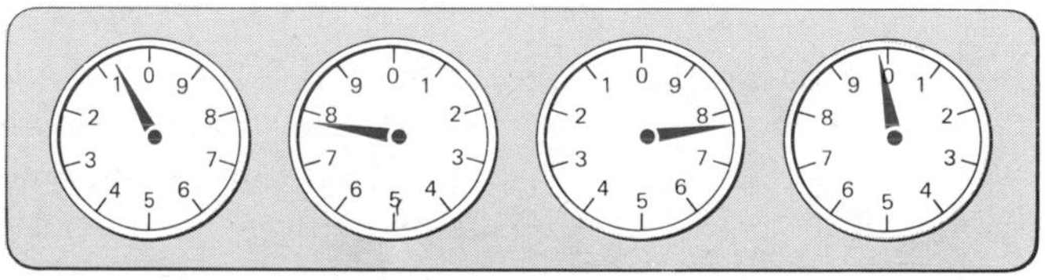

Check your answers in **36**.

---

36 1. 9741 kwh 2. 6006 kwh 3. 9024 kwh
4. 2820 kwh 5. 5472 kwh 6. 779 kwh

Natural gas use is usually measured in units of 100 cubic feet, abbreviated CCF. A common type of gas meter is shown below.

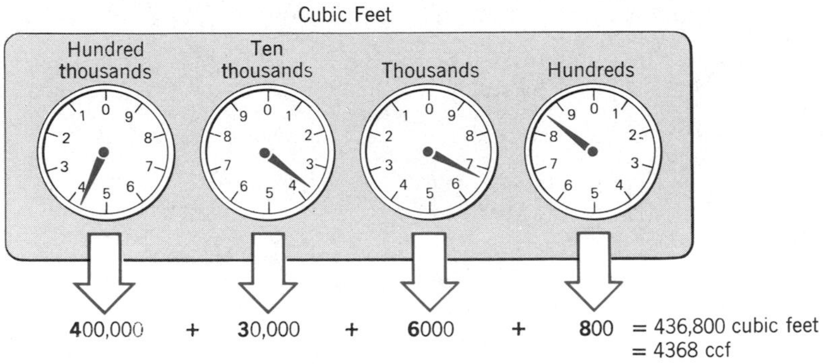

Each scale marking on the dial at the right represents 100 cubic foot of gas used or 1 CCF. One complete revolution of this dial represents 10 CCF of gas used. Again, record the smaller number when the pointer is between two scale marks.

In most models of natural gas meter a fifth dial is included and used to test the accuracy of the meter. It is not used in reading the meter.

Read the following gas meters. (Give your answer in cubic feet.)

1.

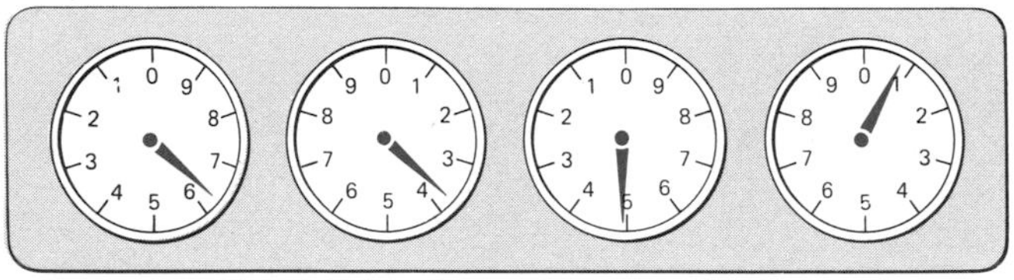

2.

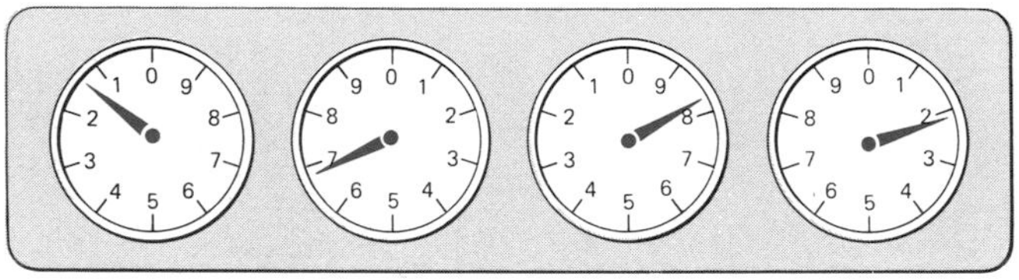

3.

4.

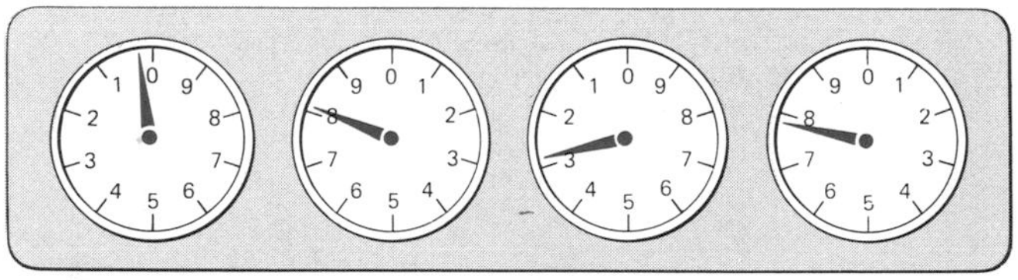

Check your readings in **37**.

---

37 1. 635,000 cubic feet 2. 168,200 cubic feet
3. 304,600 cubic feet 4. 82,700 cubic feet

There is no need to estimate the exact position of the pointer on the electric or gas meters shown above. The meters give an automatic digit-by-digit readout.

Length Measurements

Length measurement involves many different instruments that produce either fraction or decimal number readouts. Carpenters, electricians, plumbers, roofers, machinists, all use rulers, yard sticks, or other devices to make length measurements. Each instrument is made with a different scale that determines the accuracy of the measurement that can be made. The more accurate the measurement has to be, the smaller the scale divisions needed.

For example, on the carpenter's rule below, scale A is graduated in eighths of an inch. Scale B is graduated in sixteenths of an inch.

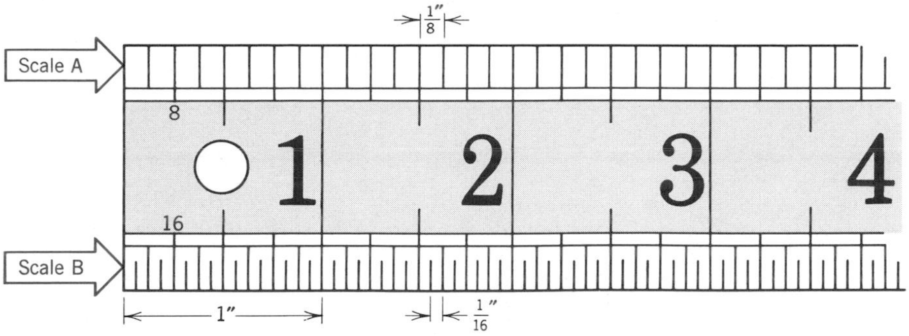

While a carpenter or roofer may need to measure no closer than $\frac{1}{8}''$ or $\frac{1}{16}''$, a machinist will often need a scale graduated in 32nds or 64ths of an inch. The finely engraved steel rule shown below is a machinist's rule.

Some machinists' rules are marked in tenths or hundredths of an inch rather than in the usual divisions of $\frac{1}{2}$, $\frac{1}{4}$, $\frac{1}{8}$, $\frac{1}{16}$, and so on.

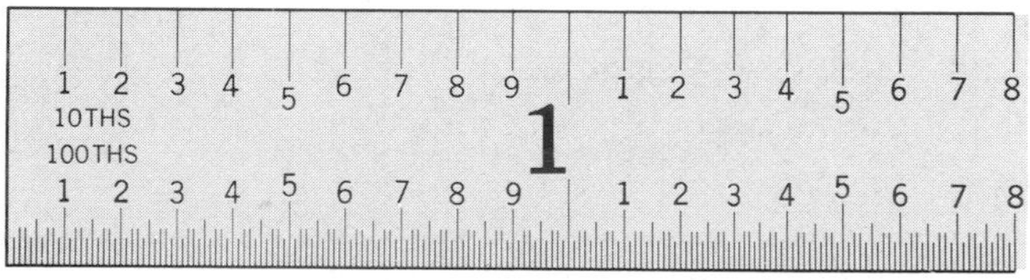

Notice that for all of these rules only a few scale markings are actually labeled with numbers. It would be easy to read a scale marked like this:

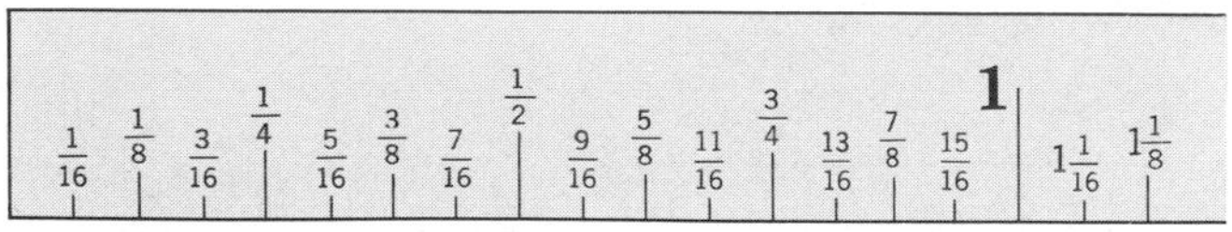

Any useful scale has divisions so small that this kind of marking is impossible. An expert in any technical field must be able to identify the graduations on the scale correctly even though the scale is *not* marked. To use any scale, first determine the size of the smallest division by counting the number of spaces in a one-inch interval. Here is an example:

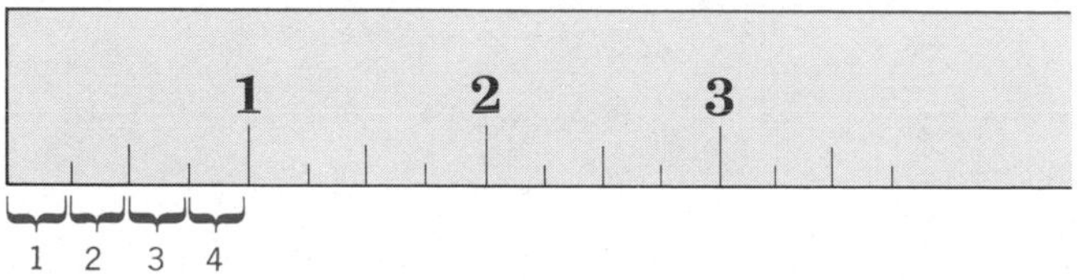

There are four spaces in the first one-inch interval. The smallest division is therefore $\frac{1}{4}''$ and we can label the scale like this:

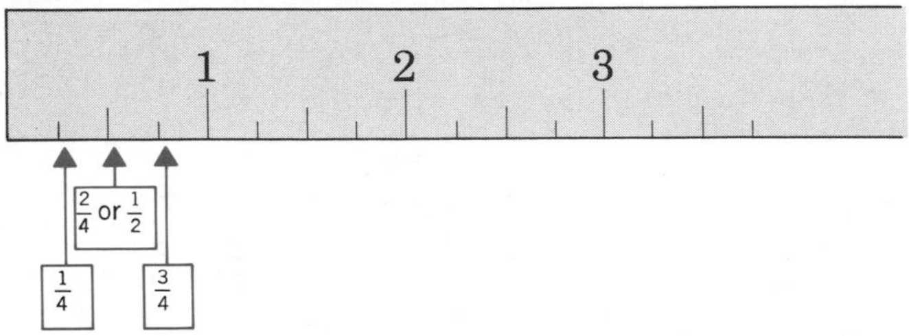

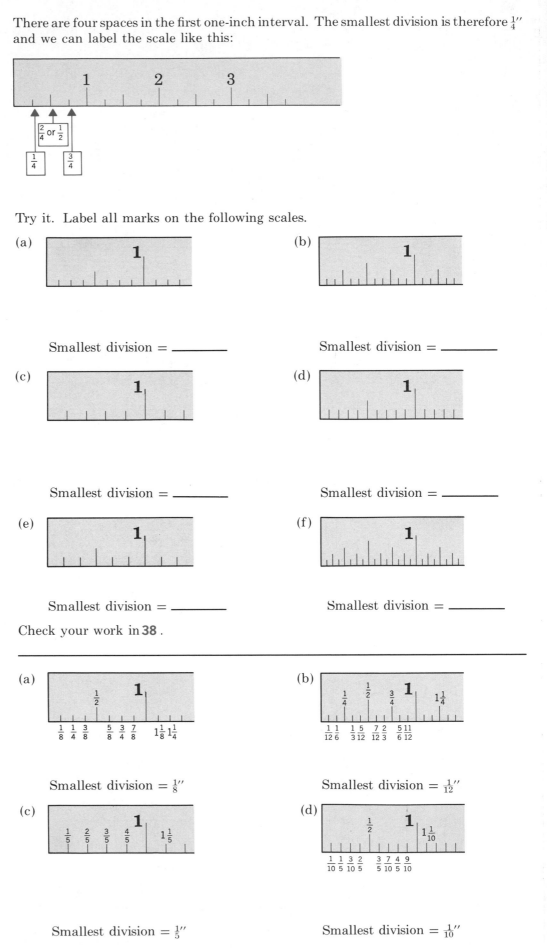

Try it. Label all marks on the following scales.

(a)

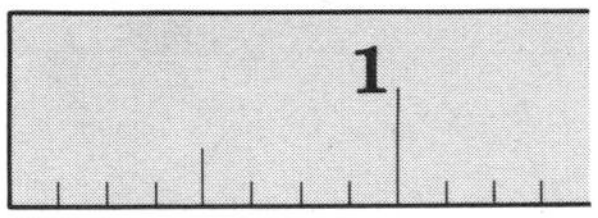

Smallest division = ______

(b)

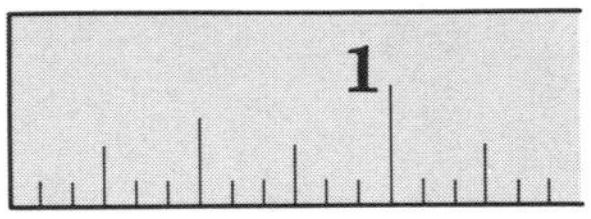

Smallest division = ______

(c)

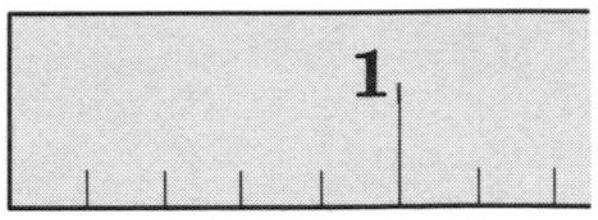

Smallest division = ______

(d)

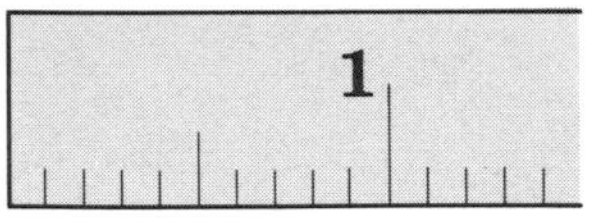

Smallest division = ______

(e)

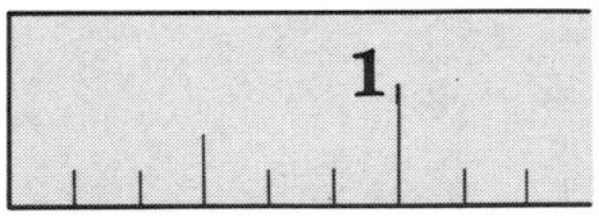

Smallest division = ______

(f)

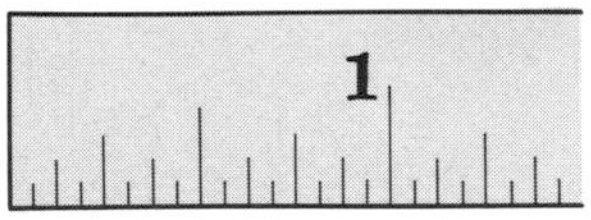

Smallest division = ______

Check your work in **38**.

---

38 (a)

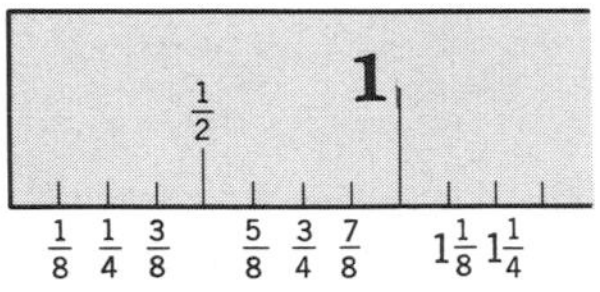

Smallest division = $\frac{1}{8}''$

(b)

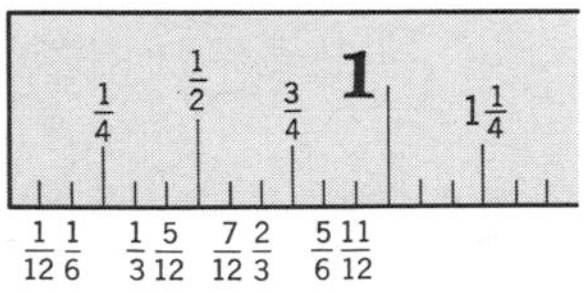

Smallest division = $\frac{1}{12}''$

(c)

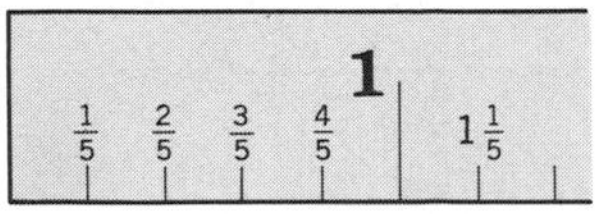

Smallest division = $\frac{1}{5}''$

(d)

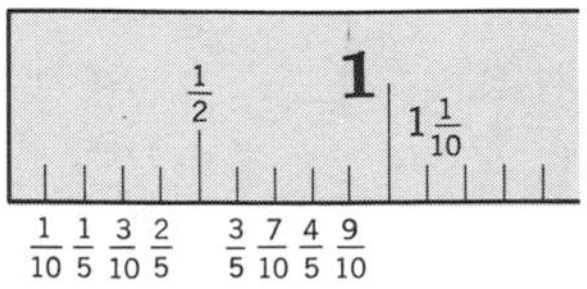

Smallest division = $\frac{1}{10}''$

(e)

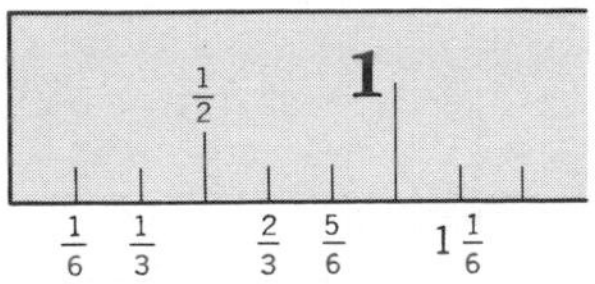

Smallest division = $\frac{1}{6}''$

(f)

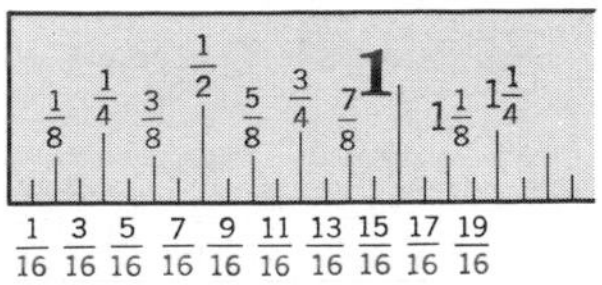

Smallest division = $\frac{1}{16}''$

To use the rule once you have mentally labeled all scale markings, first count inches, then count the numbers of smallest divisions from the last whole-inch mark.

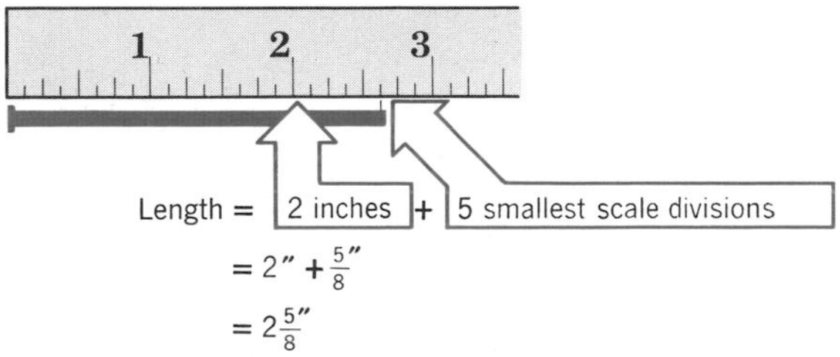

If the length of the object being measured falls between scale divisions, record the nearest scale division. For example, in this measurement

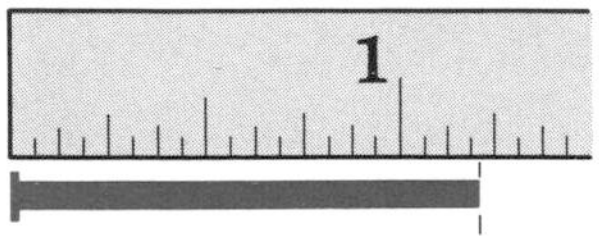

the end point of the object falls between $1\frac{3}{16}''$ and $1\frac{4}{16}''$.

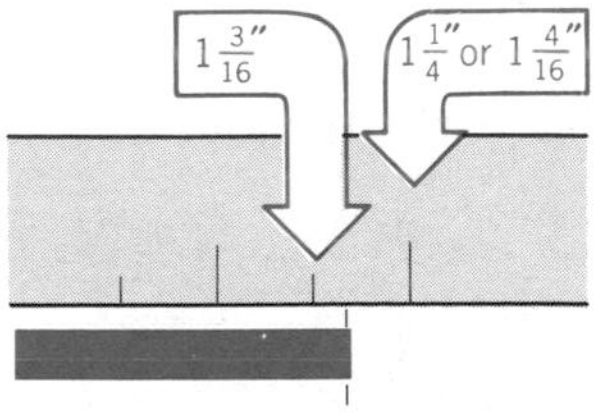

To the nearest scale division, the length is $1\frac{3}{16}''$. When a scale with $\frac{1}{16}''$ divisions is used, we usually write the measurement to the nearest $\frac{1}{16}''$, but we may estimate it to the nearest half of a scale division or $\frac{1}{32}''$.

For practice in using length rulers, find the dimensions given on the following rules.

1.

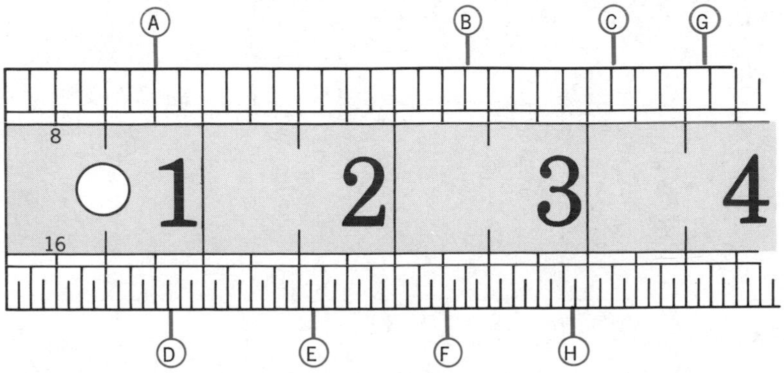

2.

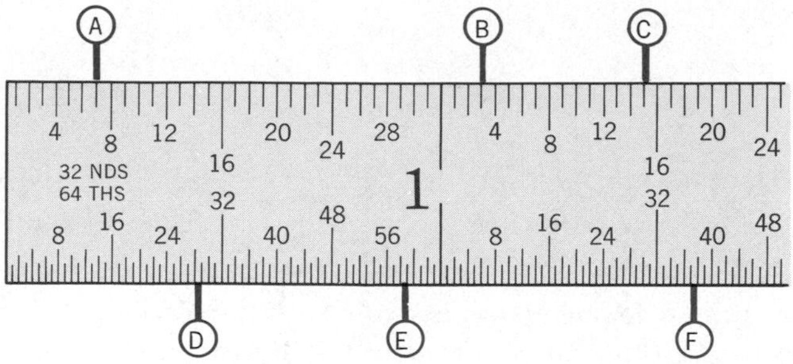

3.

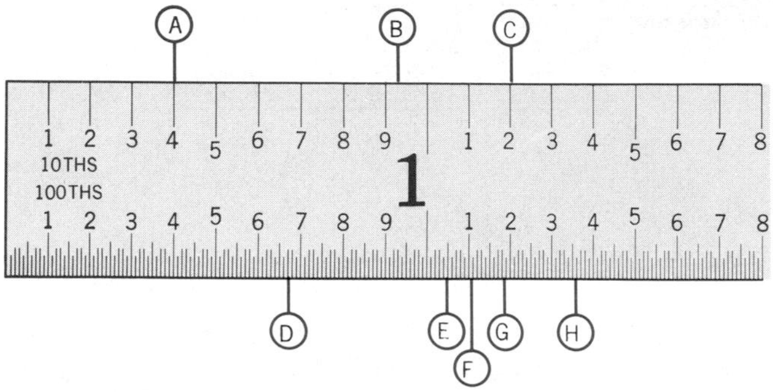

Check your answers in **39**.

---

39 1. A. $\frac{3}{4}''$ B. $2\frac{3}{8}''$ C. $3\frac{1}{8}''$ D. $\frac{13}{16}''$
E. $1\frac{9}{16}''$ F. $2\frac{1}{4}''$ G. $3\frac{5}{8}''$ H. $2\frac{15}{16}''$

2. A. $\frac{7}{32}''$ B. $1\frac{3}{32}''$ C. $1\frac{15}{32}''$
D. $\frac{28}{64}'' = \frac{7}{16}''$ E. $\frac{29}{32}''$ F. $1\frac{37}{64}''$

3. A. $\frac{4}{10}'' = 0.4''$ B. $\frac{9}{10}'' = 0.9''$ C. $1\frac{2}{10}'' = 1.2''$
D. $\frac{67}{100}'' = 0.67''$ E. $1\frac{4}{100}'' = 1.04''$ F. $1\frac{10}{100}'' = 1.10''$
G. $1\frac{18}{100}'' = 1.18''$ H. $1\frac{35}{100}'' = 1.35''$

Micrometers

A steel rule can be used to measure lengths of $\pm\frac{1}{32}''$ or $\pm\frac{1}{64}''$, or as fine as $\pm\frac{1}{100}''$. To measure lengths with greater accuracy than this, more specialized instruments are needed. A *micrometer* can be used to measure lengths to $\pm 0.001''$, one one-thousandth of an inch.

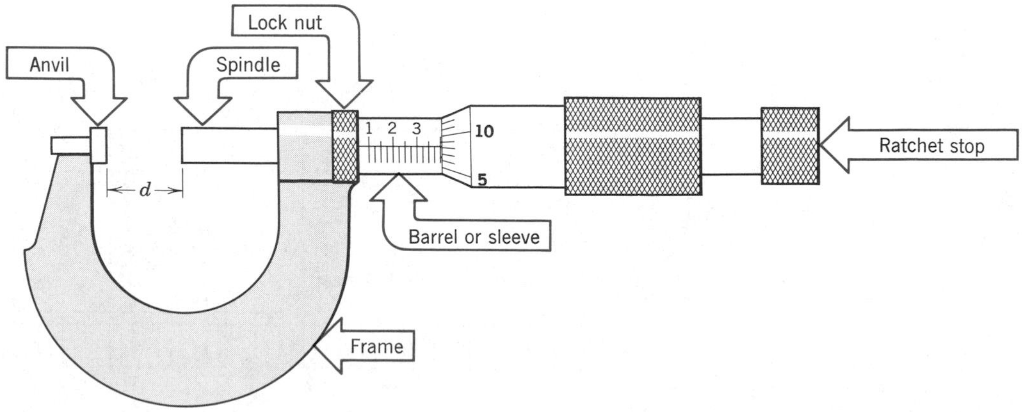

Two scales are used on a micrometer. The first scale, marked on the sleeve, records the movement of a screw machined accurately to 40 threads per inch. The smallest divisions on this scale record movements of the spindle of one-fortieth of an inch or 0.025″.

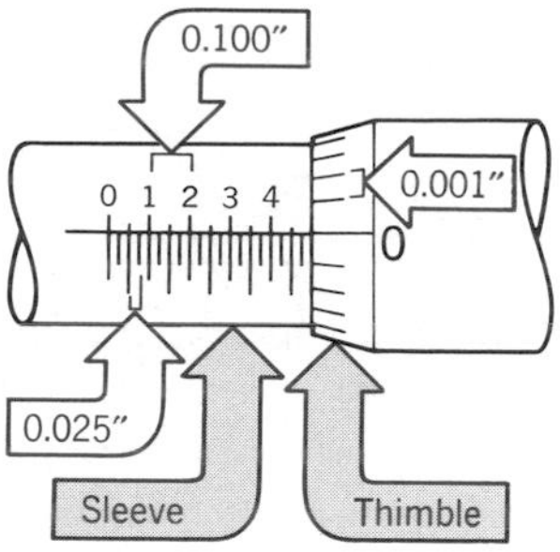

The second scale is marked on the rotating spindle. One complete turn of the thimble advances the screw one turn or $\frac{1}{40}''$. The thimble scale has 25 divisions, so that each mark on the thimble scale represents a spindle movement of $\frac{1}{25} \times \frac{1}{40}'' = \frac{1}{1000}''$ or 0.001″.

To read a micrometer, follow these steps.

Step 1 Read the largest numeral visible on the sleeve and multiply this number by 0.100″.

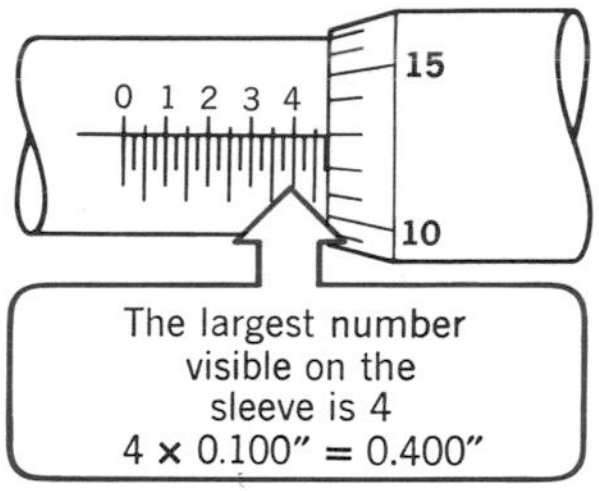

Step 2 Read the number of additional scale spaces visible on the sleeve. (This will be 0, 1, 2, or 3.) Multiply this number by 0.025″.

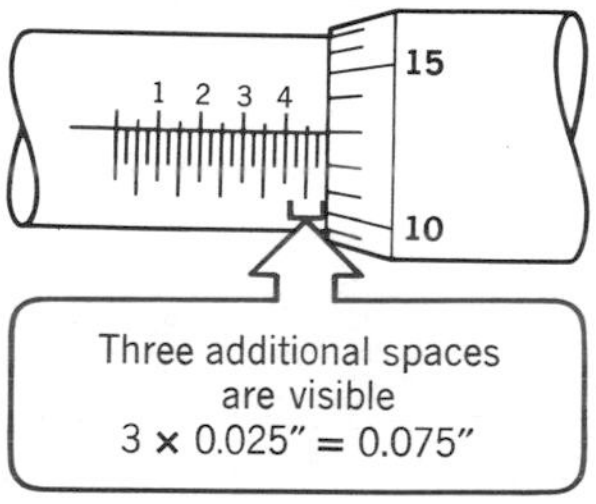

Step 3 Read the number on the thimble scale opposite the horizontal line on the sleeve and multiply this number by 0.001″.

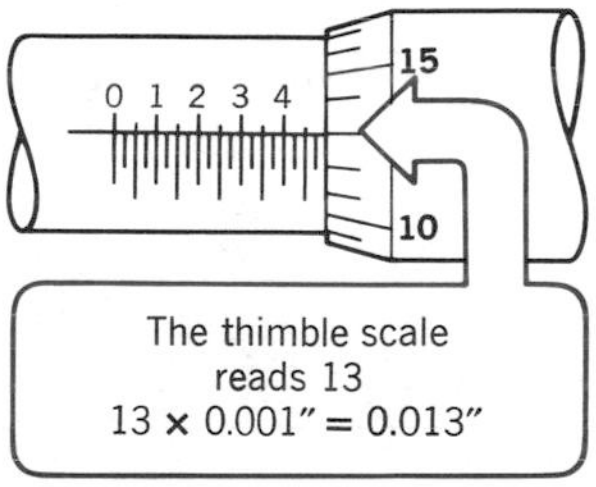

Step 4 Add these three products to find the measurement.

$$\begin{array}{r} 0.400'' \\ 0.075'' \\ +0.013'' \\ \hline 0.488'' \end{array}$$

Here are more examples.

Example 1:

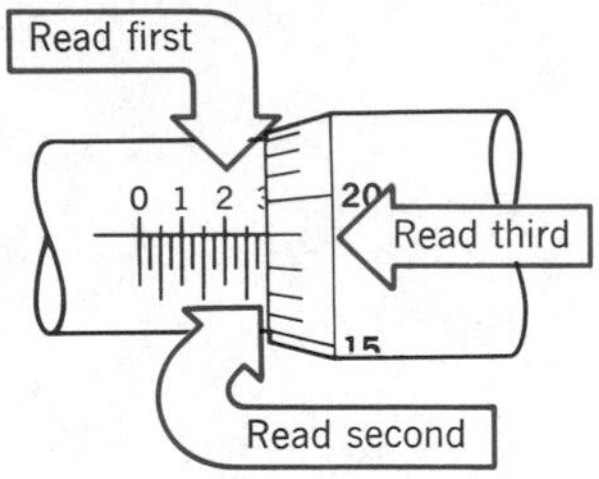

Step 1 The largest number completely visible on the sleeve is 2. $2 \times 0.100'' = 0.200''$

Step 2 Three additional spaces are visible on the sleeve. $3 \times 0.025'' = 0.075''$

Step 3 The thimble scale reads 19. $19 \times 0.001'' = 0.019''$

Step 4 Add:

$$\begin{array}{r} 0.200'' \\ 0.075'' \\ \underline{0.019''} \\ \text{length} = 0.294'' \end{array}$$

Example 2:

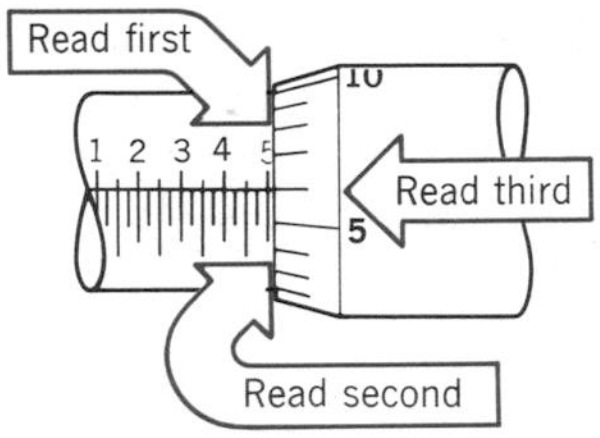

Step 1 $5 \times 0.100'' = 0.500''$

Step 2 No additional spaces are visible on the sleeve. $0 \times 0.025'' = 0.000''$

Step 3 $6 \times 0.001'' = 0.006''$

Step 4

$$\begin{array}{r} 0.500'' \\ 0.000'' \\ \underline{0.006''} \\ \text{length} = 0.506'' \end{array}$$

For practice, read the following micrometers.

1. 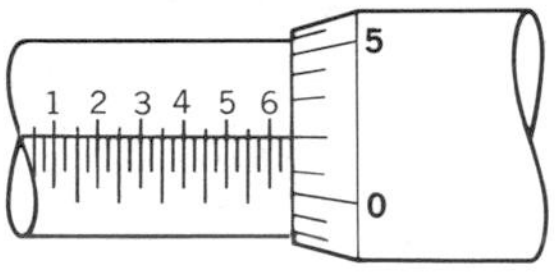

2. 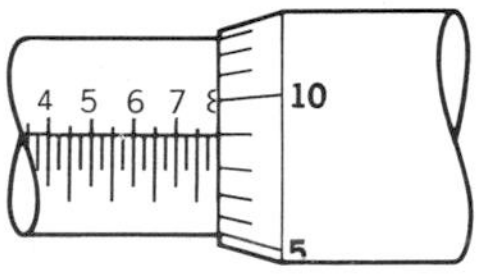

3. 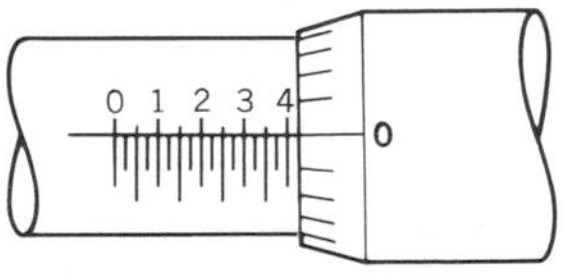

4. 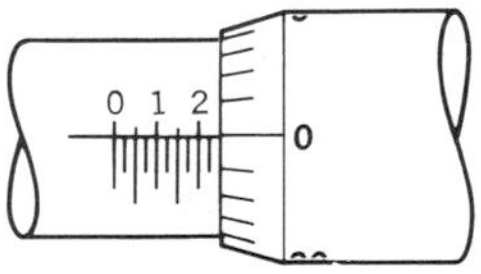

5. 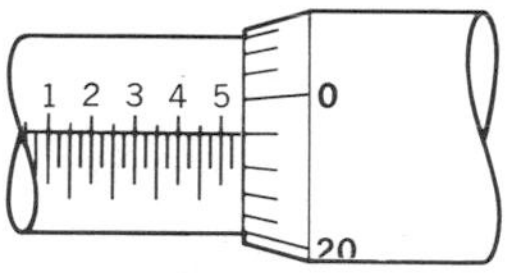

6. 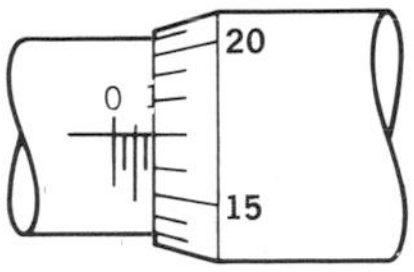

7. 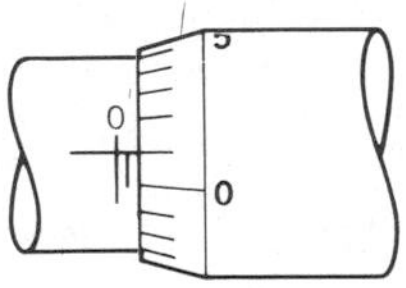

8. 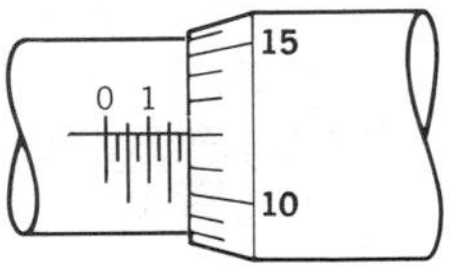

9. 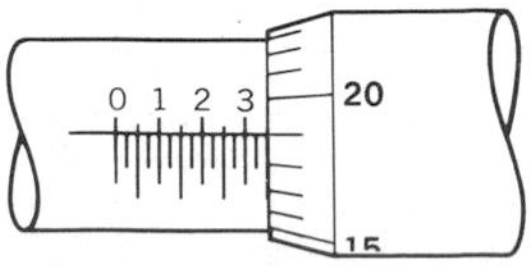

10. 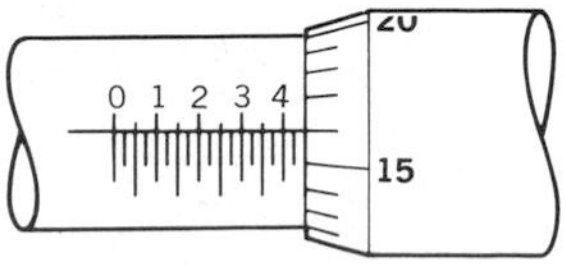

Check your micrometer readings in **40**

40 1. 0.652″ 2. 0.809″ 3. 0.425″ 4. 0.250″ 5. 0.549″
6. 0.092″ 7. 0.026″ 8. 0.187″ 9. 0.344″ 10. 0.441″

Metric micrometers are easier to read. Each small division on the sleeve represents 0.5 mm. The thimble scale is divided into 50 spaces; therefore, each division is $\frac{1}{50} \times 0.5 \text{ mm} = \frac{1}{100} \text{ mm}$ or 0.01 mm.

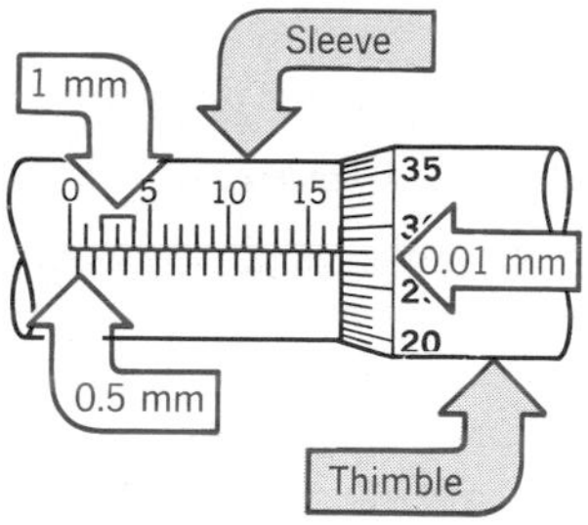

Turning the thimble scale of a metric micrometer by one scale division advances the spindle one one-hundredth of a millimeter, or 0.001 cm.

To read a metric micrometer, follow these steps.

Step 1 Read the largest mark visible on the sleeve and multiply this number by 1 mm.

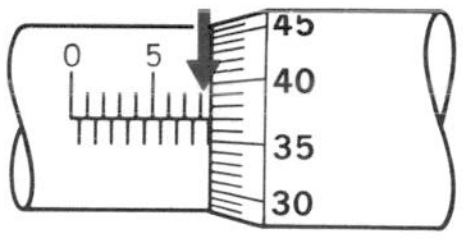

Step 2 Read the number of additional half spaces visible on the sleeve. (This number will be either 0 or 1.) Multiply this number by 0.5 mm.

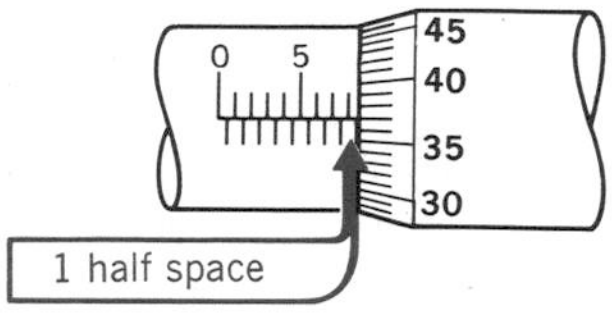

$1 \times 0.5 = 0.5$

Step 3 Read the number on the thimble scale opposite the horizontal line on the sleeve and multiply this number by 0.01 mm.

The thimble scale reads 37.
$37 \times 0.01 \text{ mm} = 0.37 \text{ mm}$

Step 4 Add these three products to find the measurement value.

8.00 mm
0.50 mm
0.37 mm
8.87 mm

Another example:

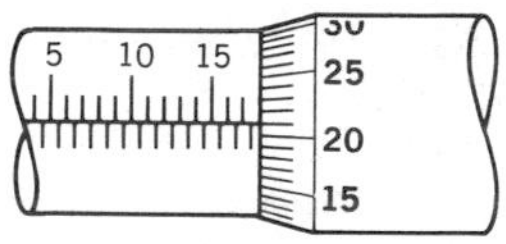

$17 \times 1 \text{ mm} = 17 \text{ mm}$
$1 \times 0.5 \text{ mm} = 0.5 \text{ mm}$
$21 \times 0.01 \text{ mm} = 0.21 \text{ mm}$
$\text{length} = 17 \text{ mm} + 0.5 \text{ mm} + 0.21 \text{ mm}$
$= 17.71 \text{ mm}$

Read these metric micrometers.

1. 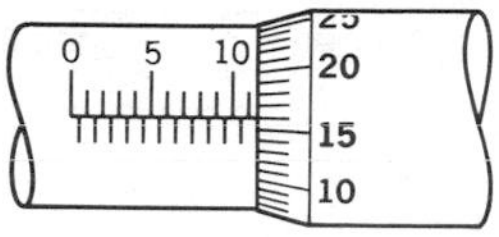

2. 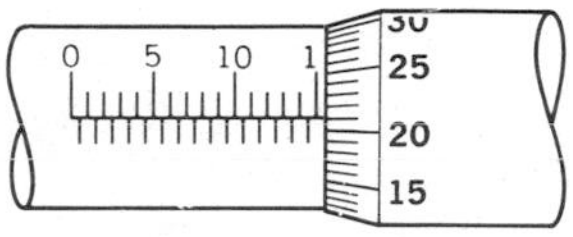

3. 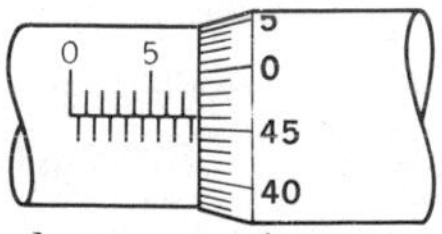

4. 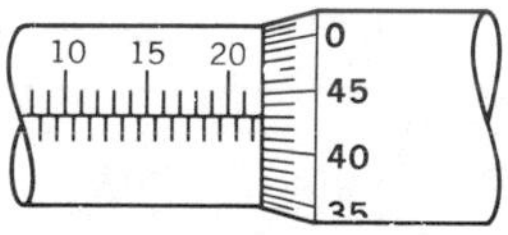

Check your micrometer readings in **41**.

41
1. 11.16 mm　　2. 15.21 mm
3. 7.96 mm　　4. 21.93 mm

When more accurate length measurements are needed, the *vernier* micrometer can be used to provide an accuracy of $\pm 0.0001''$.

Vernier Micrometers

On the vernier micrometer, a third or vernier scale is added to the sleeve. This new scale appears as a series of ten lines parallel to the axis of the sleeve above the usual scale.

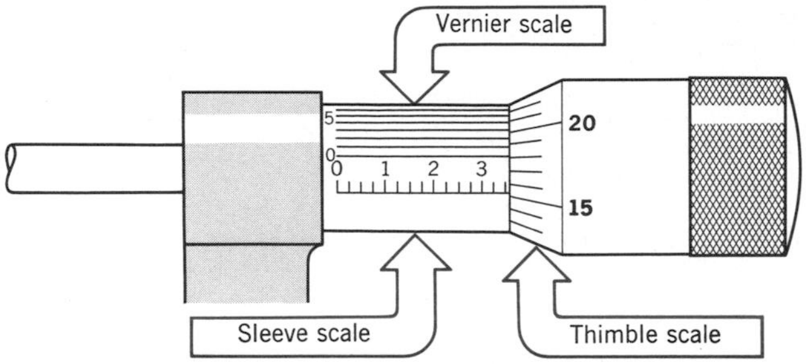

The ten vernier marks cover a distance equal to nine thimble scale divisions—as shown on this spread-out diagram of the three scales.

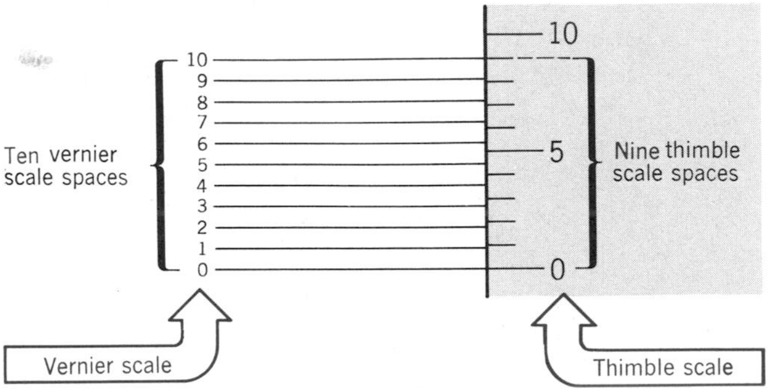

The difference between a vernier scale division and a thimble scale division is one-tenth of a thimble scale division, or $\frac{1}{10}$ of $0.001'' = 0.0001''$.

To read the vernier micrometer, follow the first three steps given for the ordinary micrometer, then

Step 4 Find the number of the line on the vernier scale that exactly lines up with any line on the thimble scale. Multiply this number by $0.0001''$ and add this amount to the other three distances.

For example, for the vernier micrometer shown,

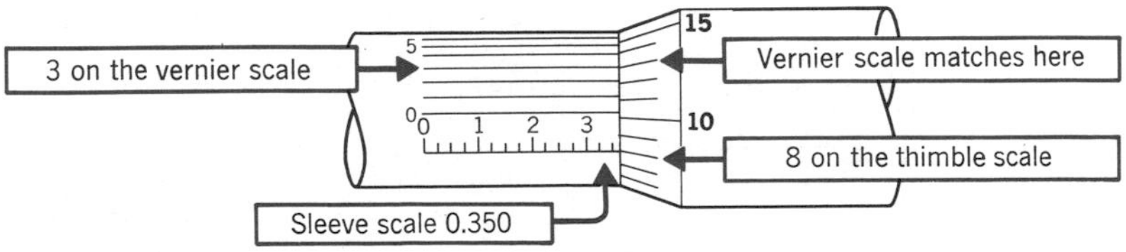

sleeve scale:	$3 \times 0.100''$	$= 0.300''$
	$2 \times 0.025''$	$= 0.050''$
thimble scale:	$8 \times 0.001''$	$= 0.008''$
vernier scale:	$3 \times 0.0001''$	$= \underline{0.0003''}$
	length	$= 0.3583''$

When adding these four lengths be careful to align the decimal digits correctly.

The vernier scale provides a way of measuring accurately the distance to an additional tenth of the finest scale division.

Now for some practice in using the vernier scale, read each of the following.

(a)

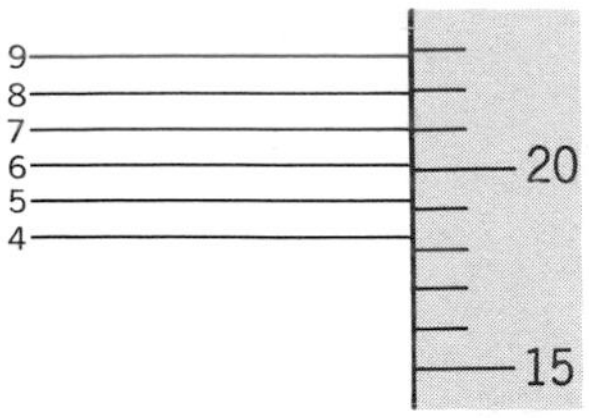

(b)

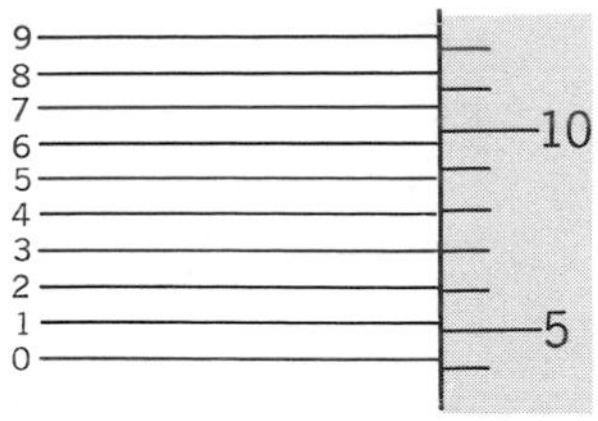

(c)

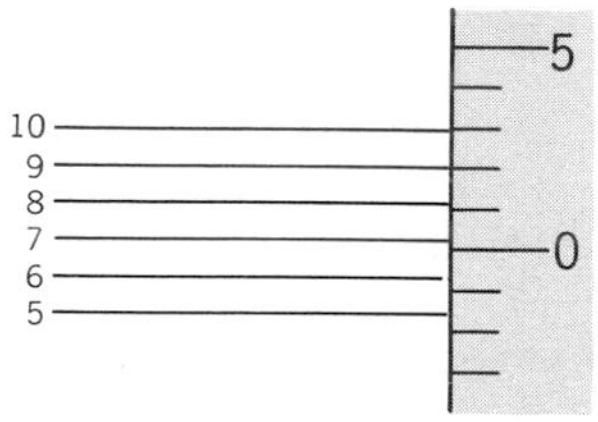

(d)

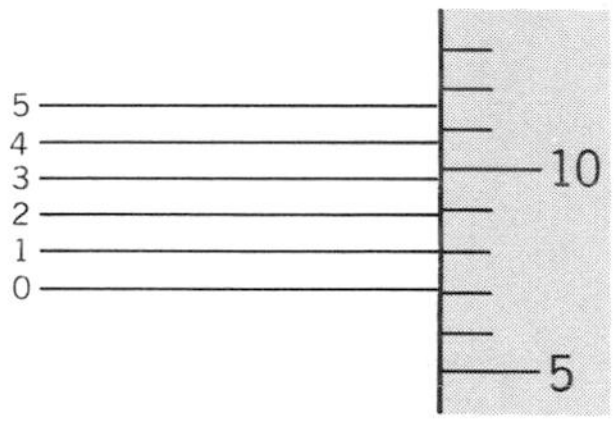

(e)

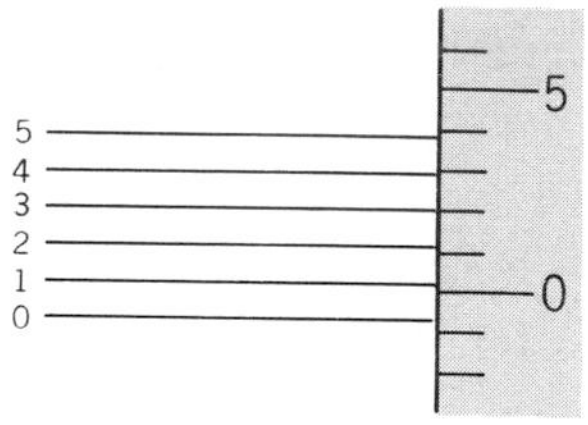

(f)

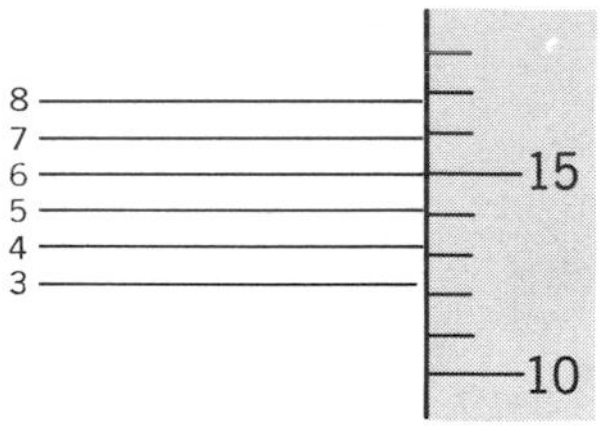

Check your vernier readings in **42**

---

42 (a) 7 (b) 3 (c) 9
(d) 1 (e) 3 (f) 6

Now read the following vernier micrometers.

(a)

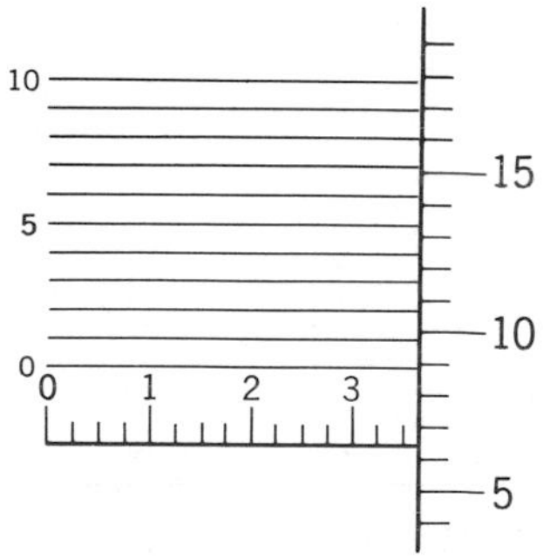

(b)

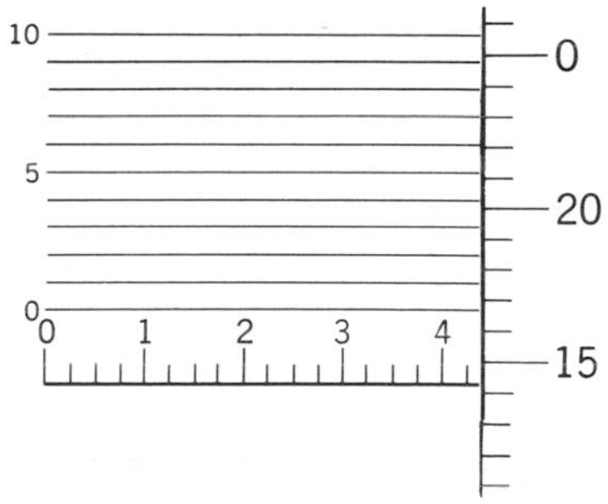

(c)

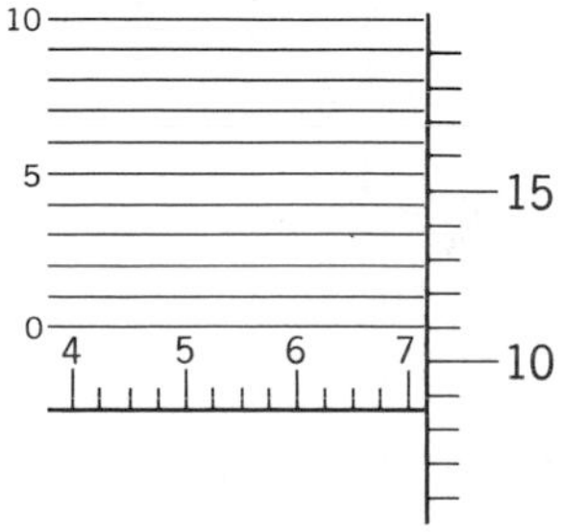

(d)

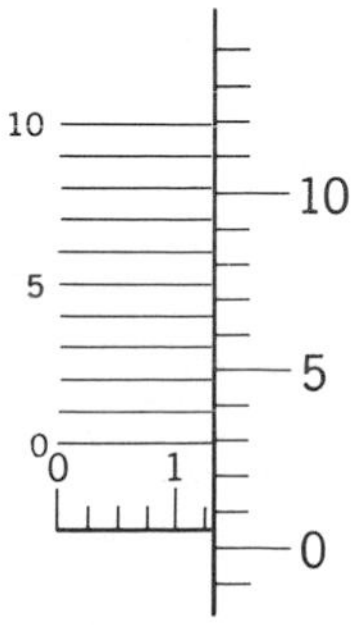

(e)

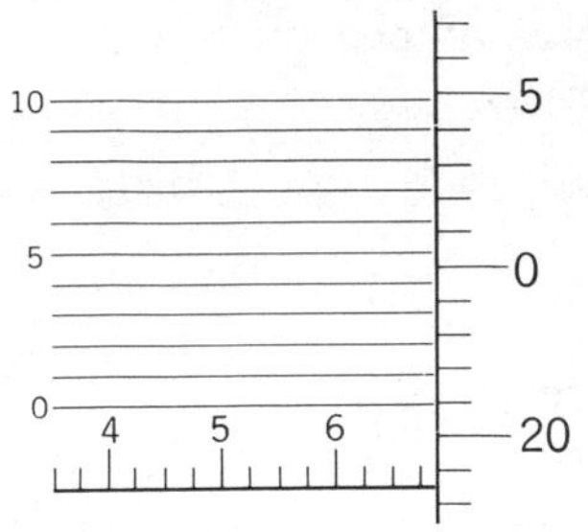

(f)

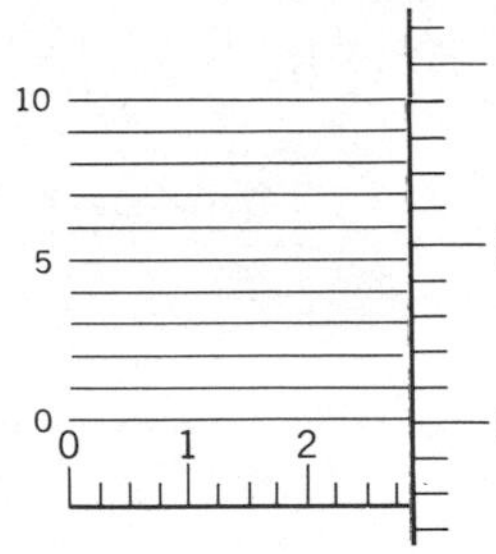

Check your micrometer readings in **43**.

---

43 (a) 0.3568″ (b) 0.4397″ (c) 0.7080″
(d) 0.1259″ (e) 0.6939″ (f) 0.2971″

Vernier Calipers

The vernier caliper is another length measuring instrument that uses the vernier principle and can measure lengths to ±0.001″.

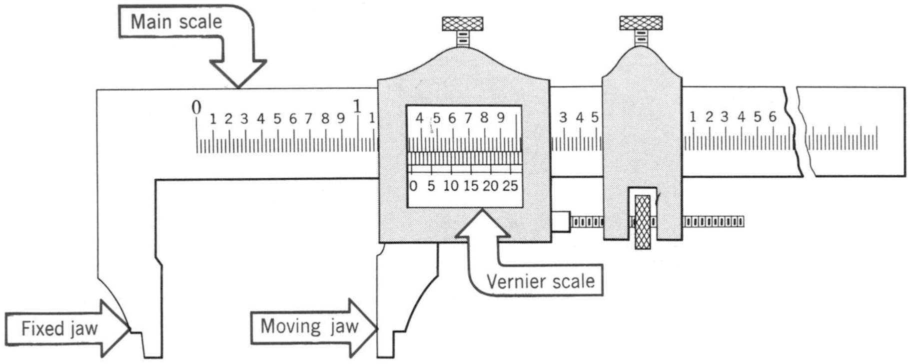

Notice that on this instrument each inch of the main scale is divided into 40 parts, so that each smallest division on the scale is 0.025″. Every fourth division or tenth of an inch is numbered. The numbered marks represent 0.100″, 0.200″, 0.300″, and so on.

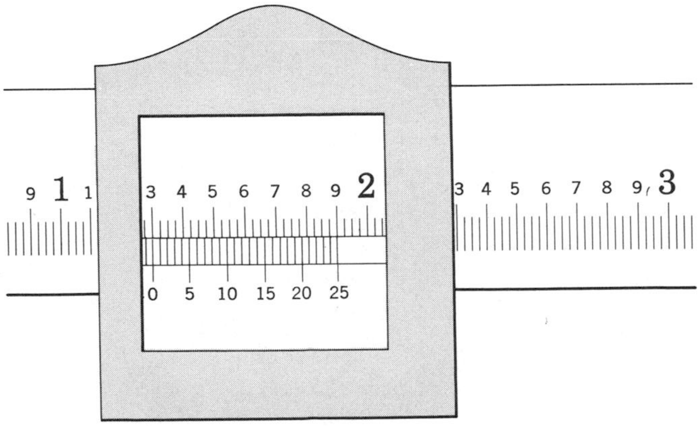

The sliding vernier scale is divided into 25 equal parts numbered by 5s. These 25 divisions on the vernier scale cover the same length as 24 divisions on the main scale.

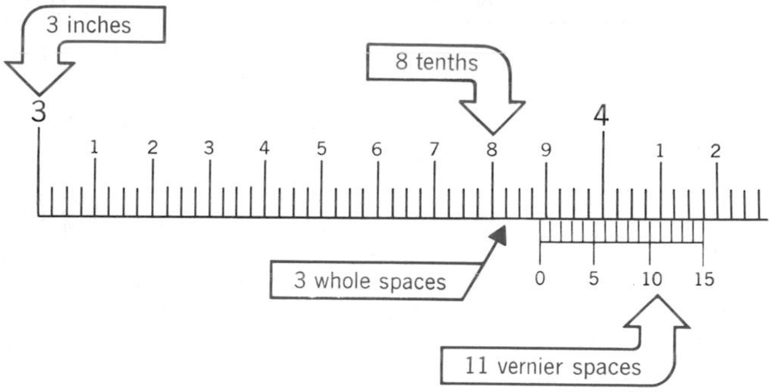

To read a vernier caliper, follow these steps.

Step 1 Read the number of whole-inch divisions on the main scale to the left of the vernier zero. This gives the number of whole inches.

Step 2 Read the number of tenths on the main scale to the left of the vernier zero. Multiply this number by 0.100″.

8 tenths
$8 \times 0.100 = 0.800''$

Step 3 Read the number of additional whole spaces on the main scale to the left of the vernier zero. Multiply this number by 0.025″.

3 spaces
$3 \times 0.025'' = 0.075''$

Step 4 On the sliding vernier scale, find the number of the line that exactly lines up with any line on the main scale. Multiply this number by 0.001″.

11 lines up exactly.
$11 \times 0.001'' = 0.011''$

Step 5 Add the four numbers.

$$\begin{array}{r} 3.000'' \\ 0.800'' \\ 0.075'' \\ +0.011'' \\ \hline 3.886'' \end{array}$$

Another example:

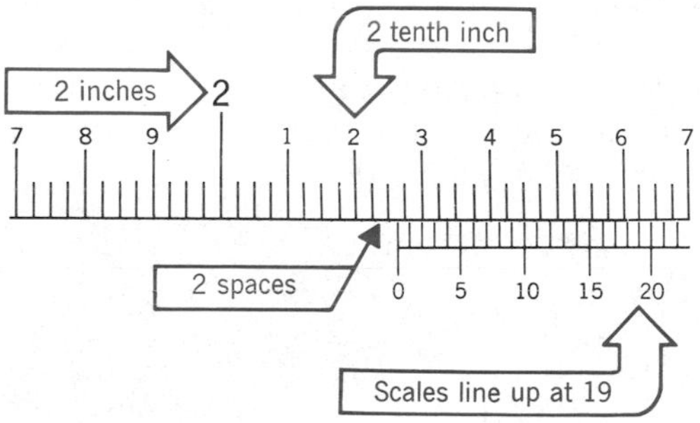

$$\begin{aligned} 2'' &= 2.000'' \\ 2 \times 0.100'' &= 0.200'' \\ 2 \times 0.025'' &= 0.050'' \\ 19 \times 0.001'' &= \underline{0.019''} \\ & \ 2.269'' \end{aligned}$$

Read the following vernier calipers.

(a) 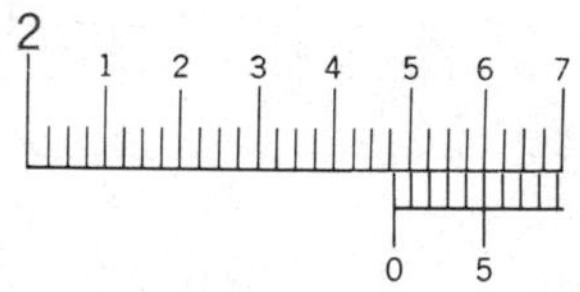

(b) 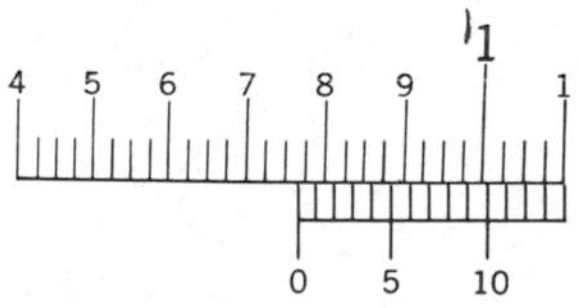

(c)
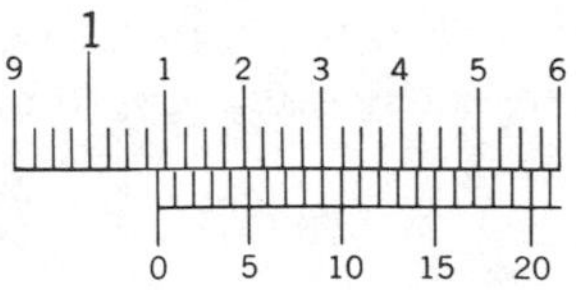

(d)
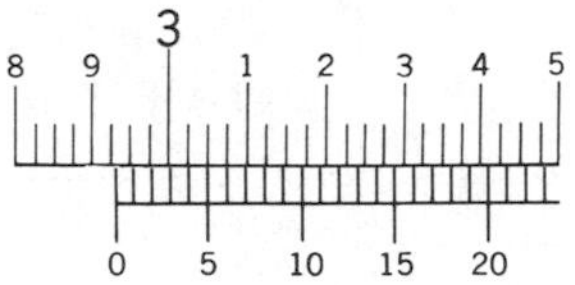

(e)
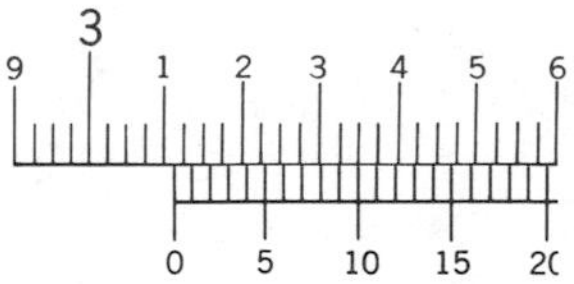

(f)
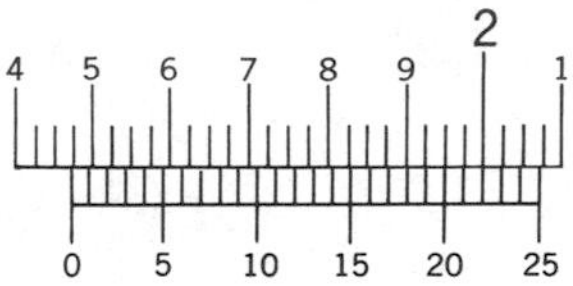

Check your caliper readings in **44**.

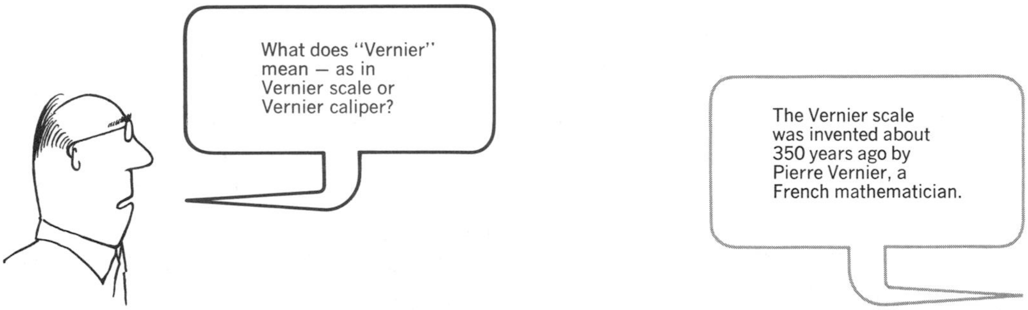

---

44 (a) 2.478″ (b) 0.763″ (c) 1.084″
(d) 2.931″ (e) 3.110″ (f) 1.469″

Protractors

A *protractor* is used to measure and draw angles. The simplest kind of protractor, shown below, is graduated in degrees with every tenth degree labeled with a number.

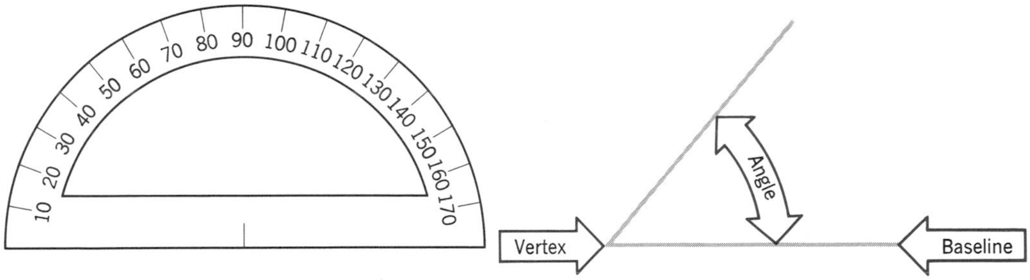

In measuring or drawing angles with a protractor, be certain that the vertex of the angle is exactly at the center of the protractor base. One side of the angle should be placed along the 0°–180° base line.

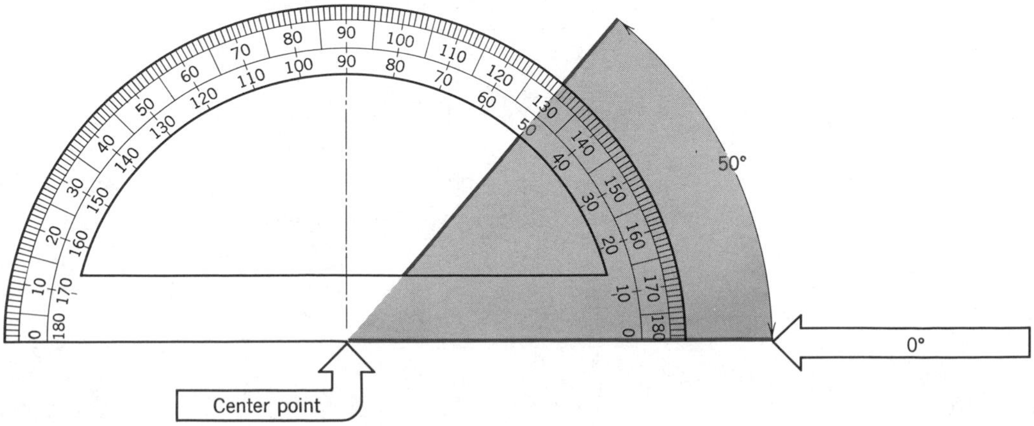

Notice that the protractor can be read from the 0° point on either the right or left side. Measuring from the right side, the angle is read as 60° on the inner scale. Measuring from the left side, the angle is 120° on the outer scale.

The *bevel protractor* is very useful in shopwork because it can be adjusted with an accuracy of ±0.5°. The angle being measured is read directly from the movable scale.

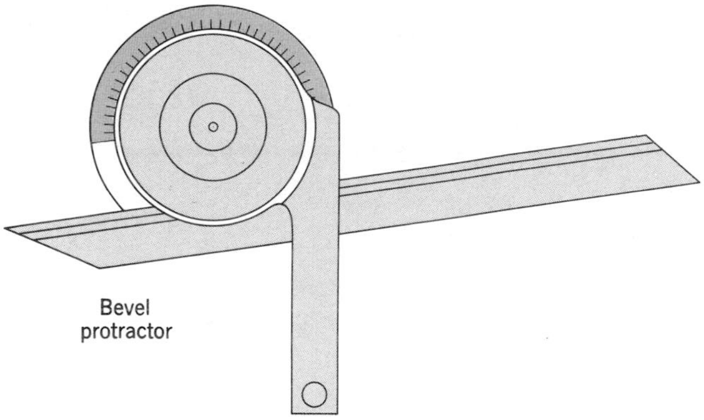

For very accurate angle measurements, the *vernier protractor* can be used with an accuracy of $\pm\frac{1}{60}$th of a degree or 1 minute, since 60 minutes = 1 degree.

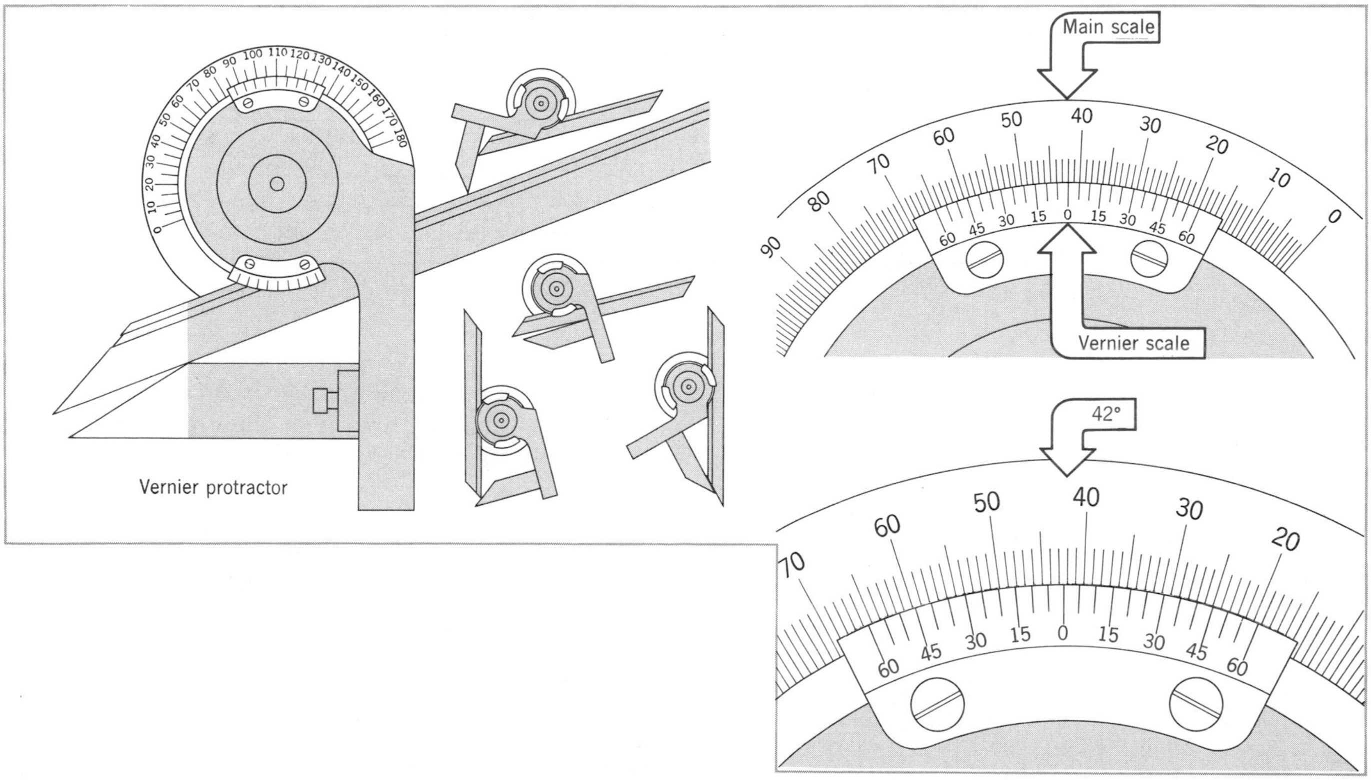

To read a vernier protractor, follow these steps.

Step 1 Read the number of whole degrees on the main scale between the zero on the main scale and the zero on the vernier scale.

42° is just to the right of the 0° mark on the vernier scale.

Step 2 Moving in the same direction, find the vernier scale line that lines up exactly with a main scale marking.

35 on the vernier scale lines up with a mark on the main scale.

$\frac{1}{60}$th of a degree = 1 minute or 1′

Therefore $\dfrac{35°}{60} = 35'$

and the vernier protractor in the example reads 42°35′.

Read the following vernier protractors.

(a)

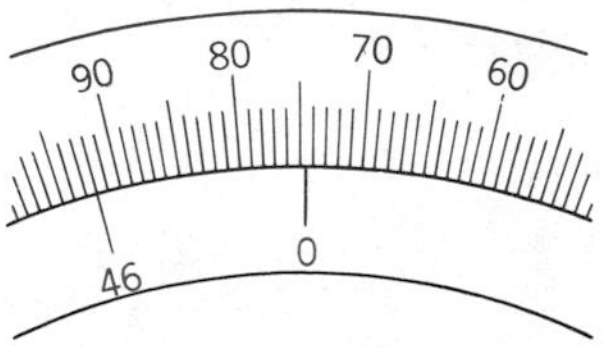

(b)

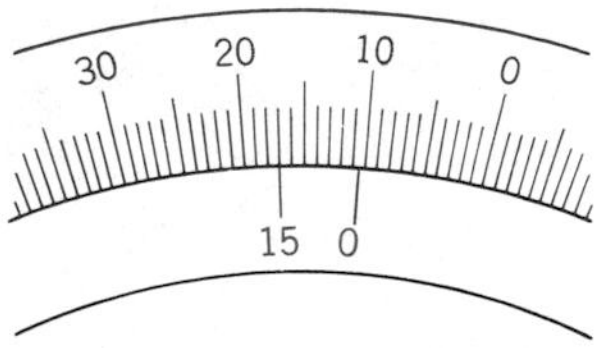

(c)

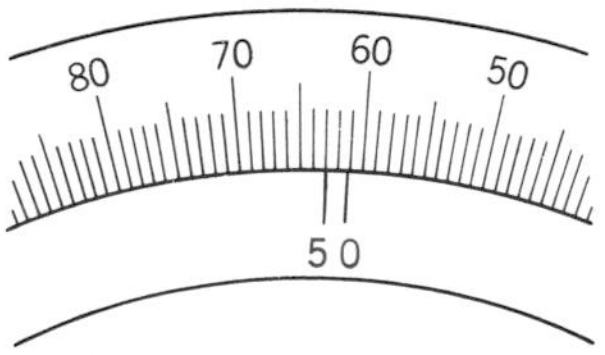

(d)

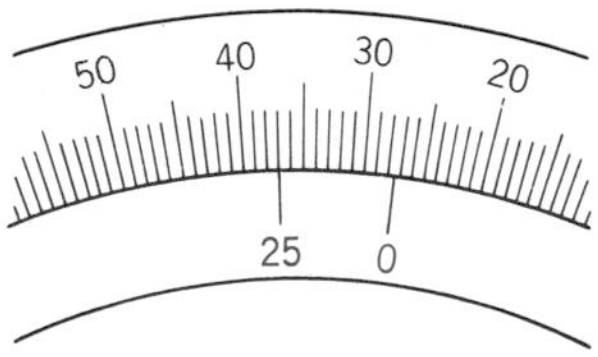

Check your angle readings in **45**.

---

45 (a) 74°46′ (b) 11°15′ (c) 61°5′ (d) 27°25′

Many measuring instruments convert the measurement into an electrical signal and then display that signal by means of a pointer and a decimal scale. Instruments of this kind, called *meters,* are used to measure electrical quantities such as voltage, current, or power, and other quantities such as air pressure, flow rates, or speed. As with length scales, every division on the meter scale is not labeled, and practice is required in order to read meters quickly and correctly.

Meters

The *range* of a meter is its full-scale or maximum reading. The *main* divisions of the scale are the numbered divisions. The *small* divisions are the smallest marked portions of a main division. For example, on the following meter scale,

The *range* is 10 volts;

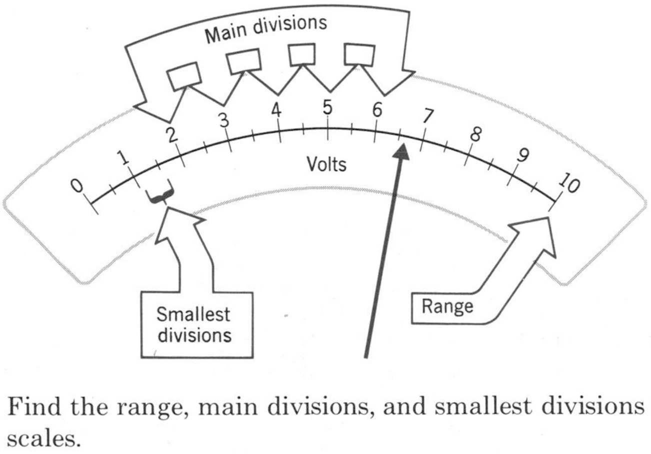

the *main* divisions are 1 volt; the *smallest* divisions are 0.5 volt.

Find the range, main divisions, and smallest divisions for each of the following meter scales.

(a) range = ________
main divisions = ________
smallest divisions = ________

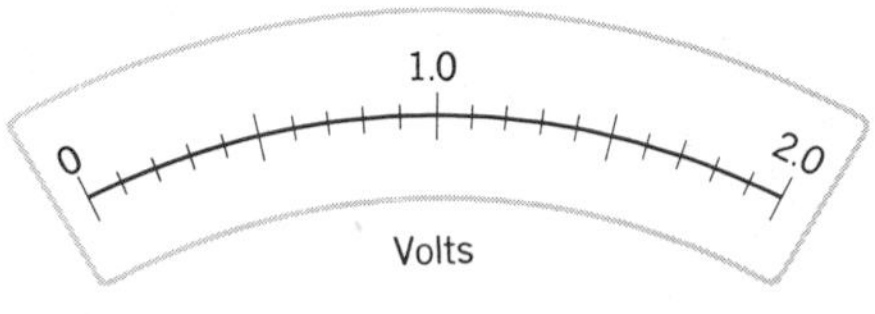

(b) range = ________
main divisions = ________
smallest divisions = ________

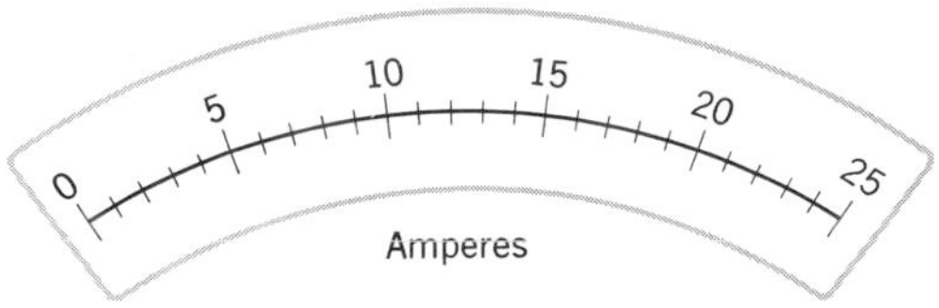

(c) range = ________
main divisions = ________
smallest divisions = ________

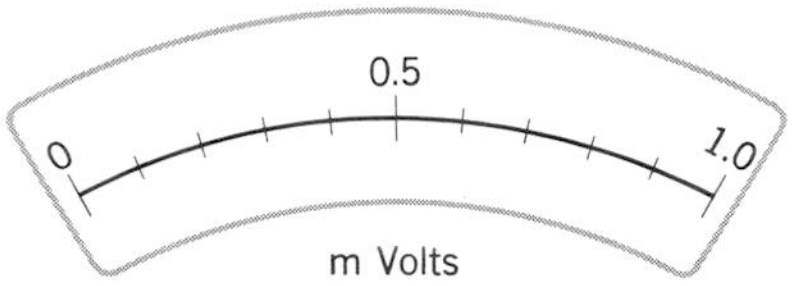

(d) range = ________
main divisions = ________
smallest divisions = ______

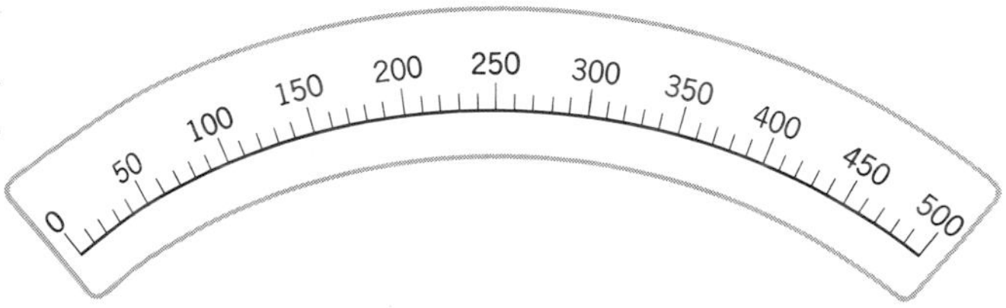

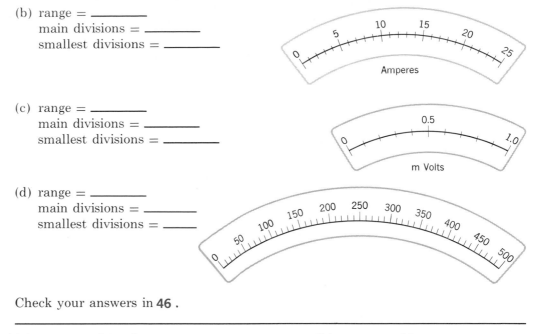

Check your answers in **46** .

---

46 (a) range = 2.0 volt
main divisions = 1.0 volt
smallest divisions = 0.1 volt

(b) range = 25 amp
main divisions = 5 amp
smallest divisions = 1 amp

(c) range = 1.0 millivolt
main divisions = 0.5 millivolt
smallest divisions = 0.1 millivolt

(d) range = 500 psi
main divisions = 50 psi
smallest divisions = 10 psi

To read a meter scale, we can either choose the small scale division nearest to the pointer, or we can estimate between scale markers. For example, this meter

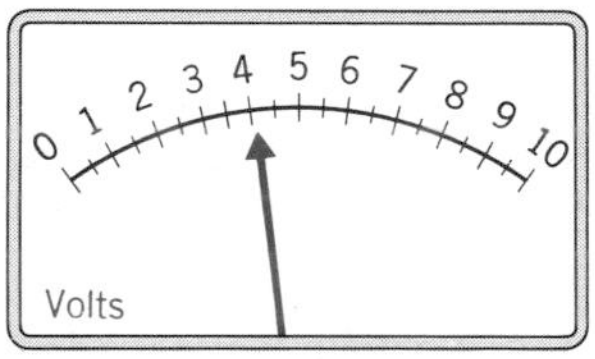

reads exactly 4.0 volts. But this meter

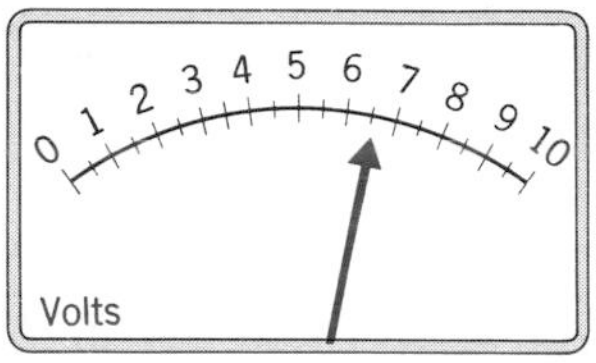

reads 6.5 volts to the nearest scale marker, or 6.6 volts if we estimate between scale markers. The smallest division is 0.5 volt.

On the following meter scale

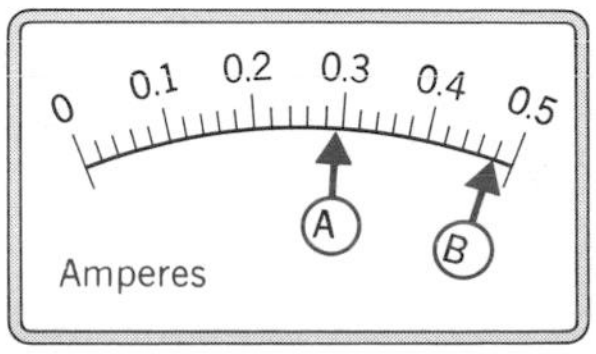

the smallest scale division is $\frac{1}{5}$ of 0.1 amp or 0.02 amp. The reading at A is therefore 0.2 plus 4 small divisions or 0.28 amp. The reading at B is 0.4 plus $3\frac{1}{2}$ small divisions or 0.47 amp.

Read the following meters. Estimate between scale divisions where necessary.

1.

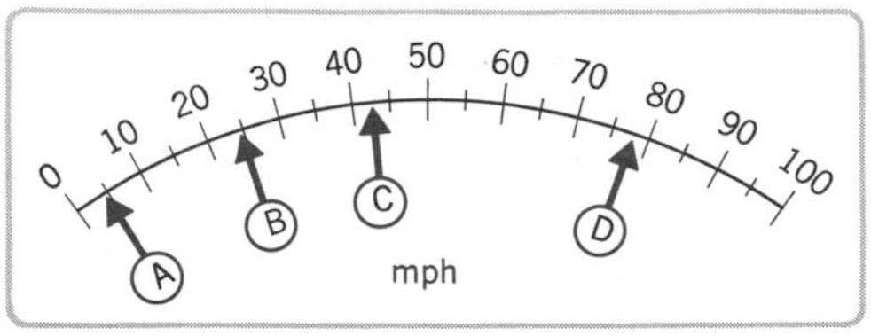

2.

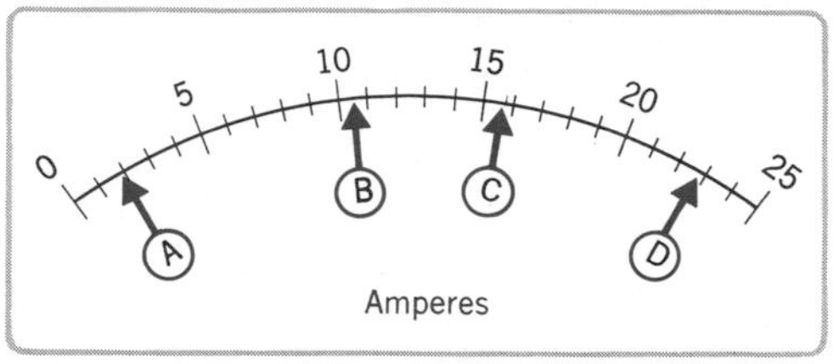

3.

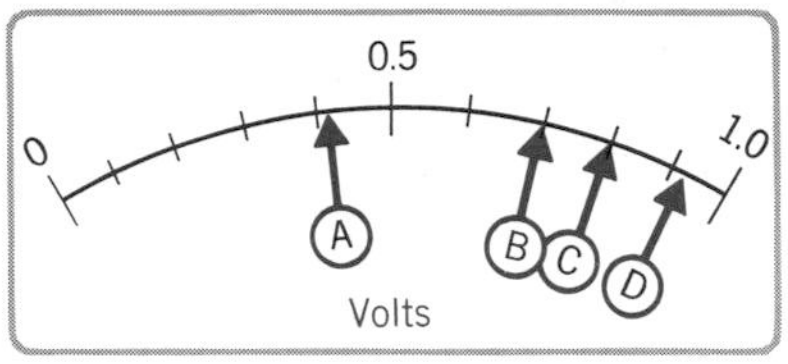

4.

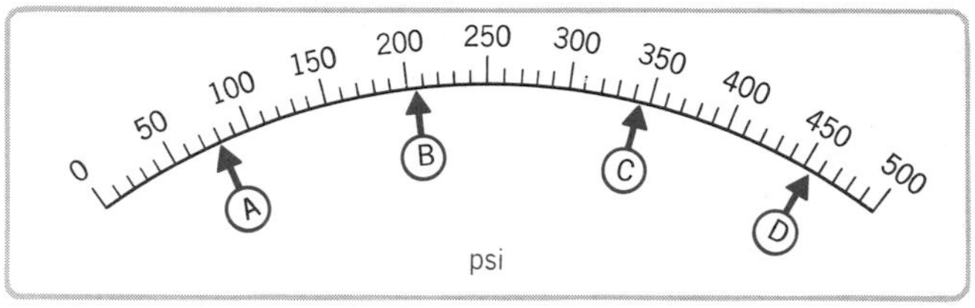

5.

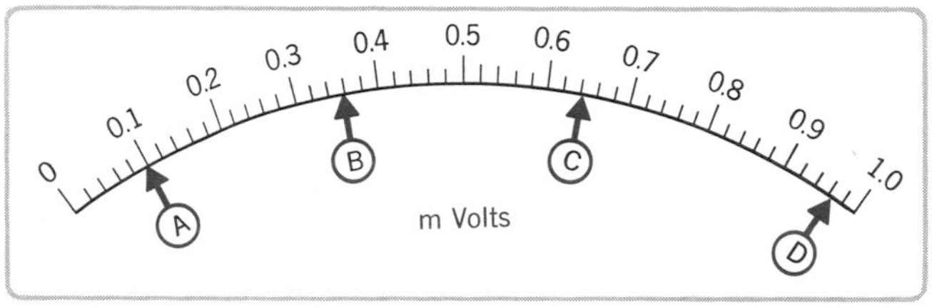

6.

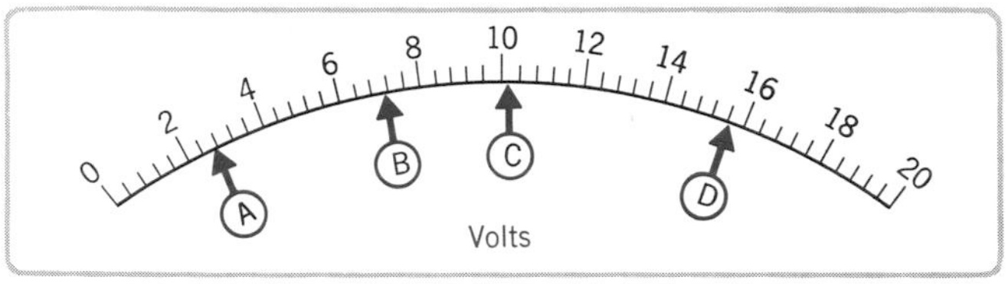

Check your meter readings in **47**

47

1.	A.	5	B. 25	C. 43	D. 78	mph
2.	A.	2	B. 10.6	C. 15.8	D. 23	amp
3.	A.	0.42	B. 0.7	C. 0.8	D. 0.93	volt
4.	A.	83	B. 209	C. 347	D. 458	psi
5.	A.	0.1	B. 0.36	C. 0.645	D. 0.965	m volt
6.	A.	2.8	B. 7.2	C. 10.2	D. 15.7	volt

Notice in meter 6 that the smallest division is $\frac{1}{5}$ of 2 volts or 0.4 volt. If your answers differ slightly from ours, the difference is probably due to the difficulty of estimating the position of the arrows on the diagrams.

Gauge Blocks

Gauge blocks are extremely precise rectangular steel blocks used for making, checking, or inspecting tools and other devices. Opposite faces of each block are ground and lapped to a high degree of flatness, so that when two blocks are fitted together they will cling tightly. Gauge blocks are sold in standard sets of from 36 to 81 pieces in precisely measured widths. These pieces may be combined to give any decimal dimension within the capacity of the set.

A standard 36-piece set would consist of blocks of the following sizes:

0.0501″	0.051″	0.050″	0.110″	0.100″	1.000″
0.0502	0.052	0.060	0.120	0.200	2.000
0.0503	0.053	0.070	0.130	0.300	
0.0504	0.054	0.080	0.140	0.400	
0.0505	0.055	0.090	0.150	0.500	
0.0506	0.056				
0.0507	0.057				
0.0508	0.058				
0.0509	0.059				

To combine gauge blocks to produce any given dimension, work from digit to digit, right to left. This procedure will produce the desired length with a minimum number of blocks.

Example: Combine gauge blocks from the set given above to equal 1.8324″.

Step 1 To get the 4 in 1.8324″ select block 0.0504″. It is the only block in the set with a 4 in the fourth decimal place.

1.8324″	The target length
−0.0504″	The first block
1.7820″	The remainder

Step 2 To get the 2 in 1.782″ select block 0.052″. It is the only block in the set with a 2 in the third decimal place.

1.782″	The new target length
−0.052″	The second block
1.730″	The remainder

Step 3 To get the 3 in 1.73″ select block 0.130″.

1.73″	The new target length
−0.13″	The third block
1.60″	The remainder

Step 4 To get 1.6 select blocks 1.000″, 0.500″, and 0.100″. The six blocks can be combined or "stacked" to total the required 1.8324″ length.

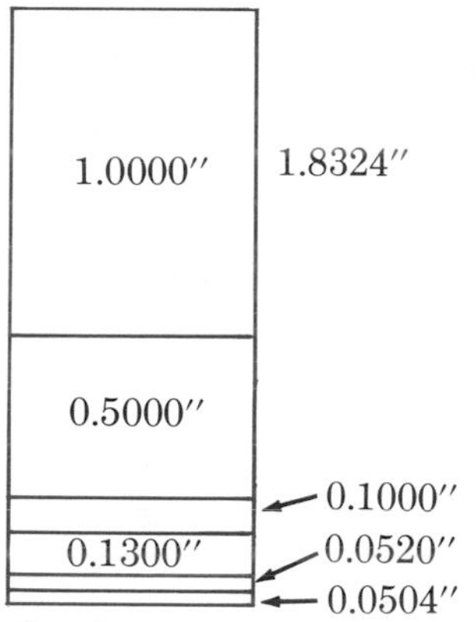

1.0000″
0.5000″
0.1000″
0.1300″
0.0520″
+0.0504″
1.8324″ the required length

Stack sets of gauge blocks that total each of the following.

(a) 2.0976″ (b) 1.8092″ (c) 3.6462″
(d) 0.7553″ (e) 1.8625″ (f) 0.7474″

Check your answers in **48**

---

48 (a) 1.000″ + 0.500″ + 0.400″ + 0.090″ + 0.057″ + 0.0506″
(b) 1.000″ + 0.500″ + 0.200″ + 0.059″ + 0.0502″
(c) 1.000″ + 2.000″ + 0.4000″ + 0.140″ + 0.056″ + 0.0502″
(d) 0.500″ + 0.100″ + 0.050″ +0.055″ + 0.0503″
(e) 1.000″ + 0.500″ + 0.200″ + 0.060″ + 0.052″ + 0.0505″
(f) 0.500″ + 0.140″ + 0.057″ + 0.0504″

Now turn to **49** for a set of practice problems designed to help you become more expert in making direct measurements.

---

49 Exercises 5-4 Direct Measurements

A. Read the following meters:

1. Electric meters. Give your answers in kwh.

(a)

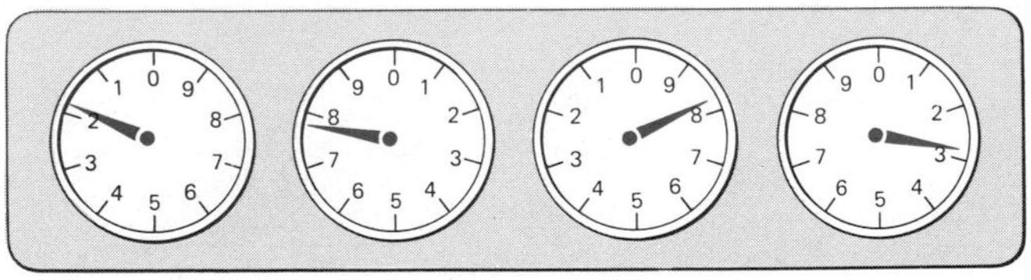

(b)

(c)

(d)

(e)

(f)

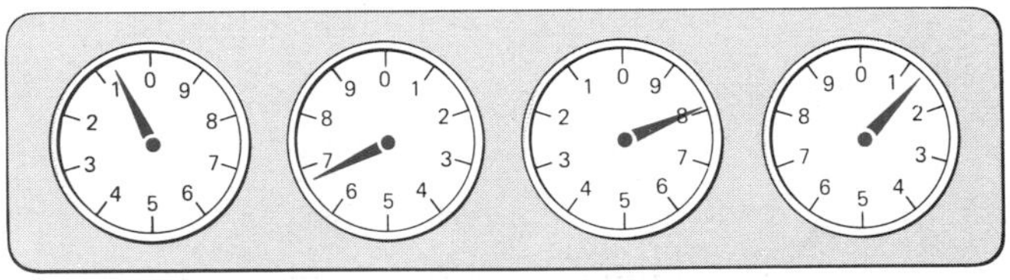

2. Gas meters. Give your answers in cubic feet.

(a)

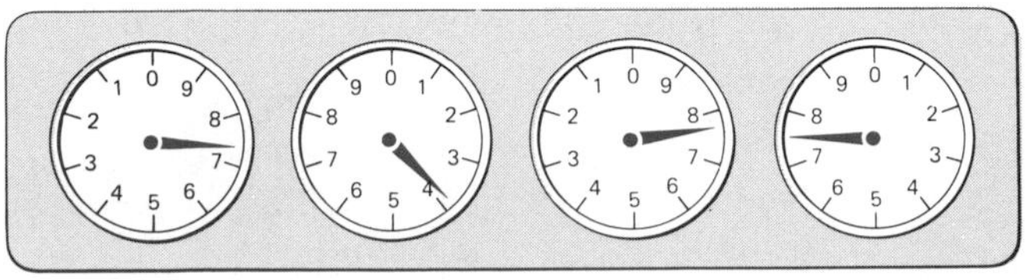

(b)

(c)

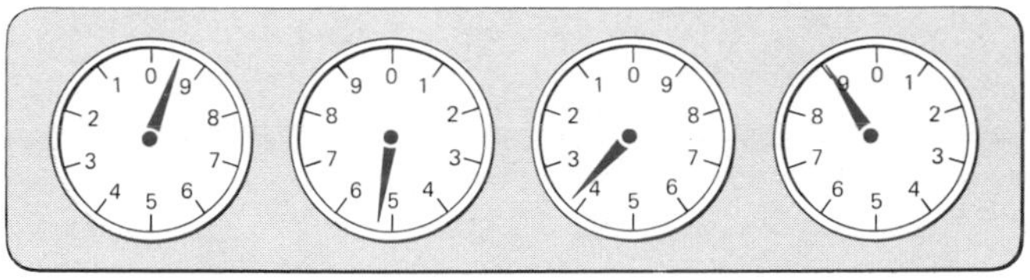

(d)

B. Find the lengths marked on the following rules:

1.

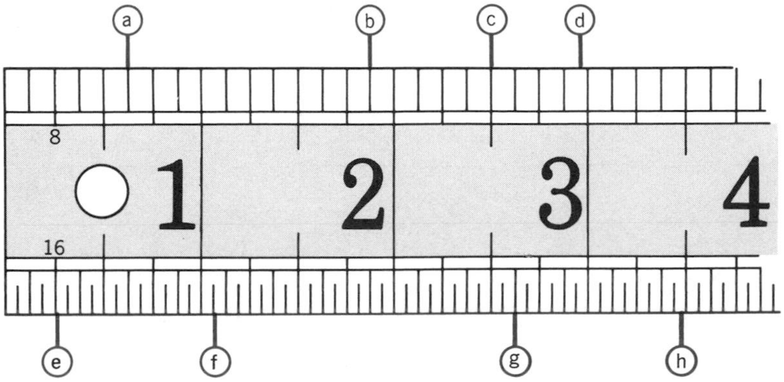

2.

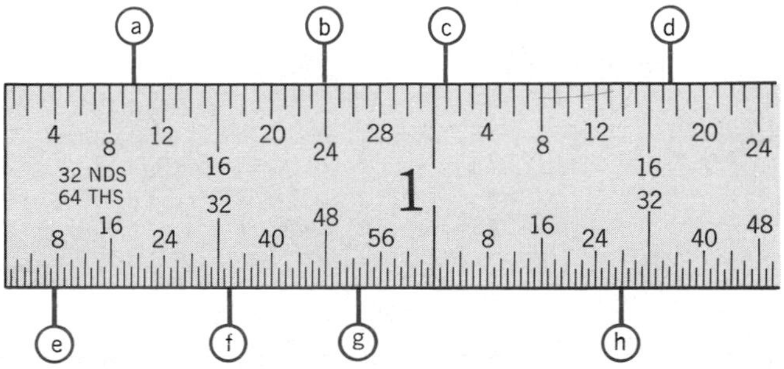

3.

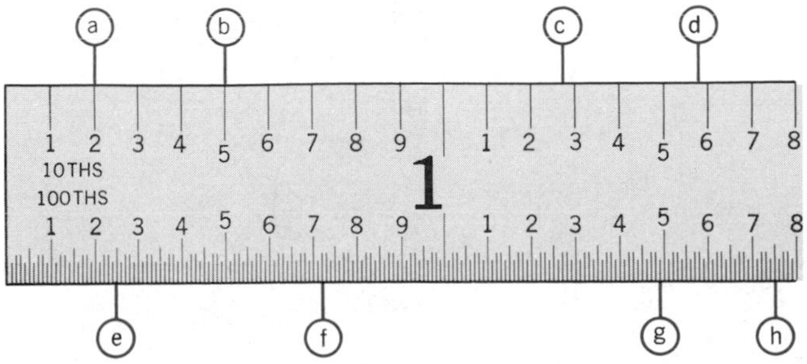

C. Read the following micrometers:

1.

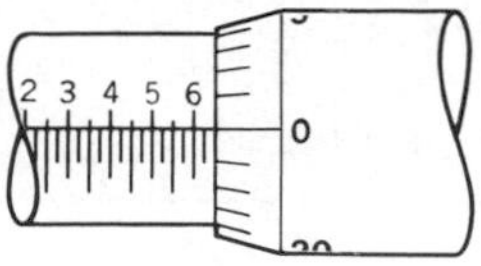

2.

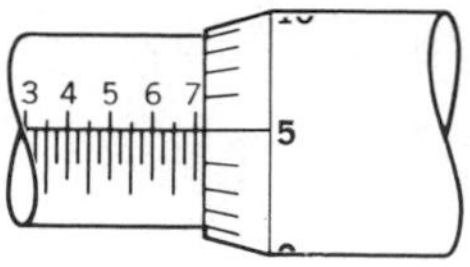

3.

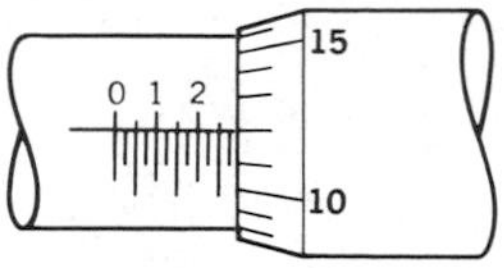

4.

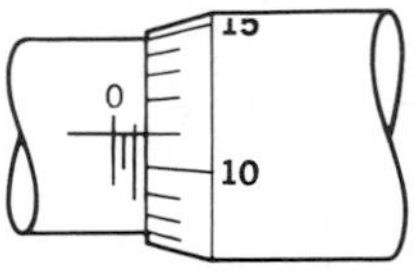

5.

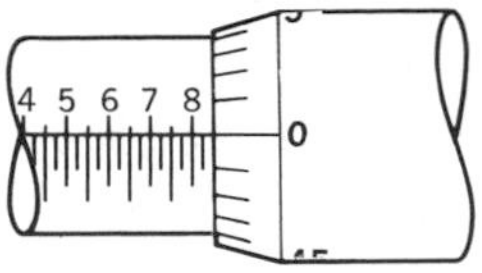

6.

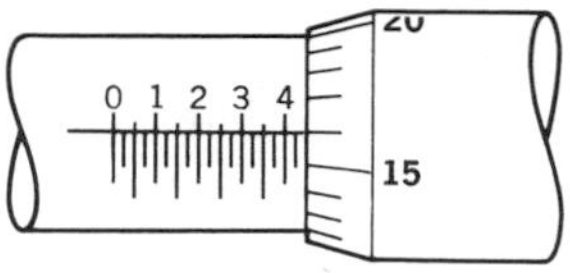

7.

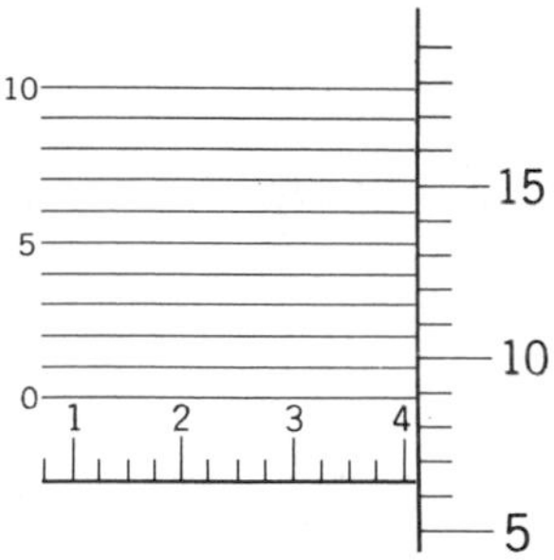

8.

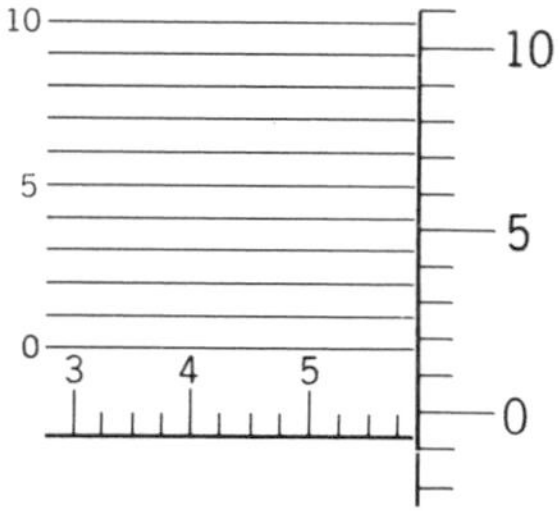

9.

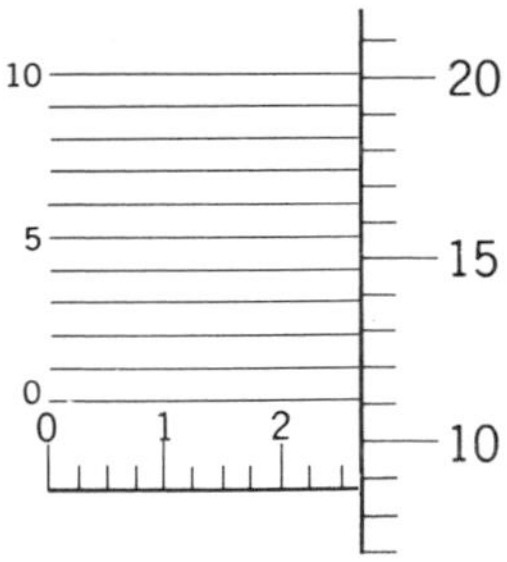

10.

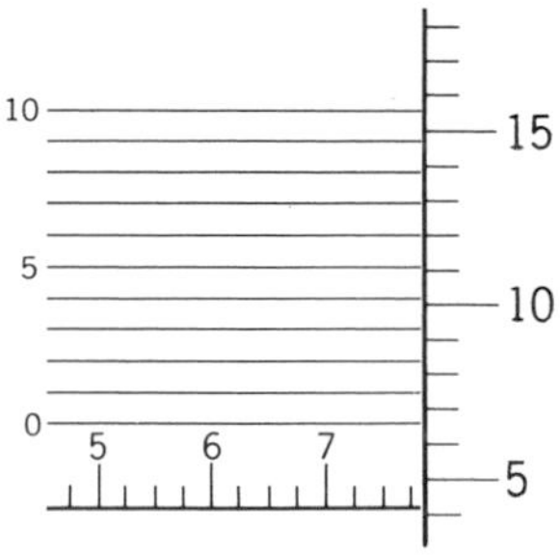

11.

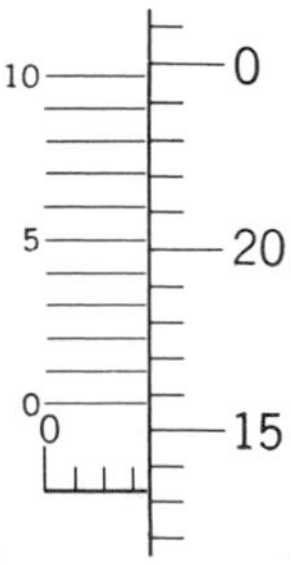

12.

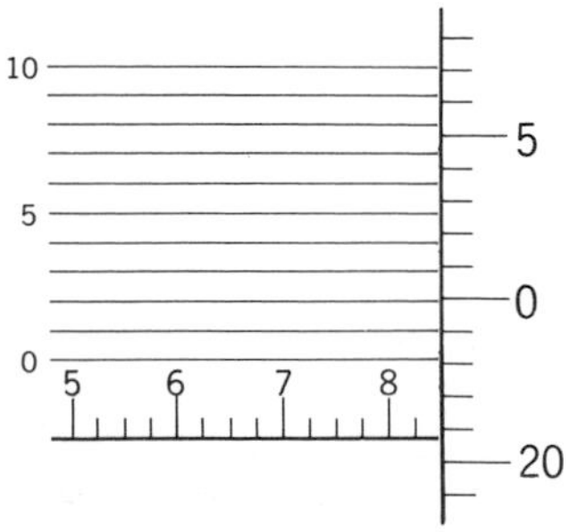

13.

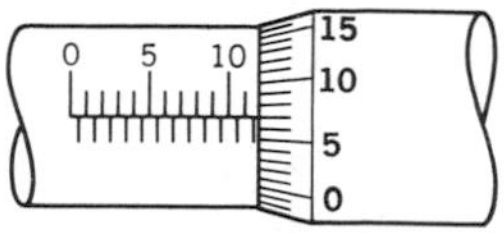

14.

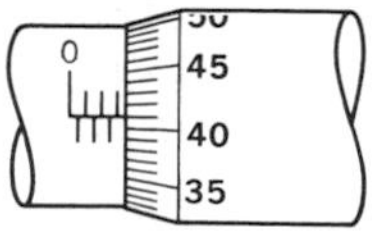

15.

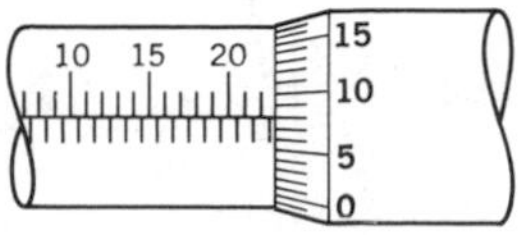

16.

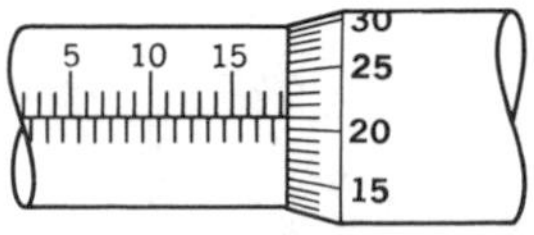

17. 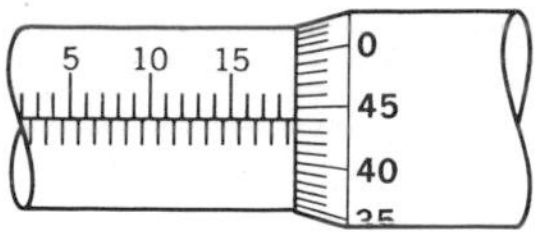

18. 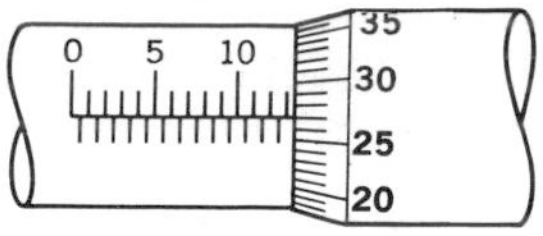

D. Read the following vernier calipers:

1. 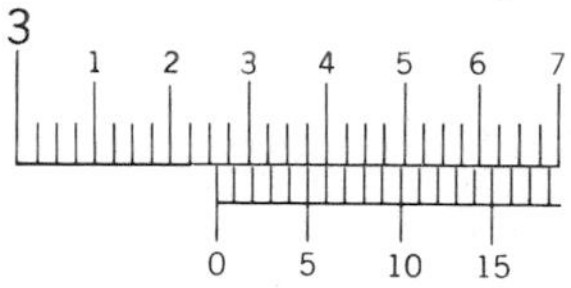

2. 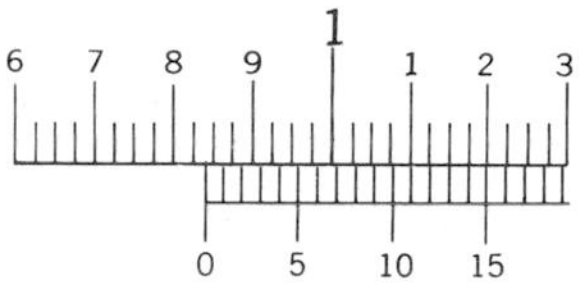

3. 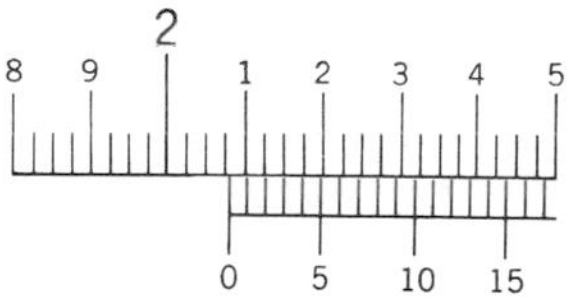

4. 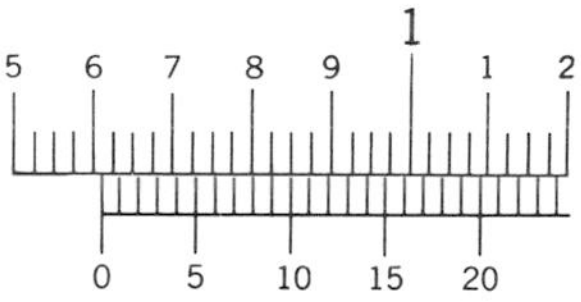

5. 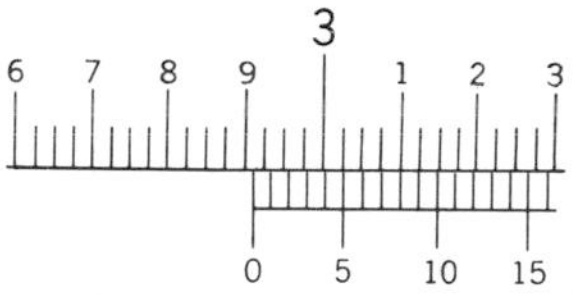

6. 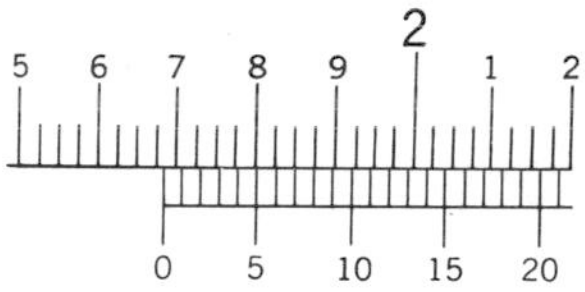

7. 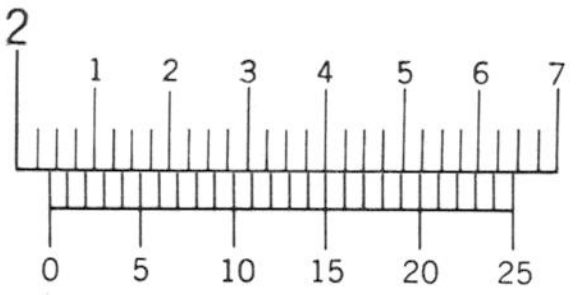

8. 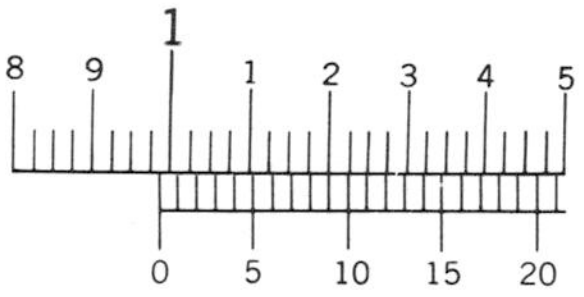

9. 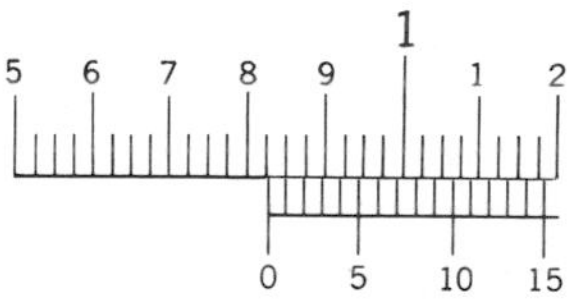

10. 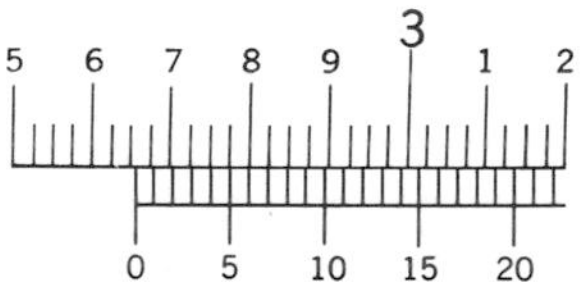

E. Read the following vernier protractors:

1. 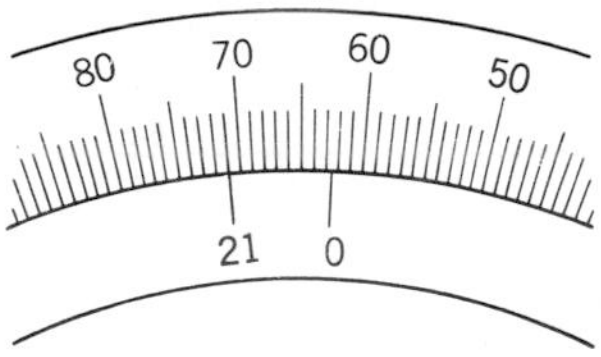

2. 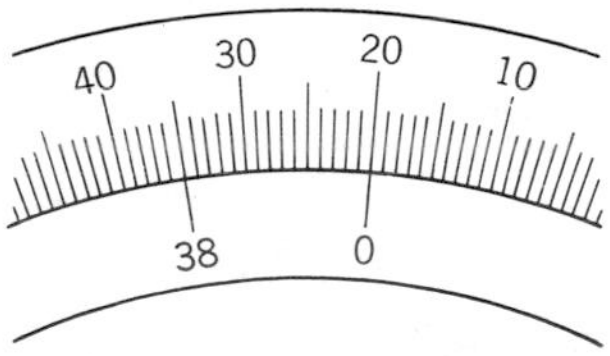

3. 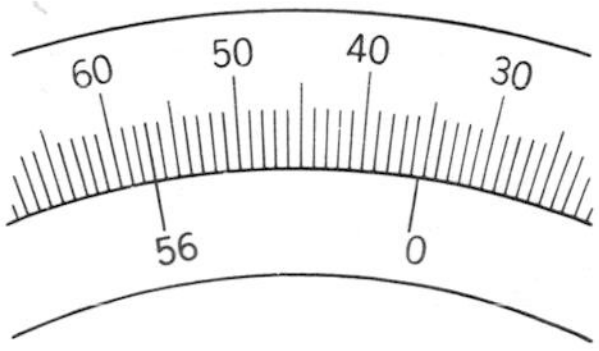

4. 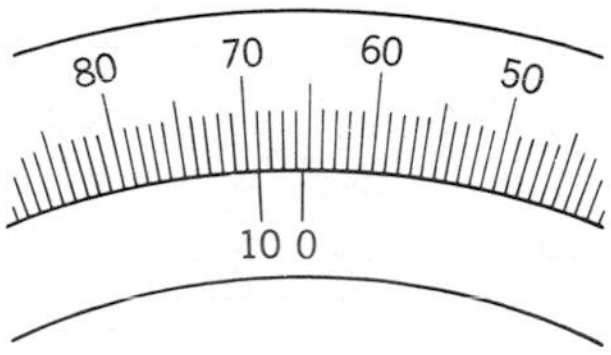

5. 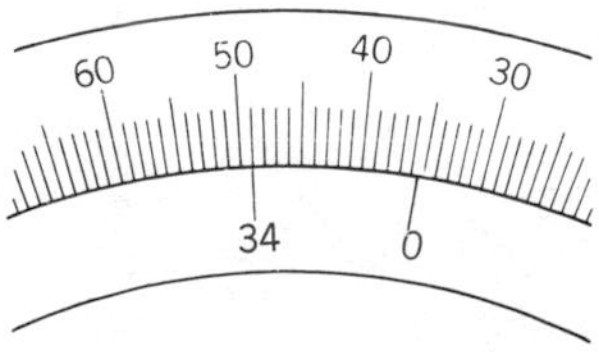

6. 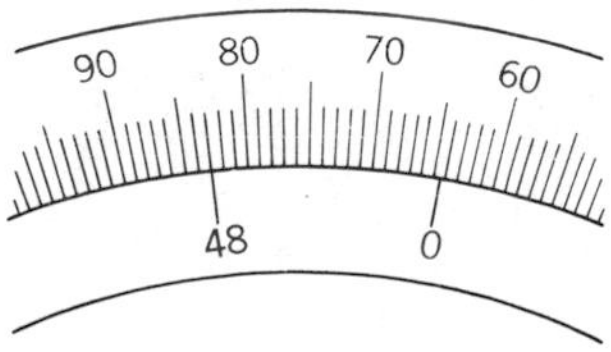

7.

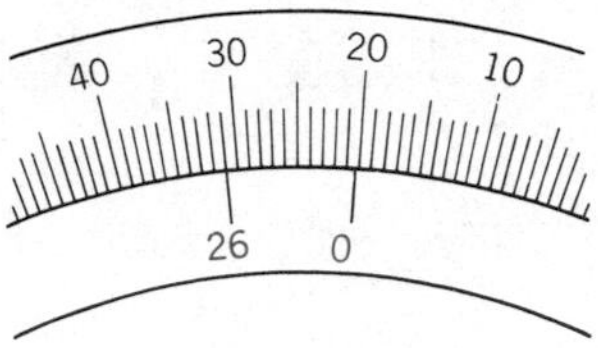

8.

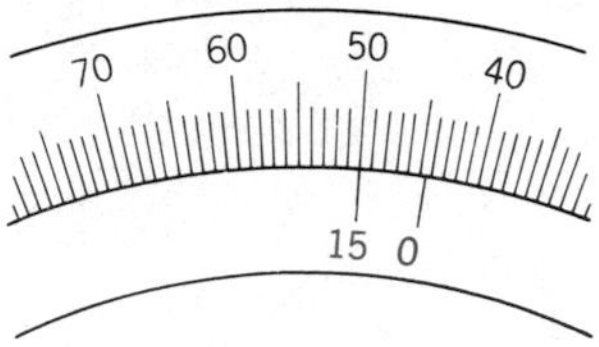

F. Find combinations of gauge blocks that produce the following lengths. (Use the 36 block set given on page 224.)

1. 2.3573″	2. 1.9467″
3. 0.4232″	4. 3.8587″
5. 1.6789″	6. 2.7588″
7. 0.1539″	8. 3.0276″
9. 2.1905″	10. 1.5066″

Now turn to frame **50** for a set of practice problems on measurement.

5 Measurement

PROBLEM SET 5

50 Measurement

Answers are on page 551.

A. Perform the following calculations with measurement numbers. Round to three decimal places when rounding is necessary.

1. 5.75 sec − 2.38 sec
2. 8 ft × 6 ft
3. 4.5 sq in. ÷ 6 in.
4. $(3.2\text{ cm})^2$
5. $7\frac{1}{2}'' + 6\frac{1}{4}'' + 3\frac{1}{8}''$
6. 9 lb 6 oz − 5 lb 8 oz
7. 765 mi ÷ 36 gal
8. 0.5 cm + 2.3 cm + 6.8 cm
9. 24.38 psi − 16.73 psi
10. $7\frac{1}{4}$ ft − 4.3 ft
11. 7.064″ − 3.195″
12. 22 psi × 18.35 sq in.
13. 8 ft 4 in. + 12 ft + 9 in.
14. $2\frac{15}{64}'' + 3\frac{1}{8}'' + 7\frac{1}{32}''$

B. Convert the following measurement numbers to new units as shown. Round to two decimal places whenever necessary.

1. 6 in. = ________ cm
2. 50 mph = ________ ft/sec
3. 22° C = ________ ° F
4. 2.6 liters = ________ qt
5. $26\frac{1}{2}$ cu ft = ________ cu in.
6. −13° F = ________ ° C
7. 14.3 kg = ________ lb
8. 123.7 psi = ________ atm
9. $14\frac{1}{2}$ mi = ________ km
10. 350 cu in. = ________ cu cm
11. 14.3 bbl = ________ cu ft
12. 160 sq ft = ________ sq m
13. 24,680 cu in. = ________ bu
14. 6.9 liters = ________ gal
15. 6.8 sq ft = ________ sq in.
16. $12\frac{1}{4}$ qt = ________ liters

C. Think metric. Choose the closest estimate.

1. Weight of a pencil	(a) 4 kg	(b) 4 g	(c) 40 g
2. Size of a house	(a) 6 sq m	(b) 200 sq m	(c) 1200 sq m
3. Winter in New York	(a) −8° C	(b) −40° C	(c) 20° C
4. Volume of a 30 gal trash can	(a) 1.5 m^3	(b) 5 m^3	(c) 50 m^3
5. Height of an oak tree	(a) 500 mm	(b) 500 cm	(c) 50 km
6. Capacity of a 50 gal water heater	(a) 25 ℓ	(b) 100 ℓ	(c) 180 ℓ

D. Read the following measuring devices.

1. Rulers.

(a)

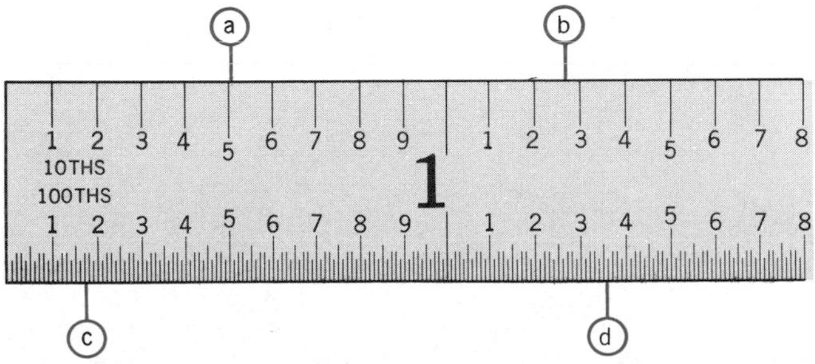

Name ________

Date ________

Course/Section ________

(b)

(c)

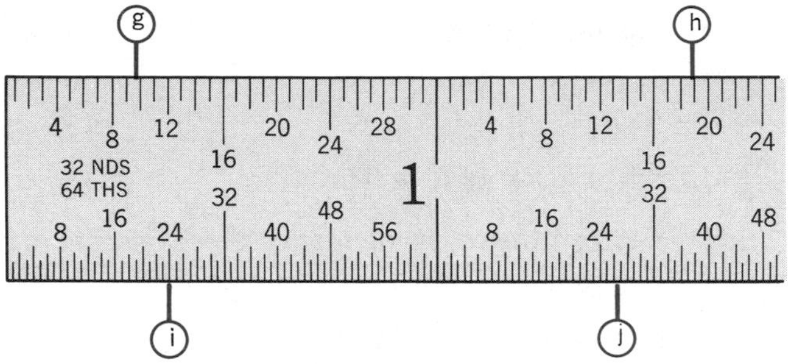

2. Electric meters. Give answers in kwh.

(a)

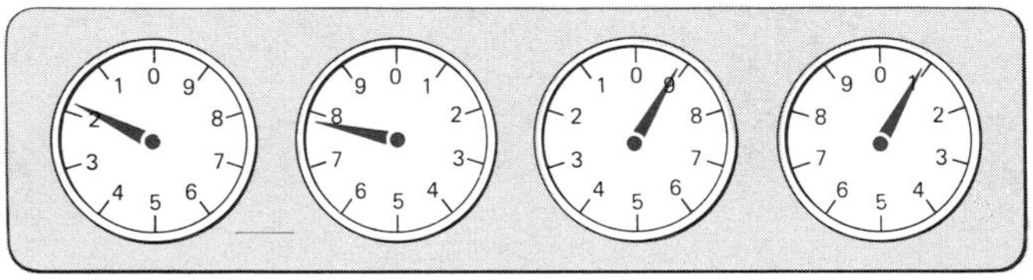

(b)

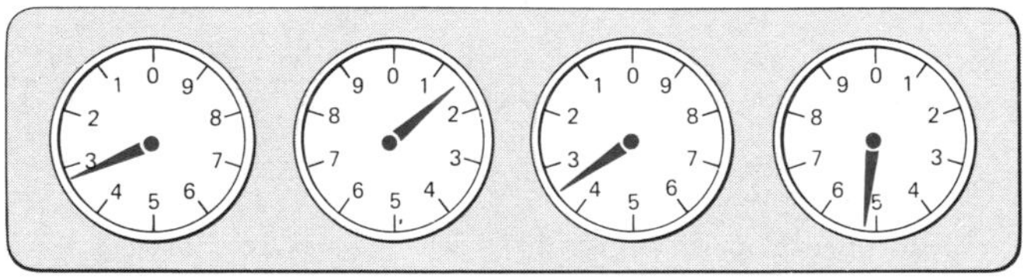

(c)

(d)

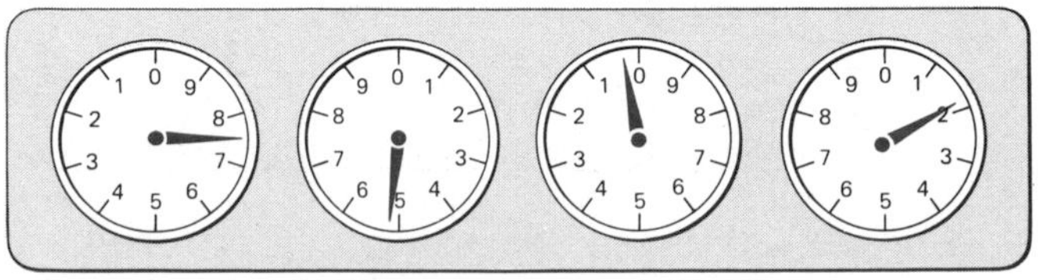

(e)

(f)

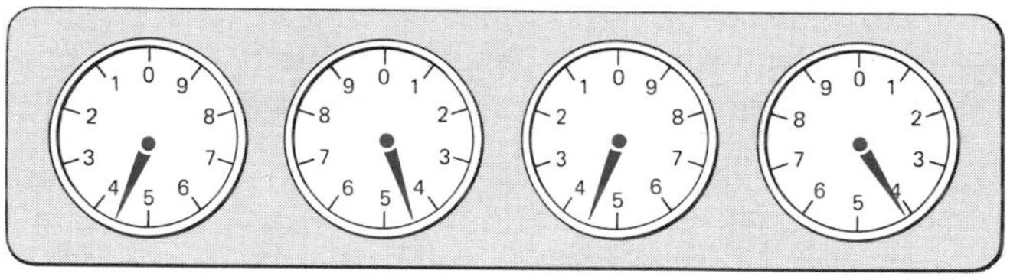

3. Gas meters. Give answers in cubic feet.

(a)

(b)

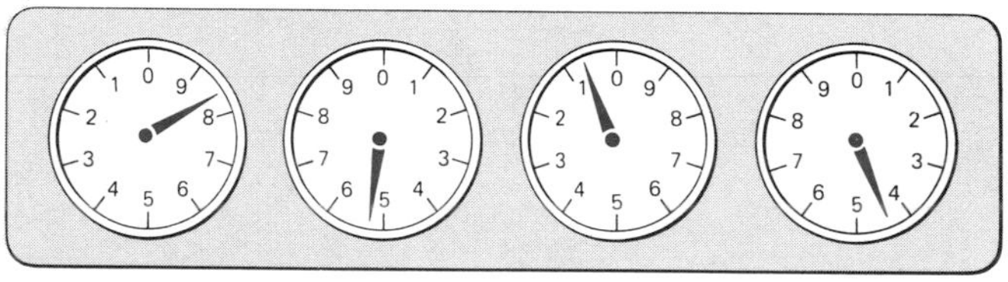

(c)

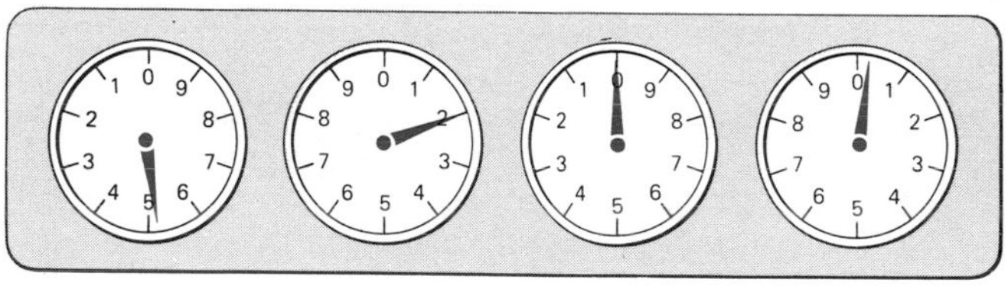

(d)

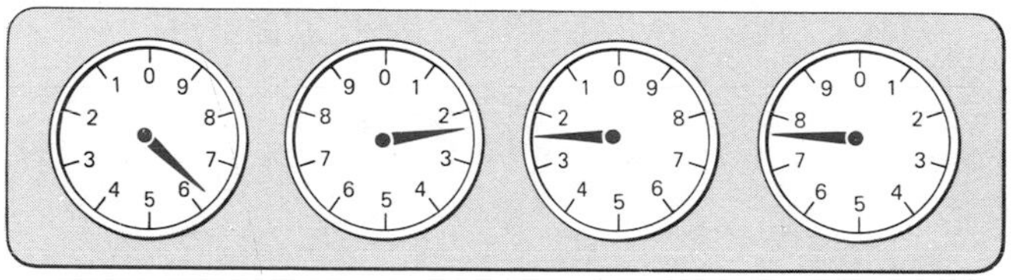

(e)

(f)

4. Micrometers. Problems (a) and (b) are English, (c) and (d) are metric, and (e) and (f) are English with a vernier scale.

(a)

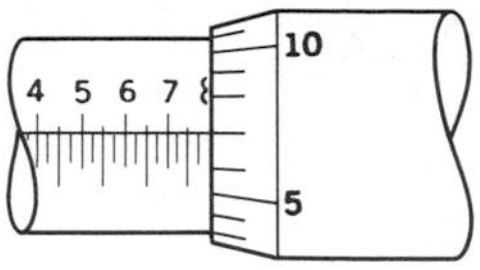

(b)

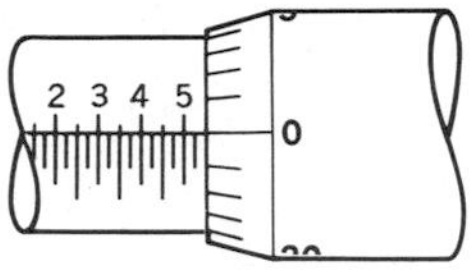

(c)

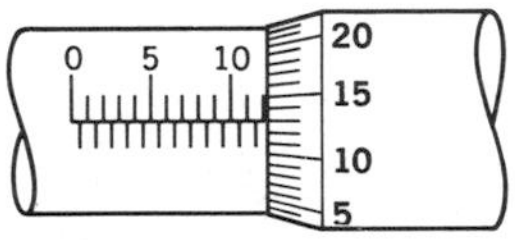

(d)

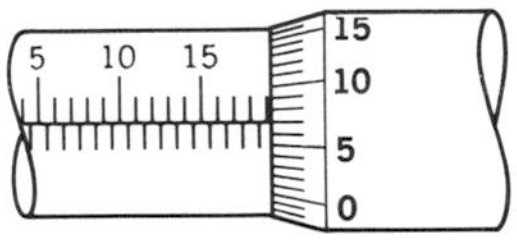

(e)

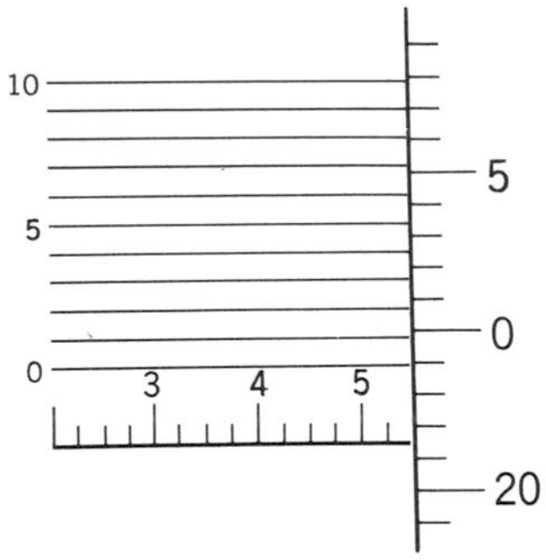

(f)

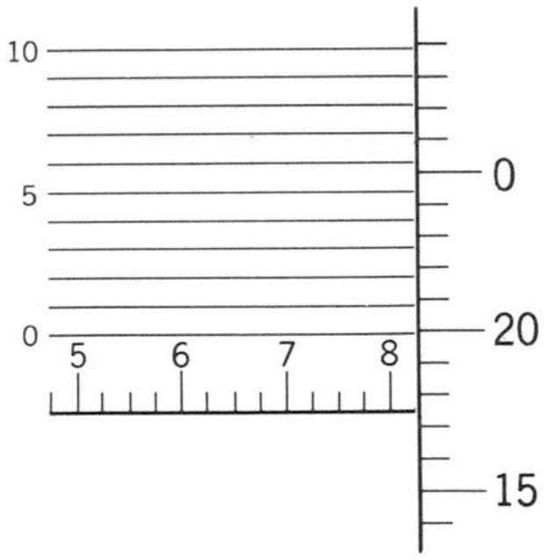

5. Vernier calipers.

(a)

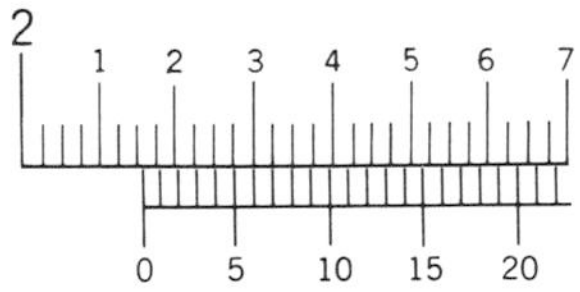

(b)

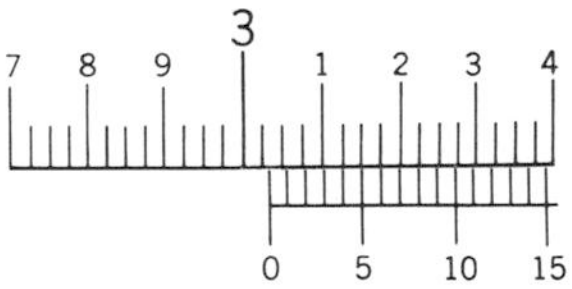

(c)

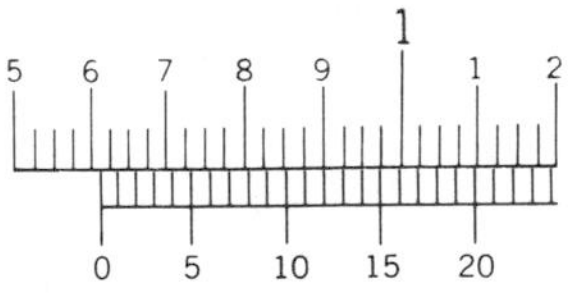

(d)

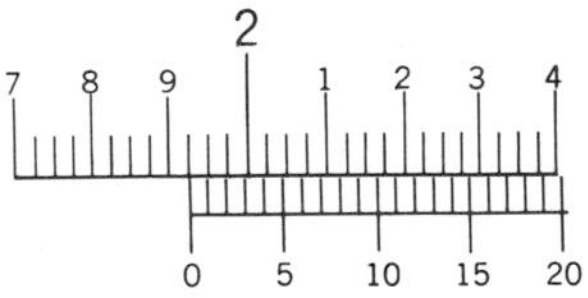

(e)

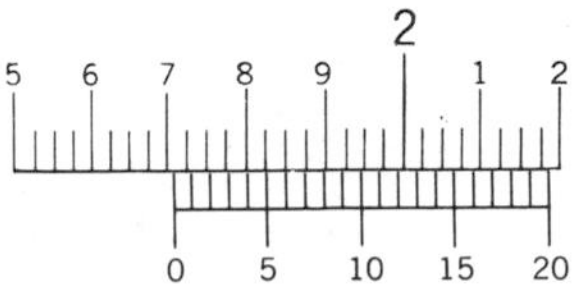

(f)

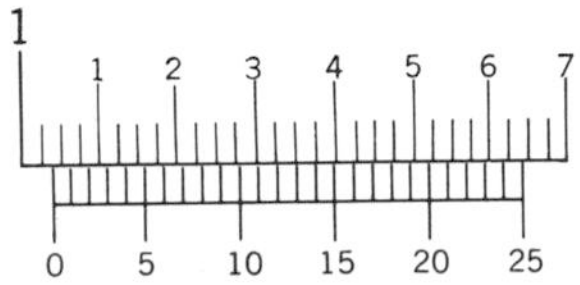

E. Practical Problems

1. A piece $6\frac{1}{32}''$ long is cut from a steel bar $28\frac{5}{64}''$ long. How much is left?
2. An American traveling in Italy notices a road sign saying 85 km/hr. What is this speed in miles per hour? (Round to the nearest mph.)
3. An automotive technician testing cars finds that a certain bumper will prevent damage up to 9 ft/sec. Convert this to miles per hour. (Round to one decimal place.)
4. The gap in the piece of steel shown in the figure should be 2.4375″ wide. Which of the gauge blocks in the set on p. 224 should be used for testing this gap?

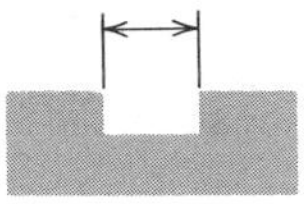

5. What size wrench, to the nearest 32nd of an inch, will fit a bolt 0.455 inch in diameter?
6. The melting point of pure iron is 1505° C. What temperature is this on the Fahrenheit scale?
7. What area of sheet metal in sq in. is needed to construct a vent with a surface area of 6.5 sq ft?
8. How many gallons of gas are needed to drive a car 350 miles if the car normally averages 18.5 mi/gal? (Round to one decimal place.)
9. A customer's gas meter read 2354 CCF last month. This month it reads:

How much gas was used this month?

10. Heated steel appears bright red at a temperature of 1650° F. What Celsius temperature does this correspond to? (Round to the nearest degree.)
11. A Datsun sports car has a 1600 cc engine. What would this be in cubic inches?
12. Find the total length of the piece of steel shown in the figure.

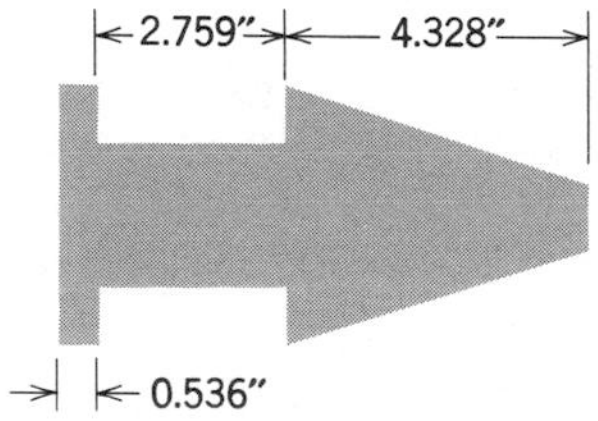

13. The Uniform Plumbing Code is being converted to the metric system. Convert the following measurements.
    (a) The amount of gas used by a furnace (136 cubic feet per hour) to cubic meters per hour.
    (b) The length of a gas line (25 feet) to meters.
    (c) The diameter of water pipe ($2\frac{1}{2}$ in.) to centimeters.
    (Round to the nearest hundredth.)
14. Wrought iron has a density of 480.0 lb/cu ft. Convert this to lb/cu in.
15. A cubic foot of water contains 7.48 gallons. What volume is this in liters?
16. What are the dimensions of a metric "two by four" (a piece of wood 2″ by 4″) in centimeters?
17. A common scale used by architects and draftsmen in making blueprints is $\frac{1}{4}'' = 1'$. What length in meters would correspond to a drawing dimension of $1\frac{1}{4}''$?

6 Basic Algebra

PREVIEW 6

Objective	Sample Problems		Where To Go for Help: Page	Frame
Upon successful completion of this unit you will be able to:				
1. Perform the basic algebraic operations.	(a) $3ax^2 + 4ax^2 - ax^2$	= ________	241	4
	(b) $2y \cdot 3ay$	= ________		
	(c) $5x - (x + 2)$	= ________		
	(d) $3x(y - 2x)$	= ________		
2. Evaluate formulas and literal expressions.	If $x = 2$, $y = 3$, $a = 5$, $b = 6$, find the value of			
	(a) $2x + y$	= ________	247	10
	(b) $\dfrac{1 + x^2 + 2a}{y}$	= ________		
	(c) $A = x^2y$	A = ________		
	(d) $T = \dfrac{2(a + b + 1)}{3x}$	T = ________		
	(e) $P = abx^2$	P = ________		
3. Translate simple English phrases and sentences into algebraic expressions and equations.	Write each phrase as an algebraic expression or equation.			
	(a) Four times the area.	________	269	28
	(b) Current squared times resistance.	________		
	(c) Efficiency is equal to 100 times the output divided by the input.	________		
	(d) Resistance is equal to 12 times the length divided by the diameter squared.	________	269	28
4. Solve linear equations in one unknown.	(a) $3x - 4 = 11$	x = ________	255	17
	(b) $2x = 18$	x = ________		
	(c) $2x + 7 = 43 - x$	x = ________		
5. Solve problems involving ratio and proportion.	If it takes a machinist 3 hr to drill 8 flanges, how many will he drill in an 8 hour day?		284	40

(Answers to these Preview problems are on page 238.)

If you are certain you can work *all* of the problems correctly, turn to page 299 for a set of practice problems. If you cannot work one or more of the previous problems, turn to the page indicated for some help. If you want to be tops in your work, turn to frame **1** and begin work there.

Answers to Preview 6

1. (a) $6ax^2$ (b) $6ay^2$ (c) $4x - 2$ (d) $3xy - 6x^2$
2. (a) 7 (b) 5 (c) 12 (d) 4 (e) 120
3. (a) 4A (b) i^2R (c) $E = \dfrac{100U}{I}$ (d) $R = \dfrac{12L}{D^2}$
4. (a) 5 (b) 9 (c) 12
5. 21

6 Basic Algebra

1 In this unit you will study algebra, but not the very formal algebra that deals with theorems, proofs, sets, and abstract problems. Instead, we will study practical or applied algebra as actually used by technical and trades workers. Let's begin with a look at the language and vocabulary of algebra.

6-1 THE LANGUAGE OF ALGEBRA

The most obvious difference between algebra and arithmetic is that in algebra letters are used to replace or to represent numbers. A mathematical statement in which letters are used to represent numbers is called a *literal* expression. Algebra is the arithmetic of literal expressions—a kind of symbolic arithmetic.

Any letters will do, but in practical algebra we use the normal lower- and uppercase letters of the English alphabet. It is helpful in practical problems to choose the letters to be used on the basis of its memory value: t for time, D for diameter, C for cost, A for area, and so on. The letter used reminds you of its meaning.

Most of the usual arithmetic symbols have the same meaning in algebra that they have in arithmetic. For example, the addition ($+$) and subtraction ($-$) signs are used in exactly the same way. However, the multiplication sign ($\times$) of arithmetic looks like the letter x and to avoid confusion we have other ways to show multiplication in algebra. The product of two algebra quantities a and b, "a times b," may be written using

a raised dot	$a \cdot b$				
parentheses	$a(b)$	or	$(a)b$	or	$(a)(b)$
or with nothing at all	ab				

Obviously this last way of showing multiplication won't do in arithmetic; we cannot write "two times four" as "24"—it looks like twenty-four. But it is a quick and easy way to write a multiplication in algebra.

Placing two quantities side by side to show multiplication is not new and it is not only an algebra gimmick; we use it every time we write 20¢ or 4 feet.

20¢ $= 20 \times 1$¢
4 feet $= 4 \times 1$ foot

Write the following multiplications using no multiplication symbols.

(a) 8 times a = ________

(b) 2 times s times t = ________

(c) 3 times x times x = ________

Check your answers in **2** .

---

2 (a) 8 times a $= 8a$
(b) 2 times s times t $= 2st$
(c) 3 times x times x $= 3x^2$

Did you notice in Problem (c) that powers of numbers are written just as in arithmetic:

$x \cdot x = x^2$
$x \cdot x \cdot x = x^3$
$x \cdot x \cdot x \cdot x = x^4$ and so on.

Parentheses ( ) are used in arithmetic to show that some complicated quantity is to be treated as a unit. For example,

$2 \cdot (13 + 14 - 6)$

means that the number 2 multiplies *all* of the quantity in the parentheses.

In exactly the same way in algebra, parentheses ( ), brackets [ ], or braces { }, are used to show that whatever is enclosed in them should be treated as a single quantity. An expression such as

$(3x^2 - 4ax + 2by^2)^2$

should be thought of as (something)^2. The expression

$(2x + 3a - 4) - (x^2 - 2a)$

should be thought of as (first quantity) − (second quantity). Parentheses are the punctuation marks of algebra. Like the period, comma, or semicolon in regular sentences, they tell you how to read an equation and get its correct meaning.

In arithmetic we would write "48 divided by 2" as

$2\overline{)48}$ or $48 \div 2$ or $\frac{48}{2}$

But the first two ways of writing division are used very seldom in algebra. Division is usually written as a fraction.

"x divided by y" is written $\dfrac{x}{y}$.

"$(2n + 1)$ divided by $(n - 1)$" is written $\dfrac{(2n+1)}{(n-1)}$ or $\dfrac{2n+1}{n-1}$.

Write the following using algebra notation.

(a) 8 times $(2a + b)$ = ________ (b) $(a + b)$ times $(a - b)$ = ________

(c) x divided by y^2 = ________ (d) $(x + 2)$ divided by $(2x - 1)$ = ________

Turn to **3** to check your answers.

---

3 (a) $8(2a + b)$ (b) $(a + b)(a - b)$

(c) $\frac{x}{y^2}$ (d) $\frac{x + 2}{2x - 1}$

The word "expression" is used very often in algebra. An *expression* is a general name for any collection of numbers and letters connected by arithmetic signs. For example,

$x + y$

$2x^2 + 4$

$3(x^2 - 2ab)$

and

$(b - 1)^2$

are all algebra expressions.

If the algebra expression has been formed by multiplying quantities, each multiplier is called a *factor* of the expression.

Expression	*Factors*
ab	a and b
$2x(x + 1)$	$2x$ and $(x + 1)$
$(R - 1)(2R + 1)$	$(R - 1)$ and $(2R + 1)$

The algebra expression could also be a sum or difference of simpler quantities or *terms*.

Expression	*Terms*	
$x + 4y$	x and $4y$ ←	The first term is x. The second term is $4y$.
$2x^2 + xy$	$2x^2$ and xy	
$A - R$	A and R	

Algebra Operations

There are three simple algebra operations you must be able to do to use algebra in practical situations. These operations are:

Operation 1. Adding and Subtracting Like Terms

Example: $2a + 5a = 7a$

Operation 2. Adding and Subtracting Expressions

Example: $(2x - y) + (x + 3y) = 3x + 2y$

Operation 3. Multiplying Simple Factors

Example: $(2ax)(3x) = 6ax^2$

Let's look at them in order. Turn to **4** to begin.

---

4 **Operation 1. Adding and Subtracting Like Terms**

Two algebra terms are said to be *like terms* if they contain exactly the same literal part. For example, the terms

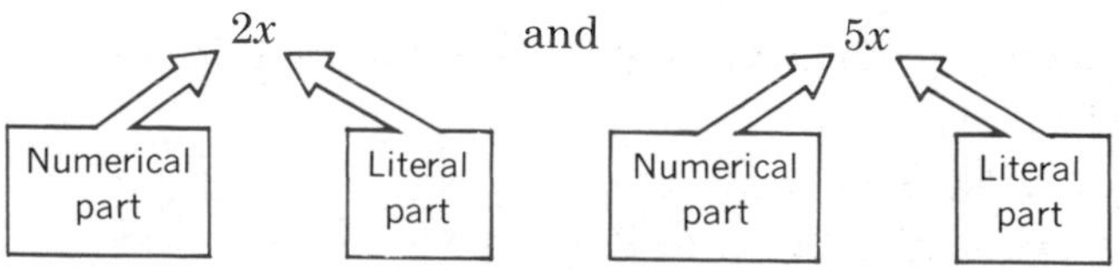

are like terms. The literal part, x, is the same for both terms.

The number multiplying the letters is called the *numerical coefficient* of the term.

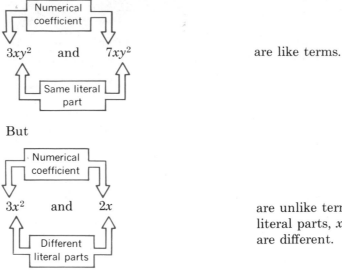

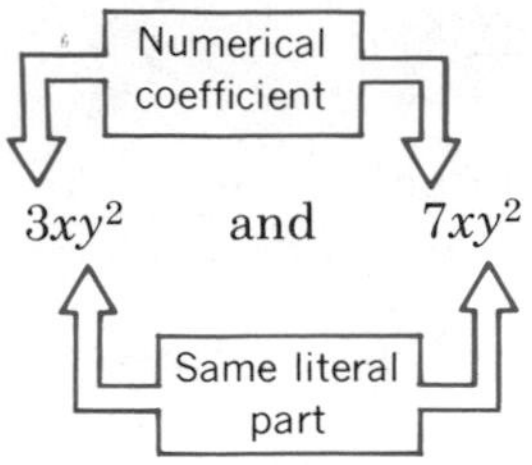

are like terms.

But

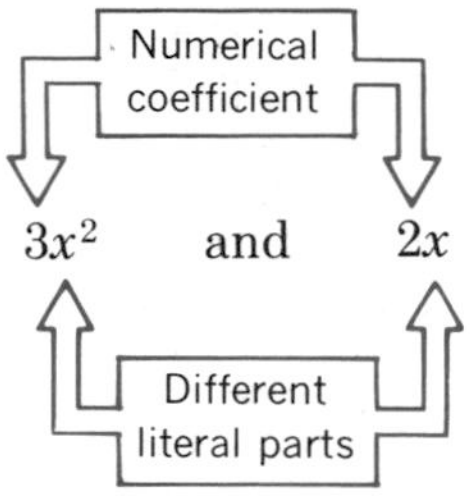

are unlike terms. The literal parts, x and x^2, are different.

To add or subtract like terms, add or subtract their numerical coefficients and keep the same literal part. For example:

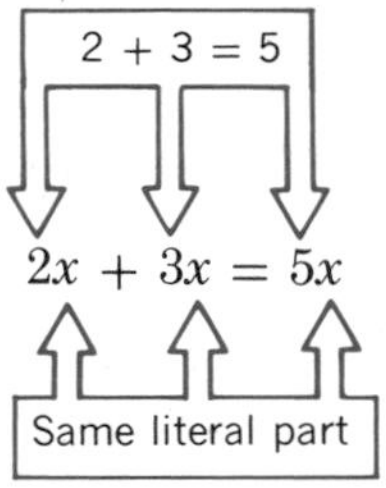

We are adding like quantities:

$$\boxed{x}+\boxed{x} \quad + \quad \boxed{x}+\boxed{x}+\boxed{x} \quad = \quad \boxed{x}+\boxed{x}+\boxed{x}+\boxed{x}+\boxed{x}$$

Another example:

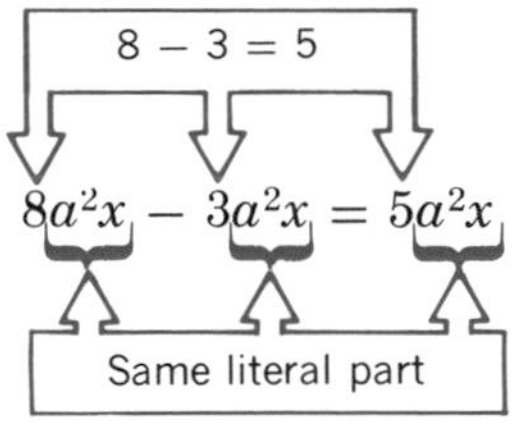

Try these problems for practice.

(a) $12d^2 + 7d^2 =$ ______ (b) $2ax - ax + 5ax =$ ______

(c) $3(y + 1) + 9(y + 1) =$ ______ (d) $8x^2 + 2xy - 2x^2 =$ ______

(e) $x + 6x - 2x =$ ______ (f) $4xy - xy + 3xy =$ ______

Turn to **5** to check your answers.

---

5 (a) $12d^2 + 7d^2 = 19d^2$
(b) $2ax - ax + 5ax = 6ax$ The term ax is equal to $1 \cdot ax$.
(c) $3(y + 1) + 9(y + 1) = 12(y + 1)$
(d) $8x^2 + 2xy - 2x^2 = 6x^2 + 2xy$

We cannot combine the x^2-term and the xy-term because the literal parts are not the same.

(e) $x + 6x - 2x = 5x$
(f) $4xy - xy + 3xy = 6xy$

In general, to simplify a series of terms being added or subtracted, first group together like terms, then add or subtract. For example,

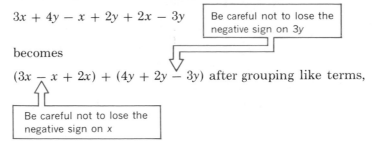

$3x + 4y - x + 2y + 2x - 3y$

Be careful not to lose the negative sign on $3y$

becomes

$(3x - x + 2x) + (4y + 2y - 3y)$ after grouping like terms,

Be careful not to lose the negative sign on x

or

$4x + 3y$ after adding and subtracting like terms.

Simplify the following expressions by adding and subtracting like terms.

(a) $5x + 4xy - 2x - 3xy$
(b) $3ab^2 + a^2b - ab^2 + 3a^2b - a^2b$
(c) $x + 2y - 3z - 2x - y + 5z - x + 2y - z$
(d) $17pq + 9ps - 6pq + ps - 6ps - pq$
(e) $4x^2 - x^2 + 2x + 2x^2 + x$

Check your answers in **6**.

---

6 (a) $(5x - 2x) + (4xy - 3xy) = 3x + xy$
(b) $(3ab^2 - ab^2) + (a^2b + 3a^2b - a^2b) = 2ab^2 + 3a^2b$
(c) $(x - 2x - x) + (2y - y + 2y) + (-3z + 5z - z) = -2x + 3y + z$
(d) $(17pq - 6pq - pq) + (9ps + ps - 6ps) = 10pq + 4ps$
(e) $(4x^2 - x^2 + 2x^2) + (2x + x) = 5x^2 + 3x$

Operation 2. Adding and Subtracting Expressions

Parentheses are used in algebra to group together terms that are to be treated as a unit. Adding and subtracting expressions usually involves working with parentheses. For example, to add

$(a + b) + (a + d)$

First, remove all parentheses: $(a + b) + (a + d)$
$= a + b + a + d$

Second, add like terms: $= a + a + b + d$
$= 2a + b + d$

Removing parentheses can be a tricky business. Remember these two rules:

Rule 1. If the parenthesis has a plus sign in front, simply remove the parentheses.

Example: $1 + (3x + y) = 1 + 3x + y$

Same sign

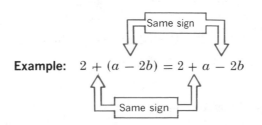

Example: $2 + (a - 2b) = 2 + a - 2b$

Same sign

Rule 2. If the parenthesis has a negative sign in front, change the sign of each term inside, then remove the parentheses.

Change sign from + to −

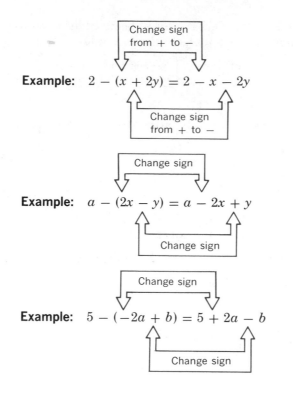

Example: $2 - (x + 2y) = 2 - x - 2y$

Change sign from + to −

Change sign

Example: $a - (2x - y) = a - 2x + y$

Change sign

Change sign

Example: $5 - (-2a + b) = 5 + 2a - b$

Change sign

Simplify the following expressions by using these two rules to remove parentheses.

(a) $x + (2y - a^2)$ (b) $4 - (x^2 - y^2)$
(c) $-(x + 1) + (y + 2)$ (d) $ab - (a - b)$
(e) $(x + y) - (x - y)$ (f) $-(-x - 2y) - (a + 2b)$
(g) $3 + (-2p - q^2)$ (h) $-(x - y) + (-3x^2 + y^2)$

Check your answers in **7**.

---

7 (a) $x + 2y - a^2$ (b) $4 - x^2 + y^2$
(c) $-x + y + 1$ (d) $ab - a + b$
(e) $2y$ (f) $x + 2y - a - 2b$
(g) $3 - 2p - q^2$ (h) $-x + y - 3x^2 + y^2$

A third rule is needed when a multiplier is in front of the parentheses.

Rule 3. If the parenthesis has a multiplier in front, multiply each term inside the parentheses by the multiplier.

Example: $+2(a + b) = +2a + 2b$

Think of this as $(+2)(a + b) = (+2)a + (+2)b$
$= +2a + 2b$

Each term inside the parentheses is multiplied by $+2$.

Example: $-2(x + y) = -2x - 2y$

Think of this as $(-2)(x + y) = (-2)x + (-2)y$

Each term inside the parentheses is multiplied by -2.

Example: $-2(a - b) = -2a + 2b$

Think of this as $(-2)a + (-2)(-b) = -2a + 2b$

Example: $-(x - y) = (-1)(x - y)$
$= (-1)(x) + (-1)(-y) = -x + y$

When you multiply negative numbers, you may need to review the arithmetic of negative numbers starting in frame **41** on page 116.

Notice that we must multiply *every* term inside the parentheses by the number outside the parentheses. Once the parentheses have been removed, you can add and subtract like terms, as explained in Operation 1.

Simplify the following expressions by removing parentheses. Use the three rules.

(a) $2(2x - 3y)$ (b) $1 - 4(x + 2y)$
(c) $a - 2(b - 2a)$ (d) $x^2 - 3(x - y)$
(e) $x - 2(-y - 2x)$ (f) $p^2 - 2(-p - 3t)$

Check your answers in **8**.

---

8 (a) $4x - 6y$ (b) $1 - 4x - 8y$
(c) $5a - 2b$ (d) $x^2 - 3x + 3y$
(e) $5x + 2y$ (f) $p^2 + 2p + 6t$

Once you can simplify expressions by removing parentheses, it is easy to add and subtract them. For example,

$(3x - y) - 2(x - 2y) = 3x - y - 2x + 4y$ Simplify by removing parentheses

$= \underbrace{3x - 2x} - \underbrace{y + 4y}$ Group like terms

$= x + 3y$ Combine like terms

Try these problems for practice:

(a) $(3y + 2) + 2(y + 1)$ (b) $(2x + 1) + 3(4 - x)$
(c) $(a + b) - (a - b)$ (d) $2(a + b) - 2(a - b)$
(e) $2(x - y) - 3(y - x)$ (f) $2(x + 1) - 3(x - 2)$
(g) $(x^2 - 2x) - 2(x - 2x^2)$ (h) $-2(3x - 5) - 4(x - 1)$

Check your answers in **9**.

---

9 (a) $5y + 4$ (b) $-x + 13$
(c) $2b$ (d) $4b$
(e) $5x - 5y$ (f) $-x + 8$
(g) $5x^2 - 4x$ (h) $-10x + 14$

Operation 3. Multiplying Simple Factors

In order to multiply two terms such as $2x$ and $3xy$, first remember that $2x$ means 2 times x. Second, recall from arithmetic that the order in which you do multiplications does not make a difference. For example, in arithmetic $2 \cdot 3 \cdot 4 = (2 \cdot 4) \cdot 3 = (3 \cdot 4) \cdot 2$ and in algebra

$a \cdot b \cdot c = (a \cdot c) \cdot b = (c \cdot b) \cdot a$ or $2 \cdot x \cdot 3 \cdot x \cdot y = 2 \cdot 3 \cdot x \cdot x \cdot y$

remember that

$x^2 = x \cdot x$
$x^3 = x \cdot x \cdot x$

and so on.

Therefore, $2x \cdot 3xy = 2 \cdot 3 \cdot x \cdot x \cdot y$
$= 6x^2y$

The following examples show how to multiply two terms.

Example 1: $a \cdot 2a = a \cdot 2 \cdot a$

$= 2 \cdot \underbrace{a \cdot a}$ Group like factors together,

$= 2 \cdot a^2$

$= 2a^2$

Example 2: $2x^2 \cdot 3xy = 2 \cdot x \cdot x \cdot 3 \cdot x \cdot y$

$$= \underbrace{2 \cdot 3} \cdot \underbrace{x \cdot x \cdot x} \cdot y \qquad \text{Group like factors together.}$$

$$= 6 \cdot x^3 \cdot y$$

$$= 6x^3y$$

Example 3: $3x^2yz \cdot 2xy = 3 \cdot x^2 \cdot y \cdot z \cdot 2 \cdot x \cdot y$

$$= \underbrace{3 \cdot 2} \cdot \underbrace{x^2 \cdot x} \cdot \underbrace{y \cdot y} \cdot z$$

$$= 6 \cdot x^3 \cdot y^2 \cdot z$$

$$= 6x^3y^2z$$

Remember to group like factors together before multiplying.

If you need to review exponents, return to frame **28** on page 108.

Now try the following problems.

(a) $x \cdot y$ = ______ (b) $2x \cdot 3x$ = ______

(c) $2x \cdot 5xy$ = ______ (d) $4a^2b \cdot 2a$ = ______

(e) $3x^2y \cdot 4xy^2$ = ______ (f) $5xyz \cdot 2ax^2$ = ______

(g) $3x \cdot 2y^2 \cdot 2x^2y$ = ______ (h) $x^2y^2 \cdot 2x \cdot y^2$ = ______

(i) $2x^2(x + 3x^2)$ = ______ (j) $-2a^2b(a^2 - 3b^2)$ = ______

The answers are in **10**.

MULTIPLYING NUMBERS IN EXPONENTIAL FORM

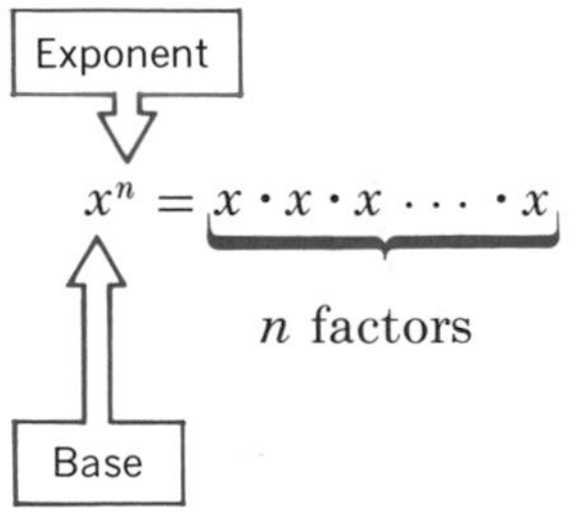

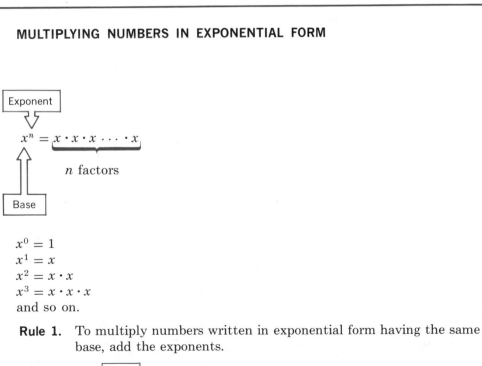

$x^0 = 1$
$x^1 = x$
$x^2 = x \cdot x$
$x^3 = x \cdot x \cdot x$
and so on.

Rule 1. To multiply numbers written in exponential form having the same base, add the exponents.

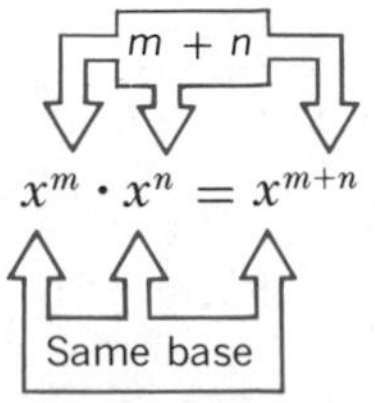

Example: $x^2 \cdot x^3 = x^{2+3}$
$= x^5$

> **Rule 2.** To divide numbers written in exponential form having the same base, subtract the exponents.
>
> $$\frac{x^m}{x^n} = x^{m-n} \qquad \textbf{Example: } \frac{x^7}{x^3} = x^{7-3} = x^4$$
>
> **Rule 3.** Use negative exponents to indicate a reciprocal.
>
> $$x^{-n} = \frac{1}{x^n} \qquad \textbf{Example: } x^{-3} = \frac{1}{x^3}$$
>
> $$\textbf{Example: } 2^{-1} = \frac{1}{2}$$

10

(a) xy	(b) $6x^2$
(c) $10x^2y$	(d) $8a^3b$
(e) $12x^3y^3$	(f) $10ax^3yz$
(g) $12x^3y^3$	(h) $2x^3y^4$
(i) $2x^3 + 6x^4$	(j) $-2a^4b + 6a^2b^3$

Were the last two problems tricky for you? Try it this way:

(i) $$\begin{aligned} 2x^2(x + 3x^2) &= (2x^2)(x) + (2x^2)(3x^2) \\ &= 2 \cdot \underbrace{x^2 \cdot x} + \underbrace{2 \cdot 3} \cdot \underbrace{x^2 \cdot x^2} \\ &= 2 \cdot x^3 + 6 \cdot x^4 \\ &= 2x^3 + 6x^4 \end{aligned}$$

(j) $$\begin{aligned} -2a^2b(a^2 - 3b^2) &= (-2a^2b)(a^2) + (-2a^2b)(-3b^2) \\ &= (-2) \cdot \underbrace{a^2 \cdot a^2} \cdot b + \underbrace{(-2) \cdot (-3)}\, a^2 \cdot \underbrace{b \cdot b^2} \\ &= -2 \cdot a^4 \cdot b + 6 \cdot a^2 \cdot b^3 \\ &= -2a^4b + 6a^2b^3 \end{aligned}$$

Evaluating Formulas

One of the most useful algebra skills for any technical or practical work involves finding the value of an algebra expression when the letters are given numerical values. A *formula* is a rule for calculating the numerical value of one quantity from the values of other quantities. The formula or rule is usually written in mathematical form because algebra gives a brief, convenient to use, and easy to remember form for the rule. Here are a few examples of rules and formulas used in the trades.

1. *Rule:* The voltage across a simple resistor is equal to the product of the current through the resistor and the value of its resistance.

   Formula: $V = iR$

2. *Rule:* The cost of setting type is equal to the total ems set in the job multiplied by the hourly wage divided by the rate at which the type is set, in ems per hour.

   Formula: $C = \dfrac{TH}{E}$

3. *Rule:* The number of standard bricks needed to build a wall is about 21 times the volume of the wall.

   Formula: $N = 21\,LWH$

Evaluating a formula or algebra expression means to find its value by substituting numbers for the letters in the expression. For example, in retail stores the following formula is used:

$M = R - C$ where M is the markup on an item,
R is the retail selling price,
and C is the original cost.

Find M if $R = \$25$ and $C = \$21$

$M =$ ________

Check your work in **11**.

---

11 $M = \$25 - \$21 = \$4$

Easy? Of course. Simply substitute the numbers for the correct letters and then do the arithmetic. A formula is a recipe for a calculation.

Try another. Automotive engineers use the following formula to calculate the horsepower rating of an engine.

$H = \dfrac{D^2N}{2.5}$ where D is the diameter of a cylinder in inches,

and N is the number of cylinders.

Find H when $D = 3\frac{1}{2}''$ and $N = 6$.

Check your work in **12**.

---

12 $H = \dfrac{(3.5)^2(6)}{2.5}$

$H = 29.4$ hp or roughly 29 horsepower.

To be certain you do it correctly, follow this two-step process:

Step 1 The numbers being substituted should be placed in parentheses, and then substituted in the formula.

Step 2 Do the arithemetic carefully *after* the numbers are substituted.

Example: Find $A = x + 2$ for $x = 5$

Step 1 $A = (5) + 2$ put 5 in parentheses

Step 2 $A = 5 + 2$ do the arithmetic

$A = 7$

Example: Find $B = x^2 - y$ for $x = 4$, $y = -3$

Step 1 $B = (4)^2 - (-3)$ put the numbers in parentheses

Step 2 $B = 16 + 3$ do the arithmetic

$B = 19$

Using parentheses in this way may seem like extra work for you, but it is the key to avoiding mistakes when evaluating formulas.

Evaluate the following formulas.

(a) $D = 2R$ for $R = 3.45''$
(b) $W = T - C$ for $T = 1420$ lb, $C = 385$ lb

(c) $P = 0.433H$ for $H = 11.4$ in.
(d) $A = bh - 2$ for $b = 1.75''$, $h = 4.2''$
(e) $P = 2(8 + x)$ for $x = -3$

Use the two-step process and check your work in **13**.

---

13 (a) $D = 2(3.45) = 6.90''$

(b) $W = (1420) - (385)$
$= 1420 - 385$
$= 1035$ lb

(c) $P = 0.433(11.4)$
$= 4.9362''$

(d) $A = (1.75)(4.2) - 2$
$= 7.35 - 2$
$= 5.35$ sq in.

(e) $P = 2(8 + (-3))$
$P = 2(8 - 3)$
$P = 2 \cdot 5$
$P = 10$

Notice how confusing problem (e) would be if you did not use parentheses around the number being substituted. A careless worker would calculate this as

$P = 2 \cdot 8 - 3$ or $16 - 3$ or 13

To be certain you know how to handle formulas that contain parentheses, find the value of each of the following number equations.

(a) $(1 + 3) + 2 =$ ______ (b) $(3 \times 2) + 5 =$ ______

(c) $4(2 + 1) =$ ______ (d) $2(1 + 4) + 3 =$ ______

(e) $1 + 2(4 + 1) =$ ______ (f) $4 + 2(5 - 2) =$ ______

Check your answers in **14**.

---

14 (a) $(1 + 3) + 2 = 4 + 2$
$= 6$

(b) $(3 \times 2) + 5 = 6 + 5$
$= 11$

(c) $4(2 + 1) = 4 \cdot 3$
$= 12$

(d) $2(1 + 4) + 3 = 2 \cdot 5 + 3$
$= 10 + 3$
$= 13$

(e) $1 + 2(4 + 1) = 1 + 2 \cdot 5$
$= 1 + 10$
$= 11$

(f) $4 + 2(5 - 2) = 4 + 2 \cdot 3$
$= 4 + 6$
$= 10$

When you evaluate formulas remember the following helpful hints.

Hint 1. Do the operations inside the parentheses first.

Example: $(2 + 3) + 4 = 5 + 4 = 9$

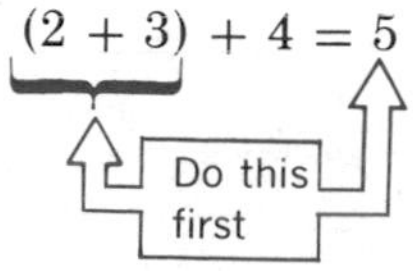

Example: $4(3 + 5) = 4 \cdot 8 = 32$

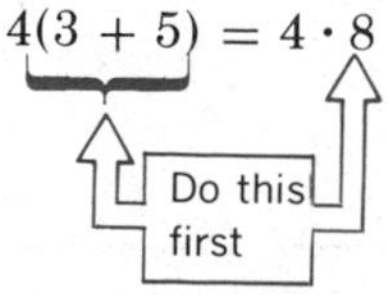

Hint 2. If the formula contains several sets of parentheses, do the calculations inside each first, then combine them.

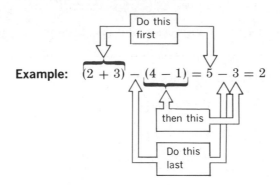

Example: $(2 + 3) - (4 - 1) = 5 - 3 = 2$

Do this first

then this

Do this last

Parentheses tell you the order in which to do the arithmetic.

Hint 3. If the formula contains a square, cube, or some other power, find the value of that factor first.

Example: $A = 3.14R^2$ — Find this value first

$A = 3.14 \cdot 3^2$ — Then multiply

$A = 3.14 \cdot 9$

$A = 28.26$

Hint 4. If the formula is a sum or difference of terms, find a numerical value for each term first, then add or subtract.

Example: $P = ab - s^2$ — Find this first

$P = 3 \cdot 4 - 2^2$ — Find this next

$= 12 - 4$

$= 8$ — Subtract last

Hint 5. Unless parentheses tell you to do otherwise, always do multiplications and divisions before adding or subtracting.

Example: $4 + 2 \cdot 6 = 4 + 12 = 16$

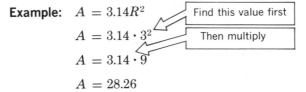

Multiply first

To summarize, when you evaluate formulas follow this order:

1. Do any operations inside parentheses.
2. Find all powers.
3. Do all multiplications or divisions.
4. Do all additions or subtractions.

Use these hints to evaluate the following formulas:

$a = 3,\ b = 4,\ c = 6$

(a) $3ab$ = ________ (b) $2a^2c$ = ________

(c) $2a^2 - b^2$ = ________ (d) $3a^2bc - ab$ = ________

(e) $2a + b$ = ________ (f) $a + (2b - c)$ = ________

(g) $2(a + b)$ = ________ (h) $(a^2 - 1) - (2 + b)$ = ________

Check your work in **15**

15 (a) $3(3)(4) = 36$

(b) $2(3)^2(6) = 2 \cdot 9 \cdot 6 = 108$

(c) $2(3)^2 - (4)^2 = 2 \cdot 9 - 16$
$= 18 - 16$
$= 2$

(d) $3(3)^2(4)(6) - (3)(4) = (3 \cdot 9 \cdot 4 \cdot 6) - (3 \cdot 4)$
$= 648 - 12$
$= 636$

(e) $2(3) + (4) = 2 \cdot 3 + 4$
$= 6 + 4$
$= 10$

(f) $(3) + (2(4) - (6)) = 3 + (2 \cdot 4 - 6)$
$= 3 + (8 - 6)$
$= 3 + 2$
$= 5$

(g) $2((3) + (4)) = 2 \cdot (3 + 4)$
$= 2 \cdot 7$
$= 14$

(h) $((3)^2 - 1) - (2 + (4)) = (9 - 1) - (2 + 4)$
$= 8 - 6$
$= 2$

Avoid the temptation to combine steps when you evaluate formulas. Take it slowly and carefully, follow our hints, and you will arrive at the correct answer. Rush through problems like these and you usually make mistakes.

Now turn to **16** for a set of practice problems in working with algebra quantities and evaluating formulas.

---

16 **Exercises 6-1 Algebra Operations**

A. Simplify by adding or subtracting like terms:

1. $3y + y + 5y$
2. $4x^2y + 5x^2y$
3. $E + 2E + 3E$
4. $ax - 5ax$
5. $9B - 2B$
6. $3m - 3m$
7. $3x^2 - 5x^2$
8. $4x + 7y + 6x + 9y$
9. $6R + 2R^2 - R$
10. $1.4A + 0.05A - 0.8A^2$
11. $x - \frac{1}{2}x - \frac{1}{4}x - \frac{1}{8}x$
12. $x + 2\frac{1}{2}x - 5\frac{1}{2}x$
13. $2 + W - 4.1W - \frac{1}{2}$
14. $q - p - 1\frac{1}{2}p$

B. Simplify by multiplying:

1. $5 \cdot 4x$
2. $(a^2)(a^3)$
3. $-3R(-2R)$
4. $3x \cdot 3x$
5. $(4x^2y)(-2xy^3)$
6. $2x \cdot 2x \cdot 2x$
7. $0.4a \cdot 1.5a$
8. $x \cdot x \cdot A \cdot x \cdot 2x \cdot A^2$
9. $\frac{1}{2}Q \cdot \frac{1}{2}Q \cdot \frac{1}{2}Q \cdot \frac{1}{2}Q$
10. $(pq^2)(\frac{1}{3}pq)(2.4p^2q)$
11. $(-2M)(3M^2)(-4M^3)$
12. $2x(3x - 1)$
13. $-2(1 - 2y)$
14. $ab(a^2 - b^2)$

C. Find the value of each of these formulas for $x = 2$, $y = 3$, $z = 4$, $R = 5$:

1. $A = 3x$
2. $D = 2R - y$
3. $T = x^2 + y^2$
4. $H = 2x + 3y - z$

5. $K = 3z - x^2$

6. $Q = 2xyz - 10$

7. $F = 2(x + y^2) - 3$

8. $W = 3(y - 1)$

9. $L = 3R - 2(y^2 - x)$

10. $A = R^2 - y^2 - xz$

11. $B = 3R - y + 1$

12. $F = 3(R - y + 1)$

D. Find the value of each of the following formulas:

1. $A = 2(x + y) - 1$ for $x = 2,\ y = 4$
2. $V = (L + W)(2L + W)$ for $L = 3.5,\ W = 5.2$
3. $I = PRT$ for $P = 150,\ R = 0.05,\ T = 2$
4. $H = 2(a^2 + b^2)$ for $a = 2,\ b = 1$
5. $T = \dfrac{(A + B)H}{2}$ for $A = 3.26,\ B = 7.15,\ H = 4.4$
6. $V = \dfrac{\pi D^2 H}{4}$ for $\pi = 3.14,\ D = 6.25,\ H = 7.2$
7. $P = \dfrac{NR(T + 273)}{V}$ for $N = 5,\ R = 0.08,\ T = 27,\ V = 3$
8. $W = D(AB - \pi R^2)H$ for $D = 8,\ A = 8,\ B = 6,\ \pi = 3.14,\ R = 2,\ H = 10$
9. $V = LWH$ for $L = 16.25,\ W = 3.1,\ H = 2.4$
10. $V = \pi R^2 A$ for $\pi = 3.14,\ R = 3.2,\ A = 0.085$

E. Practical Problems

1. The perimeter of a rectangle is given by the formula $P = 2L + 2W$, where L is the length of the rectangle and W is its width. Find P when L is $8\frac{1}{2}$ in. and W is 11 in.

2. The current in a simple electrical circuit is given by the formula $i = V/R$, where V is the voltage and R is the resistance of the circuit. Find the current in a circuit whose resistance is 10 ohms and which is connected across a 120 volt power source.

3. Find the power used in an electric light bulb, $P = i^2R$, when the current $i = 0.8$ ampere and the resistance $R = 150$ ohms. P will be in watts.

4. Find the surface area of a sphere, $A = 4\pi R^2$, when π equals roughly 3.14 and $R = 10$ cm.

5. The Fahrenheit temperature F is related to the Celsius temperature C by the formula $F = \frac{9}{5}C + 32$. Find the Fahrenheit temperature when $C = 40°$.

6. The volume of a round steel bar depends on its length L and diameter D according to the formula $V = \pi D^2L/4$. Find the volume of a bar 20 inches long and 3 inches in diameter. Use $\pi = 3.14$.

7. If D dollars is invested at p percent interest for t years, the amount A of the investment is

$$A = D\left(1 + \frac{pt}{100}\right)$$

Find A if $D = \$1000$, $p = 4\%$, and $t = 5$ years.

8. The total resistance R of two resistances a and b connected in parallel is $R = ab/(a + b)$. What is the total resistance if $a = 200$ ohms and $b = 300$ ohms?

9. The area of a trapezoid is $T = \dfrac{(A + B)H}{2}$, where A and B are the lengths of

its parallel sides and H is the height. Find the area of the trapezoid shown.

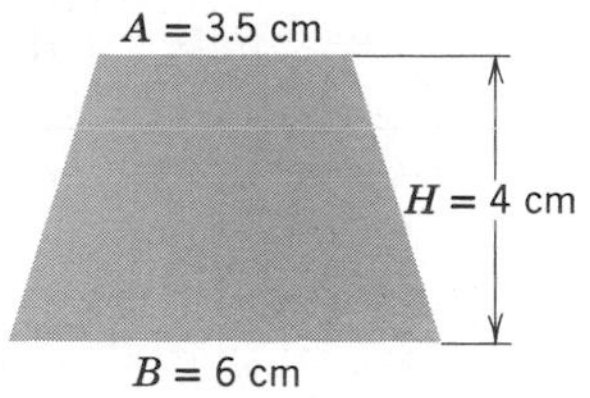

$T =$ ______

10. Find the wall thickness T of the tubing for each of the following:

	D, outside diameter	d, inside diameter
a.	2.125″	1.500″
b.	0.785″	0.548″
c.	1.400 cm	0.875 cm
d.	$\frac{15}{16}$″	$\frac{5}{8}$″

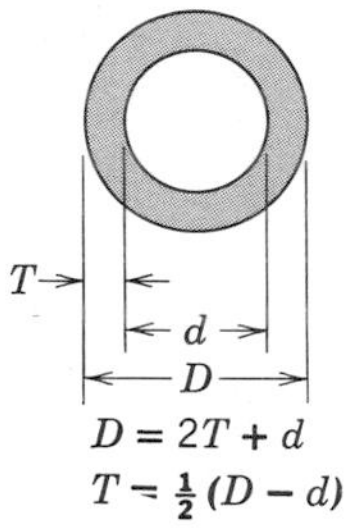

11. To make a right-angle inside bend in sheet metal, the length of sheet used is given by the formula $L = x + y + \frac{1}{2}T$. Find L when $x = 6\frac{1}{4}$″, $y = 11\frac{7}{8}$″, $T = \frac{1}{4}$″.

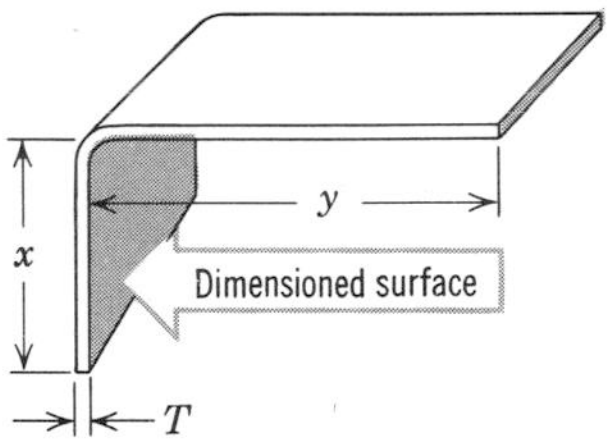

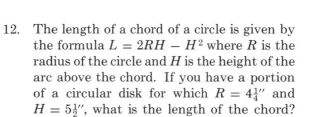

12. The length of a chord of a circle is given by the formula $L = 2RH - H^2$ where R is the radius of the circle and H is the height of the arc above the chord. If you have a portion of a circular disk for which $R = 4\frac{1}{4}$″ and $H = 5\frac{1}{2}$″, what is the length of the chord?

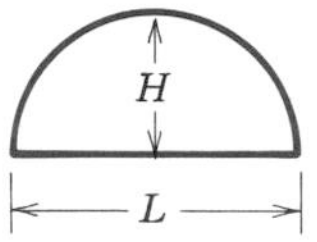

13. The rope capacity of a drum is given by the formula $L = ABC(A + D)$. How many feet of $\frac{1}{2}$″ rope can be wound on a drum where $A = 6$″, $C = 30$″, $D = 24$″, and $B = 1.05$ for $\frac{1}{2}$″ rope?

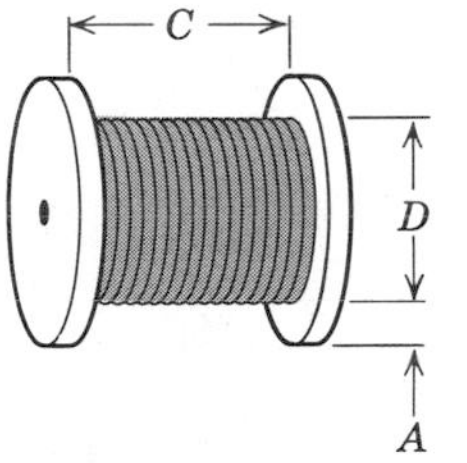

14. A millwright uses the following formula to find the required length of a pulley belt. $L = 2C + 1.57(D + d) + (D + d)/4C$. Find the length of belt needed if
$C = 36$″ between pulley centers
$D = 24$″ follower
$d = 4$″ driver

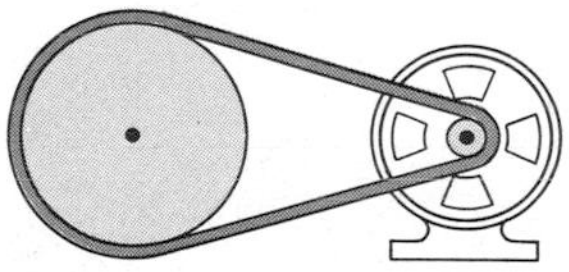

15. An electrician uses a bridge circuit to locate a ground in an underground cable several miles long. The formula

$$\frac{R_1}{L - x} = \frac{R_2}{x} \quad \text{or} \quad x = \frac{R_2 L}{R_1 + R_2}$$

is used to find x, the distance to the ground. Find x if $R_1 = 750$ ohms, $R_2 = 250$ ohms, and $L = 4000$ ft.

16. The cutting speed of a lathe is the rate, in feet per minute, that the revolving workpiece travels past the cutting edge of the tool. Machinists use the following formula to calculate cutting speed:

$$\text{Cutting speed, } C = \frac{3.1416\,DN}{12}$$

where D is the diameter of the work and N is the turning rate in rpm. Find the cutting speed if a steel shaft 3.25 inches in diameter is turned at 210 rpm. (Round to one decimal place.)

17. Find the power load P in kilowatts of an electrical circuit that takes a current I of 12 amps at a voltage V of 220 volts, if

$$P = \frac{VI}{100}$$

18. A jet engine developing T lb of thrust and driving an airplane at V mph, has a thrust horsepower, H, given approximately by the formula

$$H = \frac{TV}{375} \qquad \text{or} \qquad V = \frac{375H}{T}$$

Find the airspeed V if $H = 16{,}000$ hp and $T = 10{,}000$ lb.

19. The resistance R of a conductor is given by the formula $R = PL/A$ or $L = AR/P$

where P is the coefficient of resistivity
L is the length of conductor
A is the cross-sectional area of the conductor

Find the length in cm of #16 Nichrome wire needed to obtain a resistance of 8 ohms. For this wire $P = 0.000113$ ohm-cm and $A = 0.013\ \text{cm}^2$. (Round to the nearest centimeter.)

20. The pressure P and total force F exerted on a piston of diameter D are approximately related by the equation

$$P = \frac{1.27F}{D^2} \qquad \text{or} \qquad F = \frac{PD^2}{1.27}$$

Find the total force on a piston of diameter 3.25″ if the pressure exerted on it is 150 lb/sq in. (Round to the nearest 50 lb.)

F. Calculator Problems

1. Find the volume of a sphere of radius $R = 1.0076''$

$$\text{Volume } V = \frac{4\pi R^3}{3} \qquad \text{Use } \pi = 3.1416$$

(Round to 0.0001 cu in.)

2. How many minutes will it take a lathe to make 17 cuts each 24.5″ in length on a steel shaft if the tool feed F is 0.065 in. per revolution and the shaft turns at 163 rpm? Use the formula

$$T = \frac{LN}{FR}$$

where T = cutting time, min
N = number of cuts
L = length of cut, in.
F = tool feed rate, in./rev
R = workpiece, rpm

(Round to 0.1 minute.)

3. Find the area of each of the following circular holes using the formula $A = \frac{\pi D^2}{4}$.

(a) $D = 1.0004''$
(b) $D = \frac{7}{8}''$
(c) $D = 4.1275''$
(d) $D = 2.0605$ cm

Use $\pi = 3.1416$ and round to 0.0001 unit.

4. Suppose a steel band was placed tightly around the earth at the equator. If the temperature of the steel is raised 1° F, the metal will expand 0.000006 in. each inch. How much space would there be between the earth and the steel band if the temperature was raised 1° F? Use the formula

$$D \text{ (in ft)} = \frac{(0.000006)(\text{diameter of the earth})(5280)}{\pi}$$

where diameter of the earth = 7917 miles
$\pi = 3.1415927$

(Round to two digits.)

When you have completed these exercises, check your answers on page 552, then turn to **17** to learn how to solve algebra equations.

6-2 SOLVING EQUATIONS AND FORMULAS

17 An arithmetic equation such as $3 + 2 = 5$ means that the number named on the left $(3 + 2)$ is the same as the number named on the right (5).

An algebra equation such as $x + 3 = 7$ is a statement that the sum of some number x and 3 is equal to 7. If we choose the correct value for x, then the number $x + 3$ will be equal to 7.

x is a *variable,* a symbol that stands for a number, a blank to be filled. Many numbers might be put in the space, but only one makes the equation a true statement.

Finding the missing numbers in the following arithmetic equations:

(a) $37 + \_\_\_\_\_\_ = 58$ (b) $\_\_\_\_\_\_ - 15 = 29$

(c) $4 \times \_\_\_\_\_\_ = 52$ (d) $28 \div \_\_\_\_\_\_ = 4$

Puzzle them out, then turn to **18**.

---

18 (a) $37 + 21 = 58$ (b) $44 - 15 = 29$

(c) $4 \times 13 = 52$ (d) $28 \div 7 = 4$

We could have written these equations as

$37 + A = 58$

$B - 15 = 29$

$4C = 52$

$\frac{28}{D} = 4$

Of course any letters would do in place of A, B, C, and D in these algebra equations.

How did you solve these equations? You probably "eye-balled" them—mentally juggled the other information in the equation until you found a number that made the equation true. Solving algebra equations is very similar except that we can't "eyeball" it entirely. We need certain and systematic ways of solving the equation that will produce the correct answer quickly every time.

In this section you will learn first what a solution to an algebra equation is—how to recognize it if you stumble over it in the dark—then what a linear equation looks like, and finally how to solve linear equations.

Each value of the variable that makes an equation true is called a *solution* of the equation. For example, the solution of $x + 3 = 7$ is $x = 4$.

Check: $(4) + 3 = 7$

Another example: The solution of the equation

$$2x - 9 = 18 - 7x \qquad \text{is } x = 3$$

Check:

$$\begin{aligned} 2(3) - 9 &= 18 - 7(3) \\ 6 - 9 &= 18 - 21 \\ -3 &= -3 \end{aligned}$$

For certain equations more than one value of the variable may make the equation true. For example, the equation $x^2 + 6 = 5x$

is true for $x = 2$,

Check:

$$\begin{aligned} (2)^2 + 6 &= 5(2) \\ 4 + 6 &= 5 \cdot 2 \\ 10 &= 10 \end{aligned}$$

and it is also true for $x = 3$.

Check:

$$\begin{aligned} (3)^2 + 6 &= 5(3) \\ 9 + 6 &= 5 \cdot 3 \\ 15 &= 15 \end{aligned}$$

Use your knowledge of arithmetic to find the solution of each of the following equations.

(a) $4 + x = 11$ (b) $x - 1 = 6$
(c) $x + 2 = 9$ (d) $8 - x = 1$

Check your solution in **19**.

---

19 $x = 7$. Every equation has the same solution. If you replace the letter x by the number 7, all four equations will be true.

Equations with the exact same solution are called *equivalent* equations. The equations $2x + 7 = 13$ and $3x = 9$ are equivalent because substituting the value 3 for x makes them both true.

We say that an equation written with the variable x is *solved* if it can be put in the form

$x = \square$ where $\square$ is some number.

For example, the solution to the equation

$2x - 1 = 7$ is $x = 4$

because $2(4) - 1 = 7$
$2(4) - 1 = 7$
or $8 - 1 = 7$
is a true statement.

Equations as simple as the one above are easy to solve by guessing, but guessing is not a very happy way to do mathematics. We need some sort of rule that will enable us to rewrite the equation to be solved ($2x - 1 = 7$, for example) as an equivalent solution equation ($x = 4$).

The general rule is to treat every equation as a balance of the two sides.

$2x = 6$

$3x - 4 = 8 - x$

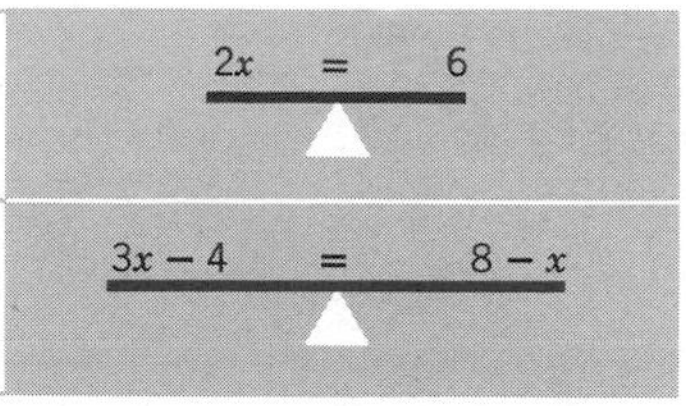

Any changes made in the equation must not disturb this balance.

Any operation performed on one side of the equation must also be performed on the other side.

Two kinds of balancing operations may be used.

Example:

1. Adding or subtracting a number on both sides of the equation does not change the balance.

and

2. Multiplying or dividing both sides of the equation by a number (but not zero) does not change the balance.

Original equation: $a = b$

$a + 2 = b + 2$

$a - 2 = b - 2$

$2 \cdot a = 2 \cdot b$

$\frac{a}{3} = \frac{b}{3}$

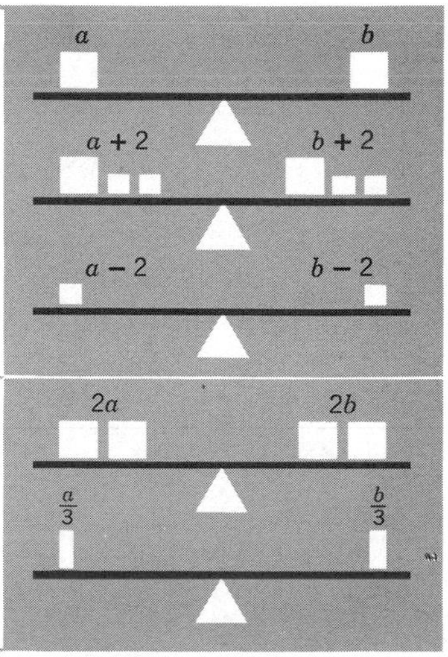

Let's work through an example.

Solve: $x - 4 = 2$.

Step 1 We want to change this equation to an equivalent equation with only x on the left, so we add 4 to each side of the equation.

$(x - 4) + 4 = (2) + 4$

Step 2 Combine terms.

$\underbrace{x - 4 + 4}_{0} = 2 + 4$

$x = 6$ Solution

Check: $(6) - 4 = 2$

$2 = 2$

Use these balancing operations to solve the equation

$8 + x = 14$

Check your work in **20**.

20 **Solve:** $8 + x = 14$

Step 1 We want to change this equation to an equivalent equation with only x on the left, so we subtract 8 from each side of the equation.

$$(8 + x) \; - 8 = (14) \; - 8$$

Step 2 Combine terms.

$$\underbrace{x + 8 \; - 8}_{0} = 14 \; - 8 \qquad (8 + x = x + 8, \quad \text{of course})$$

$x = 6$ Solution.

Check: $8 + (6) = 14$
$14 = 14$

Solve these in the same way.

(a) $x - 4 = 10$ (b) $12 + x = 27$ (c) $11 - x = 2$
(d) $x + 6 = 2$ (e) $8.4 = 3.1 + x$ (f) $4 = 1 - x$
(g) $6.7 + x = 0$ (h) $\frac{1}{4} = x - \frac{1}{2}$ (i) $2x + 6 = x$

The complete step-by-step solutions are in **21**.

---

21 (a) **Solve:** $x - 4 = 10$

$(x - 4) \; + 4 = (10) \; + 4$ Add 4 to each side.

$\underbrace{x - 4 \; + 4}_{0} = 10 \; + 4$ Combine terms.

$x = 14$ Solution.

Check: $(14) - 4 = 10$
$10 = 10$

(b) **Solve:** $12 + x = 27$

$(12 + x) \; - 12 = (27) \; - 12$ Subtract 12 from each side.

$\underbrace{x + 12 \; - 12}_{0} = 27 \; - 12$ Combine terms.
(Note that $12 + x = x + 12$.)

$x = 15$ Solution.

Check: $12 + (15) = 27$
$27 = 27$

(c) **Solve:** $11 - x = 2$

$(11 - x) \; - 11 = (2) \; - 11$ Subtract 11 from each side.

$-x + \underbrace{11 \; - 11}_{0} = 2 \; - 11$ Combine terms.

$-x = -9$
$x = 9$ Solution. Multiply each side by -1.

Check: $11 - (9) = 2$
$2 = 2$

(d) **Solve:** $x + 6 = 2$

$(x + 6) \; - 6 = (2) \; - 6$ Subtract 6 from each side.

$x + \underbrace{6 \; - 6}_{0} = 2 \; - 6$ Combine terms.

$x = -4$ Solution.

The solution is a negative number. Remember, any number, positive or negative, may be the solution of an equation.

Check: $(-4) + 6 = 2$
$2 = 2$

(e) **Solve:** $8.4 = 3.1 + x$

$(8.4) - 3.1 = (3.1 + x) - 3.1$ — Subtract 3.1 from each side.

$8.4 - 3.1 = x$ — Combine terms.

$5.3 = x$ — Decimal numbers often appear in practical problems.

or $x = 5.3$ Solution. — $5.3 = x$ is the same as $x = 5.3$.

Check: $8.4 = 3.1 + (5.3)$
$8.4 = 8.4$

(f) **Solve:** $4 = 1 - x$

$(4) - 1 = (1 - x) - 1$ — Subtract 1 from each side.

$3 = -x$

or $-x = 3$

or $x = -3$ Solution. — Always put your answer in the form $x =$ (some number).

Check: $4 = 1 - (-3)$
$4 = 1 + 3$

(g) **Solve:** $6.7 + x = 0$

$(6.7 + x) - 6.7 = 0 - 6.7$ — Subtract 6.7 from each side.

$x = -6.7$ Solution. — A negative number answer is reasonable.

Check: $6.7 + (-6.7) = 0$
$6.7 - 6.7 = 0$

(h) **Solve:** $\frac{1}{4} = x - \frac{1}{2}$

$\left(\frac{1}{4}\right) + \frac{1}{2} = \underbrace{\left(x - \frac{1}{2}\right) + \frac{1}{2}}_{0}$ — Add $\frac{1}{2}$ to each side.

$\frac{1}{4} + \frac{1}{2} = \frac{3}{4}$

$\frac{3}{4} = x$

or $x = \frac{3}{4}$ Solution.

Check: $\frac{1}{4} = \left(\frac{3}{4}\right) - \frac{1}{2}$
$\frac{1}{4} = \frac{1}{4}$

(i) **Solve:** $2x + 6 = x$

$(2x + 6) - x = \underbrace{x - x}_{0}$ — Subtract x from each side.

$x + 6 = 0$ Subtract 6 from each side.

$$x + \underbrace{6 - 6}_{0} = 0 - 6$$

$x = -6$ Solution.

Check: $2(-6) + 6 = (-6)$
$-12 + 6 = -6$
$-6 = -6$

Equations such as (i), where the variable appears on both sides, are very common in algebra. Solve them in the usual way by collecting all terms with the variable on the *same* side of the equation.

Here is a slightly different problem.

Solve: $2x = 14$

Step 1 We want to change this equation to an equivalent equation with x alone on the left, so we divide both sides by 2.

$$\frac{2x}{2} = \frac{14}{2}$$

Step 2 $x = \frac{14}{2}$ $\quad \frac{2x}{2} = \left(\frac{2}{2}\right)x = x$

$x = 7$ Solution.

Check: $2 \cdot (7) = 14$
$14 = 14$

Try this problem.

Solve: $3x = 39$

Look in **22** for the solution.

---

22 **Solve:** $3x = 39$

Step 1 $\frac{3x}{3} = \frac{39}{3}$ We want to change this equation to an equivalent equation with x alone on the left, so we divide both sides by 3.

$$\frac{3x}{3} = \left(\frac{3}{3}\right)x = x$$

Step 2 $x = \frac{39}{3}$

$x = 13$ Solution.

Check: $3 \cdot (13) = 39$
$39 = 39$

Solve these in the same way.

(a) $7x = 35$ (b) $\frac{1}{2}x = 14$

(c) $\frac{2x}{3} = 6$ (d) $3x = 0$

The step-by-step solutions are in **23**.

---

23 (a) **Solve:** $7x = 35$

Step 1 $\frac{7x}{7} = \frac{35}{7}$ Divide both sides by 7.

Step 2 $x = \frac{35}{7}$ $\qquad \frac{7x}{7} = \left(\frac{7}{7}\right)x = x$

$x = 5$ Solution.

Check: $7 \cdot (5) = 35$

$35 = 35$

(b) **Solve:** $\frac{1}{2}x = 14$

Step 1 $\left(\frac{1}{2}x\right) 2 = (14)\, 2$ Multiply both sides by 2.

Step 2 $\left(x \cdot \frac{1}{2}\right)2 = 14 \cdot 2$ $\qquad \frac{1}{2} \cdot x = x \cdot \frac{1}{2}$

$\underbrace{x \cdot \frac{1}{2} \cdot 2}_{1} = 28$ $\qquad \frac{1}{2} \cdot 2 = \frac{2}{2} = 1$

$x = 28$

Check: $\frac{1}{2} \cdot (28) = 14$

$\frac{28}{2} = 14$

$14 = 14$

(c) **Solve:** $\frac{2x}{3} = 6$

Step 1 $\left(\frac{2x}{3}\right) 3 = (6)\, 3$ Multiply both sides by 3.

$2x = 6 \cdot 3$ $\qquad \left(\frac{2x}{3}\right)3 = \frac{2 \cdot x \cdot \cancel{3}}{\cancel{3}} = 2 \cdot x$

$2x = 18$

Step 2 $\frac{2x}{2} = \frac{18}{2}$ Divide both sides by 2. $\qquad \frac{\cancel{2}x}{\cancel{2}} = x$

$x = 9$ Solution.

Check: $\frac{2 \cdot (9)}{3} = 6$

$\frac{18}{3} = 6$

$6 = 6$

(d) **Solve:** $3x = 0$

$\frac{\cancel{3}x}{\cancel{3}} = \frac{0}{3}$ Divide both sides by 3.

$x = 0$ Solution. Zero divided by any positive or negative number is still zero.

Check: $3(0) = 0$

$0 = 0$

Solving most simple algebra equations involves both kinds of operations: addition/subtraction and multiplication/division. For example,

Solve: $2x + 6 = 14$

Step 1 We want to change this equation to an equivalent equation with one x or terms with x on the left, so subtract 6 from both sides.

$$(2x + 6) - 6 = (14) - 6$$

$$\underbrace{2x + 6 - 6}_{0} = 14 - 6$$

Combine terms. (Be careful. Some careless students will add $2x$ and 6 to get $8x$! You can only add *like* terms.)

$$2x = 8$$

Now change this to an equivalent equation with only x on the left.

Step 2 $\frac{2x}{2} = \frac{8}{2}$

Divide both sides of the equation by 2.

$x = \frac{8}{2}$

$\frac{2x}{2} = x$

$x = 4$ Solution.

Check: $2 \cdot (4) + 6 = 14$
$8 + 6 = 14$
$14 = 14$

Try this one to test your understanding of the process.

Solve: $3x - 7 = 11$

Check your work in **24**.

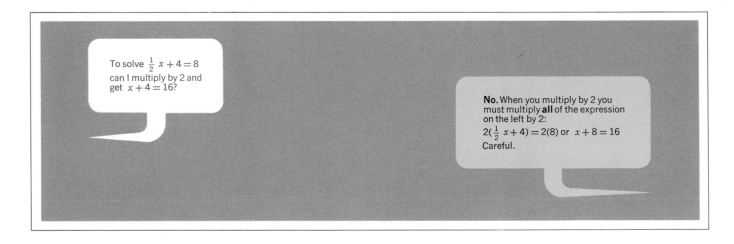

To solve $\frac{1}{2}x + 4 = 8$ can I multiply by 2 and get $x + 4 = 16$?
No. When you multiply by 2 you must multiply **all** of the expression on the left by 2:
$2(\frac{1}{2}x + 4) = 2(8)$ or $x + 8 = 16$

24 **Solve:** $3x - 7 = 11$

Step 1 $(3x - 7) + 7 = (11) + 7$

Add 7 to each side.

$$\underbrace{3x - 7 + 7}_{0} = 11 + 7$$

$$3x = 18$$

Step 2 $\frac{3x}{3} = \frac{18}{3}$

Divide both sides of the equation by 3.

$x = 6$ Solution.

Check: $3 \cdot (6) - 7 = 11$
$18 - 7 = 11$
$11 = 11$

Here are a few practice problems. Solve them as we have done above.

(a) $7x + 2 = 51$ (b) $18 - 5x = 3$ (c) $15.3 = 4x - 1.5$
(d) $2(x + 4) = 27$ (e) $x - 6 = 3x$ (f) $3(x - 2) = x + 1$

(g) $2\frac{1}{4} + \frac{1}{2}x = 3\frac{1}{2}$ (h) $2x - 9.4 = 0$ (i) $2(x - 1) = 3(x + 1)$

Our complete step-by-step solutions are in **25**.

---

25 (a) **Solve:** $7x + 2 = 51$

Change this equation to an equivalent equation with only an x-term on the left.

Step 1 $(7x + 2) - 2 = (51) - 2$ Subtract 2 from each side.

$7x \underbrace{+ 2 - 2}_{0} = 51 - 2$ Combine terms.

$7x = 49$

Step 2 $\frac{7x}{7} = \frac{49}{7}$ Divide both sides by 7.

$x = \frac{49}{7}$

$x = 7$ Solution.

Check: $7 \cdot (7) + 2 = 51$
$49 + 2 = 51$
$51 = 51$

(b) **Solve:** $18 - 5x = 3$

Step 1 $(18 - 5x) - 18 = (3) - 18$ Subtract 18 from each side.

$-5x \underbrace{+ 18 - 18}_{0} = 3 - 18$ Rearrange terms.

$-5x = -15$ Multiply both sides by -1.
$5x = 15$

Step 2 $\frac{5x}{5} = \frac{15}{5}$ Divide both sides by 5.

$x = 3$ Solution.

Check: $18 - 5 \cdot (3) = 3$
$18 - 15 = 3$
$3 = 3$

(c) **Solve:** $15.3 = 4x - 1.5$

Step 1 $(15.3) + 1.5 = (4x \underbrace{- 1.5) + 1.5}_{0}$ Add 1.5 to each side.

$16.8 = 4x$ Combine terms.
$4x = 16.8$

Step 2 $\frac{4x}{4} = \frac{16.8}{4}$ Divide both sides by 4.

$x = 4.2$ Solution.

Decimal number solutions are common in practical problems.

Check: $15.3 = 4(4.2) - 1.5$
$15.3 = 16.8 - 1.5$
$15.3 = 15.3$

(d) **Solve:** $2(x + 4) = 27$
$2x + 8 = 27$ Multiply each term inside the parentheses by 2.

$(2x + 8) - 8 = (27) - 8$	Subtract 8 from each side.
$2x = 19$	Divide each side by 2.
$x = 9\frac{1}{2}$ Solution.	

Always remove parentheses first when solving an equation.

Check: $2\left(9\frac{1}{2} + 4\right) = 27$

$2\left(13\frac{1}{2}\right) = 27$

$27 = 27$

(e) **Solve:** $x - 6 = 3x$

$(x - 6) - x = 3x - x$	Subtract x from each side, so that x-terms will appear only on the right.
$-6 = 2x$	Combine terms.
or $2x = -6$	
$x = -3$ Solution.	

Check: $(-3)-6 = 3(-3)$

$-9 = -9$

(f) **Solve:** $3(x - 2) = x + 1$

$3x - 6 = x + 1$	Multiply each term inside the parentheses by 3.
$(3x - 6) - x = (x + 1) - x$	Subtract x from each side.
$2x - 6 = 1$	Combine terms.
$(2x - 6) + 6 = (1) + 6$	Add 6 to each side.
$2x = 7$	Divide each side by 2.
$x = 3\frac{1}{2}$ Solution.	

Check: $3\left(3\frac{1}{2} - 2\right) = 3\frac{1}{2} + 1$

$3\left(1\frac{1}{2}\right) = 4\frac{1}{2}$

$4\frac{1}{2} = 4\frac{1}{2}$

(g) **Solve:** $2\frac{1}{4} + \frac{1}{2}x = 3\frac{1}{2}$

	Change all mixed numbers to fractions.
$\frac{9}{4} + \frac{x}{2} = \frac{7}{2}$	Multiply each term by 4 to eliminate fractions.
$9 + 2x = 14$	
$(9 + 2x) - 9 = (14) - 9$	Subtract 9 from each side.
$2x = 5$	Divide each side by 2.
$x = 2\frac{1}{2}$ Solution.	

Check: $2\frac{1}{4} + \frac{1}{2}\left(2\frac{1}{2}\right) = 3\frac{1}{2}$

$$2\frac{1}{4} + \frac{1}{2}\left(\frac{5}{2}\right) = 3\frac{1}{2}$$

$$2\frac{1}{4} + \frac{5}{4} = 3\frac{1}{2}$$

$$3\frac{1}{2} = 3\frac{1}{2}$$

(h) **Solve:** $2x - 9.4 = 0$

$(2x - 9.4) + 9.4 = (0) + 9.4$ — Add 9.4 to each side.

$2x = 9.4$ — Divide each side by 2

$x = 4.7$ Solution.

Check: $2(4.7) - 9.4 = 0$

$9.4 - 9.4 = 0$

(i) **Solve:** $2(x - 1) = 3(x + 1)$ — Remove parentheses by multiplying.

$2x - 2 = 3x + 3$

$(2x - 2) - 3x = (3x + 3) - 3x$ — Subtract $3x$ from each side.

$-2 - x = 3$ — Combine terms.

$(-2 - x) + 2 = (3) + 2$ — Add 2 to each side.

$-x = 5$

or

$x = -5$ Solution. — Solve for x, not $-x$.

Do you see that $-x = 5$ is the same as $x = -5$? Multiply both sides of the equation $-x = 5$ by -1 to get $x = -5$.

Check: $2(-5 - 1) = 3(-5 + 1)$

$2(-6) = 3(-4)$

$-12 = -12$

Remember:

1. Do only legal operations: add or subtract the same quantity from both sides of the equation; multiply or divide both sides of the equation by the same nonzero quantity.
2. Remove all parentheses carefully.
3. Combine like terms when they are on the same side of the equation.
4. Use legal operations to change the equation so that you have only x by itself on one side of the equation and a number on the other side of the equation.
5. Always check your answer.

Now work the following set of problems for more practice in solving algebra equations.

1. $x + 1 = 7$
2. $9 - y = 4$
3. $3 + A = 0$
4. $6.1 + x = 3.5$
5. $27 - y = 96$
6. $0.45 + q = 0.09$
7. $4 = x - 7$
8. $19 = 3 - p$
9. $12 - 2x = 4 - x$
10. $3t = 21$
11. $1/x = 0$
12. $\frac{1}{4}x = 12.1$
13. $3 + 2x = 17$
14. $5y + 3 = 28$
15. $0 = 4 - x$
16. $4 = 2x - 1$
17. $2(x - 4) = 1$
18. $16 = 3x + 25$
19. $x + 5.8 = 3x + 1.4$
20. $6x + 1 = 2x - 1$

21. $4 - (2x - 7) = 5$

22. $2x + (3 - x) = 17$

23. $1 - (4 - 2x) = -11$

24. $(x - 2) - (4 - 2x) = 12$

Check your answers in **26**

A SHORTCUT WAY TO SOLVE EQUATIONS

If you can use the balancing operations of adding/subtracting and multiplying/dividing to get an equation into the form

$ax + b = c$ where a, b, and c are any numbers, positive or negative,

then the solution of the equation is

$$x = \frac{c - b}{a}$$

Example: To solve $3x + 2 = 8$

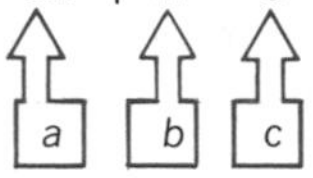

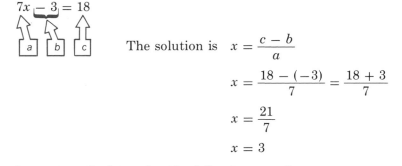

the solution is $x = \dfrac{c - b}{a}$ or $x = \dfrac{8 - 2}{3}$

$x = 2$

Example: To solve $5x - 3 = 18 - 2x$.
add $2x$ to both sides to get

$7x - 3 = 18$

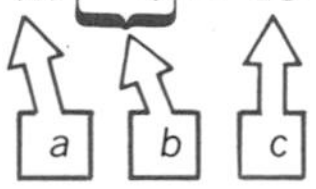

The solution is $x = \dfrac{c - b}{a}$

$$x = \frac{18 - (-3)}{7} = \frac{18 + 3}{7}$$

$$x = \frac{21}{7}$$

$x = 3$

Use this shortcut method to solve the following equations.

1. $x + 3 = 11$
2. $2x - 21 = 13$
3. $2a = 7$
4. $3R + 11 = 26$
5. $2 - x = 17$
6. $4 - 3x = x - 6$
7. $2T + 36 = -4$
8. $14 = 3y - 2$
9. $5x = x - 1$
10. $2x = 4x + 3$

The answers are on page 552.

26

1. $x = 6$	2. $y = 5$	3. $A = -3$	4. $x = -2.6$
5. $y = -69$	6. $q = -0.36$	7. $x = 11$	8. $p = -16$
9. $x = 8$	10. $t = 7$	11. $x = 0$	12. $x = 48.4$
13. $x = 7$	14. $y = 5$	15. $x = 4$	16. $x = 2\frac{1}{2}$
17. $x = 4\frac{1}{2}$	18. $x = -3$	19. $x = 2.2$	20. $x = -\frac{1}{2}$
21. $x = 3$	22. $x = 14$	23. $x = -4$	24. $x = 6$

To *solve a formula* for some letter means to rewrite the formula as an equivalent formula with that letter isolated on the left of the equals sign.

For example, the area of a triangle is given by the formula.

$A = \frac{BH}{2}$ where A is the area, B is the length of the base, and H is the height.

Solving for the base B gives the equivalent formula

$$B = \frac{2A}{H}$$

Solving for the height H gives the equivalent formula

$$H = \frac{2A}{B}$$

Solving formulas is a very important practical application of algebra. Very often a formula is not written in the form that is most useful. To use it you may need to rewrite the formula, solving it for the letter whose value you need to calculate.

To solve a formula, use the same balancing operations that you used to solve equations. You may add or subtract the same quantity on both sides of the formula and you may multiply or divide both sides of the formula by the same nonzero quantity.

For example, to solve the formula

$S = \frac{R + P}{2}$ for R

First, multiply both sides of the equation by 2 .

$$2\,S = 2 \cdot \left(\frac{R + P}{2}\right)$$

$$2S = R + P$$

Second, subtract P from both sides of the equation.

$$2S - P = R + \underbrace{P - P}_{0}$$

$$2S - P = R$$

And of course this formula can be reversed to read

$R = 2S - P$ We have solved the formula for R.

Solve the following formulas for the variable indicated.

(a) $V = \frac{3K}{T}$	for K	(b) $Q = 1 - R + T$	for R
(c) $V = \pi R^2 H - AB$	for H	(d) $Y = MX + B$	for X

Check your solutions in **27**.

---

27 (a) $V = \frac{3K}{T}$

First, multiply both sides by T to get $VT = 3K$.

Second, divide both sides by 3 to get $\frac{VT}{3} = K$

Solved for K, the formula is $K = \frac{VT}{3}$.

(b) $Q = 1 - R + T$

First, subtract T from both sides to get $Q - T = 1 - R$.
Second, subtract 1 from both sides to get $Q - T - 1 = -R$.
This is equivalent to $-R = Q - T - 1$
or $R = -Q + T + 1$ We have multiplied *all* terms by -1.
or $R = 1 - Q + T$

(c) $V = \pi R^2 H - AB$

First, add AB to both sides to get $V + AB = \pi R^2 H$.

Second, divide both sides by πR^2 to get $\dfrac{V + AB}{\pi R^2} = H$.

Notice that we divide *all* of the left side by πR^2.

Solved for H, the formula is $H = \dfrac{V + AB}{\pi R^2}$.

(d) $Y = MX + B$

First, subtract B from both sides to get $Y - B = MX$

Second, divide both sides by M to get $\dfrac{Y - B}{M} = X$.

Notice that we divide *all* of the left side by M.

Solved for X, the formula is $X = \dfrac{Y - B}{M}$.

Remember, when using the multiplication/division rule, you must multiply or divide *all* of both sides of the formula by the same quantity.

Practice solving formulas with the following problems.
Solve:

(a) $P = 2A + 3B$	for A	(b) $E = MC^2$	for M
(c) $S = \dfrac{A - RT}{1 - R}$	for R	(d) $S = \dfrac{1}{2}gt^2$	for g
(e) $P = i^2 R$	for R	(f) $I = \dfrac{V}{R + a}$	for R
(g) $A = \dfrac{2V - W}{R}$	for V	(h) $F = \dfrac{9C}{5} + 32$	for C
(i) $A = \dfrac{\pi R^2 S}{360}$	for S	(j) $P = \dfrac{t^2 dN}{3.78}$	for d
(k) $C = \dfrac{AD}{A + 12}$	for A	(l) $V = \dfrac{\pi L T^2}{6} + 2$	for L

Check your answers in **28**.

28

(a) $A = \dfrac{P - 3B}{2}$	(b) $M = \dfrac{E}{C^2}$
(c) $R = \dfrac{A - S}{T - S}$	(d) $g = \dfrac{2S}{t^2}$
(e) $R = \dfrac{P}{i^2}$	(f) $R = \dfrac{V - aI}{I}$
(g) $V = \dfrac{AR + W}{2}$	(h) $C = \dfrac{5F - 160}{9}$
(i) $S = \dfrac{360A}{\pi R^2}$	(j) $d = \dfrac{3.78P}{t^2 N}$
(k) $A = \dfrac{12C}{D - C}$	(l) $L = \dfrac{6V - 12}{\pi T^2}$

Using Square Roots in Solving Equations

The equations you learned to solve in this chapter are all *linear* equations. The variable appears only to the first power—no x^2 or x^3 terms appear in the equations. You will learn how to solve more difficult algebra equations in Chapter 10, but equations that look like

$x^2 = a$ where a is some positive number

can be solved easily.

To solve such an equation, simply take the square root of each side of the equation. The solution can be either

$x = \sqrt{a}$ or $x = -\sqrt{a}$.

Example: Solve $x^2 = 36$.

Taking square roots, $x = +6$ or $x = -6$.

There are two possible solutions, one negative and one positive. Be careful, one of them, usually the negative one, may not be a reasonable answer to a practical problem.

Example: If the cross-sectional area of a square heating duct is 75 sq in., what must be the width of the duct?

Solve $x^2 = 75$.

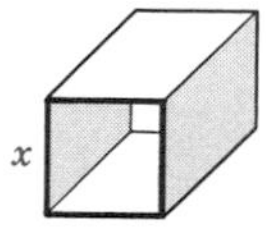

Taking the square root of each side, the positive solution is

$x = \sqrt{75}$
$x \cong 8.7$ in., rounded

If you need to review the concept of square roots, return to Section 3-6 on page 108. Square roots are easy to find with a calculator of course.

We will look at more problems like this in Chapter 10.

Translating English to Algebra

Algebra is a very useful tool for solving real problems. But in order to use it you may find it necessary to translate simple English sentences and phrases into mathematical expressions or equations. In technical work especially, the formulas to be used are often given in the form of English sentences, and they must be rewritten as algebra formulas before they can be used. For example, the statement

> "Horsepower required to overcome vehicle air resistance is equal to the cube of the vehicle speed in mph multiplied by the frontal area in square feet divided by 150,000."

translates to the formula

$$HP = \frac{MPH^3 \cdot \text{Area}}{150{,}000}$$

or

$$P = \frac{V^3A}{150{,}000} \quad \text{in algebra form}$$

In the next few pages of this chapter we will show you how to translate English statements into algebra formulas. To begin, try the following problem.

An automotive technician found the following statement in a manual:

"The pitch diameter of a cam gear is twice the diameter of the crank gear."

Translate this sentence into an algebra equation.

Check your work in **29**.

---

29 The equation is $P = 2C$ where P is the pitch diameter of the cam gear, and C is the diameter of the crank gear.

You may use any letters you wish of course, but we have chosen letters that remind you of the quantities they represent: P for *p*itch and C for *c*rank.

Notice that we write "two times C" as $2C$ as you normally do in algebra.

Certain words and phrases appear again and again in statements to be translated. They are signals alerting you to the mathematical operations to be used. Here is a handy list of the *signal words* and their mathematical translations.

SIGNAL WORDS

English Term	Math Translation	Example
Equals Is, is equal to, was, are, were The same as. . . What is left is. . . The result is. . . Gives, makes, leaves, having	$=$	$A = B$
Plus, sum of Increased by, more than	$+$	$A + B$
Minus B, subtract B Less B Decreased by B, take away B Reduced by B, diminished by B B less than A B subtracted from A Difference between A and B	$-$	$A - B$
Times, multiply, of Multiplied Product of	$\times$	AB
Divide, divided by B Quotient of	$\div$	$A \div B$ or $\frac{A}{B}$
Twice, twice as much Double	$\times 2$	$2A$
Squared Cubed		A^2 A^3

Translate the phrase "length plus 3 inches" into an algebra expression. Try it, then turn to **30** to check your answer.

---

30 **First,** make a word equation by using parentheses.

(length) (plus) (3 inches)

Second, substitute mathematical symbols.

(length) (plus) (3 inches)

↓ ↓ ↓

L $+$ 3

or $L + 3$.

Notice that signal words, such as "plus," are translated directly into math symbols. Unknown quantities are represented by letters of the alphabet, chosen to remind you of their meaning.

Translate the following phrases into math expressions:

(a) Weight divided by 13.6 ____________

(b) $6\frac{1}{4}''$ more than the width ____________

(c) One-half of the original torque ____________

(d) The sum of the two lengths ____________

(e) The voltage decreased by 5 ____________

(f) Five times the gear reduction ____________

(g) $8\frac{1}{2}''$ more than twice the height ____________

Our step-by-step translations are in **31**

---

31 (a) (weight) (divided by) (13.6)

↓ ↓ ↓

W $\div$ 13.6 or $\dfrac{W}{13.6}$

(b) ($6\frac{1}{4}''$) (more than) (the width)

↓ ↓ ↓

$6\frac{1}{4}$ $+$ W or $6\frac{1}{4} + W$

(c) (one-half) (of) (the original torque)

↓ ↓ ↓

$\frac{1}{2}$ $\times$ T or $\frac{1}{2}T$ or $\dfrac{T}{2}$

(d) (the sum of) (the two lengths)

↓ ↓

$L_1 + L_2$ or $L_1 + L_2$

(e) (the voltage) (decreased by) (5)

↓ ↓ ↓

V $-$ 5 or $V - 5$

(f) (five) (times) (the gear reduction)

↓ ↓ ↓

5 $\times$ G or $5G$

(g) ($8\frac{1}{2}''$) (more than) (twice the height)

↓ ↓ ↓ ↓

$8\frac{1}{2}$ $+$ 2 H or $8\frac{1}{2} + 2H$

Of course, any letters could be used in place of the ones used above.

So far we have translated only phrases, pieces of sentences, but complete sentences can also be translated. An English phrase translates into an algebra expression, and an English sentence translates into an algebra equation. For example, the sentence

The height of the duct	is equal to	its width	
↓	↓	↓	
translates to H	$=$	W	or $H = W$

Each word or phrase in the sentence becomes a mathematical term, letter, number, expression, or arithmetic operation sign.

Translate the following sentence into algebra form as we did above.

"The size of a drill for a tap is equal to the tap diameter minus the depth."

Check your translation in **32**.

---

32 $S = T - D$

Follow these steps when you must translate an English sentence into an algebra equation or formula.

Example:

Step 1 Cross out all unnecessary words.

~~The~~ size ~~of a drill for a tap~~ is equal to ~~the~~ tap diameter minus ~~the~~ depth.

Step 2 Make a word equation using parentheses.

(Size) (is equal to) (tap diameter) (minus) (depth)

Step 3 Substitute a letter or an arithmetic symbol for each parentheses.

S $=$ T $-$ D

Step 4 Combine and simplify.

$S = T - D$

In most formulas the units for the quantities involved must be given. In the formula above, T and D are in inches.

Translating English sentences or verbal rules into algebra formulas requires that you read the sentences very differently from the way you read stories or newspaper articles. Very few people are able to write out the math formula after reading the problem only once. You should expect to read it several times, and you'll want to read it slowly. No speed reading here!

The ideas in technical work and formulas are usually concentrated in a few key words, and you must find them. If you find a word you do not recognize, stop reading and look it up in a dictionary, textbook, or manual. It may be important. Translating and working with formulas is one of the skills you must have if you are to succeed at any technical occupation.

Here is another example of translating a verbal rule into an algebra formula:

"The electrical resistance of a length of wire is equal to the resistivity of the metal times the length of the wire divided by the square of the wire diameter."

Step 1 Eliminate all but the key words.

"~~The electrical~~ resistance ~~of a length of wire~~ is equal to ~~the~~ resistivity ~~of the metal~~ times ~~the~~ length ~~of the wire~~ divided by the square ~~of the wire~~ diameter."

Step 2 Make a word equation.

(Resistance) (is equal to) (resistivity) (times) (length) (divided by) (square of diameter)

Step 3 Substitute letters and symbols.

$$R = \frac{rL}{D^2}$$

If the resistivity r has units of ohms per inch, L and D will be in inches.

The more translations you do, the easier it gets. Translate each of the following technical statements into algebra formulas.

(a) A sheet metal worker measuring a duct cover finds that the width is $8\frac{1}{2}''$ less than the height.
(b) A 24″ length of wire is cut into two pieces so that the longer piece is five times the length of the shorter. (*Hint:* Write two separate equations.)
(c) One-quarter of a job takes $3\frac{1}{2}$ days.
(d) One-half of a coil of wire weighs $16\frac{2}{3}$ lb.
(e) Two shims are to have a combined thickness of 0.090″. The larger shim must be 3.5 times as thick as the smaller shim. (*Hint:* Write two separate equations.)
(f) The volume of an elliptical tank is approximately 0.7854 times the product of its height, length, and width.
(g) The engine speed is equal to 168 times the overall gear reduction multiplied by the speed in mph and divided by the rolling radius of the tire.
(h) The air resistance force in pounds acting against a moving vehicle is equal to 0.0025 times the square of the speed in mph times the frontal area of the vehicle.

Work carefully. Check your formulas in **33**.

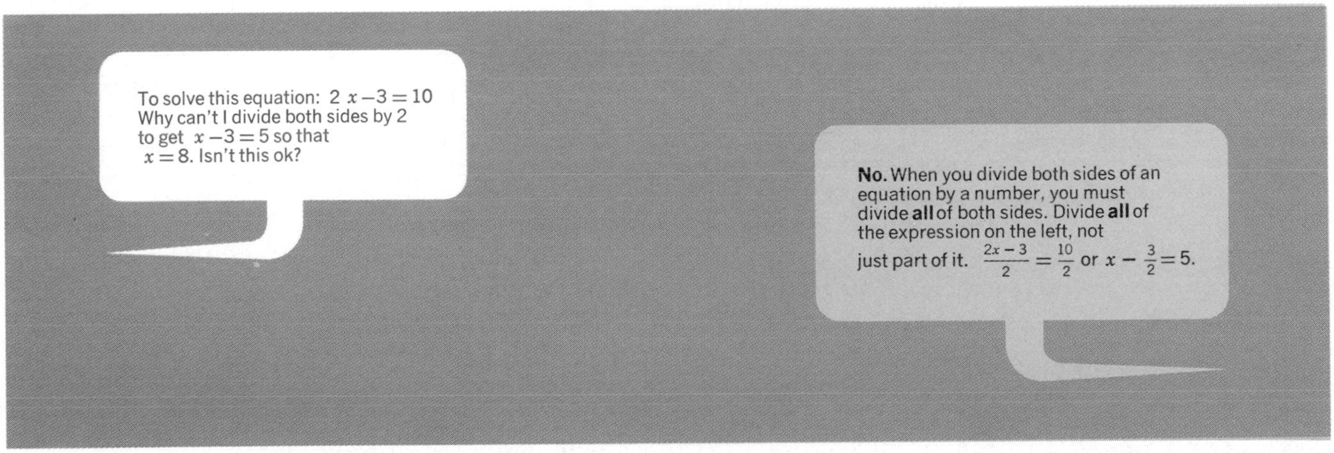
To solve this equation: $2x - 3 = 10$ Why can't I divide both sides by 2 to get $x - 3 = 5$ so that $x = 8$. Isn't this ok?
No. When you divide both sides of an equation by a number, you must divide **all** of both sides. Divide **all** of the expression on the left, not just part of it. $\frac{2x-3}{2} = \frac{10}{2}$ or $x - \frac{3}{2} = 5$.

---

33 (a) $W = H - 8\frac{1}{2}$

(b) $24 = L + S$ and $L = 5S$

(c) $\frac{1}{4}J = 3\frac{1}{2}$ or $\frac{J}{4} = 3\frac{1}{2}$

(d) $\frac{1}{2}C = 16\frac{2}{3}$ or $\frac{C}{2} = 16\frac{2}{3}$

(e) $S + L = 0.090$ and $L = 3.5S$

(f) $V = 0.7854HLW$

(g) $S = \dfrac{168 \cdot G \cdot V}{R}$ where V is in mph.

(h) $R = 0.0025 \cdot V^2 \cdot A$ where V is in mph.

Notice in Problems (b) and (e) that two equations can be written. These can be combined to form a single equation.

(b) $24 = L + S$ and $L = 5S$ give $24 = 5S + S$

(e) $S + L = 0.090$ and $L = 3.5S$ give $S + 3.5S = 0.090$

General Word Problems

Now that you can translate English phrases and sentences into algebra expressions and equations, you should be able to solve many practical word problems. For example, consider this problem:

A machinist needs to use two shims with a combined thickness of 0.084″. One shim is to be three times as thick as the other. What are the thicknesses of the two shims?

Let x = thickness of the thinner shim
Then $3x$ = thickness of the thicker shim
and the equation is

$3x + x = 0.084''$	Combine terms.
$4x = 0.084''$	Divide each side by 4.
$x = 0.021''$	**$3x = 0.063''$**

Check: $0.021'' + 3(0.021'') = 0.084''$
$0.021'' + 0.063'' = 0.084''$ which is correct.

Your turn. Use your knowledge of algebra to translate the following problem to an algebra equation and then solve it.

Two carpenters, Al and Bill, produced 42 assembly frames in one day. Al worked faster and produced 8 more than Bill. How many frames did each build?

Try it. Check your work in **34**

---

34 Let B = number of frames built by Bill
Then $B + 8$ = number of frames built by Al

and

$B + (B + 8) = 42$	Combine terms, $B + B = 2B$.
$2B + 8 = 42$	Subtract 8 from each side.
$2B = 34$	Divide both sides by 2.
$B = 17$	Bill built 17 frames.
then $B + 8 = 25$	Al built 25 frames.

Check: $17 + 25 = 42$

Practice makes perfect. Work the following set of word problems.

(a) The area of a rectangular shop floor is 400 sq ft. If the width is 16 ft, what is the length? (*Hint:* Area = Length · Width.)

(b) A 14 ft long steel rod is cut into two pieces. The longer piece is $2\frac{1}{2}$ times the length of the shorter piece. Find the length of each piece. (Ignore waste.)

(c) A carpenter wants to cut a 12 ft board into three pieces. The longest piece must have three times the length of the shortest piece and the medium-sized piece is 2 ft longer than the shortest piece. Find the actual length of each piece. (Ignore waste in cutting.)

(d) Find the dimensions of a rectangular cover plate if its length is 6 in. longer than its width and if its perimeter is 68 in. (*Hint:* Perimeter $= 2 \cdot \text{length} + 2 \cdot \text{width}$.)

(e) Mike and Jeff are partners in a small manufacturing firm. Because Mike provided more of the initial capital for the business, they have agreed that Mike's share of the profits should be $\frac{1}{4}$ greater than Jeff's. The total profit for the first quarter of this year was $17,550. How should they divide it?

When you have solved these problems, turn to **35** to check your work.

---

35 (a) Width = 16 ft Length = L

$$\text{Area} = 400 \text{ sq ft} = 16 \text{ ft} \cdot L$$

$$16L = 400 \quad \text{Divide by 16.}$$

$$L = 25 \text{ ft}$$

(b) x = shorter piece

$(2\frac{1}{2})x$ = longer piece

$$x + \left(2\frac{1}{2}\right)x = 14 \text{ ft}$$

$$\left(3\frac{1}{2}\right)x = 14 \text{ ft}$$

or $$\frac{7x}{2} = 14 \quad \text{Multiply by 2.}$$

$$7x = 28 \quad \text{Divide by 7.}$$

$$x = 4 \text{ ft}$$

The shorter piece is 4 ft long and the longer piece is 10 ft long.

(c) Let x = the shortest piece
then $3x$ = longest piece
and $x + 2$ = medium-sized piece

$x + (x + 2) + (3x) = 12$ ft Combine terms.
$5x + 2 = 12$ Subtract 2 from each side.
$5x = 10$ Divide by 5.
$x = 2$ ft The shortest piece is 2′ long.
$3x = 6$ ft The largest piece is 6′ long.
$x + 2 = 4$ ft The third piece is 4′ long.

(d) Let width = W
Then length = $W + 6$
and perimeter = 68″ = $2W + 2(W + 6)$ Remove parentheses, $2(W + 6) = 2W + 12$.

$68 = 2W + 2W + 12$ Combine terms.
or $4W + 12 = 68$ Subtract 12 from each side.
$4W = 56$ Divide by 4.
$W = 14$ in.
length $= W + 6 = 20$ in.

(e) Let Jeff's share = J
then Mike's share = $J + \frac{1}{4}J$
and

$$\$17{,}550 = J + \left(J + \frac{1}{4}J\right) \quad \text{Combine terms; remember } J = 1 \cdot J.$$

$$= 2\frac{1}{4}J \quad \text{Write the mixed number as a fraction.}$$

$$= \frac{9J}{4}$$

or $\dfrac{9J}{4} = 17{,}550$ Multiply by 4.

$9J = 70{,}200$ Divide by 9.

$J = \$7800$

Mikes share $= \$7800 + \dfrac{\$7800}{4}$

$= \$7800 + \$1950$

$= \$9750$

Now turn to **36** for a set of exercises on algebra equations and formulas.

---

36 **Exercises 6-2 Solving Equations and Formulas**

A. Solve the following equations:

1. $x + 4 = 13$
2. $23 = A + 6$
3. $17 - x = 41$
4. $z - 18 = 29$
5. $6 = a - 2\frac{1}{2}$
6. $73 + x = 11$
7. $-y - 16.01 = 8.65$
8. $11.6 - R = 3.7$
9. $-39 = 3x$
10. $-9y = 117$
11. $13a = 0.078$
12. $\dfrac{x}{3} = 7$
13. $\dfrac{z}{1.3} = 0.5$
14. $\dfrac{N}{2} = \dfrac{3}{8}$
15. $\dfrac{1}{2} + 2x = 1$
16. $3x + 16 = 46$
17. $-4a + 45 = 17$
18. $\dfrac{x}{2} + 1 = 8$
19. $-3Z + \dfrac{1}{2} = 17$
20. $2x + 6 = 0$
21. $1 = 3 - 5x$
22. $4(1 - A) = 7 + A$
23. $-5P + 18 = 3$
24. $3x + 4 = x - 8$

B. Solve the following formulas for the variable shown:

1. $S = LW$ for L
2. $A = \dfrac{1}{2}BH$ for B
3. $V = IR$ for I
4. $H = \dfrac{D - R}{2}$ for D
5. $S = \dfrac{W}{2}(A + T)$ for T
6. $V = \pi R^2 H - AB$ for H
7. $P = 2A + 2B$ for B
8. $H = \dfrac{R}{2} + 0.05$ for R
9. $T = \dfrac{RP}{R + 2}$ for R
10. $I = \dfrac{E + V}{R}$ for V

C. Translate the following into algebraic equations:

1. The height of the tank is 1.4 times its width.
2. The sum of the two weights is 167 lb.
3. The volume of a cylinder is equal to $\frac{1}{4}$ of its height times π times its diameter squared.
4. The weight in kilograms is equal to 0.454 times the weight in pounds.

5. The volume of a solid bar is equal to the product of the cross section area and the length of the bar.

6. The volume of a cone is $\frac{1}{3}$ times π times the height times the square of the radius of the base.

7. The pitch diameter D of a spur gear is equal to the number of teeth on the gear divided by the pitch.

8. The cutting time for a lathe operation is equal to the length of the cut divided by the product of the tool feed rate and the revolution rate of the workpiece.

9. The weight of a metal cylinder is approximately equal to 0.785 times the height of the cylinder times the density of the metal times the square of the diameter of the cylinder.

10. The voltage across a simple circuit is equal to the product of the resistance of the circuit and the current flowing in the circuit.

D. Practical Problems

1. The length of arc of a sector of a circle is given by the formula

$$L = \frac{2\pi R a}{360} \quad \text{where}$$

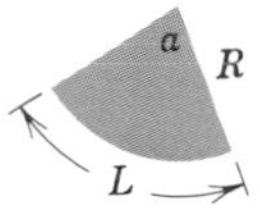

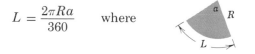

R is the radius of the circle and a is the central angle in degrees.

(a) Solve for a. (b) Solve for R.
(c) Find L when $R = 10$ in. and $a = 30°$. Use $\pi = 3.14$.

2. The area of the sector shown in Problem 1 is $A = \pi R^2 a/360$.

(a) Solve this formula for a.
(b) Find A if $R = 12$ in., $a = 45°$, $\pi = 3.14$.

3. When two resistors R_1 and R_2 are put in series with a battery giving V volts, the current through the resistors is

$$i = \frac{V}{R_1 + R_2}$$

(a) Solve for R_1.
(b) Find R_2 if $V = 100$ volts, $i = 0.4$ amp, $R_1 = 200$ ohms.

4. Machinists use a formula, known as Pomeroy's formula, to determine roughly the power required by a metal punch machine.

$$P \cong \frac{t^2 dN}{3.78}$$

where P is the power needed, in horsepower; t is the thickness of the metal being punched; d is the diameter of the hole being punched; N is the number of holes to be punched at one time.

(a) Solve this formula for N.
(b) Find the power needed to punch six 2″ diameter holes in a sheet $\frac{1}{8}$ in. thick.

5. When a gas is kept at constant temperature and the pressure on it is changed, its volume changes in accord with the pressure–volume relationship known as Boyle's Law:

$$\frac{V_1}{V_2} = \frac{P_2}{P_1}$$

where P_1 and V_1 are the beginning volume and pressure, and P_2 and V_2 are the final volume and pressure.

(a) Solve for V_1. (b) Solve for V_2.
(c) Solve for P_1. (d) Solve for P_2.
(e) Find P_1 when $V_1 = 10$ cu ft, $V_2 = 25$ cu ft, and $P_2 = 120$ psi.

6. The volume of a football is roughly $V = \pi LT^2/6$ where L is its length and T is its thickness

   (a) Solve for L.
   (b) Solve for T^2.

7. Nurses use a formula known as Young's rule to determine the amount of medicine to give a child under 12 years of age when the adult dosage is known.

$$C = \frac{AD}{A + 12}$$

   C is the child's dose; A is the age of the child in years; D is the adult dose.

   (a) Work backward and find the adult dose in terms of the child's dose. Solve for D.
   (b) Find D if $C = 0.05$ gram and $A = 7$.

8. Lens grinders use the following formula to determine the shape of a simple lens:

$$\frac{1}{F} = (n - 1)\left(\frac{1}{R} + \frac{1}{r}\right)$$

   where F is the focal length of the lens; n is its index of refraction; R and r are the radii of curvature of the two lens surfaces. Solve for n.

9. For a current transformer, $\dfrac{i_L}{i_S} = \dfrac{T_P}{T_S}$

   where i_L is the line current, i_S is the secondary

   current, T_P is the number of turns of wire in the primary coil, and T_S is the number of turns of wire in the secondary coil.

   (a) Solve for i_L.
   (b) Solve for i_S.
   (c) Find i_L when $i_S = 1.5$ amps, $T_P = 1500$, $T_S = 100$.

10. The electrical power P dissipated in a circuit is equal to the product of the current I and the voltage V, where P is in watts, I is in amps, and V is in volts.

   (a) Write an equation giving P in terms of I and V.
   (b) Solve for I.
   (c) Find V when $P = 15{,}750$ watts, $I = 42$ amps.

11. The inductance L in microhenrys of a coil constructed by a ham radio operator is given by the formula

$$L = \frac{R^2N^2}{9R + 10D}$$

   where R is the radius of the coil, D is the length of the coil, and N is the number of turns of wire in the coil.

   Find L if $R = 3$ in., $D = 6$ in., $N = 200$.

12. A sheet metal technician uses the following formula for calculating Bend Allowance, BA, in inches:

$$BA = N(0.01743R + 0.0078T)$$

   where N = number of degrees in the bend
   R = inside radius of the bend, in.
   T = thickness of the metal, in.

Find BA for each of the following situations. (You'll want to use a calculator on this one.)

	N	R	T
(a)	50°	$1\frac{1}{4}''$	0.050″
(b)	65°	0.857″	0.035″
(c)	40°	1.025″	0.0856″

When you have completed these exercises check your answers on page 552, then turn to **37** to learn about ratio and proportion.

6-3 RATIO AND PROPORTION

37 Machinists, mechanics, carpenters, and other trades workers use the ideas of ratio and proportion to solve very many technical problems. The compression ratio of an automobile engine, the gear ratio of a machine, scale drawings, the pitch of a roof, the mechanical advantage of a pulley system, and the voltage ratio in a transformer are all practical examples of the ratio concept.

Ratio

A *ratio* is a comparison of two quantities of the same kind, both expressed in the same units. For example, the steepness of a hill can be written as the ratio of its height to its horizontal extent.

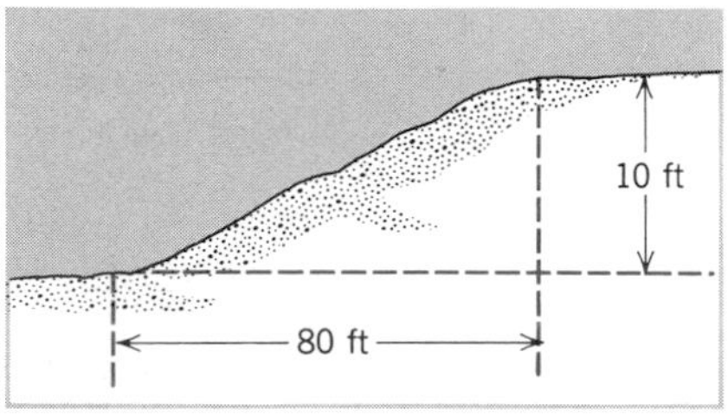

$$\text{Steepness} = \frac{10 \text{ ft}}{80 \text{ ft}} = \frac{1}{8}$$

The ratio is usually written as a fraction in lowest terms, and you would read this ratio as either "one-eighth" or "one to eight."

In the following diagram find the gear ratio of the two gears shown.

Gear ratio = ?

B has 16 teeth.
A has 64 teeth.

Gear B drives gear A.

Check your answer in **38**.

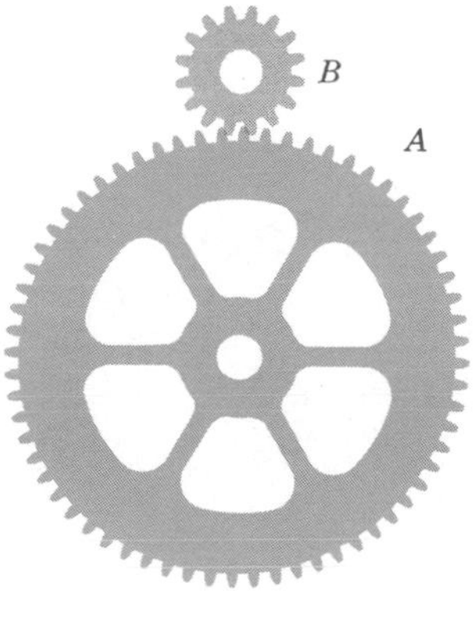

---

38 The ratio of the number of teeth on the large gear to the number of teeth on the small gear is

$$\text{Gear ratio} = \frac{64 \text{ teeth}}{16 \text{ teeth}} = \frac{4}{1}$$

Always reduce the fraction to lowest terms.

The gear ratio is 4 to 1. There are four times as many teeth on the large gear as there are on the small gear. Notice that we put the larger number on the top to calculate this gear ratio.

In general, to find a gear ratio divide the number of teeth on the *driven* gear by the number of teeth on the *driving* gear.

Here are a few important examples of the use of ratios in practical work.

Example 1: Pulley Ratios

A pulley is a device that can be used to transfer power from one system to another. A pulley system can be used to lift heavy objects in a shop or to connect a power source to a piece of machinery. The ratio of the pulley diameters will determine the relative pulley speeds.

Find the ratio of the diameter of pulley A to the diameter of pulley B in the following drawing.

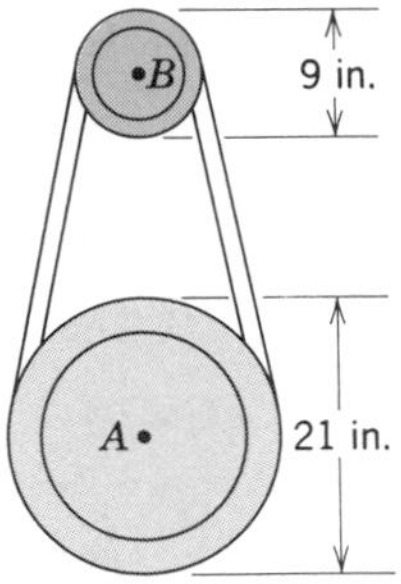

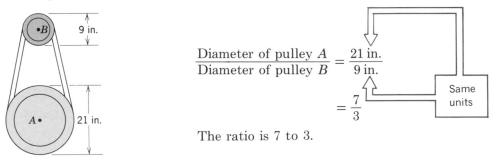

$$\frac{\text{Diameter of pulley } A}{\text{Diameter of pulley } B} = \frac{21 \text{ in.}}{9 \text{ in.}}$$

$$= \frac{7}{3}$$

Same units

The ratio is 7 to 3.

Notice that the units, inches, cancel from the ratio. A ratio is a fraction and it has no units.

Very often a colon (:) is used to write a ratio. The ratio above would be written 7:3 and this is read "seven to three."

Example 2: Pitch of a Roof

In carpentry and construction work, the *pitch* or steepness of a roof is defined as the ratio of the *rise* to the *span*.

$$\text{Pitch} = \frac{\text{rise of rafter}}{\text{span of roof}}$$

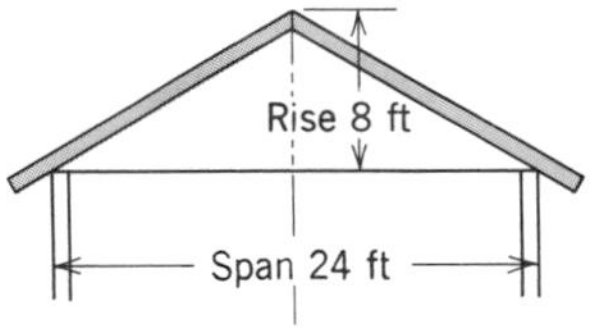

The *rise* is the vertical distance between the ridge beam and the supporting plate. The *span* is the horizontal distance between the outer faces of the supporting walls.

Find the pitch of the roof shown.

$$\text{Pitch} = \frac{8 \text{ ft rise}}{24 \text{ ft span}}$$

$$= \frac{1}{3}$$ The pitch is 1 to 3.

Example 3: Compression Ratio

In an automobile engine there is a large difference between the volume of the cylinder space when a piston is at the bottom of its stroke and when it is at the top of its stroke. This difference in volumes is called the *engine displacement.* Automotive mechanics find it very useful to talk about the compression ratio of an automobile engine. The *compression ratio* of an engine compares the volume of the cylinder at maximum expansion to the volume of the cylinder at maximum compression.

$$\text{Compression ratio} = \frac{\text{expanded volume}}{\text{compressed volume}}$$

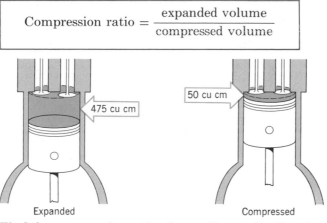

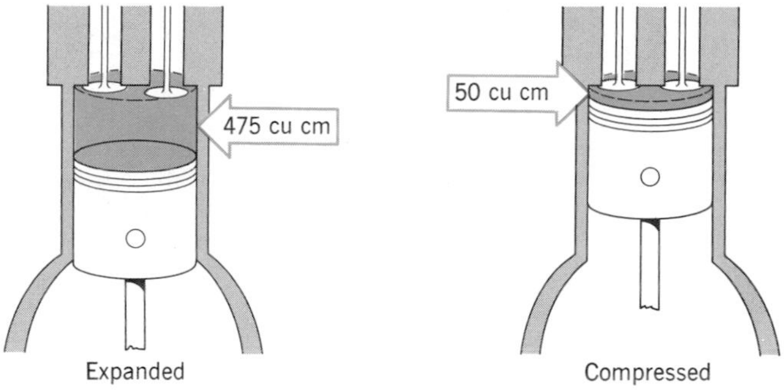

Find the compression ratio of a gasoline engine if each cylinder has a maximum volume of 475 cu cm and a minimum or compression volume of 50 cu cm.

$$\text{Compression ratio} = \frac{475 \text{ cu cm}}{50 \text{ cu cm}} = \frac{19}{2}$$

or

Compression ratio = 9.5

Compression ratios are always written so that the second number in the ratio is 1. This compression ratio would be written as $9\frac{1}{2}$ to 1.

Now, for some practice in calculating ratios, work the following problems:

1. 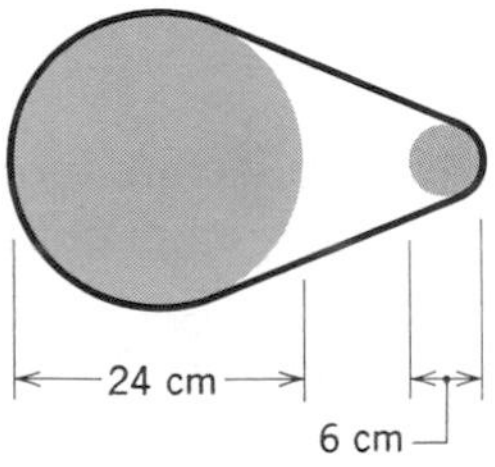

Ratio of pulley diameters =

2. 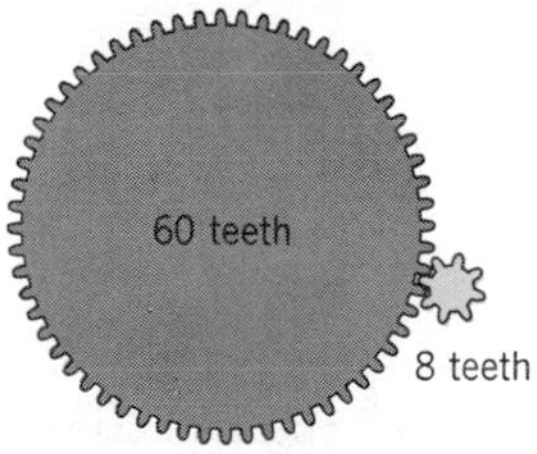

Gear ratio =

3. 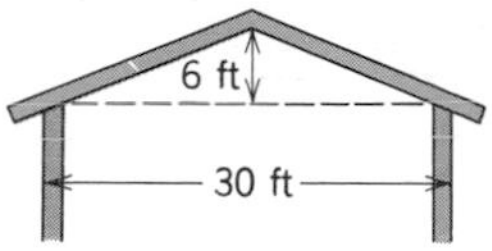

Pitch =

4. Find the rear axle ratio of a car if the ring gear has 54 teeth and the pinion gear has 18 teeth.

$$\text{Real axle ratio} = \frac{\text{number of teeth on ring gear}}{\text{number of teeth on pinion gear}}$$

5. A gasoline engine has a maximum cylinder volume of 34 cu in. and a compressed volume of 4 cu in. Find its compression ratio.

Check your answers in **39**.

---

39 1. Ratio of pulley diameters $= \dfrac{24 \text{ cm}}{6 \text{ cm}}$ or 4 to 1.

2. Gear ratio $= \dfrac{60 \text{ teeth}}{8 \text{ teeth}}$

$= \dfrac{15}{2}$ or 15 to 2.

3. Pitch $= \dfrac{6 \text{ ft rise}}{30 \text{ ft span}}$

$= \dfrac{6}{30}$ or $\dfrac{1}{5}$

The pitch is 1 to 5.

4. Rear axle ratio $= \dfrac{54 \text{ teeth}}{18 \text{ teeth}}$

$= \dfrac{54 \text{ teeth}}{18 \text{ teeth}}$ or $\dfrac{3}{1}$

The rear axle ratio is 3 to 1.

5. Compression ratio $= \dfrac{34 \text{ cu in.}}{4 \text{ cu in.}}$

$= \dfrac{17}{2}$ or $8\dfrac{1}{2}$ to 1

If you are given the value of a ratio and one of its terms, it is possible to find the other term. For example, if the pitch of a roof is supposed to be 1 to 5, and the span is 20 ft what must be the rise?

$$\text{Pitch} = \frac{\text{rise}}{\text{span}}$$

A ratio of 1 to 5 is equivalent to the fraction $\frac{1}{5}$; therefore the above equation becomes

$$\frac{1}{5} = \frac{\text{rise}}{20}$$

or

$$\frac{1}{5} = \frac{R}{20}$$

To solve for R, multiply both sides of the equation by 20.

$$\left(\frac{1}{5}\right)20 = \left(\frac{R}{\cancel{20}}\right)\cancel{20}$$

$$\frac{20}{5} = R$$

or $R = 4$ ft The rise is 4 ft.

The algebra you learned in Section 2 of this chapter will enable you to solve any ratio problem of this kind.

Try these problems.

(a) If the gear ratio on a mixing machine is 6:1 and the smaller gear has 12 teeth, how many teeth are on the larger gear?

(b) The pulley system of an assembly belt has a pulley diameter ratio of 4. If the larger pulley has a diameter of 15 in., what is the diameter of the smaller pulley?
(c) The compression ratio of a Datsun 280Z is 8.3 to 1. If the compressed volume of the cylinder is 36 cu cm, what is the expanded volume of the cylinder?
(d) On a certain construction job, concrete is made using a volume ratio of one part cement to $2\frac{1}{2}$ parts sand and 4 parts of gravel. How much sand should be mixed with 3 cu ft of cement?

Check your work in **40**.

---

40 (a) Gear ratio $= \dfrac{\text{number of teeth on large gear}}{\text{number of teeth on small gear}}$

A ratio of 6:1 is equivalent to the fraction $\dfrac{6}{1}$ or 6.

Therefore

$$6 = \frac{x}{12}$$

Multiply both sides of this equation by 12.

$$(6)12 = \left(\frac{x}{\cancel{12}}\right)\cancel{12}$$

or

$72 = x$ The large gear has 72 teeth.

(b) Pulley ratio $= \dfrac{\text{diameter of larger pulley}}{\text{diameter of smaller pulley}}$

$$4 = \frac{15\text{ in.}}{D}$$

Multiply both sides of this equation by D.

$$(4)D = \left(\frac{15}{\cancel{D}}\right)\cancel{D}$$

$$4D = 15$$

Divide both sides of this equation by 4.

$$\frac{\cancel{4}D}{\cancel{4}} = \frac{15}{4}$$

$$D = \frac{15}{4} \quad \text{or} \quad D = 3\frac{3}{4}\text{ in.}$$

(c) Compression ratio $= \dfrac{\text{expanded volume}}{\text{compressed volume}}$

$$8.3 = \frac{V}{36\text{ cu cm}}$$

Multiply both sides of the equation by 36 to get

$V = (8.3)(36)$

or

$V = 298.8$ cu cm or $V = 299$ cu cm, rounded to the nearest cu cm.

(d) Ratio of cement to sand $= \dfrac{\text{volume of cement}}{\text{volume of sand}}$

$$\frac{1}{2\frac{1}{2}} = \frac{3\text{ cu ft}}{S}$$

Multiply by S to get $\dfrac{S}{2\frac{1}{2}} = 3$.

Multiply by $2\frac{1}{2}$ to get $S = (3)(2\frac{1}{2})$

$S = 7\frac{1}{2}$ cu ft

Proportion

A *proportion* is a statement that two ratios are equal. It can be given as a sentence in words, but most often a proportion is an algebra equation.

The arithmetic equation $\dfrac{3}{5} = \dfrac{21}{35}$ is a proportion.

The algebraic equation $\dfrac{11}{4} = \dfrac{x}{5}$ is a proportion.

For the automobile shown below, the ratio of the actual length to the scale-drawing length is equal to the ratio of the actual width to the scale-drawing width.

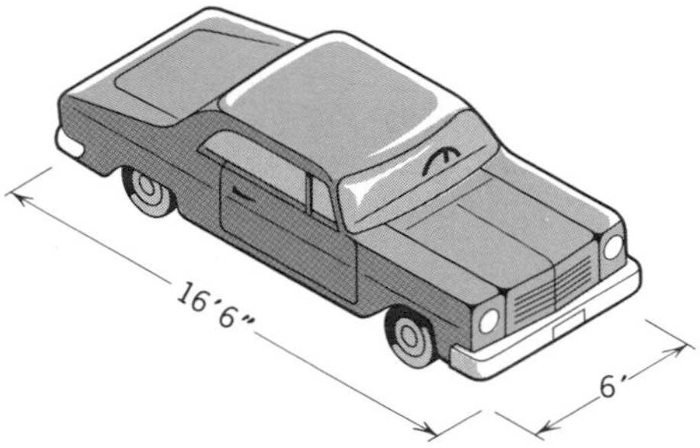

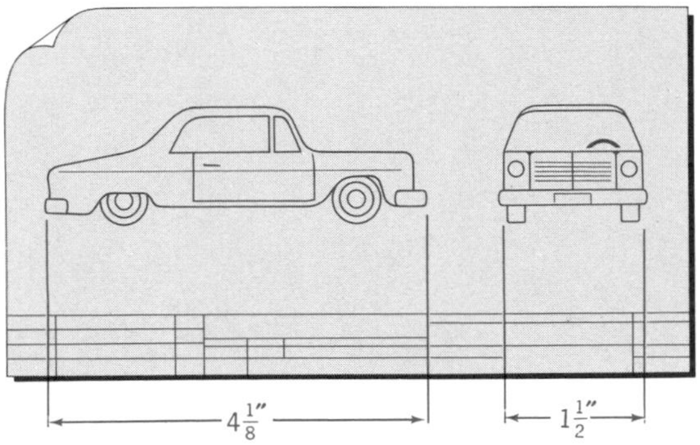

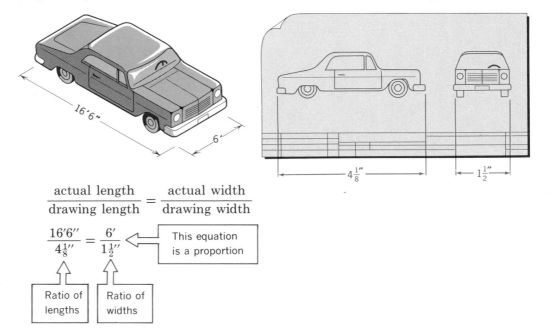

$$\frac{\text{actual length}}{\text{drawing length}} = \frac{\text{actual width}}{\text{drawing width}}$$

$$\frac{16'6''}{4\frac{1}{8}''} = \frac{6'}{1\frac{1}{2}''}$$

This equation is a proportion

Ratio of lengths

Ratio of widths

Rewrite all quantities in the same units:

$$\frac{198''}{4\frac{1}{8}''} = \frac{72''}{1\frac{1}{2}''}$$

You should notice first of all that each side of this equation is a ratio. Each side is a ratio of *like* quantities: lengths on the left and widths on the right.

Second, notice that the ratio $\dfrac{198''}{4\frac{1}{8}''}$ is equal to $\dfrac{48}{1}$.

Divide it out: $198 \div 4\frac{1}{8} = 198 \div \dfrac{33}{8}$

$= 198 \times \dfrac{8}{33}$

$= 48$

Using a Calculator

Key	Display
C	0.
1 9 8	198.
×	198.
8	8.
÷	1584.
3 3	33.
=	48.

Notice also that the ratio $\frac{72''}{1\frac{1}{2}''}$ is equal to $\frac{48}{1}$.

The common ratio $\frac{48}{1}$ is called the *scale* factor of the drawing.

The four parts of a proportion are called its *terms*.

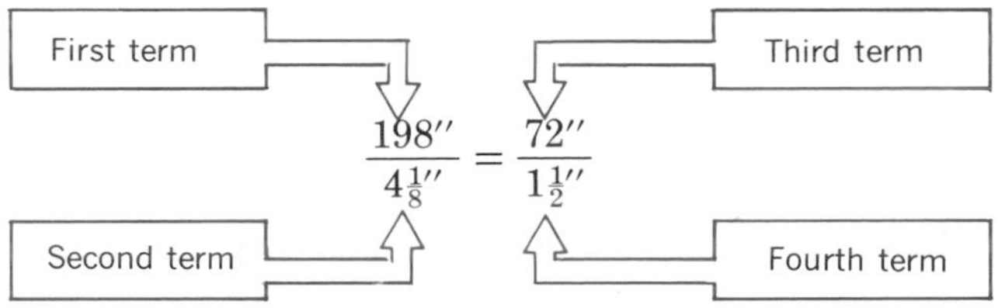

If one of the terms of the proportion is unknown, we can replace it with a letter and solve the proportion as an algebra equation. For example, suppose the actual rear bumper height of the car above is 1 ft, what would its height be on the drawing?

Set up the proportion and solve it.

Check your work in **41**.

---

41 $\frac{6'}{1\frac{1}{2}''} = \frac{1}{x}$ or $\frac{72''}{1\frac{1}{2}''} = \frac{12''}{x}$ in inches.

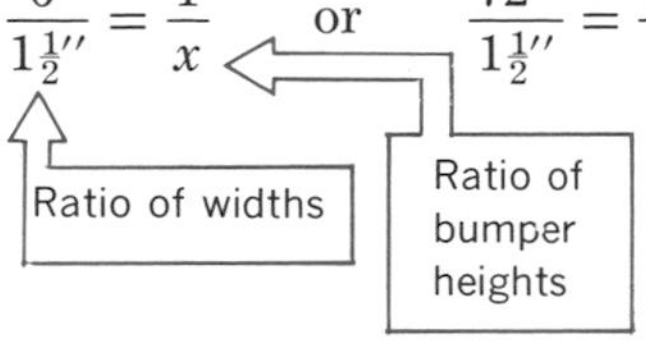

Use the methods learned earlier in this chapter to solve this equation. Multiply both equations by x.

$$\left(\frac{72}{1\frac{1}{2}}\right)x = \left(\frac{12}{\cancel{x}}\right)\cancel{x}$$

or $\frac{72x}{1\frac{1}{2}} = 12$

Now multiply both sides of this last equation by $1\frac{1}{2}$ to clear fractions:

$$\left(\frac{72x}{\cancel{1\frac{1}{2}}}\right)\cancel{1\frac{1}{2}} = (12)\,1\tfrac{1}{2}$$

$$72x = 18$$

$$x = \frac{18}{72}$$

$$x = \frac{1''}{4}$$

The bumper will be $\frac{1}{4}''$ high on the scale drawing.

Proportions can always be solved using the methods you learned for solving any linear equation, but there is also a very easy way to solve them.

> **THE CROSS-PRODUCT RULE**
>
> If $\frac{a}{b} = \frac{c}{d}$ then $ad = bc$.

The cross-products of the terms of a proportion are equal.

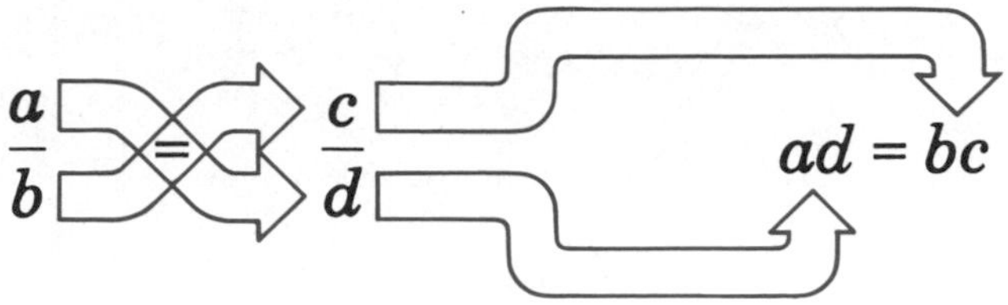

For example, if

$$\frac{2}{3} = \frac{18}{27}$$

we can use the cross-product rule to get

$$2 \cdot 27 = 3 \cdot 18$$

or $\quad 54 = 54$

In the equation above $\dfrac{72}{1\frac{1}{2}} = \dfrac{12}{x}$

Cross-multiply to get

$$72 \cdot x = 1\tfrac{1}{2} \cdot 12$$

or $\quad 72x = 18$

And this equation should be easy for you to solve.

$x = \dfrac{18}{72}$ or $\dfrac{1''}{4}$

Use the cross-product rule to solve the following proportions.

(a) $\dfrac{1}{2} = \dfrac{x}{9}$ (b) $\dfrac{y}{7} = \dfrac{3}{4}$

(c) $\dfrac{3}{z} = \dfrac{2}{5}$ (d) $\dfrac{7}{16} = \dfrac{21}{A}$

(e) $\dfrac{6}{5} = \dfrac{2}{T}$ (f) $\dfrac{21}{12} = \dfrac{R}{6}$

(g) $\dfrac{2\frac{1}{2}}{3\frac{1}{2}} = \dfrac{W}{2}$ (h) $\dfrac{0.4}{1.5} = \dfrac{12}{E}$

The answers are in **42**.

---

42 (a) $\dfrac{1}{2} = \dfrac{x}{9}$ $\quad 1 \cdot 9 = x \cdot 2$ or $9 = 2x$ $\quad x = 4\dfrac{1}{2}$

(b) $\dfrac{y}{7} = \dfrac{3}{4}$ $\quad y \cdot 4 = 3 \cdot 7$ or $4y = 21$ $\quad y = 5\dfrac{1}{4}$

(c) $\dfrac{3}{z} = \dfrac{2}{5}$ $\quad 3 \cdot 5 = 2 \cdot z$ or $2z = 15$ $\quad z = 7\dfrac{1}{2}$

(d) $\dfrac{7}{16} = \dfrac{21}{A}$ $\quad 7 \cdot A = 21 \cdot 16$ or $7A = 336$ $\quad A = 48$

(e) $T = 1\dfrac{2}{3}$ (f) $R = 10\dfrac{1}{2}$ (g) $W = 1\dfrac{3}{7}$ (h) $E = 45$

Scale Drawings

Proportion equations are found in a wide variety of practical situations. For example, when a draftsman makes a drawing of a machine part, building layout, or other large structure, he or she must *scale it down*. The drawing must represent the object accurately, but it must be small enough to fit on the paper. The draftsman reduces every dimension by some fixed ratio.

Drawings that are larger than life involve an expanded scale.

Suppose a rectangular room has a length of 18 ft and a width of 12 ft. An architectural scale drawing of this room is made so that on the drawing the length of the room is 9 in. What will be the width of the room on the drawing?

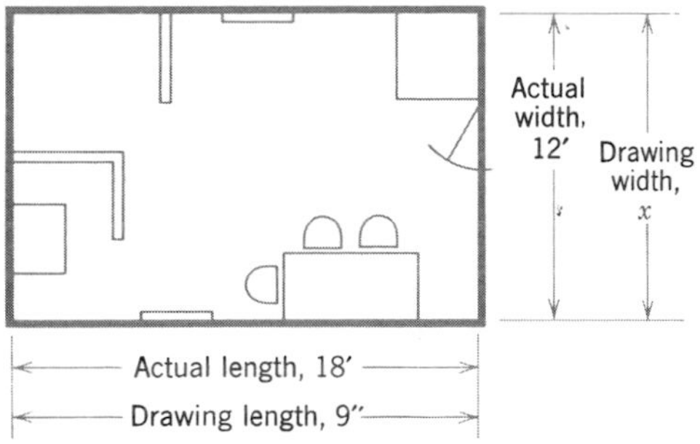

Try it. Set up a proportion equation and solve.

Check your work in **43** .

---

43 **First,** set up a ratio of lengths and a ratio of widths.

$$\text{Length ratio} = \frac{\text{actual length}}{\text{drawing length}}$$

$$= \frac{18\text{ ft}}{9\text{ in.}} = \frac{216\text{ in.}}{9\text{ in.}}$$

$$\text{Width ratio} = \frac{\text{actual width}}{\text{drawing width}}$$

$$= \frac{12\text{ ft}}{x\text{ in.}} = \frac{144\text{ in.}}{x\text{ in.}}$$

Second, write a proportion equation

$$\frac{216}{9} = \frac{144}{x}$$

Third, solve this proportion. Cross-multiply to get

$$216x = 9 \cdot 144$$
$$216x = 1296$$
$$x = 6\text{ in.}$$

The *scale ratio* of a drawing is the ratio of an actual dimension of the object to the corresponding dimension on the drawing. For the drawing shown above the scale ratio is

$$\text{Scale ratio} = \frac{18\text{ ft}}{9\text{ in.}} = \frac{216\text{ in.}}{9\text{ in.}}$$

$$= 24 \quad \text{or} \quad 24 \text{ to } 1$$

One inch on the drawing corresponds to 24 in. or 2 ft on the actual object. A draftsperson would write this as $\frac{1}{2}'' = 1'$.

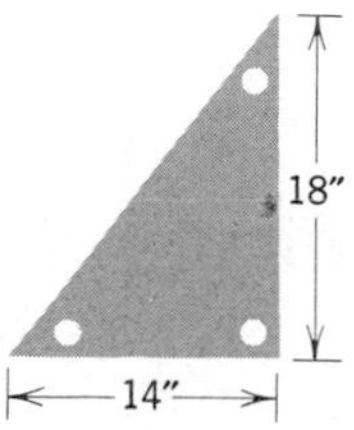

The triangular plate shown has a height of 18 in. and a base length of 14 in.

(a) What will be the corresponding height of a blueprint drawing of this plate if the base length of the drawn figure is $2\frac{3}{16}''$?
(b) Find the scale ratio of the blueprint.

Check your work in **44** .

---

44 (a) $\dfrac{18''}{x} = \dfrac{14''}{2\frac{3}{16}''}$

Cross-multiply: $(18)(2\frac{3}{16}) = 14x$

$$14x = 39\tfrac{3}{8}$$

$$x = 2\tfrac{13}{16}''$$

(b) Scale ratio $= \dfrac{18''}{2\frac{13}{16}''}$

$= 6\frac{2}{5}$ or 6.4 to 1

In general, two geometric figures that have the same shape but are not the same size are said to be *similar* figures. The blueprint drawing and the actual object are a pair of similar figures. An enlarged photograph and the smaller original are similar.

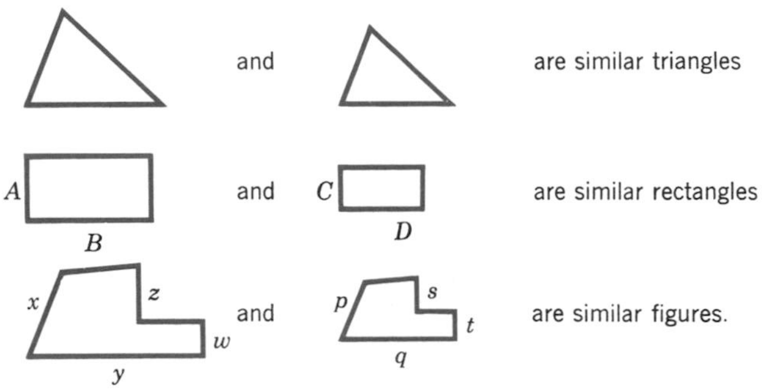

In any two similar figures, all pairs of corresponding dimensions have the same scale ratio. For example, in the rectangles above

$\dfrac{A}{C} = \dfrac{B}{D}$. In the irregular figure above,

$\dfrac{x}{p} = \dfrac{y}{q} = \dfrac{z}{s} = \dfrac{w}{t}$ and so on.

The triangles  and  are not similar.

Find the missing dimension in each of the following pairs of similar figures.

(a)

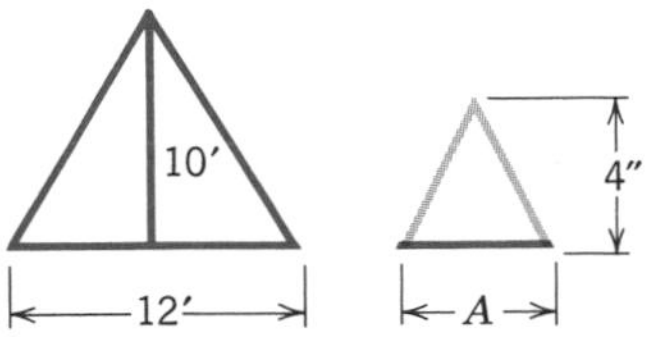

(b)

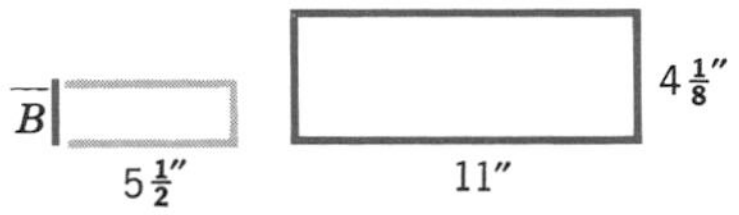

(c)

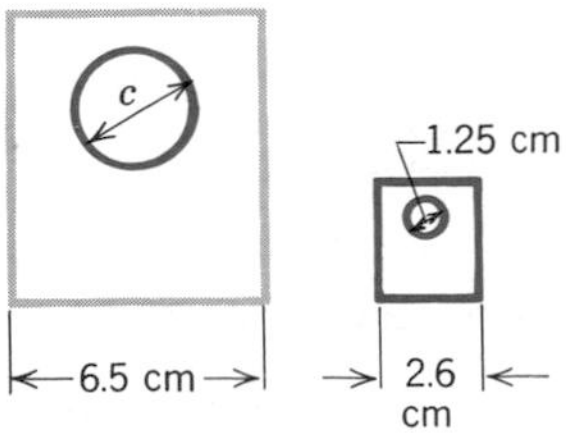

(d)

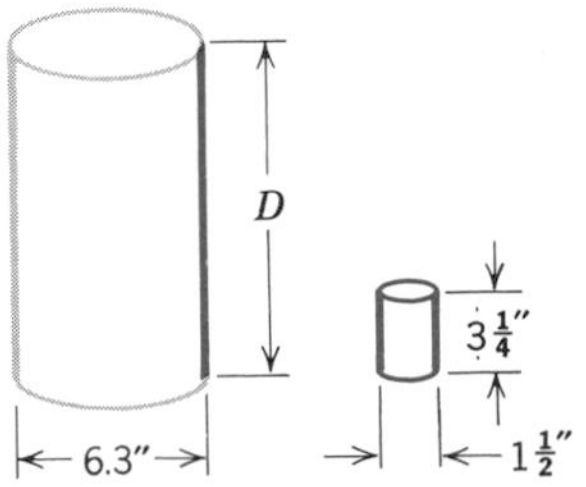

Check your answers in **45**.

45 (a) $A = 4.8''$ (b) $B = 2\frac{1}{16}''$
(c) $C = 3.125$ cm (d) $D = 13.65''$

Direct and Inverse Proportion

Many trade problems can be solved by setting up a proportion involving four related quantities. But it is important that you recognize that there are *two* kinds of proportions—direct and inverse. Two quantities are said to be *directly proportional* if an increase in one quantity leads to a proportional increase in the other quantity, or if a decrease in one leads to a decrease in the other.

> **Direct proportion:** increase ⟶ increase
> or decrease ⟶ decrease

For example, the electrical resistance of a wire is directly proportional to its length—the longer the wire, the greater the resistance. If 1 ft of nichrome heater element wire has a resistance of 1.65 ohms, what length of wire is needed to provide a resistance of 19.8 ohms?

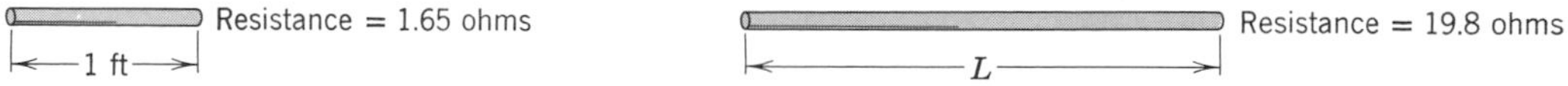

First, recognize that this problem involves a *direct* proportion. As the length of wire increases, the resistance increases proportionally.

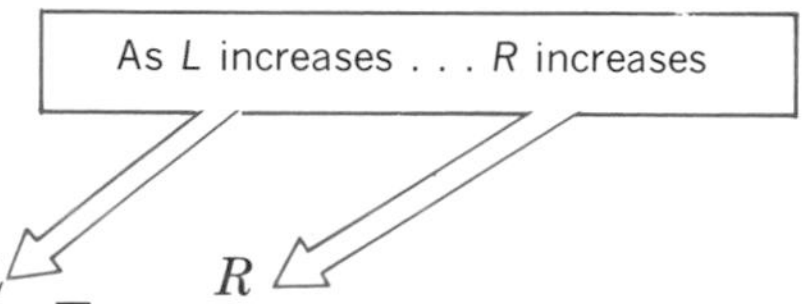

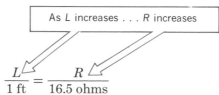

$$\frac{L}{1 \text{ ft}} = \frac{R}{16.5 \text{ ohms}}$$

Both ratios increase in size when L increases.

Second, set up a direct proportion and solve.

$$\frac{L}{1 \text{ ft}} = \frac{19.8 \text{ ohms}}{1.65 \text{ ohms}}$$

$$L = 12 \text{ ft}$$

Solve each of the following problems by setting up a direct proportion.

(a) If a Widget machine produces 88 widgets in 2 hr, how many will it produce in $3\frac{1}{2}$ hr?
(b) If one gallon of paint covers 825 sq ft, how much paint is needed to cover 2640 sq ft?
(c) What is the cost of six filters if eight filters cost $39.92?
(d) A diesel truck was driven 273 miles on 42 gallons of fuel. How much fuel is needed for a trip of 591.5 miles?
(e) A cylindrical oil tank holds 450 gallons when it is filled to its full height of 8 ft. When it contains oil to a height of 2′4″, how many gallons of oil are in the tank?

Check your solutions in **46**.

46 (a) A direct proportion—the more time spent, the more widgets produced

$$\frac{88}{x} = \frac{2\text{ hr}}{3\frac{1}{2}\text{ hr}}$$

Cross-multiply:

$2x = 308$

$x = 154$ widgets

Using a Calculator

Key	Display
C	0.
8 8	88.
×	88.
3 . 5	3.5
÷	308.
2	2.
=	154.

(b) A direct proportion—the more paint, the greater the area that can be covered—

$$\frac{1\text{ gal}}{x\text{ gal}} = \frac{825\text{ sq ft}}{2640\text{ sq ft}}$$

$x = 3.2$ gallons

(c) A direct proportion—the more you pay, the more you get—

$$\frac{6\text{ filters}}{8\text{ filters}} = \frac{x}{\$39.92}$$

$x = \$29.94$

(d) A direct proportion—the more miles you drive, the more fuel it takes—

$$\frac{273\text{ mi}}{591.5\text{ mi}} = \frac{42\text{ gal}}{x\text{ gal}}$$

$x = 91$ gallons

(e) A direct proportion—the volume is directly proportional to the height—

$$\frac{450\text{ gal}}{x\text{ gal}} = \frac{8\text{ ft}}{2\frac{1}{3}\text{ ft}}$$

$x = 131\frac{1}{4}$ gallons

Two quantities are said to be *inversely proportional* if an increase in one quantity leads to a proportional decrease in the other quantity, or if a decrease in one leads to an increase in the other.

Inverse proportion: increase → decrease
or decrease → increase

For example, the time required for a trip of a certain length is *inversely* proportional to the speed of travel. If a certain trip takes 2 hr at 50 mph, how long will it take at 60 mph?

The correct proportion equation is

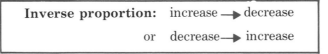

$$\frac{50\text{ mph}}{60\text{ mph}} = \frac{x\text{ hr}}{2\text{ hr}}$$

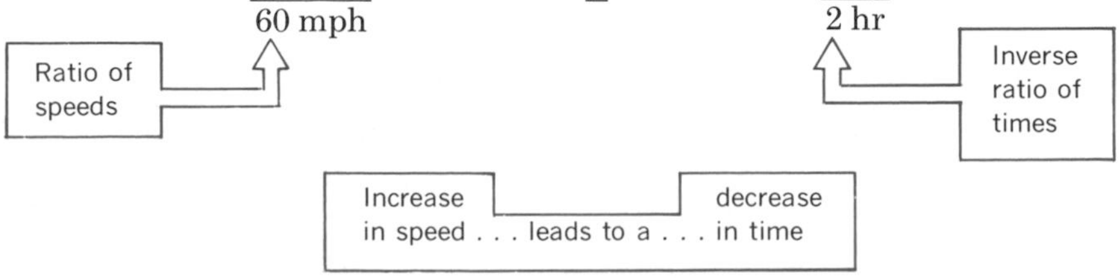

By inverting the time ratio, we have set it up so that both sides of the equation are in balance—both sides increase as speed increases.

Before attempting to solve the problem, make an estimate of the answer. We expect that the time to make the trip at 60 mph will be *less* than the time at 50 mph. The correct answer should be less than 2 hr.

Now solve it by cross-multiplying:

$$60x = 2 \cdot 50$$
$$x = 1\tfrac{2}{3}\text{ hr}$$

Remember, in a *direct* proportion

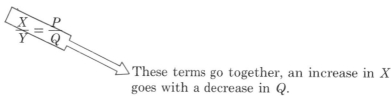

$$\frac{A}{B} = \frac{C}{D}$$

These terms go together, an increase in A goes with an increase in C.

In an *inverse* proportion

$$\frac{X}{Y} = \frac{P}{Q}$$

These terms go together, an increase in X goes with a decrease in Q.

Try this problem. In an automobile cylinder, the pressure is inversely proportional to the volume if the temperature does not change. If the volume of gas in the cylinder is 300 cu cm when the pressure is 20 psi, what is the volume when the pressure is increased to 80 psi?

Set this up as an inverse proportion and solve.

Check your work in **47**.

---

47 Pressure is inversely proportional to volume. If the pressure increases, we expect the volume to decrease.

The answer should be less than 300 cu cm.

$$\frac{P_1}{P_2} = \frac{V_2}{V_1}$$

$$\frac{20\text{ psi}}{80\text{ psi}} = \frac{V}{300\text{ cu cm}}$$

Cross-multiply:

$$80V = 20 \cdot 300$$
$$V = 75\text{ cu cm}$$

Gears and Pulleys

A particularly useful kind of inverse proportion involves the relationship between the size of a gear or pulley and the speed with which it rotates.

In the following diagram, the larger gear drives the smaller gear.

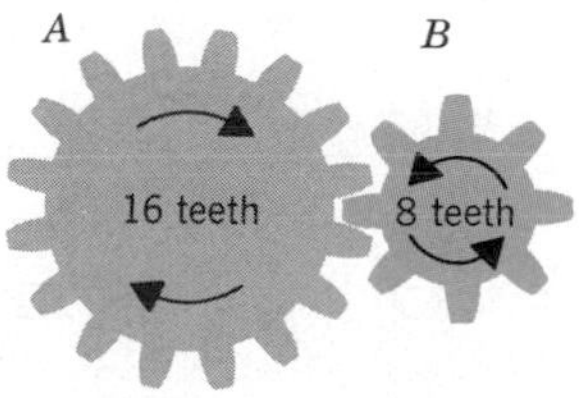

Because gear A has twice as many teeth as gear B, when A turns one turn, B will make two turns. If gear A turns at 10 turns per second, gear B will turn at 20 turns per second. The speed of the gear is inversely proportional to the number of teeth.

$$\frac{\text{speed of gear } A}{\text{speed of gear } B} = \frac{\text{teeth in gear } B}{\text{teeth in gear } A}$$

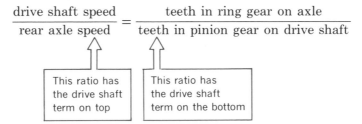

This ratio has the A term on top | This ratio has the B term on top

In this proportion, gear speed is measured in revolutions per minute, abbreviated rpm.

For the gear above, if gear A is turned by a shaft at 40 rpm, what will be the speed of gear B?

Check your work in **48**.

---

48 Because the relation is an inverse proportion, the smaller gear moves with the greater speed. We expect the speed of gear B to be faster than 40 rpm.

$$\frac{40 \text{ rpm}}{B} = \frac{8 \text{ teeth}}{16 \text{ teeth}}$$

$$B = 80 \text{ rpm}$$

On an automobile the speed of the drive shaft is converted to rear axle motion by the ring and pinion gear system.

$$\frac{\text{drive shaft speed}}{\text{rear axle speed}} = \frac{\text{teeth in ring gear on axle}}{\text{teeth in pinion gear on drive shaft}}$$

This ratio has the drive shaft term on top | This ratio has the drive shaft term on the bottom

Again gear speed is inversely proportional to the number of teeth on the gear.

If the pinion gear has 9 teeth and the ring gear has 40 teeth, what is the rear axle speed when the drive shaft turns at 1200 rpm?

Check your work in **49**.

---

49 $$\frac{1200 \text{ rpm}}{R} = \frac{40 \text{ teeth}}{9 \text{ teeth}}$$

Cross-multiply: $40R = 9 \cdot 1200$

$R = 270 \text{ rpm}$

Pulleys transfer power in much the same way as gears. For the pulley system shown, the speed of a pulley is inversely proportional to its diameter.

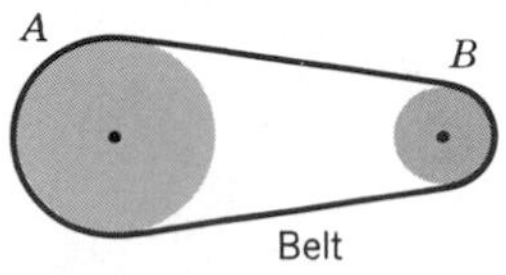

If pulley A has a diameter twice that of pulley B, then when pulley A makes one turn, pulley B will make two turns, assuming of course that there is no slippage of the belt.

$$\frac{\text{speed of pulley } A}{\text{speed of pulley } B} = \frac{\text{diameter of pulley } B}{\text{diameter of pulley } A}$$

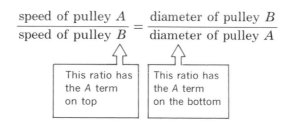

This ratio has the A term on top

This ratio has the A term on the bottom

If pulley B is 16 inches in diameter and is rotating at 240 rpm, what is the speed of pulley A if its diameter is 20 in.?

Check your work in **50**.

---

50 $\dfrac{A}{240 \text{ rpm}} = \dfrac{16 \text{ in.}}{20 \text{ in.}}$

Cross-multiply:

$$20A = 16 \cdot 240$$
$$A = 192 \text{ rpm}$$

Solve each of the following problems by setting up an inverse proportion.

(a) A 9″ pulley on a drill press rotates at 1260 rpm. It is belted to a 5″ pulley on an electric motor. Find the speed of the motor shaft.

(b) A 12 tooth gear mounted on a motor shaft drives a larger gear. The motor shaft rotates at 1450 rpm. If the speed of the large gear is to be 425 rpm, how many teeth must be on the large gear?

(c) For gases, pressure is inversely proportional to volume if the temperature does not change. If 30 cu ft of air at 15 psi is compressed to 6 cu ft, what is the new pressure?

(d) If five assembly machines can complete a given job in 3 hours, how many hours will it take for two assembly machines to do the same job?

(e) The forces and lever arm distances for a lever obey an inverse proportion.

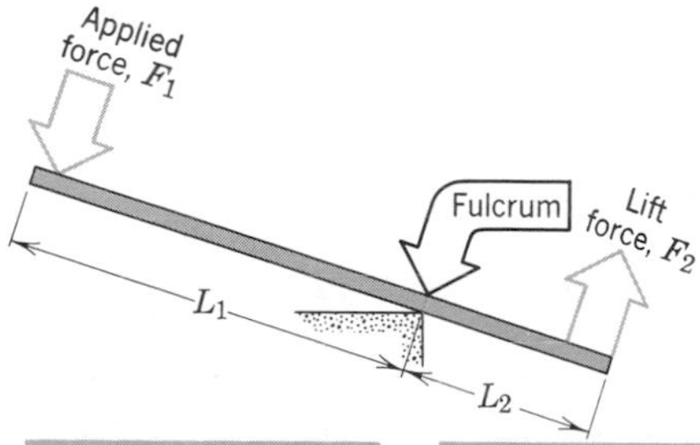

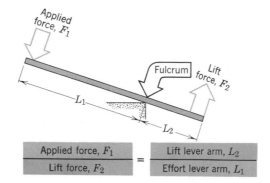

$$\frac{\text{Applied force, } F_1}{\text{Lift force, } F_2} = \frac{\text{Lift lever arm, } L_2}{\text{Effort lever arm, } L_1}$$

If a 100 lb force is applied to a 22″ crowbar pivoted 2″ from the end, what lift force is exerted?

Check your work in **51**.

---

51 (a) $\dfrac{9''}{5''} = \dfrac{x}{1260 \text{ rpm}}$ An inverse proportion: the larger pulley turns more slowly.

$$5x = 9 \cdot 1260$$
$$x = 2268 \text{ rpm}$$

(b) $\dfrac{12 \text{ teeth}}{x \text{ teeth}} = \dfrac{425 \text{ rpm}}{1450 \text{ rpm}}$ The larger gear turns more slowly.

$$425x = 12 \cdot 1450$$
$$x = 40.941...$$
$$x \cong 41$$

or 41 teeth, rounding to the nearest whole number. We can't have a part of a tooth!

(c) $\dfrac{30 \text{ cu ft}}{6 \text{ cu ft}} = \dfrac{P}{15 \text{ psi}}$

An inverse proportion: the higher the pressure the smaller the volume.

$$6P = 30 \cdot 15$$
$$P = 75 \text{ psi}$$

(d) Careful on this one! An inverse proportion should be used. The *more* machines used, the *less* the hours needed to do the job.

$$\frac{5 \text{ machines}}{2 \text{ machines}} = \frac{x \text{ hr}}{3 \text{ hr}}$$
$$2x = 15$$
$$x = 7\tfrac{1}{2} \text{ hrs}$$

Two machines will take much longer to do the job than will five machines.

(e) $\dfrac{100 \text{ lb}}{F} = \dfrac{2''}{20''}$

If the entire bar is 22″ long, and $L_2 = 2''$, then $L_1 = 20''$.

$$2F = 20 \cdot 100$$
$$F = 1000 \text{ lb}$$

Remember: $\dfrac{X_1}{X_2} = \dfrac{Y_1}{Y_2}$ In a direct proportion these are the related terms.

$\dfrac{X_1}{X_2} = \dfrac{Y_2}{Y_1}$ In an inverse proportion these are the related terms.

Now turn to 52 for a set of practice problems on ratio and proportion.

---

52 Exercises 6-3 Ratio and Proportion

A. Complete the tables given.

1.

	Teeth on Gear A	Teeth on Gear B	Gear Ratio, $\dfrac{A}{B}$
(a)	35	5	
(b)	12	7	
(c)		3	2
(d)	21		$3\frac{1}{2}$
(e)	15		1 to 3
(f)		18	1 to 2
(g)		24	2 : 3
(h)	30		3 : 5
(i)	27	18	
(j)	12	30	

2.

	Diameter of Pulley A	Diameter of Pulley B	Pulley Ratio, $\frac{A}{B}$
(a)	16″	6″	
(b)	15″	12″	
(c)		8″	2
(d)	27 cm		4.5
(e)		10 cm	4 to 1
(f)	$8\frac{1}{8}''$	$3\frac{1}{4}''$	
(g)	8.42 cm	17.56 cm	
(h)	20.41 cm		3.14 to 1
(i)		12.15 cm	1 to 2.25
(j)	4.45 cm		0.25

3.

	Rise	Span	Pitch
(a)	8′	12′	
(b)		24′	1 to 3
(c)	7′		1 to 4
(d)	14′4″	25′1″	
(e)	9′	16′9″	
(f)		20′	0.2
(g)	3′		0.15
(h)		30′6″	1:6

B. Solve these proportion equations.

1. $\frac{3}{2} = \frac{x}{8}$
2. $\frac{6}{R} = \frac{5}{72}$
3. $\frac{y}{60} = \frac{5}{3}$
4. $\frac{2}{15} = \frac{8}{H}$
5. $\frac{5}{P} = \frac{30}{7}$
6. $\frac{1}{6} = \frac{17}{x}$
7. $\frac{138}{23} = \frac{18}{x}$
8. $\frac{3.25}{1.5} = \frac{A}{0.6}$
9. $\frac{x}{34.86} = \frac{1.2}{8.3}$
10. $\frac{2\frac{1}{2}}{R} = \frac{1\frac{1}{4}}{3\frac{1}{4}}$
11. $\frac{2'6''}{4'3''} = \frac{L}{8'6''}$
12. $\frac{6.2 \text{ cm}}{x} = \frac{1.2''}{11.4''}$
13. $\frac{3'4''}{4''2''} = \frac{3.2 \text{ cm}}{x}$
14. $\frac{3\frac{1}{2}''}{W} = \frac{1.4}{0.05}$

C. Complete the following tables.

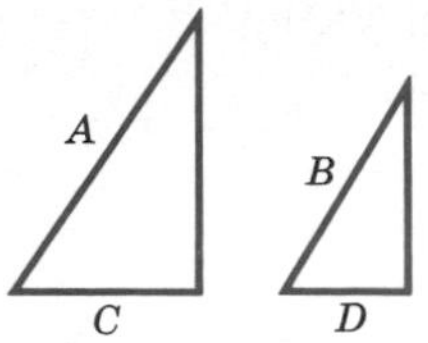

1.

	A	B	C	D
(a)	$5\frac{1}{2}''$	$1\frac{1}{4}''$	$2\frac{3}{4}''$	
(b)		23.4 cm	20.8 cm	15.6 cm

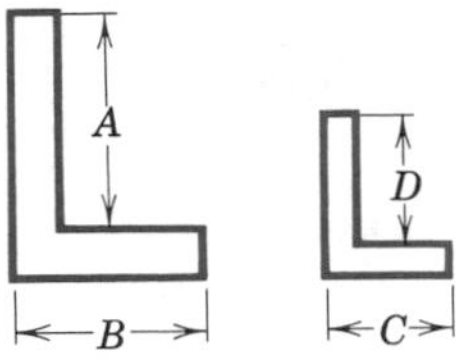

2.

	A	B	C	D
(a)	4″	5″	$3\frac{1}{2}''$	
(b)	5 ft		5 ft	2 ft

3.

	Number of Teeth on Gear 1	Number of Teeth on Gear 2	RPM of Gear 1	RPM of Gear 2
(a)	20	48	240	
(b)	25		150	420
(c)		40	160	100
(d)	32	40		1200

4.

	Diameter of Pulley 1	Diameter of Pulley 2	RPM of Pulley 1	RPM of Pulley 2
(a)	18″	24″	200	
(b)	12″		300	240
(c)		5″	400	640
(d)	14″	6″	300	

D. Practical Problems

1. A line shaft rotating at 250 rpm is connected to a 5 in. diameter pulley on a grinder. If the grinder shaft must turn at 1200 rpm, what size pulley should be attached to the line shaft?

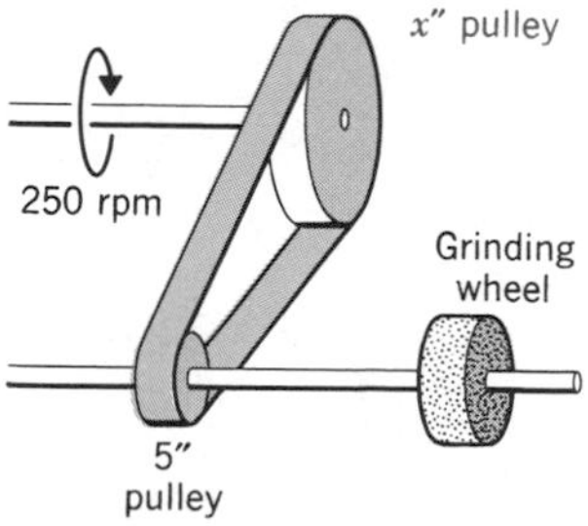

2. For a diesel engine, the combustion chamber volume at bottom dead center (BDC) is 710 cu in., and the combustion chamber volume at top dead center (TDC) is 46 cu in. What is the engine's compression ratio?

3. Horsepower developed by an engine varies directly as its displacement. How many hp will be developed by an engine with a displacement of 240 cu in. if a 380 cu in. engine of the same kind develops 220 hp?

4. If 60 gal of oil flow through a certain pipe in 16 min, how long will it take to fill a 450 gal tank using this pipe?

5. If the alternator to engine drive ratio is 2.45 to 1, what rpm will the alternator have when the engine is idling at 400 rpm? (*Hint:* Use a direct proportion.)

6. A bearing alloy is made up of 74% copper, 16% lead, and 10% tin. If 17 lb of copper are used to obtain a batch of this alloy, what weight of alloy is obtained? (Round to the nearest pound.)

7. A pair of belted pulleys have diameters of 20 in. and 16 in. If the larger pulley turns at 2000 rpm, how fast will the smaller pulley turn?

8. A 15 tooth gear on a motor shaft drives a larger gear having 36 teeth. If the motor shaft rotates at 1200 rpm, what is the speed of the larger gear?

9. The power gain of an amplifier circuit is defined as

$$\text{Power gain} = \frac{\text{output power}}{\text{input power}}$$

If the audio power amplifier circuit has an input power of 0.72 watt and a power gain of 30, what output power would be available at the speaker?

10. A 115 volt power transformer has 320 turns on the primary. If it delivers a secondary voltage of 12 volts, how many turns are on the secondary? (*Hint:* Use a direct proportion.)

E. Calculator Problems

1. The electrical resistance of a given length of wire is inversely proportional to the square of the diameter of the wire:

$$\frac{\text{resistance of wire } A}{\text{resistance of wire } B} = \frac{(\text{diameter of } B)^2}{(\text{diameter of } A)^2}$$

If a certain length of wire with diameter 34.852 mils has a resistance of 8.125 ohms, what is the resistance of the same length of the same composition wire with diameter 45.507 mils?

2. If you are paid \$138.74 for $21\frac{1}{2}$ hr of work, what amount should you be paid for 34 hours of work at this same rate of pay?

3. If the triangular plate shown is cut into 8 pieces along equally spaced dotted lines, find the height of each cut. (Round to two decimal digits.)

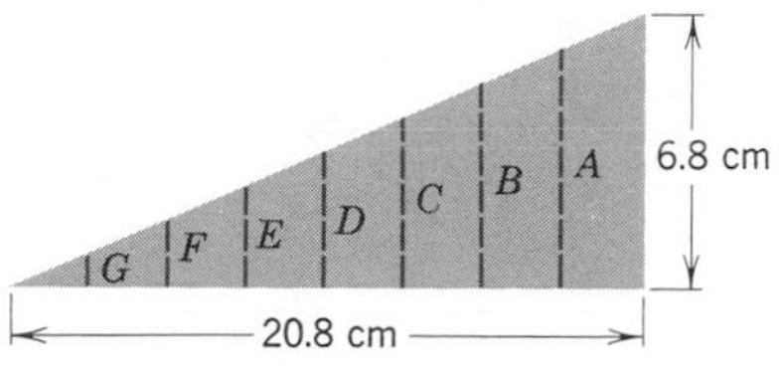

$A =$ ______ $B =$ ______

$C =$ ______ $D =$ ______

$E =$ ______ $F =$ ______

$G =$ ______

Turn to **53** for a set of practice problems over the work of Chapter 6.

6 Basic Algebra

PROBLEM SET 6

53 Answers are on page 553.

Basic Algebra

A. Simplify.

1. $4x + 6x$
2. $6y + y + 5y$
3. $3xy + 9xy$
4. $6xy^3 + 9xy^3$
5. $3\frac{1}{3}x - 1\frac{1}{4}x - \frac{5}{8}x$
6. $8v - 8v$
7. $0.27G + 0.78G - 0.65G$
8. $7y - 8y^2 + 9y$
9. $3x \cdot 7x$
10. $(4m)(2m^2)$
11. $(2xy)(-5xyz)$
12. $3x \cdot 3x \cdot 3x$
13. $3(4x - 7)$
14. $2ab(3a^2 - 5b^2)$

B. Find the value of each of the following. Round to two decimal places when necessary.

1. $L = 2W - 3$ for $W = 8$
2. $M = 3x - 5y + 4z$ for $x = 3, y = 5, z = 6$
3. $I = PRt$ for $P = 800, R = 0.06, t = 3$
4. $I = \dfrac{V}{R}$ for $V = 220, R = 0.0012$
5. $V = LWH$ for $L = 3\frac{1}{2}, W = 2\frac{1}{4}, H = 5\frac{3}{8}$
6. $N = (a + b)(a - b)$ for $a = 7, b = 12$
7. $L = \dfrac{s(P + p)}{2}$ for $s = 3.6, P = 38, p = 26$
8. $f = \dfrac{1}{8N}$ for $N = 6$
9. $t = \dfrac{D - d}{L}$ for $D = 12, d = 4, L = 2$
10. $V = \dfrac{gt^2}{2}$ for $g = 32.2, t = 4.1$

C. Solve the following equations and proportions. Round your answer to two decimal places if necessary.

1. $x + 5 = 17$
2. $m + \frac{1}{4} = \frac{3}{8}$
3. $e - 12 = 32$
4. $a - \frac{2}{3} = -\frac{5}{6}$
5. $12 = 7 - x$
6. $4\frac{1}{2} - x = -6$
7. $5x = 20$
8. $-4m = 24$
9. $\frac{1}{2}y = 16$
10. $\dfrac{M}{5} = 12$
11. $\dfrac{B}{2} = \dfrac{7}{4}$
12. $\dfrac{x}{11} = \dfrac{17}{30}$
13. $\dfrac{75}{2} = \dfrac{1500}{y}$
14. $\dfrac{32}{20} = \dfrac{E}{30}$
15. $\dfrac{F}{2.4} = \dfrac{3.6}{12}$
16. $2x + 7 = 13$
17. $-4x + 11 = 35$
18. $0.75 - 5f = 6\frac{1}{2}$
19. $0.5x - 16 = -18$
20. $5y + 8 + 3y = 24$
21. $3g - 12 = g + 8$
22. $7m - 4 = 11 - 3m$
23. $3(x - 5) = 33$
24. $7x = 0$

Date ______________

Name ______________

Course/Section ______________

Solve the following formulas for the variable shown.

25. $A = bH$ for b

26. $R = S + P$ for P

27. $P = 2L + 2W$ for L

28. $P = \frac{w}{F}$ for F

29. $S = \frac{1}{2}gt - 4$ for g

30. $V = \pi R^2H - AB$ for A

D. Practical Problems

1. In weight and balance calculations, airplane mechanics are concerned with the center of gravity of an airplane. The center of gravity may be calculated from the formula

$$CG = \frac{100(H - x)}{C}$$

where H is the distance from the datum to the empty CG, x is the distance from the datum to the leading edge of the mean aerodynamic chord (MAC), and C is the length of the MAC. CG is expressed as a percent of the MAC.

(All lengths are in inches.)

(a) Find the center of gravity if $H = 180$, $x = 155$, and $C = 80$.
(b) Solve the formula for C, and find C when $CG = 30\%$, $H = 200$, and $x = 150$.
(c) Solve the formula for H, and find H when $CG = 25\%$, $x = 125$, and $C = 60$.
(d) Solve the formula for x, and find x when $CG = 28\%$, $H = 170$, and $C = 50$.

2. Two belted pulleys have diameters of 24 in. and 10 in.
(a) Find the pulley ratio.
(b) If the larger pulley turns at 1500 rpm, how fast will the smaller pulley turn?

3. Electricians use a formula which states that the level of light in a room, in foot candles, is equal to the product of the fixture rating, in lumens, the coefficient of depreciation, and the coefficient of utilization, all divided by the area, in square feet.
(a) State this as an algebraic equation.
(b) Find the level of illumination for four fixtures rated at 2800 lumens each if the coefficient of depreciation is 0.75, the coefficient of utilization is 0.6, and the area of the room is 120 sq ft.
(c) Solve for area.
(d) Use your answer to (c) to determine the size of the room in which a level of 60 ft candles can be achieved with 10,000 lumens, given that the coefficient of depreciation is 0.8 and the coefficient of utilization is 0.5.

4. If 6 printing presses can do a certain job in $2\frac{1}{2}$ hours, how long will it take 4 presses to do the same job? (*Hint:* Use an inverse proportion.)

5. The length of a piece of sheet metal is four times its width. The perimeter of the sheet is 80 cm. Set up an equation relating these measurements and solve the equation to find the dimensions of the sheet.

6. Draftsmen usually use a scale of $\frac{1}{4}'' = 1'$ on their drawings. Find the actual lengths of the following items if their blueprint length is given.
(a) A printing press $2\frac{1}{4}''$ long.
(b) A building $7\frac{1}{2}''$ long.
(c) A bolt $\frac{1}{32}''$ long.
(d) A car $1\frac{7}{8}''$ long.

7. A certain concrete mix requires one sack of cement for every 650 lb of concrete. How many sacks of cement are needed for 3785 lb of concrete? (Remember, you cannot buy a fraction of a sack.)

8. Structural engineers have found that a good estimate of the crushing load for a square wooden pillar is given by the formula

$$L = \frac{25T^4}{H^2}$$ where L is the crushing load in tons, T is the thickness of the wood in inches, and H is the height of the post in feet.

Find the crushing load for a 6 in. thick post 12 ft high.

9. A roof has a rise of 4′ over a span of 16′. Express the pitch as a ratio.
10. An 8″ by 10″ photograph must be reduced to a width of $3\frac{1}{4}$″ to fit a space for printing. What will the reduced length be?
11. How many turns will a pinion gear having 16 teeth make if a ring gear having 48 teeth makes 120 turns?
12. Two shims must have a combined thickness of 0.048″. One shim must be twice as thick as the other. Set up an equation and solve for the thickness of each shim.
13. Find the compression ratio of a gasoline engine if each cylinder has a maximum volume of 525 cu cm and a minimum volume, or compression volume, of 50 cu cm.

E. Calculator Problems

1. If a mechanic's helper is paid $256.28 for $38\frac{1}{4}$ hours of work, how much should he receive for $56\frac{1}{2}$ hours at the same rate?
2. A pair of belted pulleys have diameters of $8\frac{3}{4}$″ and $5\frac{5}{8}$″. If the smaller pulley turns at 850 rpm, how fast will the large one turn?
3. In planning a solar energy heating system for a house, a contractor uses the formula

   $Q = 8.33GDT$ to determine the energy necessary for heating water.

   In this formula Q is the energy in BTU, G is the number of gallons heated per day, D is the number of days, and T is the temperature difference between tap water and the desired temperature of hot water. Find Q when G is 50 gallons, D is 30 days, and the water must be heated from 60° to 140°.
4. Each production employee in a plant requires an average of 25 sq ft of work area. How many employees will be able to work in an area that measures $21\frac{1}{16}$ in. by $8\frac{5}{8}$ in. on a blueprint if the scale of the drawing is $\frac{1}{32}$ in. = 1 ft.
5. Find the missing dimension in the following pair of similar figures.

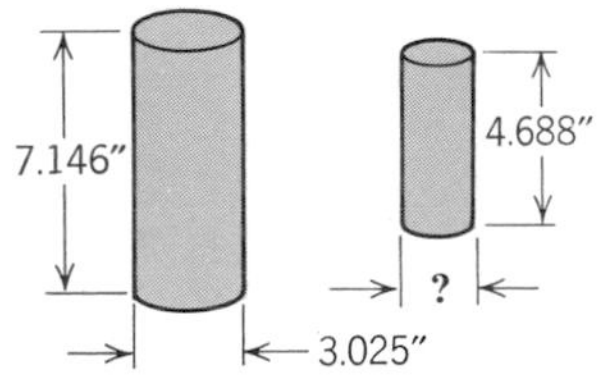

6. In a closed container, the pressure is inversely proportional to the volume when the temperature is held constant. Find the pressure of a gas compressed to 0.386 cu ft if the pressure is 12.86 psi at 2.52 cu ft.
7. The bend allowance for sheet metal is given by the formula

   $BA = N(0.01743R + 0.0078T)$ where N is the angle of the bend in degrees, R is the inside radius on the bend in inches, and T is the thickness of the metal in inches.

   Find BA if N is 47°, R is 0.725″, and T is 0.0625″.
8. To find the taper per inch of a piece of work, a machinist uses the formula

   $$T = \frac{D - d}{L}$$ where D is the diameter of the large end, d is the diameter of the small end, and L is the length.

   Find T is D = 4.1625″, d = 3.2513″, and L = 8″.
9. The modulus of elasticity of a beam is 2,650,000 at a deflection limit 360. If the modulus is directly proportional to the deflection limit, find the modulus at a deflection limit of 240.

7 Practical Geometry

PREVIEW 7

Objective	Sample Problems	Where To Go For Help: Page	Frame
Upon successful completion of this unit you will be able to:			
1. Measure angles with a protractor.	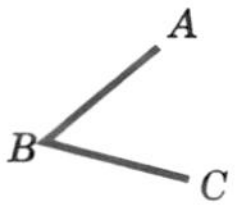 $\measuredangle ABC =$ ______	305	1
2. Use simple geometric relationships involving intersecting lines and triangles.	(a) 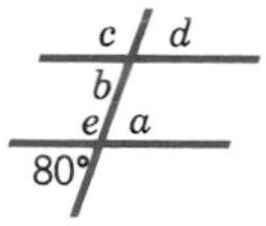 $\measuredangle a =$ ______	309	5
	$\measuredangle b =$ ______		
	$\measuredangle c =$ ______		
	$\measuredangle d =$ ______		
	$\measuredangle e =$ ______		
	(b) 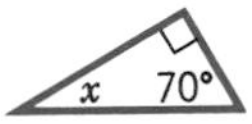 $\measuredangle x =$ ______	310	9
3. Identify polygons, including triangles (right, isosceles, equilateral), squares, rectangles, parallelograms, trapezoids, and hexagons.	(a)  ______	328	25
	(b)  ______	317	15
	(c) 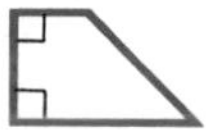 ______		
	(d) 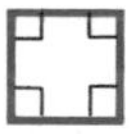 ______		
	(e)  ______		
4. Use the Pythagorean theorem.	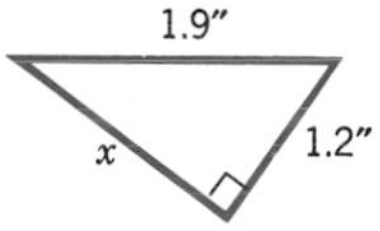 $x =$ ______ (Round to one decimal place.)	329	26
5. Find the area and perimeter of geometric figures.	(a) 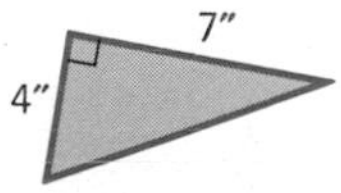 Area = ______	333	29
		323	22
	(b) 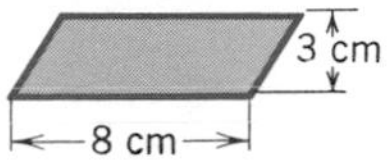 Area = ______		
	(c) 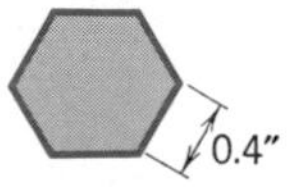 Area = ______ Perimeter = ______	337	34
	(d)  $r = 0.027''$ Area = ______ Circumference = ______	344	40

Date ______

Name ______

Course/Section ______

6. Identify solid figures, including prisms, cubes, cones, cylinders, pyramids, spheres, and frustums.

(a) 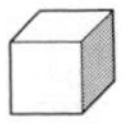 ________ 362 **49**

(b)  ________

(c) 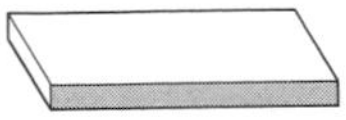 ________

7. Find the surface area and volume of solid objects.

(a) 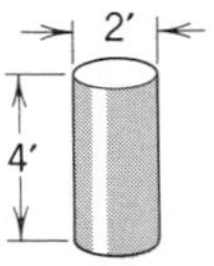

Volume = ________ 372 **57**

(b) 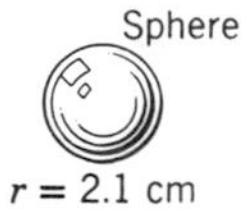

Volume = ________

Surface Area = ________ 381 **63**

(c) 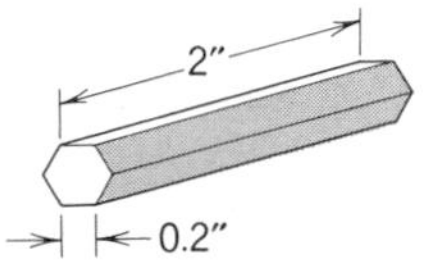

Volume = ________ 364 **50**

8. Do geometric constructions.

Construct a perpendicular to line AB at point P and a parallel to line AB through point Q. 388 **70**

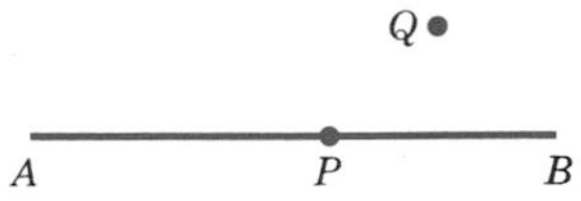

(Answers to all preview problems are at the bottom of this page.)

If you are certain you can work *all* of these problems correctly, turn to page 405 for a set of practice problems. If you cannot work one or more of the preview problems, turn to the page indicated after the problem. If you want to be certain you are successful, turn to frame **1** and begin work there.

Answers to Preview 7

1. 54° 2. (a) 80°, 80°, 100°, 80°, 100° (b) 20°
3. (a) triangle (b) parallelogram (c) trapezoid
(d) square (e) hexagon
4. 1.5 in.
5. (a) 14 sq in. (b) 24 sq cm (c) 0.416 sq in., 2.4 in.
(d) 0.0023 sq in., 0.170 in.
6. (a) cube (b) frustum of a cone (c) rectangular prism
7. (a) 12.57 cu ft (b) 38.79 cu cm, 55.42 sq cm (c) 0.208 cu in.
8. See frames **69** and **74**.

7 Practical Geometry

1 Geometry is one of the oldest branches of mathematics. It involves study of the properties of points, lines, plane surfaces, and solid figures. Ancient Egyptian engineers used these properties when they built the Pyramids, Noah used them to build the Ark, and modern trades workers use them when cutting sheet metal, installing plumbing, building cabinets, and performing countless other tasks. When they study formal geometry, students concentrate on theory and actually prove the truth of various geometric theorems. In this chapter we will concentrate on the applications of the ideas of geometry to do technical work and the trades.

7-1 ANGLE MEASUREMENT

An *angle* is a measure of the size of the opening between two intersecting lines. The point of intersection is called the *vertex,* and the lines forming the opening are called the *sides.*

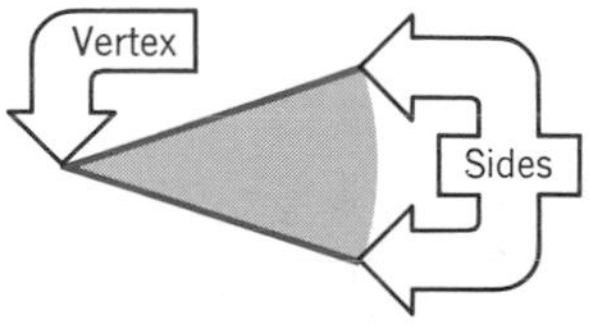

An angle may be identified in any one of the following ways:

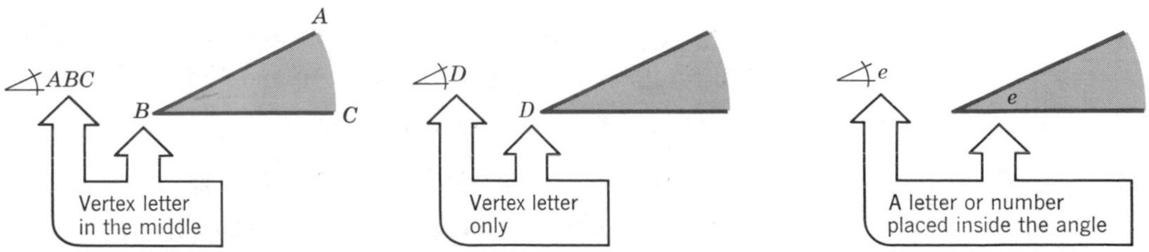

The angle symbol ∡ is simply a shorthand way to write the word "angle."

For the first angle, the middle letter is the vertex letter of the angle, while the other two letters represent points on each side. Notice that capital letters are used. The second angle is identified by the letter D outside the vertex. The third angle is named by a small letter or number placed inside the angle.

The first and third methods of naming angles are most useful when several angles are drawn with the same vertex. For example,

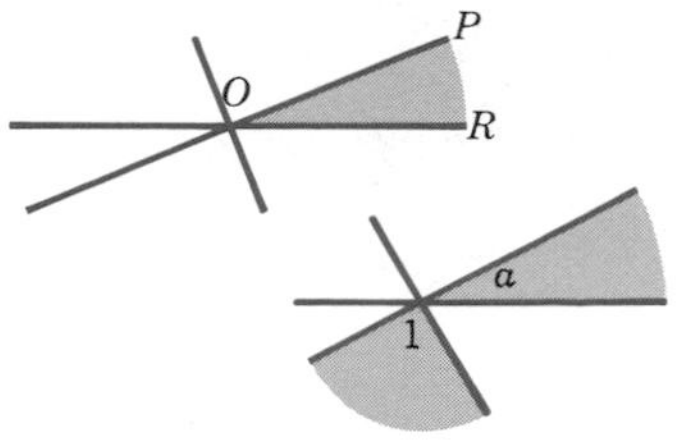

The interior of angle POR is shaded.

The interiors of angles a and 1 are shaded.

Name the angle below using the three different methods.

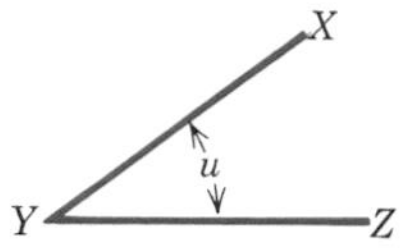

See frame **2** to check your answers.

---

2 ∡XYZ or ∡Y or ∡u

Naming or labeling an angle is the first step. Measuring it may be even more important. Angles are measured according to the size of their opening. The length of the sides of the angle is not related to the size of the angle.

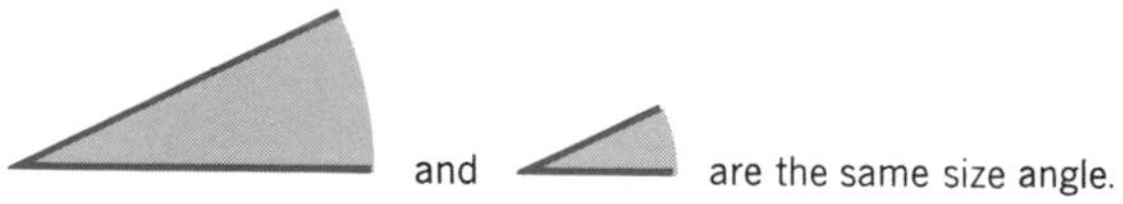

The basic unit of measurement of angle size is the *degree*. Because of traditions going back thousands of years, we define a degree like this:

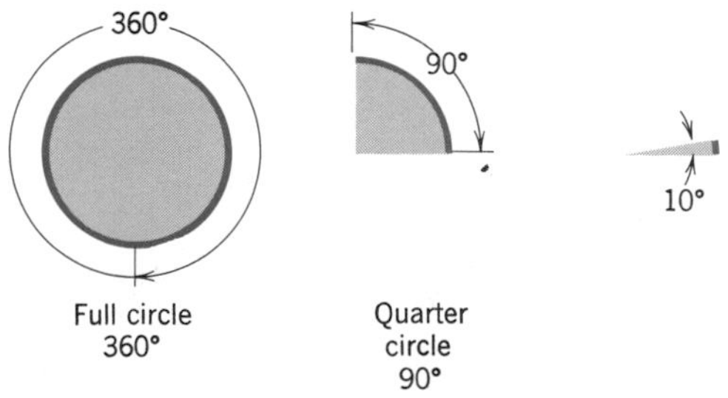

The degree can be further subdivided into finer angle units known as minutes and seconds (which are not related to the time units).

1 degree, $1° = 60$ minutes $= 60'$

1 minute, $1' = 60$ seconds $= 60''$

An angle of 62 degrees, 12 minutes, 37 seconds would be written $62°12'37''$. Most trade work requires accuracy only to the nearest degree.

As you saw in Chapter 5, page 220, a protractor is a device used to measure angles. To learn to measure angles, first look at this protractor:

A protractor that you may cut out and use is available on page 575.

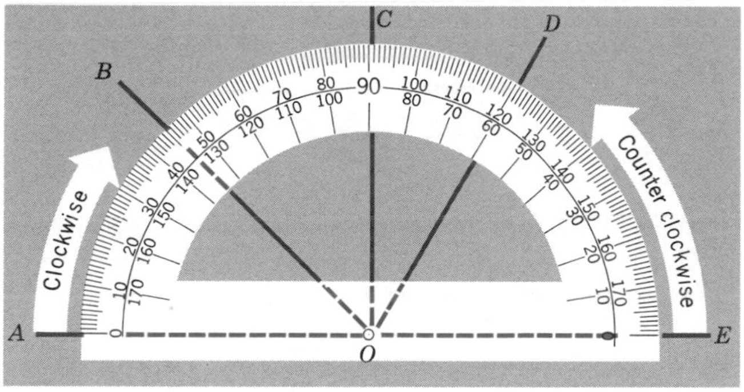

To measure an angle, place the protractor over it so that the zero-degree mark is lined up with one side of the angle, and the center mark is on the vertex. Read the measure in degrees where the other side of the angle intersects the scale of the protractor. When reading an angle clockwise, use the upper scale and, when reading counter-clockwise, use the lower scale.

In the drawing, $\measuredangle AOB$ should be read clockwise from the 0° mark on the left. $\measuredangle AOB = 45°$.

$\measuredangle EOD$ should be read counterclockwise from the 0° mark on the right. $\measuredangle EOD = 60°$.

Find (a) $\measuredangle AOD$ (b) $\measuredangle EOB$

Check your answers in **3**.

---

3 (a) $\measuredangle AOD = 120°$ Measure clockwise from the left 0° mark and read the upper scale.

(b) $\measuredangle EOB = 135°$ Measure counterclockwise from the right 0° mark and read the lower scale.

An *acute* angle is an angle *less than* 90°. Angles AOB and EOD are both less than 90°, so both are acute angles.

Acute angles: 

An *obtuse* angle is an angle *greater than* 90°. Angles AOD and EOB are both greater than 90° so both are obtuse angles.

Obtuse angles: 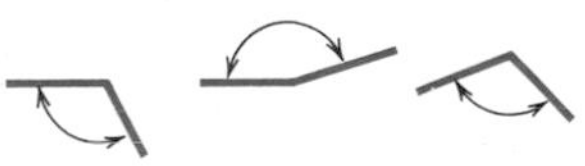

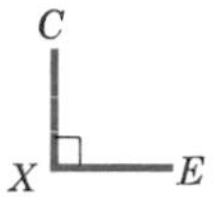

Perpendicular lines

A *right* angle is an angle exactly equal to 90°. Angle EXC in the drawing is a right angle.

Two straight lines that meet in a right or 90° angle are said to be *perpendicular*. In this drawing, CX is perpendicular to XE.

Notice that a small square is placed at the vertex of a right angle to show that the sides are perpendicular.

A *straight* angle is an angle equal to 180°.

Angle BOA shown here is a *straight* angle. 

Use your protractor to measure each of the angles given and tell whether each is acute, obtuse, right, or straight. Estimate the size of the angle before you measure.

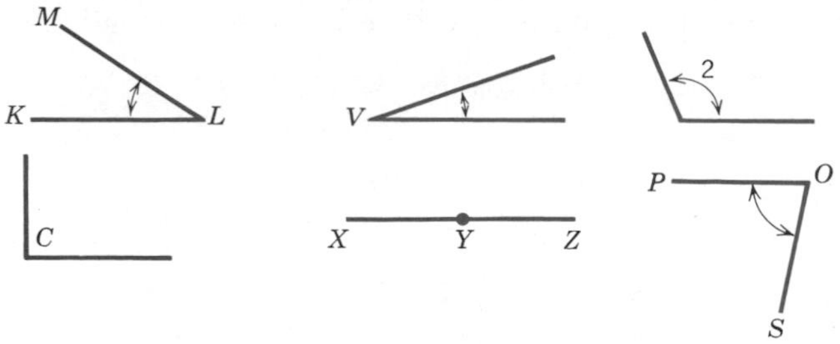

Check your answers in frame **4**.

---

4 (a) $\measuredangle KLM$ = 33°, an acute angle
(b) $\measuredangle V$ = 20°, an acute angle
(c) $\measuredangle z$ = 113°, an obtuse angle
(d) $\measuredangle c$ = 90°, a right angle
(e) $\measuredangle XYZ$ = 180°, a straight angle
(f) $\measuredangle POS$ = 80°, an acute angle

Two angles that add up to 90° are called *complementary* angles. Since 40° + 50° = 90°, the 40° angle is called the *complement* of 50°.

Two angles that add up to 180° are called *supplementary* angles. Since 120° + 60° = 180°, the 120° angle is called the *supplement* of 60°.

In addition to measuring angles that already exist, trades workers will also need to draw their own angles. To draw an angle of a given size follow these steps:

Step 1 Draw a line representing one side of the angle.

Step 2 Place the protractor over the line so that the center mark is on the vertex, and the 0° mark coincides with the other end of the line. (**Note:** We could have chosen either end of the line for the vertex.)

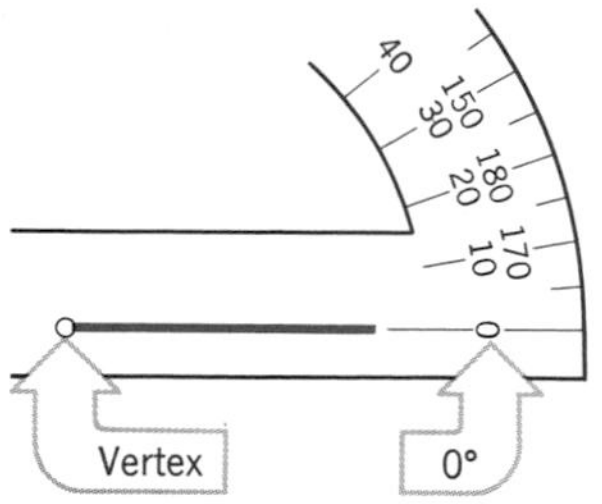

Step 3 Now place a small dot above the degree mark corresponding to the size of your angle. In this case the angle is 30°.

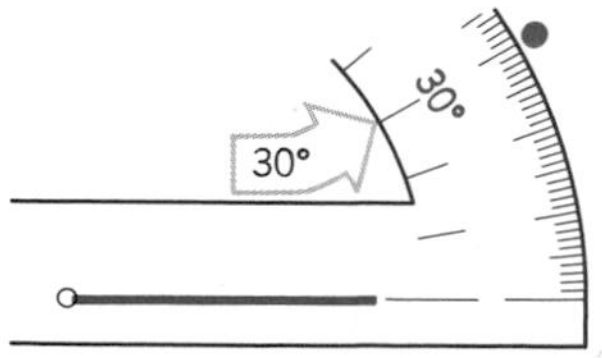

Step 4 Remove the protractor and connect the dot to the vertex.

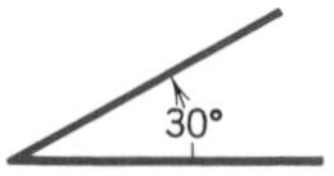

Now you try it. Use your protractor to draw the following angles:

(a) $\measuredangle A = 15°$ (b) $\measuredangle B = 115°$ (c) $\measuredangle C = 90°$

Check your work in **5**.

---

5 Here are our drawings:

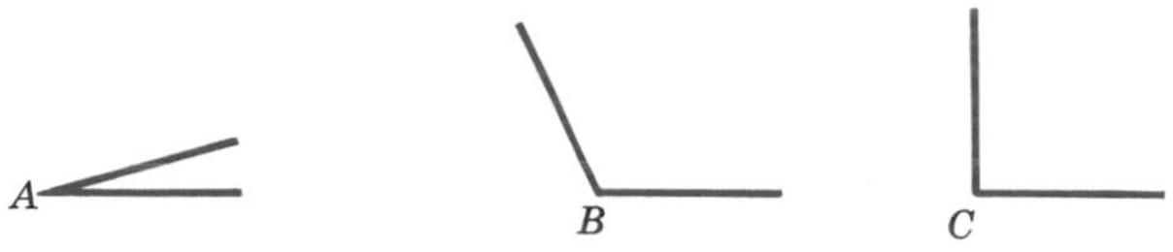

Angle Facts

There are several important geometric relationships involving angles that you should know. For example, measure the four angles created by the intersecting lines shown.

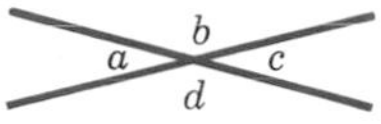

$a =$ ______ $b =$ ______ $c =$ ______ $d =$ ______

Measure them now, then check your answers in **6**.

---

6 $\measuredangle a = 30°$ $\measuredangle b = 150°$ $\measuredangle c = 30°$ $\measuredangle d = 150°$

Did you notice that opposite angles are equal?

For any pair of intersecting lines the opposite angles are called *vertical angles.* In the drawing above angles a and c are a pair of vertical angles. Angles b and d are also a pair of vertical angles.

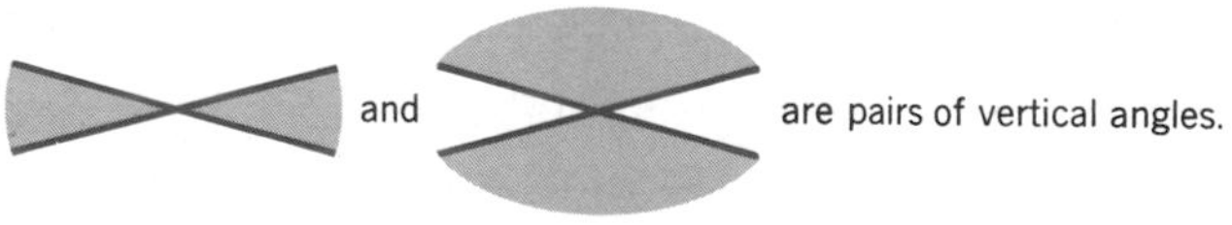

An important geometric rule is that

> When two straight lines intersect, the opposite or vertical angles are always equal.

$\measuredangle p = \measuredangle r$

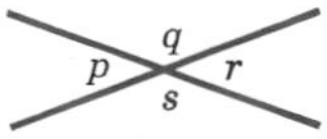

and $\measuredangle q = \measuredangle s$

If you remember that a straight angle $= 180°$, you can complete the following statements for the drawing above:

$\measuredangle p + \measuredangle q =$ ______

$\measuredangle p + \measuredangle s =$ ______

$\measuredangle s + \measuredangle r =$ ______

$\measuredangle q + \measuredangle r =$ ______

Complete these, then check your answers in **7**.

---

7 $\angle p + \angle q = 180°$

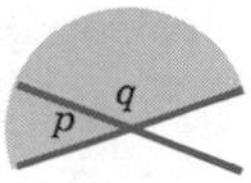

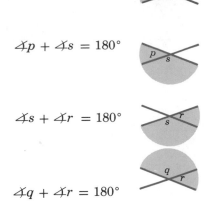

$\angle p + \angle s = 180°$

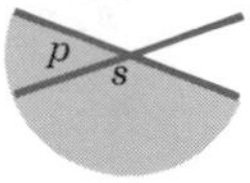

$\angle s + \angle r = 180°$

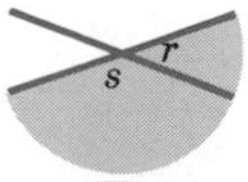

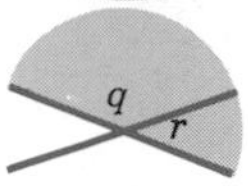

$\angle q + \angle r = 180°$

This leads to a second important geometry relation:

When two lines intersect, the adjacent angles always sum to 180°.

Use these relationships to answer the following questions:

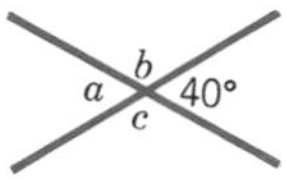

(a) $\angle a =$ ______ (b) $\angle b =$ ______ (c) $\angle c =$ ______

Check your answers in 8.

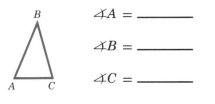

8 (a) $\angle a = 40°$ since opposite angles are equal
(b) $\angle b = 140°$ since $\angle a + \angle b = 180°$
(c) $\angle c = 140°$ since opposite angles are equal, $\angle b = \angle c$

A third important geometry relationship involves the three angles of a triangle. Use your protractor to measure each angle in the following triangle, ABC.

$\angle A =$ ______

$\angle B =$ ______

$\angle C =$ ______

Check your measurements in 9.

---

9 $\angle A = 65°$ $\angle B = 40°$ $\angle C = 75°$

Notice that the three angles of the triangle sum to 180°.

$65° + 40° + 75° = 180°$

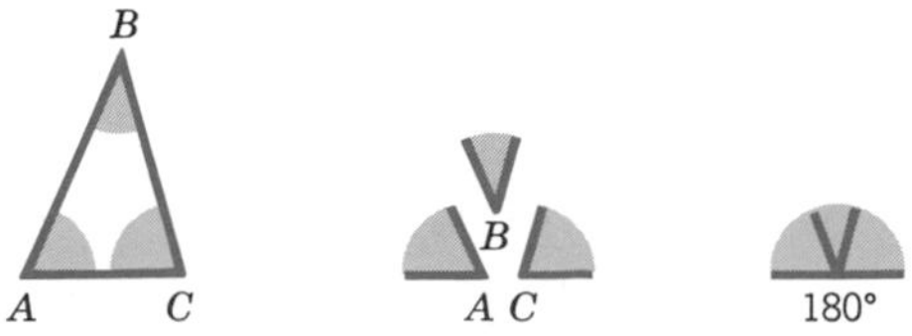

Measuring angles on many triangles of every possible size and shape leads to the third geometry relationship:

The interior angles of a triangle always add to 180°.

Use this fact to solve the following problem.

If $\measuredangle P = 50°$ and $\measuredangle Q = 85°$, find $\measuredangle R$.

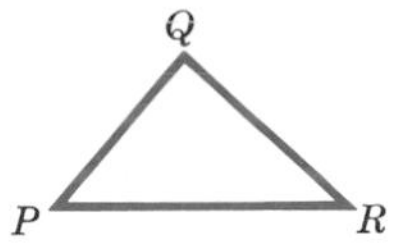

Check your answer in **10** .

---

10 By the relationship given, $\measuredangle P + \measuredangle Q + \measuredangle R = 180°$

so

$\measuredangle R = 180° - \measuredangle P - \measuredangle Q$
$\measuredangle R = 180° - 50° - 85°$
$\measuredangle R = 45°$

A fourth and final geometry fact you will find useful involves parallel lines. Two straight lines are said to be *parallel* if they always are the same distance apart and never meet. If a pair of parallel lines is cut by a third line, several angles are formed.

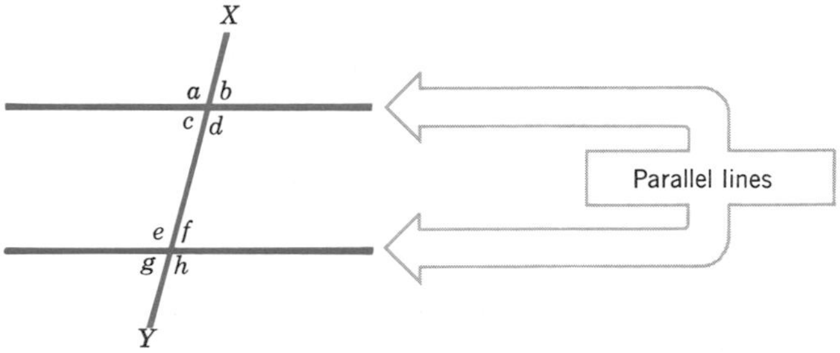

Angles c and f are called *alternate interior angles.* Angles d and e are also alternate interior angles.

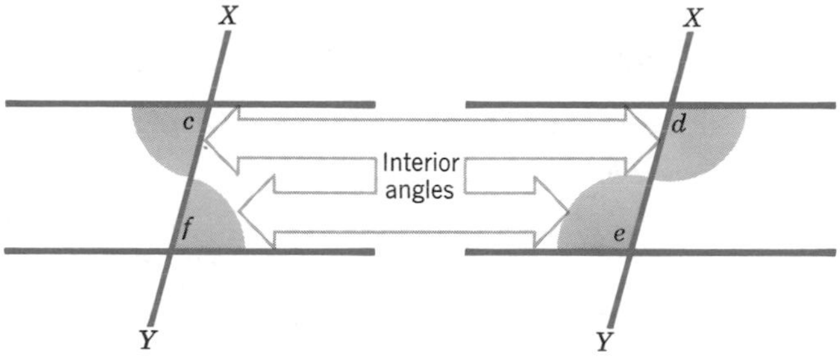

The angles are *interior* (or inside) the parallel lines and they are on *alternate* sides of the line XY. The alternate interior angles are equal.

Angles a and h are called *alternate exterior angles.*
Angles b and g are also alternate exterior angles.

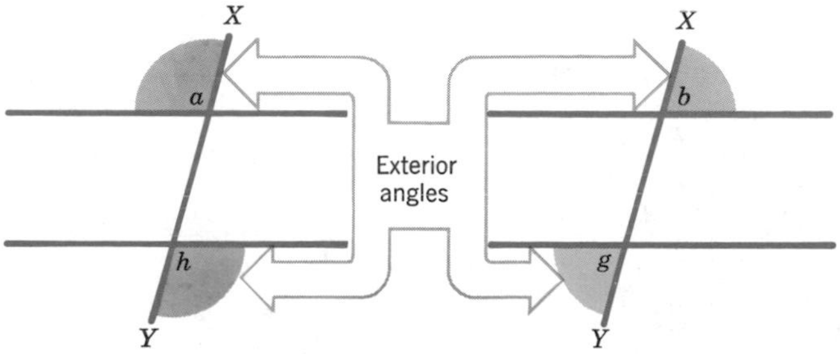

The angles are *exterior* (or outside) the parallel lines and they are on *alternate* sides of the line XY. The alternate exterior angles are equal.

The geometric rule is:

> When parallel lines are cut by a third line, the alternate exterior angles are equal and the alternate interior angles are equal.

Use this rule and the previous ones to answer the following questions:

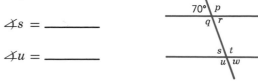

$\measuredangle p =$ ________ $\quad \measuredangle q =$ ________

$\measuredangle r =$ ________ $\quad \measuredangle s =$ ________

$\measuredangle t =$ ________ $\quad \measuredangle u =$ ________

$\measuredangle w =$ ________

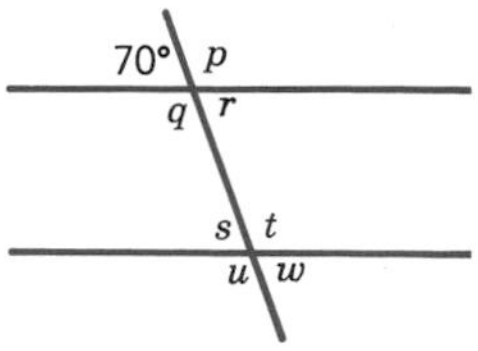

Check your work in 11.

---

11 $\measuredangle p = 110°$ since $70° + \measuredangle p = 180°$.

$\measuredangle q = 110°$ since p and q are vertical angles.

$\measuredangle r = 70°$ since it is a vertical angle to the 70° angle.

$\measuredangle s = 70°$ since $\measuredangle s = \measuredangle r$, alternate interior angles.

$\measuredangle t = 110°$ since $\measuredangle t = \measuredangle q$, alternate interior angles.

$\measuredangle u = 110°$ since $\measuredangle u = \measuredangle p$, alternate exterior angles.

$\measuredangle w = 70°$ since $\measuredangle u + \measuredangle w = 180°$.

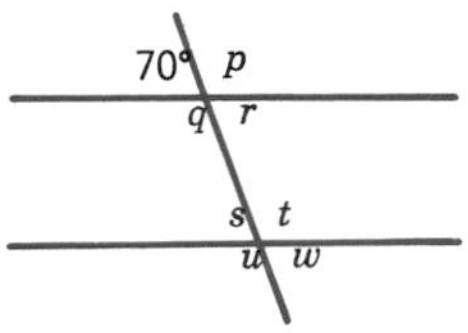

Now, ready for some problems on angle measure? Turn to 12 for practice in measuring and working with angles.

---

12 **Exercises 7-1 Angle Measurement**

A. Solve the following. If you need a protractor, cut out and use the one on page 575.

1. Name each of the following angles using the three kinds of notation shown in the text.

(a)

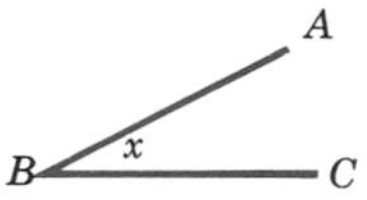

(b)

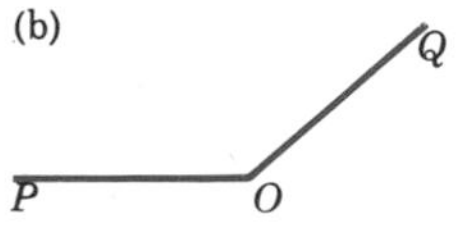

(c)

(d)

(e)

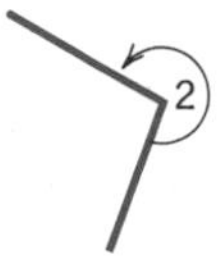

(f)

2. For this figure,

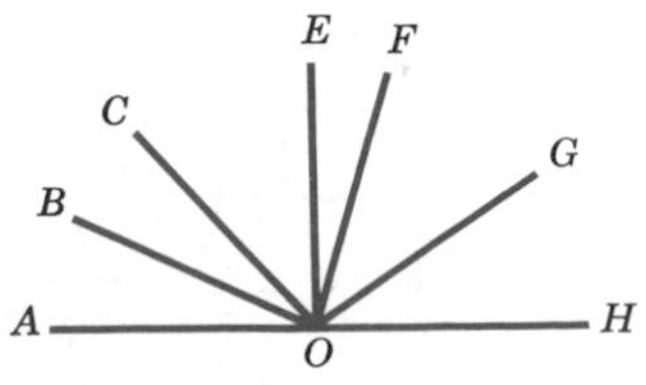

(a) Name an acute angle that has line AO as a side.
(b) Name an acute angle that has line HO as a side.
(c) Name an obtuse angle that has line AO as a side.
(d) Name a right angle.
(e) Use your protractor to measure $\measuredangle AOB$, $\measuredangle GOF$, $\measuredangle HOC$, $\measuredangle FOH$, $\measuredangle BOF$, $\measuredangle COG$.

3. Use your protractor to draw the following angles.

(a) $\measuredangle LMN = 29°$	(b) $\measuredangle 3 = 100°$
(c) $\measuredangle Y = 152°$	(d) $\measuredangle PQR = 137°$
(e) $\measuredangle t = 68°$	(f) $\measuredangle F = 48°$

4. Measure the indicated angles on the following shapes. (You will need to extend the sides of the angles to improve the accuracy of your measurements.)

(a)
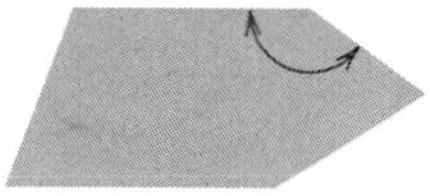
(b)

(c)
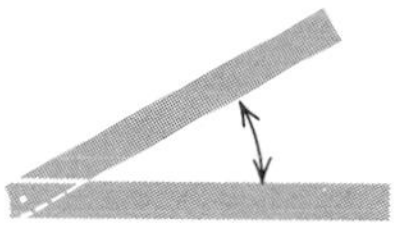
(d)
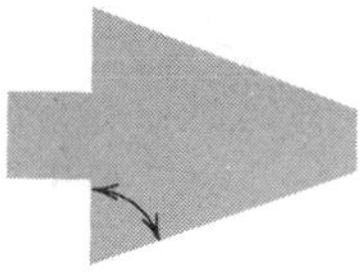
(e)

(f)
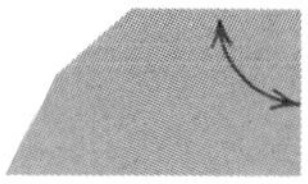
(g)
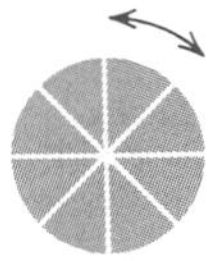
(h)
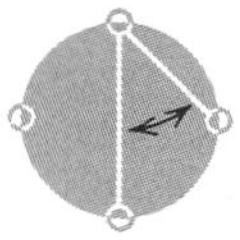
(i)
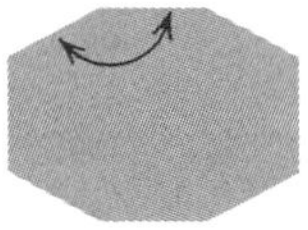

B. Use the geometry relationships to answer the following questions:

1. In each problem one angle measurement is given. Determine the others for each figure without using a protractor.

(a)
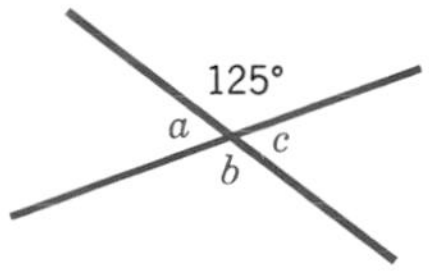

(b)
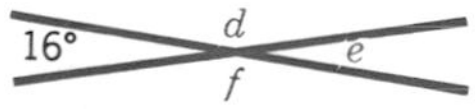

(c)
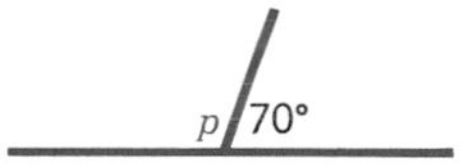

(d)
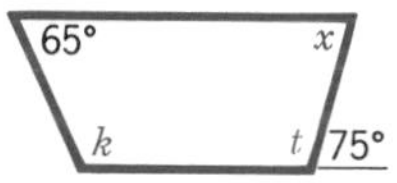

(e)
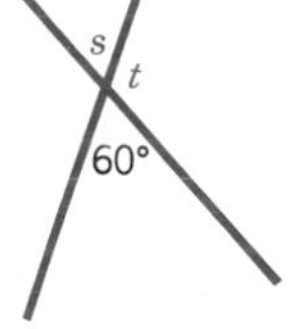

(f)
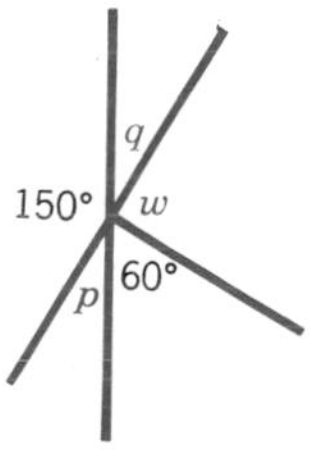

2. For each triangle shown, two angles are given. Find the third angle without using a protractor.

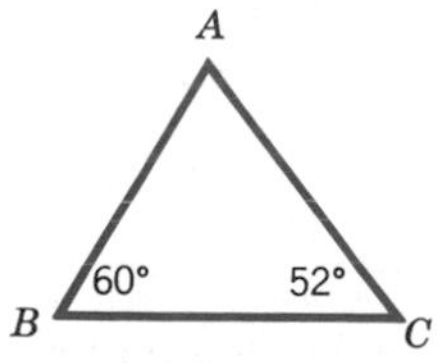

(a)

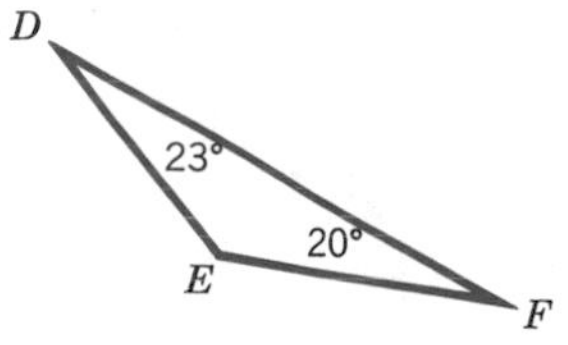

(b)

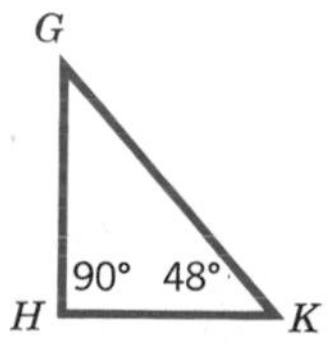

(c)

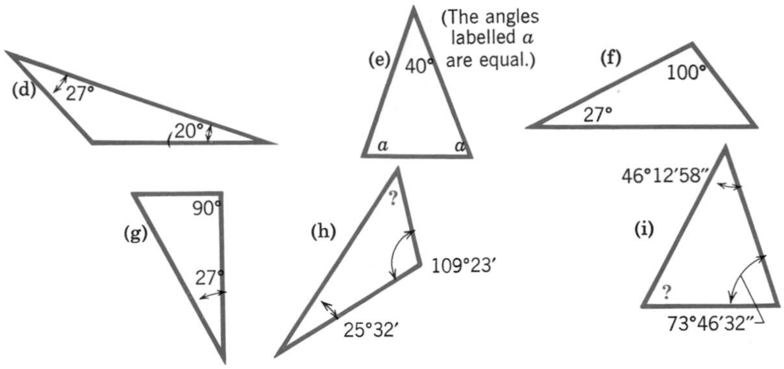

3. For each problem find the angles marked.

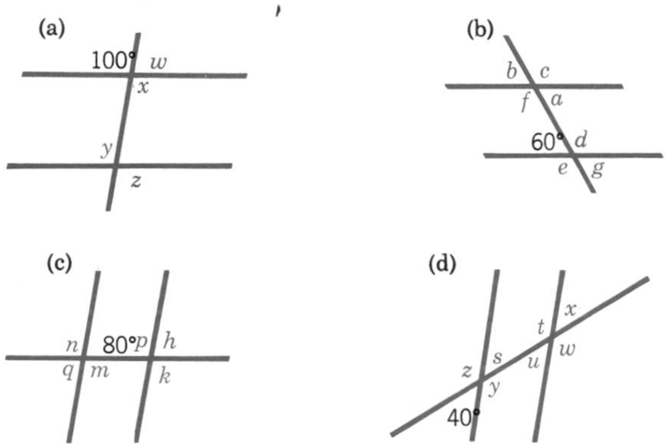

C. Applied Problems

1. A machinist must punch holes A and B in the piece of steel below by rotating the piece through a certain angle from vertex C. What is angle ACB?

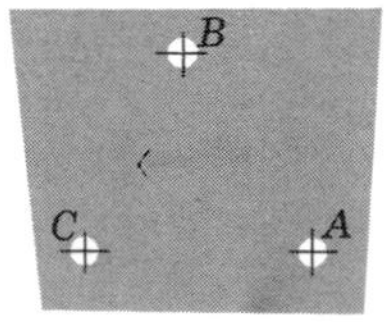

2. A plumber must connect the two pipes shown below by first selecting the proper elbows. What angle elbows does he need?

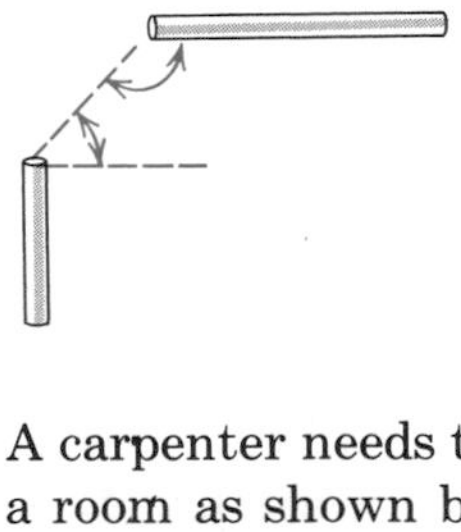

3. A carpenter needs to build a triangular hutch that will fit into the corner of a room as shown below. Measure the three angles of the hutch.

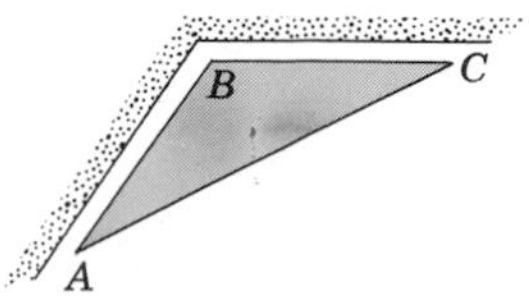

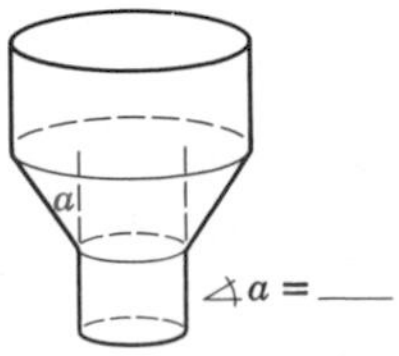

4. A sheet metal worker must connect the two vent openings shown. At what angle must the sides of the connecting pieces flare out?

5. An electrician wants to connect the two conduits shown below with a 45° elbow. How far up must he or she extend the lower conduit before they will connect at that angle? Give the answer in inches. (**Hint:** Draw the 45° connection on the upper conduit first.) Each dot represents a 1 in. extension.

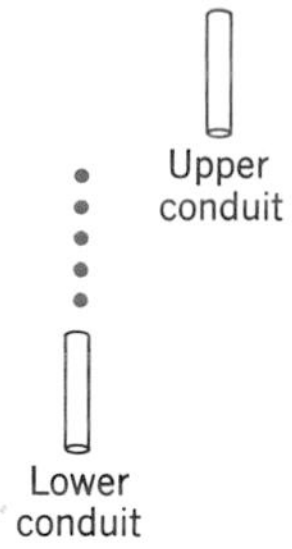

6. At what angle must the rafters be set to create the roof gable shown in the drawing?

7. A machinist receives a sketch of a part he must make as shown below:

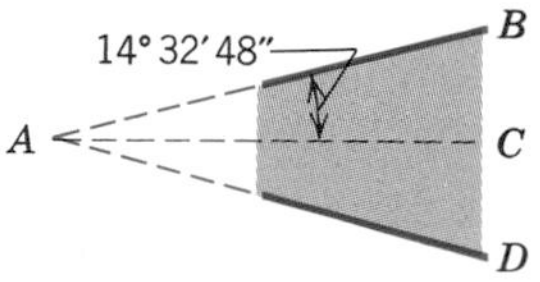

If the indicated angle BAC is precisely half of angle BAD, find $\measuredangle BAD$.

8. The drawing illustrates a cross section of a V-thread. Find the indicated angle.

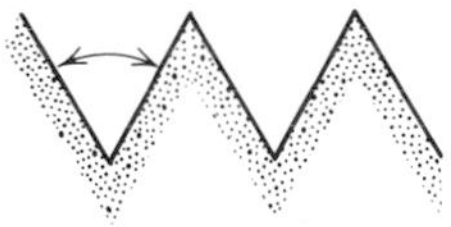

9. At what angles must drywall board be cut to create the wall shown in the figure?

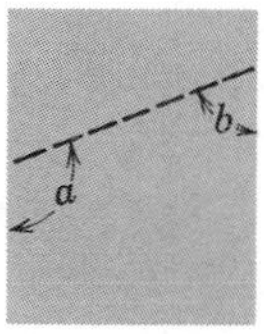

10. A carpenter wishes to make a semicircular deck by cutting 8 angular pieces of wood as shown. Measure the angle of one of the pieces. Was there another way to determine the angle measure without using a protractor?

11. A carpenter's square is placed over a board as shown in the figure. If $\measuredangle 1$ measures 74° what is the size of $\measuredangle 2$?

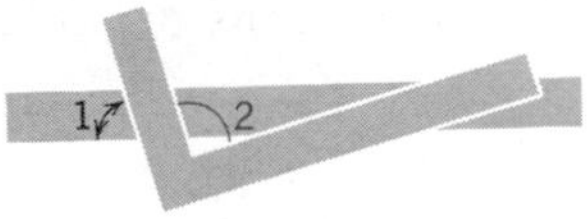

12. Board 1 must be joined to board 2 at a right angle. If $\measuredangle x$ measures 42° what must $\measuredangle y$ measure?

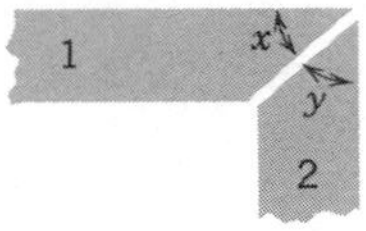

When you have finished these problems check your answers on p. 554, then continue in **13** with the study of plane figures.

7-2 PLANE FIGURES: AREA AND PERIMETER

Polygons

13 A *polygon* is a closed plane figure containing three or more angles and bounded by three or more straight sides. The word "polygon" itself means "many sides." The following figures are all polygons.

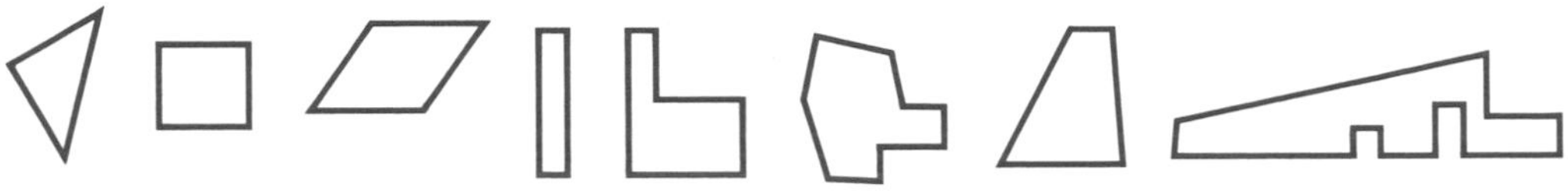

A figure with a curved side is *not* a polygon.

are not polygons

Every trades worker will find that an understanding of polygons is important. In this section you will learn how to identify the parts of a polygon, recognize the different kinds of polygons, and compute the perimeter and area of any polygon.

In the general polygon shown, each *vertex* or corner is labeled with a letter: A, B, C, D, and E. The polygon is named simply "polygon $ABCDE$"—not a very fancy name, but it will do.

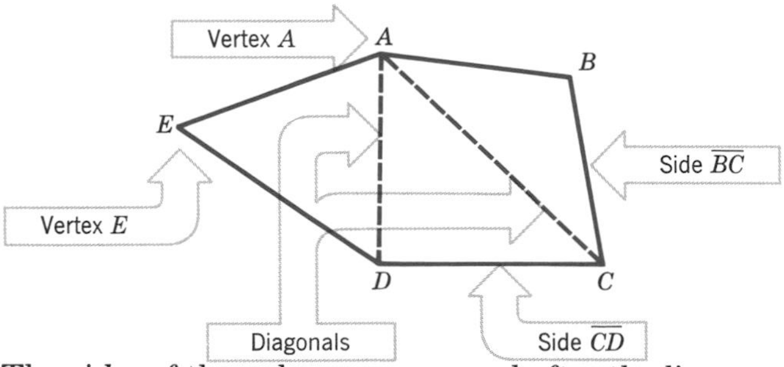

The *sides* of the polygon are named after the line segments that form each side: $\overline{AB}$, $\overline{BC}$, $\overline{CD}$, and so on. Notice that we place a bar over the letters to indicate that we are talking about a side.

The *diagonals* of the polygon are the line segments connecting nonconsecutive vertices such as $\overline{AC}$ and $\overline{AD}$. These are shown as dotted lines in the figure.

In the following polygon, identify all sides, vertices, and diagonals.

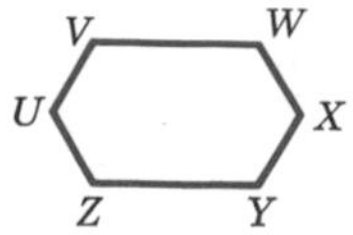

(**Hint:** There are nine diagonals. Be certain you find them all.)

See frame **14** to check your answers.

14 The sides are $\overline{UV}$, $\overline{VW}$, $\overline{WX}$, $\overline{XY}$, $\overline{YZ}$, and $\overline{ZU}$.

(Of course the letters can be reversed, side $\overline{VU}$ is the same as side $\overline{UV}$.)

The vertices are U, V, W, X, Y, and Z.

The diagonals are $\overline{UW}$, $\overline{UX}$, $\overline{UY}$, $\overline{VX}$, $\overline{VY}$, $\overline{VZ}$, $\overline{WY}$, $\overline{WZ}$, and $\overline{XZ}$.

For those who do practical work with polygons, the most important measurements are the lengths of the sides, the *perimeter,* and *area*. The perimeter of any polygon is simply the sum of the lengths of its sides. It is the distance around the outside of the polygon.

In the polygon $KLMN$ the perimeter is

$3'' + 5'' + 6'' + 4'' = 18$ inches.

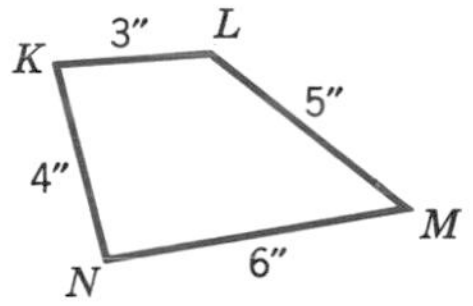

The perimeter is a length, so it has length units, in this case inches.

Find the perimeter of each of the following polygons.

(a)

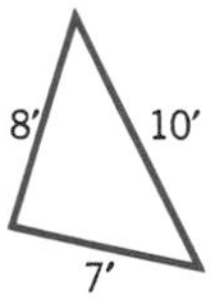

(b)

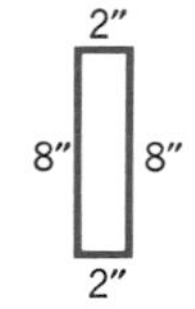

(c)

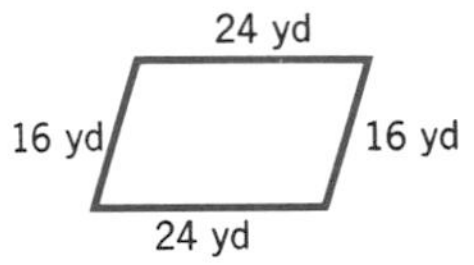

(d)

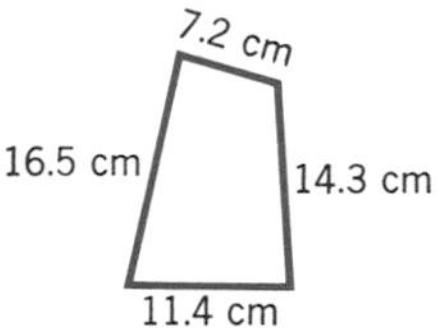

(e)

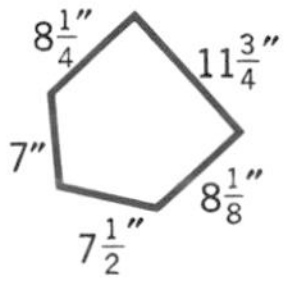

(f)

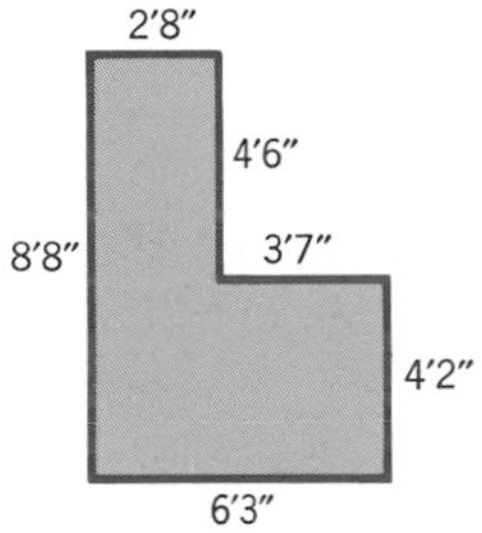

(g)

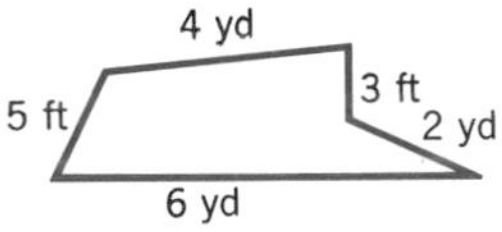

Check your answers in **15**.

---

15 (a) 25 ft. (b) 20 in. (c) 80 yd (d) 49.4 cm
(e) $42\frac{5}{8}$ in. (f) 29′10″ (g) 14 yd 2 ft or 44 ft.

Adding up the lengths of the sides will always give you the perimeter of a polygon, but handy formulas are needed to enable you to calculate the area of a polygon. Before you can use these formulas you must learn to identify the different types of polygons. Let's look at a few. First, we'll examine the *quadrilaterals* or four-sided polygons.

Figure $ABCD$ is a *parallelogram.* In a parallelogram opposite sides are parallel and equal in length.

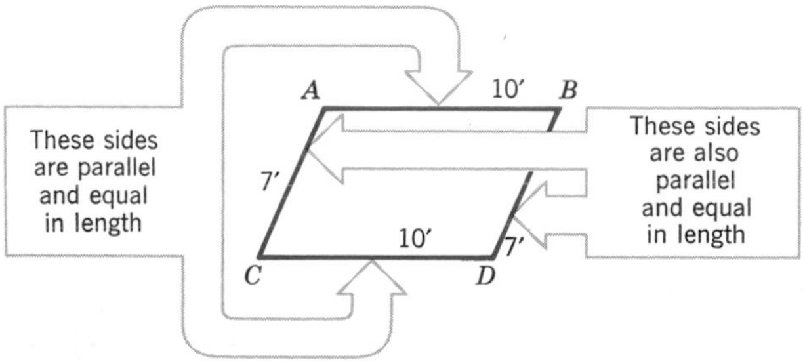

Figure *EFGH* is a *rectangle,* a parallelogram in which the four corner angles are right angles. The ⊾ symbols at the vertices indicate right angles. Since a rectangle is a parallelogram, opposite sides are parallel and equal.

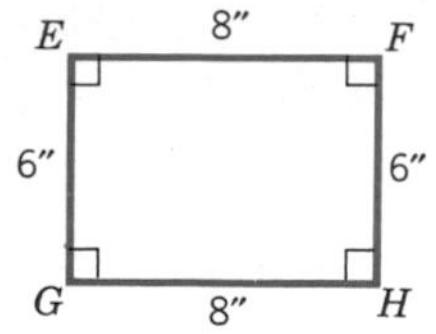

IJKL is a *square,* a rectangle in which *all* sides are the same length.

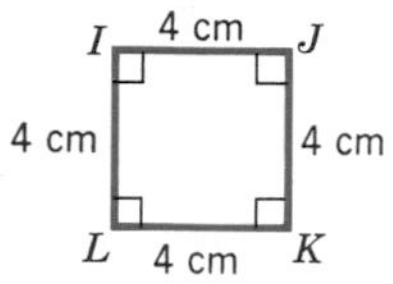

MNOP is called a *trapezoid.* A trapezoid contains two parallel sides and two nonparallel sides. *MN* and *OP* are parallel; *MP* and *NO* are not.

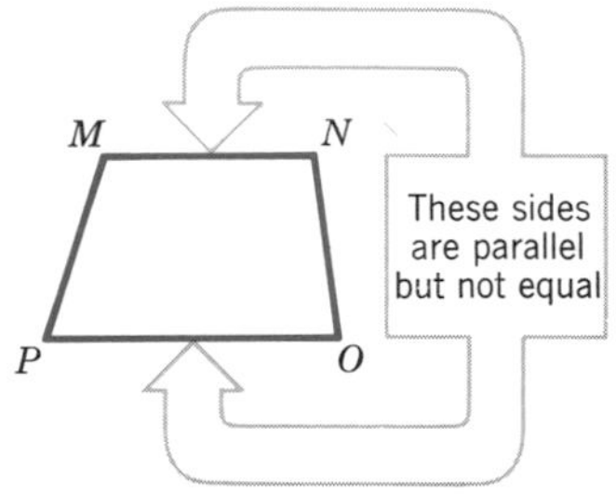

If a four-sided polygon has none of these special features—no parallel sides—we call it a *quadrilateral. QRST* is a quadrilateral.

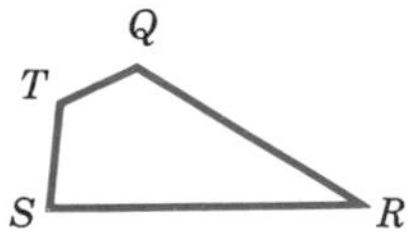

Name each of the following polygons. If more than one name fits, give the most specific name.

(a)

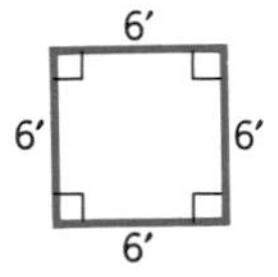

(b)

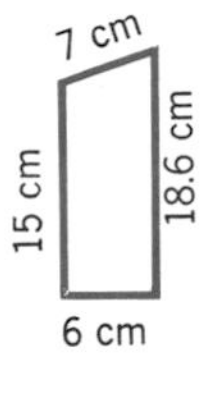

(c)

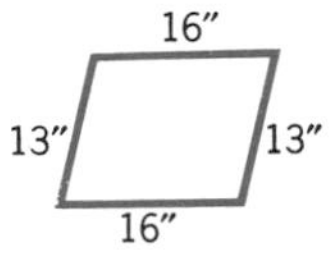

(d)

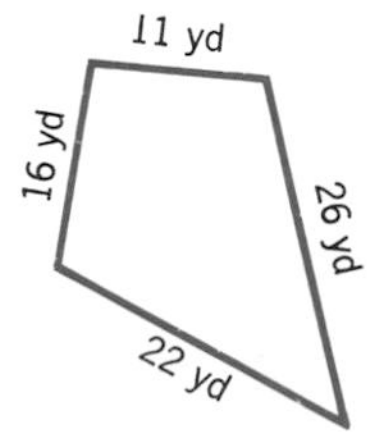

(e)

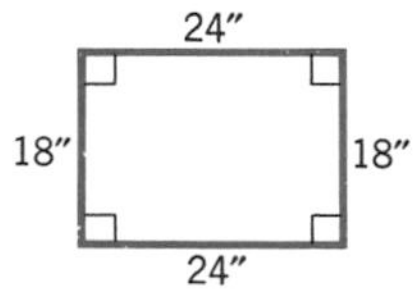

Look in **16** for the correct answers.

16 (a) Square. (It is also a rectangle and a parallelogram, but "square" is a more specific name.)
(b) Trapezoid.
(c) Parallelogram.
(d) Quadrilateral.
(e) Rectangle. (It is also a parallelogram, but "rectangle" is a more specific name.)

Once you can identify a polygon, you can use a formula to find its area. Next let's examine and use some area formulas.

Rectangles

The area of any plane figure is the number of square units of surface within the figure.

For example, in this rectangle

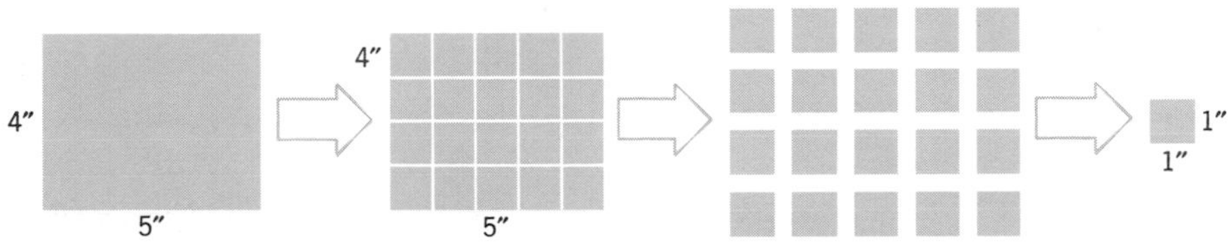

we can divide the surface into exactly 20 small squares each one inch on a side. By counting squares we can see that the area of this 4 in. by 5 in. rectangle contains 20 square inches. We abbreviate these area units as 20 square inches, 20 sq in., or 20 in.^{2}.

Of course there is no need to draw lines in this messy way. We can find the area by multiplying the two dimensions of the rectangle.

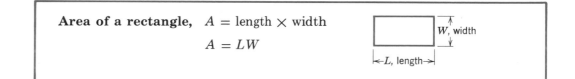

Area of a rectangle, $A = \text{length} \times \text{width}$

$A = LW$

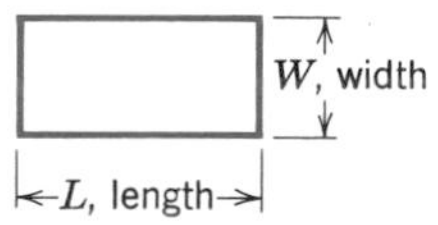

Another example:

Find the area of rectangle *EFGH*.

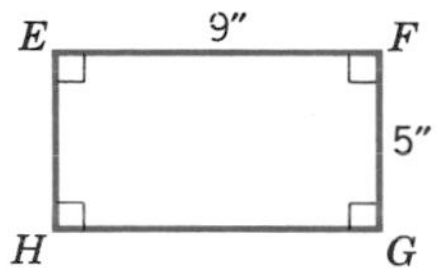

$L = 9$ in.

$W = 5$ in.

A, area $= LW = (9 \text{ in.})(5 \text{ in.})$

$A = 45$ sq in.

The formula is easy to use, but be careful. It is only valid for rectangles. If you use this formula with a different polygon, you will not get the correct answer.

Find the area of rectangle *IJKL*.

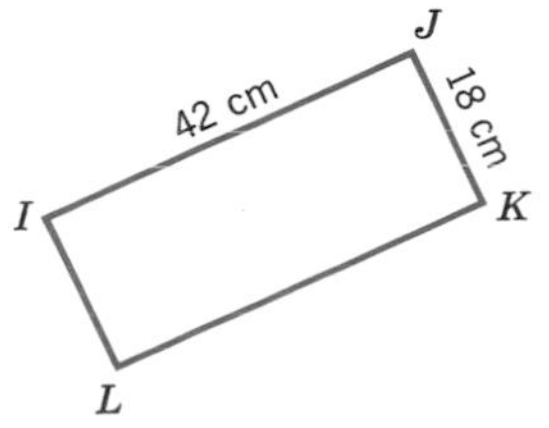

Check your work in **17**.

17 First, write the formula: $A = LW$

Second, substitute L and W: $A = (42 \text{ cm})(18 \text{ cm})$

Third, calculate the answer including the units. $A = 756 \text{ sq cm}$

Using a Calculator

Key	Display
C	0.
4 2	42.
×	42.
1 8	18.
=	756.

Of course the formula $A = LW$ may also be used to find L or W when the other quantities are known. As you learned in your study of basic algebra in Chapter 6, the following formulas are all equivalent.

$$A = LW \qquad L = \frac{A}{W} \quad \text{or} \quad W = \frac{A}{L}$$

For example, a sheet metal worker must build a heating duct in which a rectangular vent 6″ high must have the same area as a rectangular opening 8″ by 9″. Find the length of the vent.

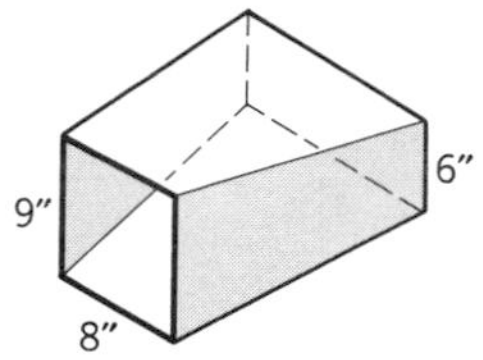

Try it, then check your work in **18**.

---

18 The area of the opening is $A = LW$

$A = (8'')(9'') = 72 \text{ sq in.}$

The length of the vent is $L = \frac{A}{W}$

$$L = \frac{72 \text{ sq in.}}{6 \text{ in.}}$$

or $L = 12 \text{ in.}$

Another formula that you may find useful allows you to calculate the perimeter of a rectangle from its length and width.

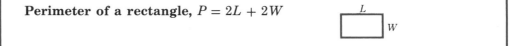

Perimeter of a rectangle, $P = 2L + 2W$

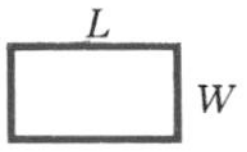

Use this formula to find the perimeter of the rectangle shown here.

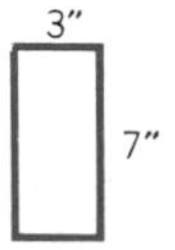

Check your work in **19**.

---

19 Perimeter $= 2L + 2W$
$= 2\,(7 \text{ in.}) + 2\,(3 \text{ in.})$
$= 14 \text{ in.} + 6 \text{ in.}$
$= 20 \text{ in.}$

Using a Calculator

Key	Display
C	0.
2	2.
×	2.
7	7.
M+	14.
2	2.
×	2.
3	3.
M+	6.
MR	20.

. . . assuming your calculator has a memory M+

For practice on using the area and perimeter formulas for rectangles, work the following problems.

1. Find the area and perimeter of each of the following rectangles.

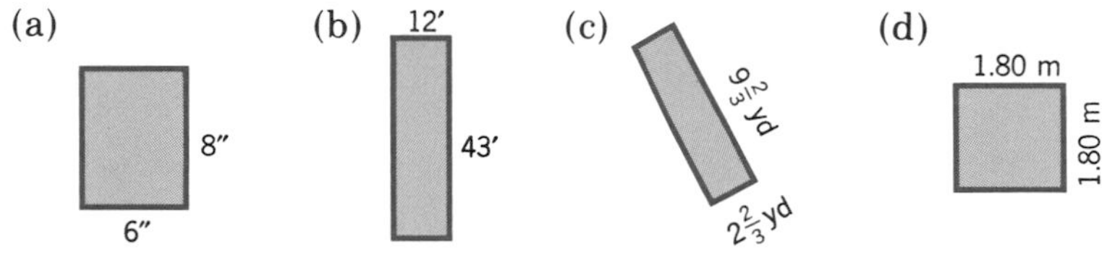

2. Find the area of a rectangle whose length is 4′6″ and whose width is 3′3″. (Give your answer in square feet in decimal form. Round to one decimal place.)
3. A rectangular opening 12″ wide must have the same total area as two smaller rectangular vents 6″ by 4″ and 8″ by 5″. What must be the height of the opening?
4. What is the cost of refinishing a wood floor in a room 35′ by 20′ at a cost of \$1.25 per square foot?
5. What is the cost of fencing a rectangular yard 45′ by 58′ at \$2.65 per foot?
6. Find the area of a rectangular opening 20″ wide and 3′ long.

Check your answers in **20**.

---

20 1. (a) 48 sq in., 28 in. (b) 516 sq ft, 110 ft.
(c) $25\frac{7}{9}$ sq yd, $24\frac{2}{3}$ yd (d) 3.24 sq m, 7.20 m
2. 14.6 sq ft
3. $5\frac{1}{3}$ in.
4. \$875
5. \$545.90
6. 720 sq in. or 5 sq ft

Squares

Did you notice that the rectangle in Problem 1(d) in frame 19 is actually a square? All sides are the same length. To save time when calculating the area or perimeter of a square, use the following formulas.

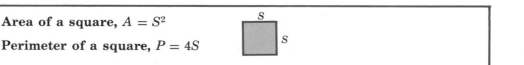

Area of a square, $A = S^2$

Perimeter of a square, $P = 4S$

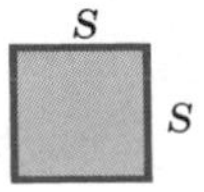

Use these formulas to find the area and perimeter of the following square.

Check your work in 21.

---

21 Area, $A = S^2$

$= (9 \text{ ft})^2$

$A = (9 \text{ ft})(9 \text{ ft}) = 81 \text{ sq ft}$

Remember S^2 means S times S and never $2S$.

Perimeter, $P = 4S$

$P = 4(9 \text{ ft})$

$P = 36 \text{ ft}$

Using a Calculator

Key	Display
C	0.
9	9.
×	9.
9	9.
=	81.
or	
C	0.
9	9.
X^2	81.

A calculation that is very useful in carpentry, sheet metal work, and many other trades involves finding the length of one side of a square given its area. The following formula will help.

Side of a square, $S = \sqrt{A}$

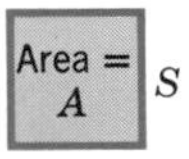

For example, if a square opening is to have an area of 64 sq in. what must be its side length?

$S = \sqrt{A}$

$S = \sqrt{64 \text{ sq in.}}$

$S = 8 \text{ in.}$

It is easiest to use a table of square roots or an electronic calculator to find square roots. If you must calculate it by hand, use the method shown on page 114.

Ready for a few problems involving squares? Try these.

1. Find the area and perimeter of each of the following squares.

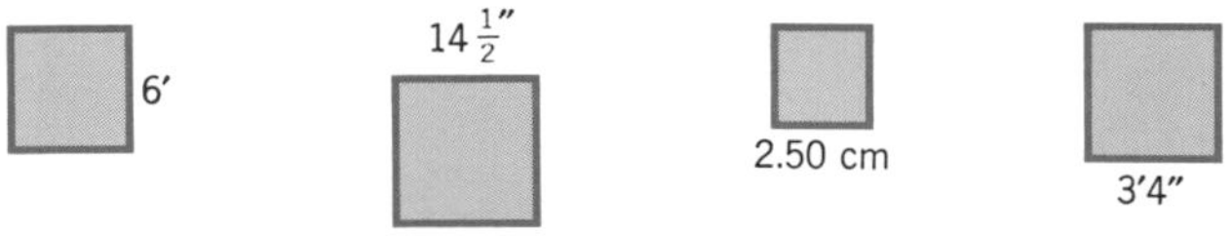

(a) (b) (c) (d)

2. Find the area of a square whose side length is $2\frac{1}{3}$ yd. (Round to two decimal places.)
3. Find the length of the side of a square whose area is 144 square meters.
4. At \$0.25 per square foot, how much will it cost to sod a square lawn 14′6″ on a side? (Round to the nearest cent.)
5. A square hot air duct must contain the same number of square inches as two rectangular ones 5″ × 8″ and 4″ × 6″. How long are the sides of the square duct?

Check your answers in 22.

---

22 1. (a) $A = 36$ sq ft; $P = 24$ ft (b) $A = 210.25$ sq in.; $P = 58$ in.
(c) $A = 6.25$ sq cm; $P = 10$ cm (d) $A = 11.11$ sq ft (rounded); $P = 13'4''$
2. 5.44 sq yd
3. 12 m
4. \$52.56
5. 8″

Parallelograms

You should recall that a parallelogram is a four-sided figure whose opposite pairs of sides are equal and parallel. Here are a few parallelograms:

To find the area of a parallelogram use the following formula:

AREA OF A PARALLELOGRAM

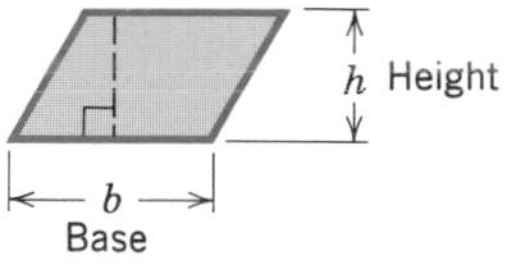

$A = bh$

Notice that we do not use only the lengths of the sides to find the area of a parallelogram. The height h is the perpendicular distance between the parallel sides at top and bottom. The height h is perpendicular to the base b.

For example, in parallelogram $ABCD$, the base is 7″ and the height is 10″. Find its area.

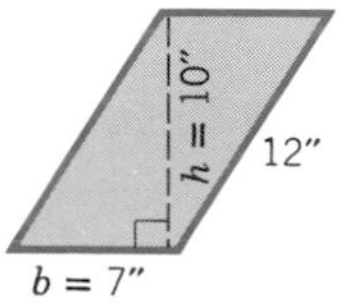

First, write the formula: $A = bh$.

Second, substitute the given values: $A = (7 \text{ in.})(10 \text{ in.})$.

Third, calculate the area including the units: $A = 70$ sq in.

Notice that we ignore the length of the slant side 12″.

Find the area of this parallelogram:

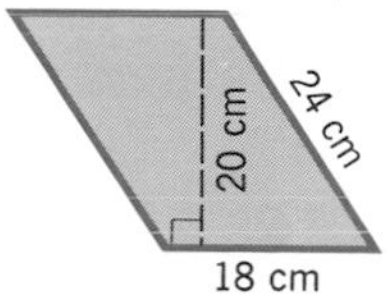

Check your work in 23.

---

23 Area $= bh$
$= (18 \text{ cm})(20 \text{ cm})$
$= 360$ sq cm

Be careful to use the correct dimensions.

This problem could also be worked using the 24 cm side as the base.

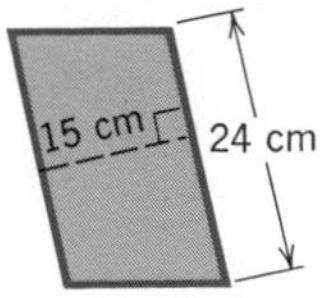

In this case the new height is 15 cm and the area is

$A = (24\text{ cm})(15\text{ cm})$
$A = 360\text{ sq cm}$

To find the perimeter of a parallelogram, simply add the lengths of the four sides.

$P = 2a + 2b$

Now try the following problems.

Find the perimeter and area for each parallelogram shown.

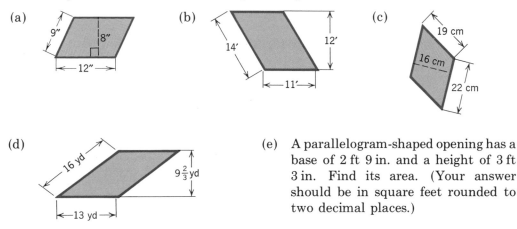

(a)

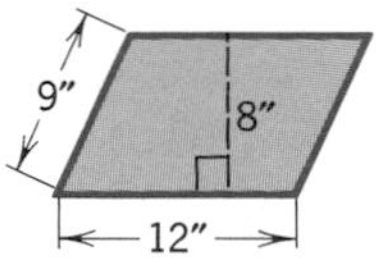

(b)

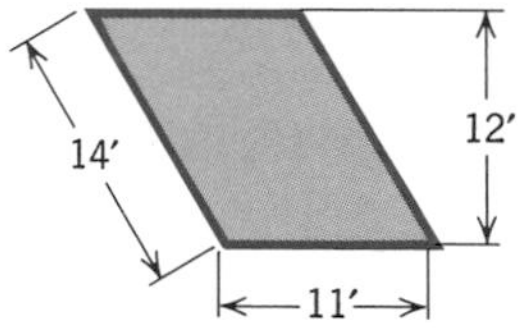

(c)

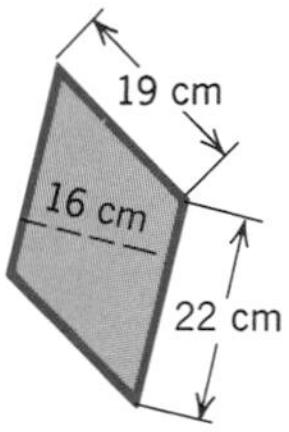

(d)

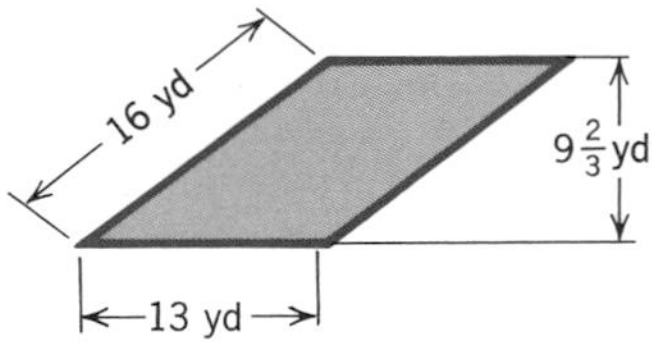

(e) A parallelogram-shaped opening has a base of 2 ft 9 in. and a height of 3 ft 3 in. Find its area. (Your answer should be in square feet rounded to two decimal places.)

Check your answers in **24**.

AREA OF A PARALLELOGRAM

You may be interested in where the formula on page 323 came from. If so, follow this explanation.

Here is a typical parallelogram. As you can see, opposite sides $\overline{AB}$ and $\overline{DC}$ are parallel, and opposite sides $\overline{AD}$ and $\overline{BC}$ are also parallel.

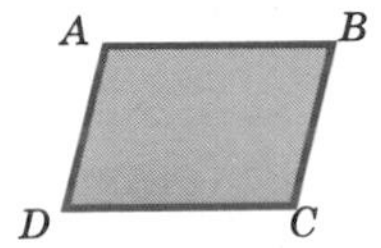

Now let's cut off a small triangle from one side of the parallelogram by drawing the perpendicular line AE.

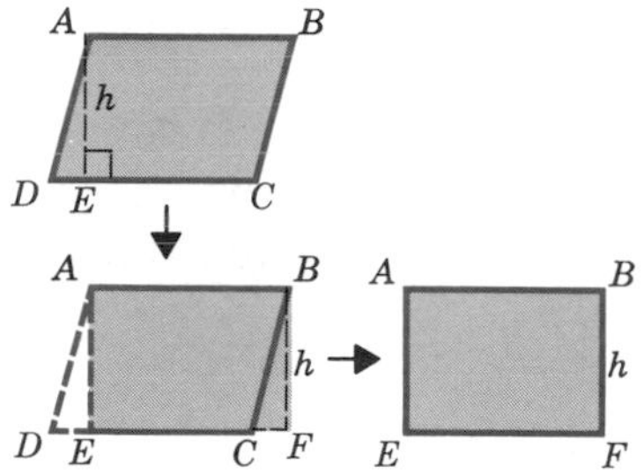

Next, reattach the triangular section to the other side of the figure, forming a rectangle.

The area of a the rectangle $ABFE$ is $A = bh$.

---

24 (a) $P = 42$ in., $A = 96$ sq in.
(b) $P = 50$ ft, $A = 132$ sq ft
(c) $P = 82$ cm, $A = 352$ sq cm
(d) $P = 58$ yd, $A = 125\frac{2}{3}$ sq yd
(e) $A = 8.94$ sq ft

Trapezoids

A trapezoid is a four-sided figure with only one pair of sides parallel. Here are a few trapezoids:

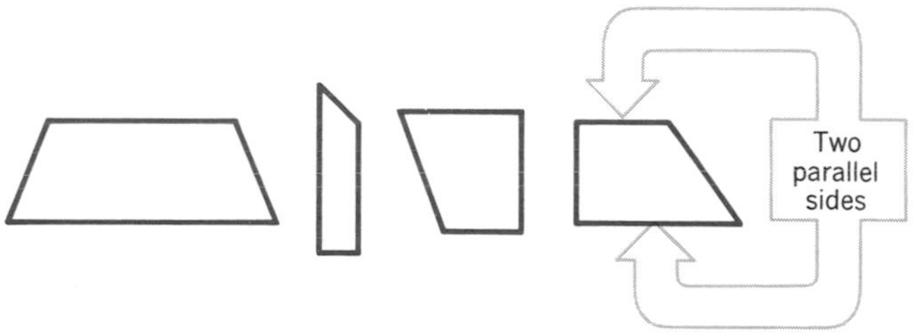

To find the area of a trapezoid use the following formula:

AREA OF A TRAPEZOID, $A = \left(\frac{b_1 + b_2}{2}\right)h$

or $\frac{h}{2}(b_1 + b_2)$

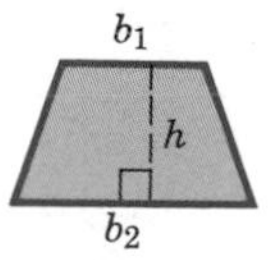

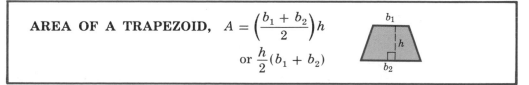

The factor $\left(\dfrac{b_1 + b_2}{2}\right)$ is the average length of the two parallel sides b_1 and b_2. The height h is the perpendicular distance between the two parallel sides.

Follow this example.
Find the area of this trapezoid-shaped metal plate.

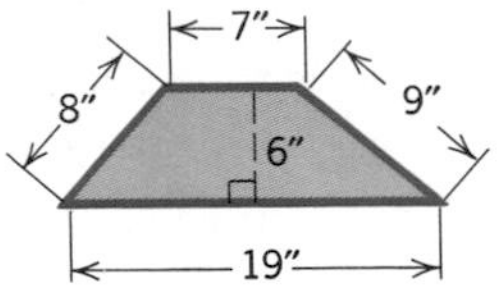

The parallel sides, b_1 and b_2, have lengths 7″ and 19″. The height h is 6″.

$A = \left(\dfrac{b_1 + b_2}{2}\right)h$

$A = \left(\dfrac{7\text{ in.} + 19\text{ in.}}{2}\right)(6\text{ in.})$ First, evaluate the quantity $\dfrac{7 + 19}{2}$.

$A = \left(\dfrac{26}{2}\text{ in.}\right)(6\text{ in.})$

$A = (13\text{ in.})(6\text{ in.})$ Now multiply (13)(6).

$A = 78\text{ sq in.}$

Using a Calculator

Key	Display
C	0.
7	7.
+	7.
1 9	19.
÷	26.
2	2.
×	13.
6	6.
=	78.

$A = \left(\dfrac{(b_1 + b_2)}{2}\right)H$

Find the area and perimeter of each of the following trapezoids.

(a)

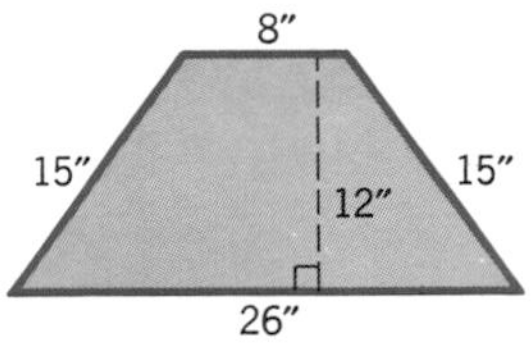

(b)

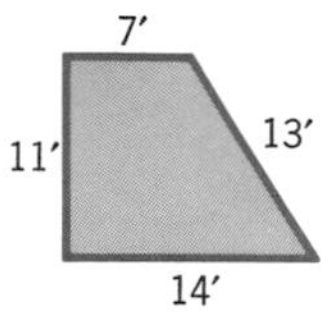

(c)

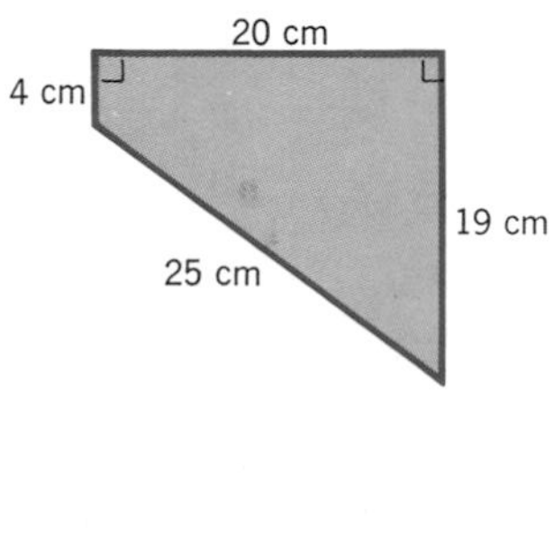

(d)

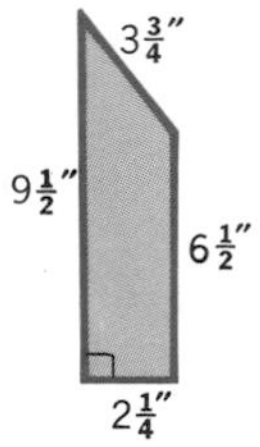

(e)

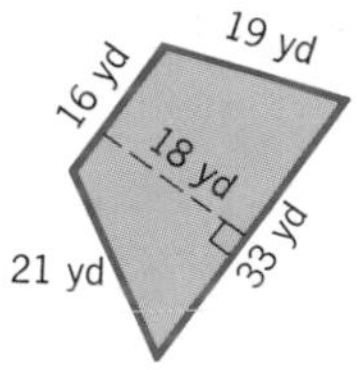

(f)

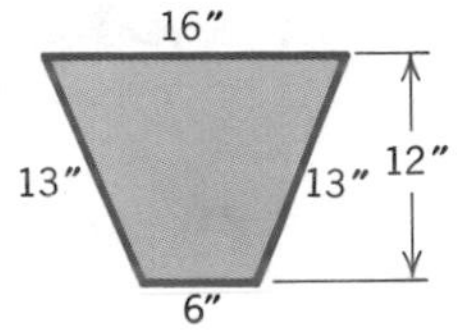

Check your answers in **25**.

What is the difference between a rectangle and a parallelogram?

A parallelogram is a four-sided figure whose opposite sides are parallel. A rectangle is a special kind of parallelogram, one in which all four interior angles are right angles.

(a) $A = \left(\frac{8'' + 26''}{2}\right)12''$ $\qquad P = 8'' + 15'' + 26'' + 15''$

$A = (17'')(12'')$ $\qquad P = 64$ in.

$A = 204$ sq in.

25 (b) $A = \left(\frac{7' + 14'}{2}\right)(11')$ $\qquad P = 11' + 7' + 13' + 14'$

$A = (10.5')(11')$ $\qquad P = 45$ ft

$A = 115.5$ sq ft

(c) $A = \left(\frac{4 \text{ cm} + 19 \text{ cm}}{2}\right)(20 \text{ cm})$ $\qquad P = 4 \text{ cm} + 20 \text{ cm} + 19 \text{ cm} + 25 \text{ cm}$

$A = (11.5 \text{ cm})(20 \text{ cm})$ $\qquad P = 68$ cm

$A = 230$ sq cm

(d) $A = \left(\frac{9\frac{1}{2}'' + 6\frac{1}{2}''}{2}\right)(2\frac{1}{4}'')$ $\qquad P = 3\frac{3}{4}'' + 6\frac{1}{2}'' + 2\frac{1}{4}'' + 9\frac{1}{2}''$

$A = (8'')(2\frac{1}{4}'')$ $\qquad P = 22''$

$A = (8'')(2.25'')$

$A = 18$ sq. in.

(e) $A = \left(\frac{16 \text{ yd} + 33 \text{ yd}}{2}\right)(18 \text{ yd})$ $\qquad P = 16 \text{ yd} + 19 \text{ yd} + 33 \text{ yd} + 21 \text{ yd}$

$A = (24.5 \text{ yd})(18 \text{ yd})$ $\qquad P = 89$ yd

$A = 441$ sq yd

(f) $A = \left(\frac{16'' + 6''}{2}\right)(12'')$ $\qquad P = 13'' + 16'' + 13'' + 6''$

$A = (11'')(12'')$ $\qquad P = 48$ in.

$A = 132$ sq in.

In a real job situation you will usually need to measure the height of the figure. Remember, the bases of a trapezoid are the parallel sides and h is the distance between them.

Triangles

A *triangle* is a polygon having three sides. It is the simplest of all plane figures.

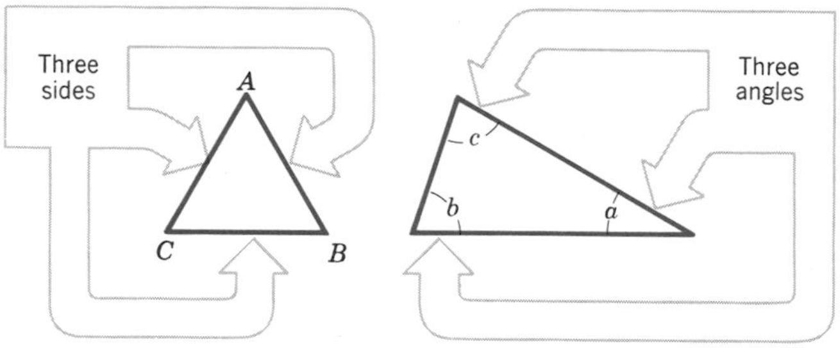

Just as there are several varieties of four-sided figures—squares, rectangles, parallelograms, and trapezoids—there are also several varieties of triangles. Fortunately, one area formula can be used with all triangles. First, you must learn to identify the many kinds of triangles that appear in practical work.

An *equilateral triangle* is one in which all three sides have the same length. An equilateral triangle is also said to be *equiangular* since, if the three sides are equal, the three angles will also be equal. In fact, since the interior angles of any triangle always add up to 180°, each angle of an equilateral triangle must equal 60°.

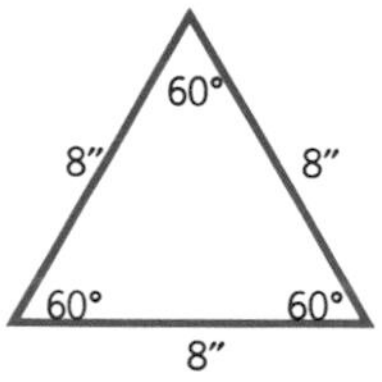

Equilateral

An *isosceles triangle* is one in which two of the three sides are equal. It is always true that the two angles opposite the equal sides are also equal.

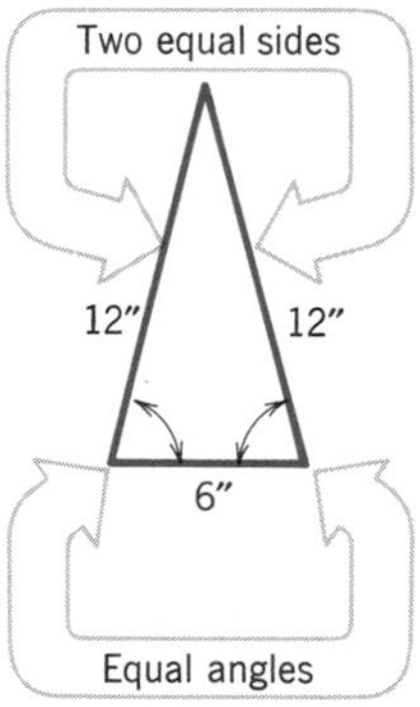

A *right triangle* contains a 90° angle. Two of the sides are perpendicular to each other. The longest side of a right triangle is always the side opposite to the right angle. This side is called the *hypotenuse* of the triangle.

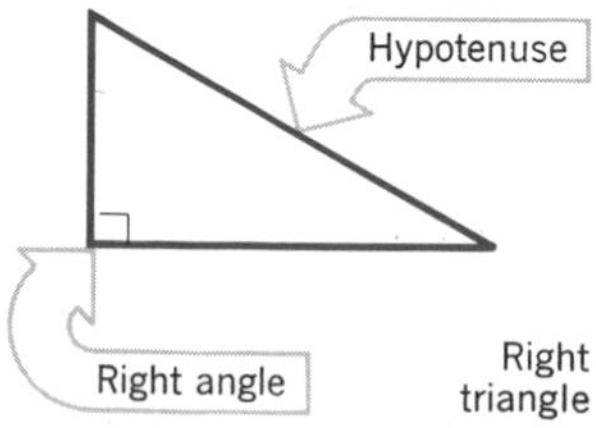

Right triangle

A *scalene triangle* is one in which no sides are equal. A right triangle could be scalene.

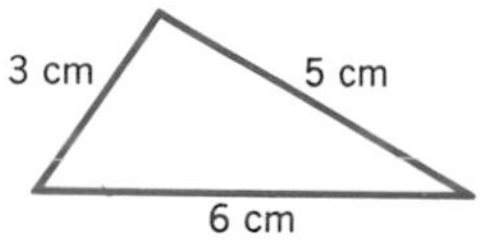

Scalene

Identify the following triangles as being equilateral, isosceles, or scalene. Also name the ones that are right triangles.

(a)

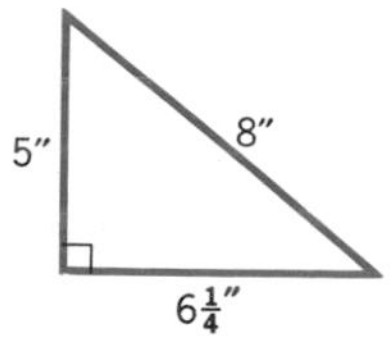

(b)

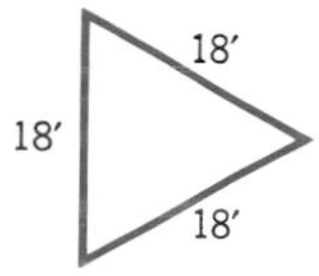

(c)

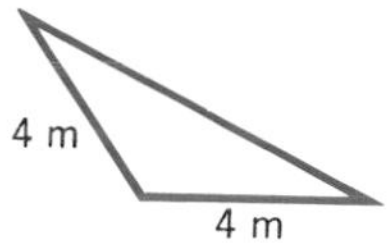

(d)

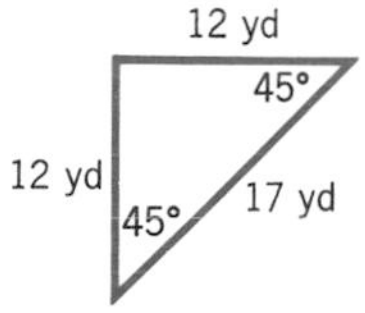

(e)

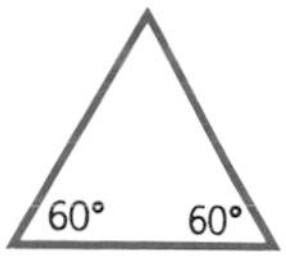

(f)

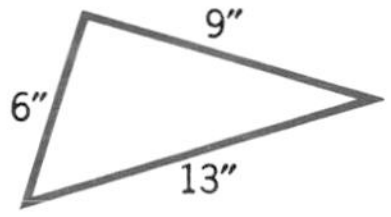

Check your answers in **26**.

---

26 (a) Scalene (b) Equilateral (c) Isosceles
(d) Isosceles (e) Equilateral (f) Scalene

(a) and (d) are also right triangles.

Did you have trouble with (d) or (e)? The answer to each depends on the fact that the interior angles of a triangle add up to 180°. In (d) the two given angles add up to 90°. Therefore the missing angle must be 90° if the three angles sum to 180°. This makes (d) a right triangle.

In (e) the two angles shown sum to 120°, so the third angle must be 180° − 120° or 60°. Because the three angles are equal we know the triangle is an equiangular triangle, and this means it is also equilateral.

Pythagorean Theorem

The Pythagorean Theorem is a rule or formula that allows us to calculate the length of one side of a right triangle when we are given the lengths of the other two sides. Although the formula is named after the ancient Greek mathematician Pythagoras, it was known to Babylonian engineers and surveyors more than a thousand years before Pythagoras lived.

PYTHAGOREAN THEOREM

For any right triangle, the square of the hypotenuse is equal to the sum of the squares of the other two sides.

$c^2 = a^2 + b^2$

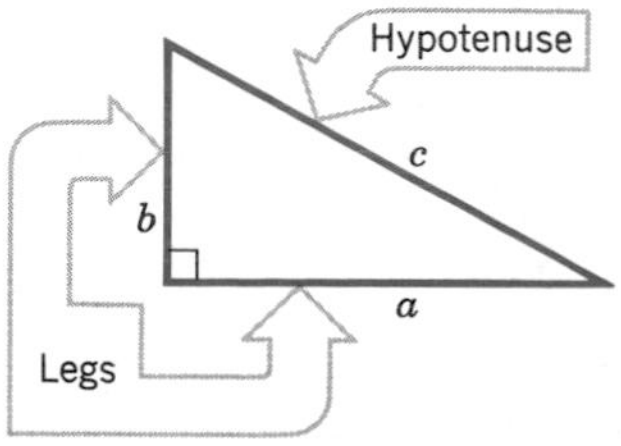

Geometrically this means that if the squares are built on the sides of the triangle, the area of the larger square is equal to the sum of the areas of the two smaller squares. For the triangle shown

$3^2 + 4^2 = 5^2$

$9 + 16 = 25$

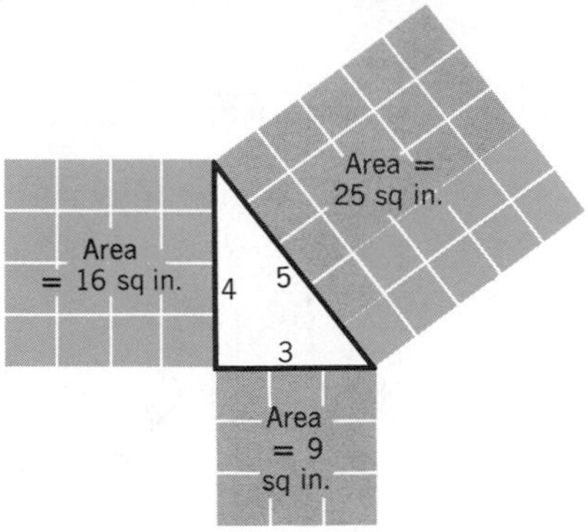

Use this formula to solve this practical problem. Find the distance d between points A and B for this rectangular plot of land.

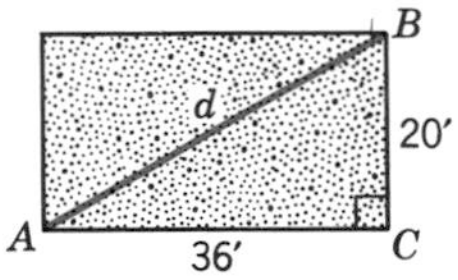

Try it. Check your work in **27**.

---

27 The Pythagorean Theorem tells us that for the right triangle ABC,

$d^2 = 36^2 + 20^2$
$d^2 = 1296 + 400$
$d^2 = 1696$

Taking the square root of both sides of the equation,

$d = 41$ ft, rounded to the nearest foot.

Using a Calculator

Key	Display
C	0.
3 6	36.
x^2	1296.
+	1296.
2 0	20.
x^2	400.
=	1696.
√	41.18252

(In some calculators you may need to store 1296 and add it to 400 later.)

This rule may be used to find any one of the three sides of a right triangle if the other two sides are given. The formula $c^2 = a^2 + b^2$ can be rewritten as

$c = \sqrt{a^2 + b^2}$ or $a = \sqrt{c^2 - b^2}$ or $b = \sqrt{c^2 - a^2}$

Be careful. This is only true for a *right* triangle—a triangle with a right or 90° angle.

Use these formulas to solve this problem.

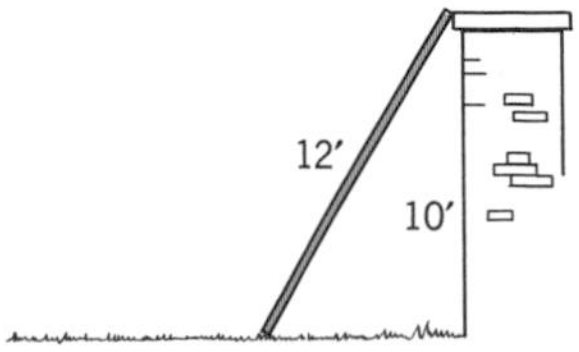

A carpenter wants to use a 12′ ladder to reach the top of a 10′ wall. How far must the base of the ladder be from the base of the wall?

Try it, then check your work in **28**.

---

28 Use the formula $a = \sqrt{c^2 - b^2}$ to get

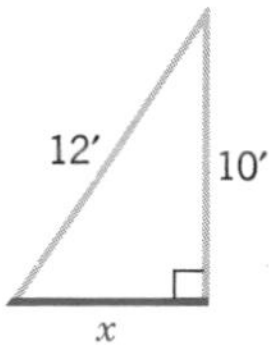

$x = \sqrt{12^2 - 10^2}$

$x = \sqrt{144 - 100}$

$x = \sqrt{44}$

$x = 6.6$ ft or 6′7″, rounded to the nearest inch

If you need help with square roots, pause here and return to page 112 for a review.

For a few right triangles, the three side lengths are all whole numbers rather than fractions or decimals. Such special triangles have been used in technical work since Egyptian surveyors used them 2,000 years ago to lay out rectangular fields for farming.

Here are a few "Pythagorean Triple" triangles:

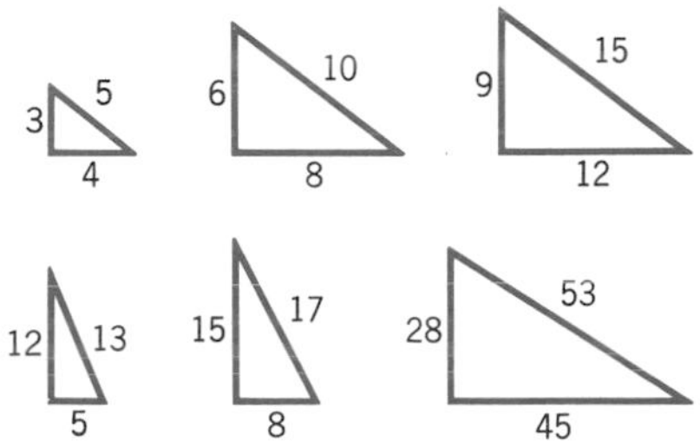

Solve the following problems, using the Pythagorean Theorem. (If your answer does not come out even, round to the nearest tenth.)

1. Find the missing side for each of the following right triangles.

(a)

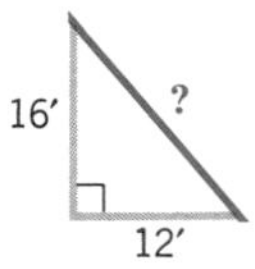

(b)

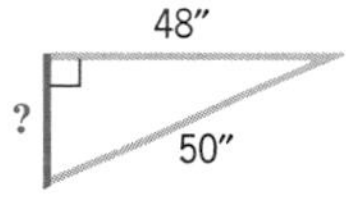

(c)

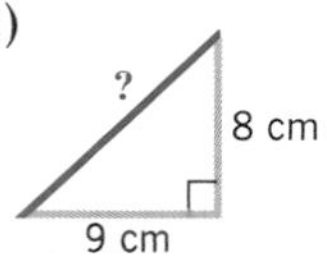

(d)

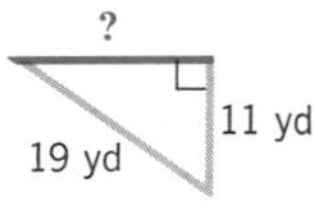

(e)

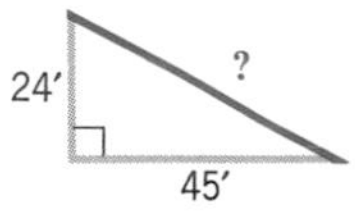

(f)

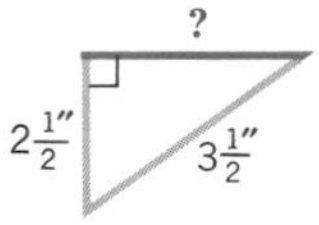

(g) Find the hypotenuse of a right triangle with legs measuring 4.6 m and 6.2 m.

2. A rectangular table measures 4′ by 6′. Find the length of a diagonal brace placed beneath the table top.

3. What is the distance between the center of two pulleys if one is placed 9″ to the left and 6″ above the other?

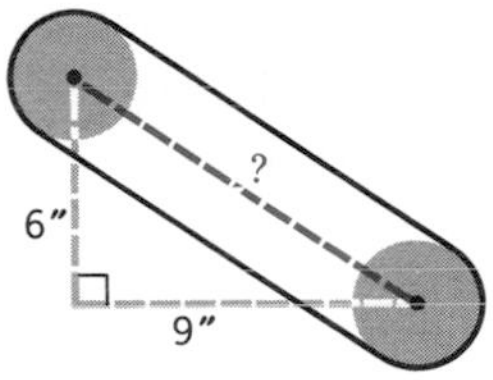

4. Find the length of the missing dimension x in the part shown in the figure.

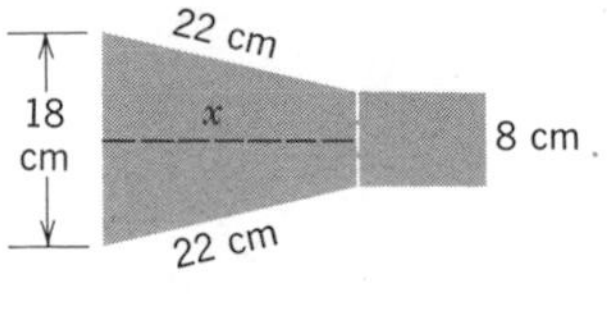

5. Find the diagonal length of the stairway shown in the drawing. Round to the nearest tenth of a foot.

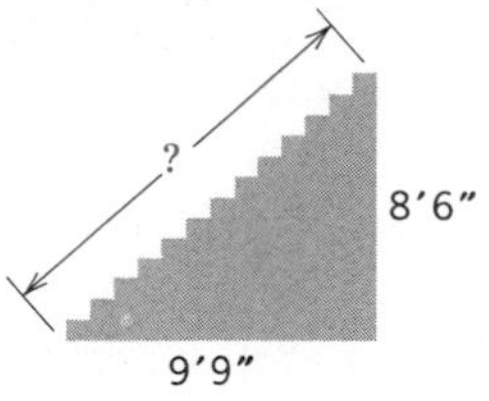

Check your answers in **29**.

SOME SPECIAL TRIANGLES

3-4-5

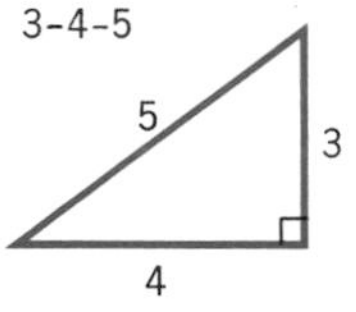

If you draw a triangle with legs 3, 4, and 5 in. long, it will always be a right triangle.

In a triangle with angles 30°, 60°, 90° the shortest side will be exactly half as long as the longest side.

30°-60°-90°

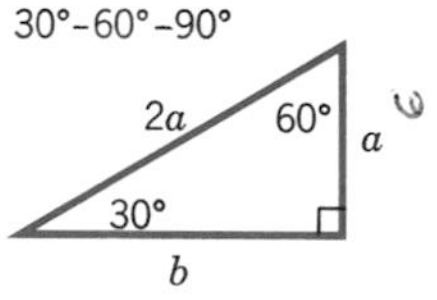

Side b is about $1.7a$.

If the two legs of a right triangle are equal, the angles will be 45°, 45° and 90°. The hypotenuse will be $\sqrt{2}$ times the length of the other sides.

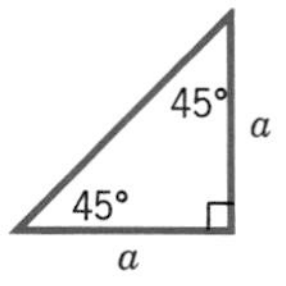

This kind of triangle is called a right-isosceles triangle.

hypotenuse $= a\sqrt{2}$ or about $1.4a$

---

29 1. (a) 20 ft (b) 14 in. (c) 12.0 cm (d) 15.5 yd
(e) 51 ft (f) 2.4 in. (g) 7.7 m
2. 7.2 ft or $7'2\frac{1}{2}''$
3. 10.8 in.
4. 21.4 cm
5. 12.9 ft or 12′11″, rounded to the nearest inch.

In Problem 4, the first difficulty is to locate the correct triangle.

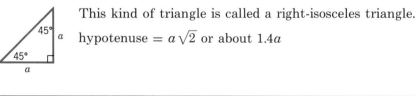

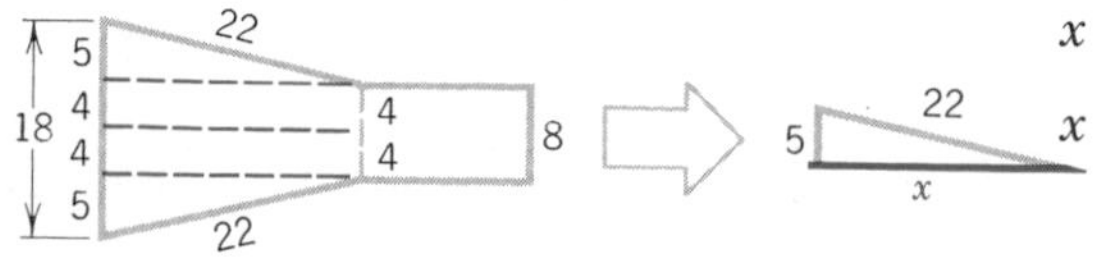

$x = \sqrt{22^2 - 5^2}$

$x = \sqrt{459}$

Area of a Triangle

The area of any triangle, no matter what its shape or size, can be found using the same simple formula.

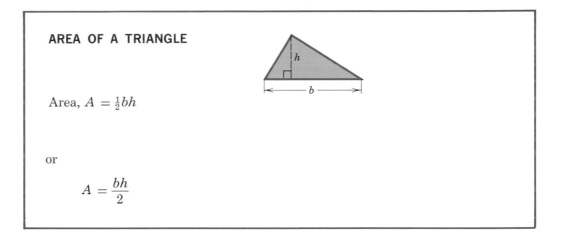

AREA OF A TRIANGLE

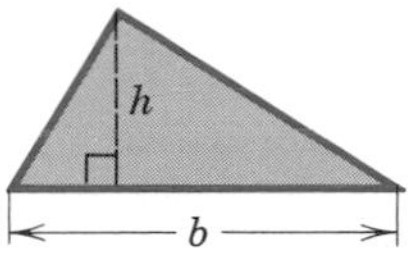

Area, $A = \frac{1}{2}bh$

or

$$A = \frac{bh}{2}$$

For example, in this triangle the base b is 13″, and the height h is 8″. Applying the formula,

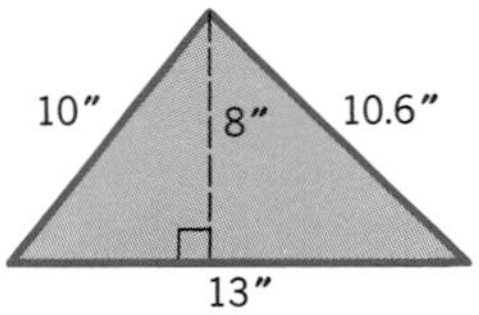

Area, $A = \frac{1}{2}bh$

$A = \frac{1}{2}(13 \text{ in.})(8 \text{ in.})$

$A = 52 \text{ sq in.}$

Of course the base used does not need to be the "bottom" of the triangle. Use any side of the triangle as the base, but be certain that the height used is perpendicular to that base. For example, in this triangle

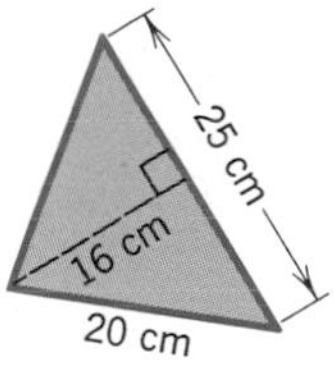

$b = 25 \text{ cm}$

$h = 16 \text{ cm}$

Area, $A = \frac{1}{2}(25 \text{ cm})(16 \text{ cm})$

$A = 200 \text{ sq cm}$

Find the area of this triangle.

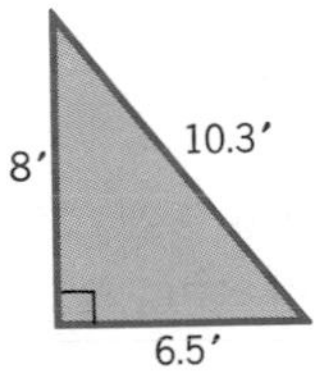

Check your work in **30**.

AREA OF A TRIANGLE

The area of any triangle is equal to one-half of its base times its height. $A = \frac{1}{2}bh$.

To learn where this formula comes from, follow this explanation:

First, for any triangle ABC,

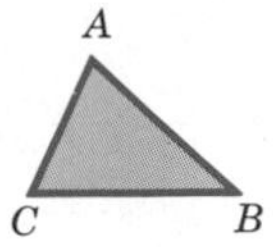

draw a second identical triangle DEF.

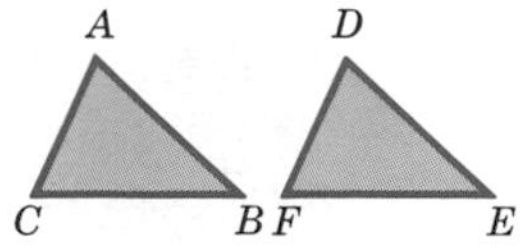

Second, flip the second triangle over and attach it to the first,

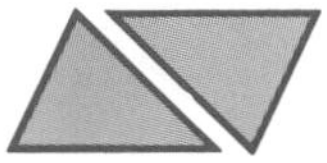

to get a parallelogram.

Finally, if the base and height of the original triangle are b and h, then these are also the base and height for the parallelogram.

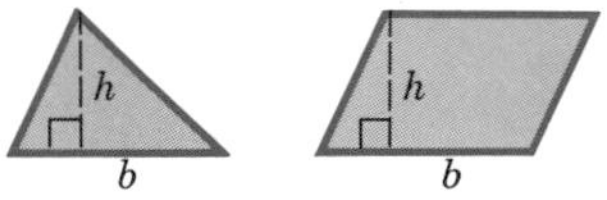

The area of the parallelogram is

$A = bh$

so the area of the original triangle must be

$A = \frac{1}{2}bh$

---

30 Base, $b = 6.5'$

Height, $h = 8'$

Area, $A = \frac{1}{2}bh$

$A = \frac{1}{2}(6.5 \text{ ft})(8 \text{ ft})$

$A = 26 \text{ sq ft}$

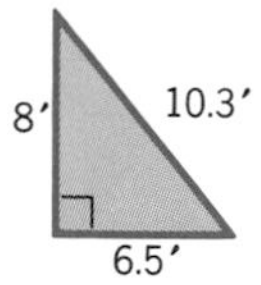

Notice that for a right triangle, the two sides meeting at the right angle can be used as the base and height.

Solve this slightly tricky problem. Find the area of the isosceles triangle shown.

(**Hint:** First use the Pythagorean Theorem to find the height, then find the area.)

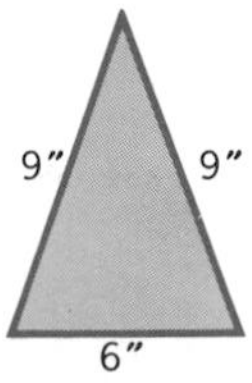

Check your work in **31**.

---

31 Draw the height perpendicular to the 6″ base. Then in the triangle on the right

$h = \sqrt{9^2 - 3^2}$

$h = \sqrt{81 - 9}$

$h = \sqrt{72}$

$h = 8.5''$, rounded

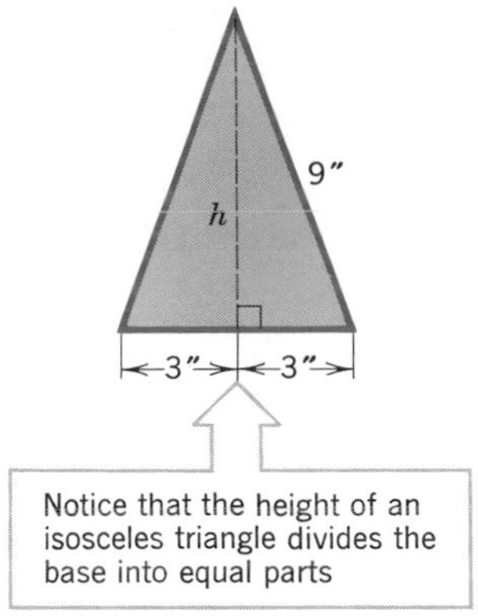

Now find the area:

$A = \frac{1}{2}bh$

$A = \frac{1}{2}(6 \text{ in.})(8.5 \text{ in.})$

$A = 25.5$ sq in.

In general, to find the area of an isosceles triangle use the following formula.

AREA OF AN ISOSCELES TRIANGLE

Area, $A = \frac{1}{2}b\sqrt{a^2 - \left(\frac{b}{2}\right)^2}$

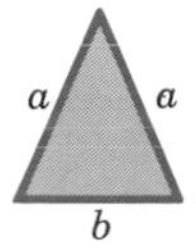

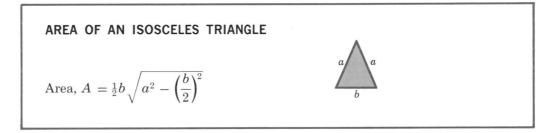

An attic wall is shaped like this: How many square feet of insulation are needed to cover the wall? (Round to the nearest sq ft.)

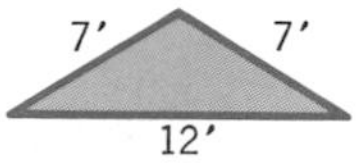

Check your work in **32**.

---

32 The triangle is isosceles since two sides are equal. Substituting $a = 7'$ and $b = 12'$ into the formula,

Area, $A = \frac{1}{2}(12 \text{ ft})\sqrt{(7 \text{ ft})^2 - \left(\frac{12 \text{ ft}}{2}\right)^2}$

$A = 6\sqrt{49 - 36}$

$A = 6\sqrt{13}$

$A = 21.63$ sq ft, or 22 sq ft, rounded.

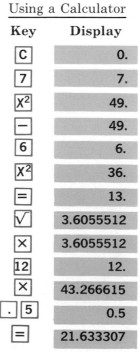

Using a Calculator

Key	Display
C	0.
7	7.
X^2	49.
−	49.
6	6.
X^2	36.
=	13.
√	3.6055512
×	3.6055512
12	12.
×	43.266615
. 5	0.5
=	21.633307

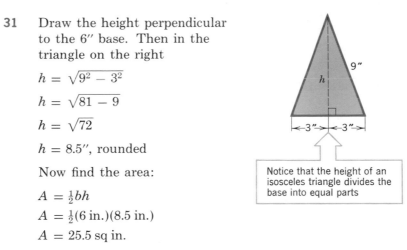

For an equilateral triangle an even simpler formula may be used.

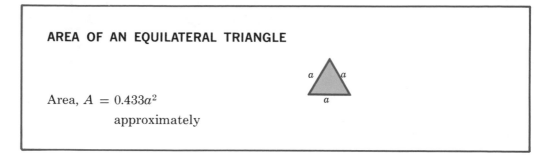

AREA OF AN EQUILATERAL TRIANGLE

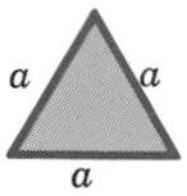

Area, $A = 0.433a^2$
approximately

What area of sheet steel is needed to make a triangular pattern where each side is 3 ft long?

Try it. Use the formula above and round to one decimal place.

Check your work in **33**.

---

33 Area, $A = 0.433a^2$ where $a = 3$ ft

$$A = 0.433(3\text{ ft})^2$$
$$A = 0.433 \cdot 9 \text{ sq ft}$$
$$A = 3.9 \text{ sq ft, rounded}$$

1. In the following problems find the area of each triangle. (Round your answer to one decimal place if it does not come out even.)

(a)

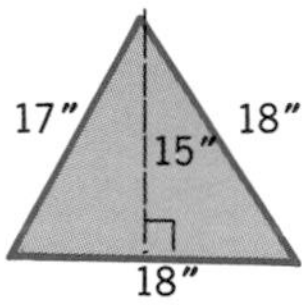

(b)

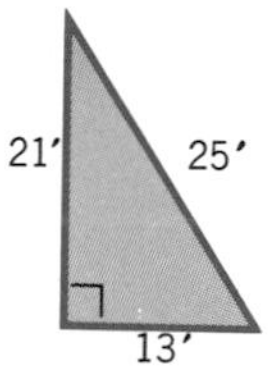

(c)

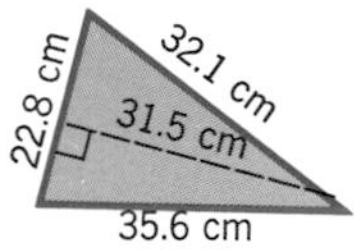

(d)

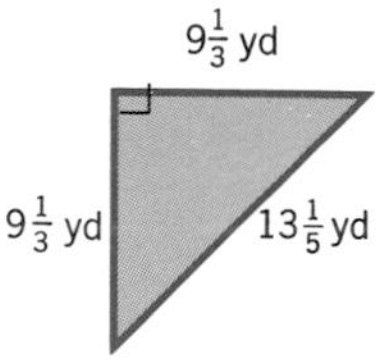

(e)

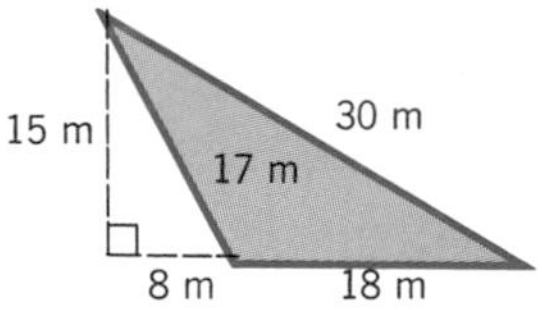

(f)

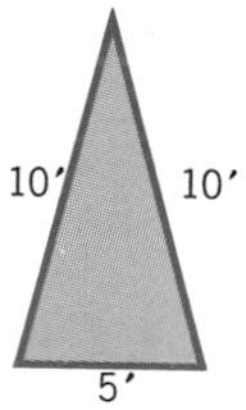

(g)

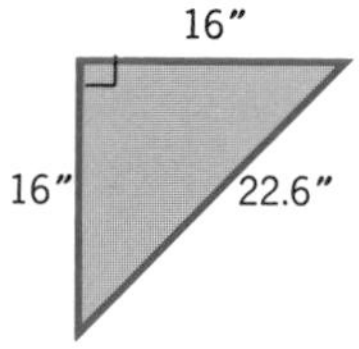

(h)

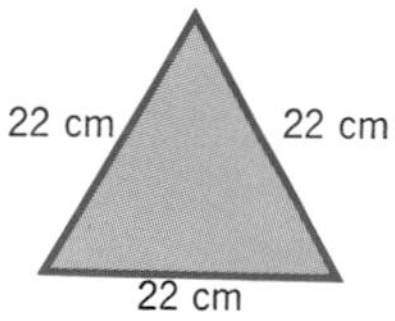

2. Find the area, to the nearest square foot, of the triangular patio deck shown.

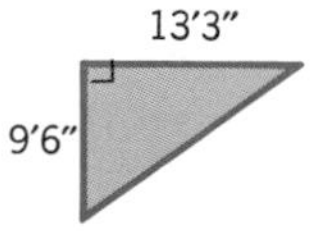

3. At a cost of 60¢ per square foot, what would it cost to tile the triangular work space shown?

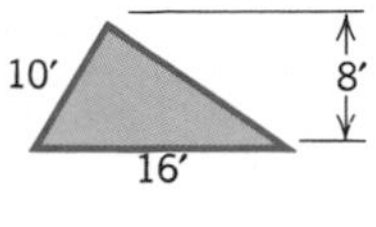

Check your answers in **34**.

HERO'S FORMULA

It is possible to find the area of any triangle from the lengths of its sides. The formula that enables us to do this was first devised almost 2,000 years ago by Hero or Heron, a Greek mathematician. It is a complicated formula, and you may feel like a hero yourself if you learn how to use it.

Area, $A = \sqrt{s(s-a)(s-b)(s-c)}$ For any triangle 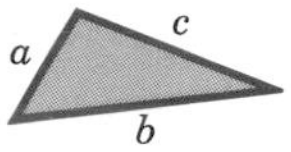

where $s = \dfrac{a+b+c}{2}$

(s stands for "semi-perimeter," or half of the perimeter.)

Example:

$a = 6''$, $b = 5''$, $c = 7''$

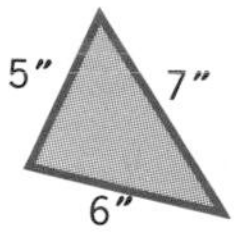

First, find s. $s = \dfrac{6+5+7}{2} = \dfrac{18}{2} = 9$

Next, substitute to find the area

$A = \sqrt{9(9-6)(9-5)(9-7)}$

$A = \sqrt{9 \cdot 3 \cdot 4 \cdot 2}$

$A = \sqrt{216}$ $= 14.7$ sq in., rounded

---

34 1. (a) 135 sq in. (b) 136.5 sq ft (c) 359.1 sq cm
(d) 43.6 sq yd (e) 135 sq m (f) 24.2 sq ft
(g) 128 sq in. (h) 209.6 sq cm

2. 63 sq ft

3. $38.40

In Problem 1 (e) you should have used 18 m as the base and *not* 26 m. $b = 18$ m, $h = 15$ m.

In Problem 3 you should have noticed that the height was given as 8 ft.

Hexagons

A polygon is a plane geometric figure with three or more sides. A *regular polygon* is one in which all sides are the same length and all angles are equal. The equilateral triangle and the square are both regular polygons.

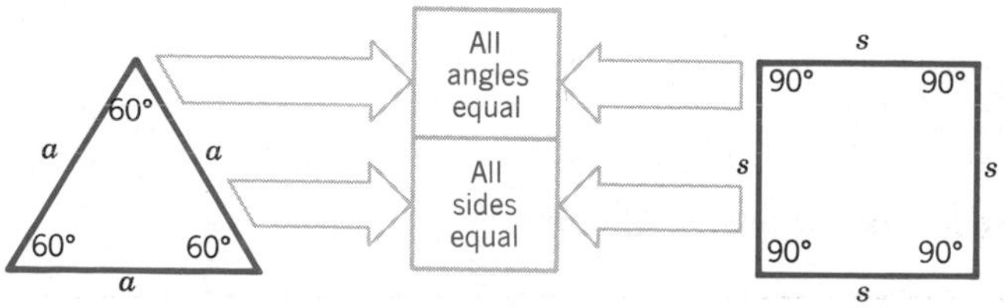

A five-sided regular polygon is called a *pentagon*.

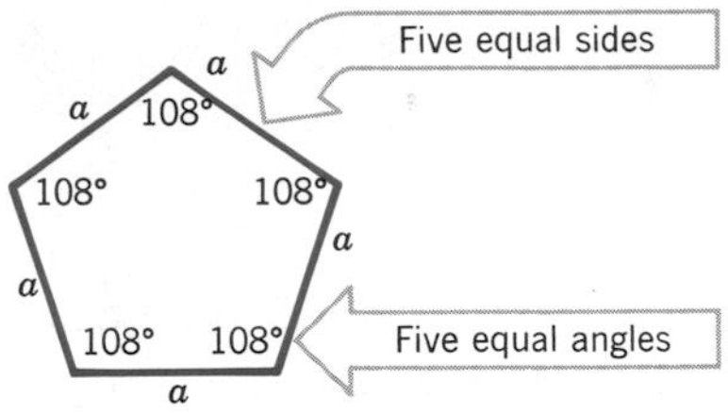

The regular *hexagon* is a six-sided polygon in which each interior angle is 120° and all sides are the same length.

Regular hexagon:
All sides equal
All angles 120°

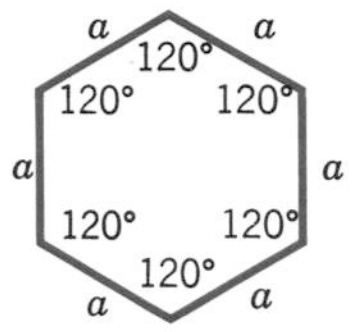

If we draw the diagonals, six equilateral triangles are formed.

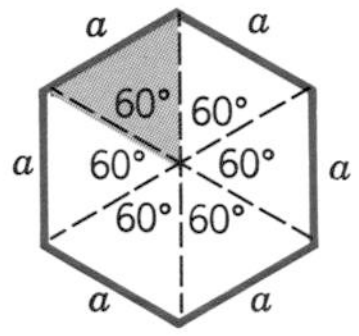

Since you already know that the area of one equilateral triangle is approximately $A = 0.433a^2$, then the area of the complete hexagon will be 6 times $0.433a^2$ or $2.598a^2$.

AREA OF A REGULAR HEXAGON

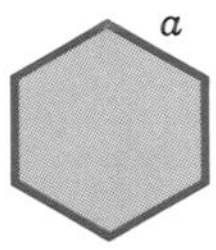

Area, $A = 2.598a^2$
approximately

For example, if $a = 6$ cm, $A = 2.598(6 \text{ cm})^2$
$A = 2.598(36 \text{ sq cm})$
$A = 93.5$ sq cm, rounded

First, square the number in parentheses.

Using a Calculator

Key	Display
C	0.
6	6.
X^2	36.
X	36.
2 . 5 9 8	2.598
=	93.528

Use this formula to find the area of a hexagonal plate 4″ on each side.

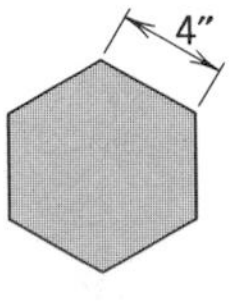

Check your work in 35.

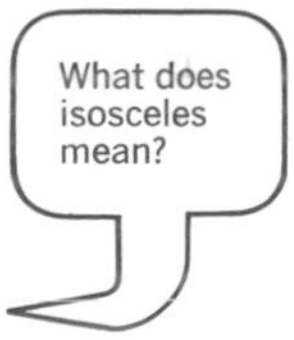
What does isosceles mean?

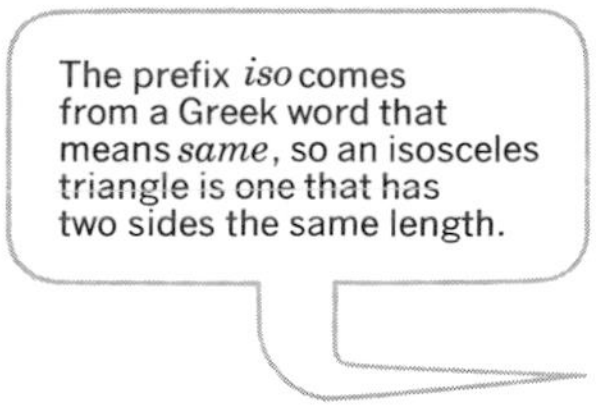
The prefix *iso* comes from a Greek word that means *same*, so an isosceles triangle is one that has two sides the same length.

35 Area, $A = 2.598(a^2)$ $\quad a = 4$ in.

$A = 2.598(4 \text{ in.})^2$

$A = 2.598(16 \text{ sq in.})$

$A = 41.6$ sq in., rounded

Because the hexagon is used so often in practical and technical work, several other formulas may be helpful to you.

A hexagon is often measured by specifying the distance across the corners or the distance across the flats.

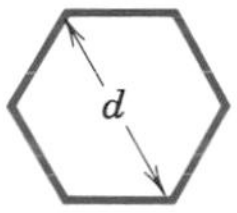
d

Distance across the corners

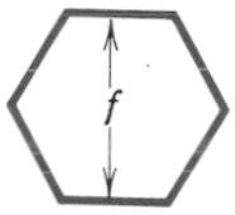
f

Distance across the flats

DIMENSIONS OF A HEXAGON

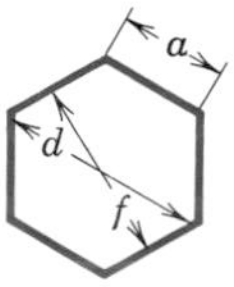
a
d
f

Distance across the corners, $d = 2a$

or $a = 0.5d$

Distance across the flats, $f = 1.732a$ approximately

or $a = 0.577f$ approximately

For example, for a piece of hexagonal bar stock where $a = \frac{1}{2}''$, the distance across the corners would be

$d = 2a$ $\quad$ or $\quad$ $d = 1$ in.

The distance across the flats would be

$f = 1.732a$ $\quad$ or $\quad$ $f = 0.866$ in.

If the cross-section of some hexagonal stock measures $\frac{3}{4}''$ across the flats, find

(a) the side length, a
(b) the distance across the corners, d
(c) the cross-sectional area.

Round to three decimal places.

Try it, then check your work in **36**.

36 $f = \frac{3}{4}''$ or 0.75 in.

(a) $a = 0.577f$
$a = 0.577(0.75 \text{ in.})$
$a = 0.433 \text{ in.}$

(b) $d = 2a$
$d = 2(0.433 \text{ in.})$
$d = 0.866 \text{ in.}$

(c) Area, $A = 2.598a^2$
$A = 2.598(0.433 \text{ in.})^2$
$A = 2.598(0.1875 \text{ sq in.})$
$A = 0.487$ sq in., approximately

Work these problems for practice.

1. Fill in the blanks with the missing dimensions of each regular hexagon. Round to the nearest thousandth if necessary.

	a	d	f	A (to two decimal places)
(a)	2 in.	______	______	______
(b)	______	$\frac{3}{4}''$	______	______
(c)	______	______	6 mm	______

2. A hex nut has a side length of $\frac{1}{4}''$. From the following list pick the smallest size wrench that will fit it.

(a) $\frac{1}{4}''$ (b) $\frac{3}{8}''$ (c) $\frac{7}{16}''$ (d) $\frac{1}{2}''$ (e) $\frac{5}{8}''$

Check your answers in **37**.

37

	a	d	f	A
1. (a)	2 in.	4 in.	3.464 in.	10.39 sq in.
(b)	$\frac{3}{8}$ in.	$\frac{3}{4}$ in.	0.650 in.	0.37 sq in.
(c)	3.462 mm	6.924 mm	6 mm	31.14 sq mm

2. $\frac{7}{16}$ in.

Let's take a closer look at Problem 2.

By drawing a picture we can see that we must find f. Since $a = \frac{1}{4}''$ we have

$f = 1.732a$
$f = 1.732(.25)$
$f = 0.433''$

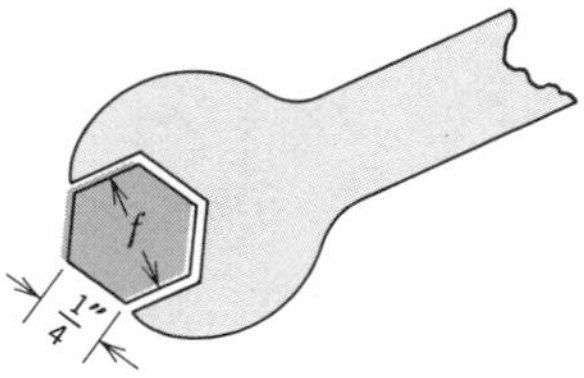

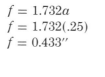

By checking a list of wrench sizes in their decimal equivalents or converting the fractions given, we find that $\frac{7}{16}''$ (0.4375″) is the smallest size that will fit the nut.

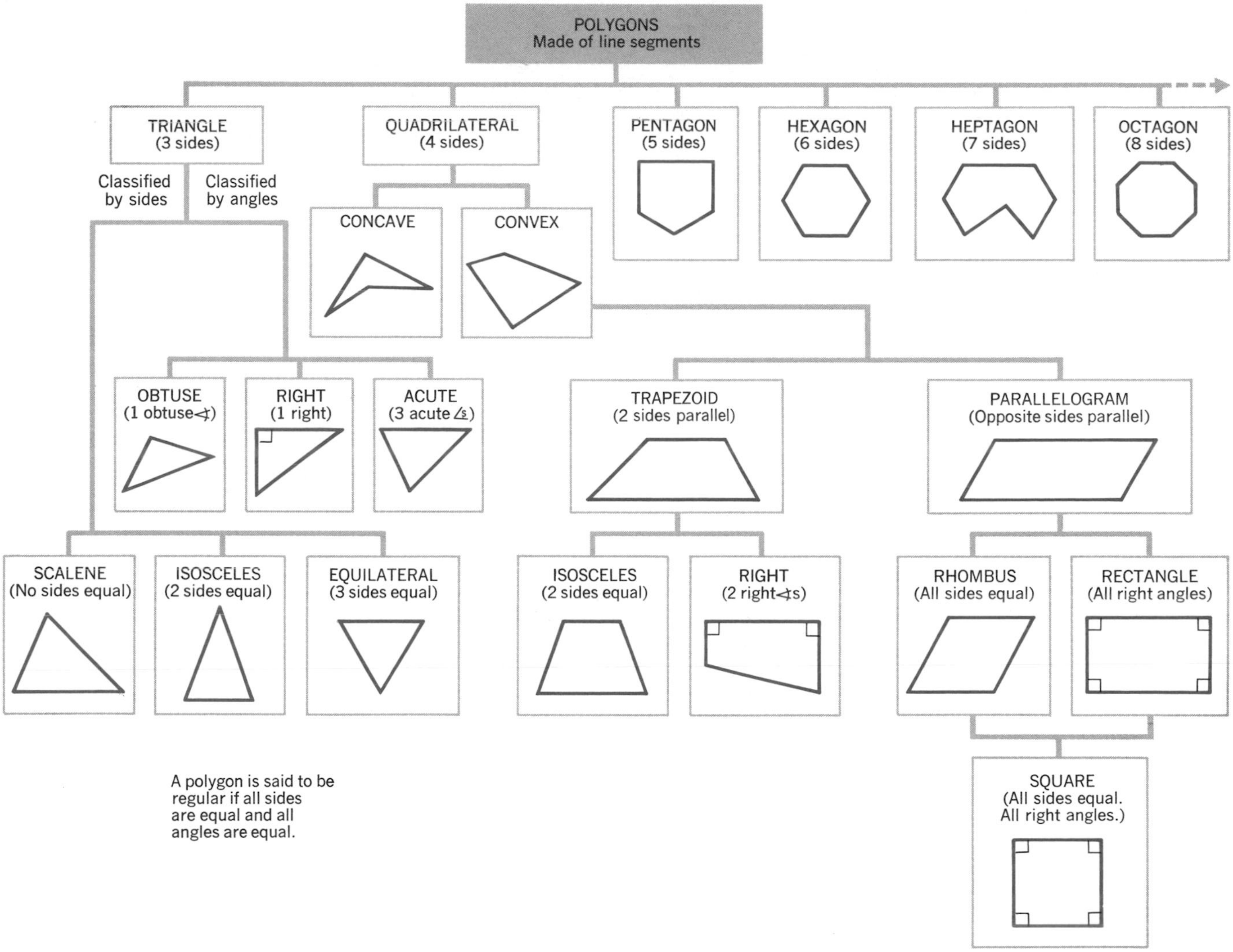

A polygon is said to be regular if all sides are equal and all angles are equal.

Irregular Polygons

Many times the shapes of polygons that appear in practical work are not the simple geometric figures that you have seen so far. The easiest way to work with irregular polygon shapes is to divide them into simpler, more familiar figures.

For example, look at this L-shaped figure:

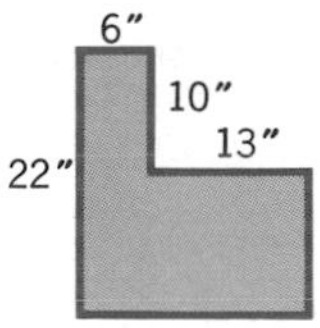

There are two ways to find the area of this shape.

Method 1: Addition

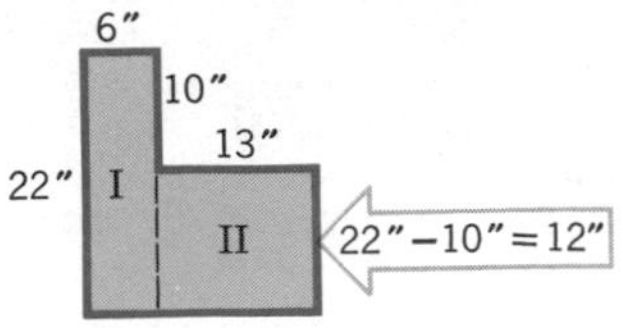

Method 2: Subtraction

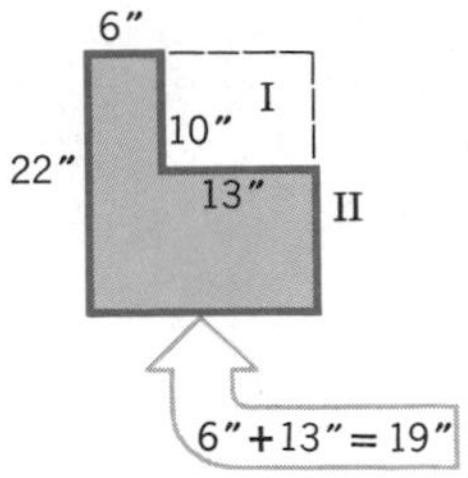

By drawing the dotted line, we have divided this figure into two rectangles, I and II. I is 22″ by 6″ for an area of 132 in^2. II is 13″ by 12″ for an area of 156 in^2. The total area is

$132 + 156 = 288$ sq in.

In this method we subtract the area of the small rectangle (I) from the area of the large rectangle (II). The area of I is

$10 \times 13 = 130$ sq in.

The area of II is

$22'' \times 19'' = 418$ sq in.
$418 - 130 = 288$ sq in.

Notice that in the addition method we had to calculate the height of rectangle II as 22″ − 10″ or 12″. In the subtraction method we had to calculate the width of rectangle II as 6″ + 13″ or 19″.

Which method you use to find the area of an irregular polygon depends on your ingenuity and on which dimensions are given.

Find the area of this shape.

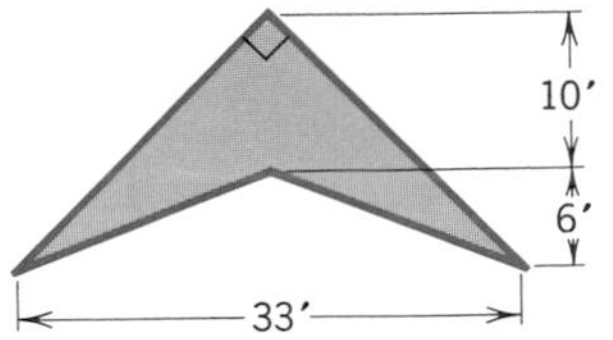

Check your work in **38**

---

38 You may at first wish to divide this into two triangles and add their areas.

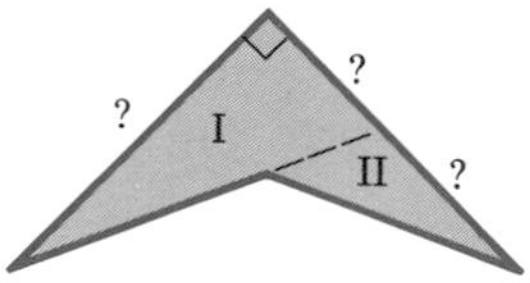

However, the dimensions provided do not allow you to do this.

or

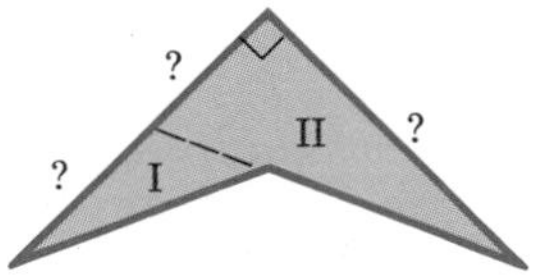

The only option here is to subtract the area of Triangle I from the area of Triangle II.

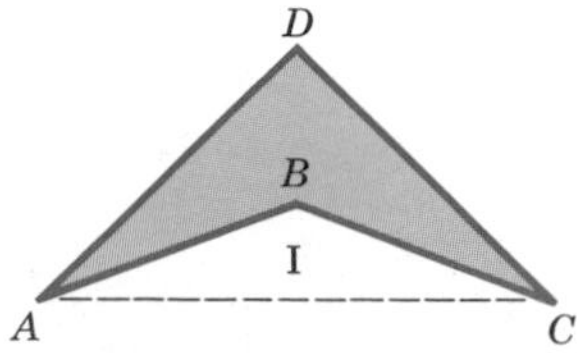

Triangle *ADC* is triangle II
Triangle *ABC* is triangle I

	Base	*Height*	*Area*
Triangle II	33 ft	16 ft	264 sq ft
Triangle I	33 ft	6 ft	− 99 sq ft
			Area = 165 sq ft

If both methods will work, pick the one that looks easiest and requires the simplest arithmetic.

Find the area of this shape.

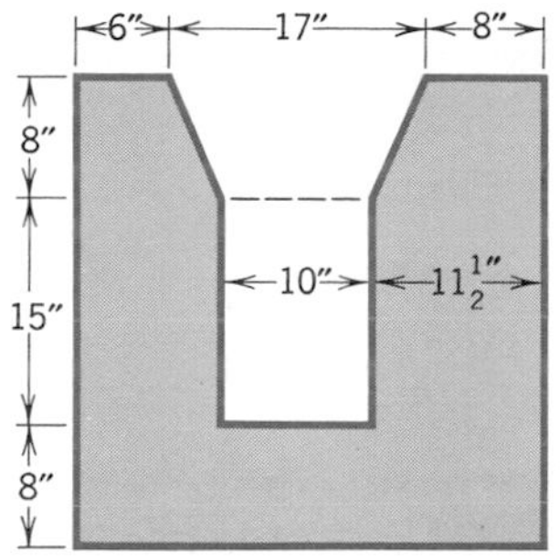

Check your work in **39**.

---

39 You might have been tempted to split the shape up into the 5 regions indicated in the figure. This will work, but there is an easier way.

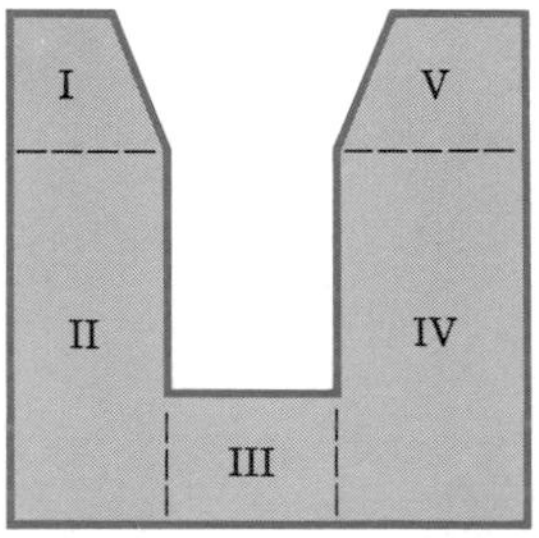

Hopefully, you thought of this alternative. Here we only need to compute three areas, then we subtract II and III from I.

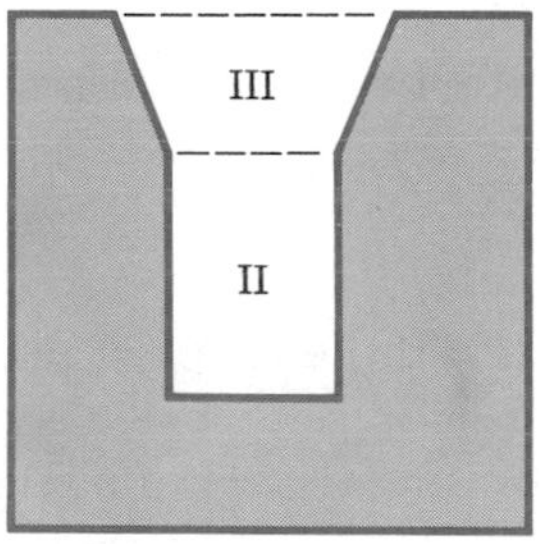

The outside rectangle is Area I.

	Base	*Height*	*Area*
Figure I	6″ + 17″ + 8″ = 31″	8″ + 15″ + 8″ = 31″	31″ × 31″ = 961 sq in.
Figure II	10″	15″	10″ × 15″ = 150 sq in.
Figure III (trapezoid)	17″ and 10″	8″	$\left(\frac{17+10}{2}\right)(8) = 108$ sq in.

The total area is 961 sq in. − 150 sq in. − 108 sq in. = 703 sq in.

When working with complex figures, organize your work carefully. Neatness will help eliminate careless mistakes.

Now, for practice, find the areas of the following shapes. Round to the nearest tenth.

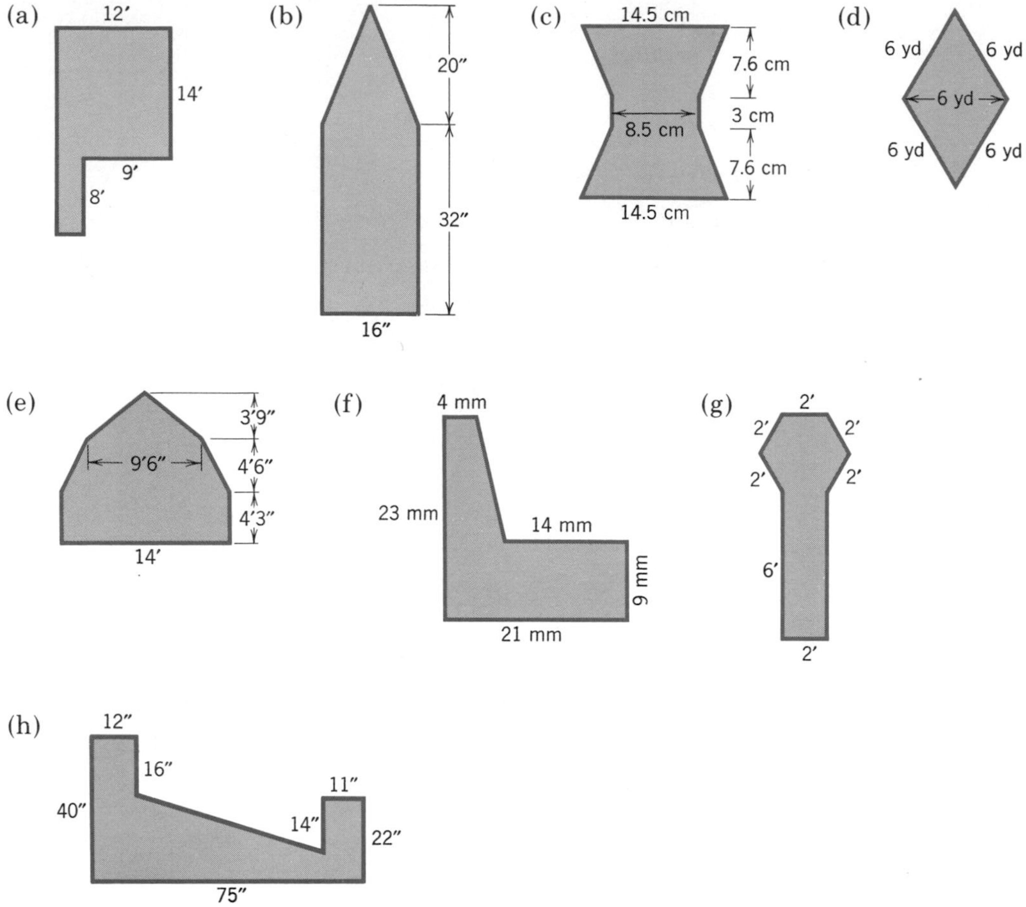

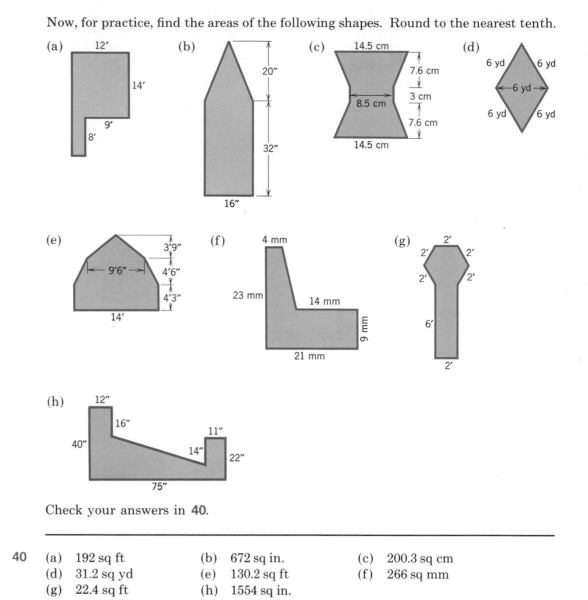

Check your answers in **40**.

---

40 (a) 192 sq ft (b) 672 sq in. (c) 200.3 sq cm
(d) 31.2 sq yd (e) 130.2 sq ft (f) 266 sq mm
(g) 22.4 sq ft (h) 1554 sq in.

In Problem (d) divide the figure into two equilateral triangles, then use the area formula $A = 0.433a^2$. Figure (g) is a hexagon plus a rectangle.

Circles

A circle is probably the most familiar and the simplest plane figure. Certainly it is the geometric figure that is most often used in practical and technical work. Mathematically, a circle is a closed curve representing the set of points some fixed distance from a given point called the *center*.

In this circle the center point is labeled O. The *radius* r is the distance from the center of the circle to the circle itself. The *diameter* d is the straight-line distance across the circle through the center point. It should be obvious from the drawing that the diameter is twice the radius.

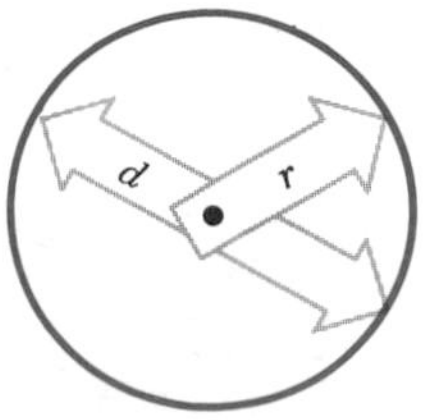

The *circumference* of a circle is the distance around it. It is a distance similar to the perimeter measure for a polygon. For all circles, the ratio of the circumference to the diameter, the distance around the circle to the distance across, is the same number.

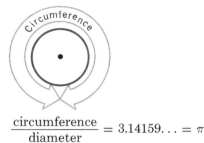

$$\frac{\text{circumference}}{\text{diameter}} = 3.14159\ldots = \pi$$

The value of this ratio has been given the label π, the Greek letter "pi." The value of π is approximately 3.14, but the number has no simple decimal form—the digits continue without ending or repeating. For most practical purposes use $\pi = 3.1416$.

The relationships between radius, diameter, and circumference can be summarized as follows:

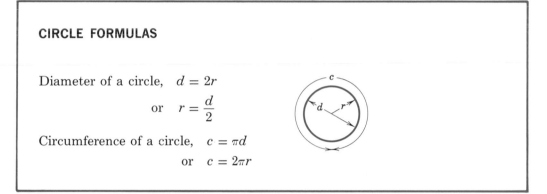

CIRCLE FORMULAS

Diameter of a circle, $d = 2r$

or $r = \dfrac{d}{2}$

Circumference of a circle, $c = \pi d$

or $c = 2\pi r$

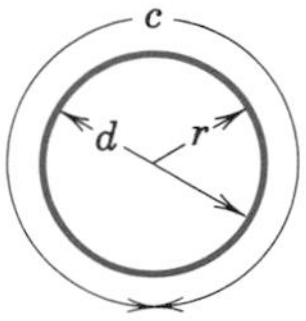

In the problems that follow, use $\pi = 3.1416$ unless you are told otherwise.

Use these formulas to find the radius and circumference of a hole with a diameter of 4 in. Round to one decimal place.

Check your work in **41**.

---

41 radius, $r = \dfrac{d}{2}$

$r = \dfrac{4 \text{ in.}}{2}$

$r = 2$ in.

circumference, $c = \pi d$

$= (3.1416)(4 \text{ in.})$

$= 12.5664 \text{ in.} \cong 12.6$ in., rounded

Use these formulas to solve the following practical problems. (Use $\pi = 3.14$ and round to two decimal places.)

1. If a hole is cut using a $\frac{3}{4}''$ diameter wood bit, what will be the radius and circumference of the hole?
2. An 8 ft strip of sheet metal is bent into a circular tube. What will be the diameter of this tube?
3. A circular redwood hot tub has a diameter of 5 ft. What length of steel band is needed to encircle the tub to hold the individual boards in position? Allow 4 in. for fastening.

4. If eight bolts are to be equally spaced around a circular steel plate at a radius of 9 in., what must be the spacing between the centers of the bolts along the curve?

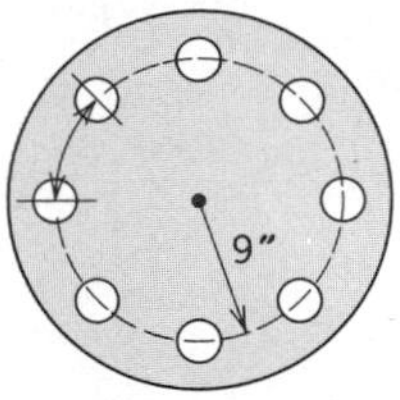
9″

5. At 85¢ per foot what will be the cost of the molding around the semicircular window shown? (Compute for the curved portion only.)

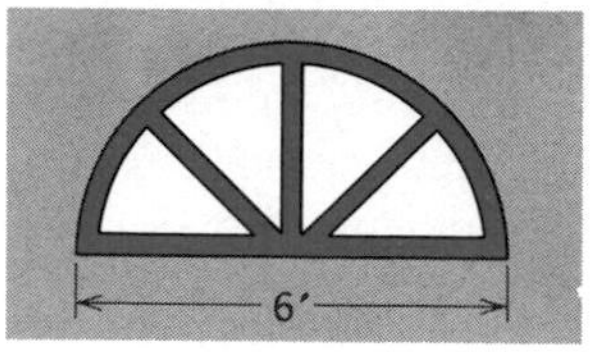
6′

Check your work in **42**.

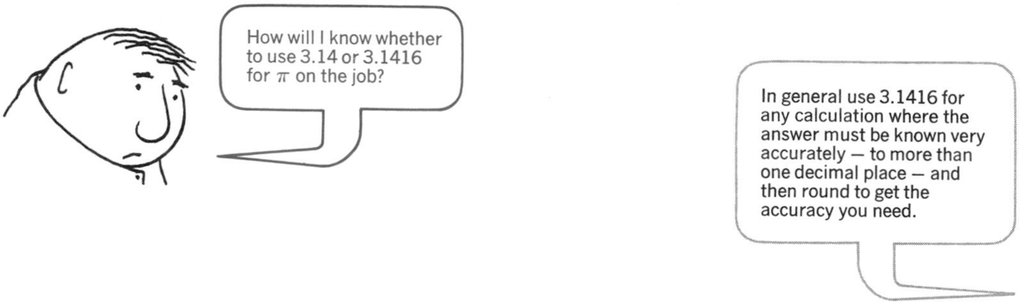
How will I know whether to use 3.14 or 3.1416 for π on the job?
In general use 3.1416 for any calculation where the answer must be known very accurately — to more than one decimal place — and then round to get the accuracy you need.

---

42 1. $r = \dfrac{d}{2}$

$r = \dfrac{\frac{3}{4}''}{2} = \dfrac{3''}{8} \qquad \dfrac{\frac{3}{4}}{2} = \dfrac{3}{4} \cdot \dfrac{1}{2} = \dfrac{3}{8}$

$c = \pi d$
$c = (3.14)(\frac{3}{4}\text{ in.})$
$c = (3.14)(0.75\text{ in.})$ Change the fraction to a decimal.
$c = 2.36\text{ in.}$

2. If $c = \pi d$ then $d = \dfrac{c}{\pi}$

and $d = \dfrac{8\text{ ft}}{3.14}$

$d = 2.55$ ft, rounded

3. $c = \pi d$
$c = (3.14)(5\text{ ft})$
$c = 15.7\text{ ft} \cong 15'8\frac{1}{2}''$
Adding 4 in., the length needed is $16'\frac{1}{2}''$
Notice that 0.7 ft is $(0.7)(12'') = 8.4''$ or about $8\frac{1}{2}''$.

4. $c = 2\pi r$
$c = 2(3.14)(9\text{ in.})$
$c = 56.52\text{ in.}$
Since eight bolts must be spaced evenly around the 9 in. circle, divide by 8.
Spacing $= 56.52 \div 8 = 7.07''$.

5. $c = \pi d$ for a complete circle
$c = (3.14)(6 \text{ ft})$
$c = 18.84 \text{ ft}$

The length of molding is half of this or 9.42 ft. The cost is $(85¢)(9.42 \text{ ft}) \simeq \$8.01$, rounded to the nearest cent.

Carpenters, plumbers, sheet metal workers, and other trades people often work with parts of circles. A common problem is to determine the length of a piece of material that contains both curved and straight segments.

Suppose we need to determine the total length of steel rod needed to form the curved piece shown. The two straight segments are no problem: we will need a total of 8″ + 6″ or 14″ of rod for these. When a rod is bent the material on the outside of the curve is stretched and the material on the inside is compressed. We must calculate the circumference for the curved section from a *neutral line* midway between the inside and outside radius.

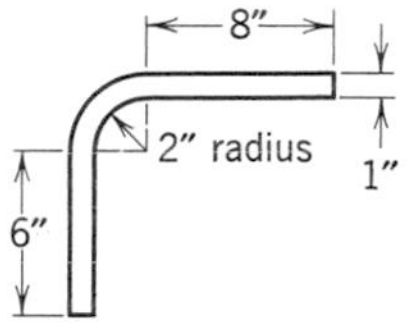

The inside radius is given as 2″. The stock is 1″ thick, so the midline or neutral line is at a radius of $2\frac{1}{2}$″. The curved section is one-quarter of a full circle, so the length of bar needed for the curved arc is

$$c = \frac{2\pi r}{4}$$ Divide the circumference by 4 to find the length of the quarter-circle arc.

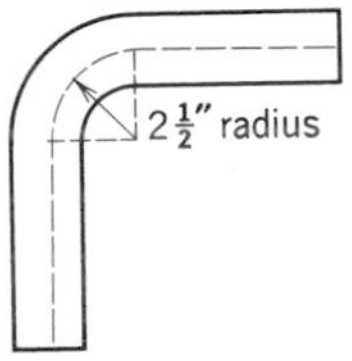

$$c = \frac{\pi r}{2}$$

$$c = \frac{(3.14)(2.5'')}{2}$$

$c = 3.93''$, rounded to two decimal places

The total length of bar needed for the piece is

$L = 3.93'' + 14''$
$L = 17.93''$

Your turn. Find the total length of ornamental iron $\frac{1}{2}$″ thick needed to bend into this shape. Use $\pi = 3.14$ and round to two decimal places.

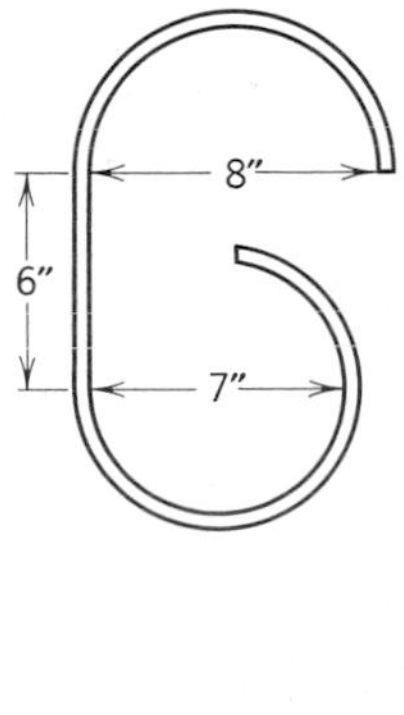

Check your work in **43**.

43 Measuring to the midline, the upper curve has a radius of $4\frac{1}{4}''$ and the lower curve has a radius of $3\frac{3}{4}''$. Calculate the lengths of the two curved pieces this way:

Upper Curve		*Lower Curve*	
(One-half circle)		(Three-fourths circle)	
$c = \dfrac{2\pi r}{2} = \pi r$	Multiply by $\frac{1}{2}$ for a half circle.	$c = \dfrac{3}{4} \cdot 2\pi r = \dfrac{3\pi r}{2}$	Multiply by $\dfrac{3}{4}$.
$c = (3.14)(4\frac{1}{4}'')$		$c = \dfrac{(3)(3.14)(3\frac{3}{4}'')}{2}$	
$c = 13.35''$		$c = 17.66''$	

The total length of straight stock needed to create this shape is $13.35'' + 17.66'' + 6'' = 37.01''$.

Find the length of stock needed to create each of the following shapes. Use 3.14 for π and round to two decimal places.

(a)
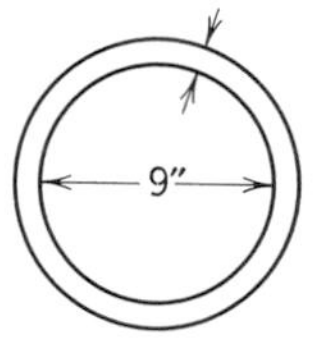

(b)
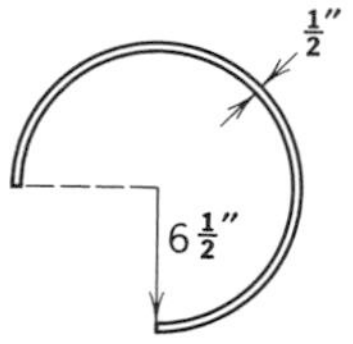

(c)
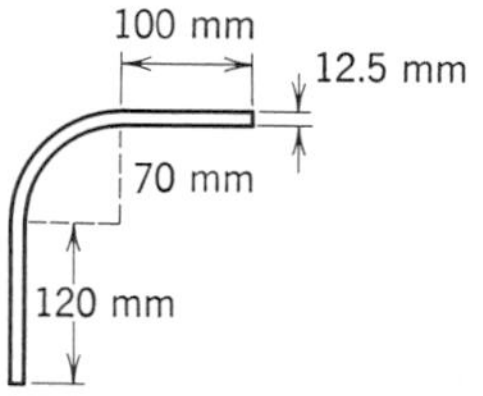

(d)
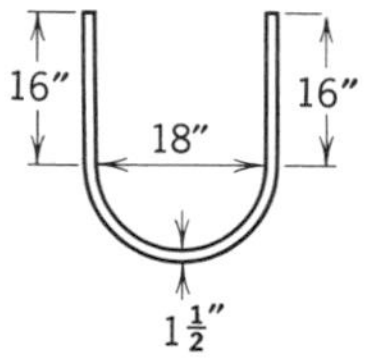

(e)
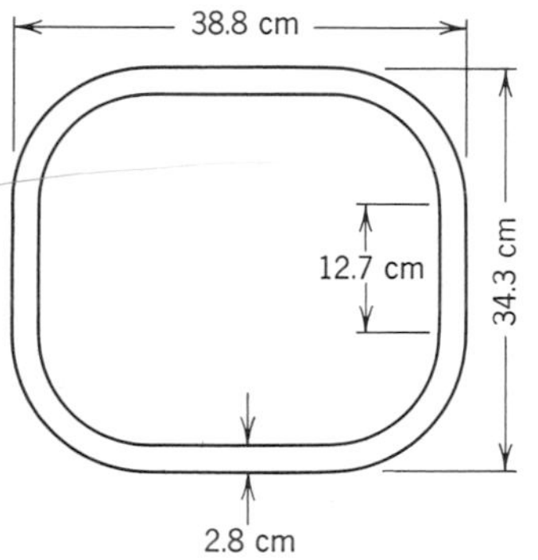

Check your answers in **44**

---

44 (a) 31.4″ (b) 31.79″ (c) 339.71 mm
(d) 62.62″ (e) 136.42 cm

On Problem (e) a bit of careful reasoning will convince you that the dimensions are as follows:

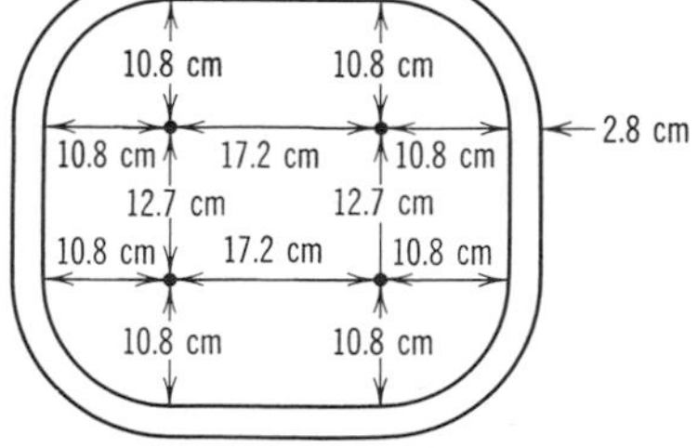

Each of the four corner arcs has radius 12.2 cm, and each is a quarter circle.

Radius $= 10.8 \text{ cm} + 1.4 \text{ cm}$
$= 12.2 \text{ cm}$

To find the *area* of a circle use one of the following formulas.

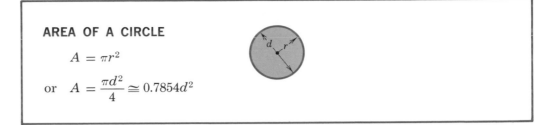

AREA OF A CIRCLE

$$A = \pi r^2$$

or $$A = \frac{\pi d^2}{4} \cong 0.7854 d^2$$

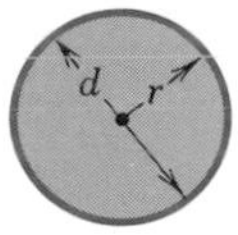

If you are curious about how these formulas are obtained, don't miss the box on p. 350.

Use the formula to find the area of a circle with a diameter of 8″.

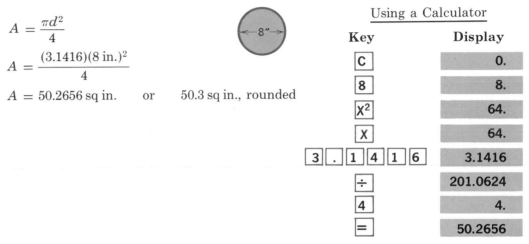

$$A = \frac{\pi d^2}{4}$$

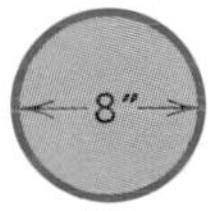

$$A = \frac{(3.1416)(8\text{ in.})^2}{4}$$

$A = 50.2656$ sq in. or 50.3 sq in., rounded

Using a Calculator

Key	Display
C	0.
8	8.
X^2	64.
X	64.
3 . 1 4 1 6	3.1416
÷	201.0624
4	4.
=	50.2656

Of course you would find exactly the same area if you used the radius $r = 4''$ in the first formula. Most people who need the area formula in their work memorize the formula $A = \pi r^2$ and calculate the radius r if they are given the diameter d.

Now for practice try these problems.

1. Find the areas of the following circles. Round as indicated.

(a)

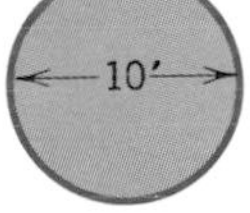

(nearest tenth)

(b)

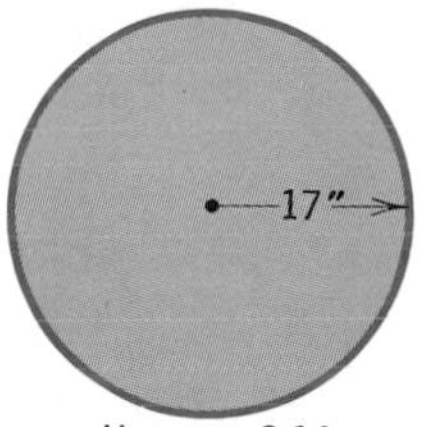

Use $\pi = 3.14$

(nearest tenth)

(c) Find the area of the head of a piston with a radius of 6.5 cm. (Round to the nearest tenth.)

(d)

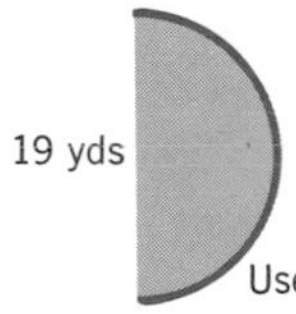

Use $\pi = 3.14$

(nearest hundredth)

(e)

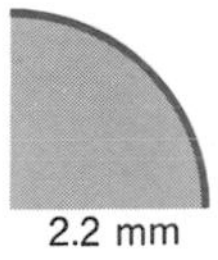

(nearest thousandth)

2. Find the number of square feet of wood needed to make a circular table 6′ in diameter. Use $\pi = 3.14$ and round to one decimal place.

3. Find the number of square inches of stained glass used in a semicircular window with a radius of 9″. Use $\pi = 3.14$ and round to one decimal place.

4. What is the area of the largest circle that can be cut out of a square piece of sheet metal 4′6″ on a side? Round to the nearest hundredth.

5. A pressure of 860 lb per square foot is exerted on the bottom of a cylindrical water tank. If the bottom has a diameter of 15′, what is the total force on the bottom? Round final answer to the nearest thousand pounds.

Check your answers in **45**.

AREA OF A CIRCLE

To find the formula for the area of a circle, first divide the circle into many pie-shaped sectors, then rearrange them like this:

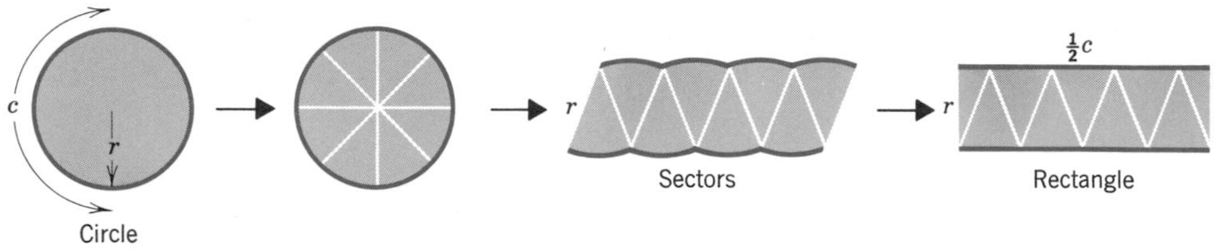

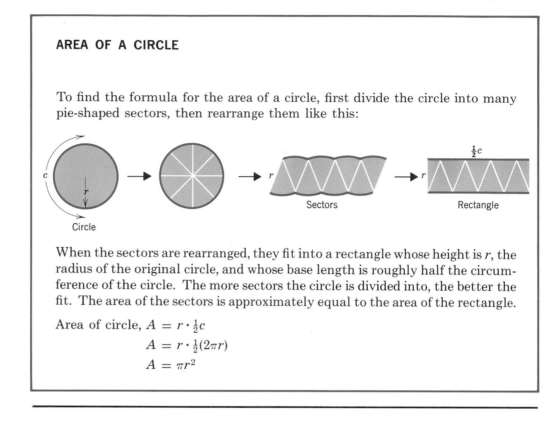

When the sectors are rearranged, they fit into a rectangle whose height is r, the radius of the original circle, and whose base length is roughly half the circumference of the circle. The more sectors the circle is divided into, the better the fit. The area of the sectors is approximately equal to the area of the rectangle.

Area of circle, $A = r \cdot \frac{1}{2}c$

$A = r \cdot \frac{1}{2}(2\pi r)$

$A = \pi r^2$

---

45 1. (a) $A = \pi r^2 = (3.1416)(5 \text{ ft})^2 = (3.1416)(25 \text{ sq ft})$
$A = 78.5$ sq ft, rounded

(b) $A = \pi r^2 = (3.14)(17 \text{ in.})^2$
$A = 907.5$ sq in.

(c) $A = \pi r^2 = (3.1416)(6.5 \text{ cm})^2$
$A = 132.7$ sq cm, rounded

(d) $A = \frac{1}{2}(\pi r^2) = \frac{1}{2}(3.14)(9.5 \text{ yd})^2$
$A = 141.69$ sq yd, rounded

(e) $A = \frac{1}{4}(\pi r^2) = \frac{1}{4}(3.1416)(2.2 \text{ mm})^2$
$A = 3.801$ sq mm, rounded

2. $A = \pi r^2 = (3.14)(3 \text{ ft})^2$
$A = 28.3$ sq ft

3. $A = \frac{1}{2}(\pi r^2) = \frac{1}{2}(3.14)(9 \text{ in.})^2$
$A = 127.2$ sq in.

4. $A = \pi r^2 = (3.1416)(2'3'')^2$
$A = (3.1416)(2.25 \text{ ft})^2 = 15.90$ sq ft

5. $A = \pi r^2 = (3.1416)(7.5 \text{ ft})^2$
$A = 176.715$ sq ft

$\text{Force} = 860 \frac{\text{lb}}{\cancel{\text{sq ft}}} \times 176.715 \cancel{\text{sq ft}}$

Force = 151,974.9 or 152,000 lb, rounded

Rings

A circular *ring* is defined as the area between two concentric circles—two circles having the same center point. A washer is a ring, so is the cross section of a pipe, a collar, or a cylinder.

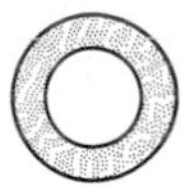

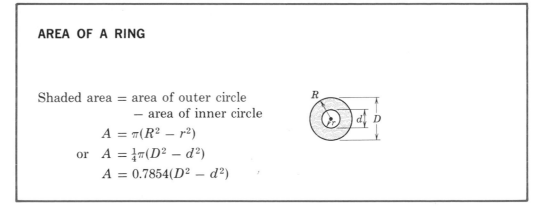

To find the area of a ring, subtract the area of the inner circle from the area of the outer circle. Use the following formula.

AREA OF A RING

Shaded area = area of outer circle − area of inner circle

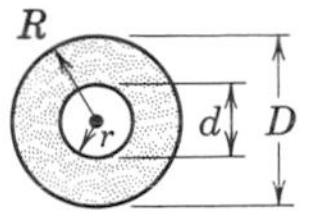

$$A = \pi(R^2 - r^2)$$

or $$A = \tfrac{1}{4}\pi(D^2 - d^2)$$

$$A = 0.7854(D^2 - d^2)$$

For example, to find the cross-sectional area of the wall of a ceramic pipe whose i.d. (inside diameter) is 8″ and whose o.d. (outside diameter) is 10″, substitute $d = 8$ in. and $D = 10$ in. into the second formula.

$A = \tfrac{1}{4}(3.1416)(10 \text{ in.}^2 - 8 \text{ in.}^2)$
$A = (0.7854)(100 \text{ sq in.} - 64 \text{ sq in.})$
$A = (0.7854)(36 \text{ sq in.})$
$A = 28.3$ sq in., rounded

Notice that $\frac{\pi}{4}$ is equal to 0.7854 rounded to four decimal places. Also notice that the diameters are *squared first* and *then* the results *subtracted.*

Try it. Find the area of a washer with an outside radius of $\frac{7}{16}''$ and a thickness of $\frac{1}{8}''$. Round to the nearest thousandth.

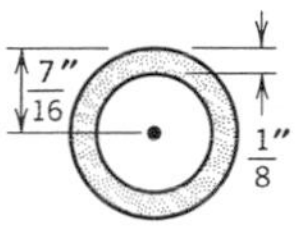

Check your work in **46**.

46 $R = \frac{7}{16}''$ $\quad r = \frac{7}{16}'' - \frac{1}{8}''$ $\quad$ or $\quad r = \frac{5}{16}''$

then

$A = \pi(R^2 - r^2)$
$A = (3.1416)[(\frac{7}{16}\text{ in.})^2 - (\frac{5}{16}\text{ in.})^2]$
$A = (3.1416)[0.4375\text{ in.})^2 - (0.3125\text{ in.})^2]$
$A = (3.1416)[0.1914\text{ sq in.} - 0.0977\text{ sq in.}]$
$A = (3.146)[0.0937\text{ sq in.}]$
$A = 0.295$ sq in., rounded

Using a Calculator

Key	Display
C	0.
MC	0.
5	5.
÷	5.
1 6	16.
=	0.3125
X^2	0.0976562
M +	0.0976562
7	7.
÷	7.
1 6	16.
=	0.4375
X^2	0.1914062
−	0.1914062
MR	0.0976562
=	0.09375
×	0.09375
3 . 1 4 1 6	3.1416
=	0.294525

Put this value in memory if your calculator has one; otherwise write it on a piece of paper.

Most people find it easiest to convert the fractions to decimals before squaring, especially if they are working with a calculator.

Solve the following problems involving rings. (Round to two decimal places.)

(a) Find the cross section area of the pipe shown in the figure. Use $\pi = 3.14$.

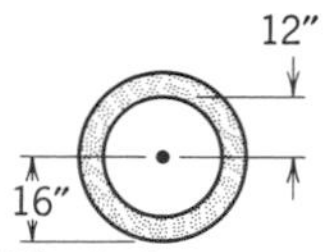

(b) Find the area of a washer with an inside diameter of $\frac{1}{2}''$ and an outside diameter of $\frac{7}{8}''$.
(c) What is the cross-sectional area of a steel collar 0.6 cm thick with an inside diameter of 9.4 cm?
(d) At 25¢ per square foot find the cost of sodding a circular lawn 6′ wide surrounding a flower garden with a radius of 20′. Use $\pi = 3.14$.
(e) For a cross section of pipe with an inside diameter of 8″, how much more material is needed to make the pipe $\frac{1}{2}''$ thick than to make it $\frac{3}{8}''$ thick?

The correct answers are in **47**.

COMBINING CIRCULAR AREAS

In many practical situations it is necessary to determine what size circle contains the same area as two or more smaller circles.

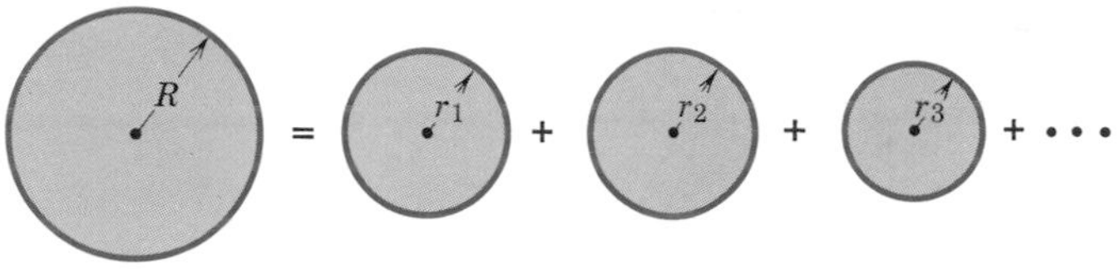

Area of larger circle = combined areas of the smaller circles

$\pi R^2 = \pi r_1^2 + \pi r_2^2 + \cdots$

Dividing all terms by π we have

$R^2 = r_1^2 + r_2^2 + \cdots$

Taking the square root of both sides of the equation

$R = \sqrt{r_1^2 + r_2^2 + \cdots}$

As an example of how to apply this formula, suppose the flow from a 4-in. radius pipe and a 6-in. radius pipe must be combined into one large pipe of area equal to the total of the smaller two. To find the radius of the larger pipe, substitute into the formula:

$R = \sqrt{r_1^2 + r_2^2}$

$R = \sqrt{(4 \text{ in.})^2 + (6 \text{ in.})^2}$

$R = \sqrt{16 \text{ sq in.} + 36 \text{ sq in.}}$

$R = \sqrt{52 \text{ sq in.}}$

$R = 7.2$ in. rounded

A very similar relationship holds true for the diameters of circles.

$D = \sqrt{d_1^2 + d_2^2 + \cdots}$

Try these problems for practice:

1. Find the radius of a steel rod whose area must equal that of four rods each of radius 2″.
2. What diameter gas line would be needed to replace two 6″ diameter lines?
3. Find the diameter of a water main with the same total area as two mains of diameter 3″ and 4″.
4. What radius vent will have the same cross-sectional area as three 6″ radius vents?
5. Find the diameter of a water pipe that has the same area as four pipes, two with 2″ diameters and two with 3″ diameters.

All answers are on p. 555.

---

47 (a) 351.68 sq in. (b) 0.40 sq in. (c) 18.85 sq cm.
(d) $216.66 (e) 3.49 sq in.

A summary of all of the formulas you have learned for plane figures appears below. Be certain you know these before you continue.

SUMMARY OF FORMULAS FOR PLANE FIGURES

Figure		*Area*	*Perimeter*
Rectangle	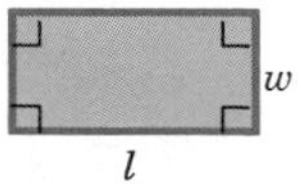	$A = LW$	$P = 2L + 2W$
Square	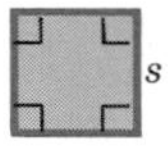	$A = S^2$	$P = 4S$
Parallelogram	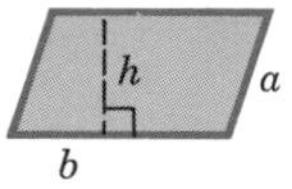	$A = bh$	$P = 2a + 2b$
Triangle	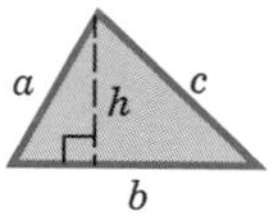	$A = \frac{1}{2}bh$ or $\frac{bh}{2}$	$P = a + b + c$
Equilateral triangle		$A = 0.433a^2$	$P = 3a$
Isosceles triangle	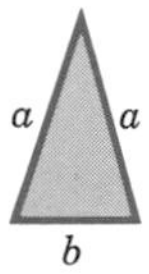	$A = \frac{b}{2}\sqrt{a^2 - \left(\frac{b}{2}\right)^2}$	$P = 2a + b$
Right triangle	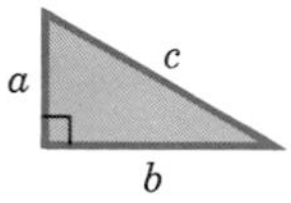	$A = \frac{1}{2}ab$	$P = a + b + c$
Trapezoid	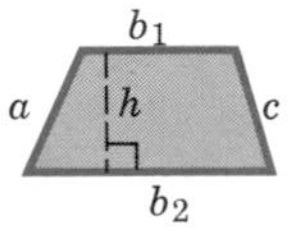	$A = \left(\frac{b_1 + b_2}{2}\right)h$ or $\frac{h}{2}(b_1 + b_2)$	$P = a + c + b_1 + b_2$
Regular hexagon	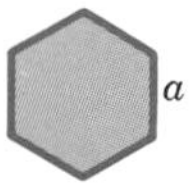	$A = 2.598a^2$	$P = 6a$
Circle	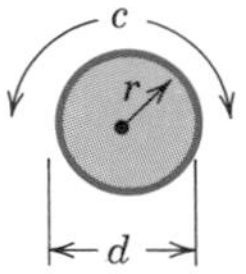	$A = \pi r^2$ $A = 0.7854d^2$	$c = \pi d$ $c = 2\pi r$
Ring	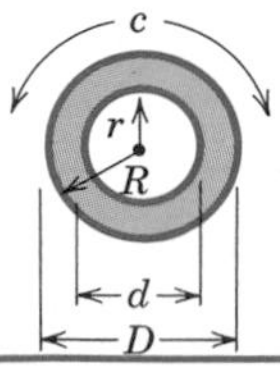	$A = \pi(R^2 - r^2)$ $A = 0.7854(D^2 - d^2)$	$c = 2\pi R$

Now turn to **48** for a set of review exercises on plane figures.

A. Identify each of the following polygons and then name all sides, vertices, and diagonals:

1.
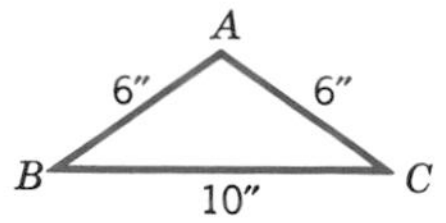

2.
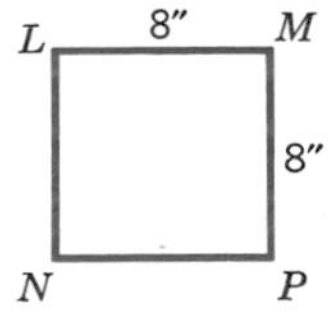

3.
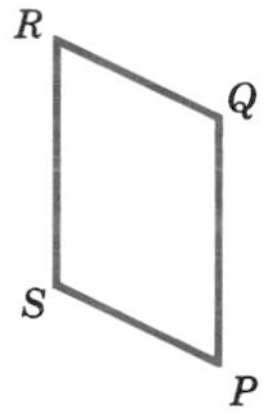

4.
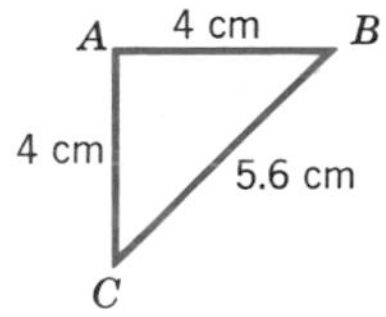

5.
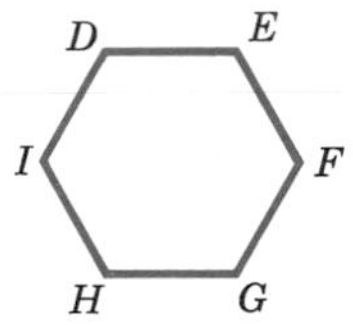

6.
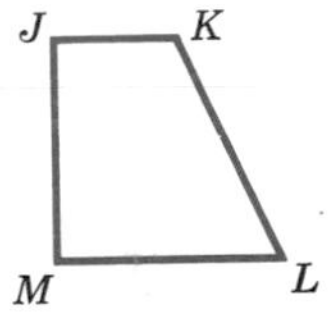

7.

8.
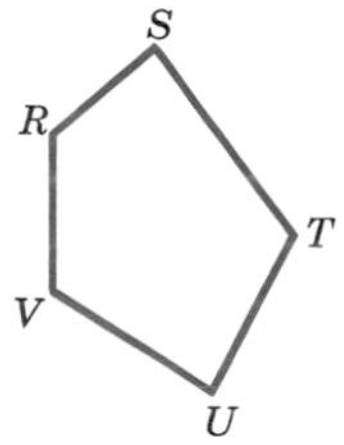

9.
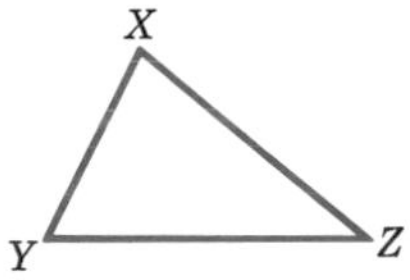

10.

11.
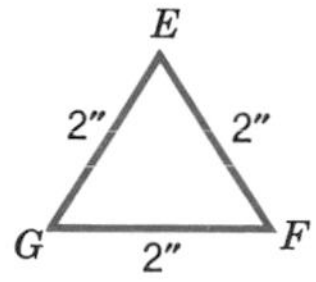

12.
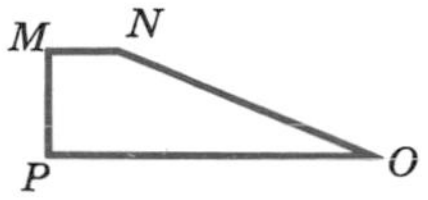

B. Find the perimeter (or circumference) and area of each figure (round to the nearest tenth if necessary):

1.

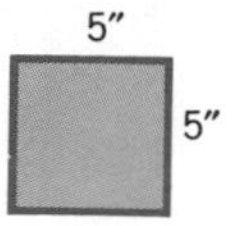

2.

3.

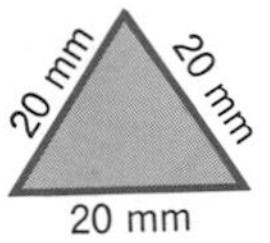

4.

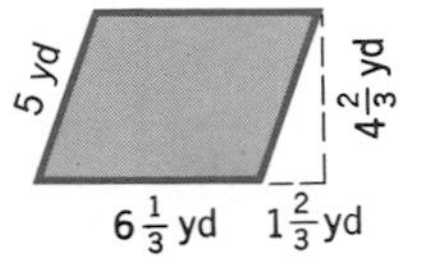

5.

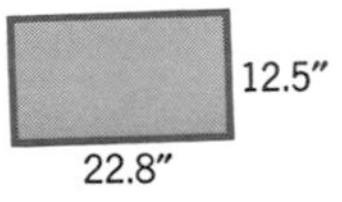

6.

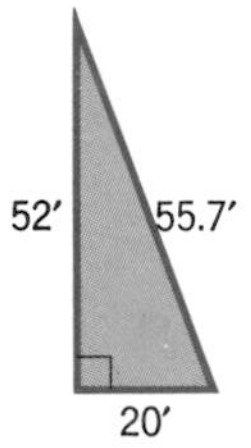

7.

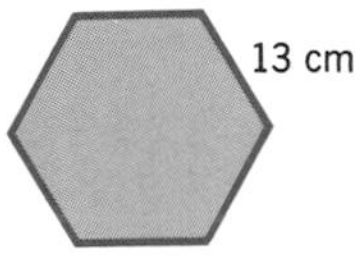

8.

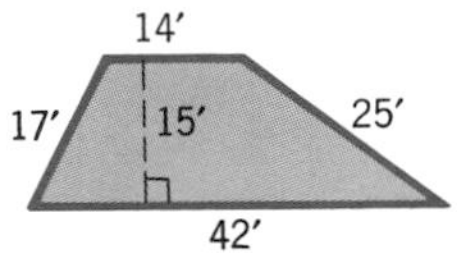

9.

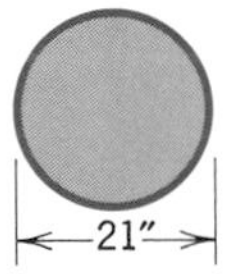

10.

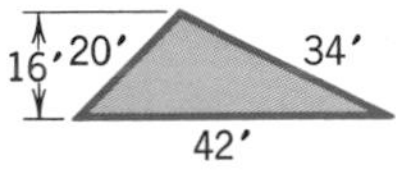

11.

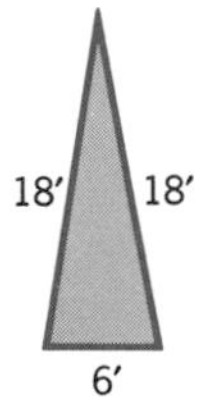

12.

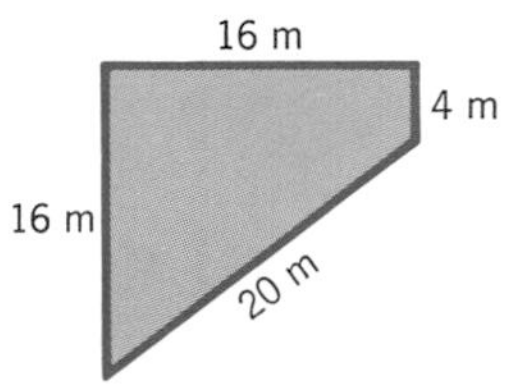

13.

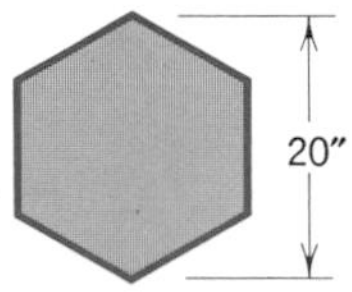

14.

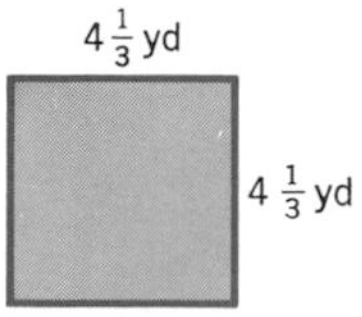

15.

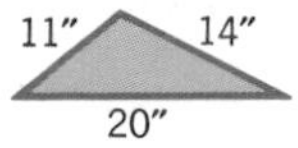

16.

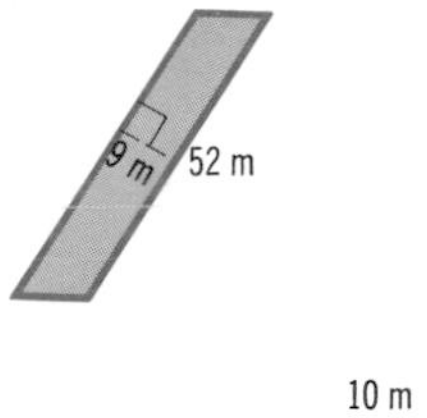

17.

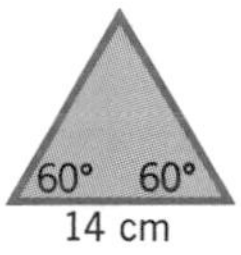

18.

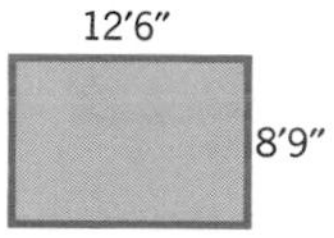

19.

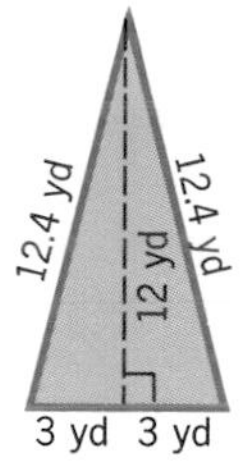

20.

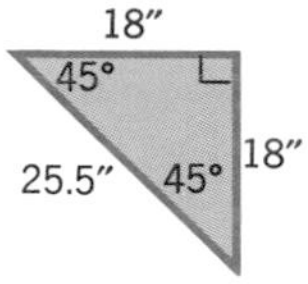

C. Find the outside perimeter (or circumference) and area of each of the following (round to the nearest tenth if necessary):

1.

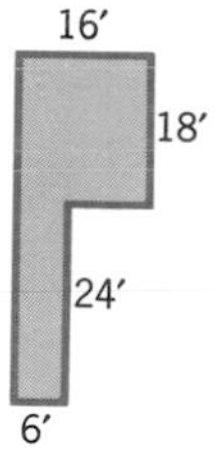

2.

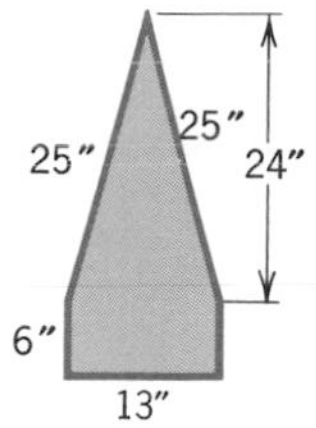

3.

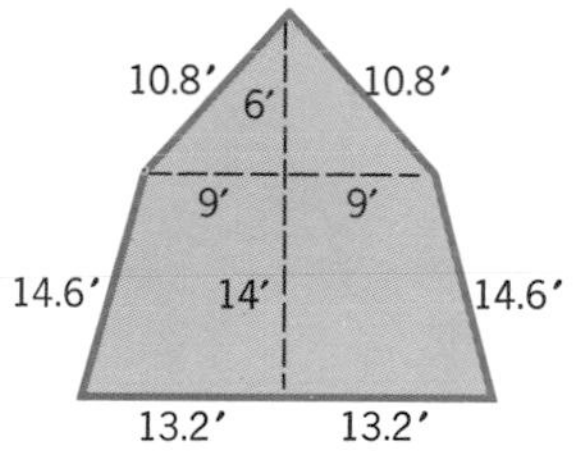

4.

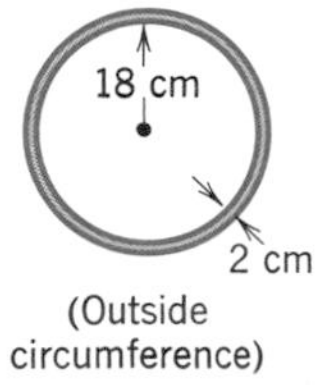

5.

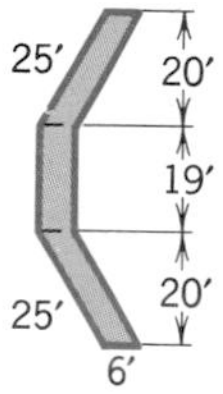

6.

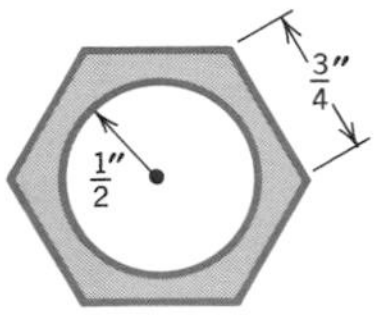

7.

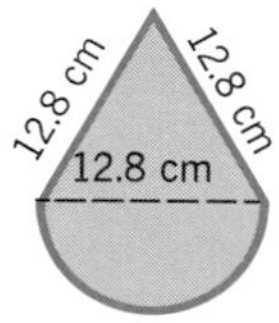

8.

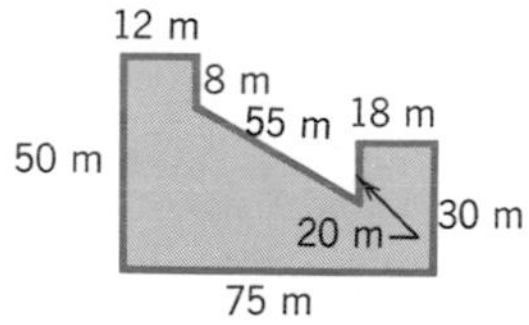

9.

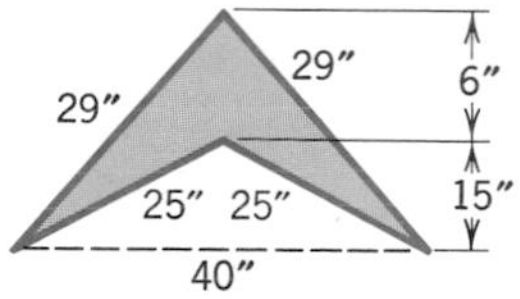

10.

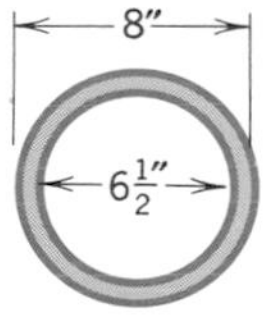

11.

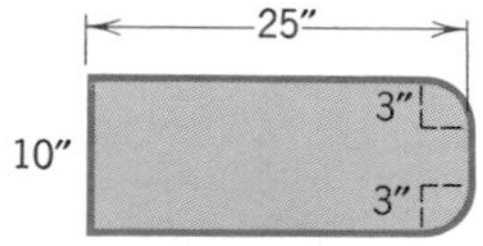

12.

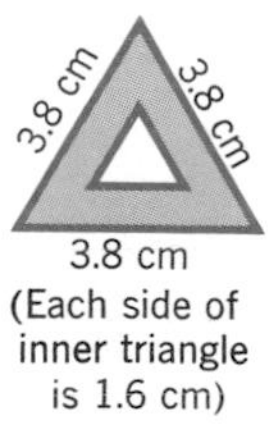

13.

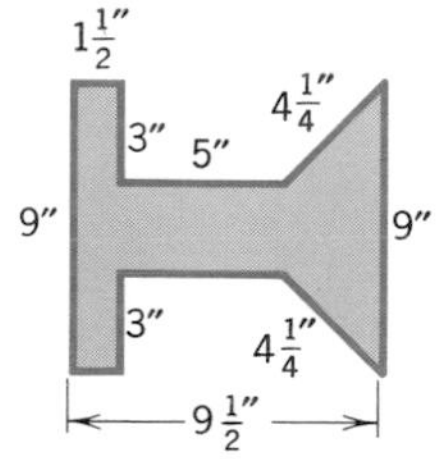

14.

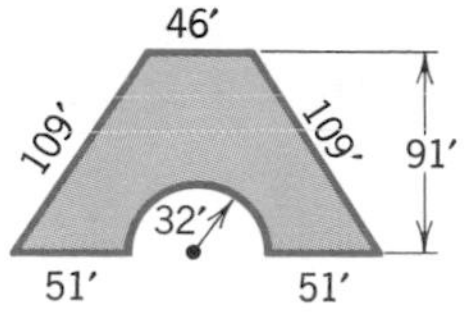

D. Find the missing dimensions of each figure shown (round to the nearest hundredth if necessary):

1. 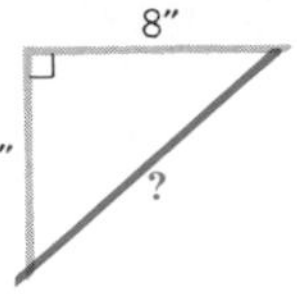

2. 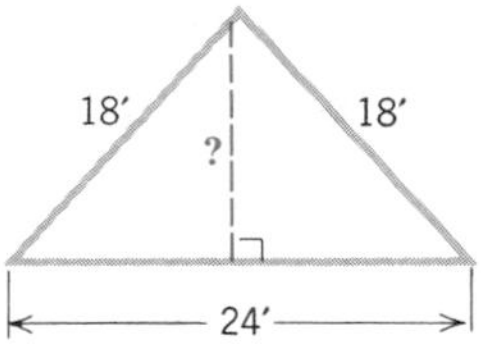

3. 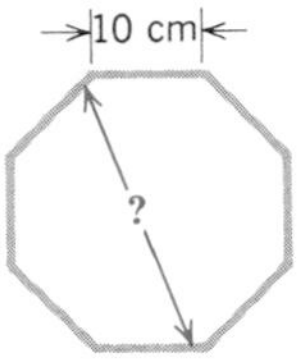

4. 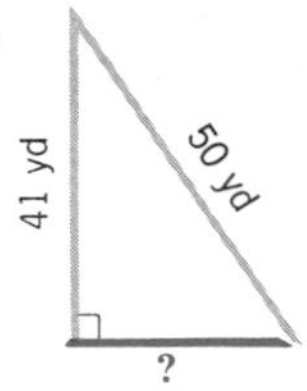

5. 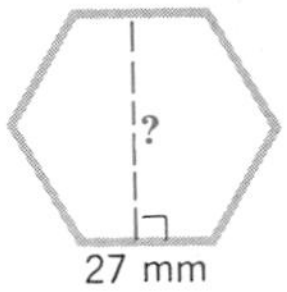

6. 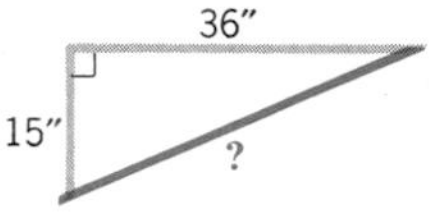

7. 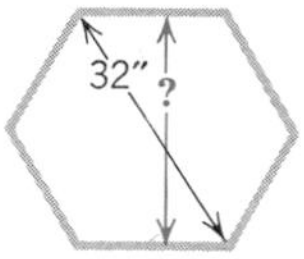

8. 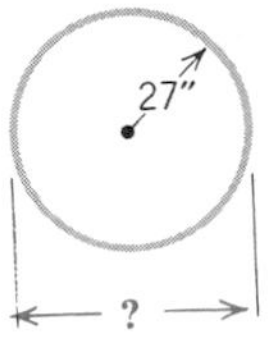

9. 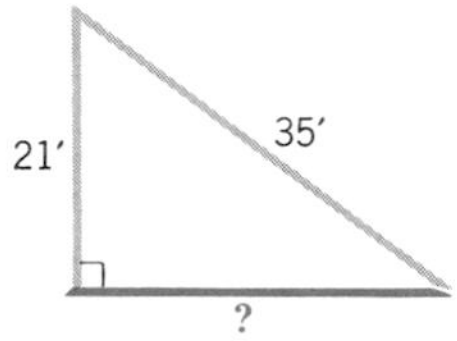

10. 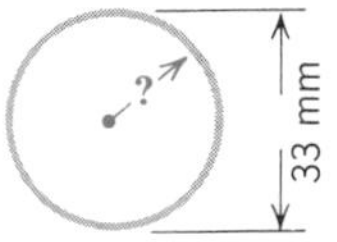

11. 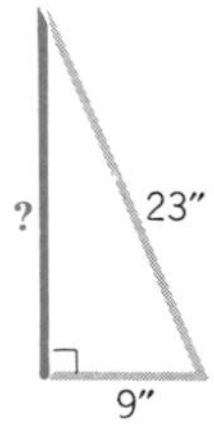

12. 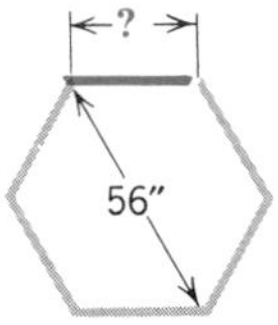

13. 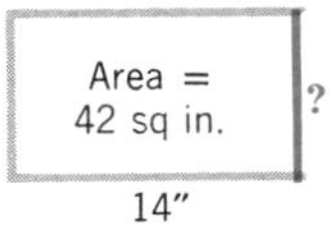

14. 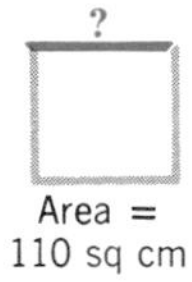

E. Practical Problems

1. What length of 1″ stock is needed to bend a piece of steel into this shape? Use $\pi = 3.14$.

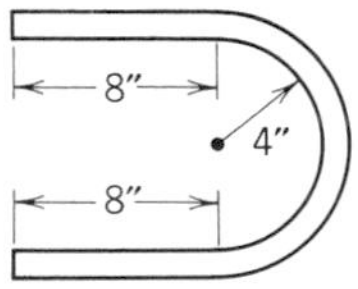

2. At 460 sq ft per gallon, how many gallons of paint are needed to cover the outside walls of a house as pictured below? (Do not count windows and doors).

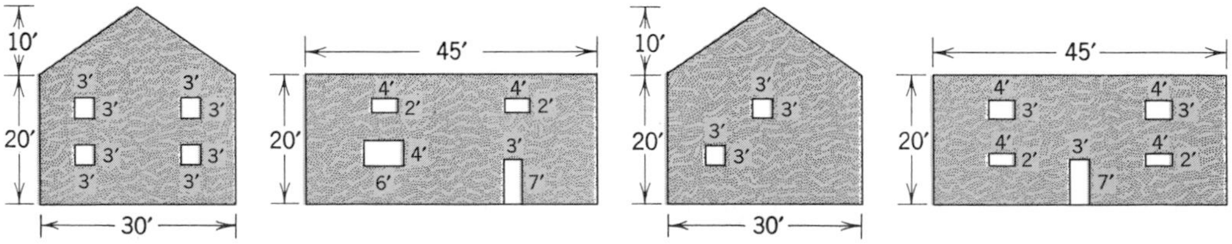

3. What diameter must a circular piece of stock be to mill a hexagonal shape with a side length of 0.3 in.?

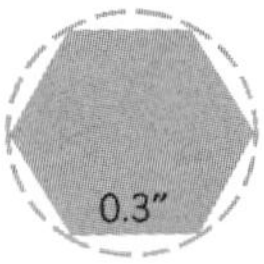

4. In Problem 3 how many square inches of stock will be wasted?

5. What is the cross-sectional area of a cement pipe 2″ thick with an inside diameter of 8′? Use $\pi = 3.14$.

6. In the piece of steel shown, $4\frac{1}{2}$″ holes are drilled at equal intervals around a 10″ diameter inner circle. Find the distance a as indicated. (**Hint:** Use the Pythagorean Theorem.) Round to the nearest thousandth.

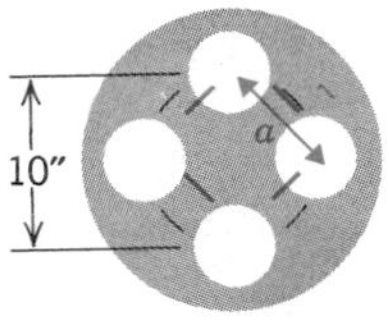

7. How many bricks will it take to build the wall shown if each brick measures $4\frac{1}{4}$″ × $8\frac{3}{4}$″ including mortar?

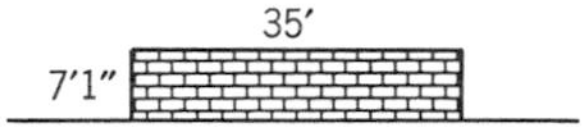

8. **How much additional cross-sectional area is there in a 12″ i.d. pipe $\frac{3}{16}$″ thick than one $\frac{1}{8}$″ thick? Round to three decimal places.**

9. What diameter round stock is needed to mill a hexagonal nut $\frac{7}{8}$″ on a side?

10. How many plants spaced every 6″ are needed to surround a circular walkway with a 25′ radius? Use $\pi = 3.14$.

11. What size diameter circular stock is needed to mill a square end 3 cm on a side?

12. How many square feet of wood are needed to build the cabinet in the figure? (Assume wood is needed for all six surfaces.)

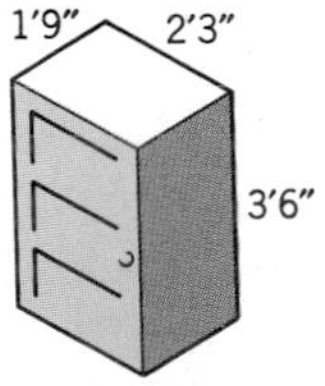

13. A four-sided vent connection must be made out of sheet metal. If each side of the vent is like the one pictured, how many total square inches of sheet metal will be used?

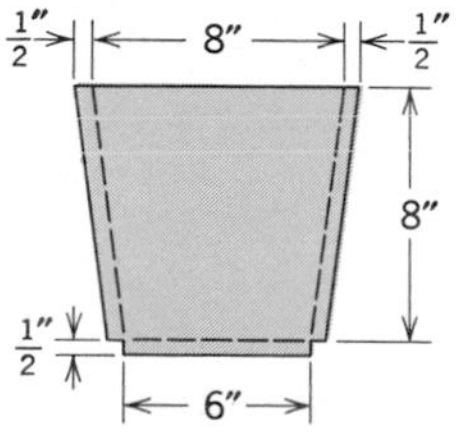

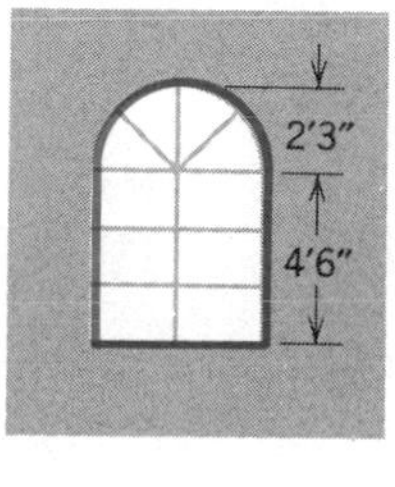

14. At $0.30 per foot, how much will it cost to put molding around the window pictured? Use $\pi = 3.14$.

15. Given the radius of the semicircular ends of the track is 32 meters, how long must each straightaway be to make a 400 meter track? Use $\pi = 3.14$.

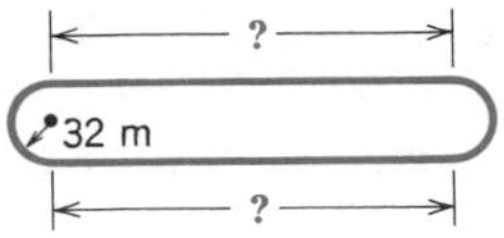

16. What is the total length of belting needed for the pulley shown?

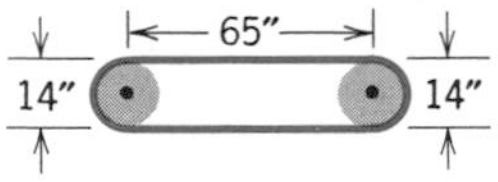

17. An electrician bends a $\frac{1}{2}''$ conduit as shown in the figure to follow the bend in a wall. Find the total length of conduit needed between A and B. Use $\pi = 3.14$.

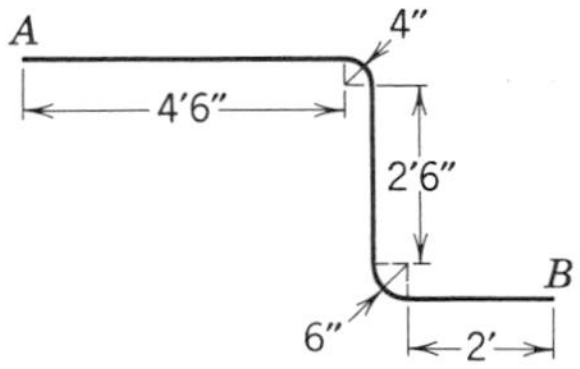

18. How many inches apart will 8 bolts be along the circumference of the metal plate shown in the figure?

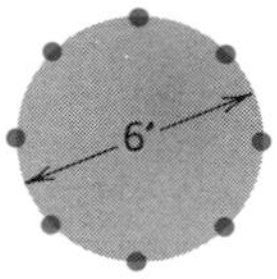

19. How many square *yards* of concrete surface are there in a walkway 8′ wide surrounding a tree if the inside diameter is 16′? Round to one decimal place.

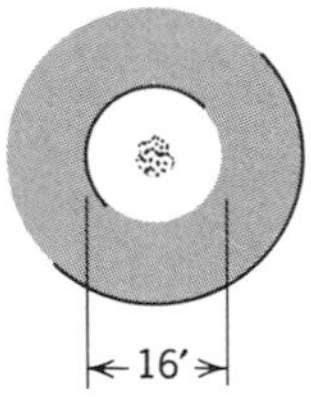

20. The aerial view of a roof produces the following design. How many squares (100 sq ft) of shingles are needed for the roof?

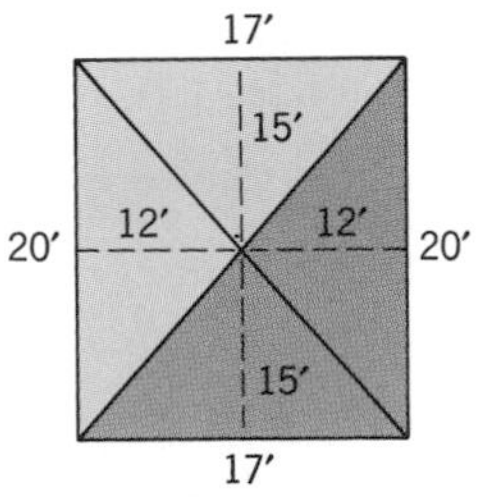

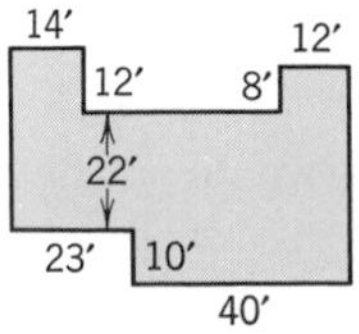

21. In Problem 20 how many feet of gutters are needed?

22. The floor plan of a house is shown in the figure. If the entire house is to be carpeted, how many square *yards* of carpet are needed? At $15.95 per square yard, how much will it cost?

23. What length of stock 3 cm thick is needed to forge the following shape? Use $\pi = 3.14$.

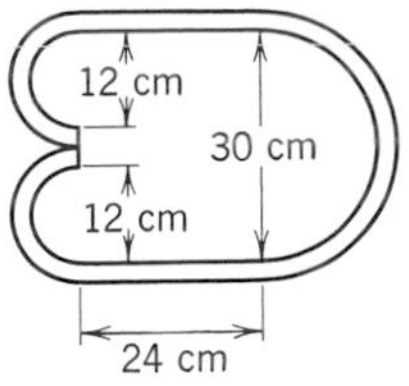

24. If a hex bolt is milled from round stock $2\frac{1}{2}''$ in diameter, what size wrench is needed to fit the bolt? Assume $\frac{1}{32}''$ extra is needed for clearance.

25. A cut is to be made in a piece of metal as indicated by the dotted line. Find the length of the cut.

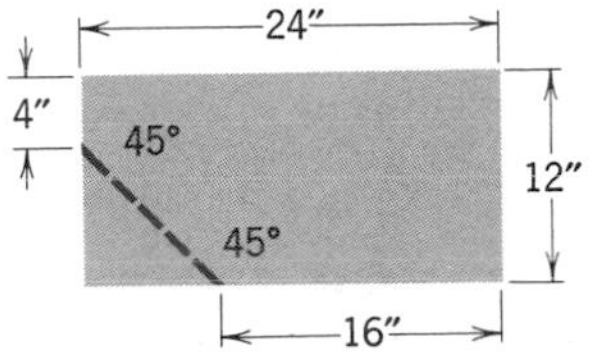

26. How many square feet of dry wall are needed for the wall shown in the figure?

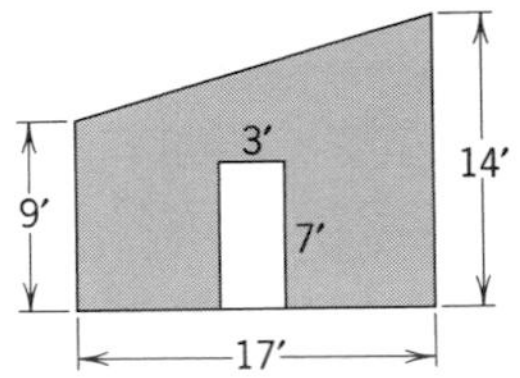

27. A circular saw with a diameter of 18 cm has 22 teeth. What is the spacing between teeth?

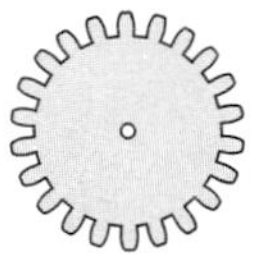

28. Allowing for a 3′ overhang, how long a rafter is needed for the gable of the house in the figure? Round to the nearest tenth.

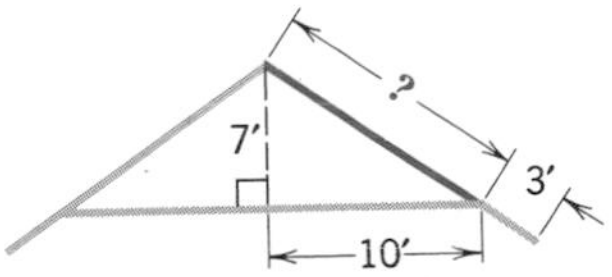

F. Calculator Problems

1. Find the area and perimeter (circumference) of each of the following figures.

(a)

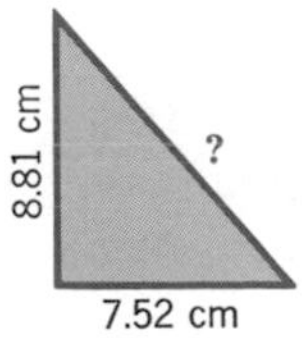

(b)

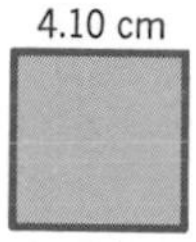

(c)

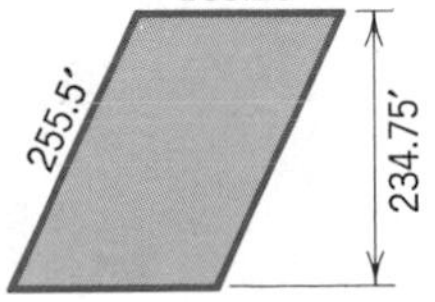

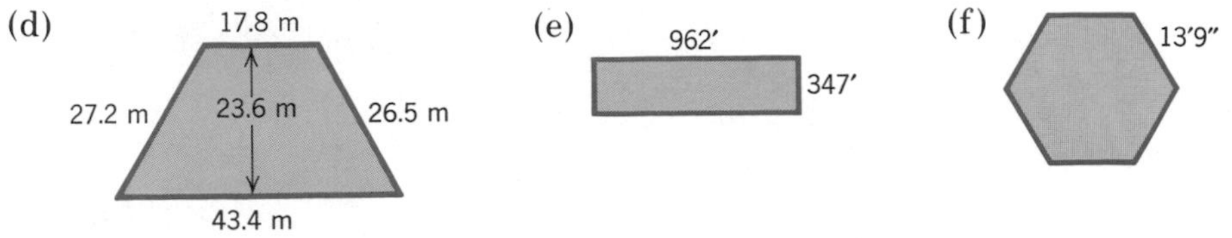

2. Find each missing dimension.

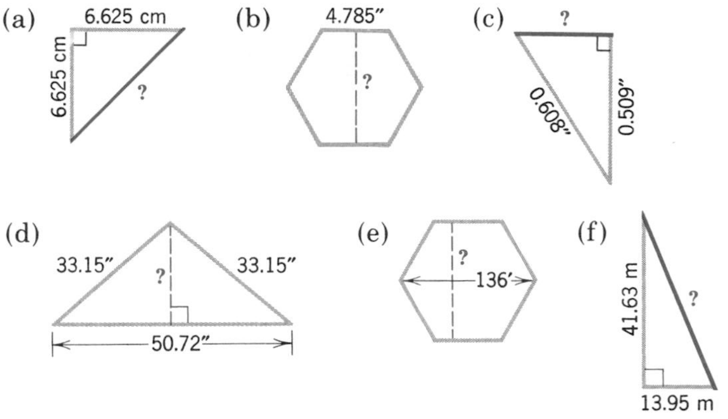

3. How much greater is the cross-sectional area of a pipe that has a 6.015 cm interior diameter and a 6.838 cm outside diameter than a pipe with the same interior diameter and a 6.575 cm outside diameter?

4. If 12 bolts are spaced equally on a steel plate 12.35′ in diameter, what is the spacing between the bolts?

5. A coil of wire has an average diameter of 12.35″. How many feet of wire are there if it contains 168 wraps?

6. Find the missing dimensions x and y on the bridge truss shown.

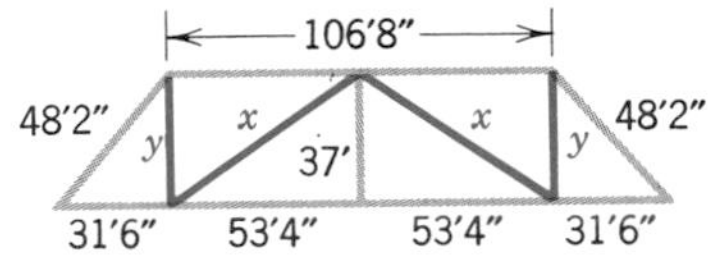

When you have finished these problems, check your answers on p. 555, then continue in **49** with the study of solid figures.

7-3 SOLID FIGURES

49 Plane figures have only two dimensions: length and width or base and height. A plane figure can be drawn exactly on a flat plane surface. Solid figures are three-dimensional: length, width, and height. Making an exact model of a solid figure requires shaping it in clay, wood, or paper. A square is a two-dimensional or plane figure; a cube is a three-dimensional or solid figure.

Trades workers encounter solid figures in the form of tanks, pipes, ducts, boxes, and buildings. They must be able to identify the solid and its component parts, and compute its surface area and volume.

Prisms

A *prism* is a solid figure having at least one pair of parallel surfaces that create a uniform cross section. The figure shows a hexagonal prism.

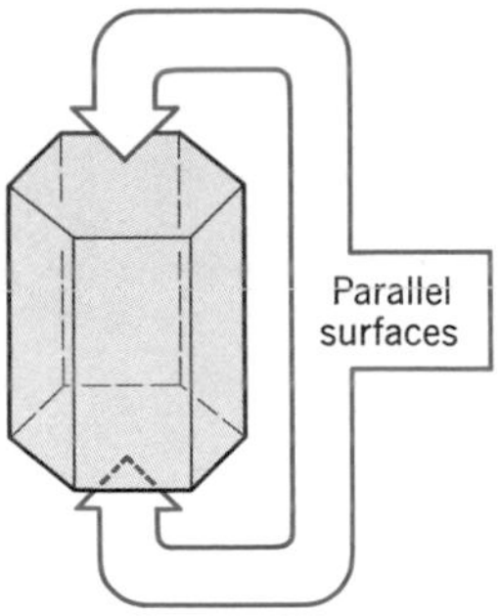

Hexagonal prism

Cutting the prism anywhere parallel to the hexagonal surfaces produces the same hexagonal cross section.

All of the polygons that form the prism are called *faces*. The faces that create the uniform cross section are called the *bases*. They give the prism its name. The others are called lateral faces.

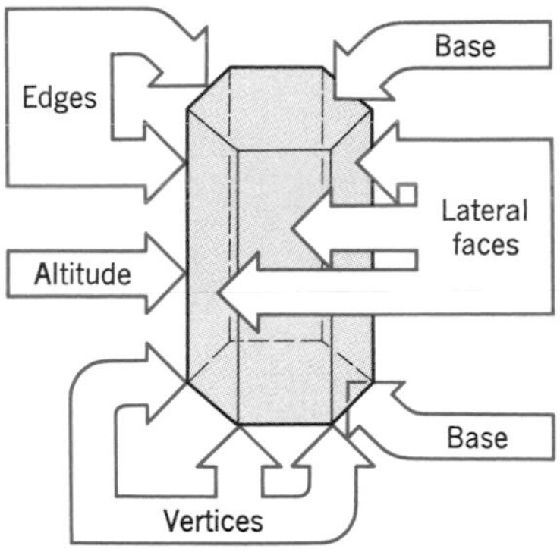

The sides of the polygons are the *edges* of the prism, and the corners are still referred to as *vertices*. The perpendicular distance between the bases is called the *altitude* of the prism.

In a *right prism* the lateral edges are perpendicular to the bases and the lateral faces are rectangles. In this section you will learn about right prisms only.

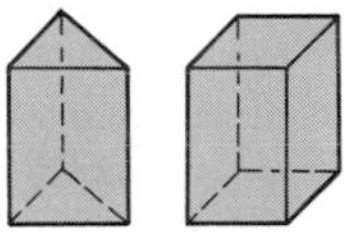

Right prisms

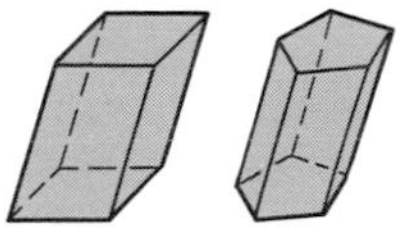

Not right prisms

The components of a prism are its faces, vertices, and edges. Count the number of faces, vertices and edges in this prism:

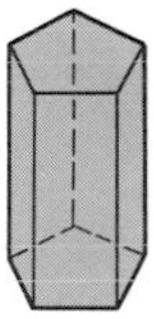

Number of faces = ________

Number of vertices = ________

Number of edges = ________

Check your answer in **50**.

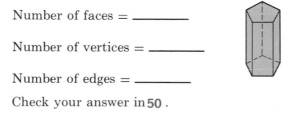

50 There are 7 faces (2 bases and 5 lateral faces), 15 edges, and 10 vertices on this prism.

Here are some examples of simple prisms.

Rectangular Prism

The most common solid in practical work is the *rectangular prism*. All opposite faces are parallel, so any pair can be chosen as bases. Every face is a rectangle. The angles at all vertices are right angles.

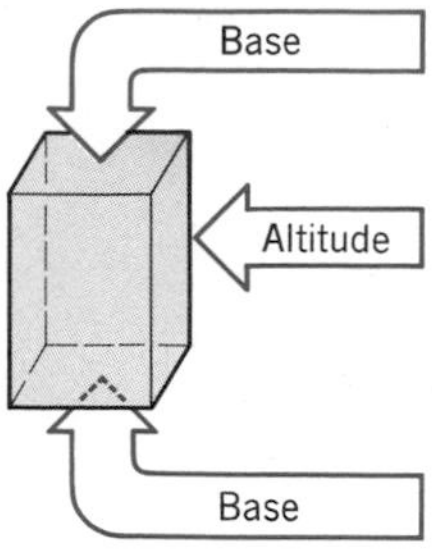

Cube

A *cube* is a rectangular prism in which all edges are the same length.

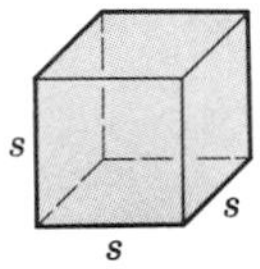

Triangular Prism

In a *triangular prism* the bases are triangular.

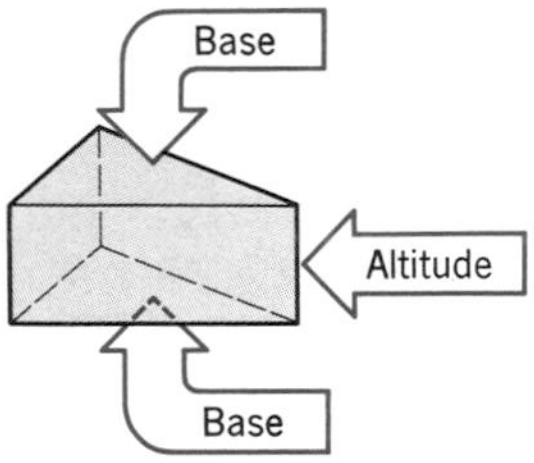

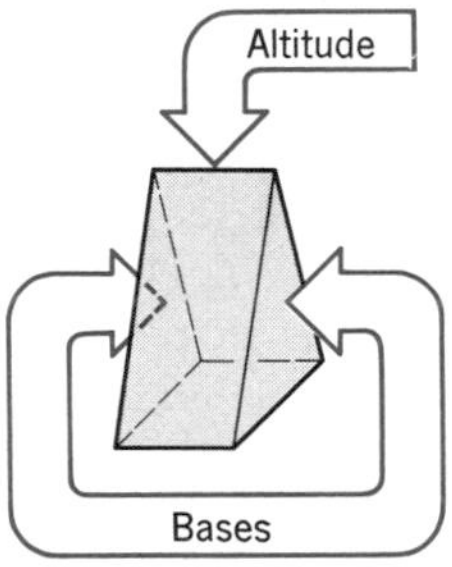

Trapezoidal Prism

In a *trapezoidal prism* the bases are identical trapezoids.

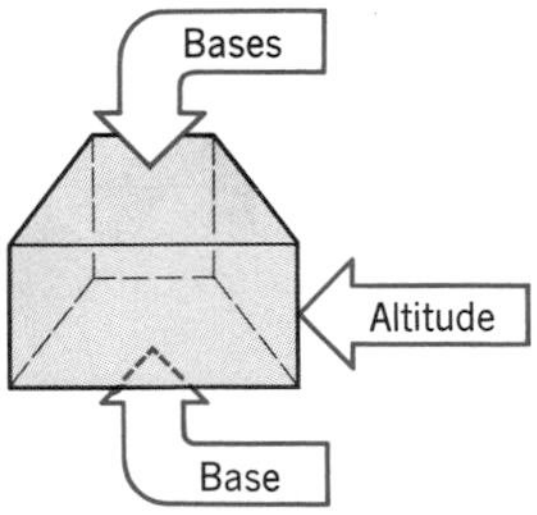

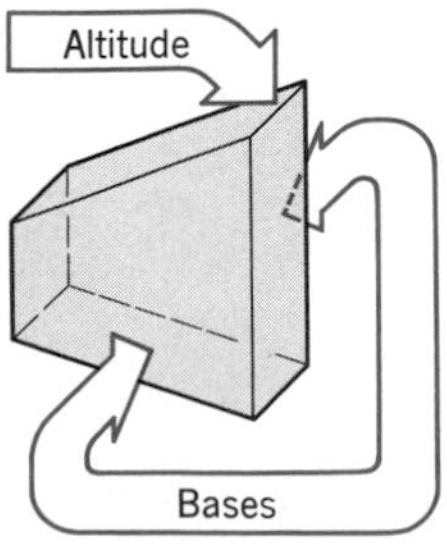

We know that the trapezoids and not the other faces are the bases because cutting the prism anywhere parallel to the trapezoid faces produces another trapezoidal prism.

Three important quantities may be calculated for any prism: the lateral surface area, the total surface area, and the volume. The following formulas apply to all right prisms.

PRISMS

Lateral surface area	$L = ph$		h = altitude
Total surface area	$S = L + 2A$		p = perimeter of the base
Volume	$V = hA$		A = area of the base
	For a right prism only		

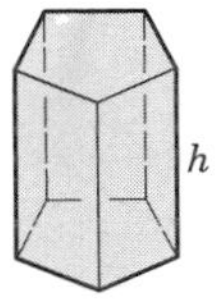

The lateral surface area L is the area of all surfaces *excluding* the two bases.

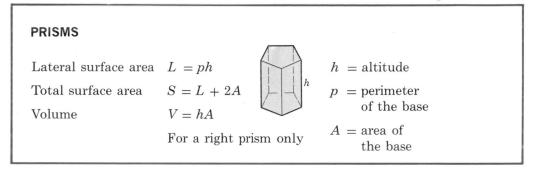

The total surface area S is the lateral surface area *plus* the area of the two bases.

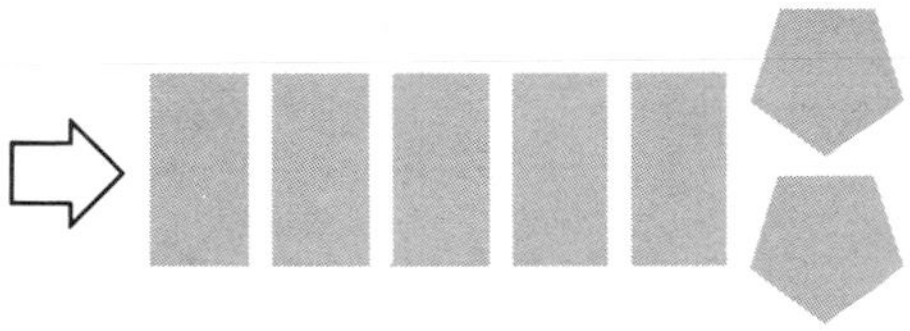

The *volume* or capacity of a prism is the total amount of space inside it. Volume is measured in cubic units. (You may want to review these units in Chapter 5, page 194).

Let's look at an example. Find the lateral surface area, total surface area, and the volume of the triangular cross-section duct shown.

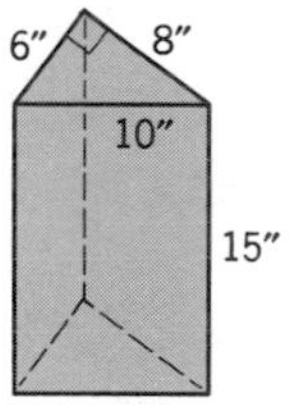

(a) The perimeter of the base is

$p = 6'' + 8'' + 10'' = 24''$ and $h = 15''$.

Therefore the lateral surface area is

$L = ph = (24 \text{ in.})(15 \text{ in.})$

$L = 360$ sq in.

(b) In this prism the bases are right triangles, so that the area of each base is

$$A = \frac{1}{2} b\, h$$

$$A = \frac{1}{2}(6 \text{ in.})(8 \text{ in.})$$

$A = 24$ sq in.

Therefore the total surface area is

$S = L + 2A$
$S = 360$ sq in. $+ 2$ (24 sq in.)
$S = 408$ sq in.

(c) The volume of the prism is

$$V = hA$$
$$V = (15 \text{ in.})(24 \text{ sq in.})$$
$$V = 360 \text{ cubic inches} = 360 \text{ cu in.}$$

Your turn. Find L, S, and V for this trapezoidal right prism.

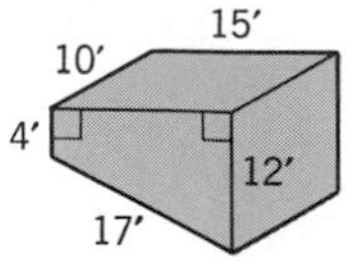

Check your work in **51**.

---

51 $h = 10$ ft and $p = 4 \text{ ft} + 15 \text{ ft} + 12 \text{ ft} + 17 \text{ ft} = 48 \text{ ft}$

The area of each trapezoidal base is

$$A = \frac{1}{2}(b_1 + b_2)h$$

$$A = \frac{1}{2}(4 \text{ ft} + 12 \text{ ft})(15 \text{ ft})$$

$$A = 120 \text{ sq ft}$$

Therefore,

$$L = ph$$
$$L = (48 \text{ ft})(10 \text{ ft})$$
$$L = 480 \text{ sq ft}$$

$$S = L + 2A$$
$$S = 480 \text{ sq ft} + 2(120 \text{ sq ft})$$
$$S = 720 \text{ sq ft}$$

$$V = hA$$
$$V = (10 \text{ ft})(120 \text{ sq ft})$$
$$V = 1200 \text{ cu ft}$$

Find the lateral surface area, total surface area, and volume for each of the following prisms.

(a)

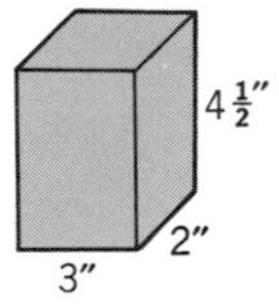

(b)

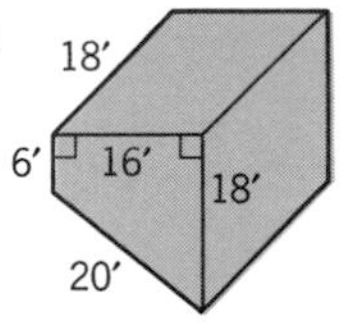

(c)

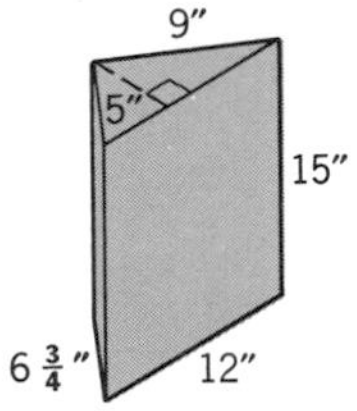

(d)

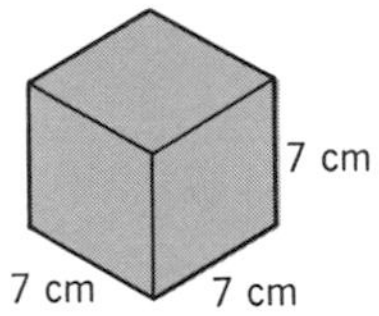

Check your work in **52**.

---

52 (a) $h = 4\frac{1}{2}$ in.

$p = 3'' + 2'' + 3'' + 2'' = 10$ in. $\quad A = (3 \text{ in.})(2 \text{ in.}) = 6$ sq in.

$$L = ph = (10 \text{ in.})(4.5 \text{ in.})$$
$$L = 45 \text{ sq in.}$$

$$S = L + 2A$$
$$S = 45 \text{ sq in.} + 2(6 \text{ sq in.})$$
$$S = 57 \text{ sq in.}$$

$$V = Ah$$
$$V = (6 \text{ sq in.})(4.5 \text{ in.})$$
$$V = 27 \text{ cu in.}$$

(b) $L = 1080$ sq ft

$S = 1464$ sq ft

$V = 3456$ cu ft

(c) $L = 416.25$ sq in.

$S = 476.25$ sq in.

$V = 450$ cu in.

(d) $L = 196$ sq cm   For a cube, $L = 4s^2$

$S = 294$ sq cm $S = 6s^2$

$V = 343$ cu cm $V = s^3$

In practical problems it is often necessary to convert volume units from cu in. or cu ft to gallons or similar units. Use the following conversion factors and set up unity fractions as shown in Chapter 5.

1 cu ft = 1728 cu in.	1 cu yd = 27 cu ft
= 7.48 gallon	1 cu in = 16.38 cu cm
= 28.3 liter	
= 0.0283 cu meter	1 gal = 231 cu in.
= 0.806 bushel	

For example, calculate the volume of water needed to fill this swimming pool in gallons. Round to the nearest hundred gallons.

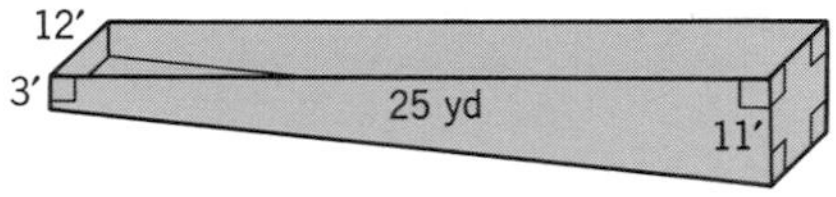

Check your calculations in 53.

---

53 $V = Ah$

$$V = \left(\frac{3' + 11'}{2}\right)(75 \text{ ft})(12 \text{ ft})$$

$$V = (7 \text{ ft})(75 \text{ ft})(12 \text{ ft})$$

$$V = 6300 \text{ cu ft}$$

In gallons

$$V = 6300 \text{ cu ft} \times \frac{7.48 \text{ gal}}{1 \text{ cu ft}}$$

$V = 47{,}124$ gal   or   47,100 gal, rounded

Using a Calculator

Key	Display
C	0.
3	3.
+	3.
1 1	11.
÷	14.
2	2.
×	7.
7 5	75.
×	525.
1 2	12.
=	6300.

In many practical problems the prisms encountered have irregular polygons for their base. For example, find the volume of the slotted bar shown. If steel weighs 0.283 lb per cu in., what would be the weight of this bar? Round to one decimal place.

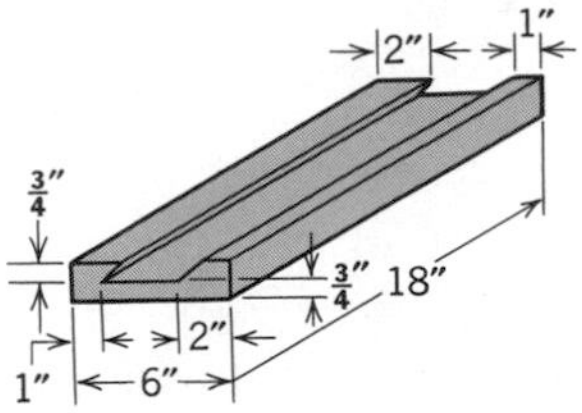

Try it. We will work it out in detail in **54**.

---

54 First, we must calculate the volume of this prism. To find this, divide the base area into more familiar shapes.

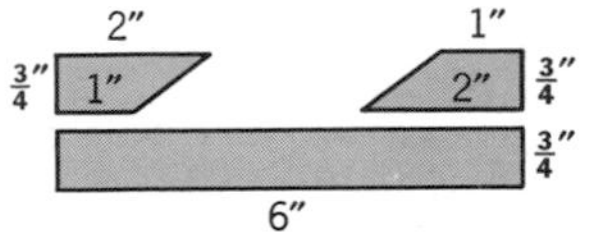

The total area of the three shapes is

$$A = \left(\frac{1'' + 2''}{2}\right)\left(\frac{3''}{4}\right) + \left(\frac{1'' + 2''}{2}\right)\left(\frac{3''}{4}\right) + (6'')\left(\frac{3''}{4}\right)$$

$A = 6.75$ sq in.

and the volume of the bar is

$V = A \cdot h$
$V = (6.75 \text{ sq in.})(18 \text{ in.})$
$V = 121.5$ cu in.

The weight of the bar is

$$W = 121.5 \text{ cu in.} \times 0.283 \frac{\text{lb}}{\text{cu in.}}$$

$W = 34.3845$ lb, or 34.4 lb, rounded

The most common source of errors in most calculations of this kind is carelessness. Organize your work neatly. Work slowly and carefully.

1. Find the volume of each of the following shapes.

(a)

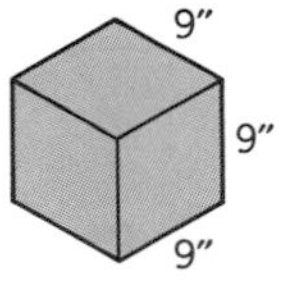

(b)

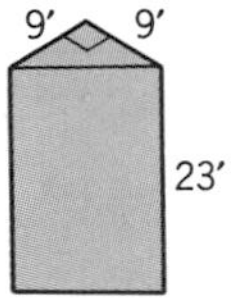

(c)

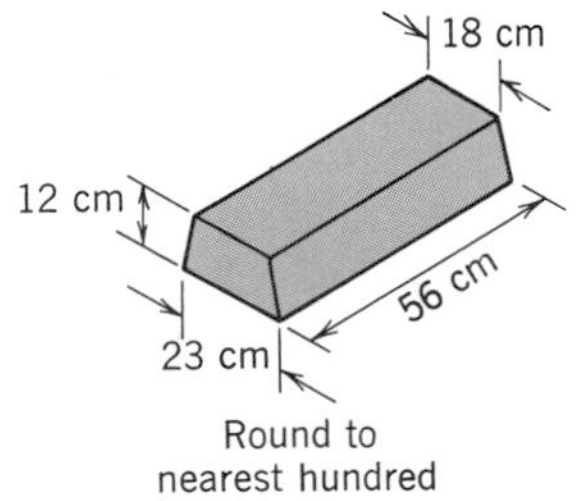

(d)

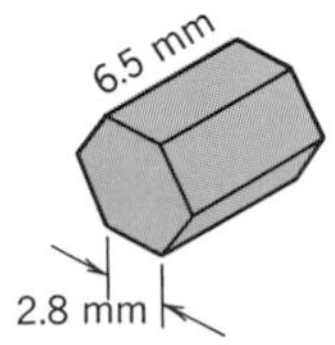

(e)

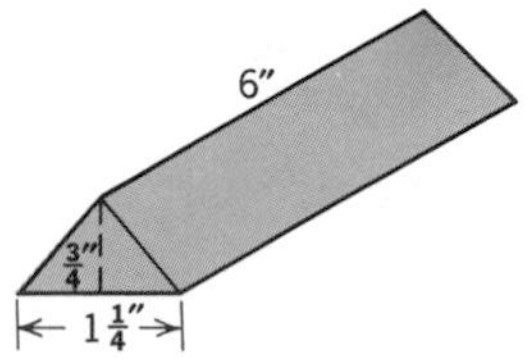

(f)

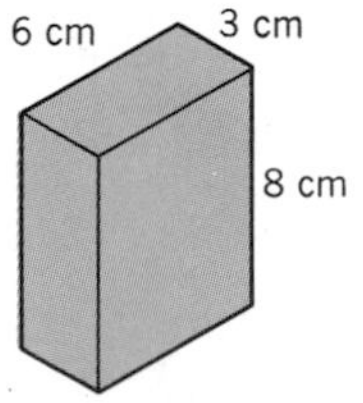

(g)

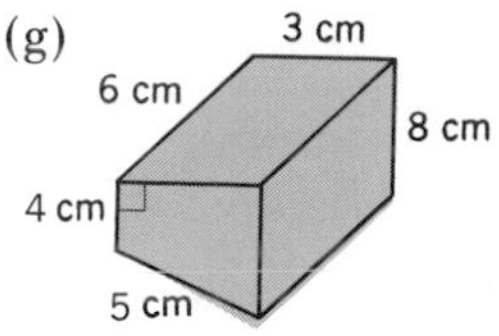

(h)

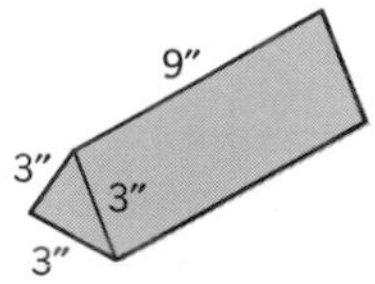

(Round to nearest cu in.)

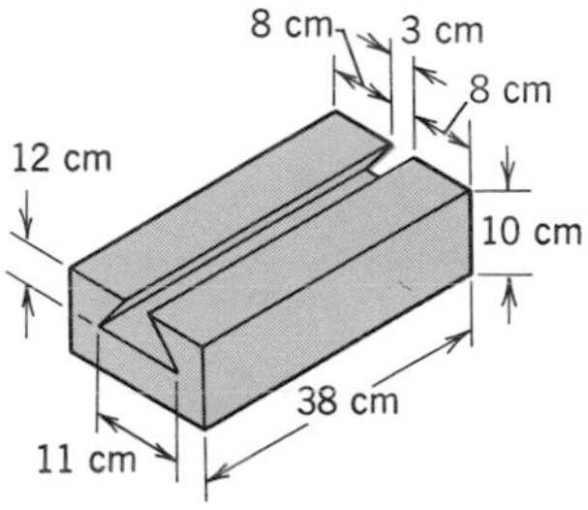

2. Find the total surface area of (a) and (d) from Problem 1.
3. Find the lateral surface area of (f), (g), and (h) from Problem 1.
4. How many cubic *yards* of concrete does it take to pour a square pillar 2 ft on a side if it is 12 ft high? (Round to one decimal place.)
5. What is the capacity to the nearest gallon of a septic tank in the shape of a rectangular prism 12′ by 16′ by 6′?
6. What is the weight of the steel *V* block in the figure at 0.0173 lb per cu cm? (Round to the nearest tenth.)
7. Find the weight of the piece of brass pictured at 0.2963 lb per cu in. (Round to the nearest tenth.)

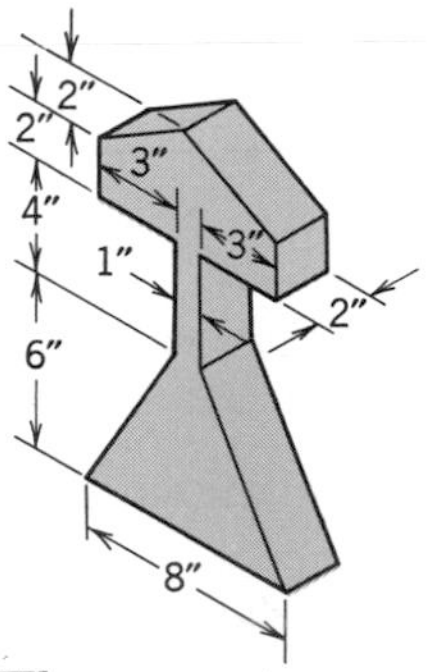

The correct answers are in **55**

---

55 1. (a) 729 cu in. (b) 931.5 cu ft (c) 13,800 cu cm
(d) 132.4 cu mm (e) 2.81 cu in. (f) 144 cu cm
(g) 108 cu cm (h) 35 cu in. (rounded)

2. 486 sq. in.; 149.9 sq mm
3. 144 sq cm; 120 sq cm; 81 sq in.
4. 1.8 cu yd
5. 8617 gal
6. 69.7 lb.
7. 30.8 lb

Problem 7 is tricky. Try separating it into four pieces like this:

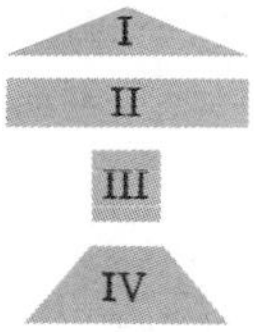

Vol. I = 14 cu in.
Vol. II = 28 cu in.
Vol. III = 8 cu in.
Vol. IV = 54 cu in.

A *pyramid* is a solid object with one base and three or more *lateral faces* that taper to a single point opposite the base. This single point is called the *apex*.

A *right pyramid* is one whose base is a regular polygon, and whose apex is centered over the base. We will examine only right pyramids here.

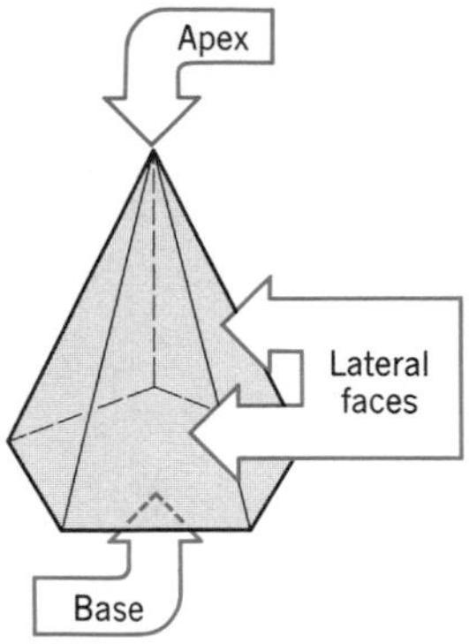

The *altitude* of a pyramid is the perpendicular distance from the apex to the base. The *slant height* of the pyramid is the height of one of the lateral faces.

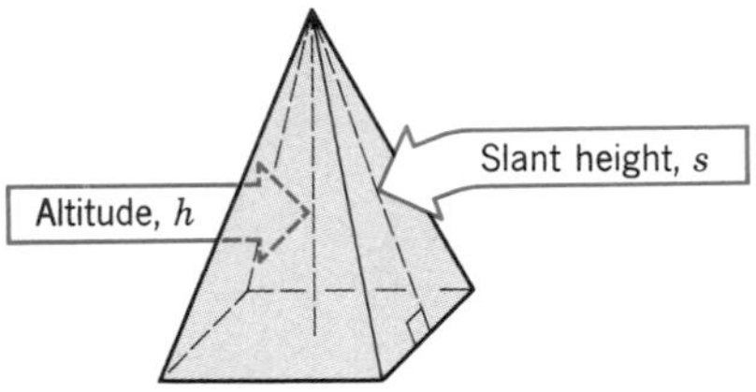

Several kinds of right pyramids can be constructed:

Square Pyramid

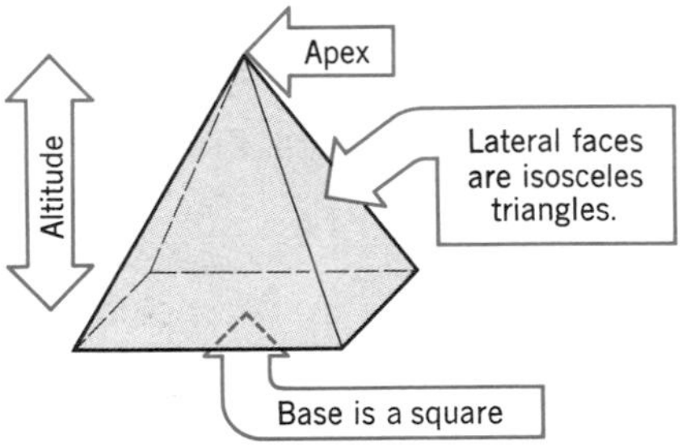

Triangular Pyramid or **Tetrahedron**

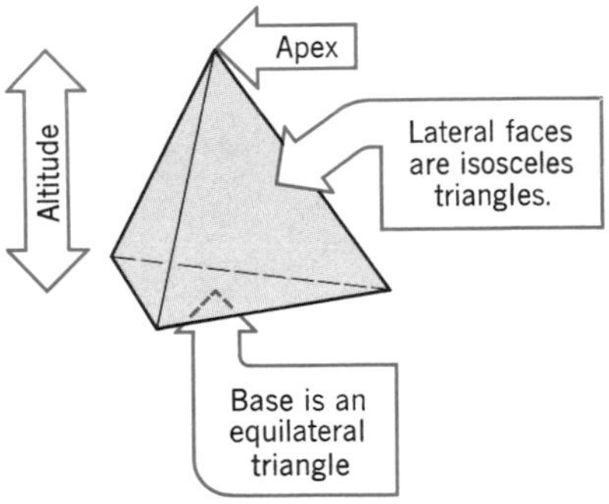

There are several very useful formulas relating to pyramids.

PYRAMIDS

Lateral surface area

$L = \frac{1}{2}ps$

Volume

$V = \frac{1}{3}Ah$

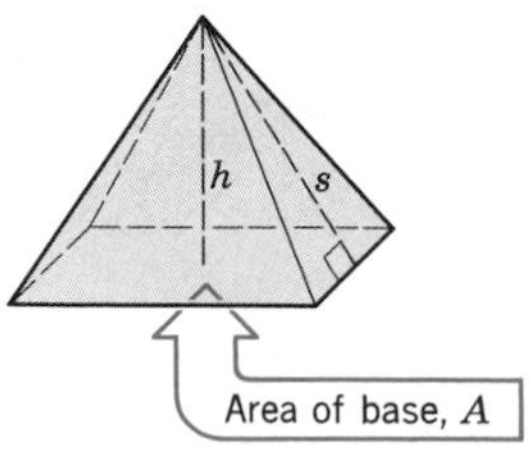

where p is the perimeter of the base,
s is the slant height of a lateral face,
A is the area of the base,
and h is the altitude.

For example, find the lateral surface area and volume of this square pyramid:

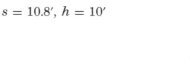

$s = 10.8'$, $h = 10'$

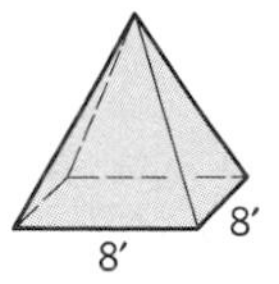

$p = 8' + 8' + 8' + 8' = 32'$ $\qquad A = (8')(8') = 64$ sq ft

The lateral surface area is

$L = \frac{1}{2}ps$
$L = \frac{1}{2}(32 \text{ ft})(10.8 \text{ ft})$
$L = 172.8$ sq ft $\qquad$ This is the area of the four triangular lateral faces.

The volume is

$V = \frac{1}{3}Ah$
$V = \frac{1}{3}(64 \text{ sq ft})(10 \text{ ft})$
$V = 213.3$ cu ft, rounded

Your turn. Find the lateral surface area and volume of this triangular prism. (Remember, the base is an equilateral triangle.) Round to the nearest whole unit.

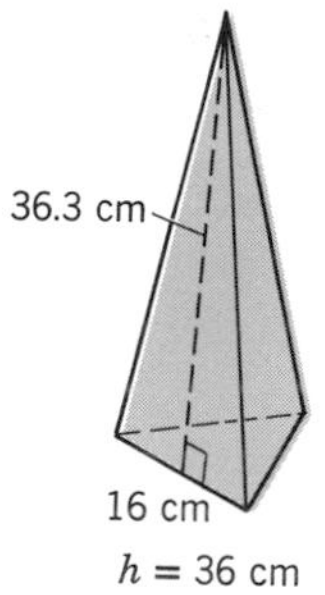

Our solution is in 56.

---

56 $p = 16 \text{ cm} + 16 \text{ cm} + 16 \text{ cm}$
$p = 48$ cm $\qquad$ perimeter of the base.
$s = 36.3$ cm

Therefore, $L = \frac{1}{2}ps$
$L = \frac{1}{2}(48 \text{ cm})(36.3 \text{ cm})$
$L = 871.2$ sq cm or 871 sq cm, rounded

$A = 0.433s^2$ for an equilateral triangle
$A = 0.433(16 \text{ cm})^2$
$A = 110.8$ sq cm $\qquad$ or $\qquad$ 111 sq cm, rounded

The volume is

$V = \frac{1}{3}Ah$
$V = \frac{1}{3}(110.8 \text{ sq cm})(36 \text{ cm})$
$V = 1330$ cu cm, rounded

More practice.

1. Find the lateral surface area and volume of each pyramid. (Round to one decimal place.)

(a) 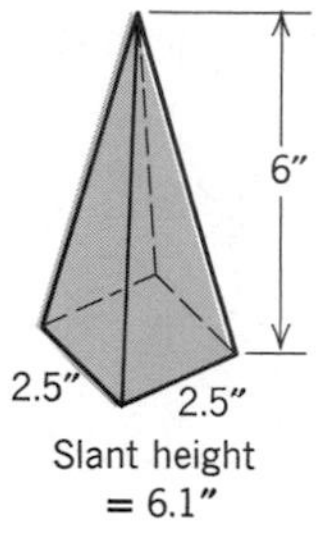

(b) 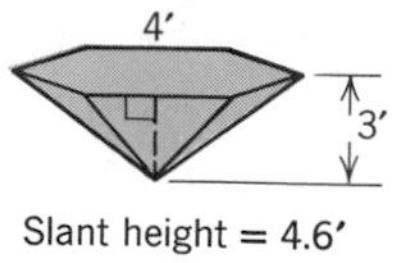

(c) 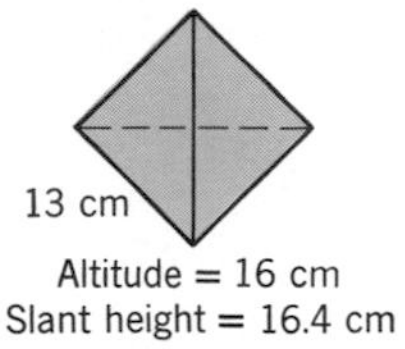

2. Find the volume of an octagonal pyramid if the area of the base is 245 sq in. and the altitude is 16″. (Round to nearest cu in.)

The correct answers are in **57**.

---

57 1. (a) 30.5 sq in.; 12.5 cu in. (b) 55.2 sq ft; 41.6 cu ft

(c) 319.8 sq cm; 390.3 cu cm

2. 1307 cu in.

Cylinders

A *cylinder* is a solid object with two identical circular bases. The *altitude* of a cylinder is the perpendicular distance between the bases.

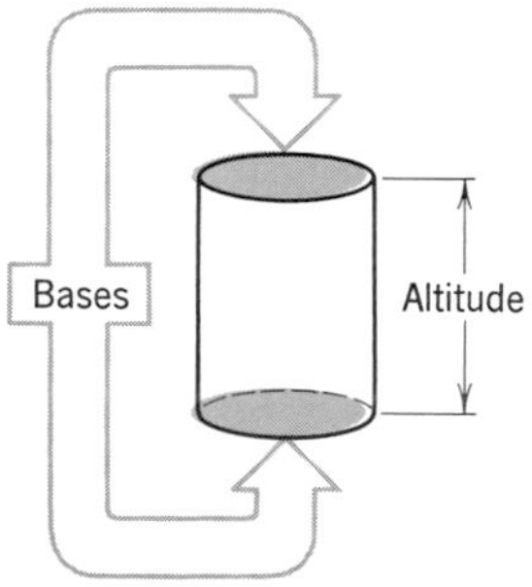

A *right* cylinder is one whose curved side walls are perpendicular to its circular base. Whenever we mention the radius, diameter, or circumference of a cylinder we are referring to those dimensions of its circular base.

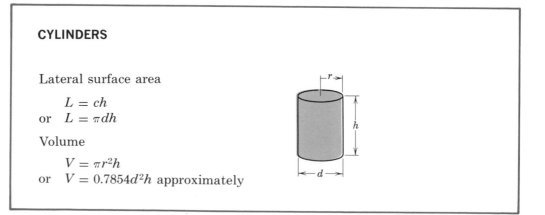

Two important formulas enable us to find the lateral surface area and volume of a cylinder.

CYLINDERS

Lateral surface area

$L = ch$

or $L = \pi dh$

Volume

$V = \pi r^2 h$

or $V = 0.7854d^2h$ approximately

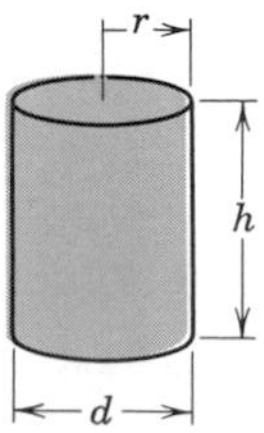

where c is the circumference of the base,
r is the radius of the base,
d is the diameter of the base,
and h is the altitude of the cylinder.

You can visualize the lateral surface area by imagining the cylinder wall flattened like this:

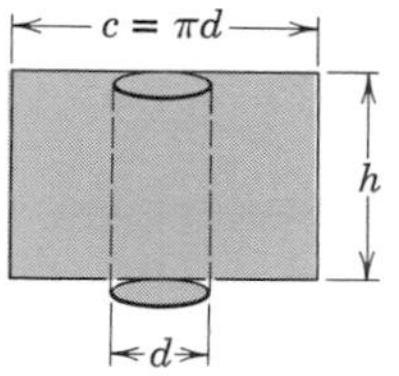

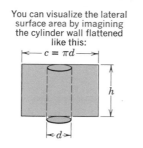

Use these formulas to find the lateral surface area and volume of a cylindrical container whose altitude is 24″ and whose diameter is 22.″

$d = 22''$ $\quad\quad\quad$ $r = 11''$ $\quad\quad\quad$ $h = 24''$

Then the lateral surface area is

$L = \pi dh$

$L = (3.1416)(22 \text{ in.})(24 \text{ in.})$

$L = 1658.8$ sq in. $\quad$ or $\quad$ 1659 sq in., rounded

and the volume is

$V = \pi r^2 h$

$V = (3.1416)(11 \text{ in.})^2(24 \text{ in.})$

$V = (3.1416)(121 \text{ sq in.})(24 \text{ in.})$

$V = 9123.2$ cu in. $\quad$ or $\quad$ 9123 cu in., rounded

Find the lateral surface area and the volume in gallons of the cylindrical storage tank shown. Round answers to the nearest unit.

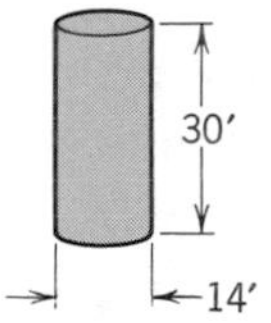

Check your work in **58**.

---

58 $\quad L = \pi dh$

$L = (3.1416)(14 \text{ ft})(30 \text{ ft})$

$L = 1319.472$ sq ft $\quad$ or $\quad$ 1319 sq ft, rounded

$V = 0.7854d^2h$ approximately

$V = (0.7854)(14 \text{ ft})^2(30 \text{ ft})$

$V = (0.7854)(196 \text{ sq ft})(30 \text{ ft})$

$V = 4618.2$ cu ft $\quad$ or $\quad$ 4618 cu ft, rounded

In gallons

$$V = 4618.2 \text{ cu ft} \times \frac{7.48 \text{ gal}}{1 \text{ cu ft}}$$

$V = 34{,}544.1$ gal $\quad$ or $\quad$ 34,544 gal, rounded

Try these problems for practice in working with cylinders.

1. Find the lateral surface area and volume of each cylinder. (Round to one decimal place.)

(a)

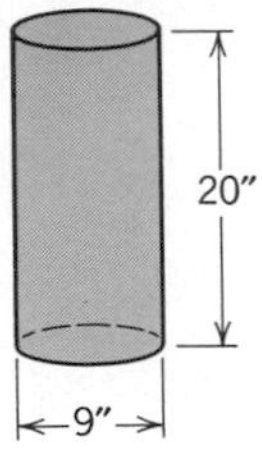

(b)

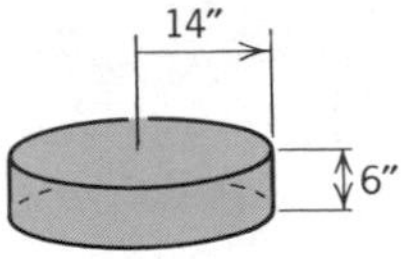

(c)

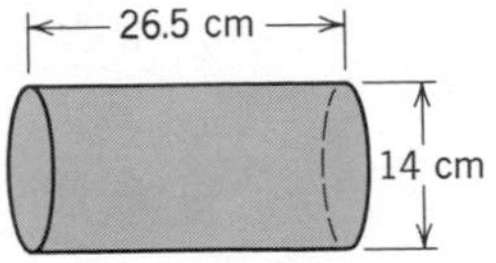

(d) A cylinder 9 yd high and 16 yd in diameter.

2. How many cubic *yards* of concrete will it take to pour a cylindrical column 14′ high with a diameter of 2′? (Round to the nearest tenth.)

3. What is the capacity in gallons of a cylindrical tank with a radius of 8′ and an altitude of 20′? (Round to the nearest gallon.)

4. A pipe 3 inches in diameter and 40′ high is filled to the top with water. If 1 cu ft of water weighs 62.4 lb, what is the weight at the base of the pipe? (Be careful of your units.)

5. The *bottom* and the *inside walls* of the cylindrical tank shown in the figure must be lined with sheet copper. How many square feet of sheet copper are needed? (Round to the nearest tenth.)

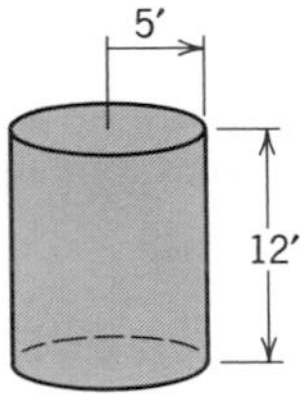

6. How many quarts of paint are needed to cover the outside walls of a cylindrical water tank 10′ in diameter and 10′ in height, if one quart covers 100 sq ft? (Assume you cannot buy a fraction of a quart.)

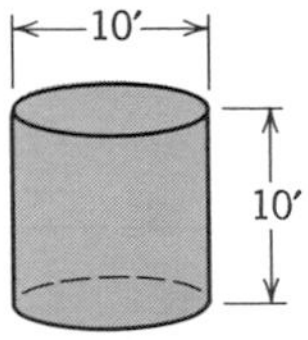

Check your answers in **59**.

59

1. (a) 565.5 sq in., 1272.3 cu in. (b) 527.8 sq in., 3694.5 cu in.
   (c) 1165.5 sq cm, 4079.4 cu cm (d) 452.4 sq yd, 1809.6 cu yd
2. 1.6 cu yd
3. 30,079 gallons, rounded
4. 122.5 lb, rounded
5. 455.5 sq ft
6. 4 qt

On Problem 4, you must change the diameter from 3 in. to 0.25 ft before substituting into the formula.

A *cone* is a pyramid-like solid figure with a circular base. The radius and diameter of a cone refer to its circular base. The *altitude* is the perpendicular distance from the apex to the base. The *slant height* is the apex to base distance along the surface of the cone.

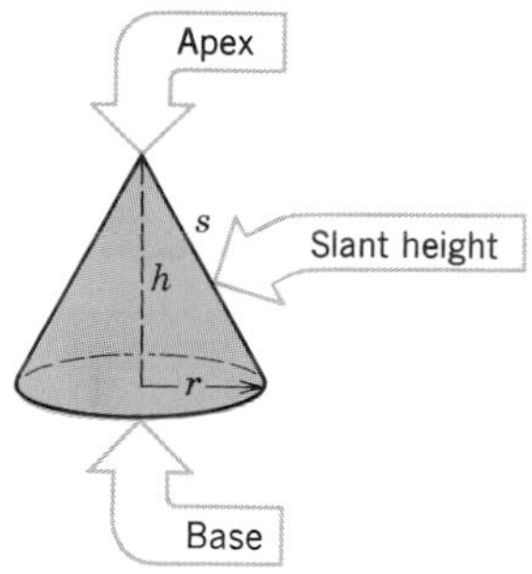

The following formulas enable us to find the lateral surface area and volume of a cone.

CONES

Lateral surface area

$L = \pi r s$

or $L = \frac{1}{2}\pi d s$

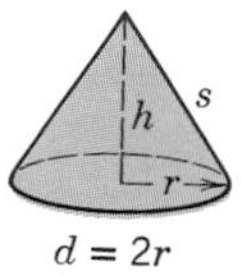

Volume

$V = \frac{1}{3}\pi r^2 h$

or $V \cong 0.2618 d^2 h$

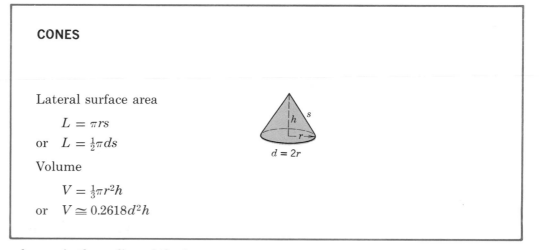

where r is the radius of the base,
h is the altitude,
and s is the slant height.

For example, find the lateral surface area and volume of the cone shown here. (Round to the nearest whole unit.)

$h = 15''$

$s = 18''$

$r = 10''$

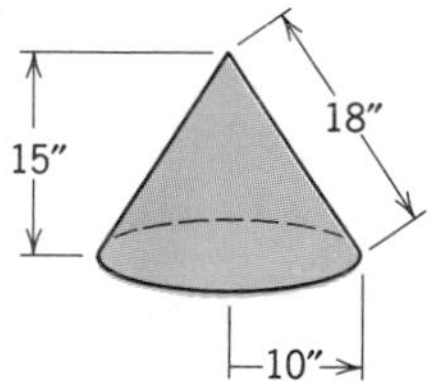

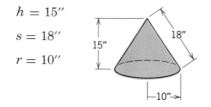

Then the lateral surface area is

$L = \pi r s$
$L = (3.1416)(10 \text{ in.})(18 \text{ in.})$
$L = 565.488$ sq in. or 565 sq in., rounded

and the volume is

$V = \frac{1}{3}\pi r^2 h$
$V = \frac{1}{3}(3.1416)(10 \text{ in.})^2(15 \text{ in.})$
$V = 1570.8$ cu in or 1571 cu in., rounded

Find the lateral surface area and volume of this cone. (Round to the nearest whole unit.)

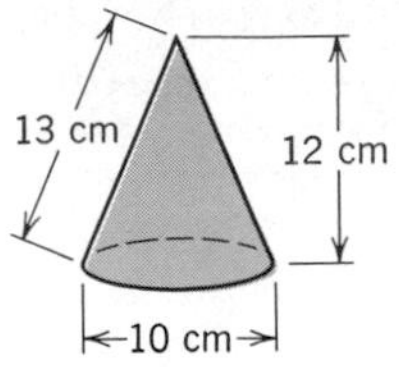

Check your work in **60**.

---

60 $h = 12$ cm

$r = 5$ cm

$s = 13$ cm

$L = \pi rs$
$L = (3.1416)(5 \text{ m})(13 \text{ cm})$
$L = 204.2$ sq cm or 204 sq cm, rounded

$V = \frac{1}{3}\pi r^2 h$
$V = \frac{1}{3}(3.1416)(5 \text{ cm})^2(12 \text{ cm})$
$V = 314.2$ cu cm or 314 cu cm, rounded

You should realize that the altitude, radius, and slant height are always related for any cone.

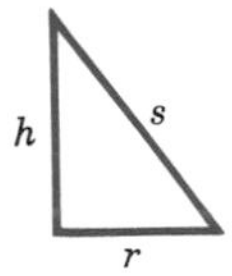

$s^2 = h^2 + r^2$

If you are given any two of these quantities, you can use this formula to find the third.

Notice that the volume of a cone is exactly one-third the volume of the cylinder that just encloses it.

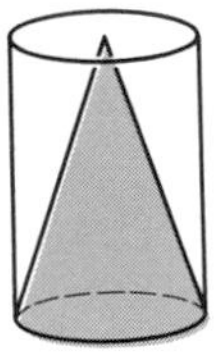

Solve the following practice problems.

1. Find the lateral surface area and volume of each of the following cones. (Round to the nearest tenth.)

   (a)

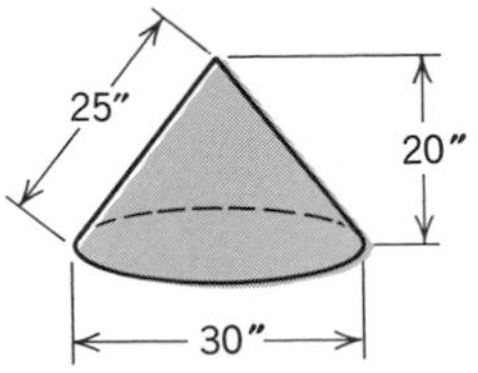

(b)

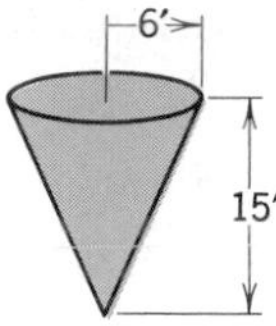

(c)

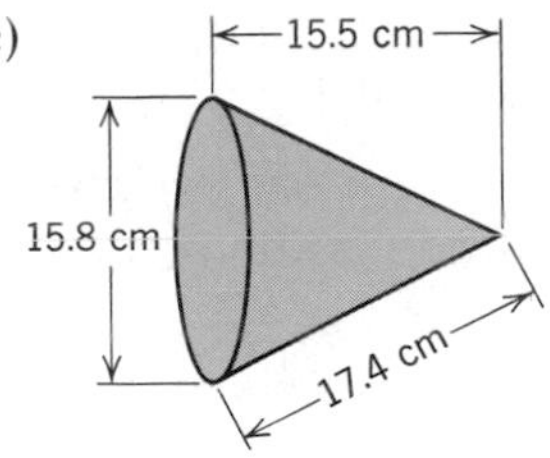

(d) A cone with radius $\frac{3}{4}''$, altitude 1″, and slant height $1\frac{1}{4}''$. (Round to two decimal places.)

2. A pile of sand dumped by a hopper is cone-shaped. If the diameter of the base is 18′3″ and the altitude is 8′6″, how many cubic feet of sand are in the pile? (Round to nearest tenth.)

3. What is the capacity in gallons of a conical drum 4′ high with a radius of 3′? (1 cubic foot = 7.48 gallons.) (Round to the nearest gallon.)

4. Find the weight of the cast iron shape shown at 0.26 lb per cubic inch. (Round to the nearest pound.)

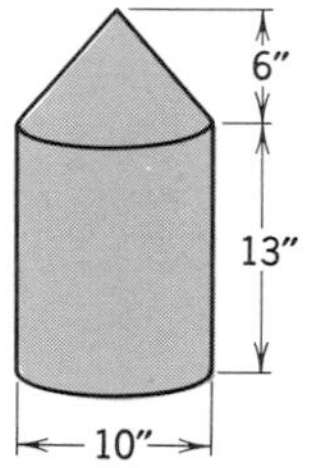

5. Find the capacity in gallons (to the nearest tenth) of a conical oil container 15″ high with a diameter of 16″ (1 gallon = 231 cu in.).

Check your answers in **61**.

---

61 1. (a) 1178.1 sq in., 4712.4 cu in.

(b) 305.4 sq ft, 565.5 cu ft $s^2 = h^2 + r^2$ $s = \sqrt{15^2 + 6^2} \cong 16.2'$

(c) 431.8 sq cm, 1013.0 cu cm

(d) 2.95 sq in., 0.59 cu in.

2. 741.2 cu ft

3. 282 gallons

4. 306 lb

5. 4.4 gallons

Frustums

The *frustum* of a pyramid or cone is the solid figure remaining after the top of the pyramid or cone is cut off parallel to the base.

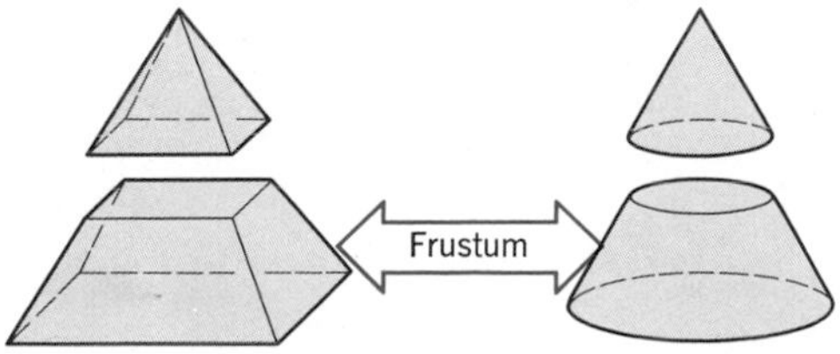

Frustum shapes appear as containers, building foundations, funnels, and as transition sections in ducts.

Every frustum has two bases, upper and lower, that are parallel and different in size. The altitude is the perpendicular distance between bases.

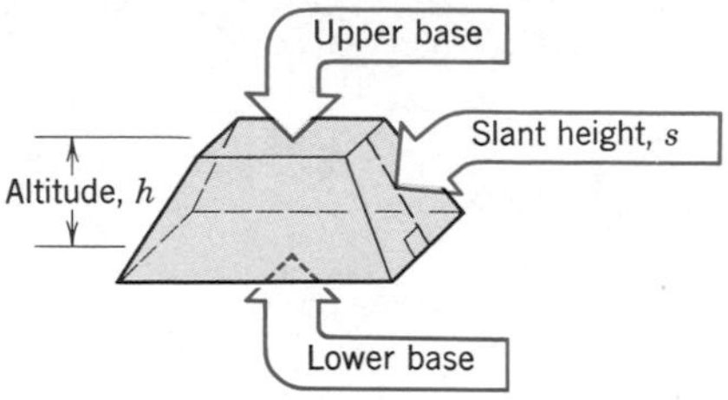

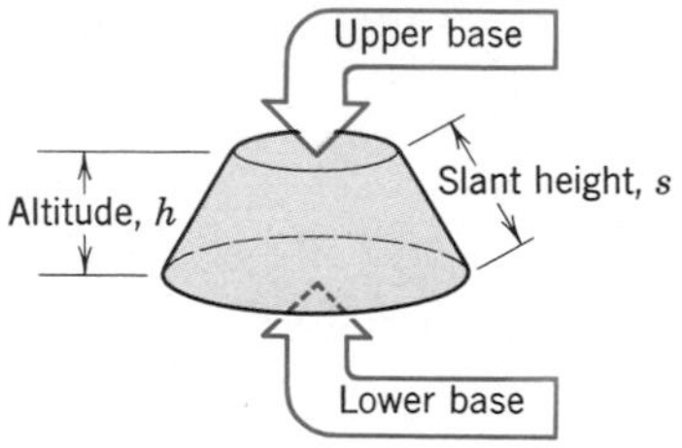

The following formulas enable you to find the lateral area and volume of any frustum, pyramid, or cone.

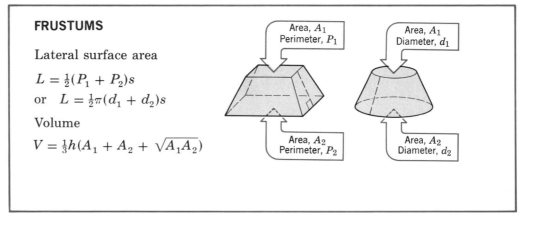

FRUSTUMS

Lateral surface area

$L = \frac{1}{2}(P_1 + P_2)s$

or $L = \frac{1}{2}\pi(d_1 + d_2)s$

Volume

$V = \frac{1}{3}h(A_1 + A_2 + \sqrt{A_1 A_2})$

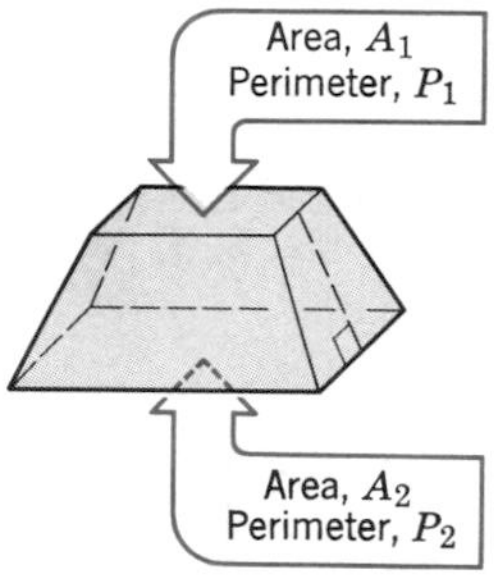

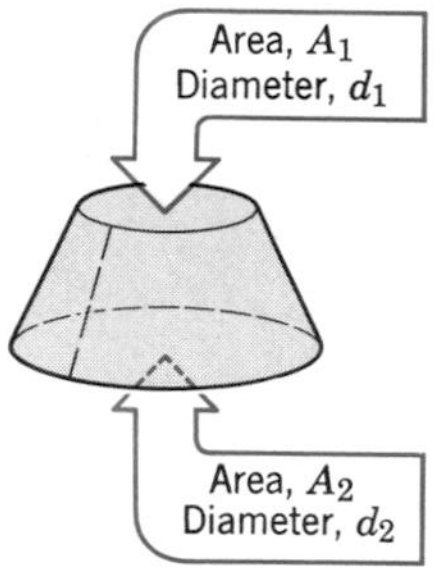

where P_1 and P_2 are the upper and lower perimeter for a pyramid frustum,
d_1 and d_2 are the upper and lower diameters for a cone frustum,
A_1 and A_2 are the upper and lower base areas for any frustum.

For example, find the lateral surface area and volume of the pyramid frustum shown.

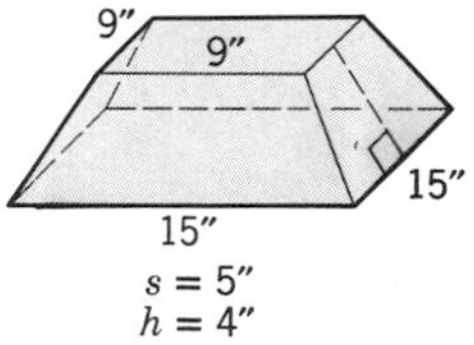

Upper perimeter, $P_1 = 9'' + 9'' + 9'' + 9''$
$= 36''$

Lower perimeter, $P_2 = 15'' + 15'' + 15'' + 15'' = 60''$

$L = \frac{1}{2}(36 \text{ in.} + 60 \text{ in.})(5 \text{ in.})$

$L = 240$ sq in.

Upper area, $A_1 = (9 \text{ in.})(9 \text{ in.}) = 81$ sq in.

Lower area, $A_2 = (15 \text{ in.})(15 \text{ in.}) = 225$ sq in.

$V = \frac{1}{3}(4 \text{ in.})(81 \text{ sq in.} + 225 \text{ sq in.} + \sqrt{81 \cdot 225})$

$V = \frac{1}{3}(4 \text{ in.})(81 + 225 + 135)$ sq in.

$V = 588$ cu in.

Using a Calculator

Key	Display	
C	0.	
8 1	81.	
×	81.	
2 2 5	225.	
=	18225.	
√	135.	
+	135.	
2 2 5	225.	
+	360.	
8 1	81.	
×	441.	
4	4.	
÷	1764.	
3	3.	
=	588.	Volume

Find the *total* surface area and volume of the cone frustum shown. Use $\pi = 3.14$ and round to one decimal place.

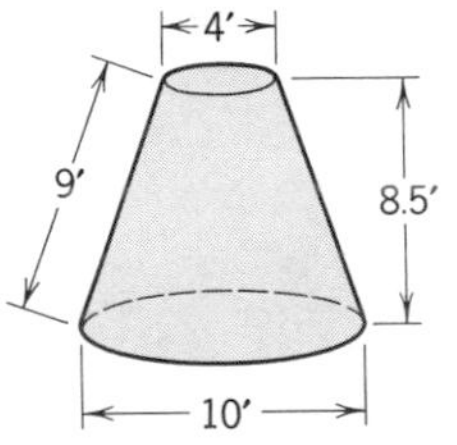

Our complete solution is in **62**.

I tried to solve the volume problem at the bottom of frame 61 on my calculator, but I couldn't figure out the correct order to push the calculator keys. What do I do now?

Follow our calculator example on that problem. Simply pushing buttons on a calculator won't solve any problem. Before you use a calculator, you must be certain you understand the formula and can do the arithmetic by hand. Use a calculator to save time, but use it only after you already know how to do the calculation.

You do the thinking and the little box will do the arithmetic.

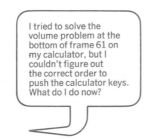

62 $d_1 = 4'$, $\quad d_2 = 10'$, $\quad s = 9'$, $\quad h = 8.5'$

The lateral surface area is

$L = \frac{1}{2}\pi(d_1 + d_2)s$
$L = \frac{1}{2}(3.14)(4\text{ ft} + 10\text{ ft})(9\text{ ft})$
$L = \frac{1}{2}(3.14)(14\text{ ft})(9\text{ ft})$
$L = 197.82\text{ sq ft}$

The total surface area is

$T = L + \pi r_1^2 + \pi r_2^2$
$T = 197.82\text{ sq ft} + (3.14)(2\text{ ft})^2 + (3.14)(5\text{ ft})^2$
$T = 288.9\text{ sq ft}$, rounded

Upper area, $A_1 = \pi r_1^2 = (3.14)(2\text{ ft})^2$
$= 12.56\text{ sq ft}$

Lower area, $A_2 = \pi r_2^2 = (3.14)(5\text{ ft})^2$
$= 78.5\text{ sq ft}$

The volume of the frustum is

$V = \frac{1}{3}h(A_1 + A_2 + \sqrt{A_1 A_2})$
$V = \frac{1}{3}(8.5\text{ ft})(12.56 + 78.5 + \sqrt{12.56 \cdot 78.5})\text{ sq ft}$
$V = \frac{1}{3}(8.5\text{ ft})(12.56 + 78.5 + 31.4)\text{ sq ft}$
$V = \frac{1}{3}(8.5\text{ ft})(122.46\text{ sq ft})$
$V = 347.0\text{ cu ft}$, rounded

Practice with these problems.

1. Find the lateral surface area and volume of each of the following frustums. (Round to the nearest whole unit and use $\pi = 3.14$.)

(a)

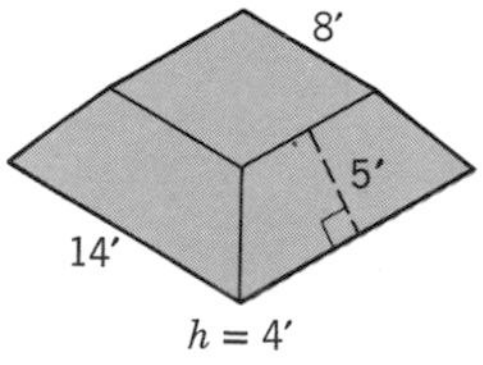

(b)

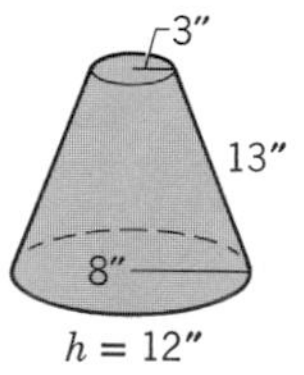

(c)

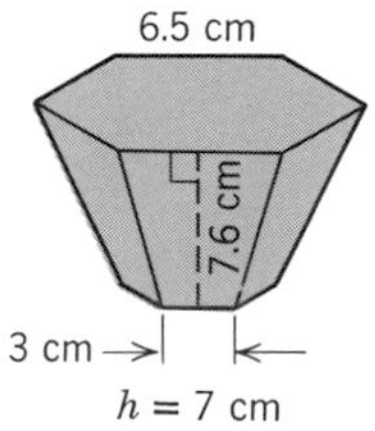

(d)

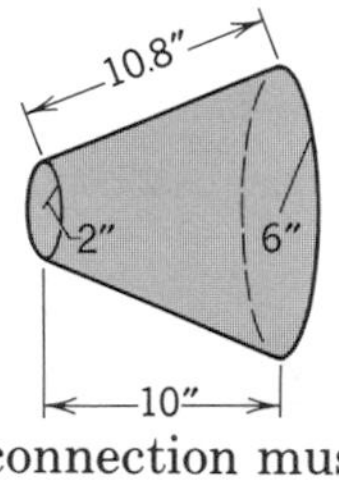

(e)

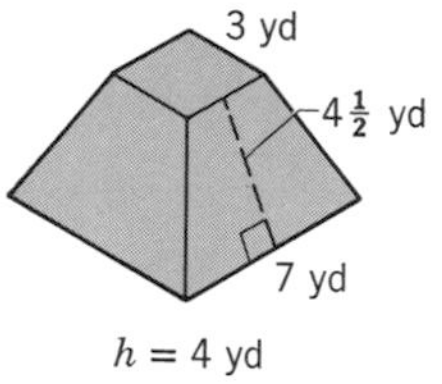

2. A connection must be made between two square vent openings in an air conditioning system. Find the total amount of sheet metal needed using the dimensions shown in the figure.

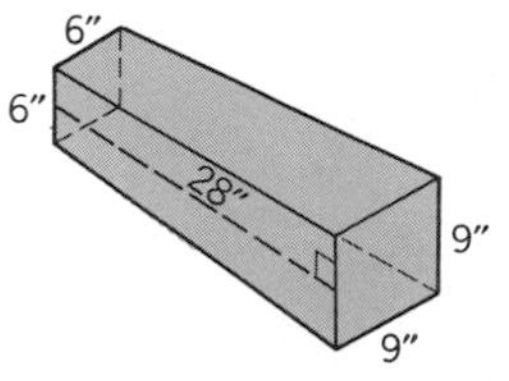

3. How many cubic *yards* of concrete are needed to pour the foundation shown in the figure? (Watch your units!) Round to the nearest hundredth.

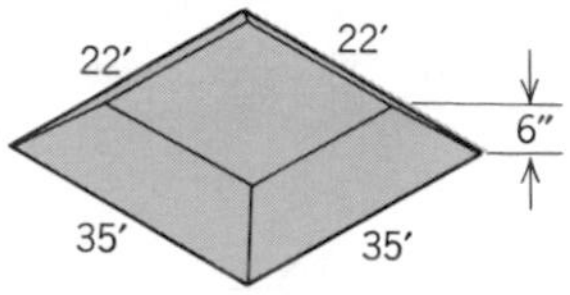

4. What is the capacity in gallons of the oil can shown in the figure? (1 gallon = 231 in.^{3}) Round to the nearest tenth.

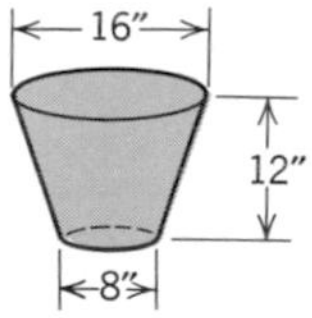

Check your answers in **63**.

---

63 1. (a) 220 sq ft, 496 cu ft

(b) 449 sq in., 1218 cu in.

(c) 217 sq cm, 429 cu cm

(d) 271 sq in., 544 cu in.

(e) 90 sq yd, 105 cu yd

2. 840 sq in.

3. 15.30 cu yd

4. 6.1 gal

Spheres

The *sphere* is the simplest of all solid geometric figures. Geometrically it is defined as the surface whose points are all equidistant from a given point called the *center*. The *radius* is the distance from the center to the surface. The *diameter* is the straight line distance across the sphere on a line through its center.

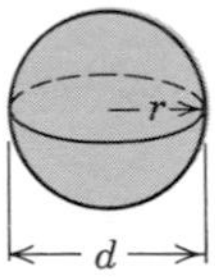

The following formulas enable you to find the surface area and volume of any sphere.

SPHERE

Surface area

$$S = 4\pi r^2$$

or $S = \pi d^2$

Volume

$$V = \frac{4\pi r^3}{3}$$

or $V = \frac{\pi d^3}{6}$

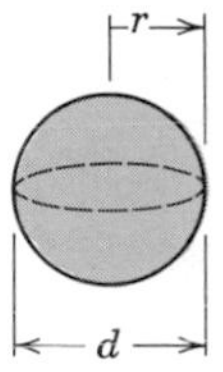

The volume formulas can be written approximately as

$V = 4.1888r^3$ and $V = 0.5236d^3$

For example, find the surface area and volume of a spherical basketball of diameter $9\frac{1}{2}$ in. Round to the nearest tenth.

The surface area is

$S = \pi d^2$
$S = (3.1416)(9.5 \text{ in.})^2$
$S = (3.1416)(90.25 \text{ sq in.})$
$S = 283.5$ sq in., rounded

The volume is

$V = \dfrac{\pi d^3}{6}$ $\quad V = \dfrac{(3.1416)(9.5 \text{ in.})^3}{6}$ $\quad V = \dfrac{(3.1416)(857.375)}{6}$ cu in.

$V = 448.9$ cu in., rounded

Your turn. Find the surface area and the volume in gallons of a spherical tank 8 ft in radius. Round to the nearest whole unit.

Check your solution in 64.

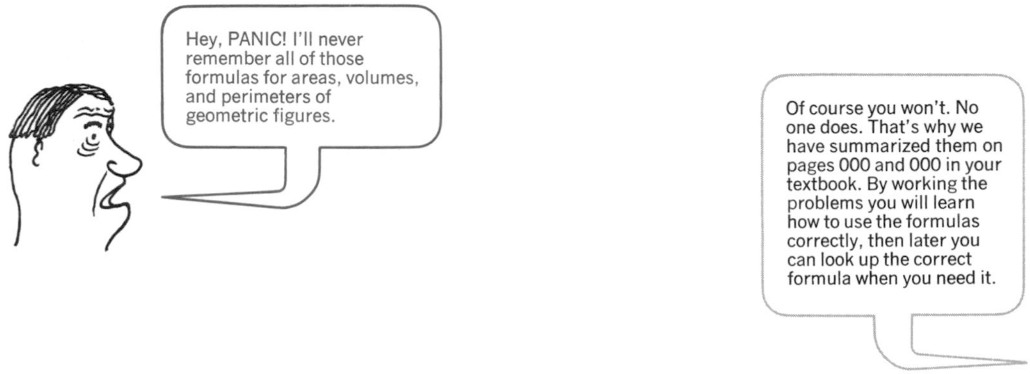
Hey, PANIC! I'll never remember all of those formulas for areas, volumes, and perimeters of geometric figures.
Of course you won't. No one does. That's why we have summarized them on pages 000 and 000 in your textbook. By working the problems you will learn how to use the formulas correctly, then later you can look up the correct formula when you need it.

---

64 The surface area is

$S = 4\pi r^2$
$S = 4(3.1416)(8 \text{ ft})^2$ Calculate 8^2 before multiplying.
$S = 4(3.1416)(64 \text{ sq ft})$
$S = 804$ sq ft, rounded

The volume is

$V = \dfrac{4\pi r^3}{3}$

$V = \dfrac{4(3.1416)(8 \text{ ft})^3}{3}$ Calculate 8^3 before multiplying.

$V = 2145$ cu ft, rounded

$V = 2145 \text{ cu ft} \times \dfrac{7.48 \text{ gallons}}{1 \text{ cu ft}}$ Convert to gallons using a unity fraction.

$V = 16{,}045$ gallons, rounded to the nearest gallon.

For practice, try the following problems.

1. Find the surface area and volume of each of the following spheres. (Round to the nearest whole unit.)

   (a) radius = 15″ (b) diameter = 22′

   (c) radius = 6.5 cm (d) diameter = 8′6″

2. How many gallons of water can be stored in a spherical tank 50″ in diameter? (Use 1 gal = 231 cu in. Round to the nearest gallon.)

3. What will be the weight of a spherical steel tank of radius 9′ when it is full of water? The steel used weighs approximately 127 lb per sq ft and water weighs 62.4 lb per cu ft. (Round to the nearest 100 lb.)

Check your answers in 65.

---

65 1. (a) 2827 sq in., 14137 cu in.

(b) 1521 sq ft, 5575 cu ft

(c) 531 sq cm, 1150 cu cm

(d) 227 sq ft, 322 cu ft

2. 283 gallons 3. 319,900 lb

In Problem 2, the volume is 65,450 cu in.

Multiply by a unity fraction to convert to gallons.

$$65{,}450 \text{ } \cancel{\text{cu in.}} \times \frac{1 \text{ gal}}{231 \text{ } \cancel{\text{cu in.}}} = \frac{65{,}450}{231} \text{ gal or } 283 \text{ gal}$$

In Problem 3 the area is 1018 sq ft. To find its weight, multiply by a unity fraction:

$$1018 \text{ } \cancel{\text{sq ft}} \times \frac{127 \text{ lb}}{1 \text{ } \cancel{\text{sq ft}}} = 129{,}300 \text{ lb, rounded}$$

The volume is 3054 cu ft or $3054 \text{ } \cancel{\text{cu ft}} \times \frac{62.4 \text{ lb}}{\cancel{\text{cu ft}}} = 190{,}600$ lb, rounded

The following table provides a handy summary of the formulas for solid figures presented in this chapter.

SUMMARY OF FORMULAS FOR SOLID FIGURES

Figure	Lateral Surface Area	Volume
Prism	$L = ph$	$V = Ah$
Pyramid	$L = \frac{ps}{2}$	$V = \frac{1}{3}Ah$
Cylinder	$L = \pi dh$ or $L = 2\pi rh$	$V = \pi r^2 h$ or $V = 0.7854d^2 h$
Cone	$L = \frac{\pi ds}{2}$ or $L = \pi rs$	$V = \frac{\pi r^2 h}{3}$ or $V = 0.2618d^2 h$
Frustum of pyramid	$L = \frac{1}{2}(P_1 + P_2)s$	$V = \frac{1}{3}h(A_1 + A_2 + \sqrt{A_1 \cdot A_2})$
Frustum of cone	$L = \frac{1}{2}\pi(d_1 + d_2)s$	$V = \frac{1}{3}h(A_1 + A_2 + \sqrt{A_1 \cdot A_2})$
Sphere	$S = 4\pi r^2$ or $S = \pi d^2$	$V = \frac{4\pi r^3}{3}$ or $V = 4.1888r^3$; $V = \frac{1}{6}\pi d^3$ or $V = 0.5236d^3$

Now turn to 66 for a set of exercises on solid geometric figures.

A. Find the lateral surface area and the volume of each of the following solids. (Round to the nearest tenth. Use $\pi = 3.14$.)

1. 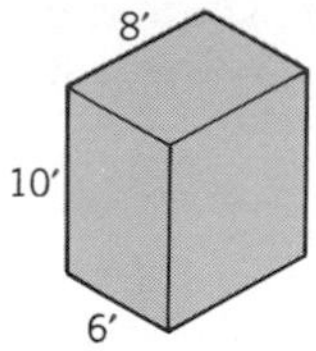

2. 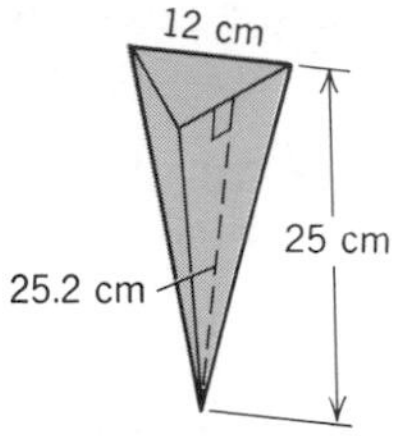

3. 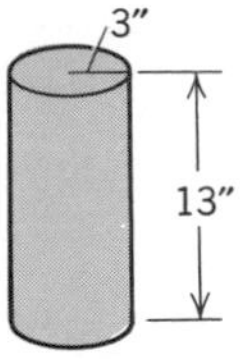

4. 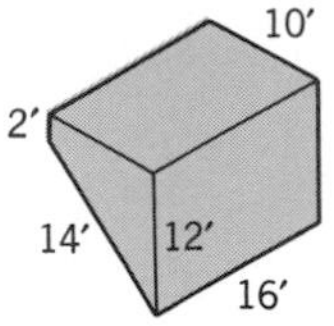

5. 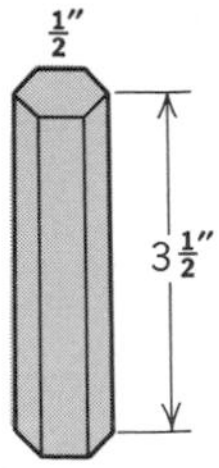

6. 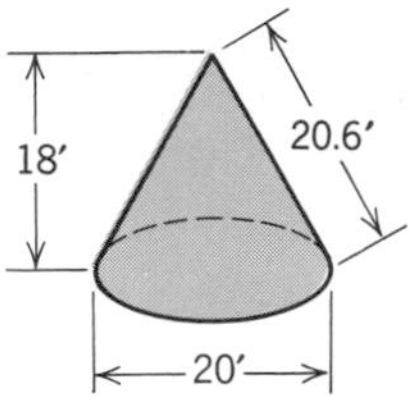

7. 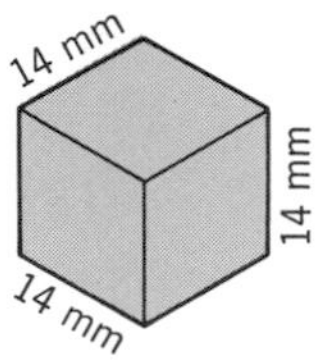

8. 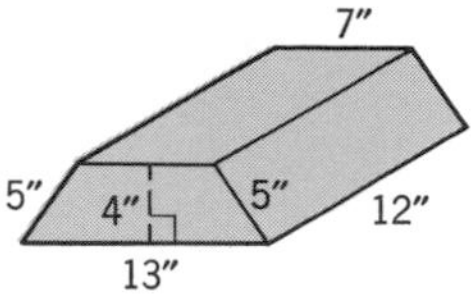

9. 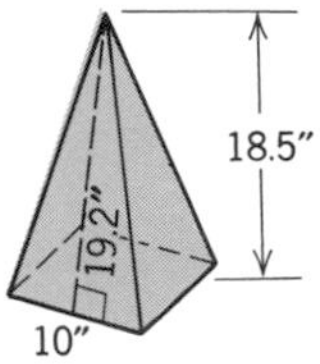

10. 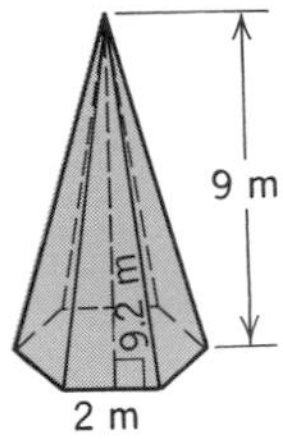

11. 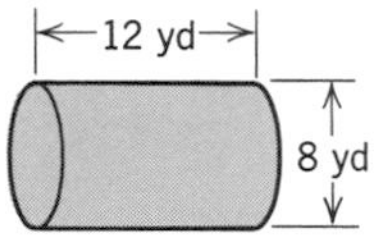

12. 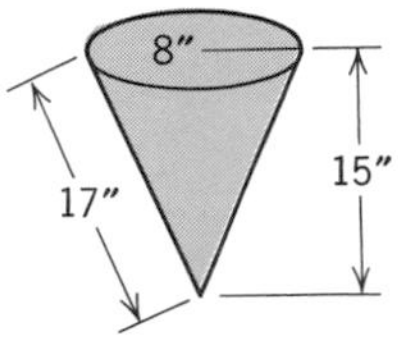

B. Find the total outside surface area and the volume of each of the following solids. (Round to the nearest tenth.):

1. 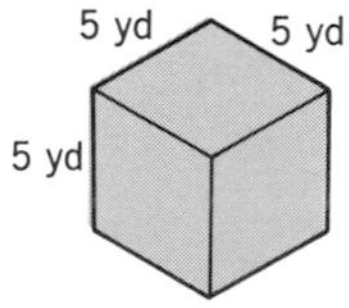

2. 

3. 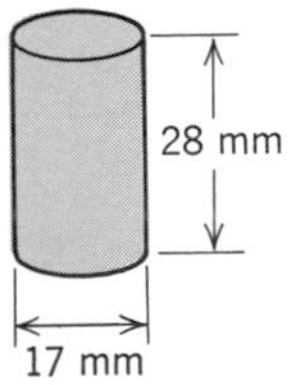

4. 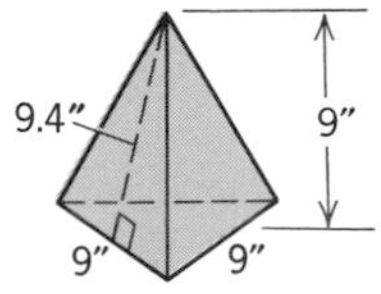

5. 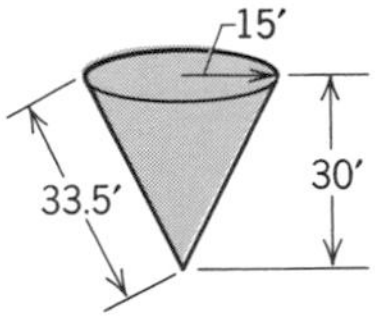

6. 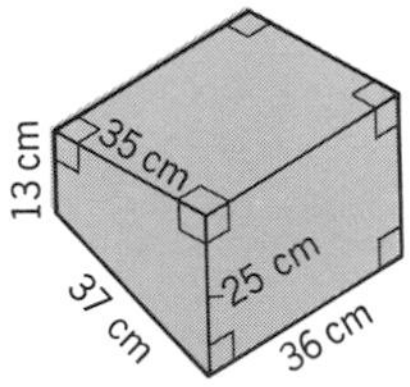

7. 

8. 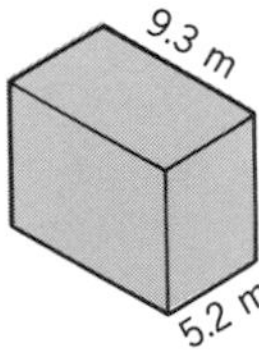

9. 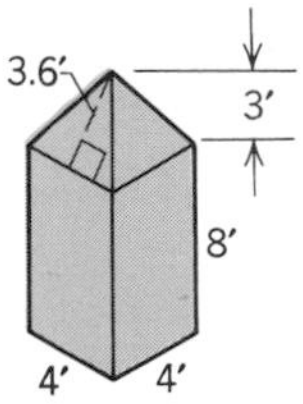

10. 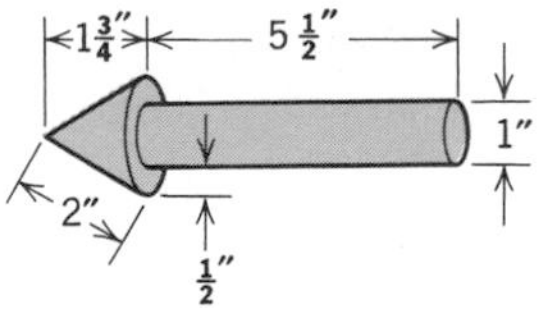

11. 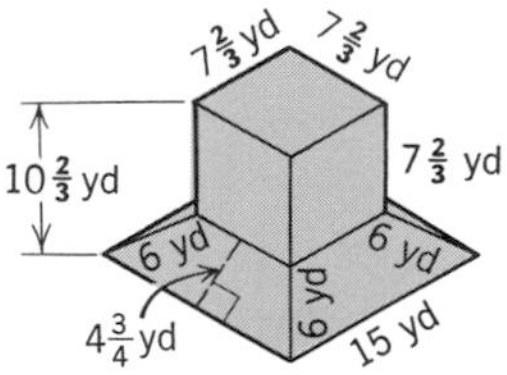

12. 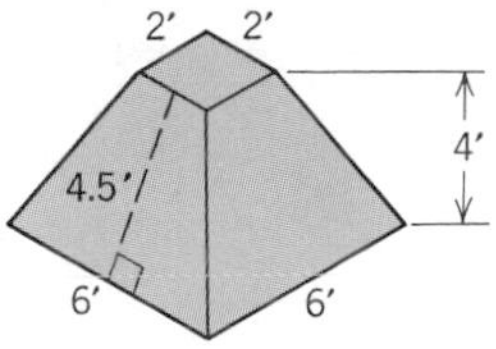

13. 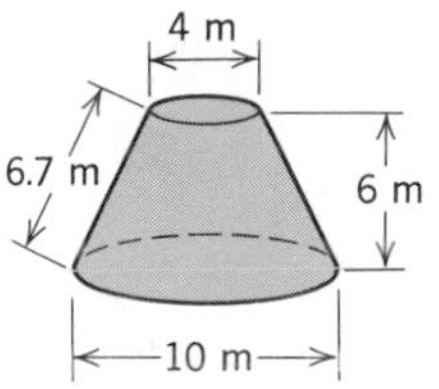

Use $\pi = 3.14$

14. 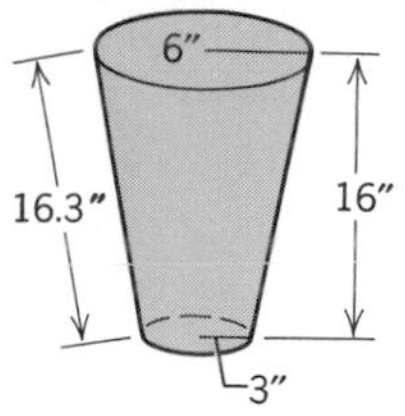

Use $\pi = 3.14$

C. Practical Problems. (Round to the nearest tenth.)

1. How many cubic *feet* of warehouse space are needed for 450 boxes 16″ by 8″ by 10″?
2. Find the weight of the piece of steel pictured. (Steel weighs 0.2833 lb per cu in.)

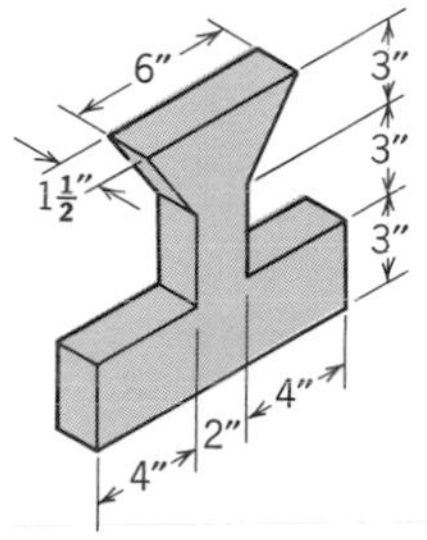

3. How many gallons of water are needed to fill a swimming pool that approximates the shape in the figure? (Round to the nearest gallon.)

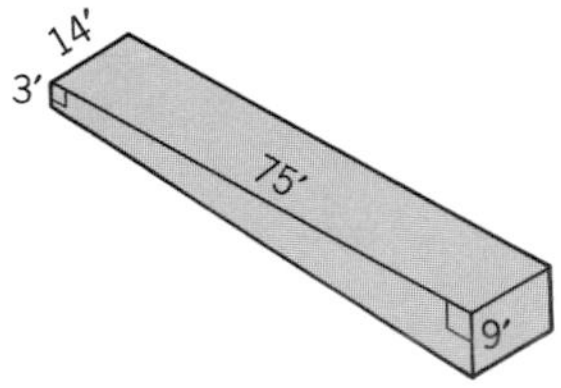

4. How many cubic meters of dirt are there in a pile, conical in shape, 10 meters in diameter and 4 meters high?
5. How many gallons of paint are needed for the outside walls of a building 26′ high by 42′ by 28′ if there are 480 sq ft of windows? One gallon covers 400 sq ft. (Assume you cannot buy a fraction of a gallon.)
6. A 4′ high cylindrical pipe has a diameter of 3″. If water weighs 62.4 lb per cu ft, what is the weight of water in the pipe when filled to the top? (**Hint:** Change the diameter to feet.)
7. Dirt must be excavated for the foundation of a building 30 yards by 15 yards to a depth of 3 yards. How many trips will it take to haul the dirt away if a truck with a capacity of 3 cu yd is used?
8. A layer of crushed rock must be spread over a circular area 23′ in diameter. How deep a layer will be obtained using 100 cubic feet of rock? (Round to the nearest hundredth of a foot.)
9. How many square centimeters of sheet metal are needed for the sides and bottom of the pail shown in the figure?

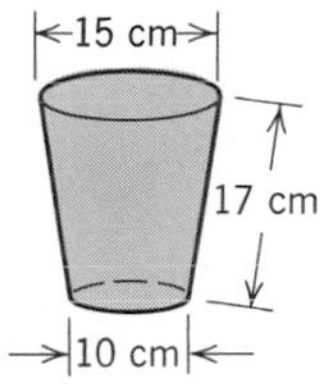

10. How many liters of water can be stored in a cylindrical tank 8 meters high with a radius of 0.5 meters? (1 m^3 contains 1000 liters.)

11. What is the weight of the piece of aluminum shown at 0.0975 lb per cu in.?

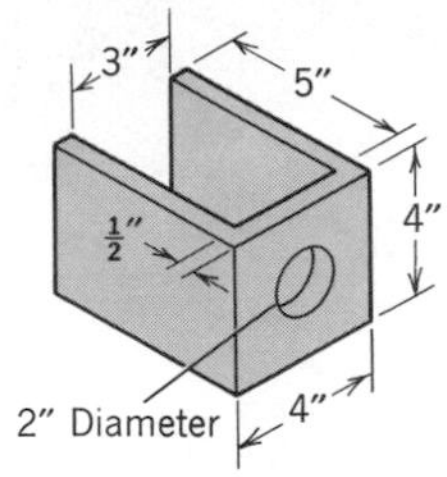

12. How many bushels will the bin in the figure hold? (1 cu ft = 1.24 bushels.)

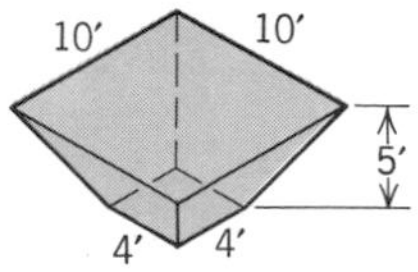

13. If brick has a density of 103 lb per cu ft, what will be the weight of 500 bricks each $3\frac{3}{4}''$ by $2\frac{1}{4}''$ by 8″?

14. Dirt cut 2′ deep from a section of land 38′ by 50′ is used to fill a section 25′ by 25′ to a depth of 3′. How many cubic *yards* of dirt are left after the fill?

15. How high should a 50 gallon tank be if it must fit into an area allowing a 16 in. diameter? (1 gallon = 231 cu in.)

16. How many square inches of sheet metal are used to make the vent transition shown?

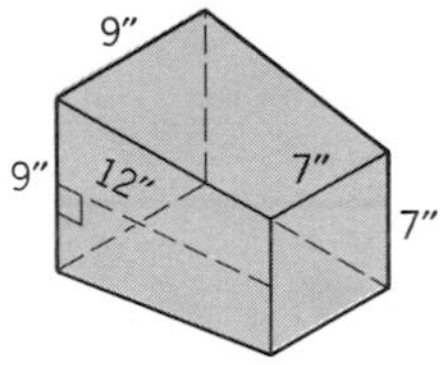

17. In a four hour percolation test, 8 ft of water has seeped out of a cylindrical trench 4 ft in diameter. The plumbing code says the soil must be able to absorb 5000 gallons in 24 hours. At this rate will the soil be up to code?

18. How many square inches of sheet copper are needed to cover the inside wall of a cylindrical tank 6′ high with a 9″ radius? (**Hint:** Change the altitude to inches.)

19. How many cubic yards of concrete are needed to pour the building foundation shown in the figure? (Write out the necessary calculation if you do not have a calculator.)

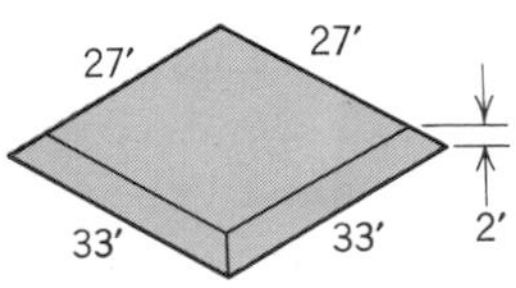

20. Find the weight of the cast iron shape in the figure. (Cast iron weighs 0.2607 lb per cu in.)

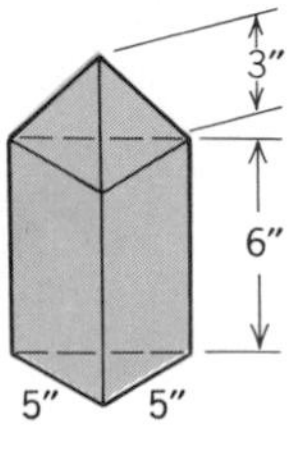

D. Calculator Problems. (Round to the nearest tenth.)

1. Find the capacity in gallons of the oil can in the figure. (1 gallon = 231 cu in. Round to the nearest tenth.)

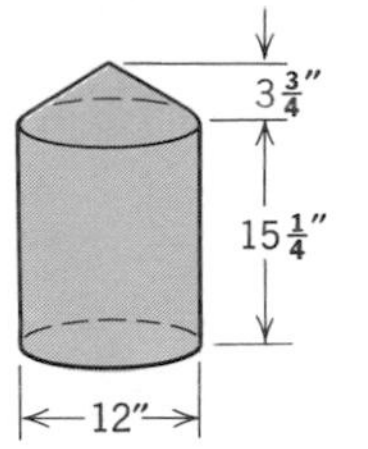

2. Find the weight the piece of brass shown. Brass weighs 0.296 lb per cu in.

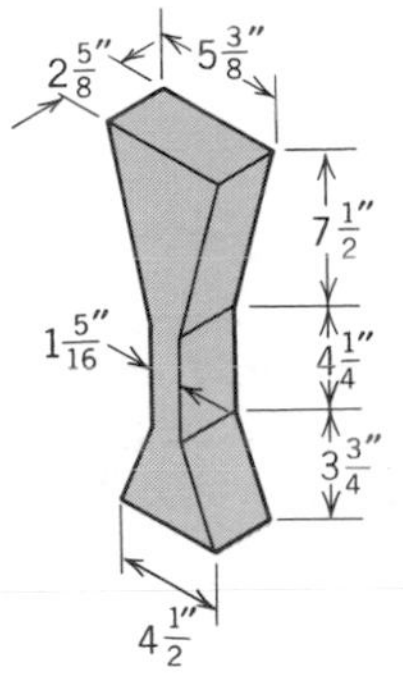

3. How many square centimeters of sheet metal will it take to make the container shown in the figure? (All measurements are in cm.)

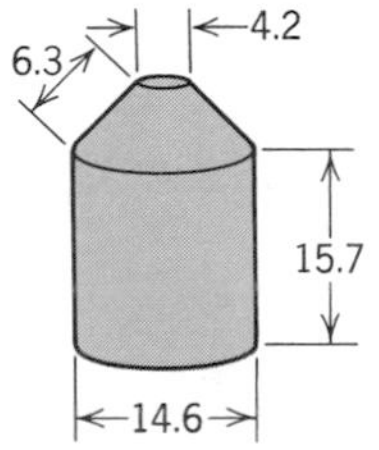

4. How high must a 40 gallon rectangular tank be if the base is a square 3′9″ on a side? (1 cu ft = 7.48 gallons.)

5. What is the weight of the taper bushing shown if it is made of steel weighing 0.2833 lb per cu inch?

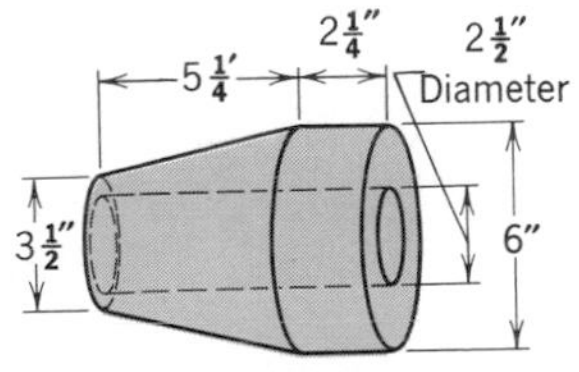

6. A septic tank has the following shape:

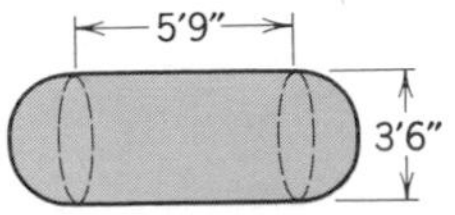

How many gallons does it hold? (1 cu ft = 7.48 gallons.)

7. Find the weight of the steel rivet shown in the figure. (Steel weighs 0.2833 lb per cu in.)

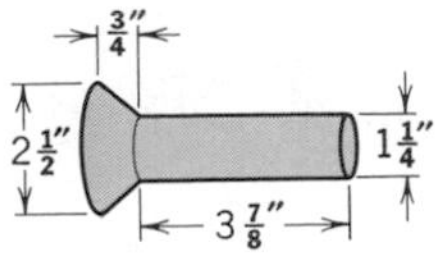

8. How many pounds of rock will you need to fill an area 25′ by 35′ to a depth of 2″ if the rock weighs 1050 lb per cubic yard?

9. How many cubic yards of concrete are needed to pour 8 cylindrical pillars 12′6″ high each with a diameter of 1′9″?

10. At a density of 0.0925 lb per cu in., calculate the weight of the aluminum piece shown.

Diameter of 4 outside holes: $\frac{1}{4}''$
Diameter of inside hole: $\frac{1}{3}''$
Thickness of piece: $\frac{1}{2}''$

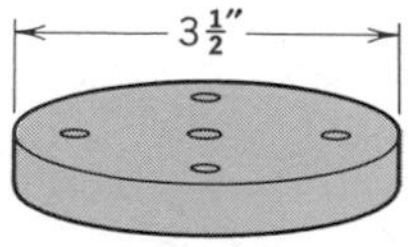

When you have completed these exercises check your answers on page 556, then turn to **67** to learn about some important geometric constructions.

7-4 GEOMETRIC CONSTRUCTIONS

67 Long before the invention of protractors, ancient Greek geometers knew how to construct angles of a given size or lines that were exactly perpendicular. They found it possible to make such geometric constructions using only a straight edge or unmarked rule, and a compass for drawing circles or arcs of circles. Many of the techniques they discovered are still used today, particularly in sheet metal work, drafting, carpentry, in machine shops, and in the construction trades. In this section we will take a step-by-step look at a few of the most useful constructions.

1. Bisecting a Line

To bisect a line means to divide it into two exactly equal parts—to cut it in half.

Example: Bisect the line segment AB. A ———— B

Step 1 Using end point A as a center and any radius greater than half the length of the line segment, draw arcs above and below the line.

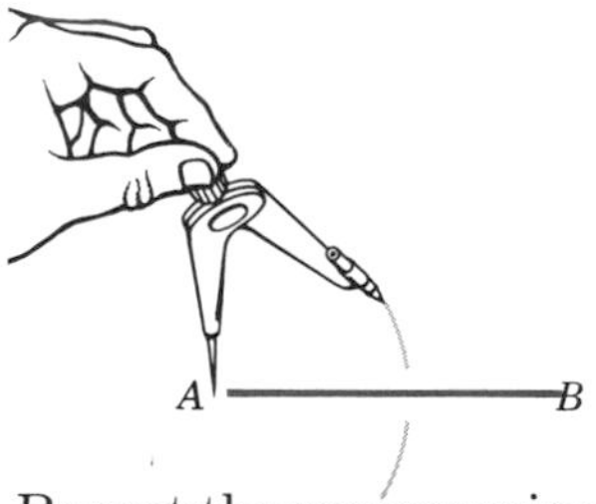

Step 2 Repeat the process using point B as the center. Do not change the radius and be sure to intersect the arcs made in Step 1. Label the points of intersection C and D.

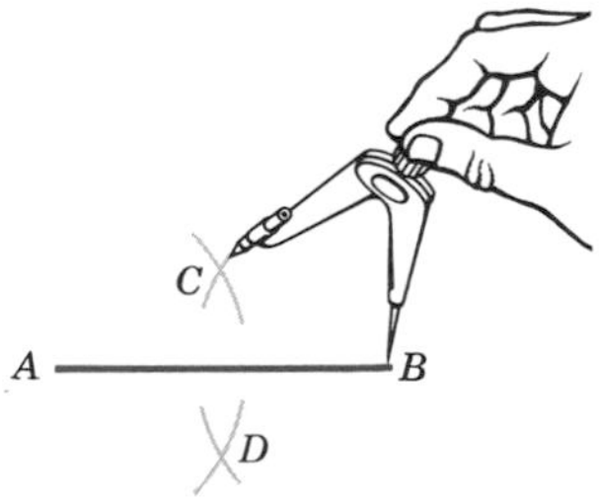

Step 3 Connect points C and D. The segment CD bisects line AB at E. Point E is the center point of line AB.

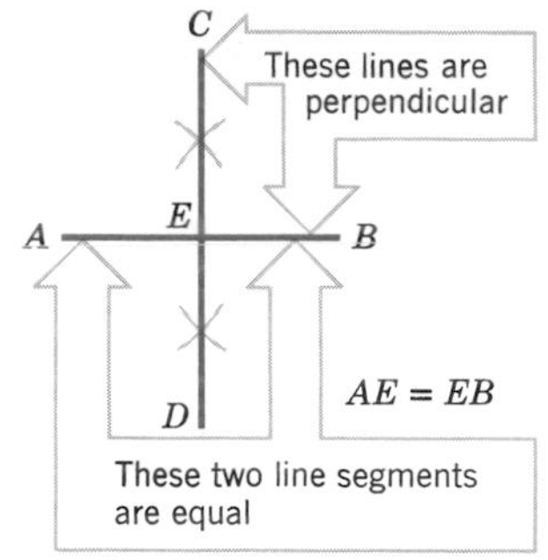

It is also true that the line CD is perpendicular to the original line AB.

Ready for a little practice? Draw three line segments of any lengths, and then bisect them using only a compass and a straightedge. Check your work with a ruler to see if the line has actually been divided into two pieces of equal length. If you have any doubts about the procedure, ask your instructor or a tutor to check your work.

Turn to **68**.

---

68 **2. Bisecting an Angle**

To bisect an angle means to draw a second angle that is exactly half of the original angle.

Example: Bisect angle X.

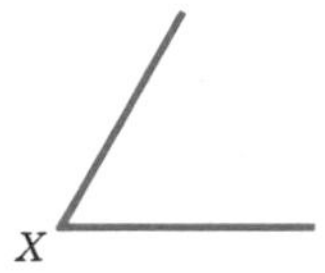

Step 1 Using X as the center and any convenient radius, draw an arc that cuts both sides of the angle, at Y and Z.

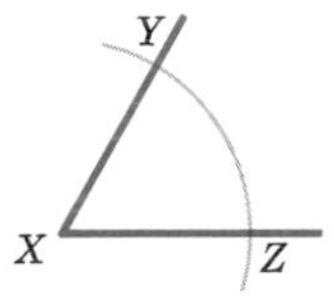

Step 2 Using Y as the center and any radius greater than half the distance between Y and Z, strike an arc above the first arc.

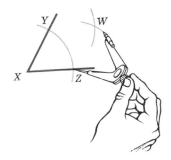

Step 3 Using Z as the center and the same radius from Step 2, draw another arc intersecting the first arc at W.

Step 4 Connect point W to point X. The line WX bisects angle X. Angles YXW and ZXW are exactly equal.

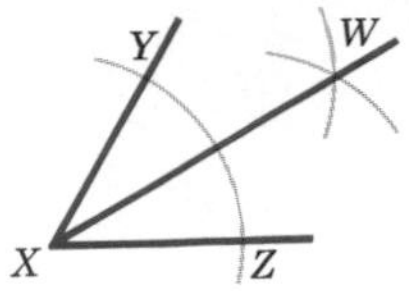

For practice bisect the following angles:

(a)

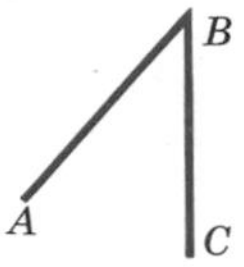

(b)

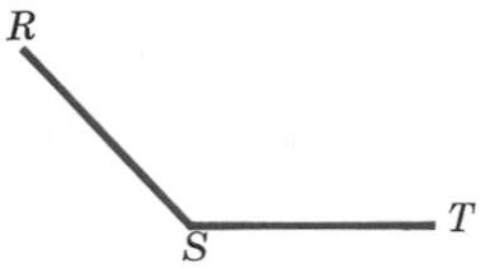

(c) Use the first construction method to draw a perpendicular to line PQ. Use the second construction to bisect one of the four right angles formed.

Check your work in 69.

---

69 (a) Angle ABD = angle CBD.

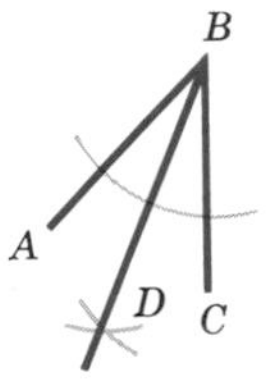

Check this by measuring them with a protractor.

(b) Angle RSW = angle TSW.

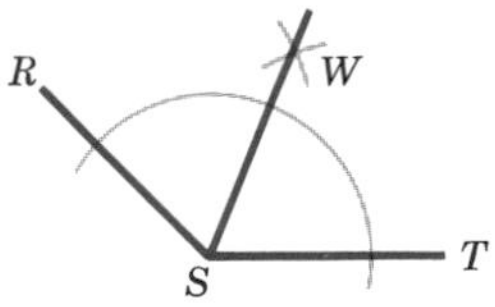

Check by measuring with a protractor.

(c) Line MN is perpendicular to line PQ. Angle MOK equals angle QOK.

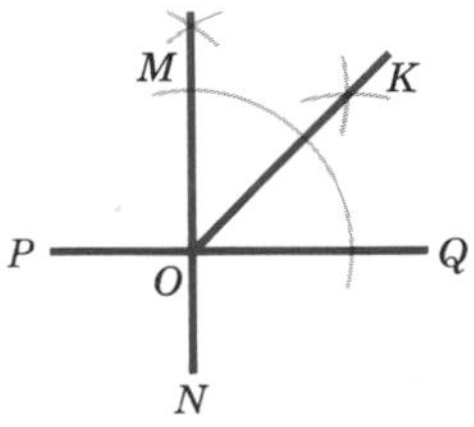

Check with a protractor.

3. Constructing a Perpendicular from a Point to a Line

Construction 1 enabled you to draw a perpendicular to any line segment at the midpoint of that line segment. This third construction enables you to draw a perpendicular to any line from any point not on the line.

Example: Construct a perpendicular from point F to line segment GH.

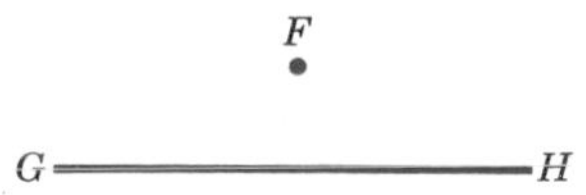

Step 1 Using F as the center, draw any arc that cuts line GH in two places, I and J.

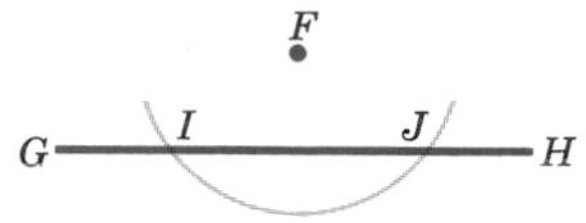

Step 2 Using I as the center, and a radius greater than half the segment IJ, draw an arc below GH.

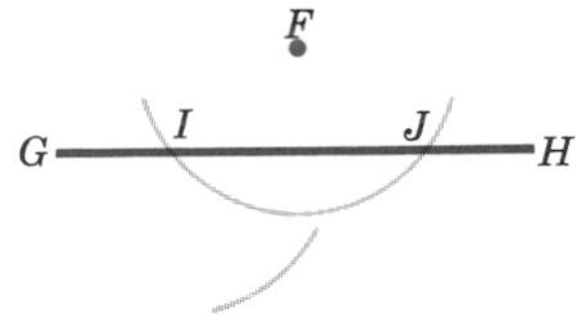

Step 3 Using J as the center and the *same* radius from Step 2, draw another arc below GH that intersects the first arc at K.

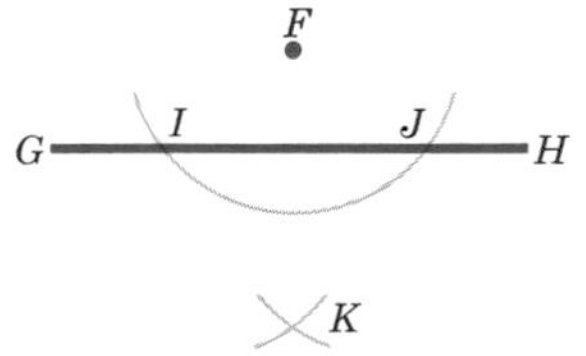

Step 4 Connect F to K. Line FK is exactly perpendicular to line GH.

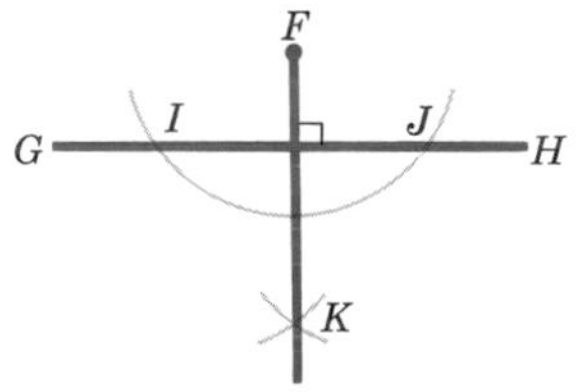

For practice, draw a line from point P perpendicular to the line AB.

(a)

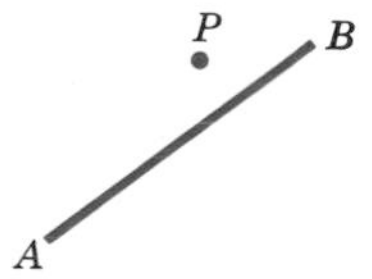

(b)

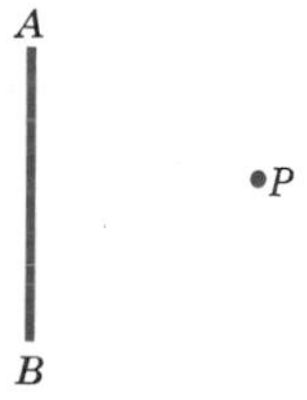

Check your work in 70.

70 (a)

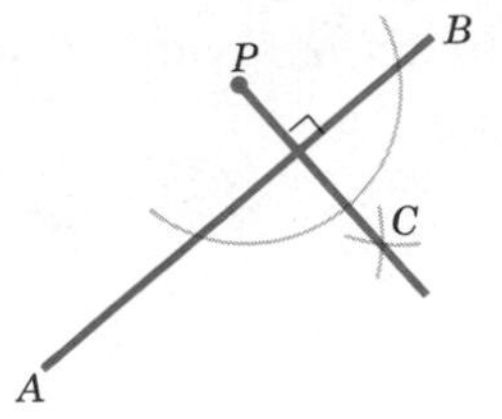

(b)

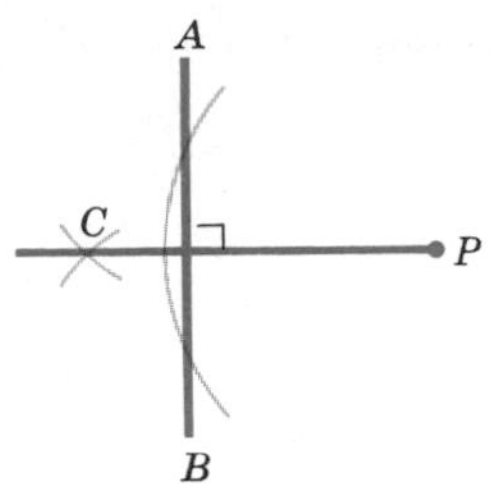

Check your drawing by measuring the angles where PC and AB intersect. Each angle should be 90°, of course.

4. Constructing a Perpendicular at a Given Point on a Line

This construction enables you to draw a perpendicular to a line at any given point on that line.

Example: Construct a perpendicular to line segment LM at point N.

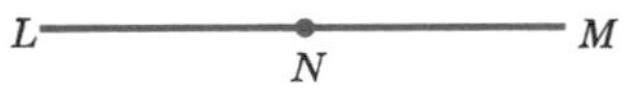

Step 1 Using N as the center and a convenient radius, draw equal arcs cutting the line LM on both sides of point N. The points where the arcs cut the line are labeled P and Q.

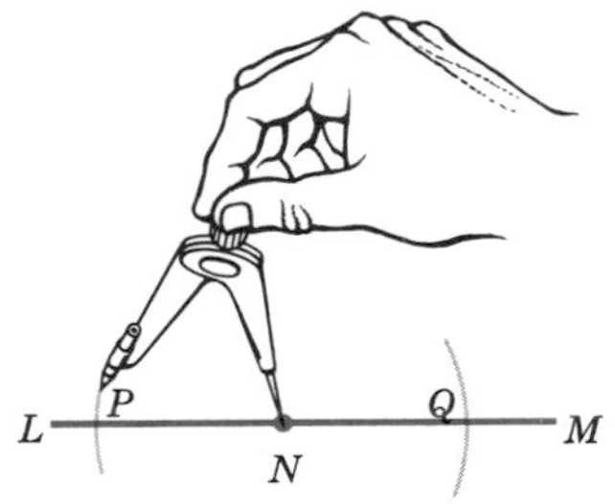

Step 2 Using P as the center and a radius greater than PN, draw an arc above (or below) the line.

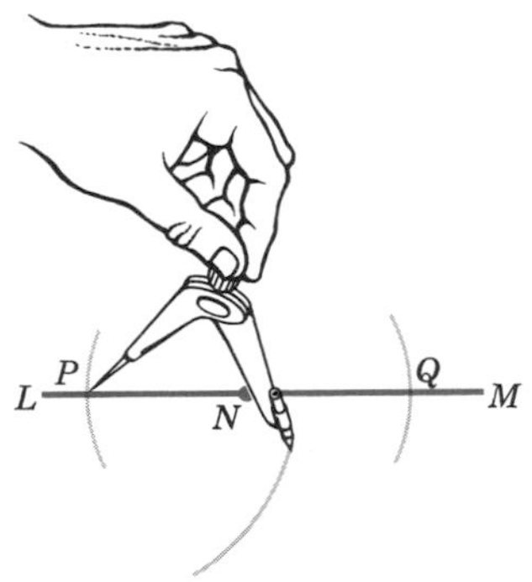

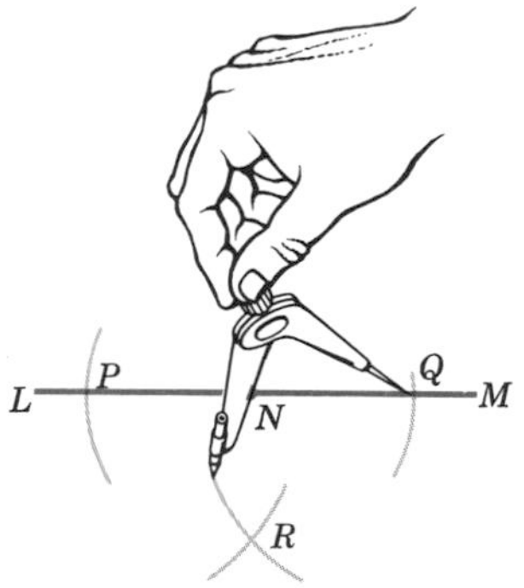

Step 3 Using Q as the center and the same radius, draw another arc that intersects the one from Step 2. The point where these arcs intersect is labeled R.

Step 4 Connect point R to point N. The line RN is perpendicular to the original line LM.

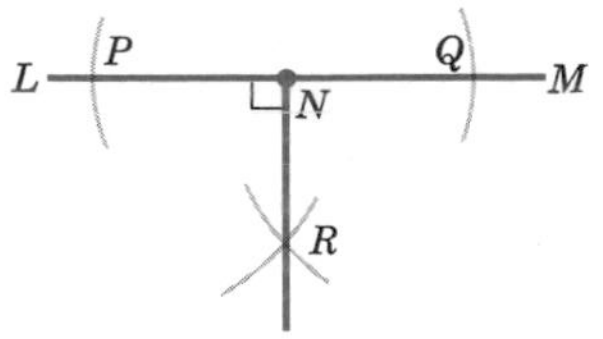

If the point to which the perpendicular must be constructed is at one end of the line, simply extend the line and use the same procedure.

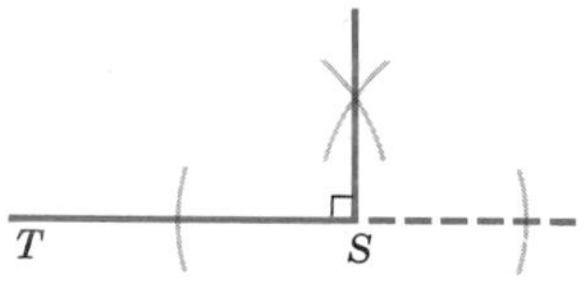

If for some reason you cannot extend the line, try this construction instead.

5. Constructing a Perpendicular to a Line at an End Point

Example: Construct a perpendicular to line XY at point X.

Step 1 Pick any point P off the line. Draw an arc as shown with radius PX. The arc should cut line XY at some point W.

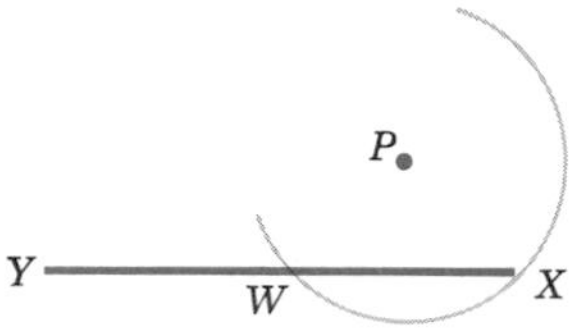

Step 2 From point W draw a line through P intersecting the arc at Z.

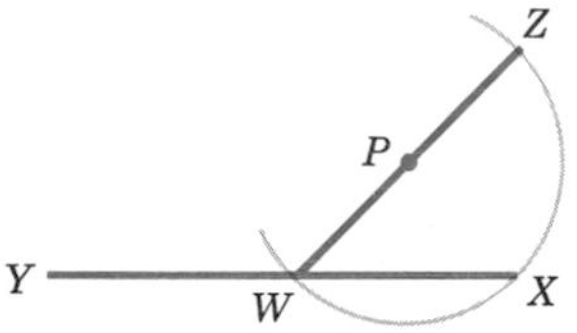

Step 3 Draw a line between points Z and X.

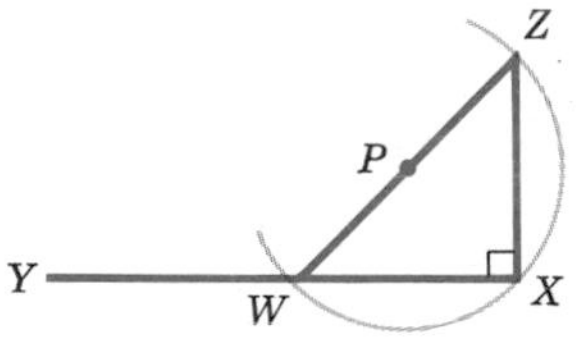

Line ZX is perpendicular to line XY.

Now use the constructions you have learned so far to do the following:

(a) Construct a square with side 2″.

(b) Construct a rectangle whose length is twice its width.

(c) Construct a 3–4–5 right triangle.

(d) Construct an isosceles right triangle.

Check your work in **71**.

71 (a) **First,** draw a line 2″ long.

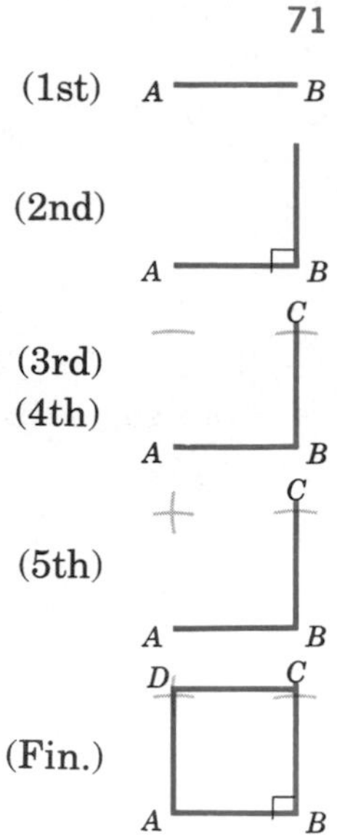

Second, draw a perpendicular to the line at one end using Construction 5.

Third, using B as a center and with radius AB, draw an arc at C.

Fourth, from A strike an arc of radius AB above A.

Fifth, with C as a center and radius AB, draw a second arc intersecting the last one. Label this point D.

Finally, connect AD and DC.

$ABCD$ is a square.

(b) Follow the procedure in Problem (a) above but make line BC equal to twice AB. Similarly, make AD equal to BC.

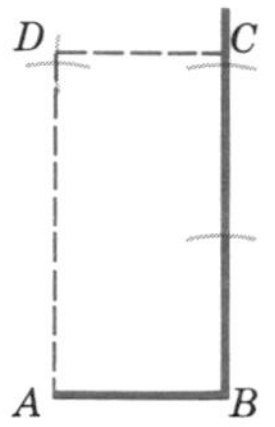

(c) **First,** use construction 3, 4, or 5 to draw a perpendicular CD to some line AB. **Second,** using some convenient radius, mark three arcs along OD and four arcs along OA.

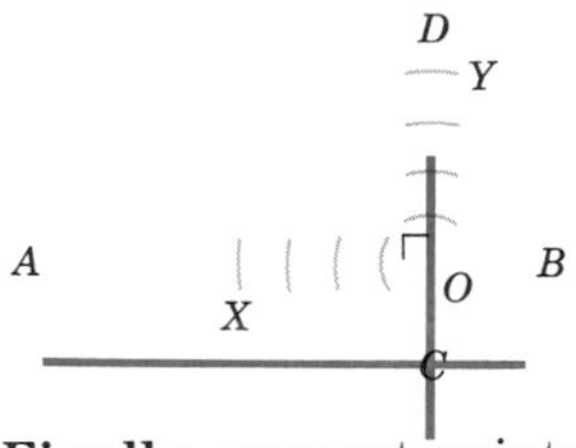

Finally, connect points X and Y. Triangle OXY is a 3–4–5 right triangle.

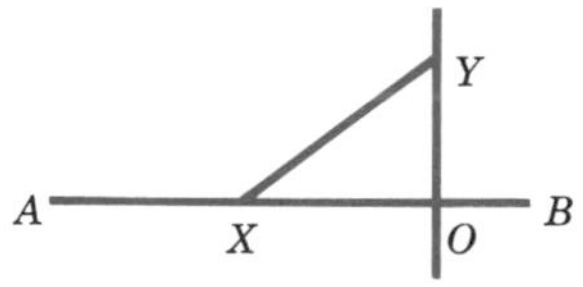

(d) **First,** use construction 3, 4, or 5 to draw a perpendicular to some given line.

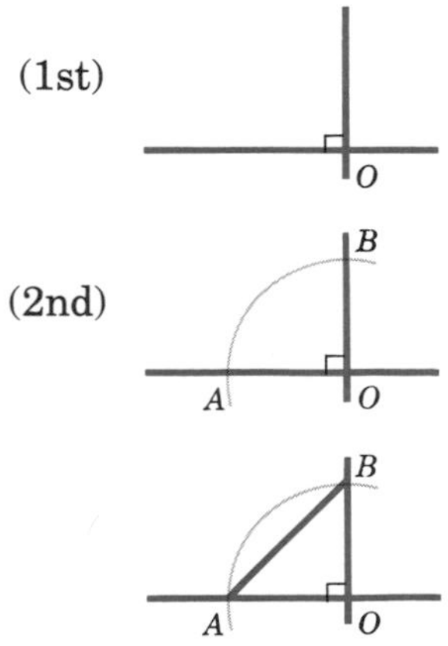

Second, using the intersection O as a center, cut both perpendicular lines with the same arc at A and B.

Triangle AOB is isosceles.
Sides AO and BO are equal.

6. Duplicating an Angle

This construction enables you to draw an angle exactly equal to some given angle, without using a protractor, of course.

Example: Construct an angle PQR equal to angle MNO.

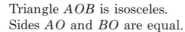

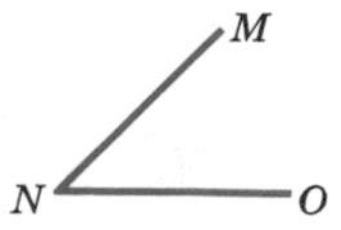

Step 1 Draw one side of *PQR* and label it. It can be of any convenient length.

Step 2 Using *N* as the center and with any convenient radius, draw an arc cutting both sides of *MNO*. Label these points *S* and *T*.

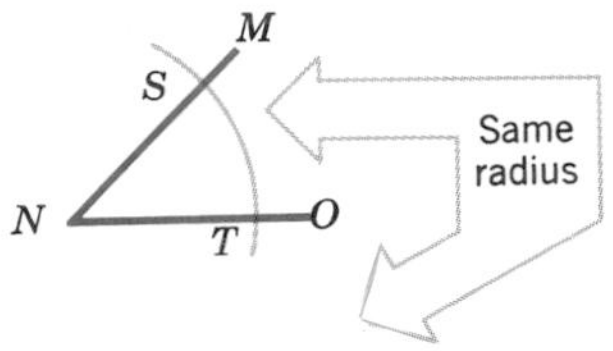

Step 3 Using the *same* radius and *Q* as the center, strike a wide arc that cuts line *QR* at point *U*.

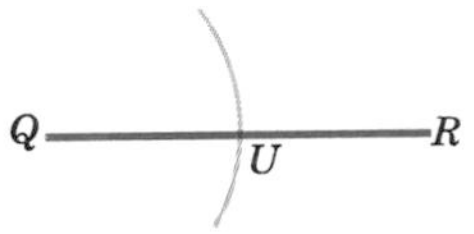

Step 4 Using *ST* as the radius, draw an arc from *U* that cuts the existing arc at *V*.

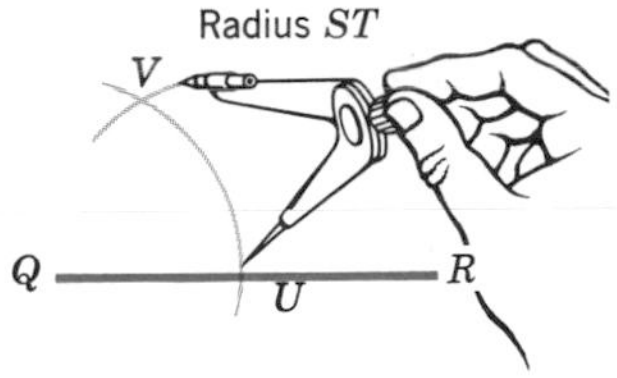

Step 5 Draw a line from point *Q* through point *V*. Then angle *PQR* is equal to the original angle *MNO*.

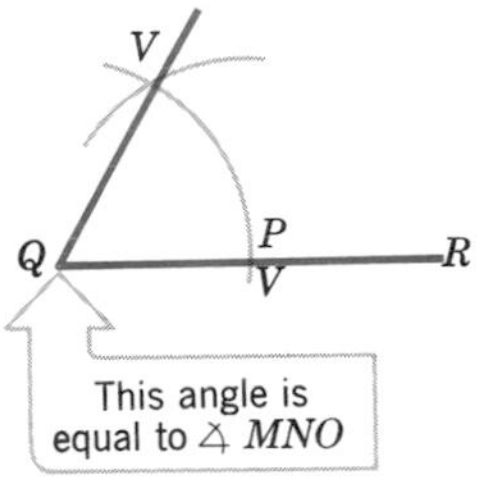

Practice this construction by duplicating the following angles. (The vertex of the duplicated angle should be at point *P*.)

(a)

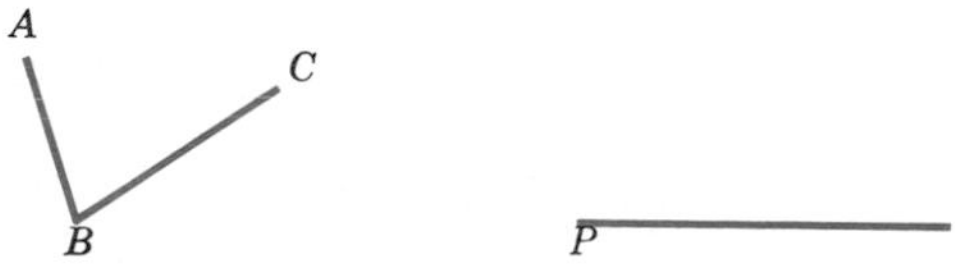

(b)

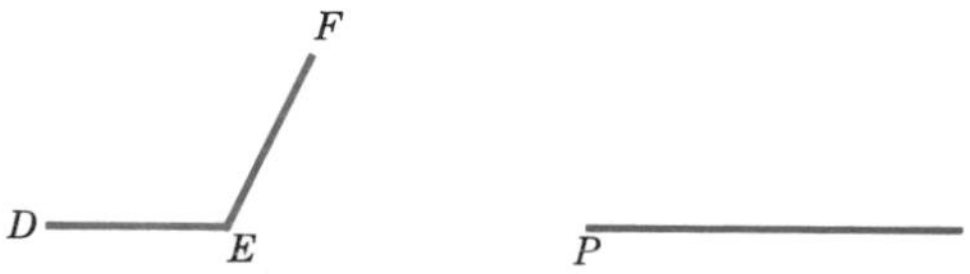

Check your work against ours in **72**.

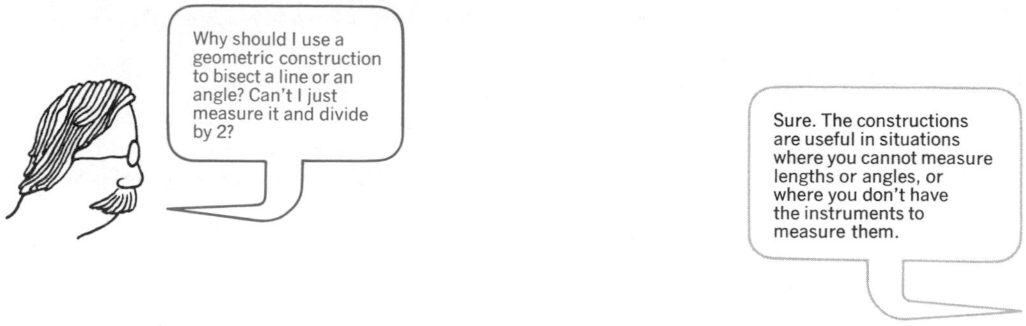

72

(a)

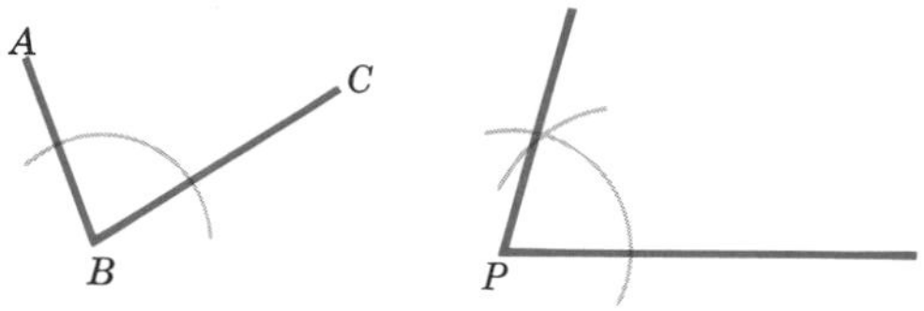

(b)

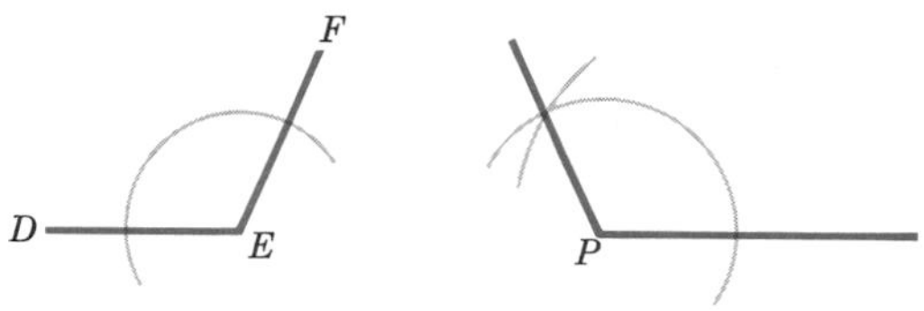

(ex.)

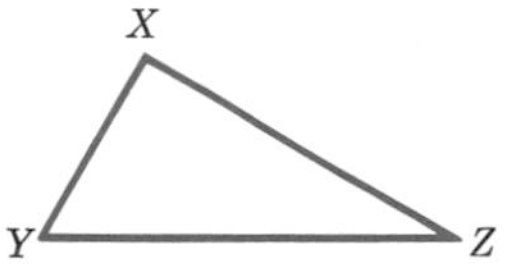

7. Duplicating a Triangle

This construction enables you to duplicate a given triangle exactly, drawing a second triangle of the same size and shape.

Example: Duplicate triangle XYZ.

(1)

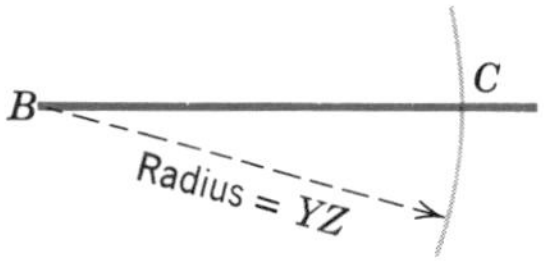

Step 1 Draw a line segment longer than YZ. From one endpoint, B, draw an arc equal to YZ cutting the line at C.

(2)

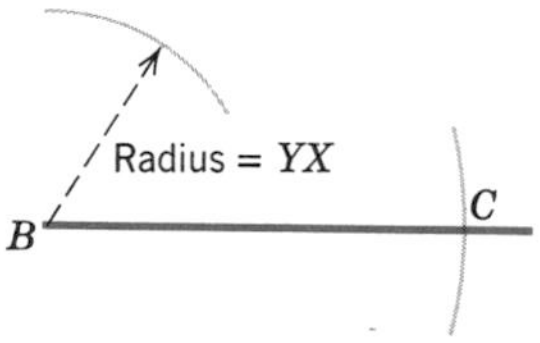

Step 2 Using B as the center and radius equal to YX, draw an arc in the direction of the third vertex of the triangle.

(3)

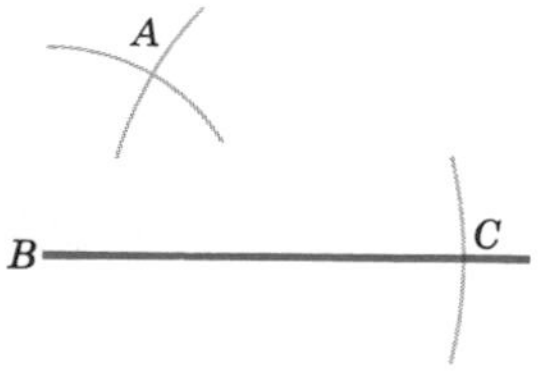

Step 3 Using C as the center and radius equal to XZ, draw an arc that intersects the last arc at A.

(4)

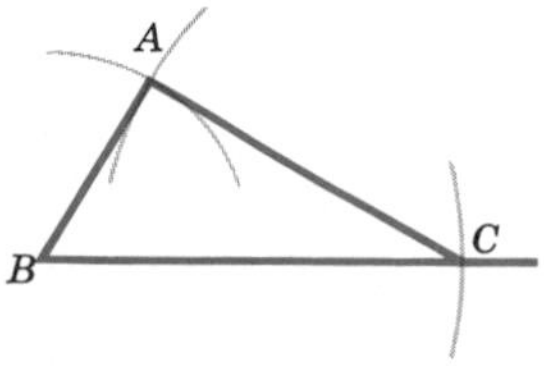

Step 4 Connect point A to point B and point C to point A. Triangle ABC is identical in both size and shape to triangle XYZ. All three angles are the same as the corresponding angles in the original triangle.

For practice, duplicate each of the following triangles, using Construction 7.

(a)

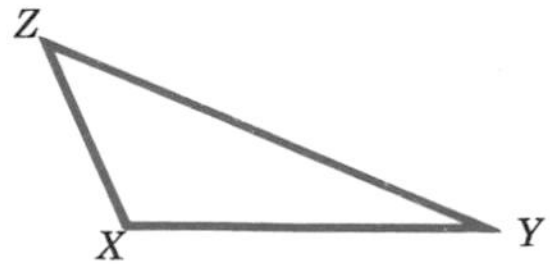

(b)

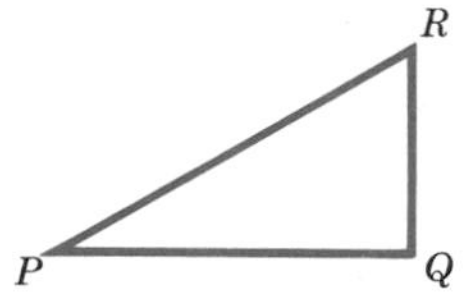

Check your work in **73**.

73 (a) (b)

8. Constructing an Equilateral Triangle

Each angle in an equilateral triangle measures 60°. Constructing an equilateral triangle enables you to draw a 60° angle without using a protractor.

Example: Construct an equilateral triangle.

Step 1 Draw a line of whatever length you choose for the side of the triangle. Label it AB.

Step 2 Using AB as a radius, draw an arc from A and a second arc from B. These will intersect at point C.

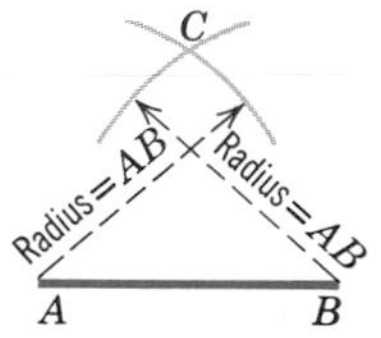

Step 3 Draw lines AC and BC. Triangle ABC is equilateral. All sides are equal in length. Each angle measures 60°.

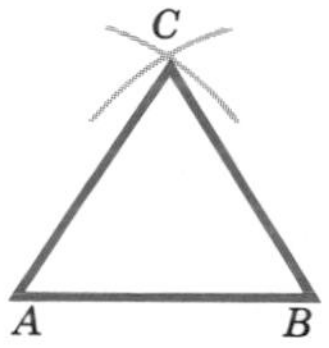

That's an easy one. For practice construct an equilateral triangle with side equal to 2″. Check your work by measuring all sides with a ruler to see if they are equal.

When this is completed turn to 74 for another construction.

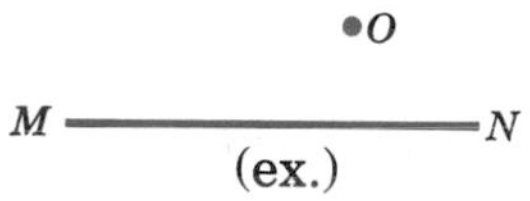

(ex.)

---

74 **9. Constructing a Parallel to a Given Line**

Many geometric figures involve parallel lines. This construction enables you to draw a line that is parallel to some given line and that goes through some given point.

Example: Construct a line parallel to MN that goes through point O.

Step 1 Using the distance MN, between points M and N, as a radius and using point O as the center draw an arc as shown.

Step 2 Next, draw a second arc with radius ON using M as the center. Label the intersection of the arcs P.

Step 3 Connect points O and P. The line OP is parallel to the line MN. If we connect points M and P and points O and N, we would have a parallelogram $MNOP$.

Practice this construction by doing the following problems.

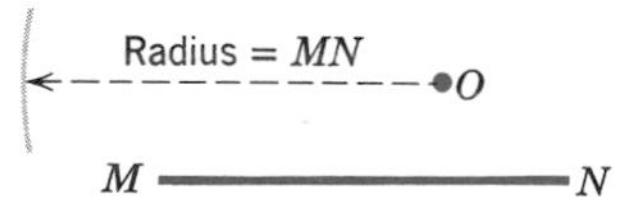

(1)

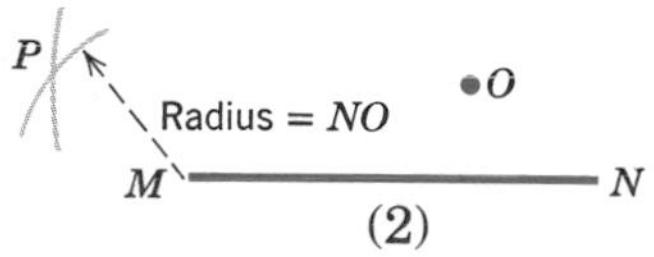

(2)

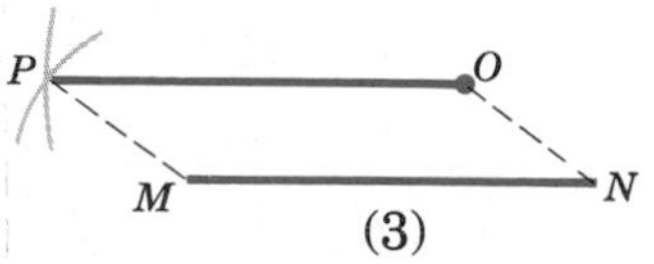

(3)

(a) Draw a line through A parallel to the base of the triangle.

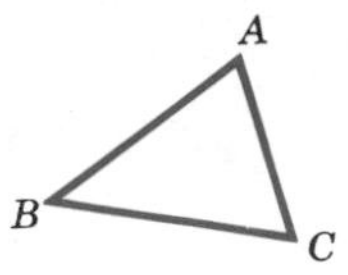

(b) Draw a line parallel to line XY and $1\frac{1}{2}$ inches from it.

Check your work in 75.

---

75 (a)

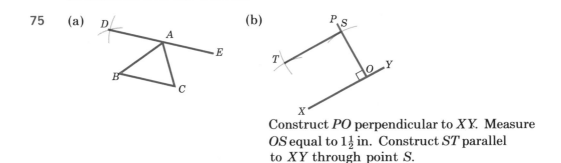

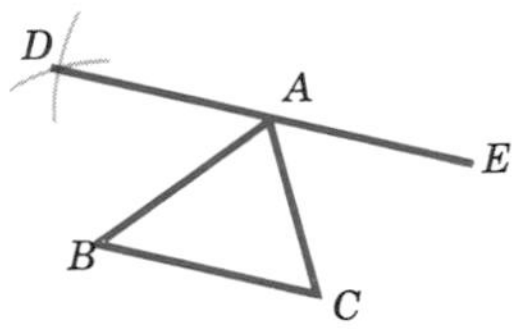

(b)

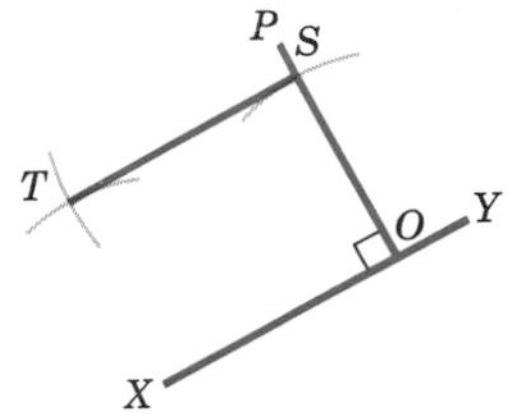

Construct PO perpendicular to XY. Measure OS equal to $1\frac{1}{2}$ in. Construct ST parallel to XY through point S.

10. Dividing a Line into a Number of Equal Parts

This construction is one of the simplest and most useful of all. It enables a sheet metal worker or carpenter to divide *any* length into *any* number of equal parts without making a measurement. For example, it will allow you to divide a 1″ length into thirds or fifths or even tenths rather than only the fourths, eighths, or sixteenths shown on a measuring rule.

Example: Divide line AB into five equal parts.

Step 1 Draw line AC of any length and at any convenient angle to AB.

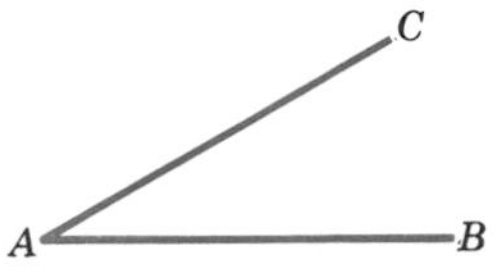

Step 2 Using any radius that will fit, mark off five arcs of equal length along AC, starting at A. Label the resulting points D, E, F, G, and H and connect the last point H to B.

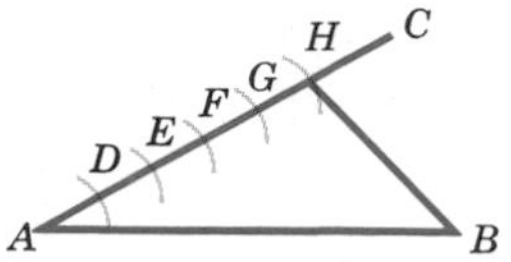

Step 3 Construct a line parallel to *BH* through point *G*. This line intersects line *AB* at point *I*. Use construction 9 to draw the parallel line.

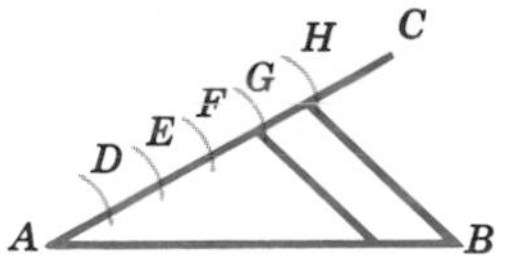

Step 4 The distance *BI* is one-fifth of the distance *AB*. To cut *AB* into five equal parts use *BI* as a radius and draw arcs at *J*, *K*, and *L* as shown.

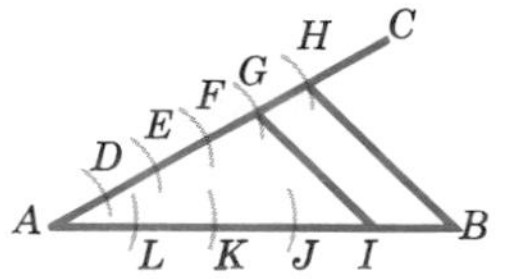

Distances *AL*, *LK*, *KJ*, *JI*, and *IB* are equal.

For practice,

(a) Divide the line below into three equal parts.

(b) Construct a line one-fifth of an inch long.

Check your work by measuring with a ruler.

Continue in 76.

76

11. Constructing a Circle of a Given Diameter

(ex)

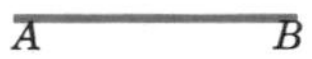

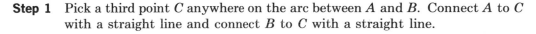

This very simple procedure allows you to draw a circle with any given diameter.

Example: Construct a circle with a diameter equal to *AB*.

Step 1 Bisect *AB*. Label the point of bisection *C*.

(1)

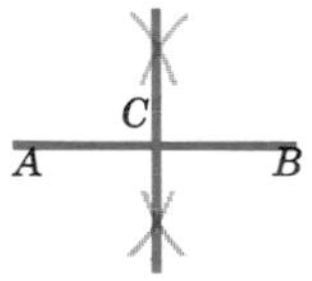

Step 2 Using radius equal to *AC* (or *CB*) construct a circle using *C* as the center.

The construction is a snap, but can you reverse the procedure?

12. Finding the Center of a Circle Given an Arc

(2)

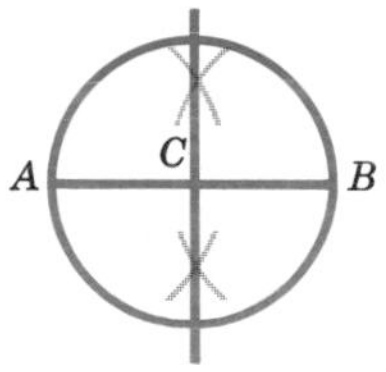

In this construction you begin with a circle or arc of a circle and locate its center point. Knowing the center point, you can find its radius or diameter.

Example: Find the center of the circle with arc $\widehat{AB}$.

Step 1 Pick a third point *C* anywhere on the arc between *A* and *B*. Connect *A* to *C* with a straight line and connect *B* to *C* with a straight line.

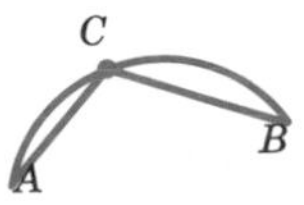

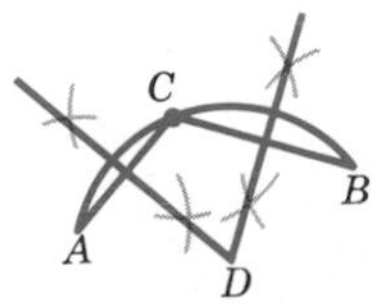

Step 2 Use construction 3 to bisect lines *AC* and *BC*. Extend the bisectors inside the arc until they meet. They meet at point *D*, the center of the circle of which arc $\widehat{AB}$ is a part.

Now you can construct the rest of the circle using DA (or DB or DC) as the radius.

Find the center of each of the following arcs.

(a)  (b) 

Check your work by using the center and radius you find to draw the complete circle. Do the arcs fit?

Continue in **77**.

---

77 Two more constructions involving circles will be useful to you.

13. Constructing a Tangent to a Circle

A *tangent* is a line that touches a circle at exactly one point.

Example: Construct a tangent to point A on the given circle with center O.

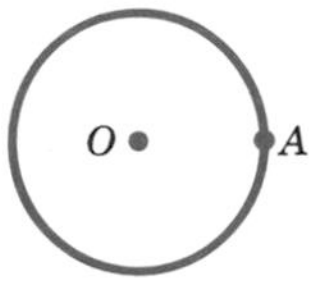

Step 1 Extend a line from O through A to X.

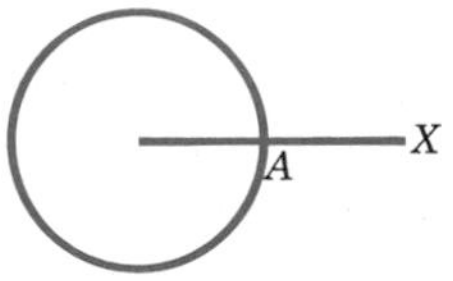

Step 2 Using construction 4, draw a perpendicular to line OX at point A. This perpendicular is tangent to the circle.

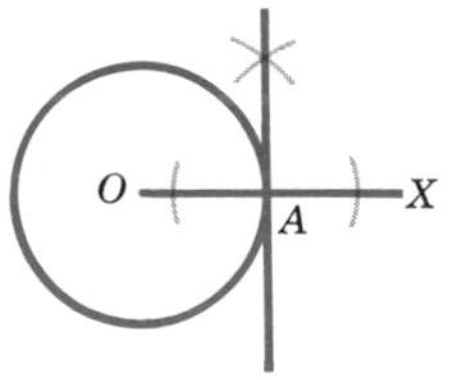

But suppose the tangent line must also go through a point outside the circle. Do it this way:

14. Construction of a Tangent to a Circle from a Point Off the Circle

Example: Construct a tangent to the circle with center X from point M.

Step 1 Draw the line MX and bisect it using construction 1. Label the midpoint P.

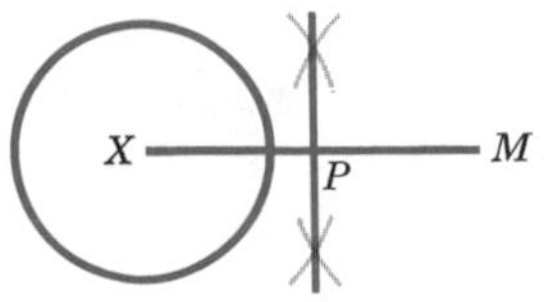

Step 2 Using P as the center and a radius equal to PX, strike an arc that cuts the circle at two points A and B.

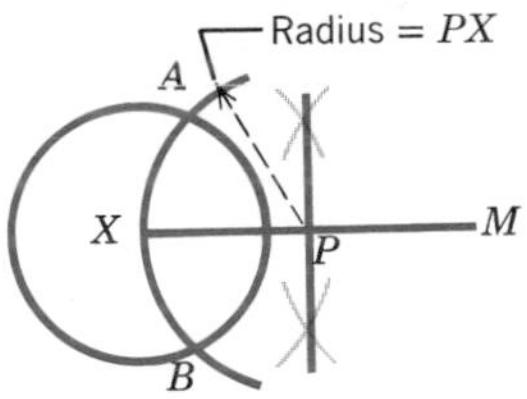

Step 3 Draw lines joining M to A and M to B. The lines MA and MB are tangent to the circle.

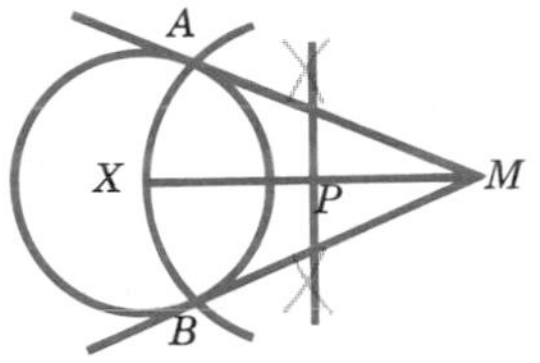

Notice that there are two lines tangent to any circle from some point outside the circle.

Practice these last two constructions by doing the following problems.

(a) Draw a tangent to this circle at point P. (The center of the circle is at O.)

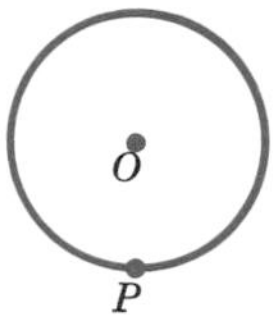

(b) Draw a tangent to this circle from point Q. (The center of the circle is at O.)

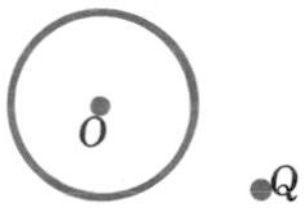

(c) Draw a tangent to this arc at point B. (**Hint:** First use Construction 12 to find the center of the arc.)

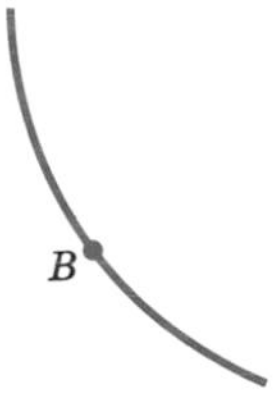

Check your constructions in **78**.

---

78 (a)

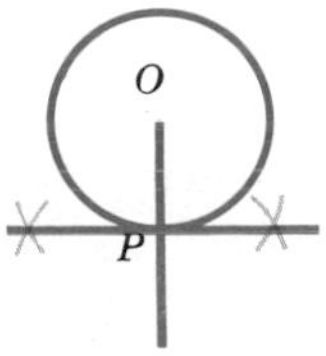

(b)

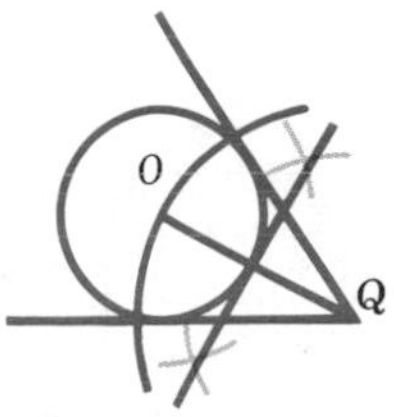

(c)

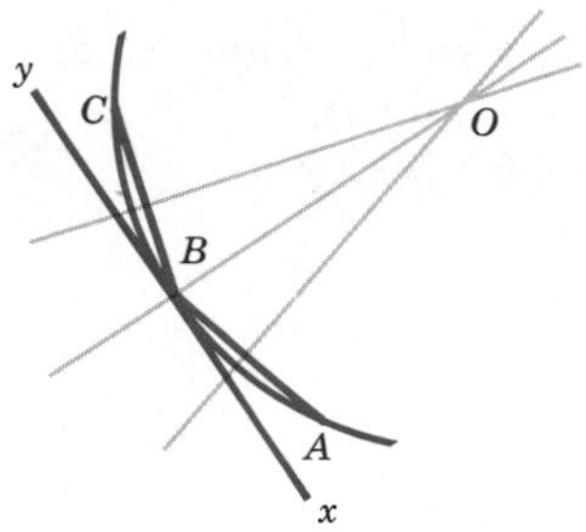

Draw lines AB and BC. Use these to find the center O. Draw line OB. Draw line XY perpendicular to OB. Then XY is tangent to the arc.

Turn to **79** for a set of exercises designed to help you become more expert at using these constructions.

---

79 **Exercises 7-4 Geometric Constructions**

Perform the following constructions. Check your own work with a ruler or protractor.

1. Draw a line segment of any length and bisect it.
2. Draw a line segment of any length and divide it into 4 equal parts.
3. Draw a line segment of any length and divide it into 7 equal parts.
4. Draw an acute angle and bisect it.
5. Draw an obtuse angle and bisect it.
6. Construct a 90° angle using any method.
7. Construct a 45° angle.
8. Draw an acute angle and duplicate it.
9. Draw an obtuse angle and duplicate it.
10. Construct a square with side length 4″.
11. Construct a rectangle such that one pair of opposite sides has three times the length of the other pair.
12. Draw a line and a point off the line. Construct a perpendicular from the point to the line.
13. Draw an irregular triangle and copy it.
14. Construct an equilateral triangle with side length $3\frac{1}{2}$″.
15. Construct a 30° angle.
16. Construct a 120° angle.
17. Construct a 135° angle.
18. Draw a line and pick a point on the line. Construct a perpendicular to the line at that point.
19. Construct a line parallel to AB through C.

20. Construct a line parallel to DE 3″ away.

21. Divide this line into 5 equal parts.

22. Use construction methods to find the center *and* complete the circle for the arc shown.

23. Construct a tangent to the circle at A.

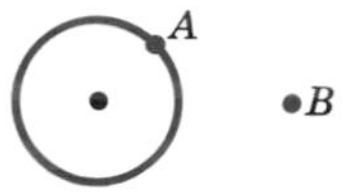

24. For the circle above, construct the two tangent lines from B to the circle.

Now go to **80** for a set of problems on practical geometry.

7 Practical Geometry

PROBLEM SET 7

80 Answers are on page 556.

Practical Geometry

A. Solve the following problems involving angles.

Name each angle and tell whether it is acute, obtuse, or right.

1. 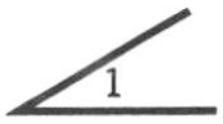

2. 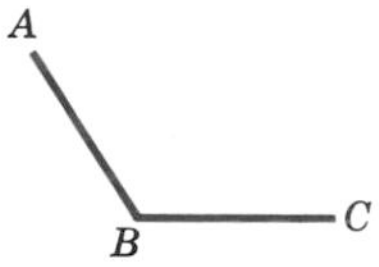

3. 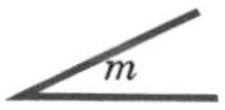

4. 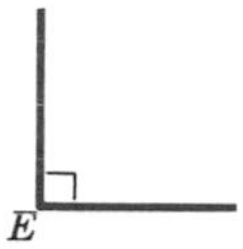

5. 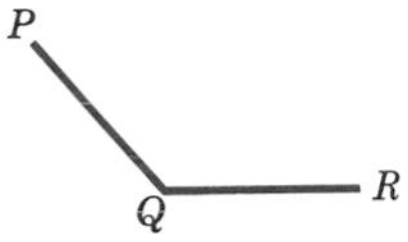

6. 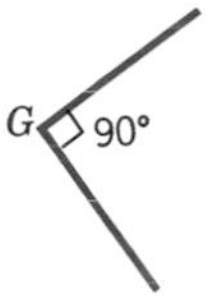

Measure the indicated angles using a protractor.

7. 

8. 

9. 

10. 

11.

12. 

13. 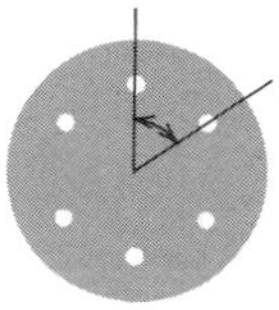

14. 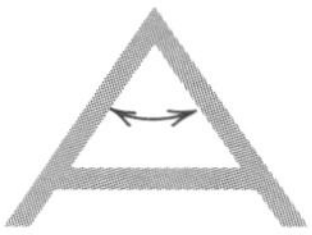

15. 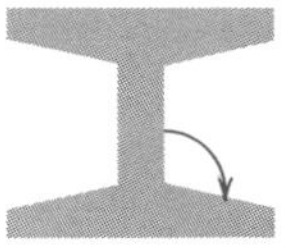

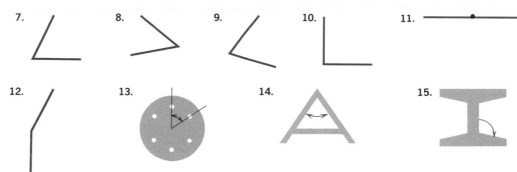

16. Use a protractor to draw angles of the following measures:
(a) 65° (b) 138° (c) 12° (d) 90°

17. Find the size of each indicated angle without using a protractor.

(a) 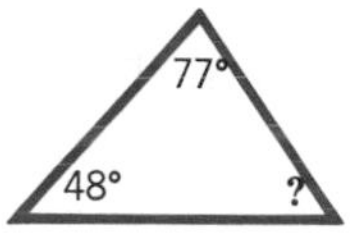

(b) 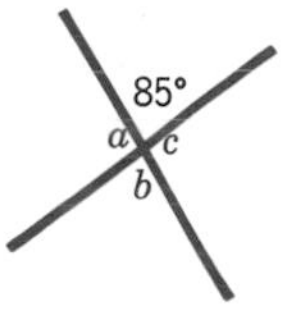

(c) 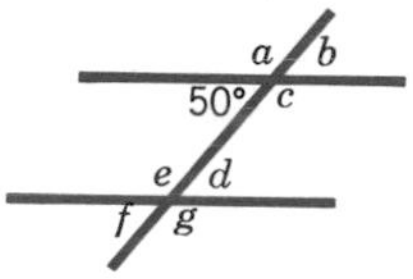

(d) 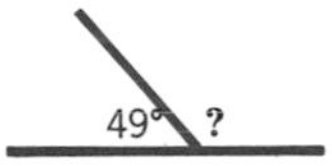

(e) 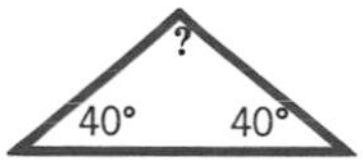

(f) 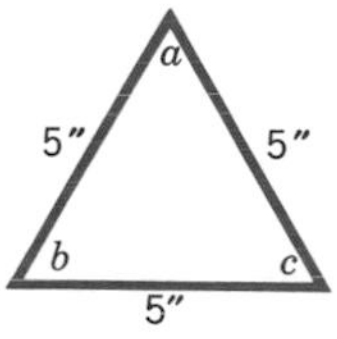

(g) 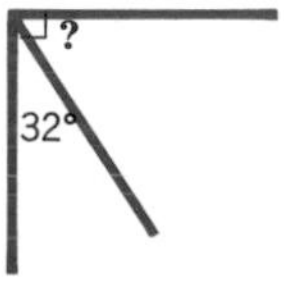

Name ____________

Date ____________

Course/Section ____________

B. **Solve the following problems involving plane figures.**

Find the perimeter (or circumference) and area of each figure. (Round to the nearest tenth.)

1. 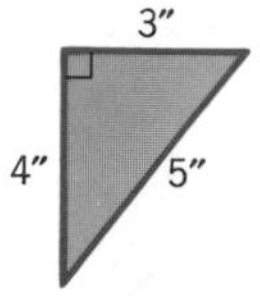

2. 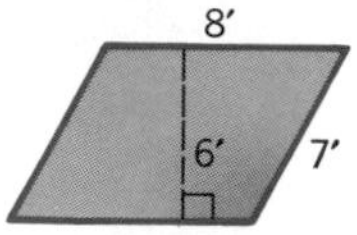

3. 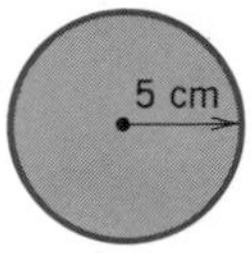

4. 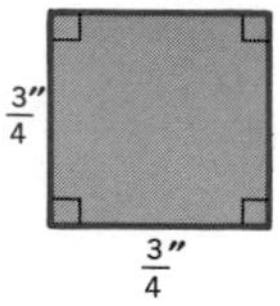

5. 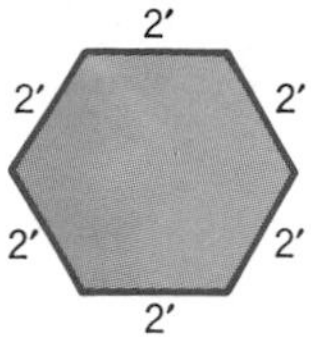

6. 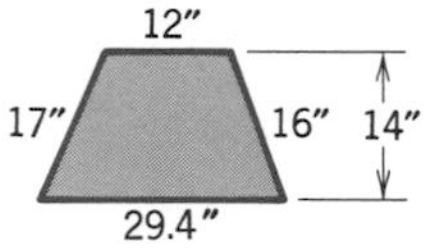

7. 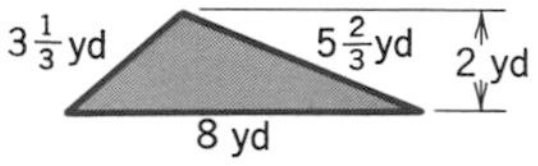

8. 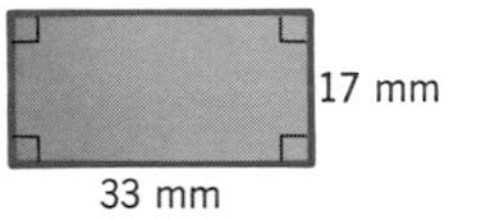

9. 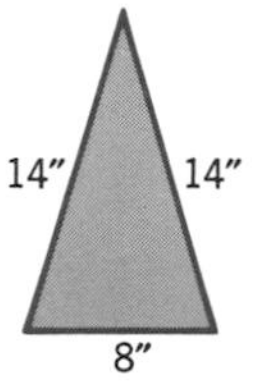

10. 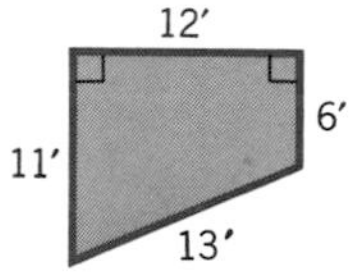

11. 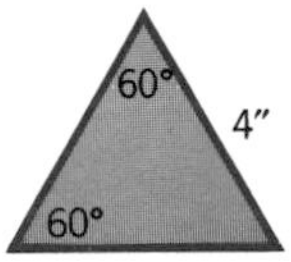

12. 

13. 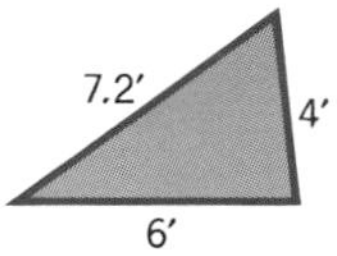

14. 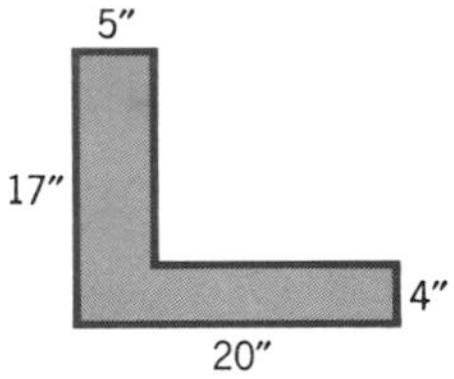

15. 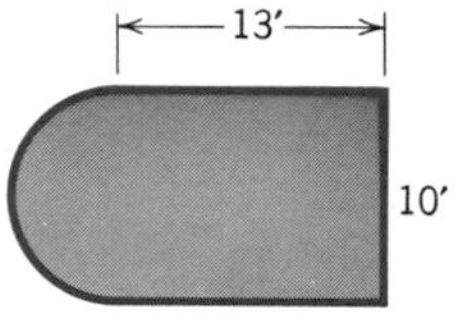

16. 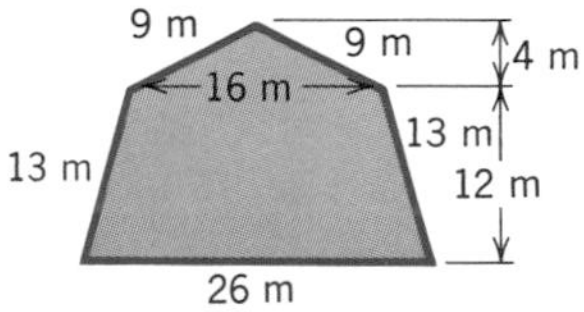

17. 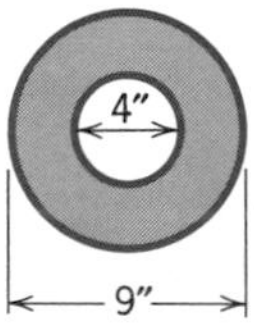

18. 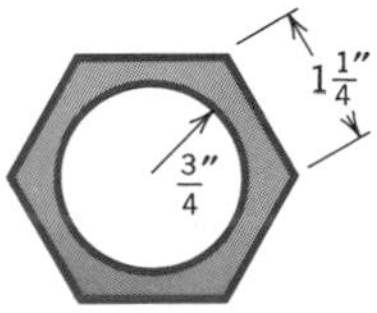

Find the missing dimensions of the following figures. (Round to the nearest hundredth if necessary.)

19. 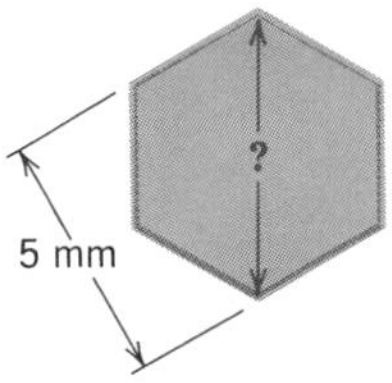

20. 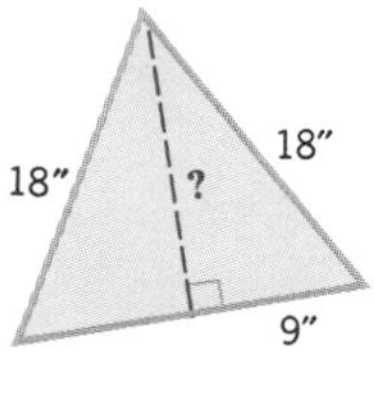

21. 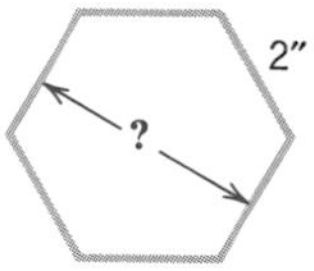

22. 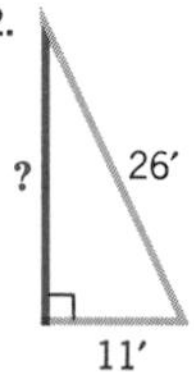

23. 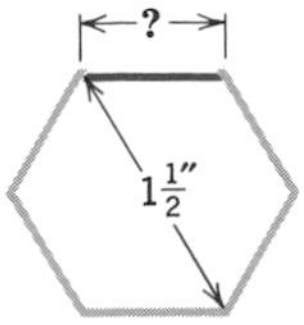

24. 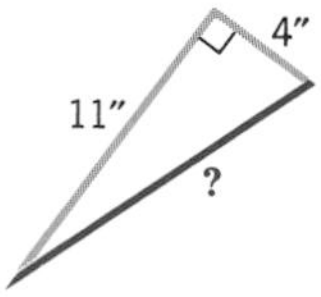

C. **Solve the following problems involving solid figures.**

Find the lateral surface area and volume of each of the following. (Round to the nearest tenth.)

1. 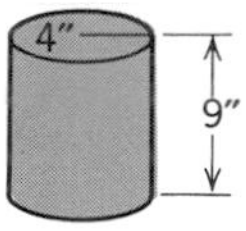

2. 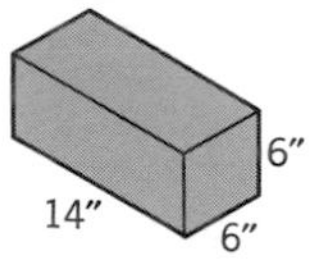

3. 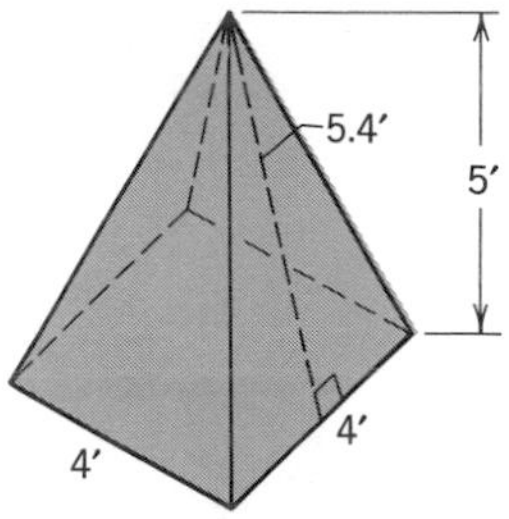

4. 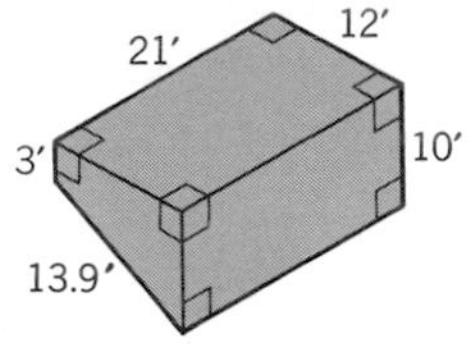

5. 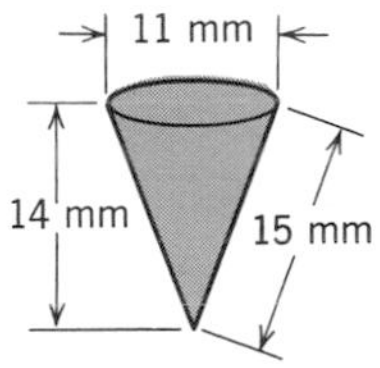

6. 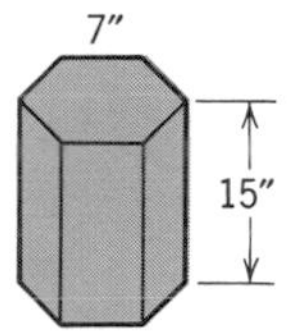

7. 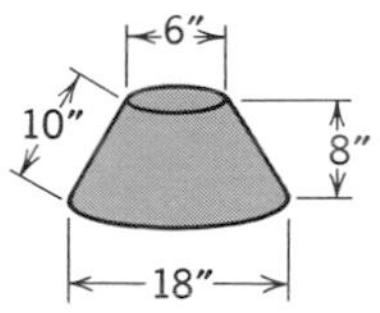

Use $\pi = 3.14$

8. 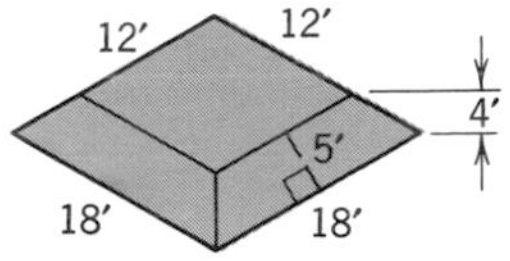

Find the total surface area and volume of each of the following. (Round to the nearest tenth.)

9. 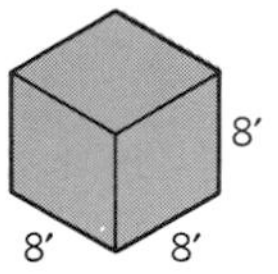

10. 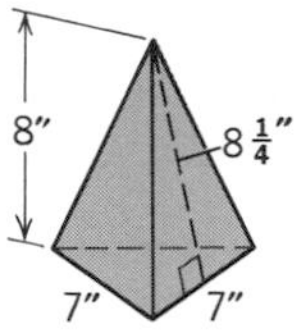

11. 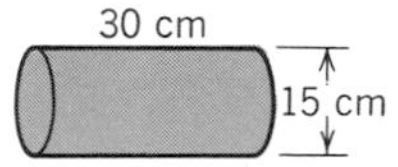

12. 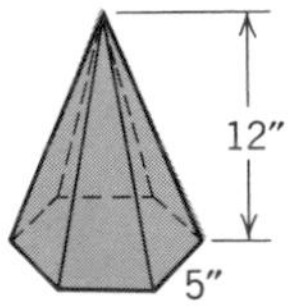

Slant height $= 12\frac{3}{4}$″

13. 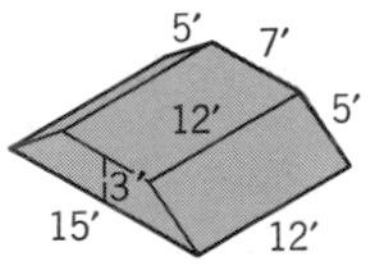

14. 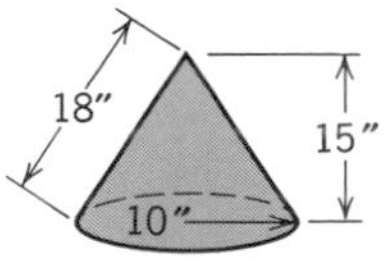

15. A sphere with a radius of 10 cm.

16. 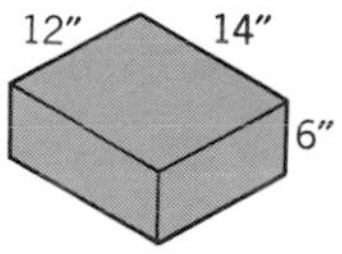

17. A sphere with a diameter of 13″.

18. 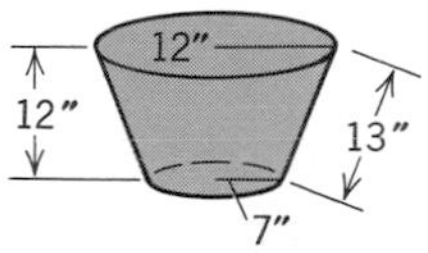

Use $\pi = 3.14$″

19. 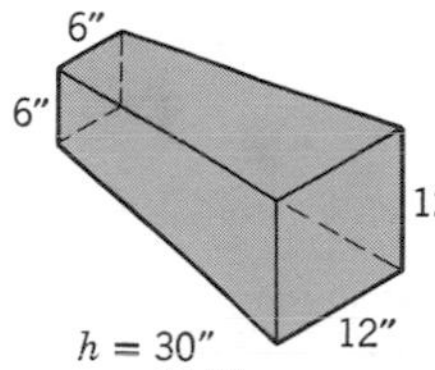

$h = 30$″
$s = 30.1$″

20. 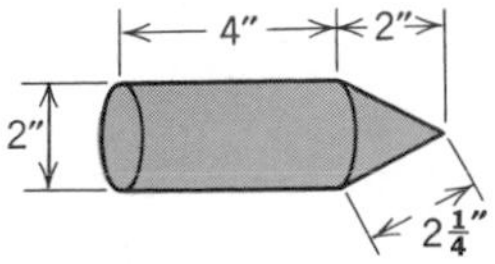

21. 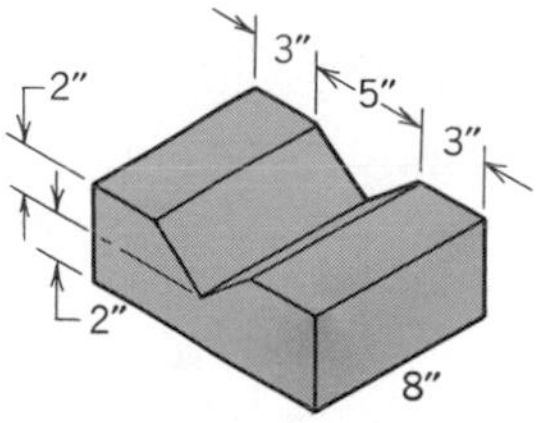

D. Perform the following constructions.

1. Copy this acute angle, and then bisect the duplicate. Check your accuracy by measuring with a protractor.

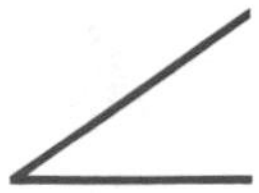

2. Bisect this line segment.

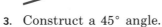

3. Construct a $45°$ angle.
4. Construct a perpendicular from the point A to the line XY.

5. Construct a square with side length $2\frac{1}{2}''$.
6. Construct an equilateral triangle with side length $3''$.
7. Construct a $30°$ angle.
8. Construct a $135°$ angle.
9. Duplicate this triangle.

10. Construct a line parallel to the line AB passing through the point P.

11. Construct a line parallel to this line and $2\frac{1}{2}''$ from it.

12. Divide this line segment into 7 equal parts.

13. Find the center of the arc AB by construction and then complete the circle.

14. Construct a line tangent to the circle at B.

15. Construct the two tangent lines from C to the circle.

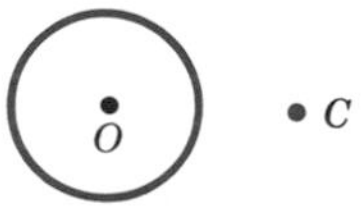

E. Practical Problems

1. What will it cost to pave a rectangular parking lot 220′ by 85′ at \$2.75 per square foot?
2. How many boxes $16'' \times 12''$ by $10''$ will fit in 1250 cubic feet of warehouse space? (1 cu ft $=$ 1728 cu in.)

3. A steel brace is used to strengthen the table leg as shown. What is the length of the brace? (Round to the nearest hundredth of an inch.)

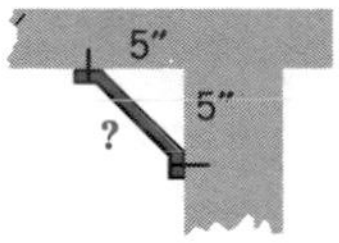

4. Holes are punched in a steel plate as indicated. Through what angle must the plate be rotated to locate the second hole?

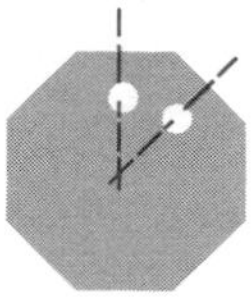

5. How many cubic *yards* of concrete are needed to pour the highway support shown in the figure? (1 cu yd = 27 cu ft.) (Round to nearest cu yd.)

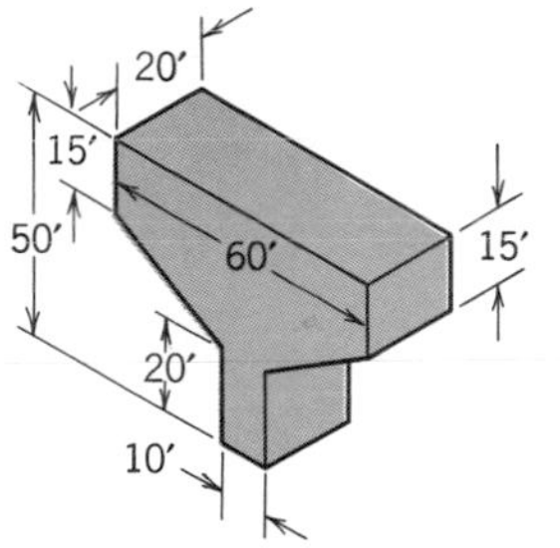

6. How many square feet of brick are needed to lay a patio in the shape shown?

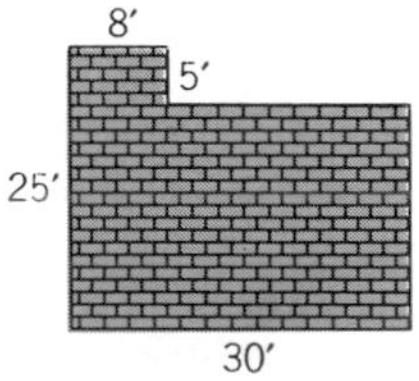

7. A coil of wire has an average diameter of 8″. How many feet of wire are there if it contains 120 turns? (Round to the nearest foot.)
8. A sheet metal worker needs to make a vent connection in the shape of the trapezoidal prism shown in the drawing. Use a protractor to find the indicated angle.

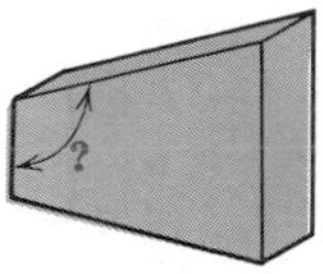

9. At $2.12 per foot, how much does it cost to fence the yard shown in the figure?

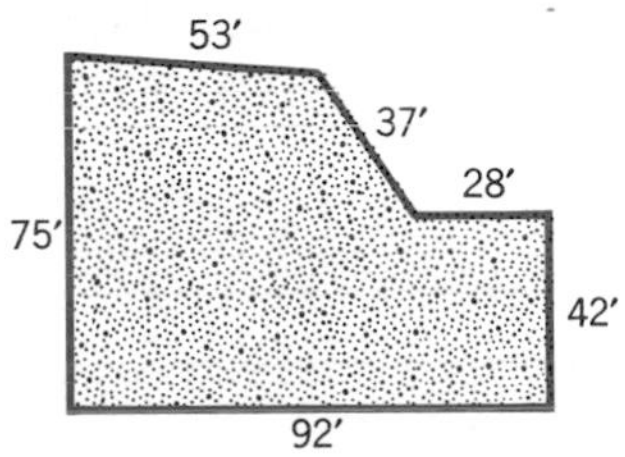

10. What is the capacity in gallons of a cylindrical water tank 3′ in diameter and 8′ high? (1 cu ft = 7.48 gal.) (Round to one decimal place.)

11. Find the length of straight stock needed to bend $\frac{1}{2}''$ steel into the shape indicated. Use $\pi = 3.14$.

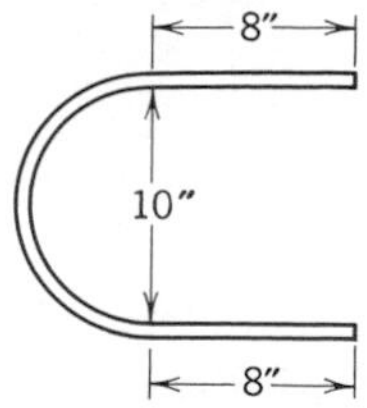

12. A hexagonal piece of steel 9″ on a side must be milled by a machinist. What diameter round stock does he need? (Round to the nearest tenth of an inch.)

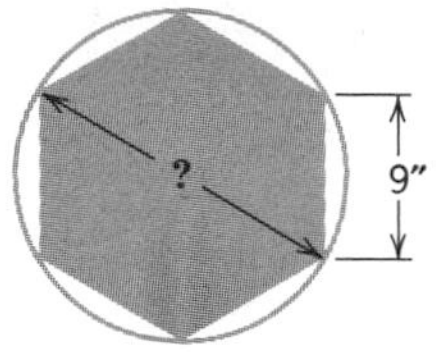

13. How much guy wire is needed to anchor a 30′ pole to a spot 20′ from its base? Allow 4″ for fastening. (Round to the nearest inch.)

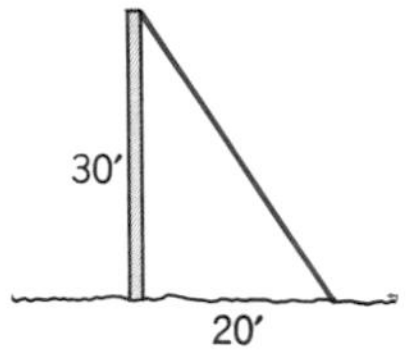

14. What is the weight of the 10 ft long steel I-beam if the density of steel is 0.2833 lb per cu in.? (Round to the nearest pound.)

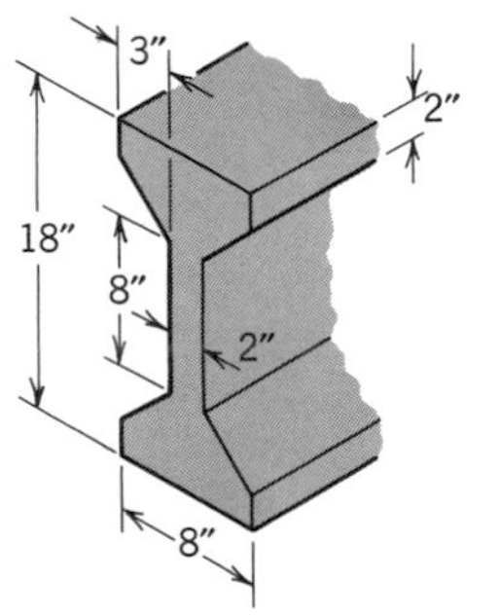

15. How many 4′ by 8′ sheets of exterior plywood must be ordered for the 4 walls of a building 20′ long, 32′ wide, and 12′ high. Assume that there are 120 sq ft of window and door space. (**Note:** You can cut the sheets to fit, but you may not order a fraction of a sheet.)
16. How many quarts of paint will it take to cover a spherical water tank 25′ in diameter if one quart covers 50 sq ft?
17. A hole must be excavated for a swimming pool in the shape shown in the figure. How many trips will be needed to haul the dirt away if the truck used has a capacity of 9 cubic meters? (Remember, a fraction of a load constitutes a trip.)

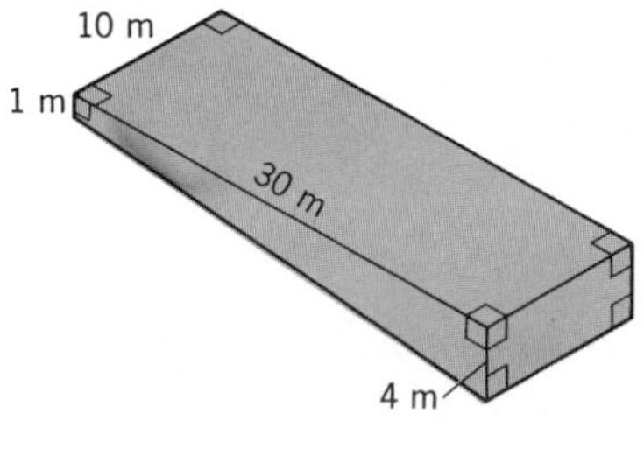

18. What must be the height of a cylindrical 750 gallon tank if it is 4′ in diameter? (1 cu ft = 7.48 gallons.) (Round to the nearest inch.)

19. Find the perimeter of the steel plate shown in the figure. (Round to the nearest tenth of an inch.)

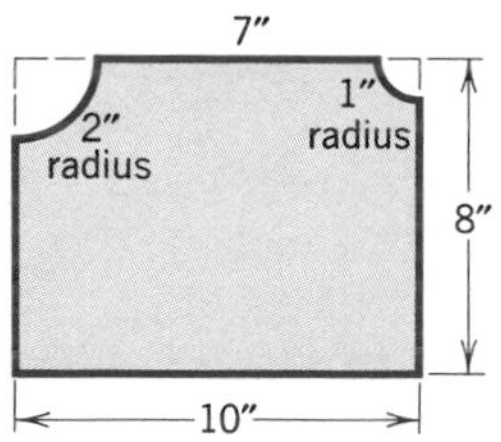

20. At what angle must sheet rock be cut to conform to the shape of the wall shown? Measure with a protractor.

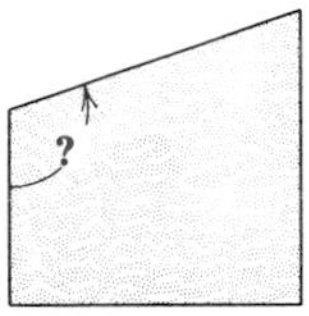

21. A triangular shape with a base of 6″ and a height of 9″ is cut from a steel plate weighing 5.1 lb per sq ft. Find the weight of the shape. (1 sq ft = 144 sq in.) (Round to the nearest ounce.)

22. How many cubic feet of propane will the tank in the figure contain? (1 cu ft = 1728 cu in.) (Round to the nearest tenth of a cu ft.)

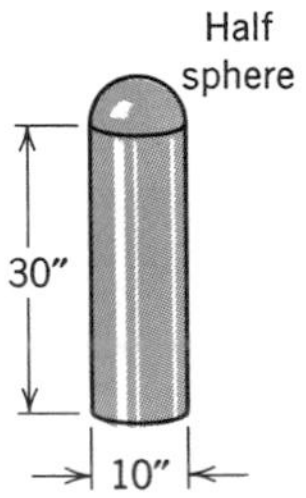

23. An air shaft is drilled from the indicated spot on the hill to the mine tunnel below. How long is the shaft to the nearest foot?

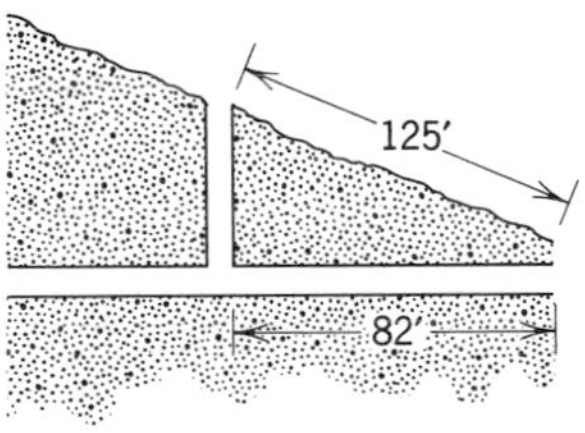

24. What diameter circular stock is needed to mill a square bolt $\frac{1}{2}$″ on a side? (Round to one-thousandth of an inch.)

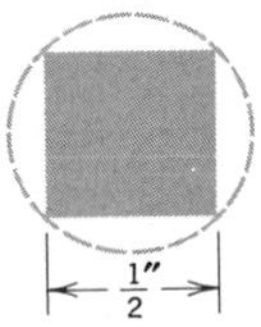

25. At a density of 42 lb per cu ft, what is the weight of fuel in the rectangular gas tank pictured?

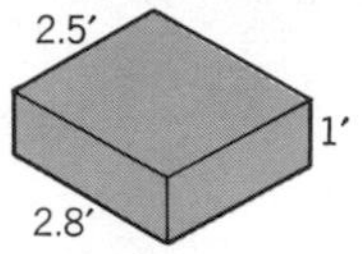

26. How many square inches of paper are needed to produce 2000 conical cups like the one in the figure? (Round to the nearest 100 sq in.)

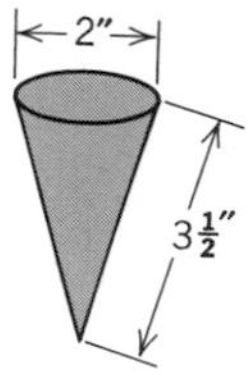

27. At 10¢ a foot how much will it cost to weatherstrip the following: 6 windows measuring 4′ by 6′; 8 windows measuring 3′ by 2′; and all sides except the bottom of two doors 3′ wide and 7′ high?

28. How many cubic yards of concrete are needed to pour the foundation shown in the figure? (Round to the nearest cu yd.)

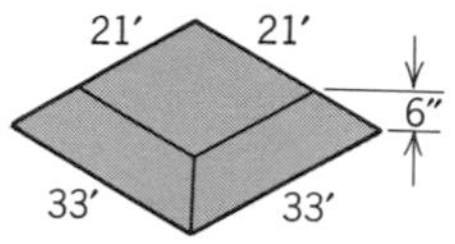

29. How many square inches of sheet metal are needed to make the bottom and sides of the pail in the figure? (Round to the nearest sq in. and use $\pi = 3.14$.)

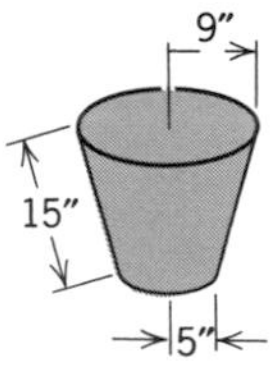

30. How many square feet of plaster board are needed to surface the wall shown? (Round to the nearest sq ft.)

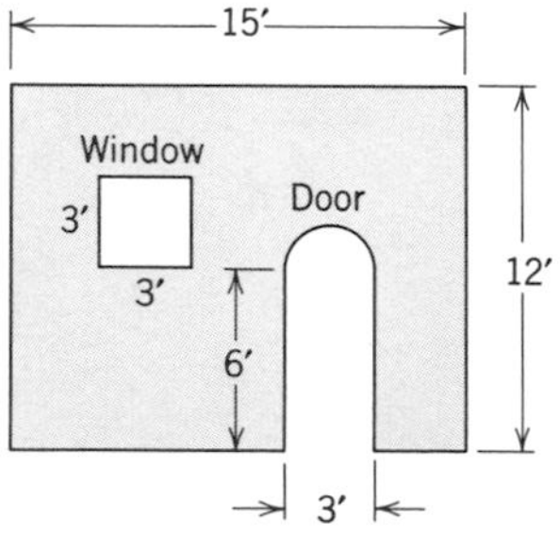

31. What length of belt is needed for the pulley shown? (Round to the nearest tenth.)

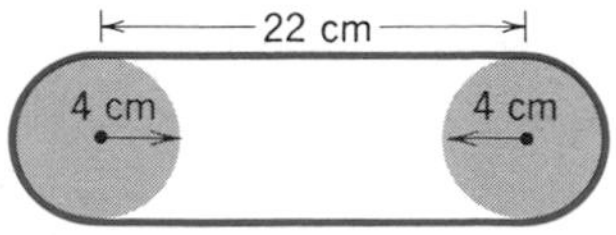

32. Electrical conduit must conform to the shape and dimensions shown. Find the total length of conduit in inches. (Use $\pi = 3.14$ and round to the nearest tenth.)

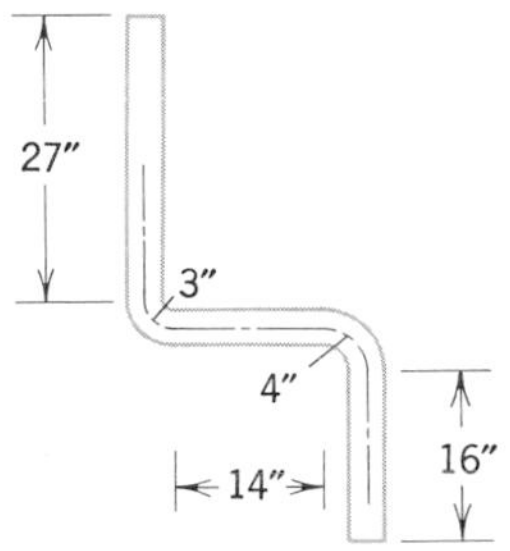

33. At \$2.75 per square foot how much will it cost to pave the circular pathway shown? (Use $\pi = 3.14$ and round to the nearest cent.)

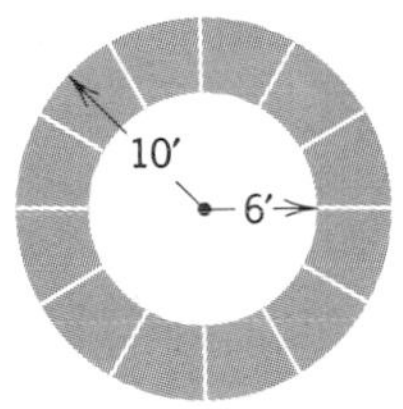

34. Find the capacity in gallons of the oil can in the figure. (1 gal = 231 cu in.) (Round to the nearest tenth of a gallon.)

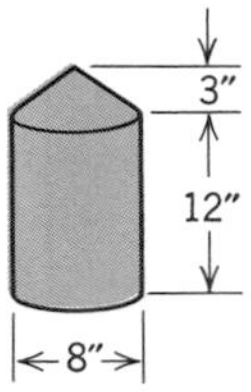

35. At 0.000017 lb per cu mm, find the weight of 1200 hex bolts, each 10 mm on a side and 5 mm thick with a hole 10 mm in diameter. (Round to the nearest tenth of a pound.)

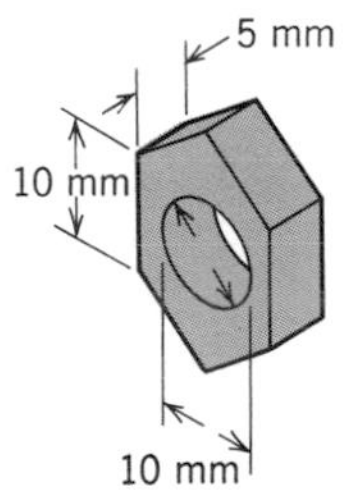

36. Find the size of ∠2 when the carpenter's square is placed over the board as shown.

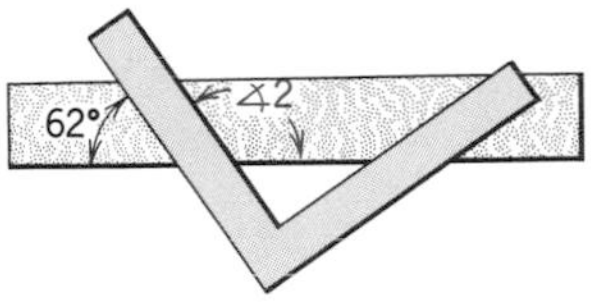

8 Graphs

PREVIEW 8

Objective

Upon successful completion of this unit you will be able to:

Sample Problems | **Where To Go For Help** (Page, Frame)

1. Read bar graphs, double bar graphs, circle graphs, and line graphs.

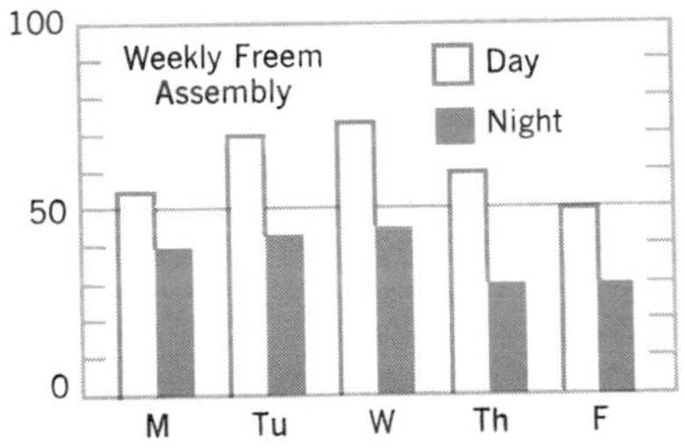

(a) Number of freems assembled by Tuesday day shift. ________ — Page 417, Frame 1

(b) Percent decrease in output from the Monday day shift to the Monday night shift. ________

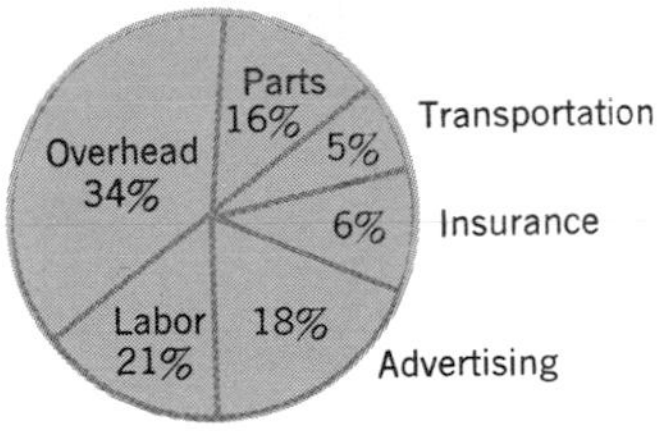

(c) If an average job costs $127.50, what part of this cost is for labor? ________ — Page 426, Frame 9

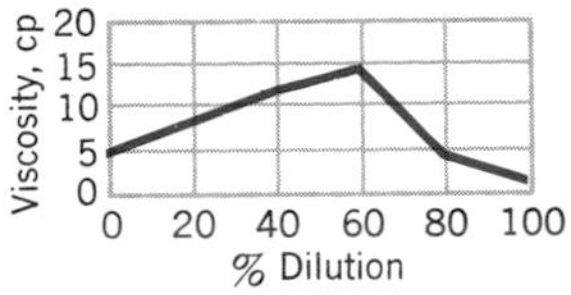

(d) What is the viscosity at 60% dilution? ________ — Page 423, Frame 5

(e) What is the approximate viscosity at 70% dilution? ________

2. Draw bar graphs, circle graphs, and line graphs from a table of data.

Work Output	Day
125	Mon.
300	Tue.
350	Wed.
310	Thu.
150	Fri.

(a) Draw a bar graph of this data. — Page 419, Frame 3

(b) Draw a line graph of this data. — Page 424, Frame 7

(c) Draw a circle graph of this data. — Page 426, Frame 9

3. Read technical graphs.

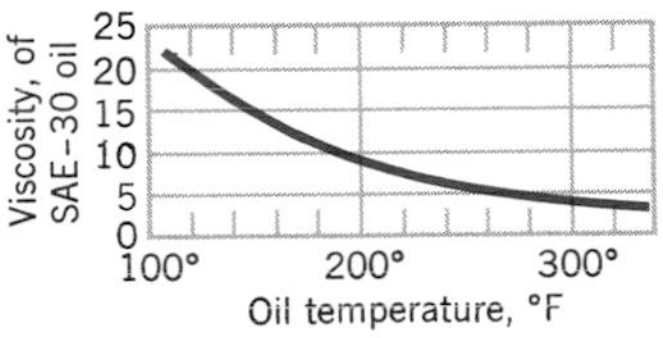

(a) What is the viscosity at 200° F? ________ — Page 435, Frame 15

(b) What is the decrease in viscosity from 180° to 250°? ________

Name ________________

Date ________________

Course/Section ________________

4. Draw technical graphs from data in tables or from formulas.

Draw graphs from each of the following. 438 **17**

(a)

t	p
0	0
2	3.8
4	7.0
6	10.5
8	14.0

(b) $D = 3A^2 + 4$ 441 **18**

(The answers to all preview problems are on the bottom of this page.)

If you are certain you can work *all* of these problems correctly, turn to page 447 for a set of practice problems. If you cannot work one or more of the preview problems, turn to the page indicated after the problem. If you want to be certain you are successful at this, turn to frame **1** and begin work there.

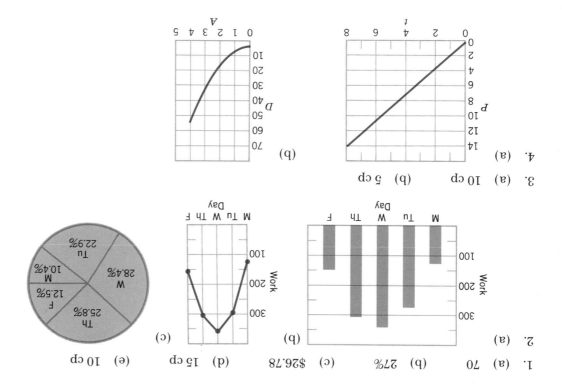

Answers to Preview 8

1. (a) 70 (b) 27% (c) $26.78 (d) 15 cp (e) 10 cp

2. (a)

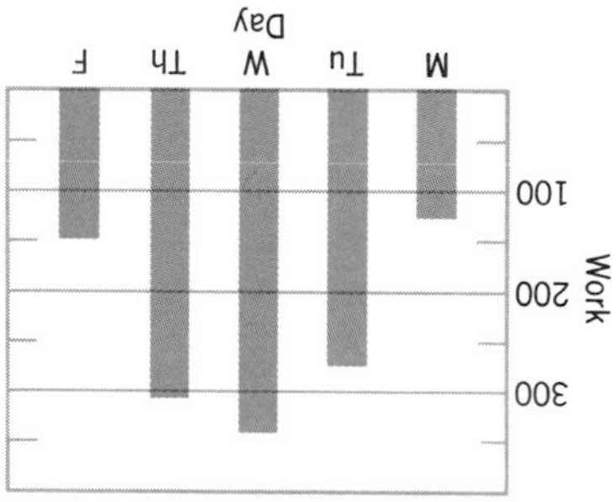

(b)

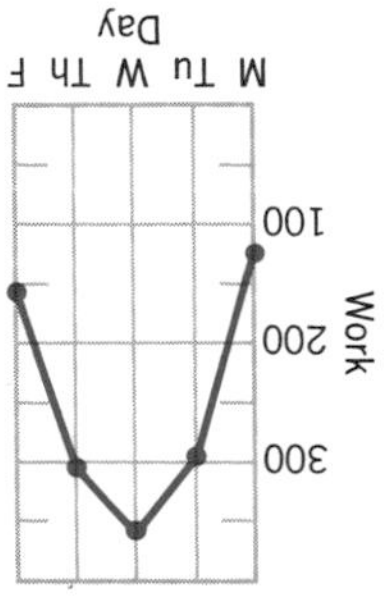

(c)

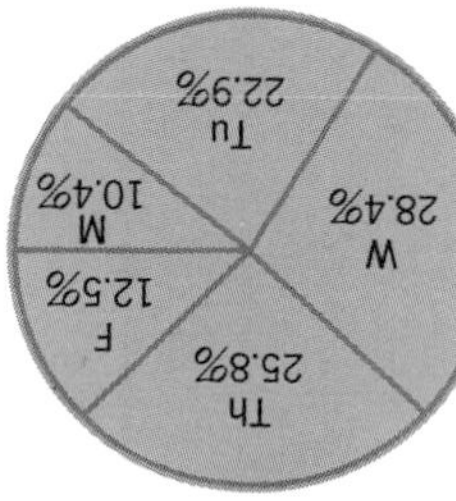

3. (a) 10 cp (b) 5 cp

4. (a)

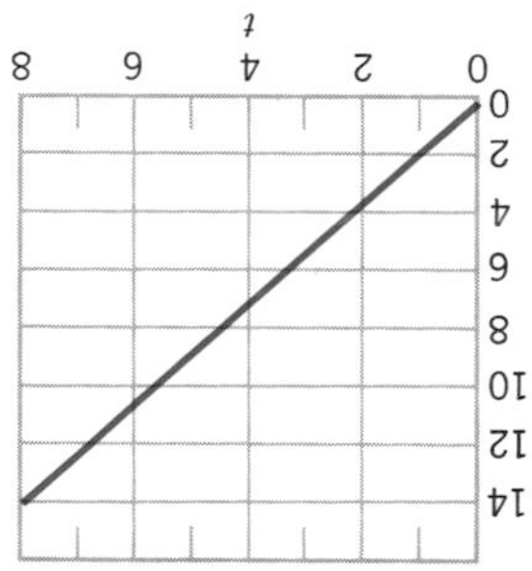

(b)

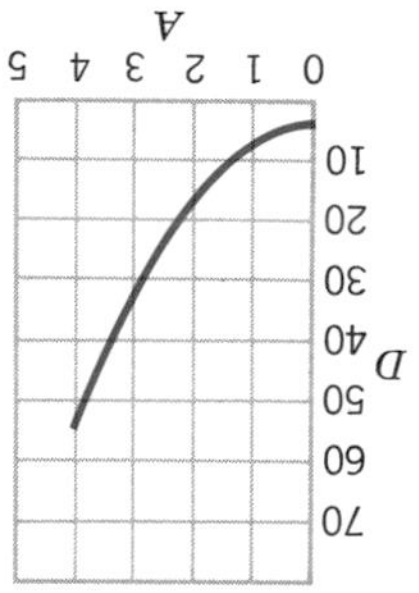

8 Graphs

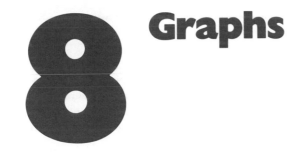

A graph is a pictorial display of information. By means of a graph a jumble of confusing data can be made to reveal significant trends and meaningful comparisons. A graph can be used to show the relationship between two quantities in a formula and thereby eliminate the need for long calculations. Many trades workers and technicians find it necessary to read the graphs that appear in handbooks, technical reports, or instruction manuals. A few find it necessary to make graphs using their own data. In this chapter you will learn how to read graphs and how to make them.

8-1 GRAPHICAL INFORMATION

Bar Graphs

1 A **bar graph** is used to display and compare the sizes of different but related quantities.

For example, the graph shown gives price comparisons for different types of plywood. The grades of plywood are listed along the bottom or *horizontal* axis of the graph. The cost per sheet in dollars is given along the side or *vertical axis.*

Comparative Cost of $\frac{3}{4}''$ Plywood

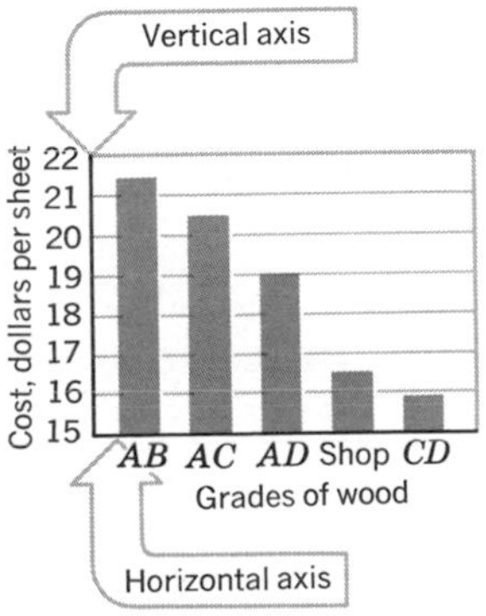

Notice that the cost scale goes from \$15 to \$22; no type costs more than \$22 or less than \$15. Also notice that the scale must increase uniformly in a bar graph. In this graph each vertical unit represents an increase of \$1.

To find the cost of a particular type of plywood, locate the top of its bar. Move straight across to the vertical scale, and read the price. If the price falls between two numbers, estimate the fraction of a dollar. For example, CD plywood, the least expensive grade, costs approximately \$15.90 per sheet, since the top of its bar falls just below \$16.

Use this graph to answer these questions.

(a) What is the cost of a sheet of "shop grade" $\frac{3}{4}''$ plywood?
(b) What is the cost of three sheets of grade AC $\frac{3}{4}''$ plywood?
(c) What is the difference in cost between a sheet of AB grade plywood and a sheet of AD grade plywood?

Check your answers in **2**.

---

2 (a) \$16.50
(b) AC grade costs \$20.50 per sheet. Three sheets will cost \$61.50.
(c) AB grade costs \$21.40. AD grade costs \$19.00. The difference is \$2.40.

In a **double bar graph** side-by-side comparisons are made by two related quantities. For example, in this double bar graph the costs of two different types of insurance are compared.

Average Annual Premium for \$1000 of Life Insurance

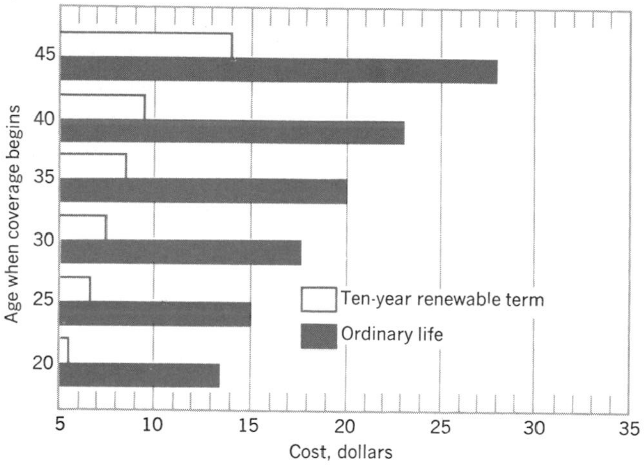

This graph differs from the first type in that the categories of comparison are along the side, but of course it can be drawn either way.

On this graph the white bars represent ten-year renewable term insurance, and the shaded bars represent ordinary life insurance. To read the cost, follow the right edge of the bar down to the horizontal scale. For example, for coverage beginning at age 30, ten-year renewable term costs about \$7.50 per year for \$1000 coverage. \$1000 of ordinary life at the same age costs about \$17.70 per year.

Use this graph to answer the following questions:

(a) How much would \$10,000 of ordinary life insurance cost at age 35?
(b) How much would \$3000 of ten-year renewable term insurance cost at age 40?
(c) What is the yearly difference in cost between ordinary life and ten-year renewable term for a \$5000 policy at age 25?

Check your answers in **3**.

---

3 (a) $\$20.00 \times 10 = \$200.00$ (c) $\$15.00 - \$6.50 = \$8.50$
(b) $\$9.50 \times 3 = \$28.50$ $\$8.50 \times 5 = \$42.50$

Bar graphs are easy to read, and it is rather easy to draw a bar graph to illustrate numerical data. For example, suppose you want to display the following data in a bar graph showing radio and television sales in five stores.

SALES OF RADIOS AND TELEVISIONS AT FIVE APPLIANCE STORES

Store	Radios	Televisions
Ace	140	65
Wilson's	172	130
Martin's	185	200
XXX	195	285
Shop-Rite	190	375

To draw a bar graph, follow these steps:

Step 1 Decide what type of bar graph to use. Because we are comparing two different items, we should use a double bar graph. The bars can be placed either horizontally or vertically—you get to choose. In this case let's make the bars horizontal.

Step 2 Choose a suitable spacing for the vertical (side) axis and a suitable scale for the horizontal (bottom) axis. Label each axis. When you label the vertical or side axis it should read in the normal way.

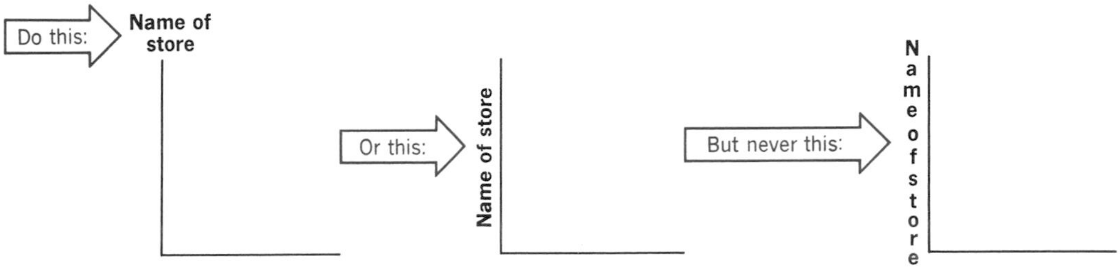

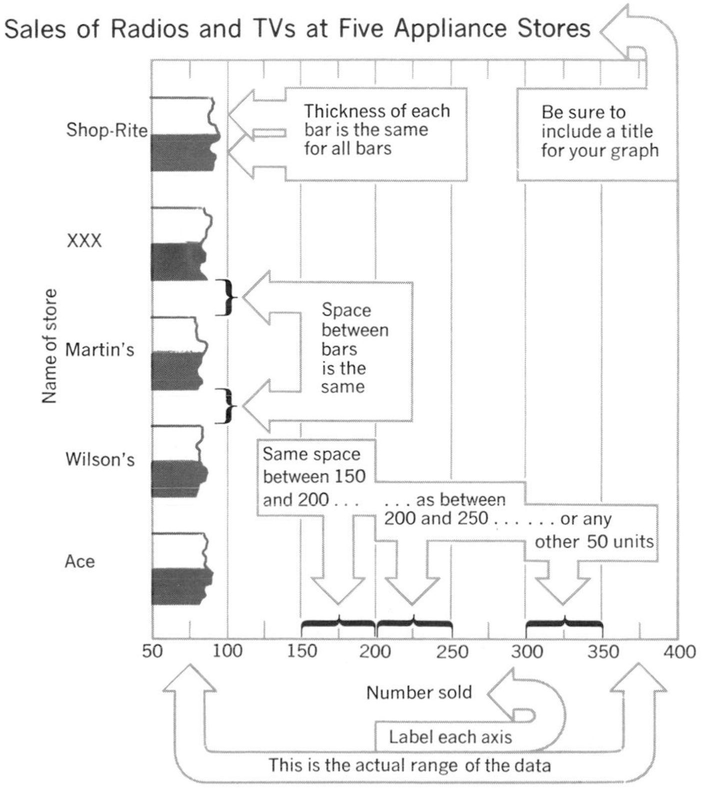

Notice that the numbers on the bottom axis are evenly spaced and cover the entire range of numbers in the data, in this case from 65 to 375.

Also notice that we arranged the stores in order of amount of sales. The biggest seller, Shop-Rite, is at the top, and the smallest seller, "Ace," is at the bottom. This is not necessary, but it makes your graph easier to read.

Step 3 Use a straightedge to mark the length of each bar according to the data given. Round the numbers if necessary. Drawing your bar graph on graph paper will make the process easier.

The final bar graph will look like this:

Sales of Radios and TVs at Five Appliance Stores

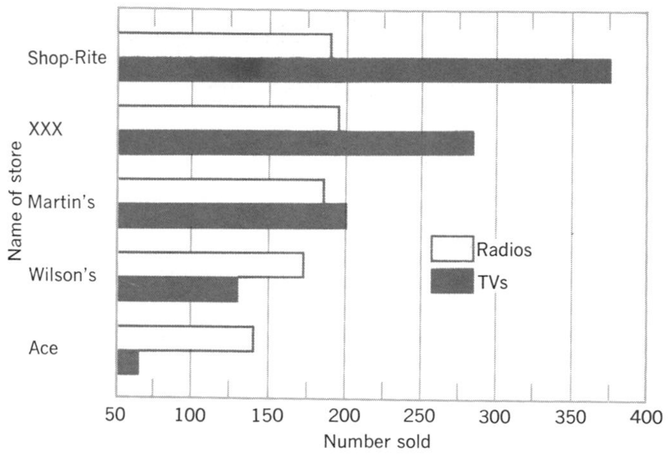

Use the following data to make a bar graph.

AVERAGE ULTIMATE COMPRESSION STRENGTH OF COMMON MATERIALS

Material	Compression Strength (lb. per sq. in.)
Hard bricks	12,000
Light red bricks	1,000
Portland cement	3,000
Portland concrete	1,000
Granite	19,000
Limestone and sandstone	9,000
Trap rock	20,000
Slate	14,000

Compare your graph with ours in **4**.

---

4 Average Ultimate Compression Strength of Common Materials

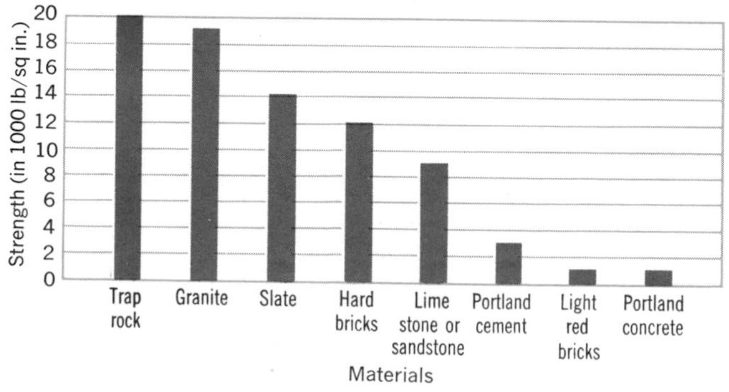

Don't worry if your graph does not look exactly like this one. The only requirement is that all of the data be displayed clearly and accurately.

Notice that the vertical scale has units of *thousands* of lb per sq in. We drew it this way to save space.

Test your understanding of bar graphs by answering the following questions.

Energy Consumption of Gas Appliances

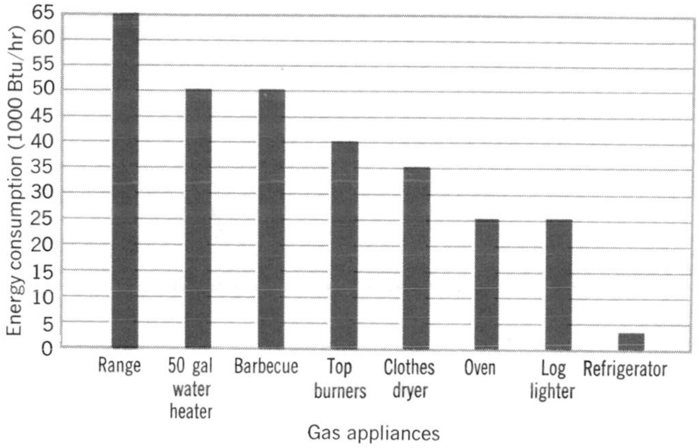

1. (a) Which appliance uses the most energy in an hour?
   (b) Which appliance uses the least energy in an hour?
   (c) How many BTU/hr does a gas barbecue use?
   (d) How many BTU/*day* (24 hr) would a 50-gal water heater use?
   (e) Is there any difference between the energy consumption of the range and that of the top burners plus oven?
   (f) How many BTUs are used by a log lighter in 15 minutes?
   (g) What is the difference in energy consumption between a 50-gal water heater and a clothes dryer?

2. Draw a bar graph displaying the following information.

WORLD'S LARGEST REFLECTOR TELESCOPES (AS OF 1980)

Observatory	Size (inches)
Hale	200
Kitt Peak	158
Cerro Tololo	158
La Silla	150
Siding Spring	140
Lick	120
McDonald	107
Crimean	104

3. Draw a bar graph that will display the following data:

MINIMUM RADIUS OF CONDUIT BENDS

	Radius (Inches)	
Size of Conduit	Conductors with Lead Sheath	Conductors without Lead Sheath
$\frac{1}{2}''$	6	4
$1''$	11	6
$1\frac{1}{2}''$	16	10
$2''$	21	12
$2\frac{1}{2}''$	25	15
$3''$	31	18
$3\frac{1}{2}''$	36	21
$4''$	40	24

Check your answers and graphs in **5**.

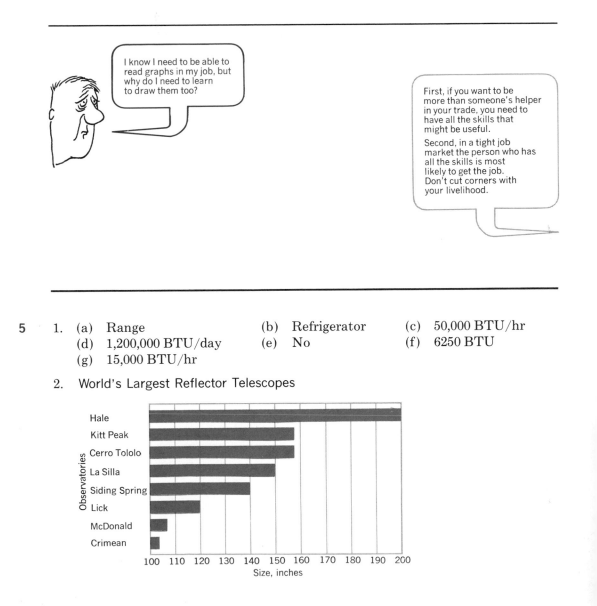

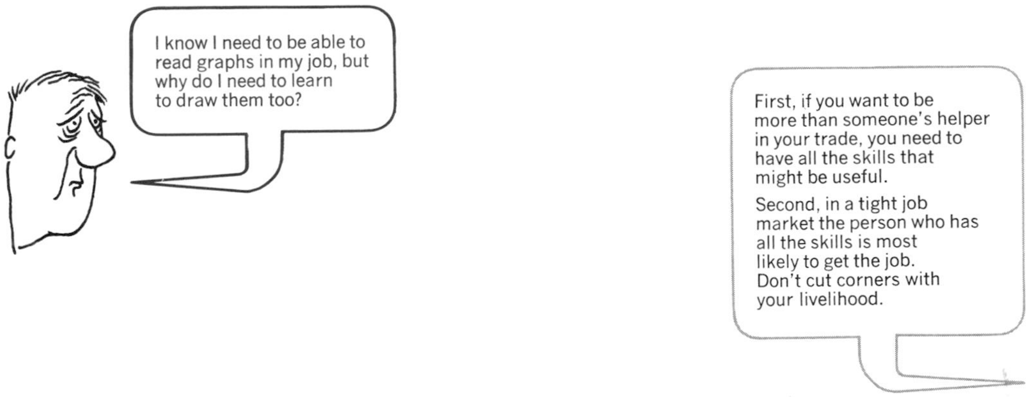
I know I need to be able to read graphs in my job, but why do I need to learn to draw them too?
First, if you want to be more than someone's helper in your trade, you need to have all the skills that might be useful.
Second, in a tight job market the person who has all the skills is most likely to get the job. Don't cut corners with your livelihood.

5 1. (a) Range (b) Refrigerator (c) 50,000 BTU/hr
(d) 1,200,000 BTU/day (e) No (f) 6250 BTU
(g) 15,000 BTU/hr

2. World's Largest Reflector Telescopes

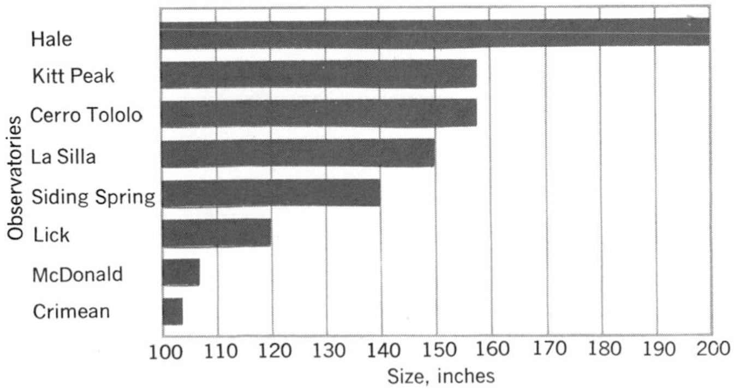
Hale
Kitt Peak
Cerro Tololo
La Silla
Siding Spring
Lick
McDonald
Crimean
Observatories
100 110 120 130 140 150 160 170 180 190 200
Size, inches

3. Minimum Radius of Conduit Bends

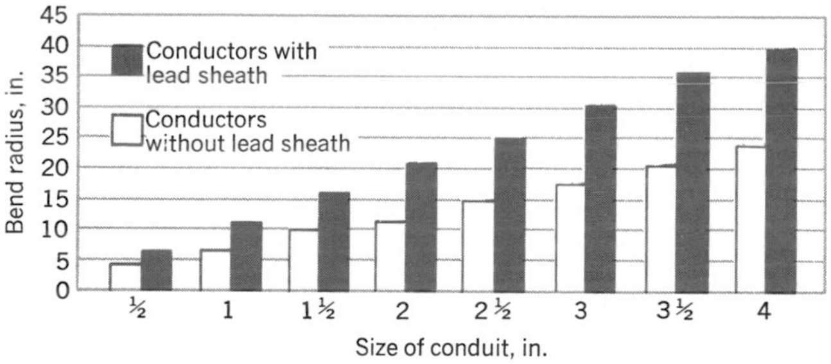

Line Graphs

A line graph is a display that shows the continuous change in quantity, usually over a period of time.

The graph shown is an example of a broken-line graph. Each dot represents the population of Great Britain at 50-year intervals from 1700 to 1950. The dots are connected to emphasize that the growth in population is continuous.

Population of Great Britain, 1700–1950

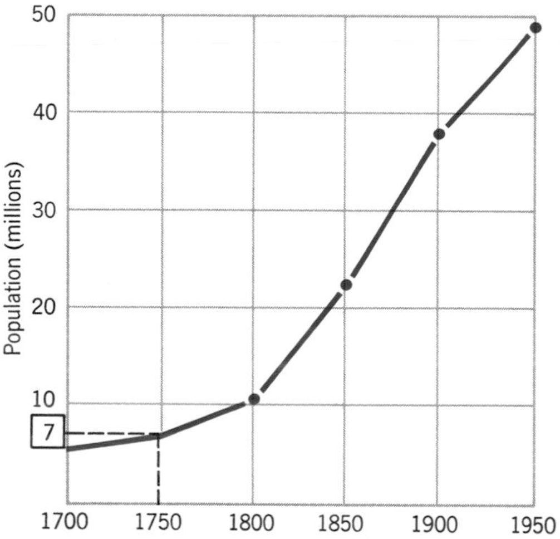

Notice that the numbers on both axes are evenly spaced. Also notice that the time axis is the horizontal axis—this is a convention that should always be followed.

The times plotted on the horizontal axis are spaced every 50 years in order to make it easy to estimate population numbers between points with accuracy. In general, choose points spaced 1, 2, 5, 10—or some multiple of these units—apart.

To read a line graph find the desired time along the horizontal axis and follow the line on the grid upwards until you reach the graph line. Then move horizontally to the left and read the value on the vertical axis. For example, the population in Britain in 1750 was approximately 7 million. Look at the graph to see how we arrived at this answer.

What was the population of Great Britain in 1850?

Check your answer in **6**.

---

6 Locate 1850 on the horizontal axis, move vertically up to the graph line and horizontally left to the axis to about 22 million or 22,000,000. We estimate the answer as 22,000,000.

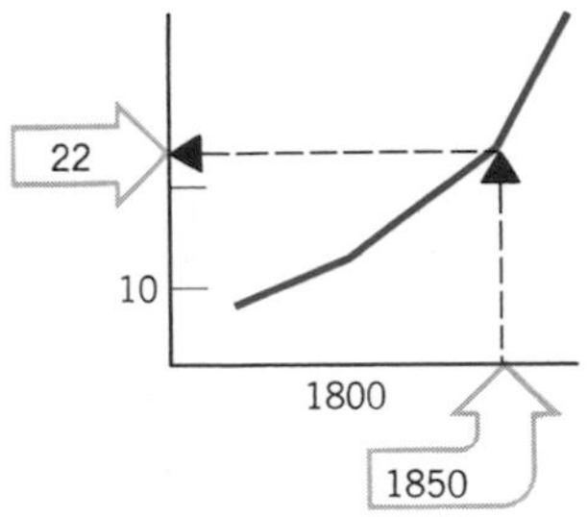

Use the graph to answer the following questions.

(a) What was the population of Great Britain in 1880?
(b) When did the population break the 20 million mark?
(c) By how much did the population increase between 1900 and 1950?
Check your answers in **7**.

---

7 (a) 32,000,000 (b) 1840 (c) 11,000,000

Notice in Problem (b) that we have estimated the answer from the straight-line portion of the graph.

Constructing a broken-line graph is similar to drawing a bar graph. For example, to construct a broken-line graph of the following data, follow the steps outlined.

COMPARATIVE PRICES OF CORN, 1969-1974

Year	Price per Bushel
1969	$1.16
1970	1.33
1971	1.08
1972	1.57
1973	2.55
1974	2.95

Step 1 Decide on a suitable spacing or scale for each of the axes. The scale markings must be evenly spaced. Time must be plotted along the horizontal axis.

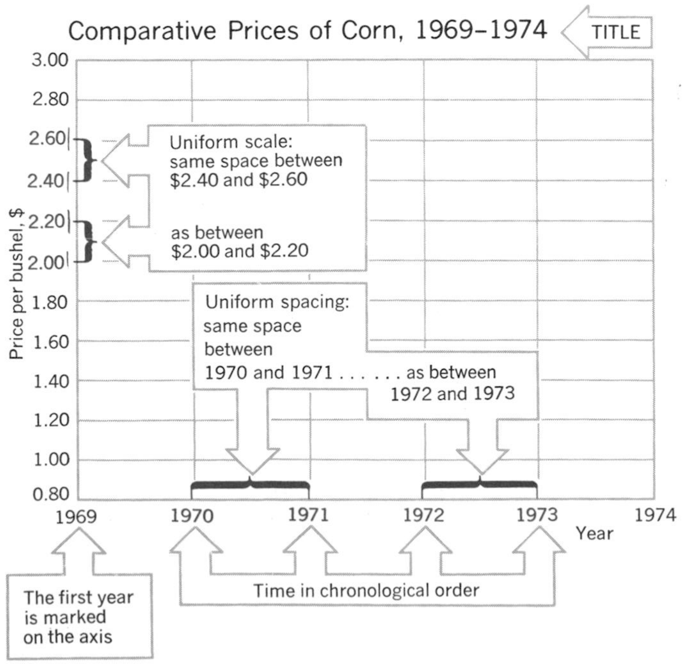

Notice that we place the year numbers each directly *on* a graph line and not between lines.

Step 2 Mark dots representing the yearly prices. This is called *plotting* the graph points. For each pair of numbers, locate the year on the horizontal scale and the price for that year on the vertical scale. Draw imaginary lines up from the year and horizontally from the price. Place a dot where these lines intersect.

For example, for 1970 $1.33

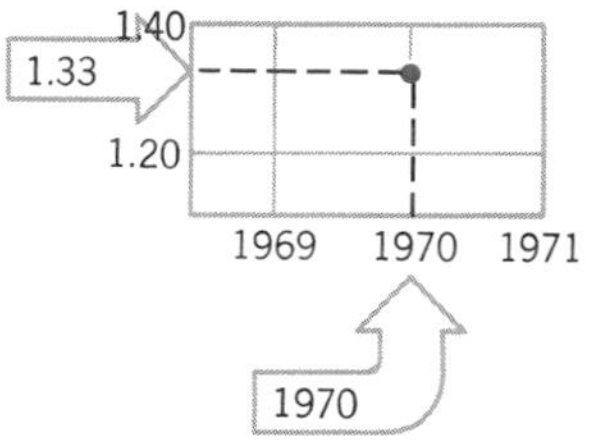

Step 3 After all points are plotted on the graph, connect adjacent points with straight-line segments.

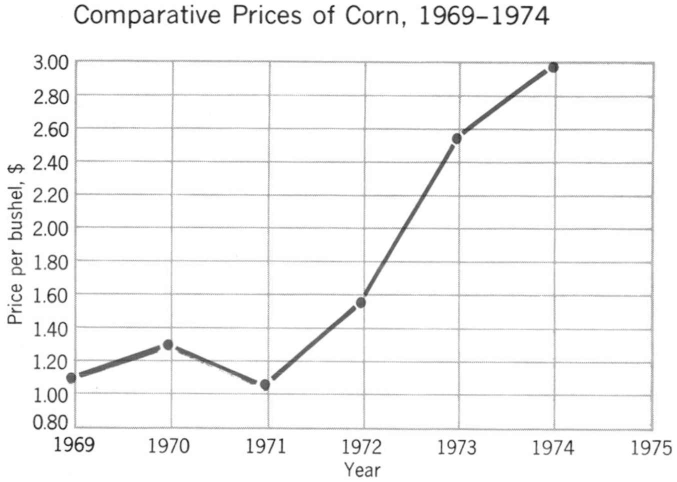

Plot the following data on this same graph.

COMPARATIVE PRICES OF BARLEY, 1969–1974

Year	Price per Bushel
1969	$0.89
1970	0.97
1971	0.99
1972	1.21
1973	2.13
1974	2.72

Check your graph in **8**.

8 Comparative Prices of Corn and Barley, 1969–1974

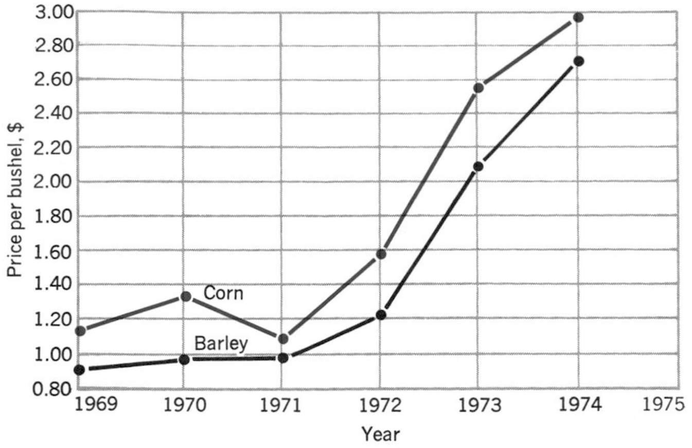

For more practice draw a broken-line graph displaying the following data.

AVERAGE MONTHLY RAINFALL IN TWO U.S. CITIES

	Rainwater, Washington	Mildew, Georgia
January	19.3 cm	4.7 cm
February	13.7	4.3
March	14.6	4.7
April	8.3	4.9
May	4.9	4.3
June	3.8	5.5
July	1.1	6.6
August	1.3	5.7
September	5.3	4.6
October	12.3	3.4
November	18.3	3.6
December	18.1	4.7

Check your graph in **9**.

---

9 Average Rainfall in Two U.S. Cities

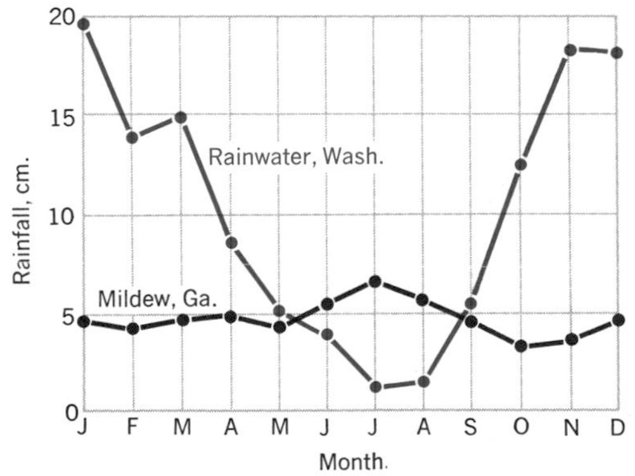

Circle Graphs

A **circle graph** or "pie graph" is used to show what fraction of the whole is represented by its different parts. For example, the circle graph shown here gives the distribution of expenses for a typical American family. The area of the circle represents the total budget of the family, and the percents add up to 100%. Each category of expense takes a proportional share of the circle area. Since clothing represents 8% of the budget, it also represents 8% of the circle or a pie-shaped sector of 29°.

Family Budget

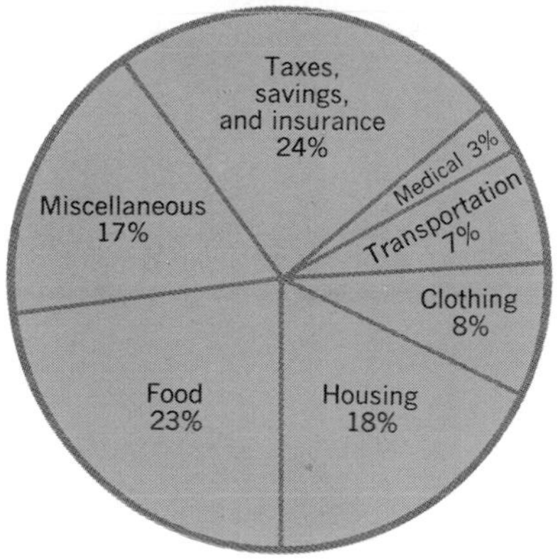

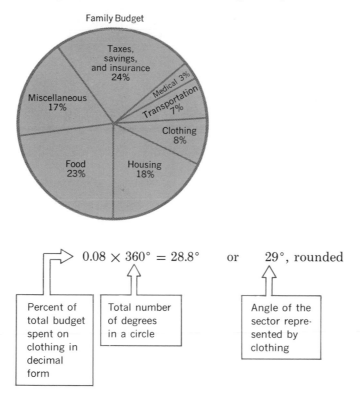

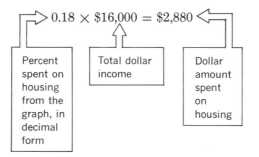

$0.08 \times 360° = 28.8°$ or $29°$, rounded

- Percent of total budget spent on clothing in decimal form
- Total number of degrees in a circle
- Angle of the sector represented by clothing

Use this same sort of calculation to determine the angle of the sector represented by food.

Check your answer in **10**.

---

10 $0.23 \times 360° = 82.8°$ or $83°$, rounded

Now suppose we were told that the family represented by this pie graph had a net income of $16,000. What actual dollar amount did they spend on housing?

$0.18 \times \$16,000 = \$2,880$

- Percent spent on housing from the graph, in decimal form
- Total dollar income
- Dollar amount spent on housing

Use this circle graph to answer the following questions:

(a) If the total income was $16,000, what dollar amount was spent on transportation?
(b) How much money went to taxes, savings, and insurance?
(c) If their net income rose to $18,000 the following year, and if the percents remained the same, what would be the increase in their food expenditures?

Look in **11** for the answers.

---

11 (a) $1120 (b) $3840 (c) $460

Making a circle graph also involves percent calculations. If the percents are already given, as in this table, the procedure is easy.

Continent	Percent of Earth Occupied
Asia	30%
Africa	20%
North America	16%
South America	12%
Antartica	10%
Europe	7%
Australia	5%

Step 1 Convert each percent to a decimal and multiply each by 360°. This tells you the number of degrees of the circle each category will occupy.

Asia	$0.30 \times 360° = 108°$
Africa	$0.20 \times 360° = 72°$
North America	$0.16 \times 360° = 57.6°$
South America	$0.12 \times 360° = 43.2°$
Antarctica	$0.10 \times 360° = 36°$
Europe	$0.07 \times 360° = 25.2°$
Australia	$0.05 \times 360° = 18°$
	360°

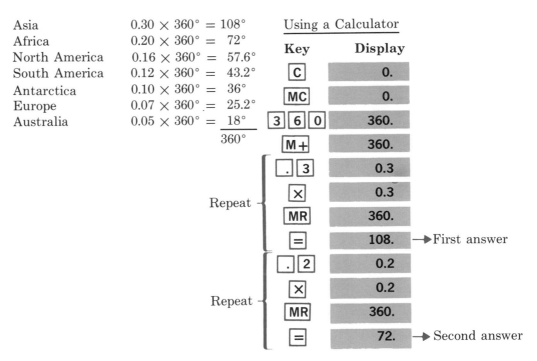

Using a Calculator

	Key	Display	
	C	0.	
	MC	0.	
	3 6 0	360.	
	M+	360.	
Repeat	. 3	0.3	
	×	0.3	
	MR	360.	
	=	108.	→ First answer
Repeat	. 2	0.2	
	×	0.2	
	MR	360.	
	=	72.	→ Second answer

Step 2 Most protractors are accurate only to the nearest degree; therefore all angles from Step 1 should be rounded to the nearest degree.

Asia	108°
Africa	72°
North America	58°
South America	43°
Antarctica	36°
Europe	25°
Australia	18°

The total angle should remain 360°. If the total is 359° round up the answer with the largest fraction less than 0.5. If the total is 361° round down the answer with the smallest fraction greater than 0.5. In this example the angles total 360°.

Step 3 Draw a circle and use a protractor to draw pie-shaped sectors corresponding to each angle. Label the areas by name and percent. Title the graph.

Land Area of the Earth by Continents

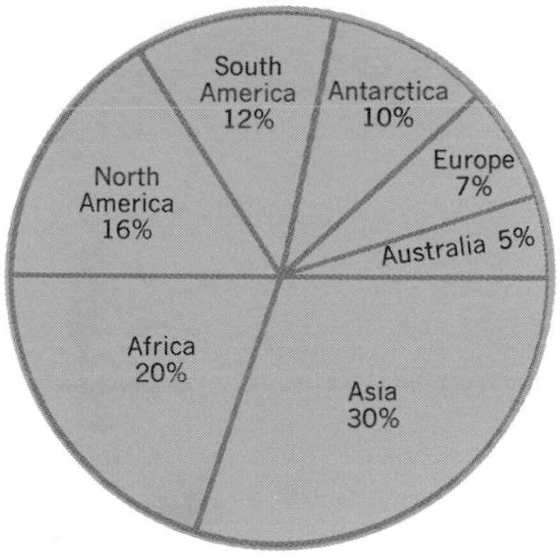

Ready to construct a circle graph on your own? Use the following data to draw a circle graph.

INCOME OF COLLEGE GRADUATES

Income Bracket	Percent of People
0 to $2,999	4%
$3000 to $5999	7%
$6000 to $9999	13%
$10,000 to $14,999	26%
$15,000 and up	50%

Compare your graph to ours in **12**.

12

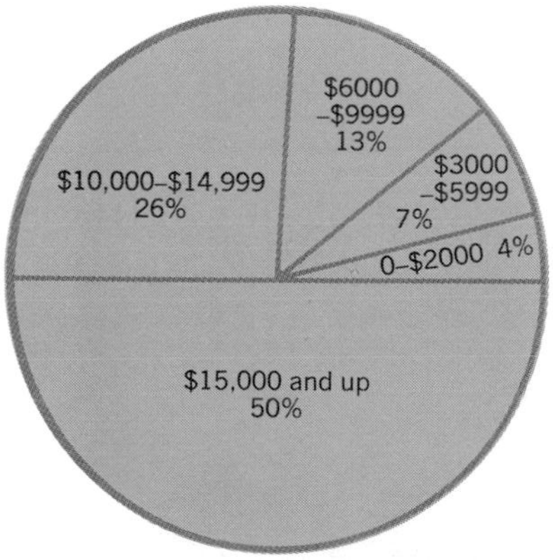

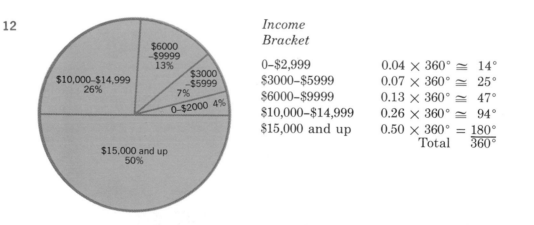

Income Bracket		
0–$2,999	$0.04 \times 360° \cong$	14°
$3000–$5999	$0.07 \times 360° \cong$	25°
$6000–$9999	$0.13 \times 360° \cong$	47°
$10,000–$14,999	$0.26 \times 360° \cong$	94°
$15,000 and up	$0.50 \times 360° =$	180°
	Total	360°

If the percents are not already given, it is necessary that you calculate them in order to draw a circle graph. For example, Joe Smith's 1979 property tax bill was $1955. This money was spent by the county as follows:

County indebtedness	$ 515
Fire protection/Police	$ 191
Schools	$1030
Sanitation	$ 109
Water	$ 72
Flood control	$ 30
Transportation	$ 8
Total	$1955

In this case rewrite each dollar amount as a percent of the total tax, then calculate the circle graph angle. For example, the first item, "county indebtedness," becomes

$\frac{\$\ 515}{\$1955} = 0.2634$ or 26.3%

Divide each amount by the total

$0.2634 \times 360° = 94.8°$

Multiply each decimal fraction by 360°

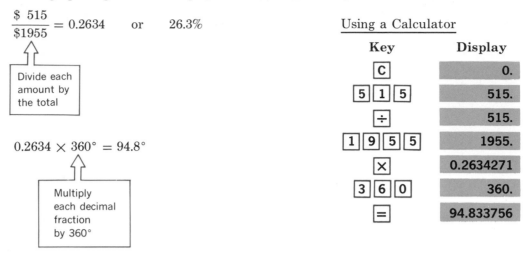

Using a Calculator

Key	Display
C	0.
5 1 5	515.
÷	515.
1 9 5 5	1955.
×	0.2634271
3 6 0	360.
=	94.833756

Continue this process for the other quantities on this list. Find the angle corresponding to each dollar amount, and construct a circle graph.

Check your work in **13**.

13

	Amount	Percent	Angle	Angle, Rounded
County indebtedness	\$ 515	26.3%	94.8°	95°
Fire protection/Police	191	9.8	35.3	35
Schools	1030	52.7	189.7	190
Sanitation	109	5.6	20.2	20
Water	72	3.7	13.3	13
Flood control	30	1.5	5.4	5
Transportation	8	0.4	1.4	2°
			Total	360°

Notice that we rounded 1.4° to 2° in order to get a total of 360°. The final circle graph will look like this:

Property Tax Expenditures

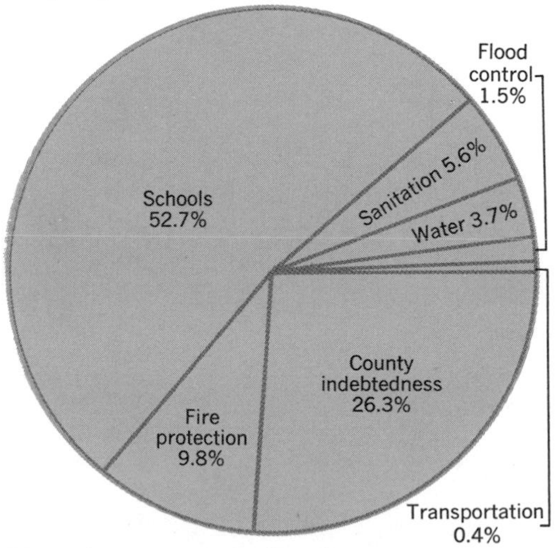

Now for some practice in using and constructing bar graphs, line graphs, and circle graphs, turn to **14** for a set of problems.

A. Reading Graphs

1. Answer the following questions in reference to the bar graphs in the figure.

Social Security Disability Insurance

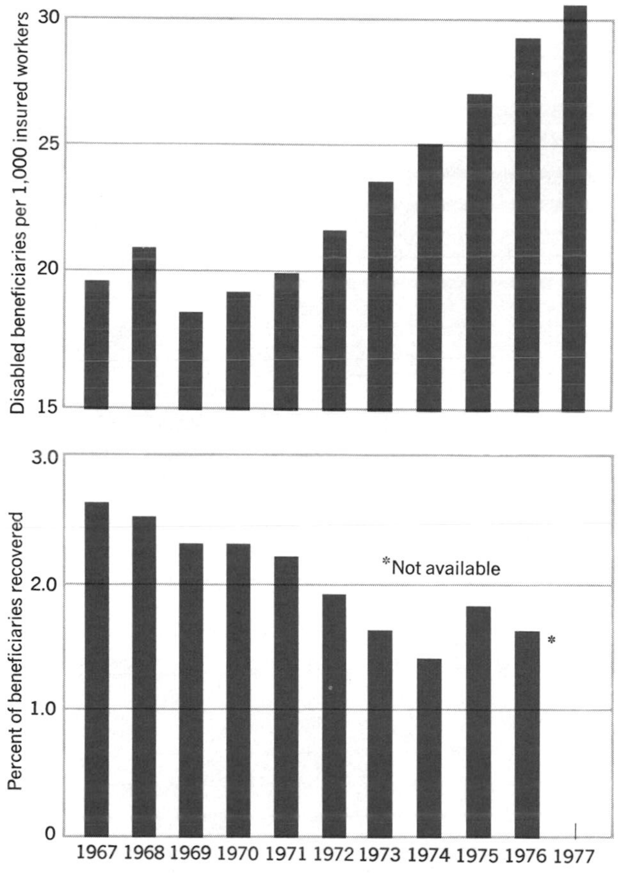

(a) In 1967, approximately how many disabled beneficiaries were there per 1000 insured workers?

(b) In 1976, approximately what percent of beneficiaries recovered?

(c) Did the number of disabled per 1000 ever decline? If so, when?

(d) If there were 2,500,000 disabled beneficiaries in 1976, approximately how many recovered?

(e) If there were 80 million insured workers in 1974, approximately how many of these were disabled?

2. Answer the following questions using the line graph shown.

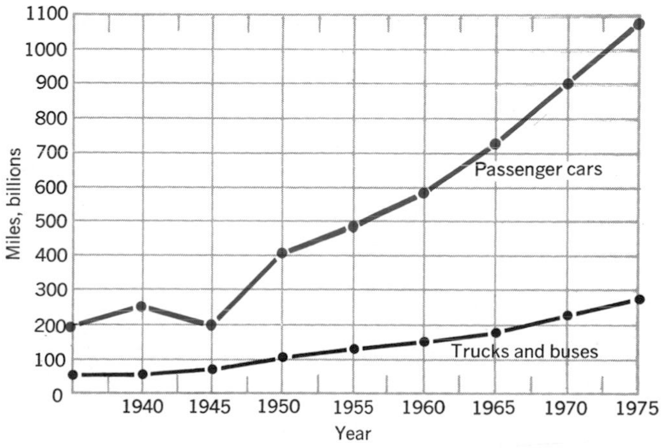

(a) How many miles were traveled by trucks and buses in 1950?
(b) How many miles were traveled by all vehicles in 1960?
(c) By how many miles did passenger car use increase from 1965 to 1975?

3. Answer the following questions using the circle graph.

Expenses of the Zzap Electrical Co.

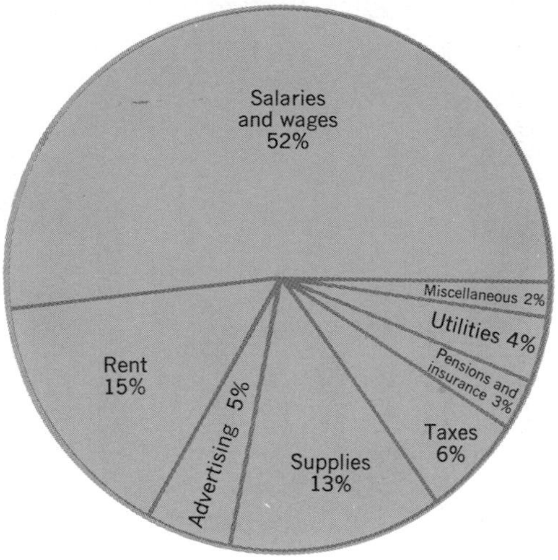

(a) What percent of Zzap's expenses were allotted to rent?
(b) What percent was spent on pension·and insurance?
(c) If the total expenses were $420,000, how much was spent on taxes? On rent?

4. Use the circle graph to answer the following questions.

Assets of the Allbrite Corp.

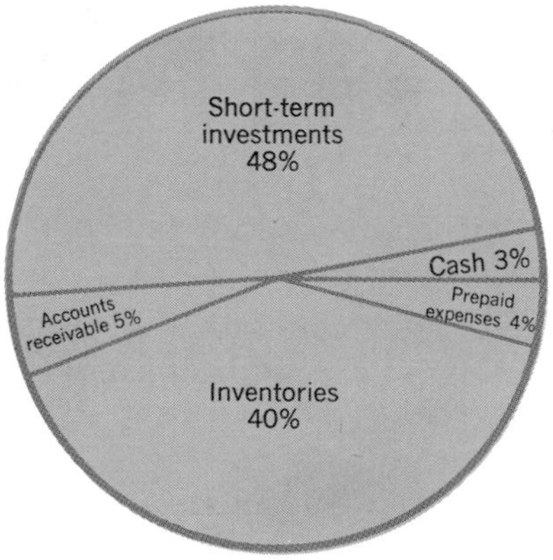

(a) What percent of the assets of the Allbrite Corporation are in the form of cash?
(b) If the total assets of the company are $90 million, how much do they have in inventory?
(c) If the total assets declined to $75 million and they wished to maintain the same percentage of short-term investments, how much money should they invest?
(d) If they had $2,500,000 in prepaid expenses, how much would their total assets be?

B. Draw the indicated graphs.

1. Make bar graphs of the following sets of data.

(a) **ABSORPTION CAPACITIES OF FIVE TYPICAL SOILS**

Type of Soil	Required Leaching Area (Square Feet)
Coarse sand	20
Gravel	20
Fine sand	25
Sandy loam	40
Sandy clay	40
Clay with large amount of sand or gravel	60
Clay with small amount of sand or gravel	90

(b) **AVERAGE MODULUS OF ELASTICITY OF VARIOUS MATERIALS**

Material	Tension (million psi)
Aluminum	10
Brass	15.5
Bronze	15
Copper	15
Iron	15
Steel	30
Zinc	12
Wood	1.5
Concrete	3

(c) **DISTRIBUTION OF PASSENGER CAR TRIPS**

Length of Trip	Percent of Total Trips	Percent of Vehicle Miles
Under 5 miles	59.6%	13.2%
5–9 miles	19.9%	15.4%
10–19 miles	12.3%	19.4%
20–49 miles	6.1%	21.2%
50 miles +	2.1%	30.8%

(d) **AMOUNT OF STEEL USED BY WILLIAMS MANUFACTURING COMPANY**

Month	Amount (in Thousands of Pounds)
January	85
February	103
March	132
April	138
May	121
June	100
July	75
August	112
September	106
October	135
November	158
December	127

2. Draw line graphs using the following sets of data.

(a) **AUTO FACTORY SALES IN THE UNITED STATES**

Year	Sales (in Thousands)
1900	4
1905	24
1910	181
1915	896
1920	1906
1925	3735
1930	3787
1935	3273
1940	3717
1945	70
1950	6665
1955	7920
1960	6674
1965	9305
1970	6546

(b) **HIGH AND LOW TEMPERATURES DURING A WEEK IN HELENA, MONTANA**

Day	High (°F)	Low (°F)
Monday	42	25
Tuesday	37	9
Wednesday	28	−3
Thursday	24	12
Friday	33	16
Saturday	45	28
Sunday	40	21

(c) **NEW BUSINESS INCORPORATIONS: 1976**

Month	New Incorporations
January	30,100
February	29,500
March	29,000
April	30,800
May	28,000
June	31,600
July	29,400
August	32,200
September	32,000
October	32,300
November	33,500
December	33,900

(d) **SALES OF CONSTRUCTION MATERIAL: SMYTH BUILDING SUPPLIES**

Year	Wood (× $1000)		Masonry (× $1000)
1971	243		112
1972	289		131
1973	325		133
1974	296	(000's omitted)	106
1975	288		92
1976	307		94
1977	412		89

3. Draw circle graphs illustrating the following sets of data.

(a) *California's Present Sources of Electricity* (*1977*)

Geothermal	1%
Nuclear	4%
Coal	7%
Gas	21%
Hydroelectric	24%
Oil	43%

(b) *Family Budget: Robbins Family*

Food	$330
Housing	$425
Utilities	$126
Transportation	$189
Clothing	$ 85
Entertainment	$ 50
Savings	$100
Other expenses	$ 70

(c) *Housing Situation: Pine Valley*

Own home	16,500
Rent home	2,700
Own condominium	3,800
Rent condominium	1,600
Own apartment	650
Rent apartment	9,600
Other	1,200

(d) *Government Budget: Center City*

Education	$17,000,000
Police and fire	$ 8,000,000
Health and sanitation	$11,000,000
Pensions	$ 6,000,000
Debt service	$15,000,000
Miscellaneous	$32,000,000

Check your answers on page 557 and then proceed to **15** to learn about the use of graphs in technology.

8-2 GRAPHS IN TECHNOLOGY

15 In this section we will study graphs as they are used in technology. A technologist often finds it necessary to use graphs of actual formulas or experimental results. In such graphs, a definite and continuous relationship exists between two quantities, and the graph shows this relationship as a straight line or a smooth curve. With this kind of graph, values can be read anywhere along the line or curve rather than at certain points as in a broken-line graph.

Surface Speed of a Grinding Wheel

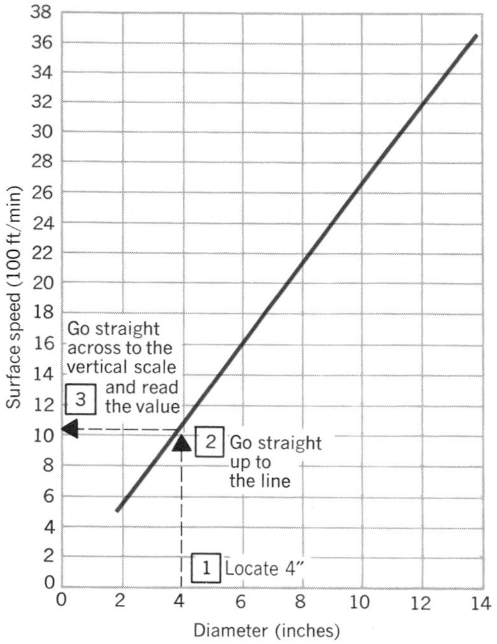

For example, the graph in this figure shows how the surface speed of a grinding wheel is related to its diameter when it is turning at 1000 rpm.

The possible diameters of the wheel are given along the horizontal axis. Of course, any diameter may be used, not only the axis labels 2″, 4″, 6″, and so on.

The surface speed in hundreds of feet per minute is plotted along the vertical axis. The graph tells you that the surface speed increases in direct proportion as the diameter of the wheel increases—the bigger the wheel, the faster the rim moves.

The dotted lines on the graph show how to find the surface speed for a grinding wheel 4 inches in diameter. First, locate 4″ along the horizontal axis. Second, follow the grid straight up until you reach the graph line. Third, follow a horizontal line across to the vertical axis and read the speed value. In this case the surface speed is just over 1000 feet per minute. Notice we add two zeros to the value on the scale because it is given in hundreds.

Use this graph to answer the following questions:

(a) Find the surface speed for a wheel with a 10″ diameter rotating at 1000 rpm.
(b) What is the surface speed for a wheel $3\frac{1}{2}$″ in diameter rotating at 1000 rpm?
(c) What diameter wheel is needed to produce a surface speed of 1700 ft/min?

Check your answers in **16**.

---

16 (a) 2600 ft/min
(b) 900 ft/min
(c) $6\frac{1}{2}$ in.

Even though the number $3\frac{1}{2}$ was not written on the horizontal scale, it is easy to locate.

In Problem (c) work backwards: Find 17 on the vertical axis, move horizontally to the graph line then down to the horizontal axis. The point at $6\frac{1}{2}$″ corresponds to 17 on the vertical axis.

Surface Speed of a Grinding Wheel

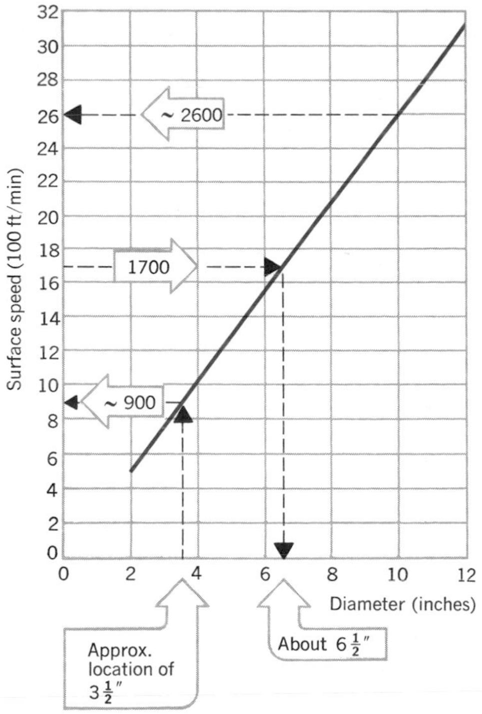

Many technical and scientific quantities are so related that the graph line is curved rather than straight. For example, the graph shown gives the relationship between viscosity and temperature for various weight of oil. To save space, curves representing seven different weights of oil have been included on the same graph.

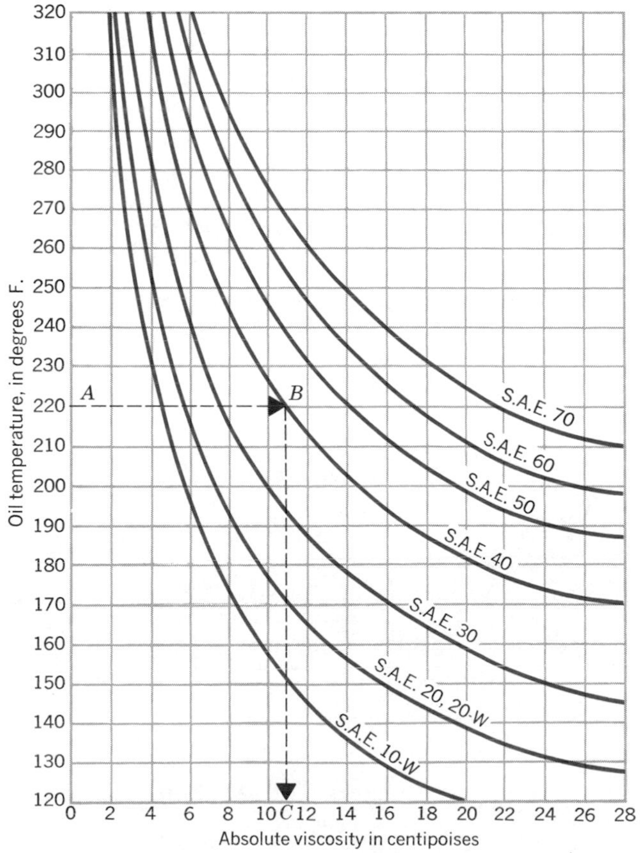

Viscosity vs. temperature — SAE oils.

To read a graph like this, follow the same procedures you used for straight line graph. For example, suppose you want to know the viscosity of SAE 40 oil at 220° F. The dotted lines show how it is done. First, locate 220° on the vertical or temperature scale at point A. Second, follow the grid line straight across to the right until you meet the curve labeled SAE 40 at point B. Third, from point B move down to the horizontal axis and read the viscosity value. The viscosity is about 11 centipoise.

Use the graph on page 437 to answer the following questions.

(a) What is the viscosity of SAE 30 at 180°?
(b) How much does the viscosity of SAE 10 decrease as the temperature rises from 130° to 200°?
(c) At what temperature will SAE 70 oil have a viscosity of 8 centipoise?

Check your answers in **17**.

---

17 (a) About 13.5 centipoise
(b) About 10 centipoise
(c) About 295°

Although the graphs used in technical work usually appear in handbooks and technical manuals, it is sometimes necessary for the technician to plot graphs from measurements or formulas. For example, this table gives the results of a calibration measurement on a spring.

Elongation, L (in.)	0	0.05	1.0	1.5	2.0	2.5	3.5	5.5	7.0	7.0
Applied Force, F (lb)	0	1	2	3	4	5	6	7	8	9

These points may be plotted on a graph in exactly the same way as a broken-line graph was drawn in Section 8-1.

Step 1 Draw and label the horizontal and vertical axes. Be sure to include the correct units. In the graph of an experiment or series of measurements, always put the *cause* on the horizontal axis and the *effect* on the vertical axis. In this case the applied force caused the spring to stretch.

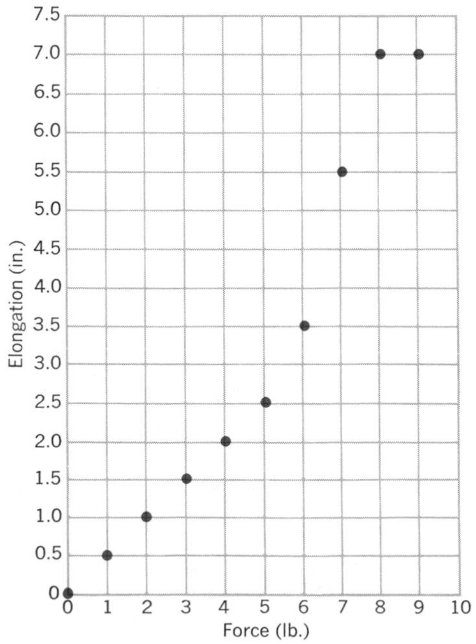

Step 2 Plot the points. Find each quantity on its axis and place a dot at the intersection of its grid lines.

For example, a force of 2 lb produces an elongation of 1.0 in.

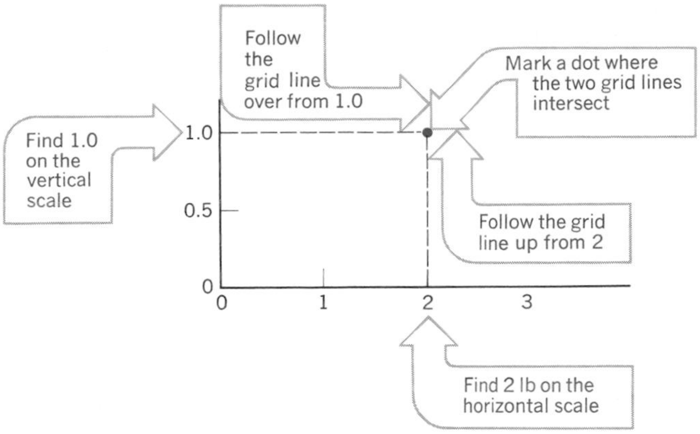

Step 3 Connect the points with a uniform straight line or curve. In this case we have both. The graph is a straight line between 0 and 5 lb, then it curves upward from 5 to 8 lb until flattening out completely. This is typical of a stretched spring. The spring will elongate uniformly until reaching its elastic limit, then it will deform rapidly, and finally break. The elastic limit of this spring was 2.5 in., and it broke at 7.0 in. By graphing the results, we were able to analyze the behavior in a way that was not obvious from the data itself.

Force Elongation of a Spring

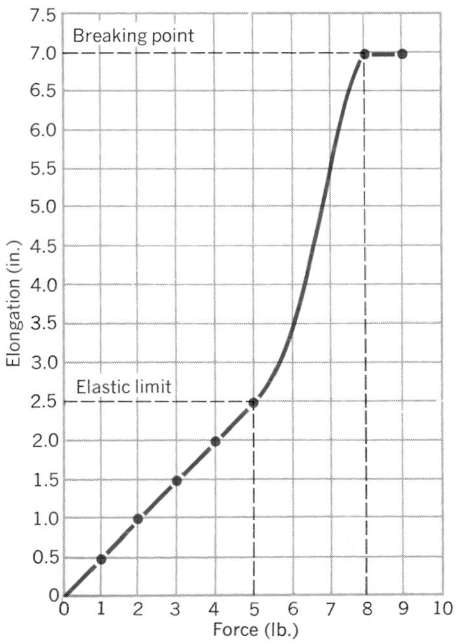

Practice graphing by plotting the points in the following tables of data.

(a) The following data shows the energy (in BTU) necessary to heat a certain object from 15° F to 210° F. Plot the points and connect with a smooth line or curve:

Heat Energy, BTU	750	1500	2250	3000	5250	7500
Temperature, °F	15°	75°	120°	165°	195°	210°

(b) The surface energy and surface tension of water vary with the temperature. Plot both sets of data on the same graph. **Note:** You can use the same scale for energy and tension even though the units are different.

Temperature (°C)	Surface Energy $\left(\frac{\text{erg}}{\text{cm}^2}\right)$	Surface Tension $\left(\frac{\text{dyne}}{\text{cm}^2}\right)$
0	144	75
50	143	65
100	141	55
150	137	45
200	131	35
250	124	25
300	112	15
350	82	5
374	0	0

Check your graphs in **18**

What is the difference between a broken-line graph and a graph in science and technology?

For a scientific graph, the quantities plotted along the horizontal and vertical axes are continuous. Even though we plot only a few points, we can draw a smooth, unbroken line or curve.

---

18 (a) Heat Energy vs. Temperature

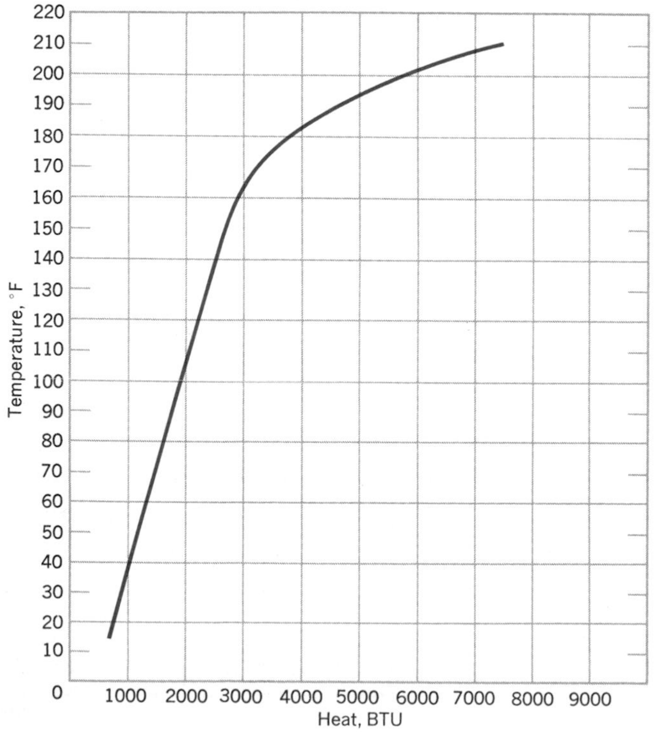

(b) Surface Energy and Surface Tension of Water

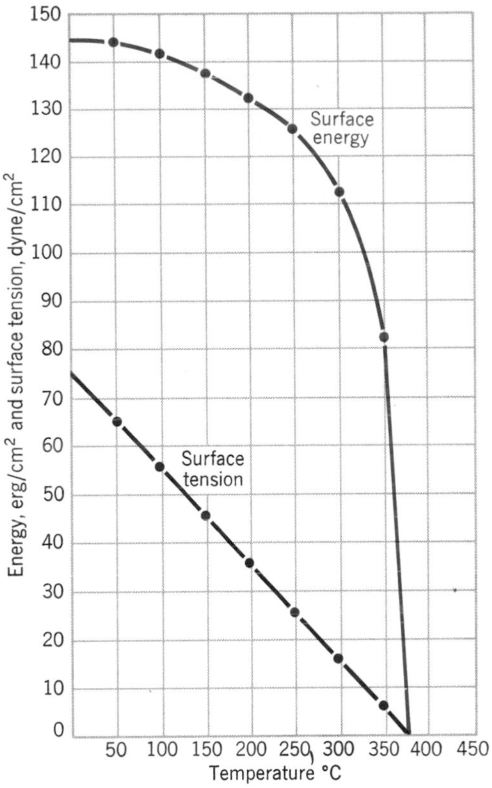

Graphing Formulas

Another important skill in technology is the ability to draw a graph of a formula. A graph is a picture of the information contained in a formula. Once you have a graph of a formula, you can simply read values from the graph rather than evaluate the formula each time you need to use it.

For example, the formula $D = 55t$ gives the distance D in miles traveled by an automobile moving at 55 mph for t hours. To make a graph of this formula, follow these steps.

Step 1 Make up a table or chart of values for D and t. Choose several values for t over the desired range.

Distance, D (mi)				
Time, t (hr)	1	5	9	12

Step 2 Calculate the value of D that corresponds to each chosen value of t.

For $t = 1$, $D = 55(1) = 55$ mi
For $t = 5$, $D = 55(5) = 275$ mi
For $t = 9$, $D = 55(9) = 495$ mi
For $t = 12$, $D = 55(12) = 660$ mi

Enter these values of D into the table.

Distance, D (mi)	55	275	495	660
Time, t (hr)	1	5	9	12

Step 3 Construct a graph of these values.

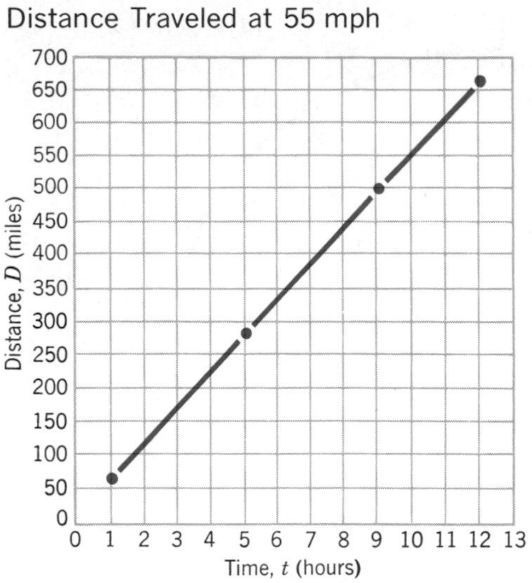

Once a graph of the formula is available, it can be used in place of the formula to answer many questions without using algebra. For example, use this graph to answer the following questions:

(a) How far will this automobile travel in $7\frac{1}{2}$ hr?
(b) What time will be required to travel 200 miles?

Check your answers in **19** .

---

19 (a) About 410 mi (b) About $3\frac{1}{2}$ hr

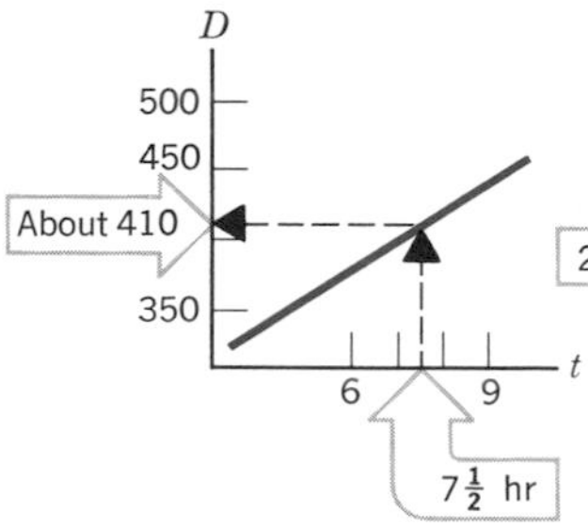

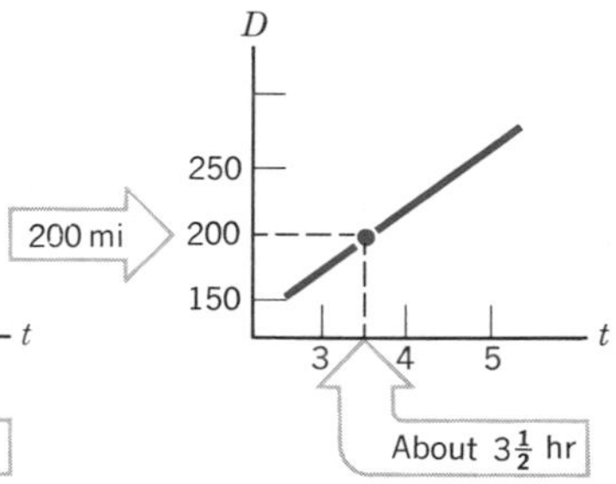

The answers are not exact, but they are very close to the values that you would get by doing the actual calculations, and the process of reading the graph is very simple and quick.

For practice in graphing, draw graphs of the following formulas.

(a) In a simple electric circuit the power generated is equal to voltage times current. If P is the power in watts and I is the current in amperes, the formula

$$P = 220I$$

enables you to calculate the power developed in a 220-volt circuit. Graph this formula over a range from one to ten amps.

(b) The conversion formula from the Celsius to the Fahrenheit temperature scale is $F = \frac{9}{5}C + 32$. Graph this formula over a range from 0° to 100° C. (*Hint:* It will be easier if you choose values for C that are multiples of 5.)

(c) The formula for the area of a circle is approximately $A = 3.14R^2$, where R is the radius. Graph this formula over a range of radii from 2 to 12 in.

Check your graphs in **20** .

20 (a) Power in a 220-Volt Circuit

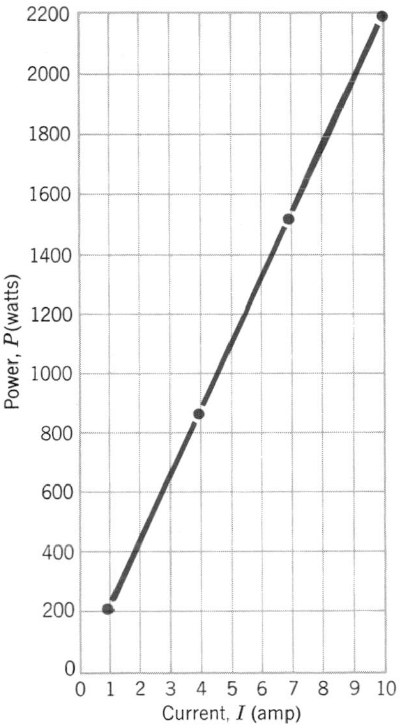

P	220	880	1540	2200
I	1	4	7	10

(b) Temperature Conversion

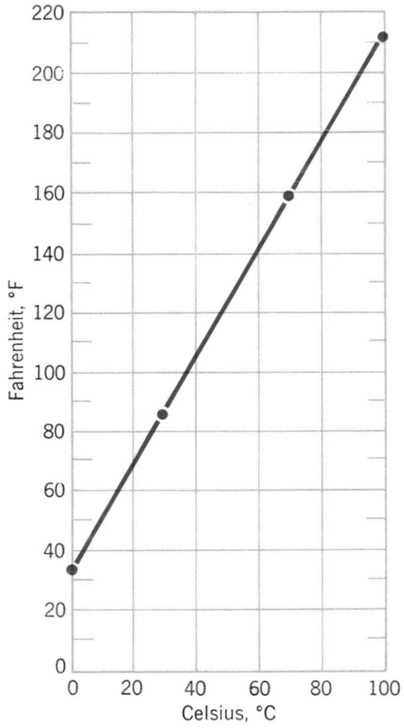

F	32	86	158	212
C	0	30	70	100

(c) Area of a Circle

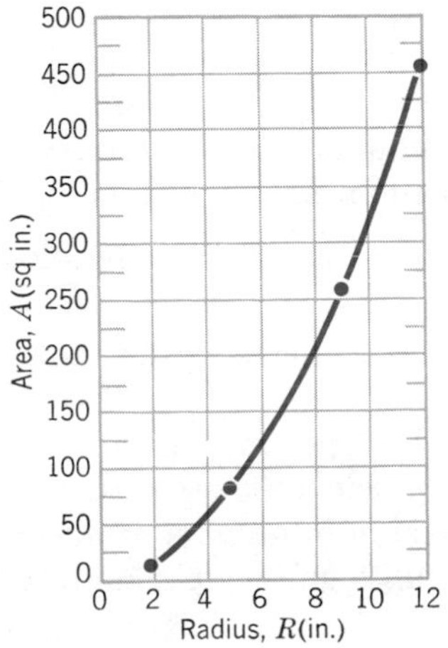

A	13	79	254	452
R	2	5	9	12

Notice that we used four points for each graph. At least two points are necessary for a straight-line graph and many more may be needed for a curve-line graph. You may have used different points from ours, but your graphs should look the same.

Now turn to **21** for a set of problems on technical graphs.

---

21 **Exercises 8-2 Graphs in Technology**

A. Interpreting Graphs

1. Use the graphs in **20** to answer the following questions.

   (a) As current increases, power ________.

   (b) 60° C = ________° F.

   (c) The area of a circle with a radius of 7″ is roughly ________.

   (d) How many amps of current will produce 1400 watts of power at 220 volts?

   (e) 50° F = ________° C.

   (f) What is the approximate radius of a circle with an area of 200 square inches?

   (g) How many watts of power are produced by a current of 9 amps at 220 volts?

   (h) What is the area of a circle with a radius of 11″?

   (i) 25° C = ________° F.

   (j) What is the radius of a circle with an area of 275 in.^{2}?

2. Use this graph to answer the following questions.

Demand Load Curves for Water Systems

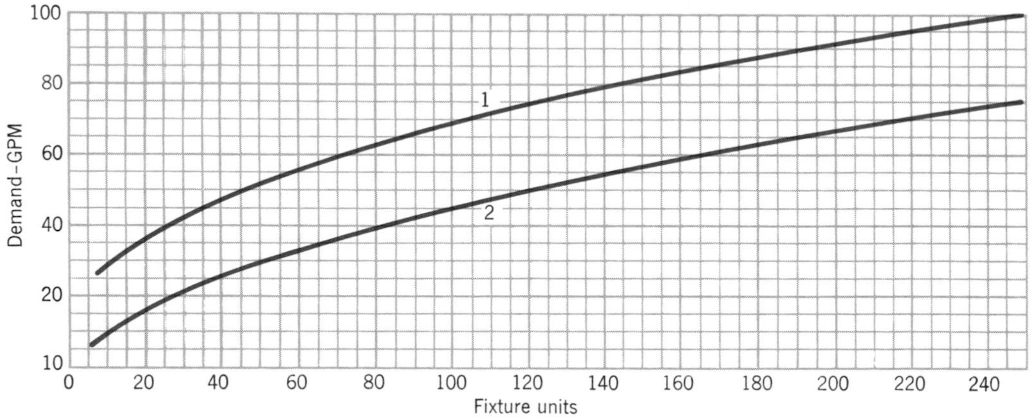

No. 1 for system predominantly of flush valves.
No. 2 for system predominantly of flush tanks.

(a) In general, systems with flush valves demand ____________ (more, less) water than ones with flush tanks.

(b) What is the demand for 20 fixture units in a predominantly flush valve system?

(c) How many gallons per minute are required by 190 fixture units in a predominantly flush tank system?

(d) What is the maximum number of fixture units you could accommodate with a supply of 45 gallons per minute in a predominantly flush valve system?

(e) What is the maximum number of fixture units you could accommodate with a supply of 25 gallons per minute in a predominantly flush tank system?

B. Construct graphs from the following sets of data.

1. In a stress analysis the stretch of a copper wire was measured as different weights were hung from the wire:

Load (lb)	Stretch (inches)
0	0
2	0.02
4	0.04
6	0.06
8	0.08
10	0.10
12	0.12
14	0.63

2. The measurements below show the growth of a carrot plant. Draw the graph as a continuous curve.

Height, H (cm)	0	1.5	4.0	6.0	6.5
Time, t (weeks)	0	1	2	3	4

3. The following measurements were made on an object sliding along a smooth, level, nearly frictionless surface.

Distance, d (meters)	0	0.9	1.8	2.7	3.6	5.4	7.2
Time, t (seconds)	0	1	2	3	4	6	8

4. Plot the following data as two curves on the same graph. Use the same vertical scale to represent horsepower and percent efficiency.

Current (Amperes)	Horsepower	Percent Efficiency
50	14	85
75	20	87
100	27	85
125	34	83
150	39	82
175	44	81
200	49	80

5. The following table shows how the voltage of a battery and the specific gravity of the battery fluid are related.

Voltage, V (volts)	2.00	1.98	1.96	1.92	1.88	1.84	1.80	1.72	1.50
Specific gravity, S	1.300	1.295	1.285	1.280	1.275	1.270	1.265	1.253	1.250

6. For relatively low altitudes, air temperature decreases as altitude increases. The following table shows some typical measurements of temperature with altitude.

Altitude, H (ft)	Sea Level	1000	2000	3000	4000	5000	6000	7000	8000
Temperature, T (F°)	85°	83°	80°	75°	65°	60°	48°	35°	10°

C. Graph the following formulas:

1. The resistance in ohms of a copper wire L ft long and 40 mils in diameter is given by the formula

   $R = 0.0065\ L$

   Graph for values of L from 100 to 1000 ft.

2. The approximate number of standard bricks needed to build a wall of volume V is given by the formula $B = 21V$, if a $\frac{1}{4}''$ bed joint is used. Graph for values of V from 0 to 500 cu ft.

3. The power dissipated in a DC circuit with a 15 ohm total resistance is given by the formula $P = 15I^2$. (P is the power in watts and I is the current in amps.)

   Graph for values of I from 0 to 5 amps.

4. The current in a two-phase, 3-wire, 120 volt circuit is related to the power by the formula $P = 216I$. Plot P vs. I from $I = 0$ to 10 amps.

5. The lift in pounds on a certain type of airplane wing is given by the formula

   $L = 0.083V^2$

   where V is the speed of air flow over the wing in ft/sec. Plot L vs. V from $V = 50$ ft/sec to 400 ft/sec.

6. Graph each of the following algebra formulas:

   (a) $y = 2x - 1$ from $x = 0$ to $x = 6$
   (b) $y = x^2 - 1$ from $x = 0$ to $x = 5$
   (c) $y = 4 - x$ from $x = 0$ to $x = 5$
   (d) $y = 2x^2 - x$ from $x = 0$ to $x = 5$

Check your answers on page 561, then turn to **22** for a set of problems on all aspects of graphing.

8 Graphs

PROBLEM SET 8

Graphs

22 Answers are on page 566.

A. Answer the questions based on the graphs pictured.
Questions 1 to 5 refer to Graph I.

Graph I

Minimum Size of Circuit Wires

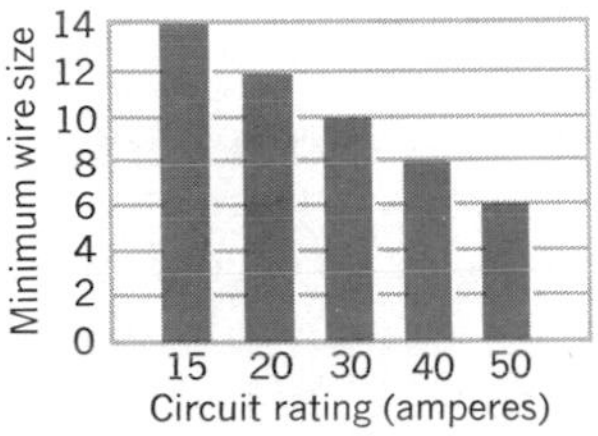

1. In general, as amps increase, the necessary wire size ______________. (increases, decreases)
2. What is the minimum size of wire needed for a circuit rating of 30 amps?
3. What is the minimum wire size needed for a circuit rating of 15 amps?
4. What circuit ratings could a wire size of 8 be used for?
5. What circuit ratings could a size 14 wire be used for?

Questions 6 to 15 refer to Graph II.

Graph II

Employee Production at Miller Printing Co.

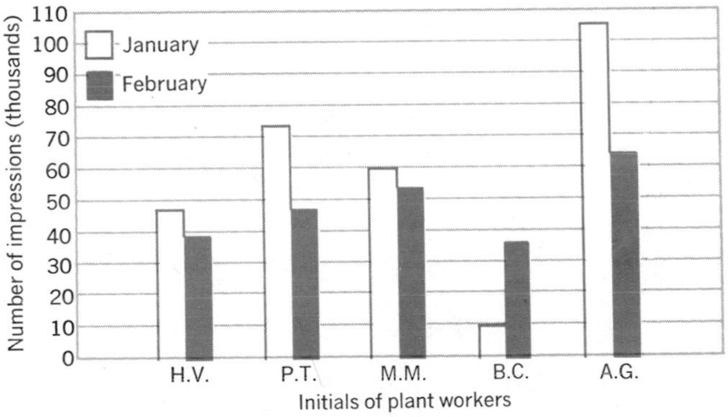

6. What was HV's output in February?
7. What was P.T.'s output for January?
8. Who produced the most in January?
9. Who produced the least in January?
10. Who produced the most in February?
11. Who produced the least in February?
12. Which workers produced more in February than January?
13. Which workers produced less in February than January?
14. Determine the total production for January.
15. Determine the total production for February.

Name ______________

Date ______________

Course/Section ______________

Problems 16 to 22 refer to Graph III.

Graph III

Earnings Per Share of F.B.G. Corp.

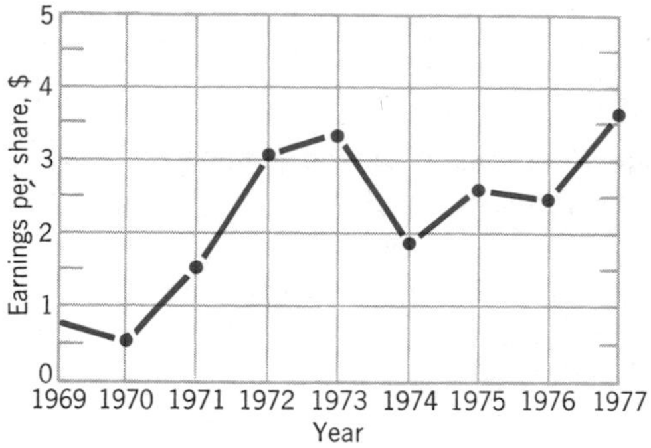

16. When were the highest earnings reported by the company? What were they?
17. When were earnings the lowest? What were they?
18. When did the biggest increase in earnings occur? By how much did they increase?
19. When did the biggest drop in earnings occur? By how much did they drop?
20. If the company paid a dividend equal to 50% of its earnings in 1975, what did this dividend amount to?
21. If 8,000,000 shares of F.B.G. stock existed in 1973, how much did the company earn altogether?
22. If 6,500,000 shares of F.B.G. stock existed in 1970, how much did the company earn?

Problems 23 to 30 refer to Graph IV.

Graph IV

Projected vs. Actual Sales

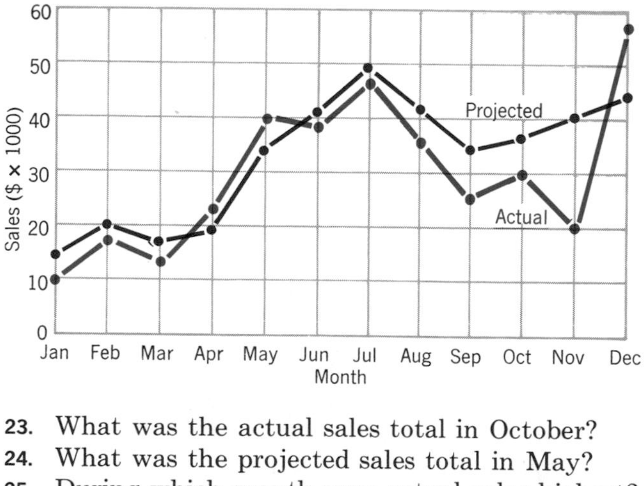

23. What was the actual sales total in October?
24. What was the projected sales total in May?
25. During which month were actual sales highest?
26. During which month were projected sales highest?
27. During which month were actual sales lowest?
28. During which month were projected sales lowest?
29. During which months were actual sales lower than projected sales?
30. During which month was the gap between actual and projected sales the largest? What was this difference?

Problems 31 to 35 refer to Graph V.

Graph V

Composition of Marine Bronze

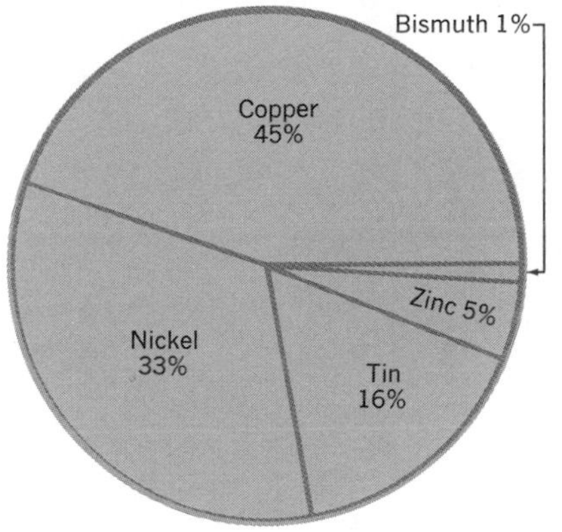

31. What percent of marine bronze is composed of tin?
32. What percent of this alloy is made up of quantities other than copper?
33. Without measuring, calculate how many degrees of the circle should have gone to nickel.
34. How many ounces of zinc are there in 50 lb of marine bronze?
35. How many grams of bismuth are there in 73 grams of this alloy?

Problems 36 to 41 refer to Graph VI.

Graph VI

Grade Horsepower Requirements (65,000 lbs. GCW)

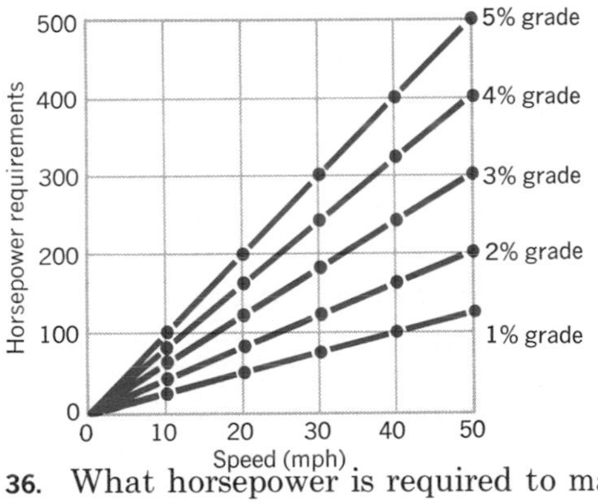

36. What horsepower is required to make a 4% grade at 30 mph?
37. What horsepower is required to make a 2% grade at 50 mph?
38. How fast will 250 horsepower get the vehicle over a 3% grade?
39. How fast will 125 horsepower carry the vehicle over a 1% grade?
40. How much more horsepower is required at 35 mph over a 5% grade than over a 3% grade?
41. How much more horsepower is required at 30 mph over a 4% grade than over a 1% grade?

Questions 42 to 46 refer to Graph VII on page 450.

42. In general, as the diameter of the wire increases, resistance ____________________.
(increases, decreases)
43. What is the resistance of 3000 ft of wire 30 mils in diameter?
44. What is the resistance of 1000 ft of wire 65 mils in diameter?
45. What diameter wire has a resistance of 20 ohms of 1000 ft?
46. What diameter wire has a resistance of 220 ohms for 4000 ft?

Graph VII

Electrical Resistance of Copper Wire

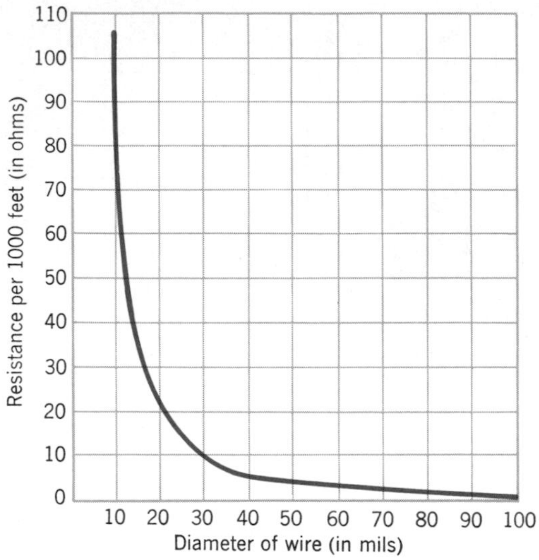

B. Construct graphs from the data given.

1. Construct a bar graph based on this data.

FIRE RESISTANCE RATINGS OF VARIOUS PLYWOODS

Type of Wood	Fire Resistant Rating
A: $\frac{1}{4}''$ plywood	10
B: $\frac{1}{2}''$ plywood	25
C: $\frac{3}{4}''$ tongue and groove	20
D: $\frac{1}{2}''$ gypsum wallboard	40
E: $\frac{1}{2}''$ gypsum—two layers	90
F: $\frac{5}{8}''$ gypsum wallboard	60

2. Construct a double bar graph based on this data.

OUTPUT OF BRICKLAYERS FOR THE ABC CONSTRUCTION CO.

	Bricks Laid	
Initials of Employee	March	April
R.M.	1725	1550
H.H.	1350	1485
A.C.	890	1620
C.T.	1830	1950
W.F.	1175	1150
S.D.	2125	1875

3. Construct a broken-line graph based on this data.

Day	High Temperature
Monday	75°
Tuesday	72°
Wednesday	81°
Thursday	77°
Friday	66°
Saturday	64°
Sunday	68°

4. Construct a double broken-line graph based on this data.

Year	Earnings per Share	Dividends per Share
1972	\$1.12	\$0.28
1973	\$1.27	0.32
1974	\$1.75	0.40
1975	\$1.92	0.40
1976	\$2.16	0.45
1977	\$2.85	0.55

5. Construct a circle graph based on this data.

Metal	Alloy Composition
Copper	43%
Aluminum	29%
Zinc	3%
Tin	2%
Lead	23%

6. Use this data to construct a circle graph.

WORKER EXPERIENCE AT EVANS TOOL AND DYE

Length of Service	Number of Workers
20 years or more	9
15–19	7
10–14	14
5–9	20
1–4	33
Less than 1 year	5

7. Construct a straight line or curve graph with the following information.

AMOUNT OF GAS SUPPLIED BY DIFFERENT LENGTHS OF $\frac{3}{4}''$ PIPE

Length, ft	Volume, Cubic Feet/Hour
10	360
20	245
30	198
40	169
50	150
60	135
70	123
80	115
90	108
100	102

8. Construct a double straight line or continuous curve graph based on the following data.

CUTTING DEPTH VS. NOZZLE PRESSURE OF PVC AND ACRYLIC PIPE

PVC

Cutting depth (inches)	0.06	0.18	0.54	1.62
Nozzle pressure (psi)	10	20	40	80

ACRYLIC

Cutting depth (inches)	0.06	0.18	0.54	0.90
Nozzle pressure (psi)	15	30	60	80

C. Graph the following formulas and then answer the questions below each one:

1. The power dissipated in a certain electronic circuit is given by:

   $P = 275I$ (P is the power in watts, and I is the current in amperes.)

   Range of graph: $I = 1$ to $I = 10$ amp.

   (a) How many amps are required to produce a power of 1375 watts?
   (b) How many watts are produced by current of 7 amps?

   (c) As current increases, power ________________.
   (increases, decreases)

2. In a machine shop the output of a man finishing an average of 25 pieces of work per day is $W = 25d$. (W is the output in total pieces, and d is the number of days of work.)

   Range of graph: $d = 1$ to $d = 28$ days.

   (a) How many pieces of work can this man finish in 27 days?
   (b) How long would it take him to finish 375 pieces?

3. The total simple interest earned on $1000 invested at 6% is $I = 60t$. (I is the interest earned, and t is the time in years.)

   Range of graph: $t = 1$ year to 15 years.

   (a) How much interest would be earned in 7 years?
   (b) How long would it take to earn $600 interest?

4. The area of a square is given by the formula $A = s^2$ where A is the area in square cm, and s is the side length in cm.

Range of graph: $s = 1$ cm to $s = 25$ cm

(a) What is the length of a side of a square having an area of 400 cm^2?
(b) What is the area of a square with a side length of 13 cm?

5. The horsepower produced by an 8 cylinder engine can be calculated from the formula

$H = 3.2D^2$

(where H is the horsepower, and D is the diameter of the cylinder in inches.)

Range of graph: $D = 3''$ to $D = 10''$

(a) What horsepower is produced by an engine with $3\frac{1}{2}''$ cylinders?
(b) What size cylinders are needed to produce 200 horsepower?

6. The power dissipated in a circuit is roughly proportional to the square of the current flowing through the circuit. The formula relating power and current for a certain circuit is

$P = 5.4I^2$

(P is the power in watts, and I is the current in amps.)

Graph this formula from $I = 1$ amp to $I = 6$ amps.

(a) What power is dissipated in a circuit carrying 4 amps?
(b) What current is needed to produce 20 watts?

9 Triangle Trigonometry

PREVIEW 9

Objective | **Sample Problems** | **Where To Go For Help**

Upon successful completion of this unit you will be able to:

Objective	Sample Problems		Page	Frame
1. Determine whether or not a figure is a right triangle, given certain information.	(a) 		457	1
	(b) 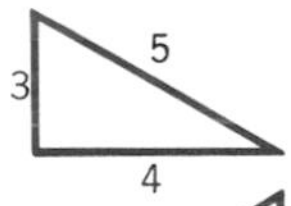		464	11
	(c) 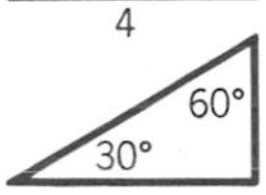		463	10
	(d) 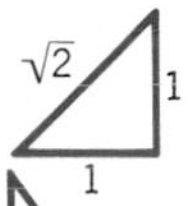		463	9
	(e) 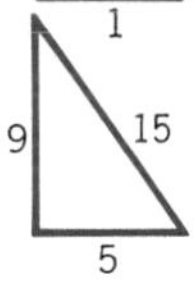		462	8
2. Find the values of trig ratios from a table.	(a) sin 26°	= ________	467	13
	(b) cos 84°	= ________	469	16
	(c) tan 43°20′	= ________	470	18
	(Round to three decimal places.)			
3. Find the angle when given the value of a trig ratio.	Find $\angle x$ to the nearest minute:			
	(a) $\sin x = 0.242$	$\angle x =$ ________	473	22
	(b) $\cos x = 0.549$	$\angle x =$ ________	474	
	(c) $\tan x = 3.821$	$\angle x =$ ________		
4. Solve problems involving triangles.	(a) 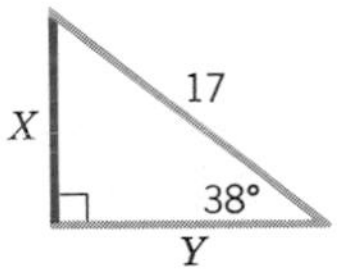	$X =$ ________ $Y =$ ________	478	30
	(b) 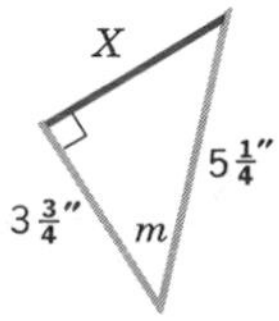	$\angle m =$ ________ $X =$ ________	479	31

Name ________

Date ________

Course/Section ________

(c) In a pipe-fitting job, what is the run if the offset is $22\frac{1}{2}°$ and the length of set is $16\frac{1}{2}''$?

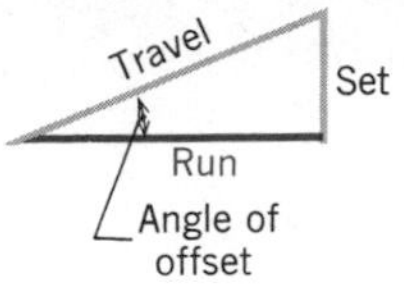

run = ____________

(d) Find the angle of taper below:

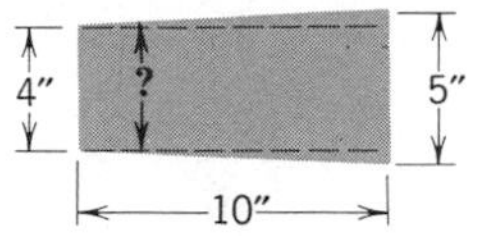

angle = ____________

(Answers to these preview problems are at the bottom of this page.)

If you are certain you can work all of these problems correctly, turn to page 487 for a set of practice problems. If you cannot work one or more of the preview problems, turn to the page indicated for help. If you want to be tops in your work, turn to frame **1** and begin work there.

Answers to Preview 9

1. (a) Yes (b) Yes (c) Yes (d) Yes (e) No
2. (a) 0.438 (b) 0.105 (c) 0.944
3. (a) 14°0′ (b) 56°43′ (c) 75°19′
4. (a) $X \approx 10.47$ $Y \approx 13.40$ (b) $\measuredangle m = 44°25'$ $X \approx 3.67''$
 (c) $\sim 39.8''$ (d) $\sim 5°46'$

9 Triangle Trigonometry

"Not the Bermuda triangle?"

9-1 ANGLES AND TRIANGLES

The word *trigonometry* means simply "triangle measurement." We know that the ancient Egyptian engineers and architects of 4000 years ago used practical trigonometry in building the pyramids. By 140 B.C. the Greek mathematician Hipparchus had made trigonometry a part of formal mathematics and taught it as astronomy. In this chapter we will look at only the simple practical trigonometry used in electronics, drafting, machine tool technology, aviation mechanics, and similar technical work.

1 In Chapter 7 you learned some of the vocabulary of angles and triangles. To be certain you remember this information, try this short quiz.

1. Find the value of each of the angles shown.

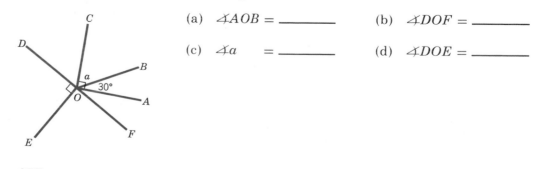

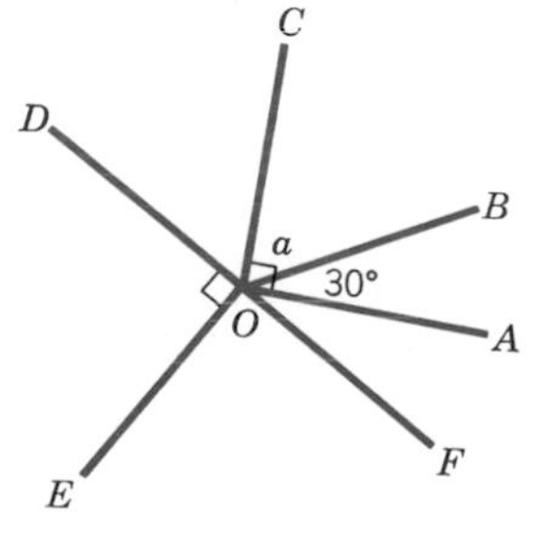

(a) $\measuredangle AOB =$ _______ (b) $\measuredangle DOF =$ _______

(c) $\measuredangle a =$ _______ (d) $\measuredangle DOE =$ _______

2. Label each of these angles with the correct name: acute angle, obtuse angle, straight angle, right angle.

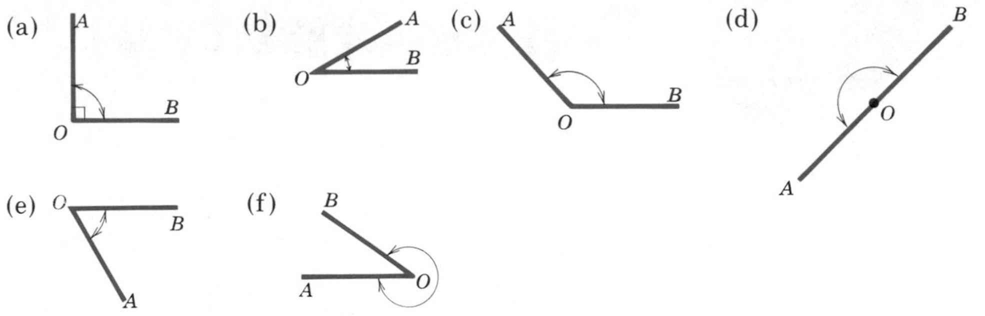

Check your answers in **2**.

---

2 1. (a) $\measuredangle AOB = 30°$ (b) $\measuredangle DOF = 180°$ (c) $\measuredangle a = 60°$ (d) $\measuredangle DOE = 90°$ (The symbol ∟ means "right angle" or 90°.)

2. (a) right angle (b) acute angle (c) obtuse angle (d) straight angle (e) acute angle (f) obtuse angle

If you missed any of these, you should return to frame **1** on page 305 for a quick review; otherwise continue here.

In most practical work angles are measured in degrees and fractions of a degree. Smaller units have been defined as follows:

60 minutes = 1 degree abbreviated, $60' = 1°$
60 seconds = 1 minute abbreviated, $60'' = 1'$

For most technical purposes angles can be rounded to the nearest minute.

Often in your work — especially if you use a hand calculator — you will need to convert an angle measured in decimal degrees to its equivalent in degrees and minutes. For example,

$17\tfrac{1}{2}° = 17° + \tfrac{1}{2}°$
$= 17° + \tfrac{1}{2}(60')$ Multiply the fractional part by $60'$ since $1° = 60'$.
$= 17°30'$

Write this angle to the nearest minute:

$36.25° =$ ________

Check your answer in **3**.

---

3 $36.25° = 36° + 0.25°$
$= 36° + (0.25)(60')$ Multiply the decimal part by $60'$.
$= 36°15'$

The reverse procedure is also useful. For example, to convert the angle $72°6'$ to decimal form,

$72°6' = 72° + 6'$

$= 72° + \left(\dfrac{6'}{60'}\right)$

Divide the minutes part by $60'$ since $1° = 60'$.

$= 72° + 0.1°$
$= 72.1°$

For practice in working with angles, rewrite each of the following angles as shown.

1. Write in degrees and minutes.

(a) $24\frac{1}{3}^\circ$ = ______ (b) $64\frac{3}{4}^\circ$ = ______

(c) 15.3° = ______ (d) 38.6° = ______

(e) $46\frac{3}{8}^\circ$ = ______ (f) 124.67° = ______

2. Write in decimal form.

(a) 10°48′ = ______ (b) 96°9′ = ______

(c) 57°36′ = ______ (d) 168°54′ = ______

(e) 33′ = ______ (f) 1°12′ = ______

Check your answers in **4**.

RADIAN MEASURE

In some scientific and technical work you may find that angles are sometimes measured or described in a new angle unit called the *radian*. By definition,

$$1 \text{ radian} = \frac{180^\circ}{\pi} \text{ degrees or about } 57.296^\circ$$

One radian is the angle at the center of a circle that corresponds to an arc exactly one radius in length.

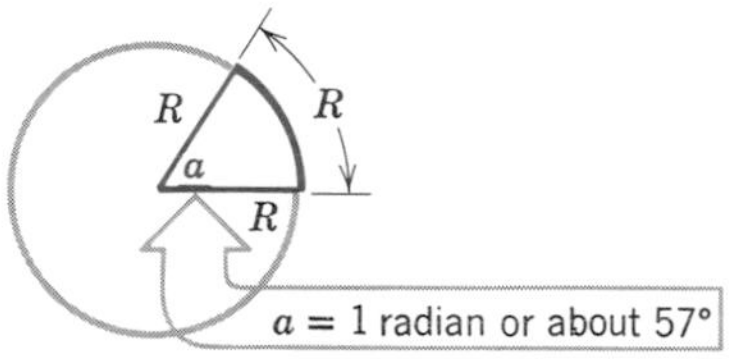

To understand where the definition comes from, set up a proportion:

$$\frac{\text{angle } a}{\text{total angle in the circle}} = \frac{\text{arc length}}{\text{circumference}}$$

$$\frac{a}{360} = \frac{R}{2\pi R}$$

$$a = \frac{360^\circ \not{R}}{2\pi \not{R}}$$

$$a = \frac{180^\circ}{\pi} \qquad \frac{180^\circ}{3.1416} \quad \text{is about } 57.296^\circ$$

To convert angles from degrees to radian measure, simply multiply by 0.0175 and round as needed. For example,

43° = 43 × 0.0175
43° = 0.7525 radians
43° ≅ 0.75 radians

To convert angles from radians to degrees, multiply by 57.296 and round. For example,

1.2 radians = 1.2 × 57.296
1.2 radians = 68.7552°
1.2 rad ≅ 68.76° or 68°46′

Try these problems for practice in using radian measure.

1. Convert to radians. (Round to two decimal places.)

   (a) 10° (b) 35° (c) 90° (d) 120°

2. Convert to degrees. (Round to the nearest minute.)

   (a) 0.3 rad (b) 1.5 rad (c) 0.8 rad (d) 0.05 rad

Check your answers on p. 570.

4

1.	(a) 24°20′	(b) 64°45′	(c) 15°18′
	(d) 38°36′	(e) 46°22$\frac{1}{2}$′	(f) 124°40′
2.	(a) 10.8°	(b) 96.15°	(c) 57.6°
	(d) 168.9°	(e) 0.55°	(f) 1.2°

Trigonometry is triangle measurement, and, to keep it as simple as possible, we can begin by studying only *right* triangles. You should remember from Chapter 7 that a right triangle is one that contains a right angle, a 90° angle. All of the following are right triangles.

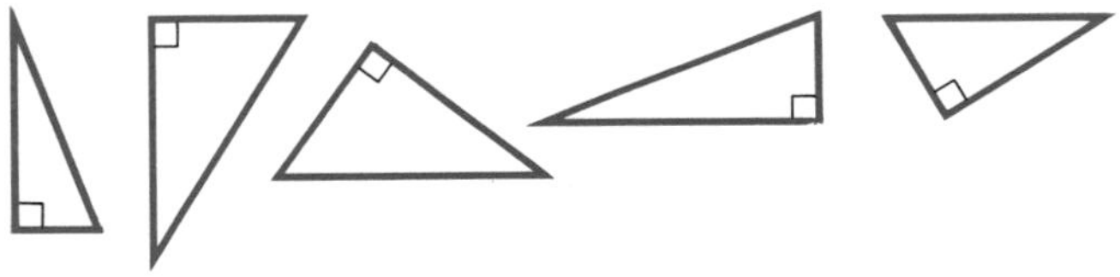

In each triangle the right angle is marked.

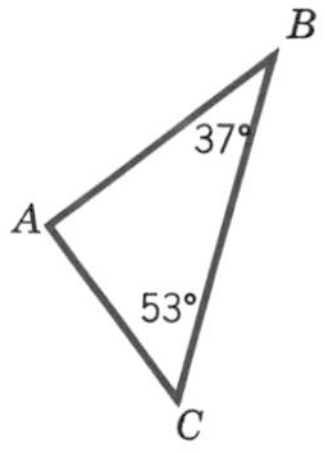

Is triangle ABC a right triangle?

The answer is in **5**.

---

5 You should remember that the sum of the three angles of any triangle is 180°. For triangle ABC two of the angles total 37° + 53° or 90°; therefore angle A must equal 180° − 90° or 90°. The triangle *is* a right triangle.

To make your study of triangle trigonometry even easier, we will always work with triangles that are in a *standard position*. For this collection of right triangles

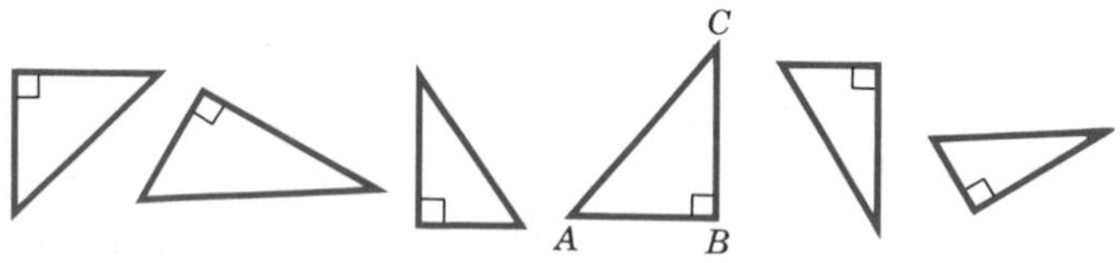

only triangle ABC is in standard position.

Place the triangle so that the right angle is on the right side, one side (AB) is horizontal, and the other side (BC) is vertical. The longest side (AC) will slope up from left to right.

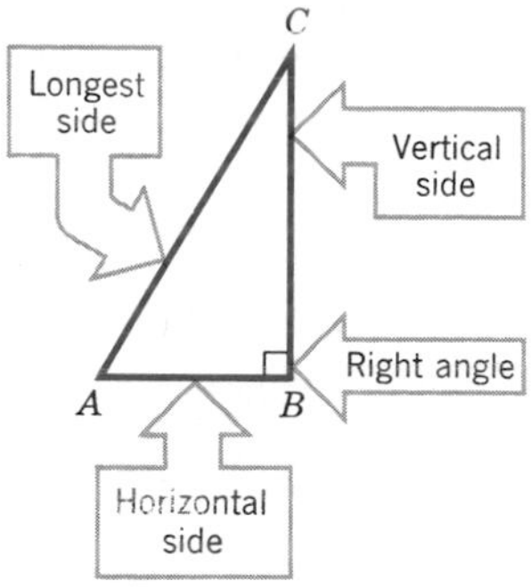

Try it. Rotate the following triangle so that it is in standard position.

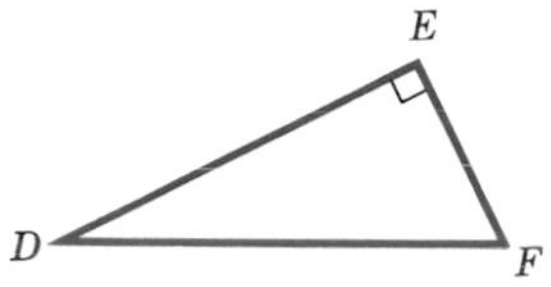

Check it in **6**.

6

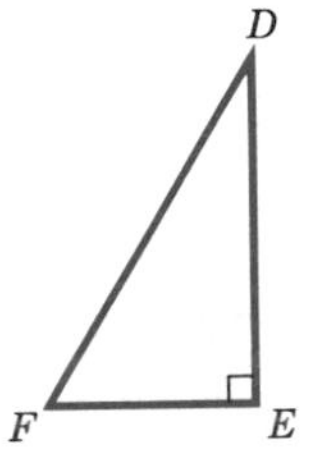

Once we have placed the right triangle in this position, we can name the sides in a way that has also become standard.

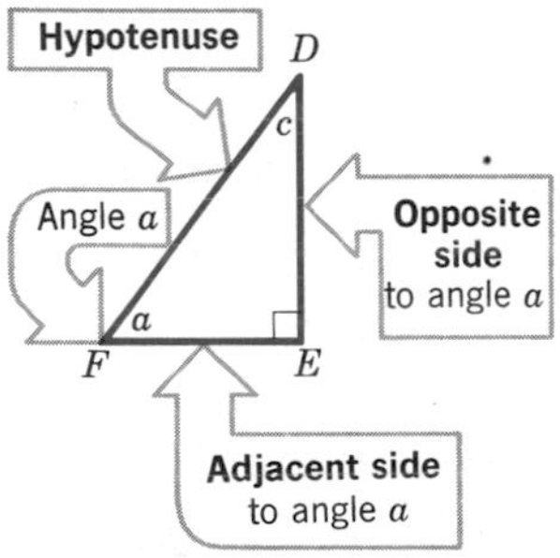

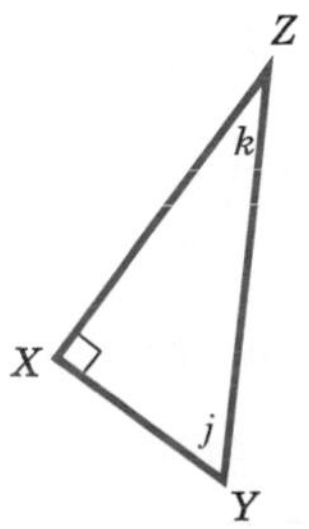

The longest side DF, the side opposite the right angle, is called the *hypotenuse.*

The side DE is called the *opposite side* for angle a. The side EF is called the *adjacent side* for angle a.

For angle c, side DE is its adjacent side, and side EF is its opposite side.

Check your understanding of this vocabulary by putting the triangle in standard position and by completing the following statements.

(a) The hypotenuse is ________ .

(b) The opposite side for angle Y is ________.

(c) The adjacent side for angle Y is ______.

(d) The opposite side for angle Z is ______.

(e) The adjacent side for angle Z is ______.

Check your answers in **7**.

7

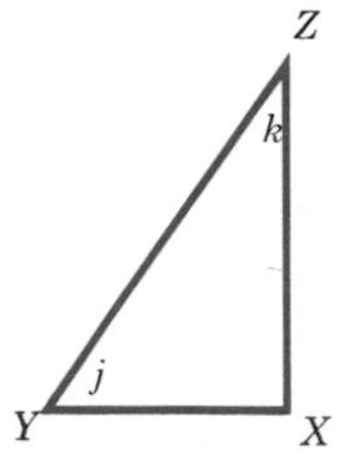

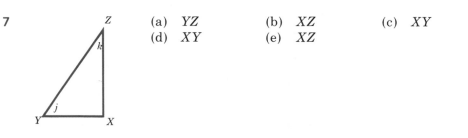

(a) YZ (b) XZ (c) XY
(d) XY (e) XZ

Of course, these labels are only used with right triangles.

For each of the following right triangles, place the triangle in standard position, mark the hypotenuse with the letter H, mark the side adjacent to the given angle with the letter A, and mark the side opposite to the given angle with the letter B.

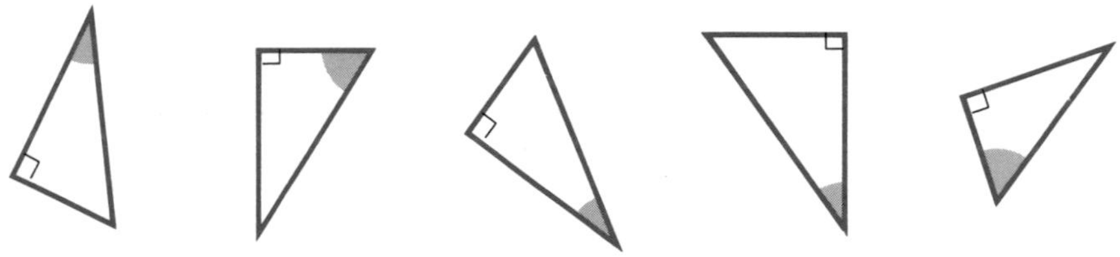

(a) (b) (c) (d) (e)

Check your work in **8**.

8

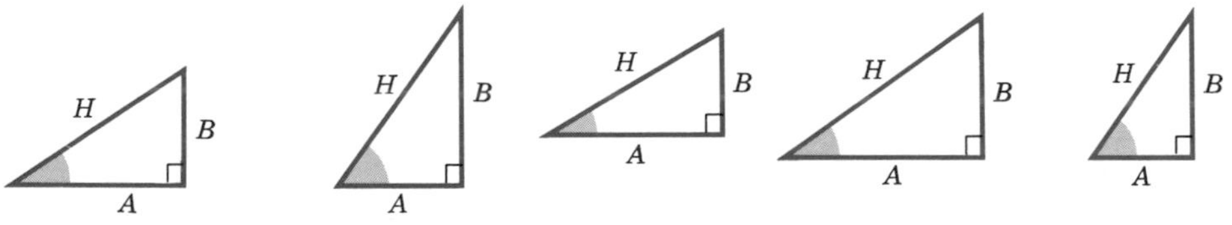

(a) (b) (c) (d) (e)

You should also recall from Chapter 7 that the *Pythagorean Theorem* is an equation relating the lengths of the sides of any right triangle.

PYTHAGOREAN THEOREM

$$H^2 = A^2 + B^2$$

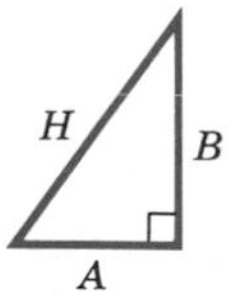

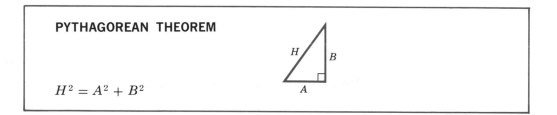

This equation is true for every right triangle. For example, the triangle

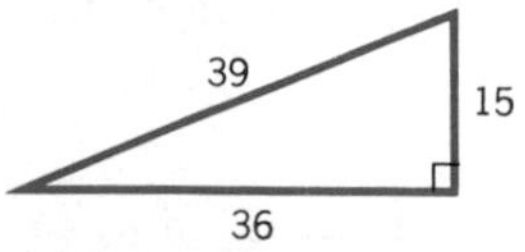

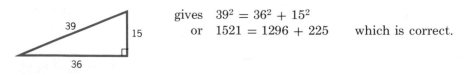

gives $39^2 = 36^2 + 15^2$
or $1521 = 1296 + 225$ which is correct.

For practice, use this rule to find the missing side for each of the following triangles:

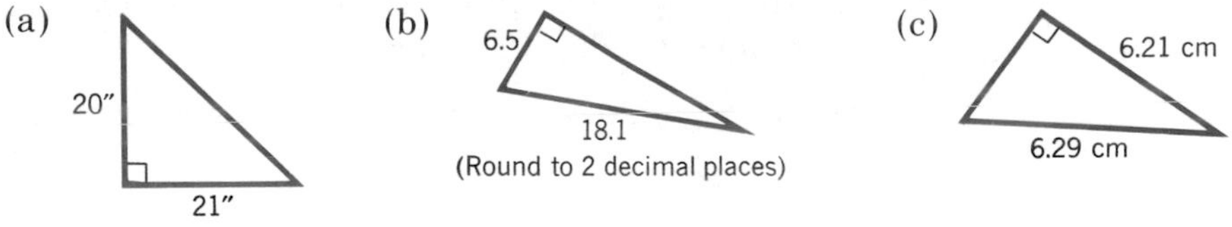

Check your answers in **9**.

---

9 (a)

$$H^2 = 21^2 + 20^2$$
$$H^2 = 441 + 400$$
$$H^2 = 841$$
$$H = 29''$$

(b)

$$18.1^2 = B^2 + 6.5^2$$
$$327.61 = B^2 + 42.25$$
$$B^2 = 327.61 - 42.25$$
$$B^2 = 285.36$$
$$B \cong 16.89$$

(c)

$$6.29^2 = 6.21^2 + A^2$$
$$39.5641 = 38.5641 + A^2$$
$$A^2 = 39.5641 - 38.5641$$
$$A^2 = 1.0000$$
$$A = 1.00 \text{ cm}$$

Three very special right triangles are used very often in practical work. Here they are:

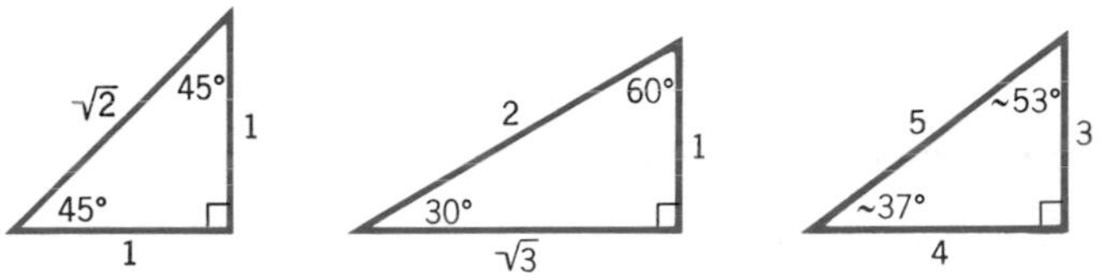

The first triangle is a *45° right triangle*. Its angles are 45°, 45°, and 90°, and it is formed by the diagonal and two sides of a square. Notice that it is an isosceles triangle—the two shorter sides are the same length.

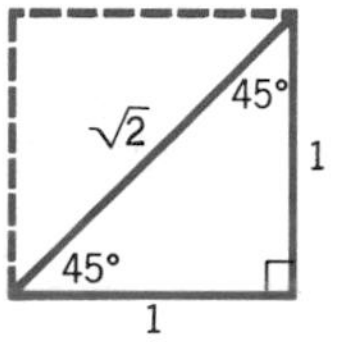

Our triangle has the shorter sides one unit long, but we can draw a 45° right triangle with any length sides. If we increase the length of one side the others increase in proportion, while the angles stay the same size.

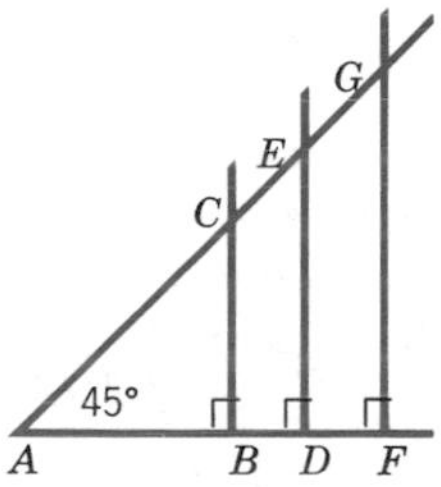

Triangle ABC is a 45° right triangle.
Triangle ADE is also a 45° right triangle.
Triangle AFG is also a 45° right triangle, . . . and so on.

If $AD = 2''$ (a) How long is DE?
(b) How long is AE?

Look for the answers in **10**.

---

10 (a) $DE = 2$ in.

The sides of a 45° right triangle opposite the 45° angles are always equal in length. If AD is 2 inches long, DE is also 2 inches long.

(b) $AE = 2\sqrt{2}$ in. or about 2.8 in.

The hypotenuse of a 45° right triangle is $\sqrt{2}$ times the length of one of the smaller sides.

No matter what the size of the triangle, if it is a right triangle and if one angle is 45°, then the third angle is also 45°. The two smaller sides are equal, and the hypotenuse is $\sqrt{2} \simeq 1.4$ times the length of a side.

The second of our special triangles is the 30°–60°right triangle.

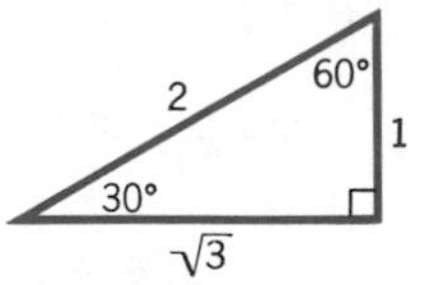

In a 30°–60° right triangle the length of the side opposite the 30° angle, the smallest side, is exactly one-half the length of the hypotenuse. The length of the third side is $\sqrt{3}$ times the length of the smaller side.

In this triangle, if $a = 30°$ and $A = 2$ inches, calculate

(a) $H =$
(b) $B =$
(c) $b =$

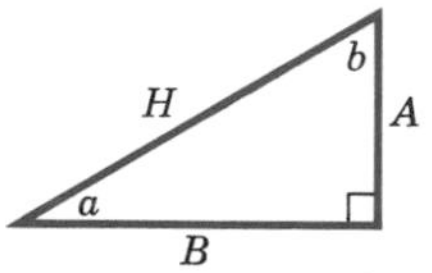

Compare your work with ours in **11**.

---

11 (a) $H = 4$ in. (b) $B = 2\sqrt{3}$ in. or about 3.5 in. (c) $b = 60°$

Again, we may form any number of 30°–60° right triangles, and in each triangle the hypotenuse is twice the length of the smallest side. The third side is $\sqrt{3}$ or about 1.7 times the length of the smallest side. When you find a right triangle with one angle equal to 30° or 60°, you automatically know a lot about it.

The third of our special triangles is the 3–4–5 triangle that was discussed in Chapter 7. If a triangle is found to have sides of length 3, 4, and 5 units (any units—in., ft, cm), it must be a right triangle. The 3–4–5 triangle is the smallest right triangle with sides whose lengths are whole numbers.

Construction workers or surveyors can use this triangle to set up right angles.

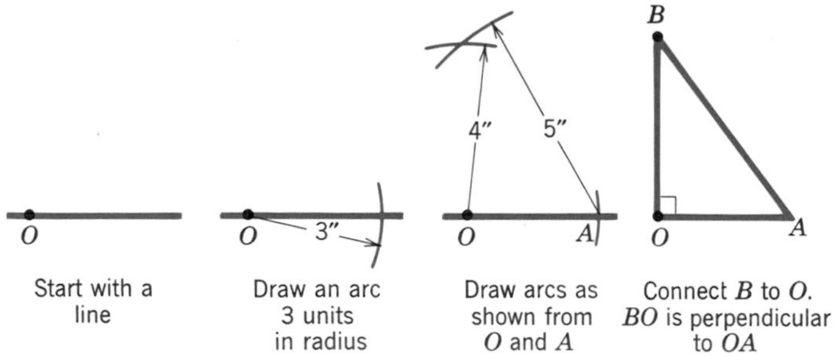

Notice that in any 3–4–5 triangle the acute angles are approximately 37° and 53°. The smallest angle is always opposite the smallest side.

You will find these three triangles appearing often in triangle trigonometry. Now, for some practice in working with angles and triangles, turn to **12**.

---

12 **Exercises 9-1 Angles and Triangles**

A. Label the shaded angles as acute, obtuse, right, or straight.

1. 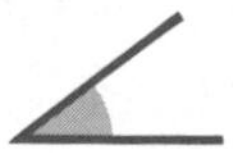 2. 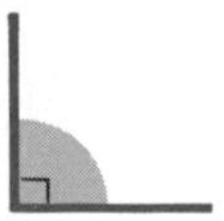 3. 

4. 

5. 
6. 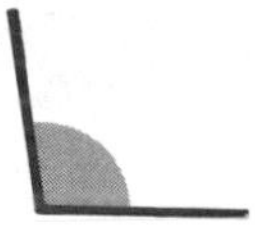
7. 

8. 
9. 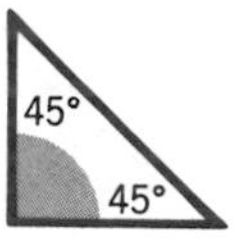

10. 

B. Write the following angles in degrees and minutes. (Round to the nearest minute.)

1. $36\frac{1}{4}°$	2. $73\frac{3}{5}°$	3. 65.45°
4. $17\frac{1}{2}°$	5. $47\frac{3}{8}°$	6. 16.11°
7. 24.75°	8. 185.8°	9. 1.054°
10. 216.425°	11. 9.65°	12. 80.60°

C. Solve the following problems involving triangles.

Place each of the following right triangles in standard position. Place an *H* near the hypotenuse, an *A* near the side adjacent to angle *a*, and a *B* near the side opposite angle *a*.

1. 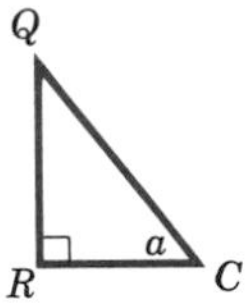

2. 

3. 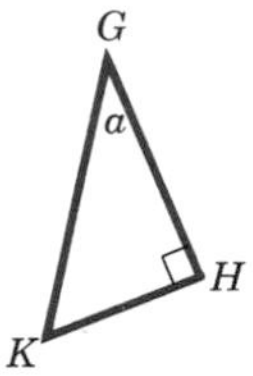

4. 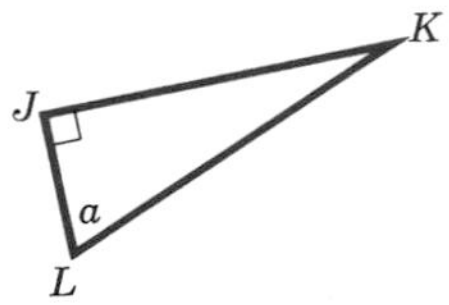

5. 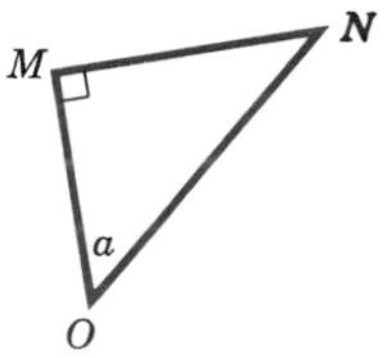

6. 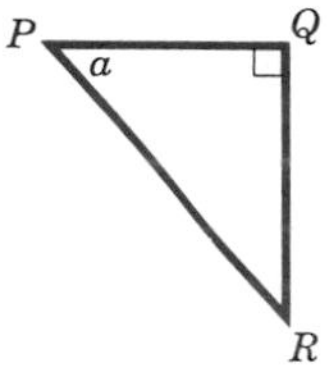

7. 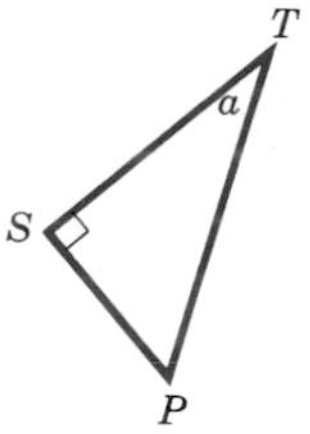

8. 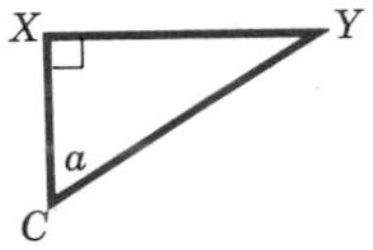

9. 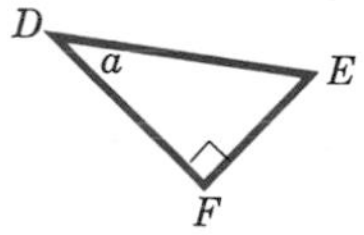

10. 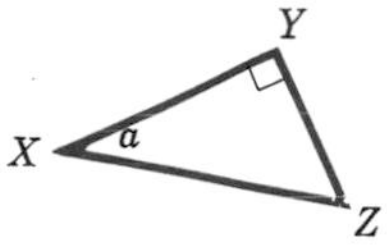

Find the lengths of the missing sides below using the Pythagorean Theorem. (Round to one decimal place.)

11. 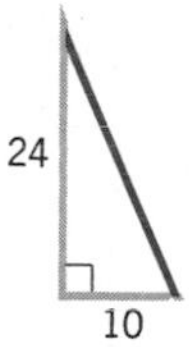

12. 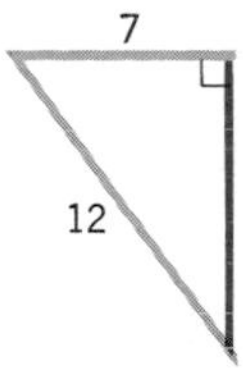

13. 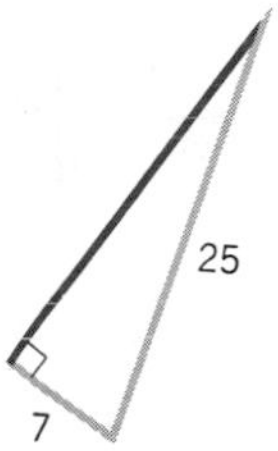

14. 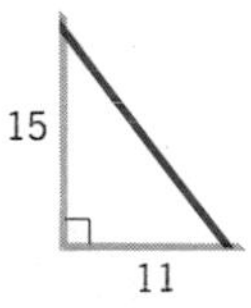

15. 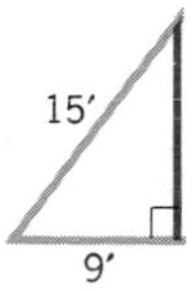

16. 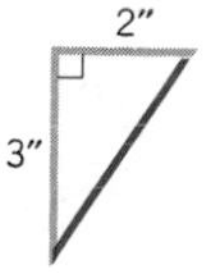

17. 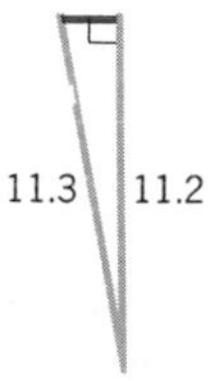

18. 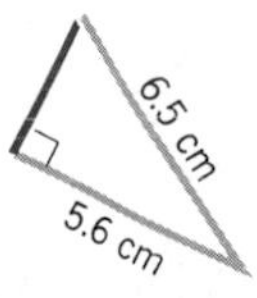

19. 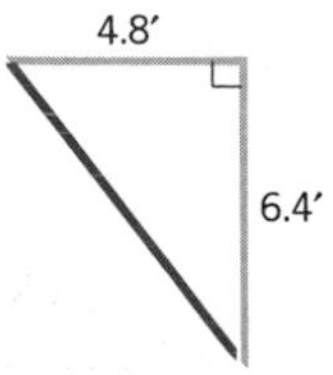

20. 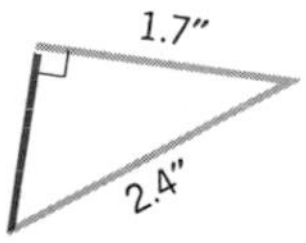

Each of the following right triangles is an example of one of the three "special triangles" described in this chapter. Find the quantities indicated without using the Pythagorean Theorem.

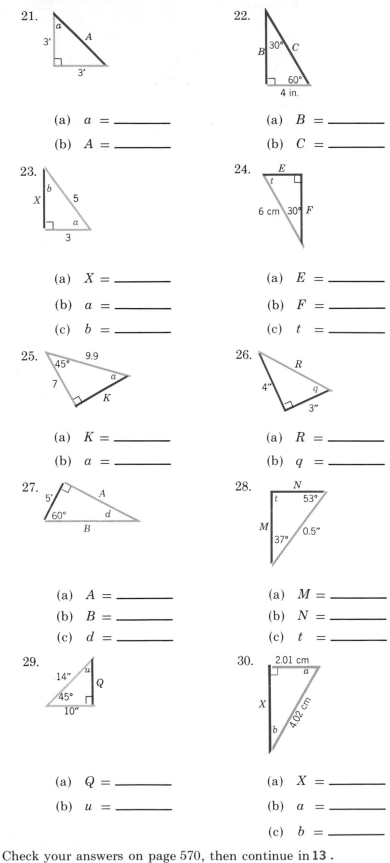

21.

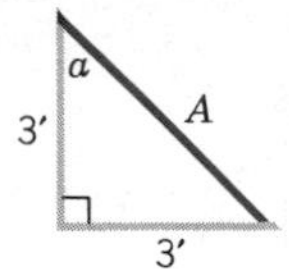

(a) a = ________

(b) A = ________

22.

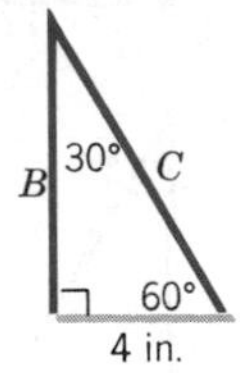

(a) B = ________

(b) C = ________

23.

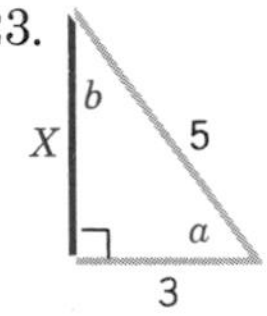

(a) X = ________

(b) a = ________

(c) b = ________

24.

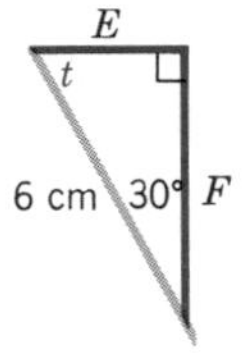

(a) E = ________

(b) F = ________

(c) t = ________

25.

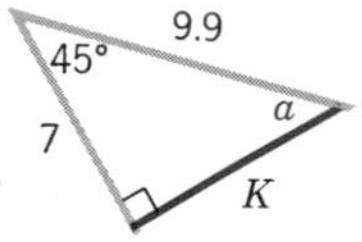

(a) K = ________

(b) a = ________

26.

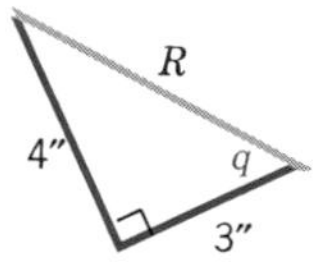

(a) R = ________

(b) q = ________

27.

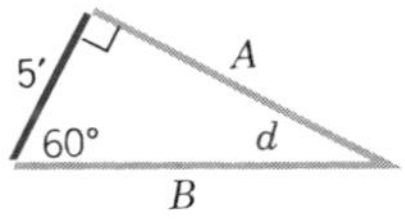

(a) A = ________

(b) B = ________

(c) d = ________

28.

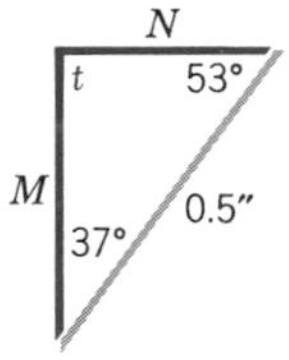

(a) M = ________

(b) N = ________

(c) t = ________

29.

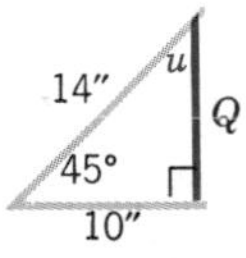

(a) Q = ________

(b) u = ________

30.

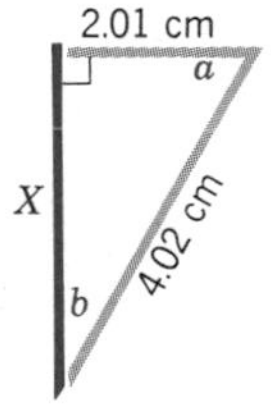

(a) X = ________

(b) a = ________

(c) b = ________

Check your answers on page 570, then continue in **13** .

13 In Section 9-1 we showed that in *all* 45° right triangles, the adjacent side is equal in length to the opposite side, and the hypotenuse is roughly 1.4 times this length. Similarly, in *all* 30°–60° right triangles, the hypotenuse is twice the length of the side opposite the 30° angle, and the adjacent side is about 1.7 times this length.

The key to understanding trigonometry is to realize that in *every* right triangle there is a fixed relationship connecting the lengths of the adjacent side, opposite side, hypotenuse, and the angle that characterizes the triangle.

For example, in all right triangles that contain the angle 20°,

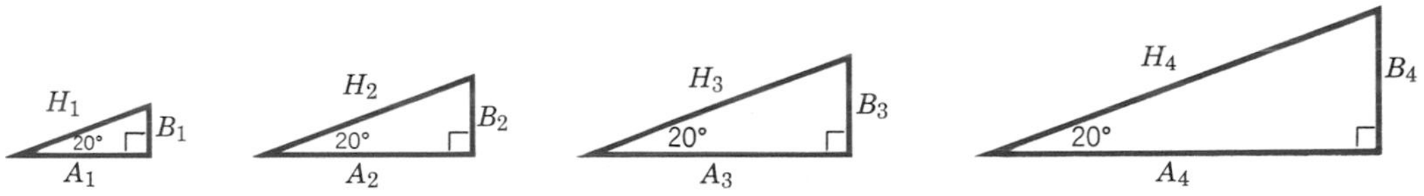

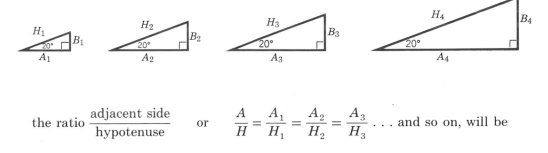

the ratio $\dfrac{\text{adjacent side}}{\text{hypotenuse}}$ or $\dfrac{A}{H} = \dfrac{A_1}{H_1} = \dfrac{A_2}{H_2} = \dfrac{A_3}{H_3}$ . . . and so on, will be

approximately 0.9397.

This kind of ratio of the side lengths of a right triangle is called a *trigonometric ratio*. It is possible to write down six of these ratios, but we will use only three here. Each ratio is given a special name.

For the triangle

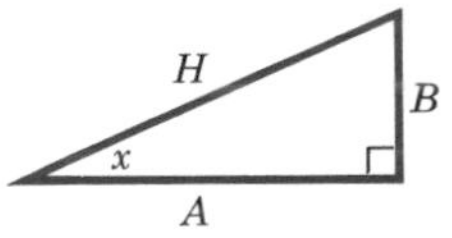

the ratio ***sin x*** is defined as

$$\sin x = \frac{\text{opposite side}}{\text{hypotenuse}} = \frac{B}{H}$$

(Pronounce "sin" as "sign" to rhyme with "dine" not with "sin." *Sin* is an abbreviation for *sine*.)

For example, in the triangle

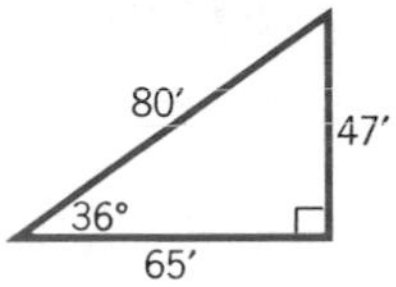

$$\sin 36° = \frac{47'}{80'}$$

$\sin 36° = 0.588$ approximately

Use this triangle to find sin 60°. (Round to three decimal places.)

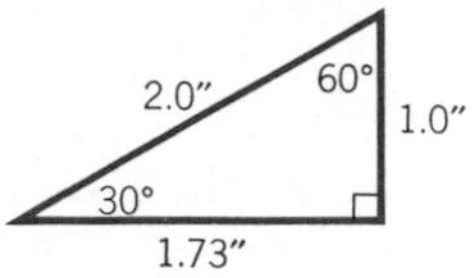

sin 60° = ________

Check your work in **14** .

---

14 In standard position

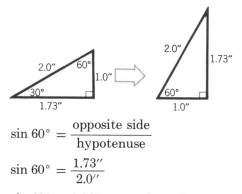

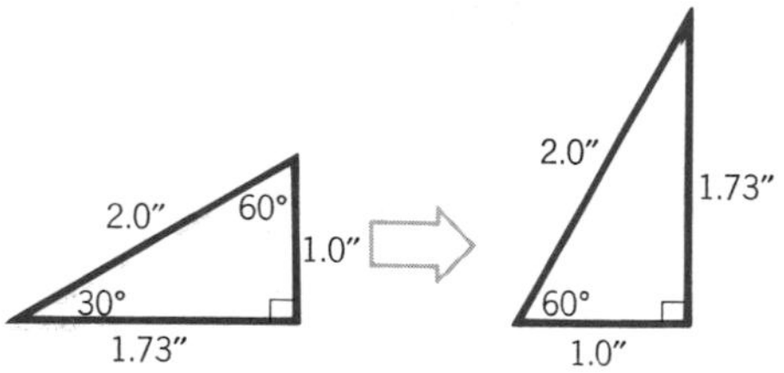

$$\sin 60^\circ = \frac{\text{opposite side}}{\text{hypotenuse}}$$

$$\sin 60^\circ = \frac{1.73''}{2.0''}$$

sin 60° = 0.865 approximately

Every possible angle x will have some number sin x associated with it, and we can calculate the value of sin x from any right triangle that contains the angle x.

For example, sin 40° can be calculated from this triangle.

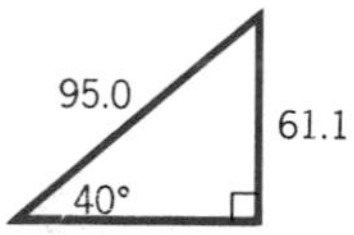

$$\sin 40^\circ = \frac{61.1}{95.0}$$

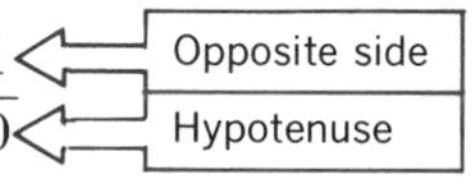

$\cong 0.643$ rounded

What is the value of sin 0°?

(*Hint:* Construct a right triangle with a very small angle and guess.)

Check your answer in **15** .

---

15 sin 0° = 0 In this triangle

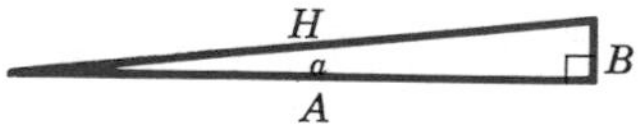

side B is much smaller than H, therefore the number $\frac{B}{H}$ is very small.

As angle a decreases to zero, side B gets smaller and smaller, and sin a get closer and closer to zero.

In a similar way you can show that sin 90° = 1. (Try it.)

Now, remember those three special triangles from Section 9-1.

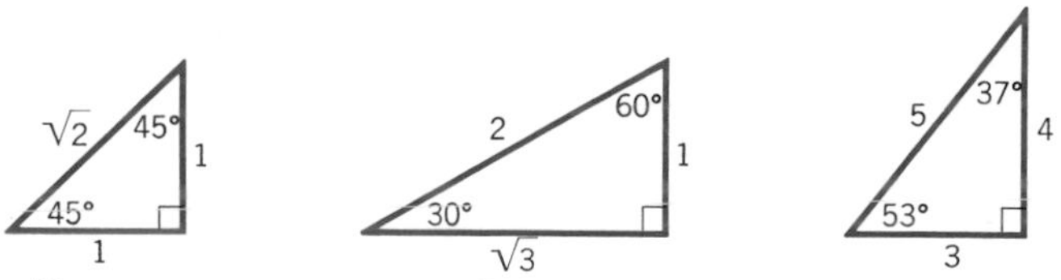

$\sqrt{2} \cong 1.414$ $\sqrt{3} \cong 1.732$

Use these drawings to calculate the following values of $\sin x$. (Round to the nearest hundredth.)

x	30°	37°	45°	53°	60°
$\sin x$					

Check your answers in **16**.

16

x	30°	37°	45°	53°	60°
$\sin x$	0.50	0.60	0.71	0.80	0.87

Sin x is a trigonometric *ratio* or trigonometric *function* associated with the angle x. The value of $\sin x$ can never be greater than 1.

A second useful trigonometric ratio is the *cosine* of angle x, abbreviated *cos* x. This ratio is defined as

$$\cos x = \frac{\text{adjacent side}}{\text{hypotenuse}} = \frac{A}{H}$$

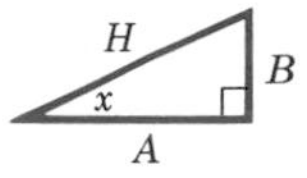

For example, in the triangle

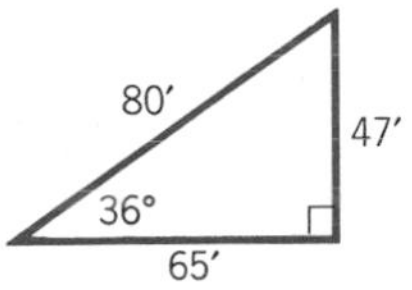

$$\cos 36° = \frac{65'}{80'}$$

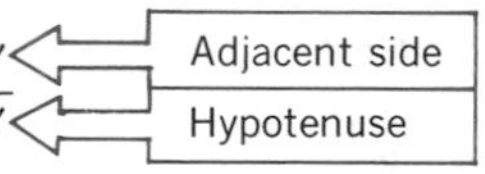

$\cos 36° = 0.81$ approximately

Use the following triangle to find $\cos 60°$.

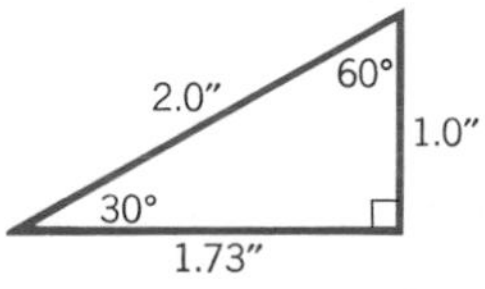

Check your work in **17**.

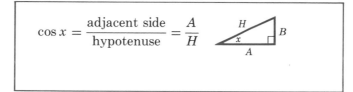

17

In standard form

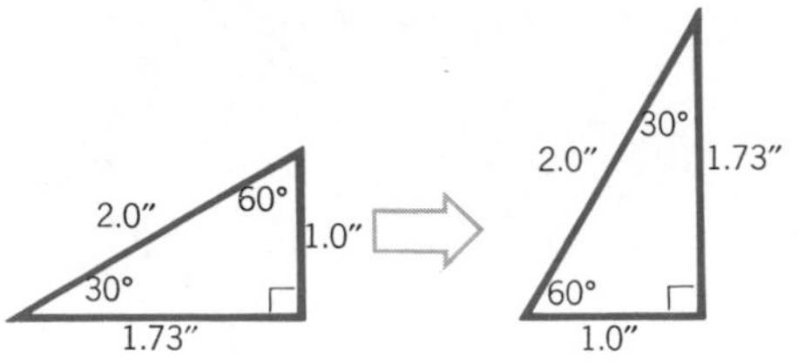

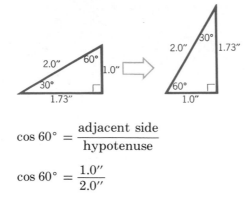

$$\cos 60° = \frac{\text{adjacent side}}{\text{hypotenuse}}$$

$$\cos 60° = \frac{1.0''}{2.0''}$$

$$\cos 60° = 0.50$$

Now use the drawing of the three special triangles in **9** on page 463 to help you to complete this table. (Round to two decimal places.)

x	0°	30°	37°	45°	53°	60°	90°
cos x	1						1
sin x	0						0

We have entered the values for 0° and 90°.

Check your table against ours in **18**.

18

x	0°	30°	37°	45°	53°	60°	90°
cos x	1	0.87	0.80	0.71	0.60	0.50	0
sin x	0	0.50	0.60	0.71	0.80	0.87	1

A third trigonometric ratio may also be defined. The ratio *tangent x,* abbreviated *tan x,* is defined as

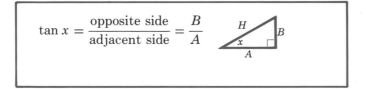

$$\tan x = \frac{\text{opposite side}}{\text{adjacent side}} = \frac{B}{A}$$

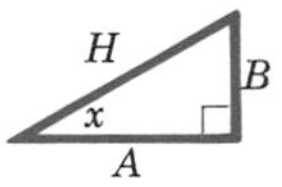

For example, in the triangle

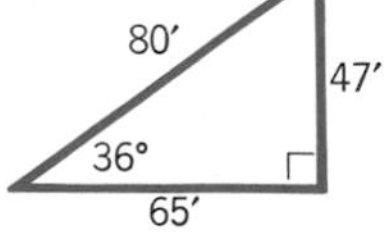

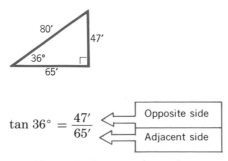

$$\tan 36° = \frac{47'}{65'}$$

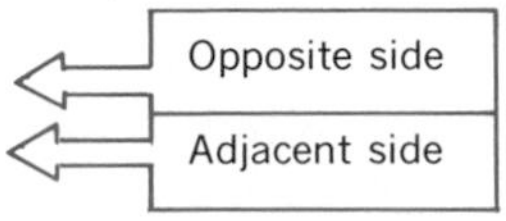

$\tan 36° = 0.72$ approximately

Use the following triangle to find tan 30°.

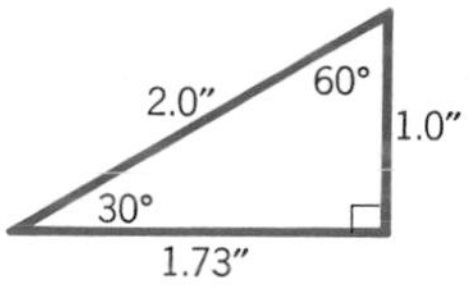

Check your work in **19**.

19

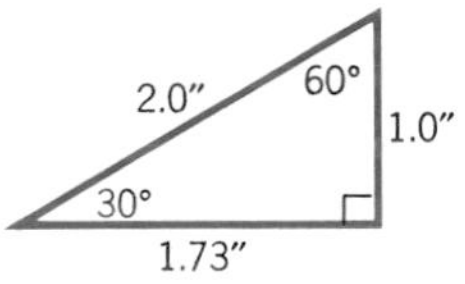

$$\tan 30° = \frac{\text{opposite side}}{\text{adjacent side}}$$

$$\tan 30° = \frac{1.0''}{1.73''}$$

$$\tan 30° = 0.58 \text{ approximately}$$

Complete this table for tan x using the three special triangles. (Round to two decimal places.)

x	0°	30°	37°	45°	53°	60°	90°
tan x							

(tan x for x = 90° is very tricky.)

Check your table against ours in **20**.

20

x	0°	30°	37°	45°	53°	60°	90°
tan x	0	0.58	0.75	1.00	1.33	1.73	not defined

To find tan 90° draw a triangle with x very large.

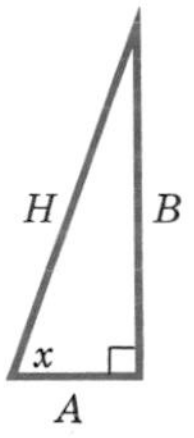

As angle x gets larger and larger,
the opposite side B gets longer and longer.

The ratio $\tan x = \dfrac{B}{A}$ gets larger and larger. As x approaches 90°, tan x becomes infinitely large, and we have no number large enough to name it.

You have been asked to calculate these "trig" ratios to help you get a feel for how the ratios are defined and to encourage you to memorize a few values. In actual practical applications of trigonometry, the values of the ratios are found by looking them up in a "Table of Trigonometric Functions" or "Trig Table" or by using an electronic calculator with built-in trigonometric function keys.

A table of trigonometric ratios to three decimal place is given on page 540. Here is a portion of that table:

TABLE OF TRIGONOMETRIC FUNCTIONS

Angle	Sine	Cosine	Tangent
0°	0.000	1.000	0.000
1°	.018	1.000	.018
2°	.035	.999	.035
3°	.052	.999	.052
4°	.070	.998	.070
5°	.087	.996	.088
6°	.105	.995	.105
7°	.122	.993	.123
8°	.139	.999	.141
9°	.156	.988	.158
10°	.174	.985	.176
11°	.191	.982	.194
12°	.208	.978	.213

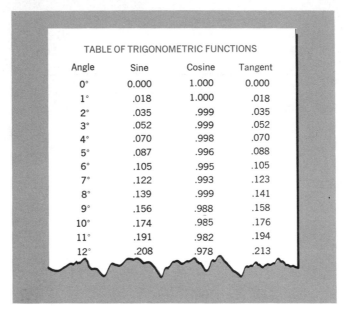

Use this table to find

(a) cos 4° (b) tan 9° (c) sin 7°

Check your answers in **21**.

---

21 (a) cos 4° = 0.998

(b) tan 9° = 0.158

(c) sin 7° = 0.122

To use the table, first locate the angle in the left column. Second, move to the right to the column representing the appropriate trig ratio. For example, to find cos 27°,

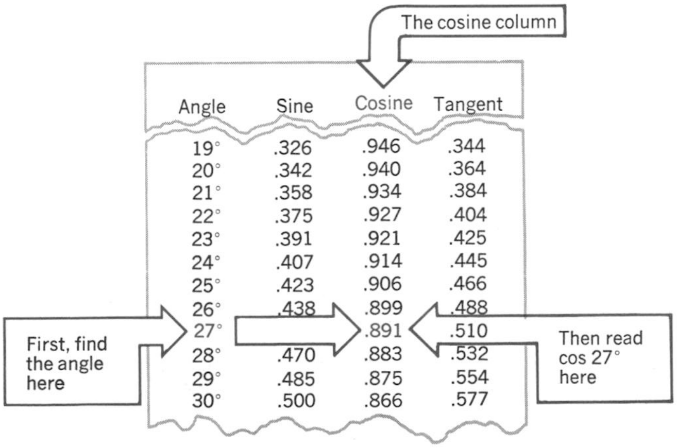

Angle	Sine	Cosine	Tangent
19°	.326	.946	.344
20°	.342	.940	.364
21°	.358	.934	.384
22°	.375	.927	.404
23°	.391	.921	.425
24°	.407	.914	.445
25°	.423	.906	.466
26°	.438	.899	.488
27°		.891	.510
28°	.470	.883	.532
29°	.485	.875	.554
30°	.500	.866	.577

cos 27° = 0.891

Use the table on page 540 to find the following trigonometric ratios.

(a) cos 87° (b) sin 61° (c) tan 43°

(d) sin 21° (e) tan 77° (f) cos 55°
(g) tan 19° (h) cos 32° (i) sin 84°

Check your answers in **22**.

---

22 (a) 0.052 (b) 0.875 (c) 0.933
(d) 0.358 (e) 4.331 (f) 0.574
(g) 0.344 (h) 0.848 (i) 0.995

In some applications of trigonometry it may happen that we know the value of the ratio number and we need to work backward to find the angle associated with it. For example, find the angle whose cosine is 0.819.

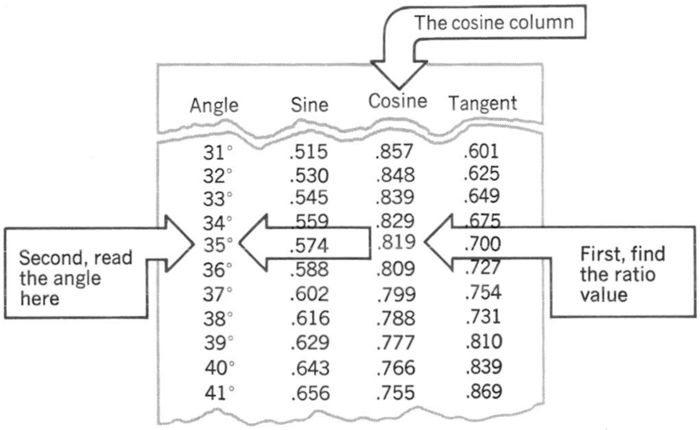

Use the table on page 540 to find

(a) the angle whose tangent is 0.404
(b) the angle whose sine is 0.695
(c) the angle whose cosine is 0.407

Check your answers in **23**.

---

23 (a) tan 22° = 0.404
(b) sin 44° = 0.695
(c) cos 66° = 0.407

If you use an electronic calculator to find the values of the trigonometric ratios, you can find the value for any angle, or fraction of an angle, quickly and easily. The tables we have been using contain no values for fractions of a degree. To include fractions of a degree would require a much larger table. But we can *estimate* or *interpolate* the value of a ratio for fractions of a degree.

Use the table to find sin 22.5°.

Try it. Then turn to **24** for a careful explanation.

---

24 22.5° is halfway between 22° and 23°, so sin 22.5° should be about halfway between sin 22° and sin 23°. From the table

sin 22° = 0.375
sin 23° = 0.391

Therefore sin 22.5° = 0.383 since 0.383 is midway between 0.375 and 0.391. This is an estimate, but it is quite accurate enough for any practical work. The process of calculating an intermediate value is called *interpolation*.

In general, follow these steps to estimate the value of a trigonometric ratio for a fractional angle.

Step 1 Convert the angle from degrees and minutes to a fraction.

Example:

$22.5° \quad = 22\frac{1}{2}°$

$38°40' = 38\frac{2}{3}°$ $\qquad 40' = \frac{40}{60} = \frac{2}{3}°$

$76°15' = 76\frac{1}{4}°$ $\qquad 15' = \frac{15}{60} = \frac{1}{4}°$

Step 2 Find the values of the trigonometric ratio for the whole number degrees on either side of the angle given.

Example: To find $\sin 76°15' = \sin 76\frac{1}{4}°$

$\sin 76° = 0.970$

$\sin 77° = 0.974$

$\sin 76\frac{1}{4}°$ will be between 0.970 and 0.974

Step 3 Multiply the degree fraction by the difference in the whole number values, then add that result to the smaller sine value.

Example:

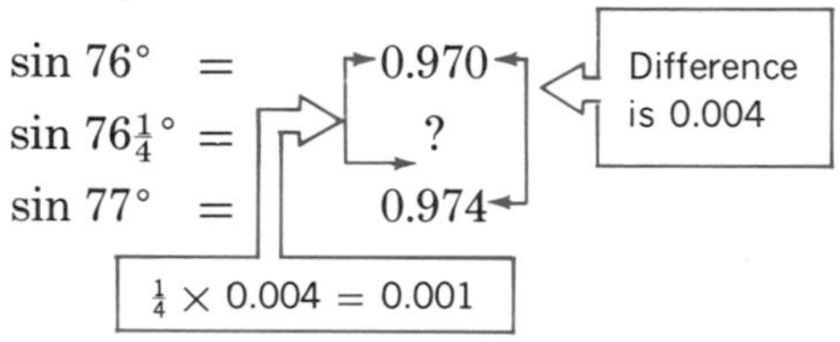

So $\sin 76\frac{1}{4}° = 0.970 + 0.001$

$\sin 76\frac{1}{4}° = 0.971$

Your turn. Find $\cos 63°20'$.

Check your work in **25**.

---

25 **Step 1** $63°20' = 63\frac{1}{3}°$ $\qquad 20' = \frac{20}{60} = \frac{1}{3}°$

Step 2 $\cos 63° = 0.454$

$\cos 64° = 0.438$

$\cos 63\frac{1}{3}°$ is between 0.454 and 0.438

Step 3

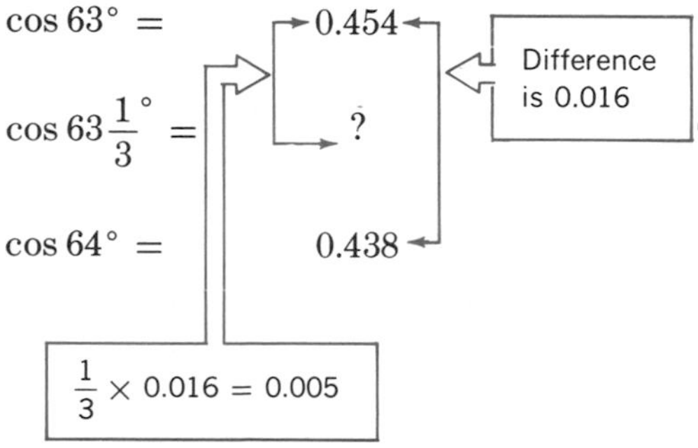

So $\cos 63\frac{1}{3}° = 0.454 - 0.005$

or $\cos 63\frac{1}{3}° = 0.449$

Notice that in this case we must *subtract* 0.005 in order to get an answer between 0.454 and 0.438.

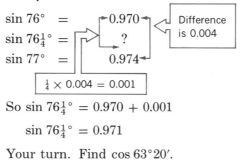

Find the following trigonometric ratio values using this estimation procedure.

(a) cos 27°40′ (b) sin 72.75 (c) tan 46°15′
(d) sin 38°20′ (e) tan 15°30′ (f) cos 81.2°

Check your answers in **26**.

TRIGONOMETRIC RATIOS FOR ANGLES BETWEEN 90° AND 180°

To find values of the sine, cosine, and tangent ratios for angles greater than 90° use these formulas.

$$\cos(90° + a) = -\sin a$$
$$\sin(90° + a) = \cos a$$
$$\tan(90° + a) = -\tan(90° - a)$$

For example,

$$\cos 120° = \cos(90° + 30°)$$
$$= -\sin 30°$$

Break the angle down to 90° plus some smaller angle, and use the smaller angle in the table

$\cos 120° = -0.500$ since $\sin 30° = 0.500$ in the table

Here is another example.

$\tan 147° = \tan(90° + 57°)$
$= -\tan(90° - 57°)$ from the formula above
$= -\tan(33°)$
$= -0.649$ $\tan 33° = 0.649$ in the table

Find these ratios using the formulas above.

(a) cos 110° (b) sin 163° (c) tan 125°
(d) sin 132° (e) tan 171° (f) cos 144°

Check your answers on page 451.

---

26 (a) 0.886 (b) 0.955 (c) 1.045
(d) 0.620 (e) 0.277 (f) 0.153

If you need to find the angle corresponding to a trigonometric value that is between two of the values on the table, follow a reverse procedure.

Step 1 Find the values in the table on either side of the given value.

Example: Find the angle whose cosine is 0.826.

$\cos 34° = 0.829$
$\cos 35° = 0.819$

The angle is between 34° and 35°

Step 2 Find the differences between the table values and between the unknown angle value and the table value for the smaller angle.

Example:

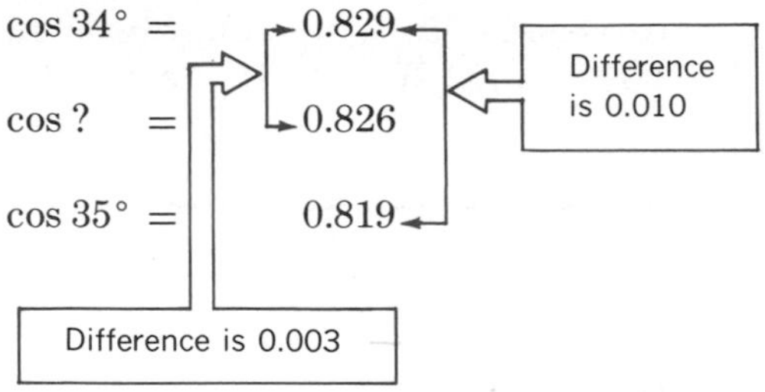

Step 3 Use these differences to find the angle fraction.

Example:

$$\frac{0.003}{0.010} = \frac{3}{10}$$

The angle is $\frac{3}{10}$ of the way between 34° and 35°, or 34.3°.

cos 34.3° = 0.826

The cosine value 0.826 corresponds to an angle of 34.3°.

Use this procedure to determine the angle whose sine is 0.464.

Check your work in **27**.

What is an "arc sin"?

The arc sin is the inverse of the sine and is often abbreviated $\sin^{-1}$. The arc sin of a number means that you are working backward from that number to get the angle. For example, if $\sin 30° = 0.5$, then $\arc \sin 0.5 = 30°$. That notation appears in many advanced math books and on some calculators.

---

27 **Step 1** The ratio value 0.464 lies between

sin 27° = 0.454
sin 28° = 0.469

The angle is between 27° and 28°.

Step 2

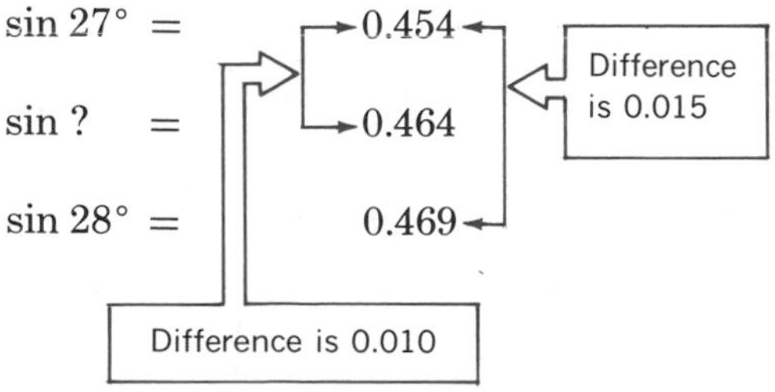

Step 3 The angle is $\frac{0.010}{0.015}$ or $\frac{2}{3}$ of a degree from 27° or at $27\frac{2}{3}°$. The sine value 0.464 corresponds to an angle of $27\frac{2}{3}°$ or 27°40′. (Because the table is accurate only

to three decimal places, the answer found using an electronic calculator will be slightly different, in this case about 27°38.7′.)

Use this procedure to find angle x for each of the following:

(a) $\sin x = 0.814$	(b) $\tan x = 0.988$	(c) $\cos x = 0.949$
(d) $\tan x = 2.08$	(e) $\sin x = 0.262$	(f) $\cos x = 0.387$

(Round to nearest minute.)

Check your answers in **28**.

---

28 (a) 54°30′ (b) 44°39′ (c) 18°24′
(d) 64°19′ (e) 15°11′ (f) 67°15′

Finding trig ratios and angles is a fairly complex task using a table of trigonometric functions. However, it is easy if you use a calculator with built-in trig functions.

Now turn to **29** for a set of practice problems on trigonometric ratios.

---

29 **Exercises 9-2 Trigonometric Ratios**

A. For each of the following triangles *calculate* the indicated trigonometric ratios of the given angle. (Round to nearest hundredth.)

1.
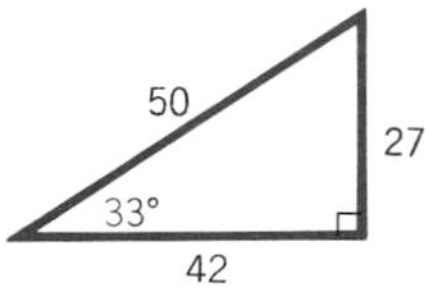

sin 33°

2.
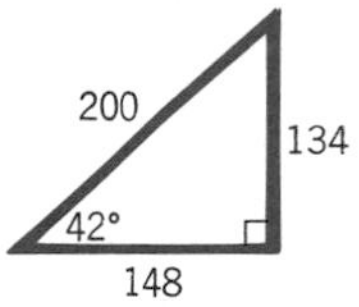

cos 42°

3.
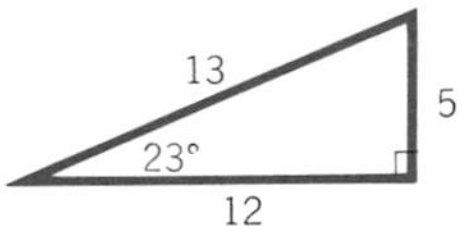

tan 23°

4.
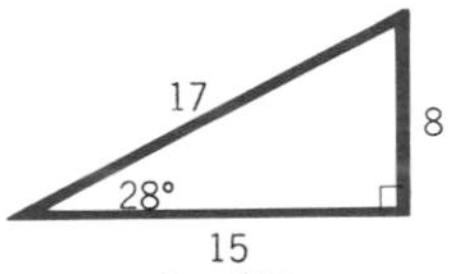

tan 28°

5.
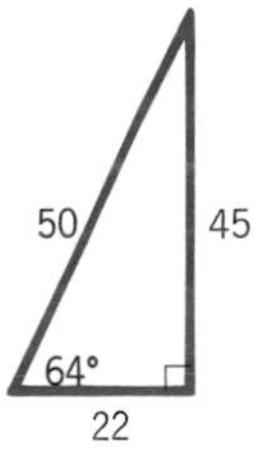

cos 64°

6.
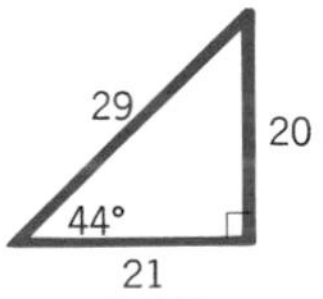

sin 44°

B. Find each of the following trig values using the table on page 540. (Round to three decimal places.)

1. sin 27°	2. cos 38°
3. tan 12°	4. sin 86°
5. cos 79°	6. tan 6°
7. cos 87°	8. cos 6°30′
9. tan 50°20′	10. sin 75°40′
11. cos 41.25°	12. cos 81°25′

C. Find the angle x. (Round to the nearest minute.)

1. $\sin x = 0.934$	2. $\cos x = 0.819$
3. $\tan x = 2.05$	4. $\sin x = 0.087$
5. $\cos x = 0.242$	6. $\tan x = 0.404$
7. $\sin x = 0.168$	8. $\cos x = 0.346$

Check your answers on page 571, then turn to **30** to learn how to use trig ratios to solve problems.

---

30 In the first two sections of this chapter you learned about angles, triangles, and the trigonometric ratios that relate the angles to the side lengths in right triangles. Now it's time to look at a few of the possible applications of these trig ratios.

If you know the length of one of the three sides of a right triangle and the size of either of the two acute angles, it is always possible to determine the other side lengths and the third angle. For example, suppose you need to measure the distance between two points A and B on opposite sides of a river. You have a tape measure and a protractor but no way to stretch the tape measure across the river. How can you measure the distance AB?

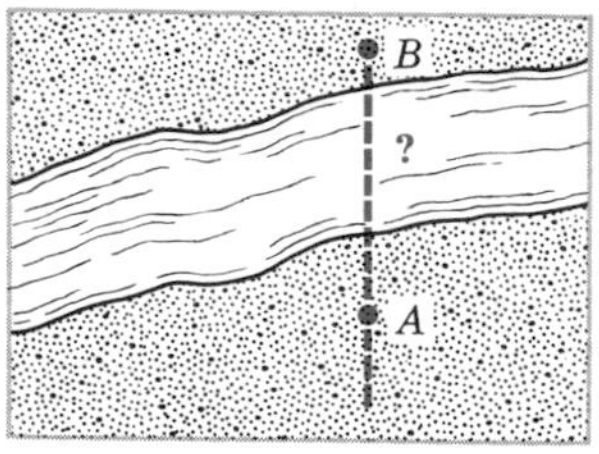

Try it this way:

1. Choose a point C in line with points A and B. If you stand at C, A and B will appear to "line up."

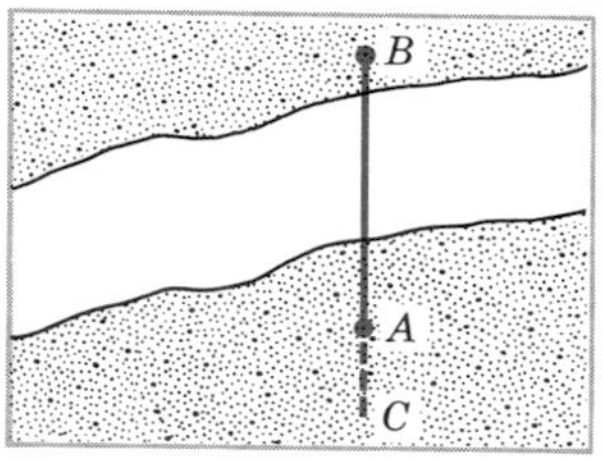

2. Construct a perpendicular to line BC at A and pick any point D on this perpendicular.

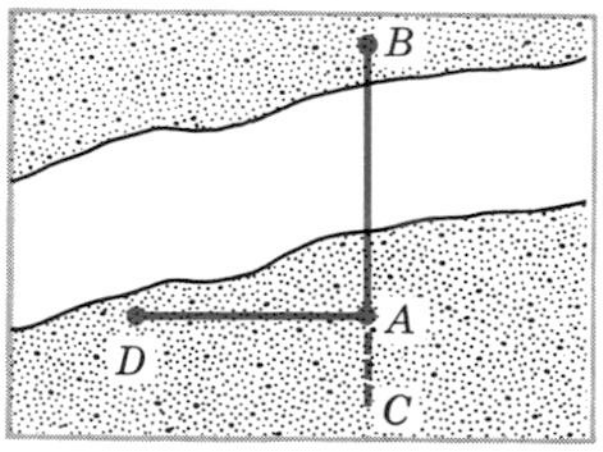

3. Find a point E where B and D appear to "line up." Draw DE and extend DE to F.

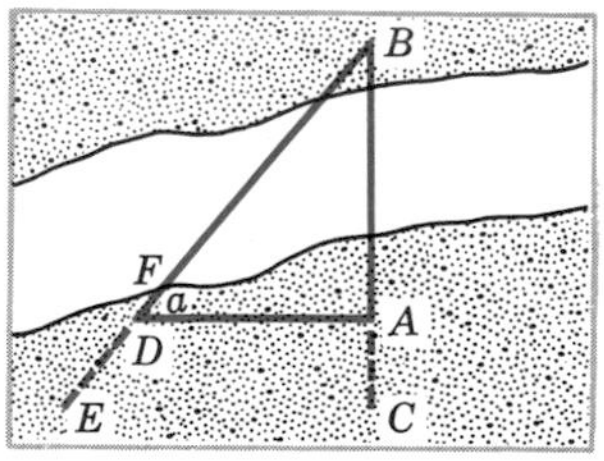

4. Triangle ADB is a right triangle. You can measure AD and angle a, and, using the trigonometric ratios, you can calculate AB. Suppose $AD = 100$ ft and $\measuredangle a = 49°$. Find AB.

Try it. Check your work in **31**.

31

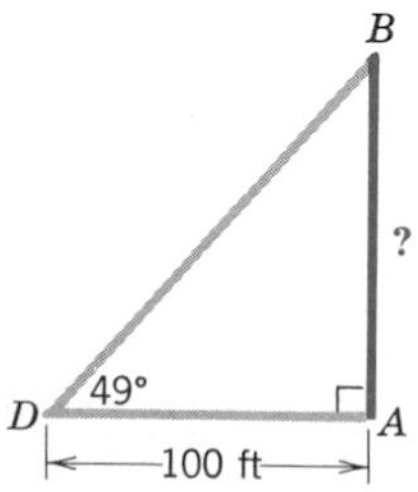

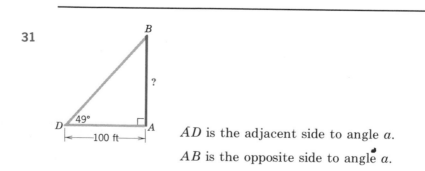

AD is the adjacent side to angle a.

AB is the opposite side to angle a.

The trig ratio that relates the opposite to the adjacent side is the tangent.

$$\tan a = \frac{\text{opposite side}}{\text{adjacent side}}$$

$$\tan a = \frac{AB}{AD}$$

$$\tan 49^\circ = \frac{AB}{100 \text{ ft}}$$

Multiply both sides of this equation by 100 to get 100 tan 49° $= AB$.

Use a trigonometric table or calculator to find that tan 49° = 1.150 approximately.

Therefore

$AB = 100 \times 1.150$
$AB = 115$ ft, rounded to the nearest foot

Using trig ratios, you have managed to calculate the distance from A to B without ever coming near point B.

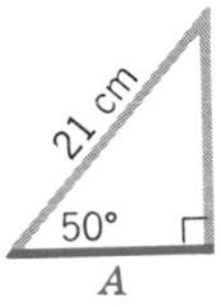

Try another example: Find the length of side A in the triangle shown.

Step 1 Decide which trig ratio is the appropriate one to use. In this case, we are given the hypotenuse (21 cm) and we need to find the adjacent side (A). The trig ratio that relates the adjacent side and the hypotenuse is the cosine.

$$\text{cosine } a = \frac{\text{adjacent side}}{\text{hypotenuse}}$$

Step 2 Write the cosine of the given angle in terms of the sides of the given right triangle.

$$\cos 50^\circ = \frac{A}{21 \text{ cm}}$$

Step 3 Solve for the unknown quantity. In this case multiply both sides of the equation by 21. Find cos 50° from a table or with a calculator.

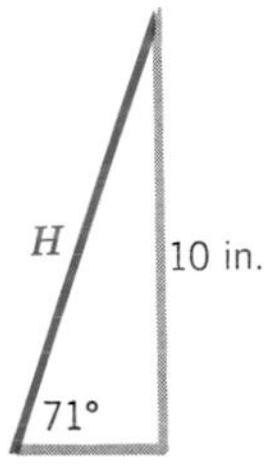

$A = 21 \cdot \cos 50^\circ$
$A = 21 \cdot 0.643$
$A = 13.503$ cm or 13.5 cm, rounded

Your turn. Find the length of the hypotenuse H in the triangle at the left. Round to the nearest tenth.

Work it out using the three steps shown above, then turn to **32** to check your answer.

---

32 **Step 1** We are given the side opposite to the angle 71°, and we need to find the hypotenuse. The trig ratio relating the opposite side and the hypotenuse is the sine.

$$\sin a = \frac{\text{opposite side}}{\text{hypotenuse}}$$

Step 2 $\sin 71^\circ = \dfrac{10 \text{ in.}}{H}$

Step 3 Solve for H. $H \cdot \sin 71^\circ = 10$ in.

or $H = \dfrac{10 \text{ in}}{\sin 71^\circ}$

Find sin 71° in a table or using a calculator.
sin 71° = 0.946

$$H = \frac{10 \text{ in.}}{0.946}$$

$H = 10.6$ in., rounded

Here are a few problems for practice in solving right triangles. Round to the nearest tenth.

(a)

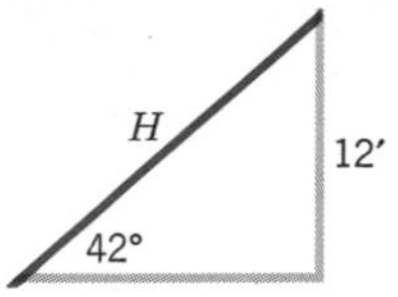

Find H

(b)

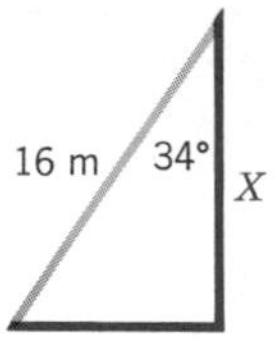

Find X

(c)

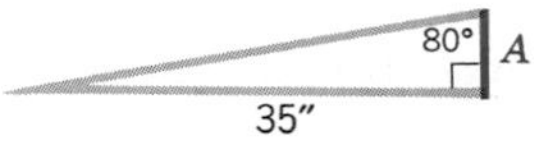

Find A

(d)

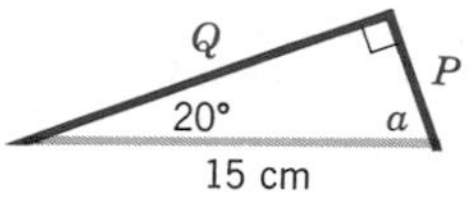

Find P, Q, and a

Check your work in **33**.

---

33 (a) $\sin 42° = \dfrac{12 \text{ ft}}{H}$

$H = \dfrac{12}{\sin 42°}$ $\quad \sin 42° = 0.669$

$H = \dfrac{12}{0.669}$

$H = 17.9$ ft, rounded

(b) $\cos 34° = \dfrac{X}{16 \text{ m}}$

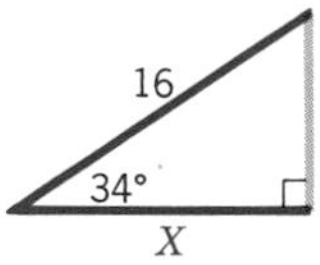

$X = 16 \cdot \cos 34°$ $\quad \cos 34° = 0.829$

$X = (16)(0.829)$

$X = 13.3$ m, rounded

(c) $\tan 80° = \dfrac{35''}{A}$

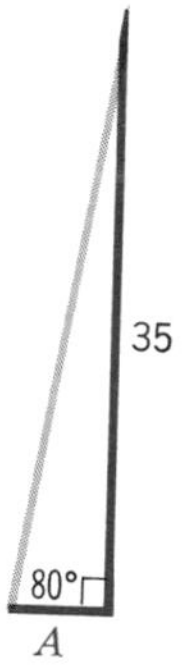

$A = \dfrac{35''}{\tan 80°}$ $\quad \tan 80° = 5.671$

$A = \dfrac{35}{5.671}$

$A = 6.2''$, rounded

(d) $\sin 20° = \dfrac{P}{15 \text{ cm}}$

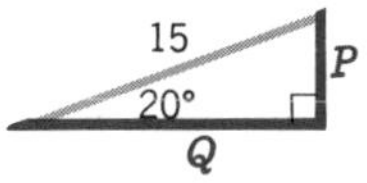

$P = 15 \cdot \sin 20°$ $\quad \sin 20° = 0.342$

$P = (15)(0.342)$

$P = 5.1$ cm, rounded

$\cos 20° = \dfrac{Q}{15 \text{ cm}}$

$Q = 15 \cdot \cos 20°$ $\quad \cos 20° = 0.940$

$Q = (15)(0.940)$

$Q = 14.1$ cm, rounded

$a = 90° - 20° = 70°$

On Problems (b), (c), and (d) redraw the triangle so that it is in standard position as shown.

Practical problems are usually a bit more difficult to set up, but they are solved with the same three-step approach. Try the following practical problems.

(a) Find the angle a in the casting shown: Round to the nearest minute.

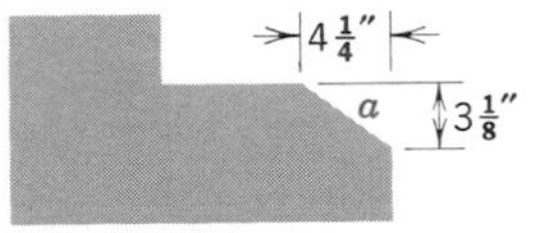

(b) In a pipe-fitting job, what is the length of set if the offset angle is $22\frac{1}{2}°$ and the travel is $32\frac{1}{8}''$? Round to the nearest tenth.

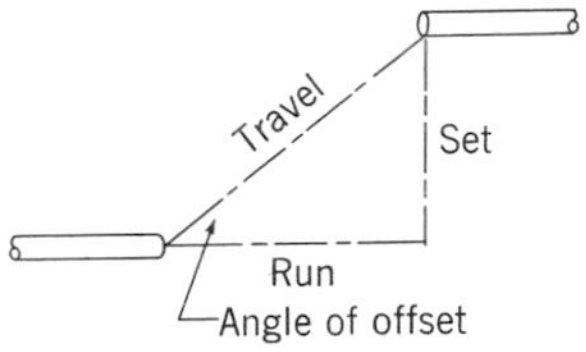

(c) Find the angle of taper, x.

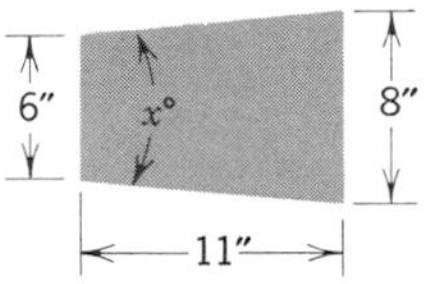

(d) Find the width, w, of the v slot shown. (Round to 0.001 cm.)

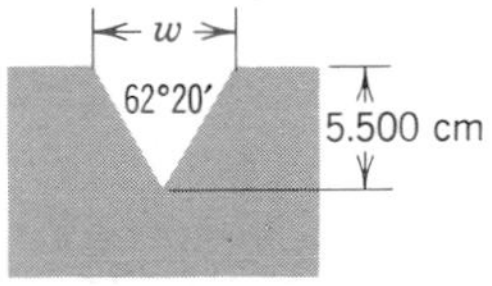

Our step-by-step solutions are in **34**.

---

34 (a) In this problem the angle a is unknown. The information given allows us to calculate tan a:

$$\tan a = \frac{\text{opposite side}}{\text{adjacent side}} = \frac{3\frac{1}{8}''}{4\frac{1}{4}''}$$

$$\tan a \cong 0.735$$

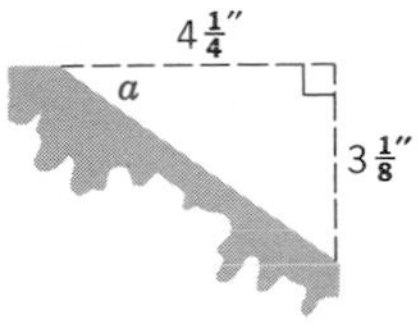

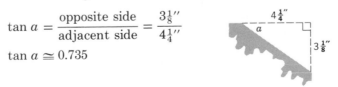

Now look down the "tan" column in the table until you find this value. Tan 36° is 0.727 and tan 37° is 0.754, therefore angle a is $\frac{0.735 - 0.727}{0.754 - 0.727}$ or $\frac{8}{27}$ of the way between 36° and 37°. Since $\frac{8}{27} \times \frac{60'}{1} \cong 18'$

$$a \cong 36°18'$$

(b) In this triangle the opposite side is unknown and the hypotenuse is given. Use the formula for sine:

$$\sin 22\tfrac{1}{2}^\circ = \frac{\text{opposite side}}{\text{hypotenuse}}$$

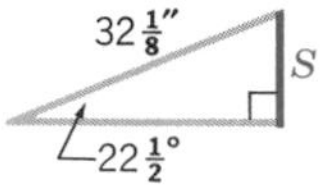

Then substitute, $0.383 = \dfrac{S}{32\frac{1}{8}''}$

And solve, $S = (32\tfrac{1}{8}'')(0.383)$
$S = 12.3''$

(c) Notice that angle x, the *taper angle,* is twice as large as angle m. We can find angle m from triangle ABC.

$$\tan m = \frac{\text{opposite side}}{\text{adjacent side}} = \frac{AC}{BC}$$

$$\tan m = \frac{1''}{11''}$$

$$\tan m = 0.091$$

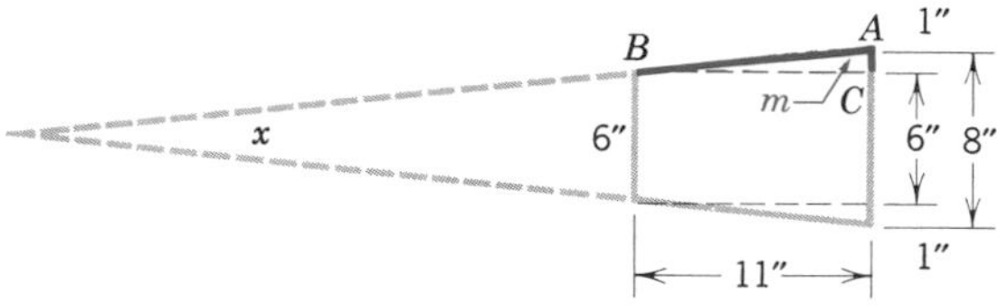

and angle $m = 5^\circ 11'$, rounded to the nearest minute

So that angle $x = 10^\circ 22'$

(d) The V slot creates an isosceles triangle whose height is 5.500 cm. The height divides the triangle into two identical right triangles as shown in the figure.

To find y,

$$\tan 31^\circ 10' = \frac{y}{5.500}$$

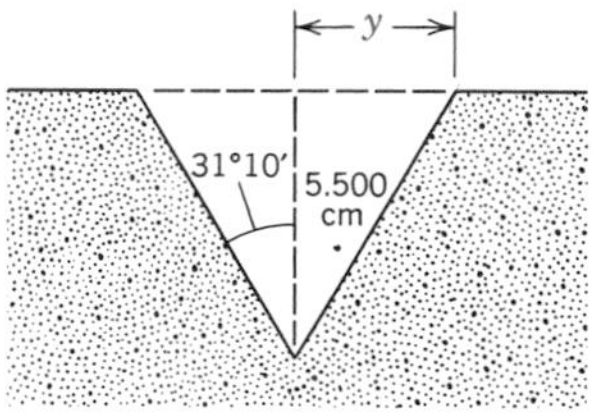

or

$y = (5.500)(\tan 31^\circ 10')$
$y = (5.500)(0.605)$
$y = 3.328$ cm

$w = 2y$
$w = 6.656$ cm

Now turn to **35** for a set of problems on solving right triangles using trigonometric ratios.

Exercises 9-3 Solving Right Triangles

35 A. Find the missing quantities as indicated (degrees to the nearest minute, sides to the nearest hundredth).

1.

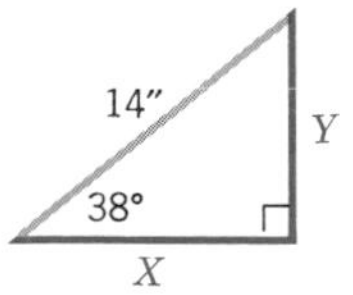

$X =$ ______

$Y =$ ______

2.

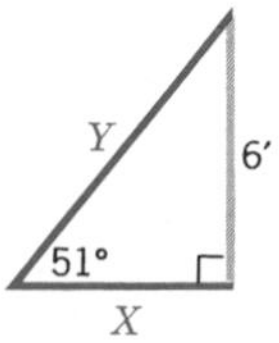

$X =$ ______

$Y =$ ______

3.

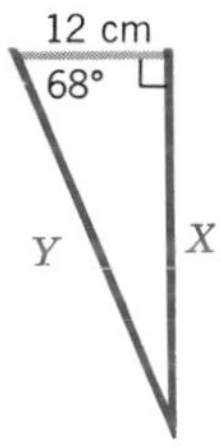

$X =$ ________

$Y =$ ________

4.

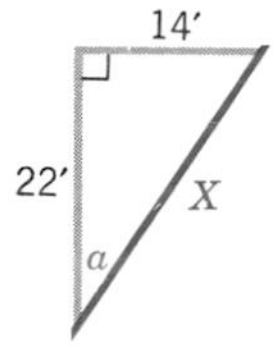

$\measuredangle a =$ ________

$X =$ ________

5.

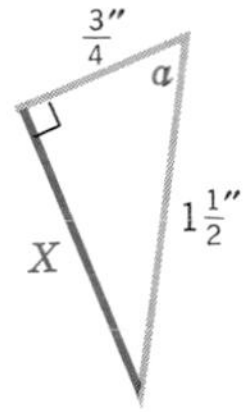

$\measuredangle a =$ ________

$X =$ ________

6.

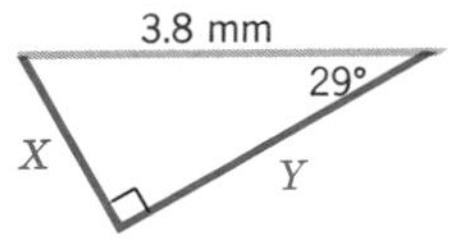

$X =$ ________

$Y =$ ________

7.

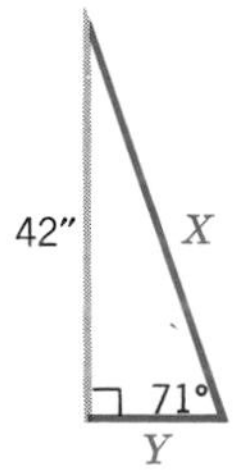

$X =$ ________

$Y =$ ________

8.

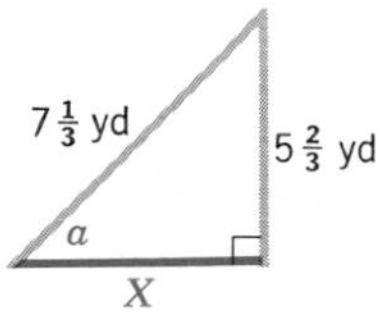

$\measuredangle a =$ ________

$X =$ ________

9.

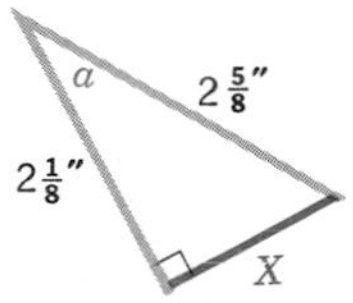

$\measuredangle a =$ ________

$X =$ ________

10.

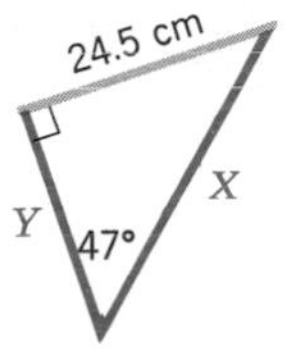

$X =$ ________

$Y =$ ________

B. Practical Problems. (Round all angles to the nearest minute and all lengths to the nearest hundredth.)

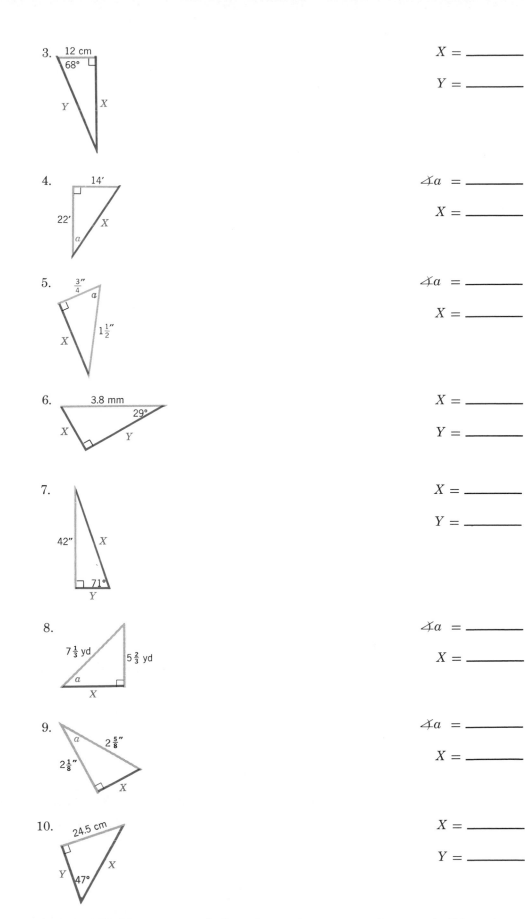

1. The most efficient operating angle for a certain conveyor belt is 31°. If the parts must be moved a vertical distance of 16′, what length of conveyor is needed?

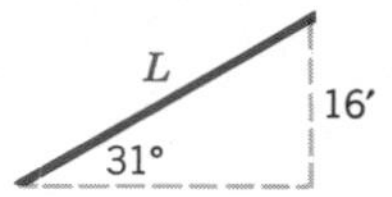

2. Find the angle m in the casting shown.

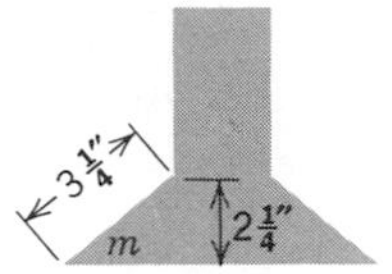

3. A pipe fitter must connect a pipeline to a tank as shown. The run from the pipeline to the tank is 62′, while the set is 38′.
   (a) How long is the connection?
   (b) Will the pipe fitter be able to use standard pipe fittings (i.e., $22\frac{1}{2}$°, 30°, 45°, 60°, or 90°)?

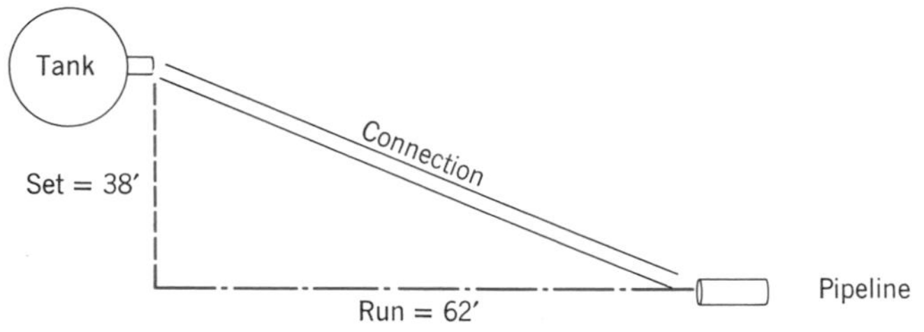

4. A helicopter, flying directly over a fishing boat at an altitude of 1200 ft, spots an ocean liner at a 15° angle of depression. How far from the boat is the liner? (Round to the nearest foot.)

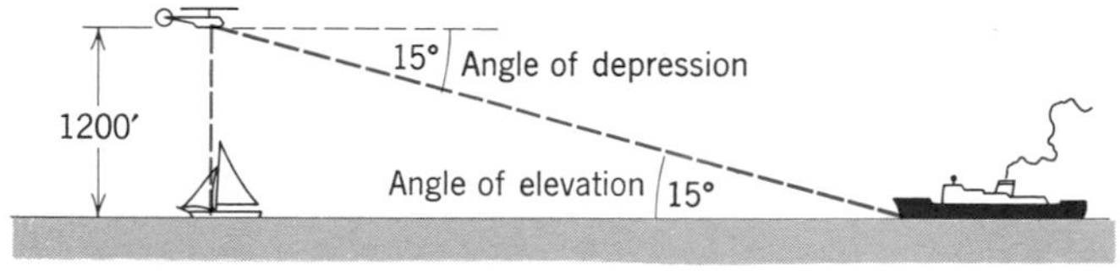

5. Find the angle of taper on the piece of work below if it is equal to twice m.

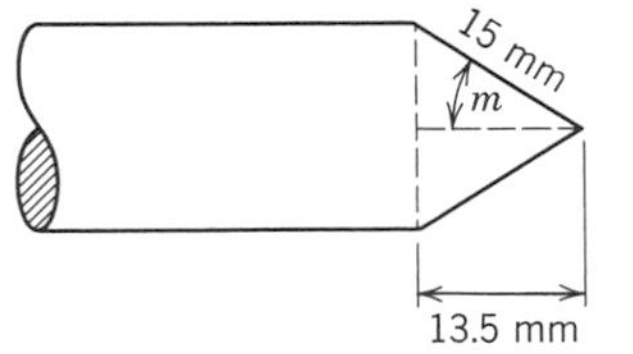

6. A road has a rise of 6′ in 80′. What is the gradient angle of the road?

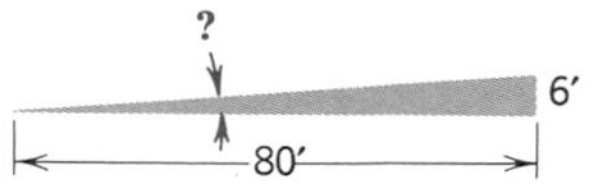

7. Find the length of the rafter below.

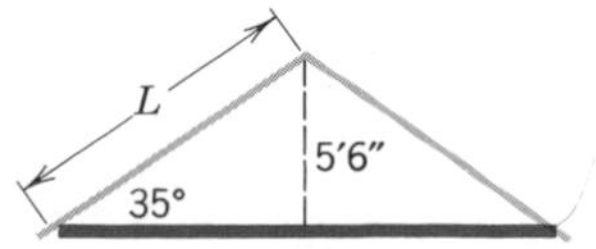

8. Three holes are drilled into a steel plate. Find the distances A and B as shown in the figure.

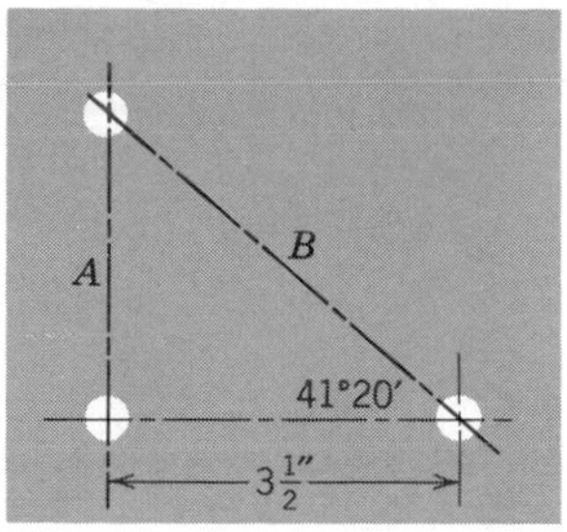

9. Find the depth of cut x needed for the V-slot shown. (Assume that the V-slot is symmetric.)

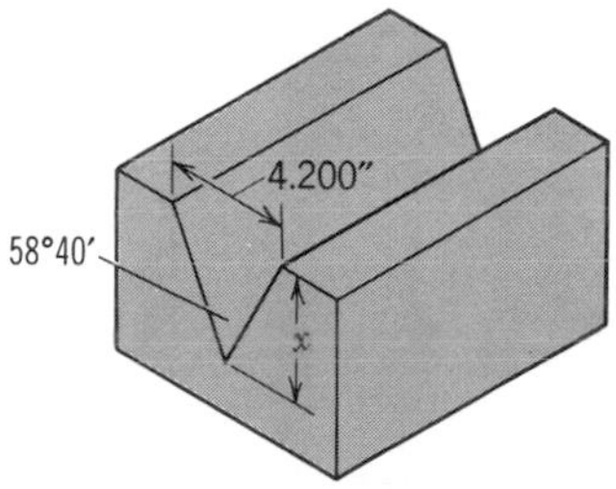

10. A 20-ft ladder leans against a building at an 80° angle with the ground. Will it reach a window 17′ above the ground?

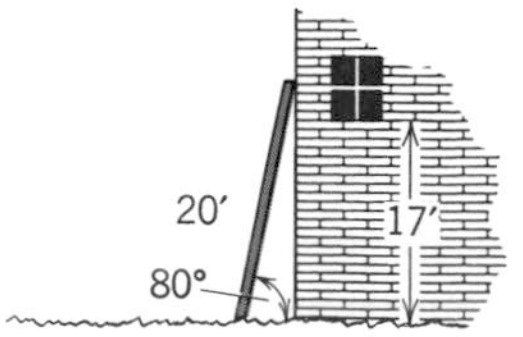

C. Calculator Problems

1. Find the missing dimensions as indicated.

(a)

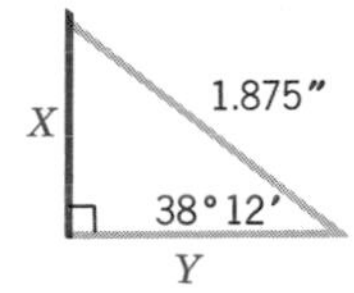

$X =$ ________

$Y =$ ________

(b)

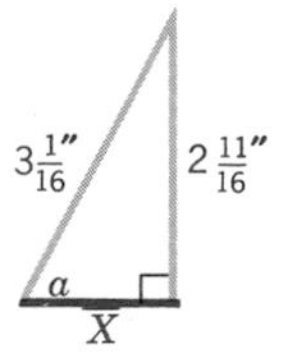

$\measuredangle a =$ ________

$X =$ ________

(c)

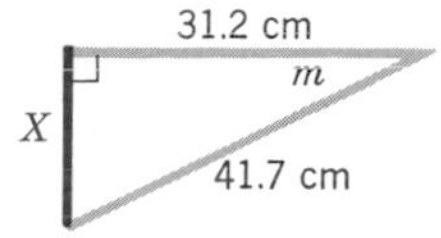

$\measuredangle m =$ ________

$X =$ ________

(d)

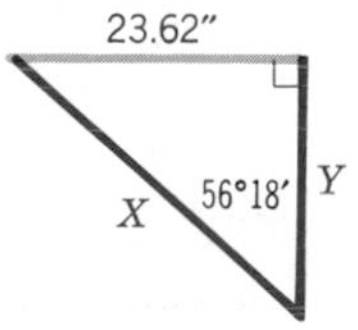

$X =$ ________

$Y =$ ________

(e)

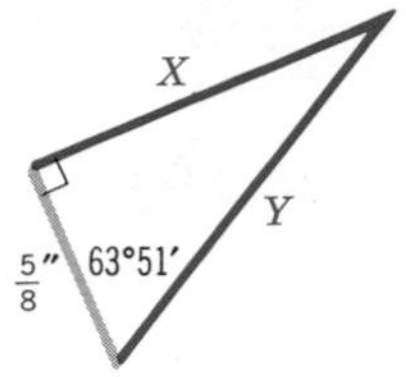

$X =$ ______

$Y =$ ______

(f)

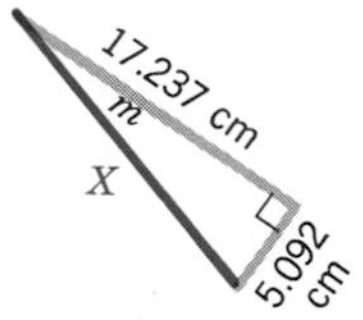

$\measuredangle m =$ ______

$X =$ ______

2. Solve the following problems.

   (a) Find angle a in the figure.

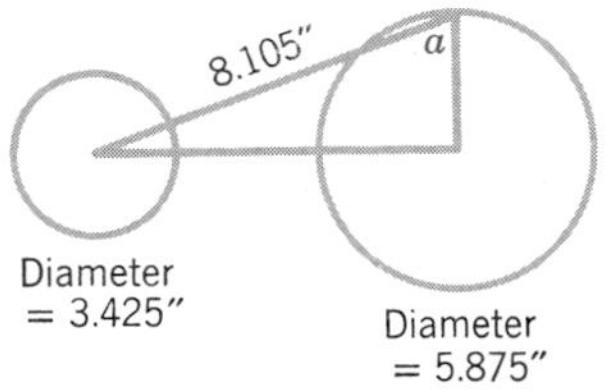

   (b) Find the distance B in the rivet.

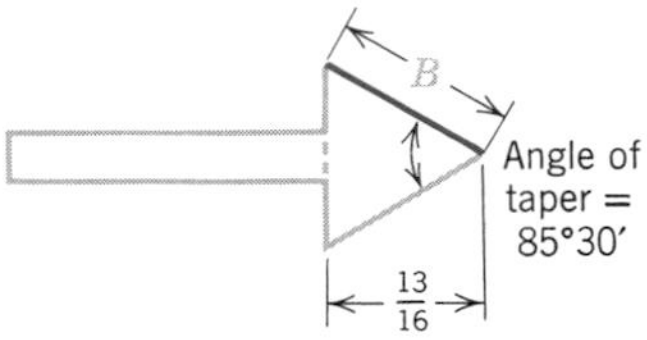

   (c) A road has a slope of 2°25′. Find the rise in 6500′ of horizontal run.

   (d) Find the missing dimension d in the flathead screw.

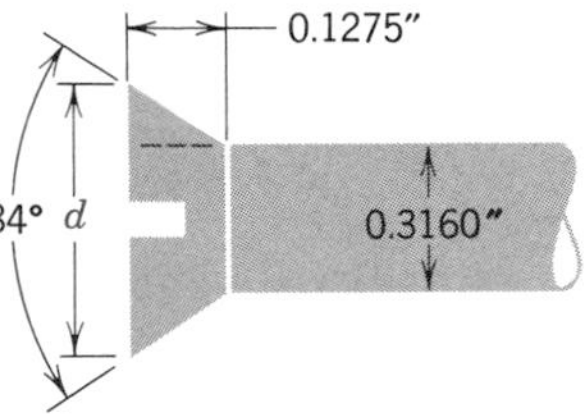

   (e) What total length of conduit is needed for sections A and B of the pipe connection shown?

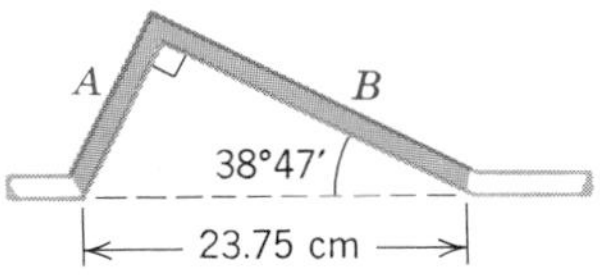

   (f) Find the included angle m of the taper below:

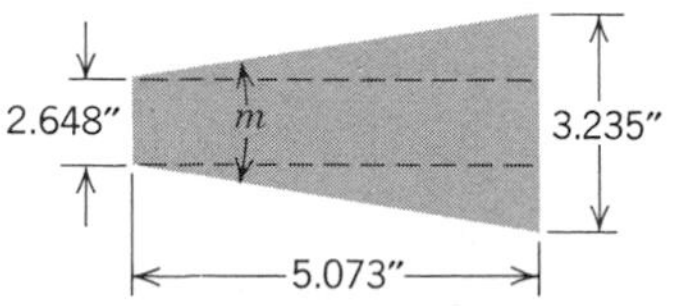

Turn to page 482 to check your answers to these problems. Then go to **36** for a set of problems covering triangle trigonometry.

9 Triangle Trigonometry

PROBLEM SET 9 Answers are on page 571.

36 **A. Which of the following are right triangles?**

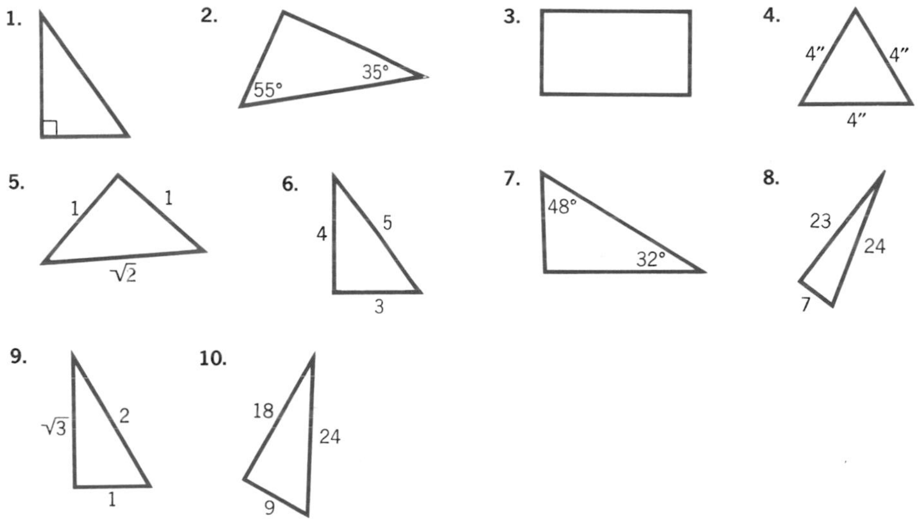

B. Find each of the following trig values using a trigonometric table or a calculator.

1. $\cos 67°$
2. $\tan 81°$
3. $\sin 4°$
4. $\cos 63°10'$
5. $\tan 35°45'$
6. $\sin 29°12'$

Use a trigonometric table or a calculator to find x to the nearest minute.

7. $\sin x = 0.242$
8. $\tan x = 1.54$
9. $\sin x = 0.927$
10. $\cos x = 0.309$
11. $\tan x = 0.194$
12. $\cos x = 0.549$
13. $\tan x = 0.823$
14. $\sin x = 0.672$

C. Find the missing dimensions in the right triangles as indicated. (Round all distances to one decimal place, and all angles to the nearest degree.)

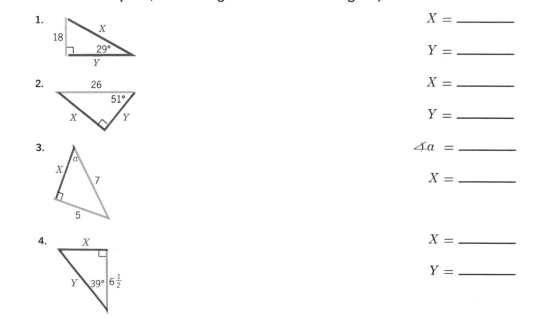

1.
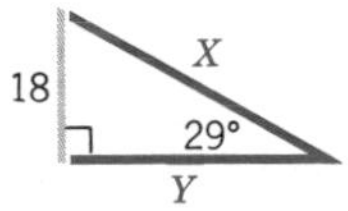

$X =$ ________

$Y =$ ________

2.
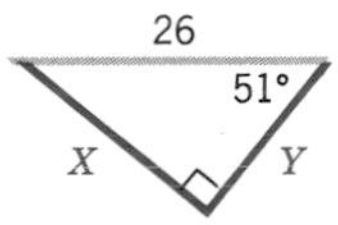

$X =$ ________

$Y =$ ________

3.
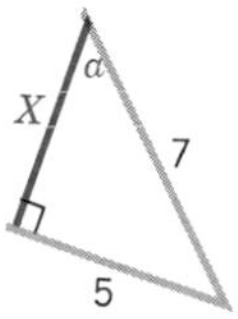

$\angle a =$ ________

$X =$ ________

4.
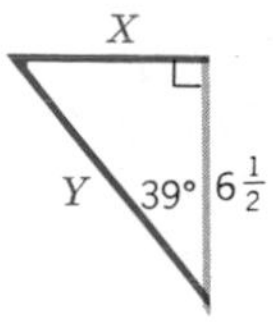

$X =$ ________

$Y =$ ________

Name

Date

Course/Section

5. 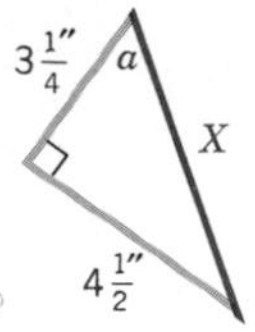

$\measuredangle a$ = ______

X = ______

6. 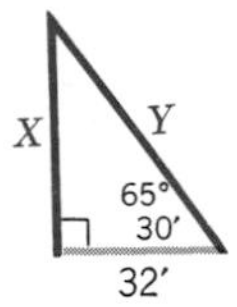

X = ______

Y = ______

7. 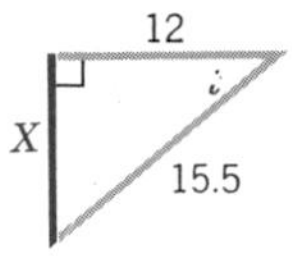

$\measuredangle t$ = ______

X = ______

D. Practical Problems

Round all angles to the nearest minute, and all distances to one decimal place, unless told otherwise.

1. What height of gage blocks is required to set an angle of 7°15′ for 10-inch plots?

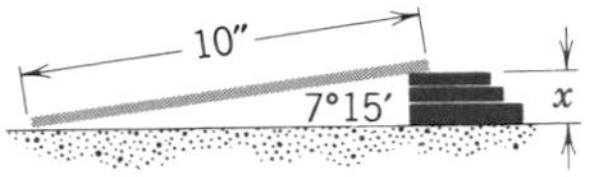

2. The destination of an airplane is due north, and the plane flies at 120 mph. There is a crosswind of 20 mph from the west. What heading angle a should the plane take? What is its relative ground speed v?

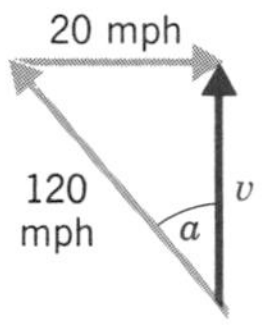

3. Six holes are spaced evenly around a 4-inch diameter circle. Find the distance, d, between the holes.

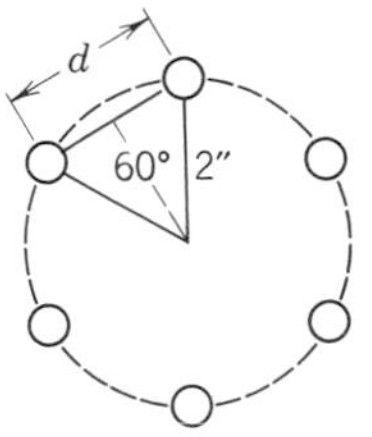

(*Hint:* The dotted line creates two identical right triangles.)

4. Find the angle of taper of the shaft in the figure.

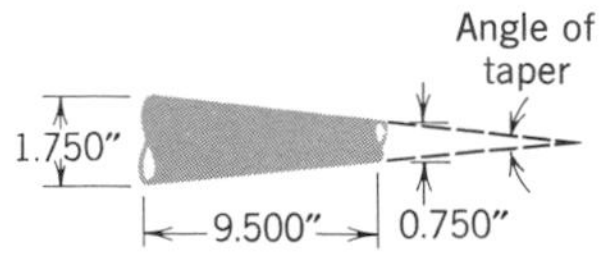

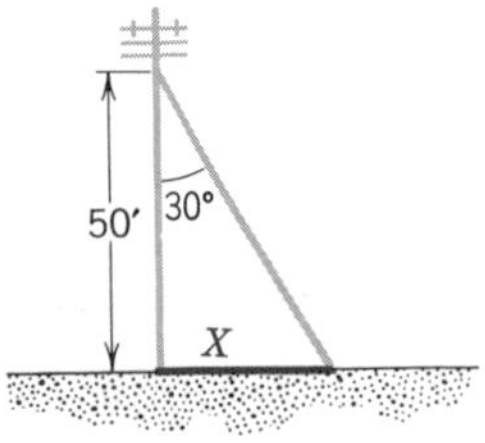

5. A TV technician installs a 50′ antenna on a flat roof. Safety regulations require a minimum angle of 30° between mast and guy wires.
   (a) Find the minimum value of X.
   (b) Find the length of the guy wires.

6. A bridge approach is 12 ft high. The maximum slope allowed is 15%—that is, $\frac{15}{100}$.
   (a) What is the length of the approach?
   (b) What is the angle a of the approach?

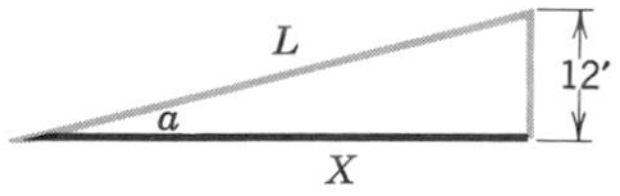

$\left(\textbf{Hint: } \frac{15}{100} = \frac{12}{x}\right)$

7. The *helix angle* of a screw is the angle at which the thread is cut. The *lead* is the distance advanced by one turn of the screw. The circumference refers to the circumference of the screw as given by $C = \pi d$. The following formula applies:

$$\text{Tangent of helix angle} = \frac{\text{lead}}{\text{circumference}}$$

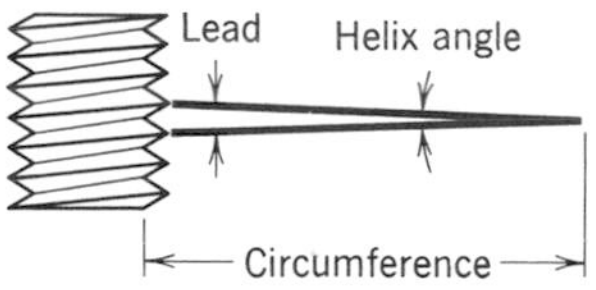

   (a) Find the helix angle for a 2″ diameter screw if the lead is $\frac{1}{8}$″.
   (b) Find the lead of a 3″ diameter screw if the helix angle is 2°.

8. Find the lengths of the beams x and y in the bridge truss below.

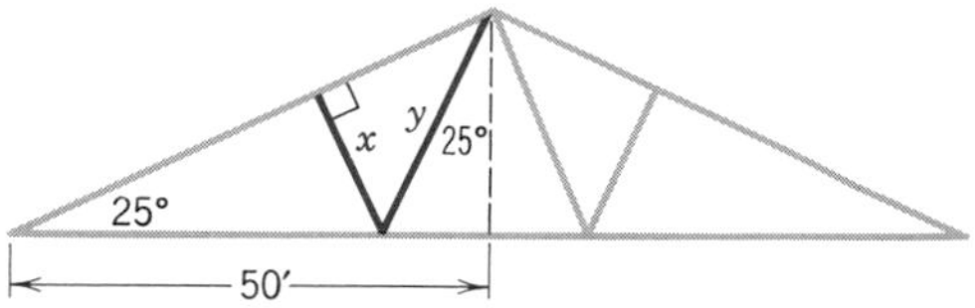

9. Find the width X of the v thread below. (Round to the nearest thousandth.)

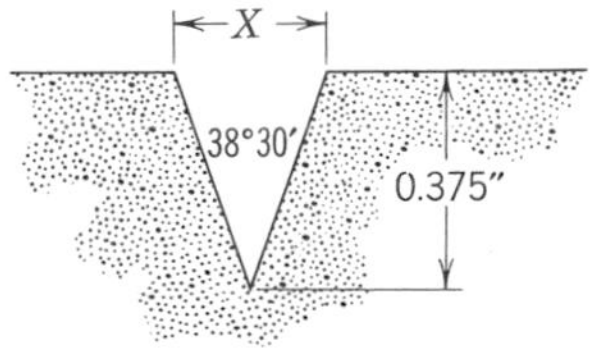

10. Find the head angle a of the screw shown.

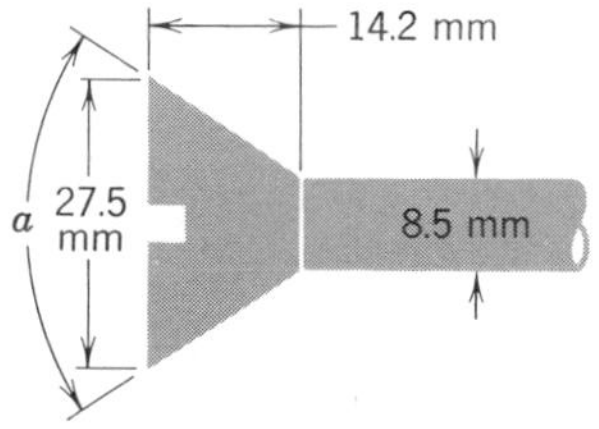

10 Advanced Algebra

PREVIEW 10

Objective	Sample Problems	Where To Go For Help: Page	Frame
Upon successful completion of this unit you will be able to:			
1. Solve a system of two linear equations in two variables.	(a) $5x + 2y = 20$ $3x - 2y = 12$	493	1
	(b) $x = 3y$ $2y - 4x = 30$		
	(c) $3x - 5y = 7$ $6x - 10y = 14$		
	(d) $2y = 5x - 7$ $10x - 4y = 9$		
2. Solve word problems involving systems of equations in two variables.	(a) The sum of two numbers is 35. Twice the smaller decreased by the larger is 10. Find them.	507	14
	(b) The perimeter of a rectangular vent is 28 inches. The length of the vent is two inches greater than its width. Find the dimensions.		
	(c) A mixture of 800 screws costs \$60. If the mixture consists of one type costing 5¢ each and another type costing 9¢ each, how many of each kind are there?		
3. Solve quadratic equations.	(a) $x^2 = 16$	513	20
	(b) $x^2 - 7x = 0$		
	(c) $x^2 - 5x = 14$		
	(d) $3x^2 + 2x - 16 = 0$		
	(e) $2x^2 + 3x + 11 = 0$		

Name

Date

Course/Section

4. Solve word problems involving quadratic equations.	(a) Find the side length of a square opening whose area is 36 cm^2.	520	**29**
	(b) Find the radius of a circular pipe whose cross-sectional area is 220 sq in.		
	(c) In the formula $P = RI^2$, find I (in amps) if $P = 2500$ watts and $R = 15$ ohms.		
	(d) The length of a rectangular vent is three inches longer than its width. Find the dimensions if the cross-sectional area is 61.75 sq in.		

(Answers to these preview problems are on the bottom of this page.)

If you are certain you can work *all* of the problems correctly, turn to page 525 for a set of practice problems. If you cannot work one or more of the preview problems, turn to the page indicated for some help. If you want to be tops in your work, turn to frame **1** and begin work there.

Answers to Preview 10

1. (a) (4, 0) (b) (−9, −3) (c) dependent (d) inconsistent
2. (a) 20 and 15 (b) 8″ and 6″ (c) 300 of the 5¢ screws, and 500 of the 9¢ screws
3. (a) $x = 4$ or $x = -4$ (b) $x = 0$ or $x = 7$
   (c) $x = 7$ or $x = -2$ (d) $x = -2\frac{2}{3}$ or $x = 2$ (e) no solution
4. (a) 6 cm (b) ~8.4″ (c) 12.9 amp (d) 6.5″ by 9.5″

10 Advanced Algebra

Reprinted by permission of Jefferson Communications, Inc., Vienna, Virginia

10-1 SYSTEMS OF EQUATIONS

1 This chapter is designed for students who are interested in highly technical occupations. We will explain how to solve systems of linear equations and how to solve quadratic equations. Before beginning, you may want to return to Chapter 6 to review the basic algebra operations explained there. When you have had the review you need, return here and continue.

Before you can solve a system of two equations in two unknowns, you must be able to solve a single linear equation in one unknown. Let's review what you learned in Chapter 6 about solving linear equations.

A *solution* to an equation such as $3x - 4 = 1$ is a number that we may use to replace x in order to make the equation a true statement. To find such a number, we *solve* the equation by changing it to an *equivalent* equation with only x on the left. The process goes like this:

Solve $3x - 4 = 11$

Step 1 Add 4 to both sides of the equation

$$(3x \underbrace{- 4) + 4}_{0} = \underbrace{(11) + 4}_{15}$$

$$3x = 15$$

Step 2 Divide both sides of the equation by 3.

$$\frac{\not{3}x}{\not{3}} = \frac{15}{3}$$

$$x = 5$$

This is the solution and you can check it by substituting 5 for x in the original equation.

$3(5) - 4 = 11$
$15 - 4 = 11$ which is true.

Solve each of the following equations and check your answer.

(a) $\frac{3x}{2} = 9$ (b) $17 - x = 12$

(c) $2x + 7 = 3$ (d) $3(2x + 5) = 4x + 17$

The correct answers are in **2**.

---

2 (a) 6 (b) 5 (c) -2 (d) 1

A *system of equations* is a set of equations with a common solution. For example, the pair of equations

$2x + y = 11$
$4y - x = 8$

has the common solution $x = 4$, $y = 3$.

This pair of numbers will make each equation a true statement. If we substitute 4 for x and 3 for y, the first equation becomes

$2(4) + 3 = 8 + 3$ or 11

and the second equation becomes

$4(3) - 4 = 12 - 4$ or 8

The single pair of numbers (4, 3) satisfies *both* equations.

By substituting, show that the numbers $x = 2$, $y = -5$ give the solution to the pair of equations

$5x - y = 15$
$x + 2y = -8$

Check your work in **3**.

---

3 The first equation is

$5(2) - (-5) = 15$

$10 + 5 = 15$, which is correct.

The second equation is

$(2) + 2(-5) = -8$

$2 - 10 = -8$, which is also correct.

Solution by Substitution

In this chapter we will show you two different methods of solving a system of two linear equations in two unknowns. The first method is called the method of *substitution*. For example, to solve the pair of equations

$y = 3 - x$
$3x + y = 11$

follow these steps.

Step 1 *Solve* the first equation for x or y and substitute this expression into the second equation.

The first equation is already solved for y:

$y = 3 - x$

Substituting this expression for y in the second equation,

$3x + y = 11$ becomes
$3x + (3 - x) = 11$

Substitute this for y

Step 2 *Solve* the resulting equation.

$3x + (3 - x) = 11$	becomes
$2x + 3 = 11$	Subtract 3 from each side.
$2x = 8$	Divide each side by 2.
$x = 4$	

Step 3 *Substitute* this value of x into the first equation and find a value for y.

$y = 3 - x$
$y = 3 - 4$
$y = -1$

The solution is $x = 4$, $y = -1$ or $(4, -1)$

x value

y value

Step 4 *Check* your solution by substituting it back into the second equation.

$3x + y = 11$ becomes
$3(4) + (-1) = 11$
$12 - 1 = 11$ which is correct.

Try it. Use this substitution procedure to solve the system of equations

$x - 2y = 3$
$2x - 3y = 7$

Check your work in **4**.

---

4 **Step 1** *Solve* the first equation for x by adding $2y$ to both sides of the equation.

$x - 2y + 2y = 3 + 2y$

$x = 3 + 2y$

Substitute this expression for x in the second equation.

$2x - 3y = 7$ becomes

$2(3 + 2y) - 3y = 7$

Step 2 *Solve:*

$6 + 4y - 3y = 7$	Simplify by combining the y-terms.
$6 + y = 7$	Subtract 6 from each side.
$y = 1$	

Step 3 *Substitute* this value of y in the first equation to find x.

$x - 2y = 3$ becomes
$x - 2(1) = 3$
$x - 2 = 3$ Add 2 to each side.
$x = 5$

The solution is $x = 5$, $y = 1$ or $(5, 1)$.

Step 4 *Check* the solution in the second equation.

$2x - 3y = 7$ becomes
$2(5) - 3(1) = 7$
$10 - 3 = 7$ which is correct.

Of course, it does not matter which variable, x or y, we solve for on Step 1, or which equation we use on Step 3. For example, in the pair of equations

$2x + 3y = 22$
$x - y = 1$

the simplest procedure is to solve the *second* equation for x to get

$x = 1 + y$ and substitute this expression for x into the first equation.

Solve this set of equations.

Check your work in **5**.

---

5 $2x + 3y = 22$
$x - y = 1$

Step 1 From the second equation, $x = 1 + y$.

When we substitute into the first equation,

$2x + 3y = 22$ becomes

$2(1 + y) + 3y = 22$

Step 2 *Solve:*

$2 + 2y + 3y = 22$ Combine terms.
$2 + 5y = 22$ Subtract 2 from each side.
$5y = 20$ Divide each side by 5
$y = 4$

Step 3 *Substitute* 4 for y in the second equation.

$x - y = 1$ becomes
$x - (4) = 1$

or

$x = 5$ The solution is $x = 5$, $y = 4$.

Step 4 *Check* the solution by substituting these values into the first equation.

$2x + 3y = 22$ becomes
$2(5) + 3(4) = 22$
$10 + 12 = 22$ which is correct.

Solve the following systems of equations by using the substitution method.

(a) $x = 1 + y$
$2y + x = 7$

(b) $3x + y = 1$
$y + 5x = 9$

(c) $x - 3y = 4$
$3y + 2x = -1$

(d) $y + 2x = 1$
$3y + 5x = 1$

(e) $x - y = 2$
$y + x = 1$

(f) $y = 4x$
$2y - 6x = 0$

Check your answers in **6**.

---

6 (a) $x = 3, y = 2$ (b) $x = 4, y = -11$ (c) $x = 1, y = -1$
(d) $x = 2, y = -3$ (e) $x = 1\frac{1}{2}, y = -\frac{1}{2}$ (f) $x = 0, y = 0$

So far we have looked only at systems of equations with a single solution—one pair of numbers. Such a system of equations is called a *consistent* system. However, it is possible for a system of equations to have no solution at all or to have very many solutions.

For example, the system of equations

$$y + 3x = 5$$
$$2y + 6x = 10$$

has *no unique* solution. If you solve for y in the first equation

$$y = 5 - 3x$$

and substitute this expression into the second equation

$$2y + 6x = 10$$

or $\quad 2(5 - 3x) + 6x = 10$

This resulting equation simplifies to

$$10 = 10$$

There is no way of solving to get a unique value of x or y.

A system of equations that does not have a single unique number-pair solution is said to be *dependent*. The two equations are essentially the same. For the system shown in this example, the second equation is simply twice the first equation. There are infinitely many pairs of numbers that will satisfy this pair of equations. For example,

$x = 0, y = 5$
$x = 1, y = 2$
$x = 2, y = -1$
$x = 3, y = -4$

and so on.

If a system of equations is such that our efforts to solve it produce a false statement, the equations are said to be *inconsistent*. For example, the pair of equations

$y - 1 = 2x$
$2y - 4x = 7$ is inconsistent.

If we solve the first equation for y

$$y = 2x + 1$$

and substitute this expression into the second equation,

$2y - 4x = 7$ becomes

$$2(2x + 1) - 4x = 7$$

$4x + 2 - 4x = 7$
or $\quad 2 = 7$ which is false.

All of the variables have dropped out of the equation, and we are left with an incorrect statement. The original pair of equations is said to be inconsistent, and it has no solution.

Try solving the following systems of equations.

(a) $2x - y = 5$
$2y - 4x = 3$

(b) $3x - y = 5$
$6x - 10 = 2y$

Check your work in **7**.

7 (a) Solve the first equation for y.

$y = 2x - 5$

Substitute this into the second equation.

$$2y - 4x = 3 \quad \text{becomes}$$
$$2(2x - 5) - 4x = 3$$
$$4x - 10 - 4x = 3$$
$$-10 = 3$$

This is impossible. There is no solution for this system of equations. The equations are inconsistent.

(b) Solve the first equation for y.

$y = 3x - 5$

Substitute this into the second equation.

$$6x - 10 = 2y \quad \text{becomes}$$
$$6x - 10 = 2(3x - 5)$$
$$6x - 10 = 6x - 10$$
$$0 = 0$$

This is true, but all of the variables have dropped out and we cannot get a single unique solution. The equations are dependent.

Solution by Elimination

The second method for solving a system of equations is called the *method of elimination.* When it is difficult or "messy" to solve one of the two equations for either x or y, the method of elimination may be the simplest way to solve the system of equations. For example, in the system of equations

$$2x + 3y = 7$$
$$4x - 3y = 5$$

neither equation can be solved for x or y without introducing fractions that are difficult to work with. But we can simply add these equations and the y-terms will be eliminated.

$$2x + 3y = 7$$
$$4x - 3y = 5 \quad \text{Add like terms.}$$
$$6x + 0 = 12$$
$$6x = 12$$
$$x = 2$$

Now substitute this value of x back into either one of the original equations to obtain a value for y.

The first equation

$$2x + 3y = 7 \quad \text{becomes}$$
$$2(2) + 3y = 7$$
$$4 + 3y = 7 \quad \text{Subtract 4 from each side.}$$
$$3y = 3 \quad \text{Divide each side by 3.}$$
$$y = 1$$

The solution is $x = 2$, $y = 1$, or $(2, 1)$.

Check the solution by substituting it back into the second equation.

$$4x - 3y = 5$$
$$4(2) - 3(1) = 5$$
$$8 - 3 = 5 \quad \text{which is correct.}$$

Try it. Solve the following system of equations by adding.

$2x - y = 3$
$y + x = 9$

Check your work in **8**.

---

8 $2x - y = 3$
$y + x = 9$

becomes

$2x - y = 3$
$x + y = 9$ by rearranging the terms in the second equation.

Add to get

$$\begin{aligned} 3x + 0 &= 12 \\ 3x &= 12 \\ x &= 4 \end{aligned}$$

Substitute 4 for x in the first equation.

$2x - y = 3$	
$2(4) - y = 3$	Simplify.
$8 - y = 3$	Subtract 8 from each side.
$-y = 3 - 8$	Combine terms.
$-y = -5$	Multiply both sides by -1.
$y = 5$	

The solution is $x = 4$, $y = 5$, or $(4, 5)$.

It is important to rearrange the terms in the equations so that the x and y terms appear in the same order in both equations.

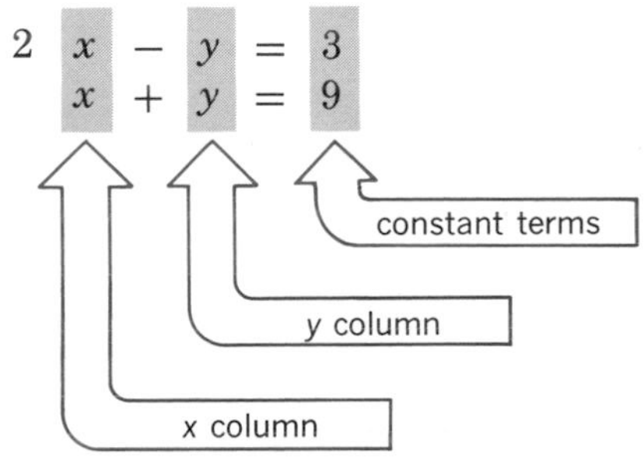

Solve the following systems of equations by this process of elimination.

(a) $x + 5y = 17$, $-x + 3y = 7$

(b) $x + y = 16$, $x - y = 4$

(c) $3x - y = -5$, $y - 5x = 9$

(d) $3x - 4y = 30$, $4y + 3x = -6$

(e) $6x - y = 5$, $y - x = -5$

(f) $\frac{1}{2}x + 2y = 10$, $y + 1 = \frac{1}{2}x$

Our step-by-step solutions are in **9**.

---

9 (a) **Solve:**

$$\begin{array}{r} x + 5y = 17 \\ -x + 3y = 7 \\ \hline 0 + 8y = 24 \\ 8y = 24 \\ y = 3 \end{array}$$

adding like terms.

Substitute 3 for y in the first equation.

$$x + 5(3) = 17$$
$$x + 15 = 17$$
$$x = 2$$ This solution is $x = 2$, $y = 3$, or (2, 3).

Check:
$$-x + 3y = 7$$
$$-(2) + 3(3) = 7$$
$$-2 + 9 = 7$$ which is correct.

(b) **Solve:**

$$\begin{array}{r} x + y = 16 \\ x - y = 4 \\ \hline 2x + 0 = 20 \end{array}$$ adding like terms.
$$2x = 20$$
$$x = 10$$

Substitute 10 for x in the first equation.

$$(10) + y = 16$$
$$y = 6$$ The solution is $x = 10$, $y = 6$, or (10, 6).

Check:
$$x - y = 4$$
$$(10) - (6) = 4$$
$$10 - 6 = 4$$ which is correct.

(c) **Solve:**

$$3x - y = -5$$
$$y - 5x = 9$$

Rearrange the order of the terms in the second equation.

$$\begin{array}{r} 3x - y = -5 \\ -5x + y = 9 \\ \hline -2x + 0 = 4 \end{array}$$ adding like terms.
$$-2x = 4$$
$$x = -2$$

Substitute -2 for x in the first equation.

$$3(-2) - y = -5$$
$$-6 - y = -5$$
$$-y = -5 + 6 = 1$$
$$y = -1$$ The solution is $x = -2$, $y = -1$, or $(-2, -1)$.

Check:
$$y - 5x = 9$$
$$(-1) - 5(-2) = 9$$
$$-1 + 10 = 9$$ which is correct.

(d) **Solve:**

$$3x - 4y = 30$$
$$4y + 3x = -6$$

Rearrange the order of terms in the second equation.

$$\begin{array}{r} 3x - 4y = 30 \\ 3x + 4y = -6 \\ \hline 6x + 0 = 24 \end{array}$$ adding like terms.
$$6x = 24$$
$$x = 4$$

Substitute 4 for x in the first equation.

$$\begin{aligned} 3(4) - 4y &= 30 \\ 12 - 4y &= 30 \\ -4y &= 18 \\ y &= -4\tfrac{1}{2} \end{aligned}$$

The solution is $x = 4$, $y = -4\frac{1}{2}$, or $(4, -4\frac{1}{2})$.

Check:

$$\begin{aligned} 4y + 3x &= -6 \\ 4(-4\tfrac{1}{2}) + 3(4) &= -6 \\ -18 + 12 &= -6 \end{aligned}$$

which is correct.

(e) **Solve:**

$$\begin{aligned} 6x - y &= 5 \\ y - x &= -5 \end{aligned}$$

Rearrange terms in the second equation.

$$\begin{aligned} 6x - y &= 5 \\ -x + y &= -5 \\ \hline 5x + 0 &= 0 \quad \text{adding like terms.} \\ 5x &= 0 \\ x &= 0 \end{aligned}$$

Substitute 0 for x in the first equation.

$$\begin{aligned} 6(0) - y &= 5 \\ 0 - y &= 5 \\ -y &= 5 \\ y &= -5 \end{aligned}$$

The solution is $x = 0$, $y = -5$, or $(0, -5)$.

Check:

$$\begin{aligned} y - x &= -5 \\ (-5) - (0) &= -5 \\ -5 &= -5 \end{aligned}$$

(f) **Solve:**

$$\begin{aligned} \tfrac{1}{2}x + 2y &= 10 \\ y + 1 &= \tfrac{1}{2}x \end{aligned}$$

Rearrange terms in the second equation so that they are in the same order as in the first equation. Subtract 1 from both sides, then subtract $\frac{1}{2}x$ from both sides.

$$\begin{aligned} \tfrac{1}{2}x + 2y &= 10 \\ -\tfrac{1}{2}x + y &= -1 \\ \hline 0 + 3y &= 9 \quad \text{adding like terms} \\ 3y &= 9 \\ y &= 3 \end{aligned}$$

Substitute 3 for y in the first equation.

$$\begin{aligned} \tfrac{1}{2}x + 2(3) &= 10 \\ \tfrac{1}{2}x + 6 &= 10 \\ \tfrac{1}{2}x &= 4 \\ x &= 8 \end{aligned}$$

The solution is $x = 8$, $y = 3$, or $(8, 3)$.

Check:

$$\begin{aligned} y + 1 &= \tfrac{1}{2}x \\ (3) + 1 &= \tfrac{1}{2}(8) \\ 3 + 1 &= 4 \end{aligned}$$

which is correct.

With some systems of equations neither x nor y can be eliminated by simply adding like terms. For example, in the system

$$3x + y = 17$$
$$x + y = 7$$

adding like terms will not eliminate either variable. To solve this system of equations, multiply all terms of the second equation by -1 so that

$x + y = 7$ becomes $-x - y = -7$

and the system of equations becomes

$$3x + y = 17$$
$$-x - y = -7$$

The system of equations may now be solved by adding the like terms as before.

$$\begin{aligned} 3x + y &= 17 \\ -x - y &= -7 \\ \hline 2x + 0 &= 10 \quad \text{adding like terms} \\ 2x &= 10 \\ x &= 5 \end{aligned}$$

Substitute 5 for x in the first equation.

$$\begin{aligned} 3(5) + y &= 17 \\ 15 + y &= 17 \\ y &= 17 - 15 = 2 \end{aligned}$$

The solution is $x = 5$, $y = 2$, or $(5, 2)$.

Check:

$$\begin{aligned} x + y &= 7 \\ (5) + (2) &= 7 \\ 5 + 2 &= 7 \end{aligned}$$

which is correct.

Use this "multiply and add" procedure to solve the following system of equations.

$$2x + 7y = 29$$
$$2x + y = 11$$

Check your solution in **10** .

10

$$2x + 7y = 29$$
$$2x + y = 11$$

Multiply all terms in the second equation by -1.

$$\begin{aligned} 2x + 7y &= 29 \\ -2x - y &= -11 \\ \hline 0 + 6y &= 18 \quad \text{adding like terms.} \\ 6y &= 18 \\ y &= 3 \end{aligned}$$

Substitute 3 for y in the first equation.

$$\begin{aligned} 2x + 7(3) &= 29 \\ 2x + 21 &= 29 \\ 2x &= 8 \\ x &= 4 \end{aligned}$$

The solution is $x = 4$, $y = 3$, or $(4, 3)$.

Check the solution by substituting in the original equations.

$$\begin{aligned} 2x + y &= 11 \\ 2(4) + (3) &= 11 \\ 8 + 3 &= 11 \end{aligned} \qquad \begin{aligned} 2x + 7y &= 29 \\ 2(4) + 7(3) &= 29 \\ 8 + 21 &= 29 \end{aligned}$$

Solving by the multiply and add procedure may involve multiplying by constants other than -1, of course. For example, use this method to solve the pair of simultaneous equations.

$2x + 4y = 26$
$3x - 2y = 7$

Look for our solution in **11**.

---

11 $2x + 4y = 26$
$3x - 2y = 7$

First, look at these equations carefully. Notice that the y terms can be eliminated easily if we multiply all terms in the second equation by 2.

The y column $\begin{array}{|c|}\hline +4y \\ -2y \\ \hline\end{array}$ becomes $\begin{array}{|c|}\hline +4y \\ -4y \\ \hline\end{array}$ when we multiply by 2 .

sum = 0

The second equation becomes $2\,(3x) - 2\,(2y) = 2\,(7)$

or $6x - 4y = 14$

and the system of equations is converted to the equivalent system

$$\begin{aligned} 2x + 4y &= 26 \\ \underline{6x - 4y} &\underline{= 14} \\ 8x + 0 &= 40 \quad \text{adding like terms.} \\ 8x &= 40 \\ x &= 5 \end{aligned}$$

Substitute 5 for x in the first equation.

$$\begin{aligned} 2(5) + 4y &= 26 \\ 10 + 4y &= 26 \\ 4y &= 16 \\ y &= 4 \end{aligned}$$

The solution is $x = 5$, $y = 4$, or $(5, 4)$.

Check the solution by substituting it back into the original equations.

Try these problems to make certain you understand this procedure.

(a) $3x - 2y = 14$
$5x - 2y = 22$

(b) $5x + 6y = 14$
$3x - 2y = -14$

(c) $5x - y = 1$
$2y + 3x = 11$

(d) $-x - 2y = 1$
$19 - 2x = -3y$

Our solutions are in **12**.

I solved the equations $x + 2y = 5$ and $8x + y = -5$ by adding to get rid of the 5s, and I got $9x + 3y = 0$. What do I do now?
You goofed it. In solving a system of equations by elimination, you must work to eliminate one of the variables x or y. For example, in this case try multiplying each term in the second equation by -2 and then add them.

12 (a) **Solve:**

$$3x - 2y = 14$$
$$5x - 2y = 22$$

Multiply each term in the first equation by -1.

$$(-1)(3x) - (-1)(2y) = (-1)(14)$$
$$-3x + 2y = -14$$

The system of equations is therefore

$$\begin{array}{r} -3x + 2y = -14 \\ 5x - 2y = 22 \\ \hline 2x + 0 = 8 \end{array}$$ adding like terms.

$$2x = 8$$
$$x = 4$$

Substitute 4 for x in the first equation.

$$3(4) - 2y = 14$$
$$12 - 2y = 14$$
$$-2y = 2$$
$$-y = 1$$
$$y = -1$$ The solution is $x = 4$, $y = -1$, or $(4, -1)$.

Check:

$$3x - 2y = 14 \qquad\qquad 5x - 2y = 22$$
$$3(4) - 2(-1) = 14 \qquad\qquad 5(4) - 2(-1) = 22$$
$$12 + 2 = 14 \qquad\qquad 20 + 2 = 22$$
$$14 = 14 \qquad\qquad 22 = 22$$

(b) **Solve:**

$$5x + 6y = 14$$
$$3x - 2y = -14$$

Multiply each term in the second equation by 3.

$$(3)(3x) - (3)(2y) = (3)(-14)$$
$$9x - 6y = -42$$

The system of equations is now

$$\begin{array}{r} 5x + 6y = 14 \\ 9x - 6y = -42 \\ \hline 14x + 0 = -28 \end{array}$$ adding like terms.

$$14x = -28$$
$$x = -2$$

Substitute -2 for x in the first equation.

$$5(-2) + 6y = 14$$
$$-10 + 6y = 14$$
$$6y = 24$$
$$y = 4$$ The solution is $x = -2$, $y = 4$, or $(-2, 4)$.

Be certain to check your solution.

(c) **Solve:**

$$5x - y = 1$$
$$2y + 3x = 11$$

Rearrange to put the terms in the second equation in the same order as they are in the first equation.

$$5x - y = 1$$
$$3x + 2y = 11$$

Multiply each term in the first equation by 2.

$$\begin{array}{r} 10x - 2y = 2 \\ 3x + 2y = 11 \\ \hline 13x + 0 \;\; = 13 \\ 13x = 13 \\ x = 1 \end{array} \qquad \text{adding like terms.}$$

Substitute 1 for x in the first equation.

$$\begin{aligned} 5(1) - y &= 1 \\ 5 - y &= 1 \\ -y &= 1 - 5 = -4 \\ y &= 4 \end{aligned} \qquad \text{The solution is } x = 1,\ y = 4, \text{ or } (1, 4).$$

Check it.

(d) **Solve:**

$$\begin{aligned} -x - 2y &= 1 \\ 19 - 2x &= -3y \end{aligned}$$

Rearrange the terms in the second equation in the same order as they are in the first equation.

$$\begin{aligned} -x - 2y &= 1 \\ -2x + 3y &= -19 \end{aligned}$$

Multiply each term in the first equation by -2.

$$\begin{aligned} (-2)(-x) - (-2)(2y) &= (-2)(1) \\ 2x + 4y &= -2 \end{aligned}$$

The system of equations is now

$$\begin{array}{r} 2x + 4y = -2 \\ -2x + 3y = -19 \\ \hline 0 \;\; + 7y = -21 \\ 7y = -21 \\ y = -3 \end{array} \qquad \text{adding like terms.}$$

Substitute -3 for y in the first equation of the original problem.

$$\begin{aligned} -x - 2(-3) &= 1 \\ -x + 6 &= 1 \\ -x &= 1 - 6 = -5 \\ x &= 5 \end{aligned} \qquad \text{The solution is } x = 5,\ y = -3, \text{ or } (5, -3).$$

Check your solution.

If you examine the system of equations

$$\begin{aligned} 3x + 2y &= 7 \\ 4x - 3y &= -2 \end{aligned}$$

you will find that there is no single integer we can use as a multiplier that will allow us to eliminate one of the variables when the equations are added. Instead we must convert each equation separately to an equivalent equation, so that when the new equations are added one of the variables is eliminated. For example, with the system of equations above, if we wish to eliminate the y variable, we must multiply the first equation by 3 and the second equation by 2.

First equation: $\boxed{3x + 2y = 7}$ → Multiply by 3 → $\boxed{9x + 6y = 21}$

Second equation: $\boxed{4x - 3y = -2}$ → Multiply by 2 → $\boxed{8x - 6y = -4}$

The new system of equations is

$$\begin{array}{rl} 9x + 6y &= 21 \\ 8x - 6y &= -4 \\ \hline 17x + 0 &= 17 \quad \text{adding like terms.} \\ 17x &= 17 \\ x &= 1 \end{array}$$

Substitute 1 for x in the original first equation.

$$\begin{aligned} 3x + 2y &= 7 \\ 3(1) + 2y &= 7 \\ 3 + 2y &= 7 \\ 2y &= 4 \\ y &= 2 \end{aligned}$$

The solution is $x = 1$, $y = 2$, or $(1, 2)$.

Check this solution by substituting it back into the original pair of equations.

Check:

$$\begin{aligned} 3x + 2y &= 7 \\ 3(1) + 2(2) &= 7 \\ 3 + 4 &= 7 \\ 7 &= 7 \end{aligned} \qquad \begin{aligned} 4x - 3y &= -2 \\ 4(1) - 3(2) &= -2 \\ 4 - 6 &= -2 \\ -2 &= -2 \end{aligned}$$

Use this same procedure to solve the following system of equations:

$$\begin{aligned} 2x - 5y &= 9 \\ 3x + 4y &= 2 \end{aligned}$$

Check your work in **13**.

---

13 We can eliminate x from the two equations as follows:

First equation: $2x - 5y = 9$ → Multiply by -3 → $-6x + 15y = -27$

Second equation: $3x + 4y = 2$ → Multiply by 2 → $6x + 8y = 4$

The new system of equations is:

$$\begin{array}{rl} -6x + 15y &= -27 \\ 6x + 8y &= 4 \\ \hline 0 + 23y &= -23 \quad \text{adding like terms.} \\ 23y &= -23 \\ y &= -1 \end{array}$$

Substitute -1 for y in the first original equation.

$$\begin{aligned} 2x - 5y &= 9 \\ 2x - 5(-1) &= 9 \\ 2x + 5 &= 9 \\ 2x &= 4 \\ x &= 2 \end{aligned}$$

The solution is $x = 2$, $y = -1$, or $(2, -1)$.

Of course, we could have chosen to eliminate the y-variable and we would have arrived at the same solution. Try it.

When you are ready to continue, practice your new skills by solving the following systems of equations.

(a) $2x + 2y = 4$
$5x + 7y = 18$

(b) $5x + 2y = 11$
$6x - 3y = 24$

(c) $3x + 2y = 10$
$2x = 5y - 25$

(d) $-7x - 13 = 2y$
$3y + 4x = 0$

When you have solved these systems of equations, check your answers in **14** .

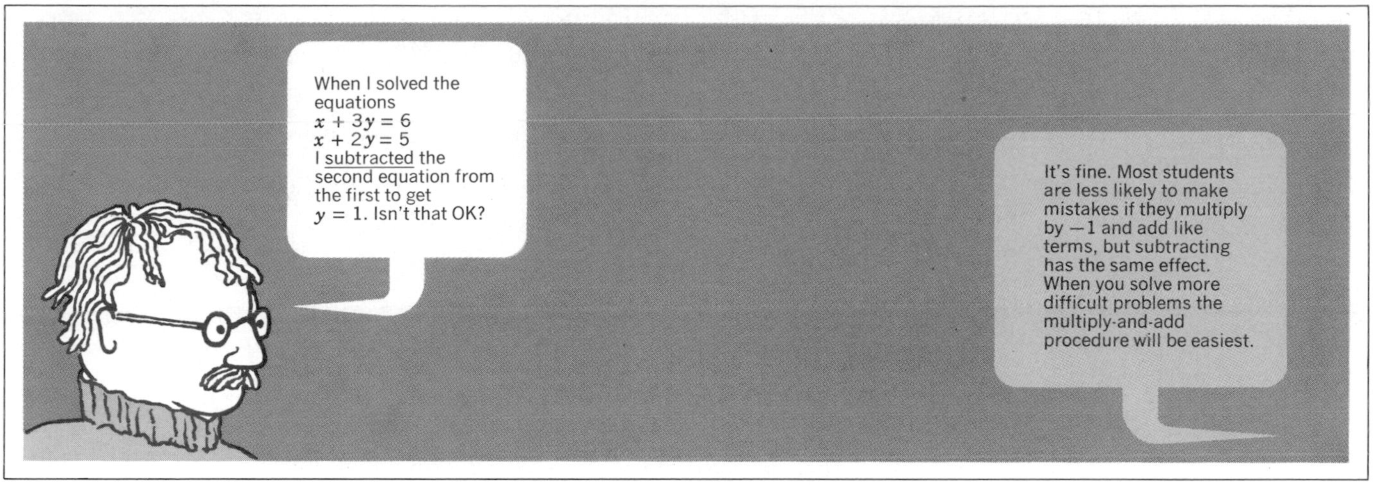

14 (a) $x = -2,\ y = 4$ (b) $x = 3,\ y = -2$

(c) $x = 0,\ y = 5$ (d) $x = -3,\ y = 4$

In (a) multiply the first equation by -5 and the second equation by 2.

In (b) multiply the first equation by 3 and the second equation by 2.

In (c) rearrange the terms of the second equation, then multiply the first equation by -2 and the second equation by 3.

In (d) rearrange the terms of the first equation to agree with the second equation, then multiply the first equation by 3 and the second equation by 2.

Word Problems

In many practical situations not only must you be able to solve a system of equations, you must also be able to write the equations in the first place. You must be able to set up and solve word problems. In Chapter 6 we listed some "signal words" and showed how to translate English sentences and phrases to mathematical equations and expressions. If you need to review the material on word problems in Chapter 6, turn to page 269 now; otherwise continue here.

Translate the following sentence into *two* equations.

"The sum of two numbers is 26 and their difference is 2." (Let x and y represent the two numbers.)

Check your work in **15**.

15 The sum of two numbers is 26

$x + y \quad = 26$

. . . their difference is 2.

$x - y \quad = 2$

The two equations are

$x + y = 26$
$x - y = 2$

Solve this pair of equations using the elimination method. Check your answer in **16**.

16 The solution is $x = 14$, $y = 12$. Check the solution by seeing if it fits the original problem. The sum of these numbers is 26 ($14 + 12 = 26$) and their difference is 2 ($14 - 12 = 2$).

Ready for another word problem? Translate and solve this one:

"The difference of two numbers is 14, and the larger number is three more than twice the smaller number."

Our step-by-step solution is in **17**.

---

17 The first phrase in the sentence should be translated as

"The difference of two numbers is 14 . . ."

$$L - S = 14$$

The larger number is L; the smaller number is S; the difference must be $L - S$.

and the second phrase should be translated as

". . . the larger number is three more than twice the smaller. . . ."

$$L = 3 + 2S$$

The system of simultaneous equations is

$$L - S = 14$$
$$L = 3 + 2S$$

To solve this system of equations, substitute the value of L from the second equation into the first equation. Then the first equation becomes

$$(3 + 2S) - S = 14$$

or $\quad 3 + S = 14$

$$S = 11$$

Now substitute this value of S into the first equation to find L.

$$L - (11) = 14$$

$L = 25$ The solution is $L = 25$, $S = 11$.

Check the solution by substituting it back into both of the original equations. Never neglect to check your answer.

Translating word problems into systems of equations is a very valuable and very practical algebra skill. Translate each of the following problems into a system of equations.

(a) The total value of an order of nuts and bolts is \$1.40. The nuts cost 5¢ each and the bolts cost 10¢ each. If the number of bolts is four more than twice the number of nuts, how many of each are there? (*Hint:* Keep all money values in cents to avoid decimals.)

(b) The perimeter of a rectangular lot is 350′. The length of the lot is 10 ft more than twice the width. Find the dimensions of the lot.

(c) A materials yard wishes to make a 900 cubic foot mixture of two different types of rock. One type of rock costs \$1.40 per cubic foot and the other costs \$2.60 per cubic foot. If the cost of the mixture is to run \$1530, how many cubic feet of each should go into the mixture?

(d) A lab technician wishes to mix a 5% salt solution and a 15% salt solution to obtain 4 liters of a 12% salt solution. How many liters of each solution must be added?

Check your work in **18**.

---

18 (a) In problems of this kind it is often very helpful to first set up a table:

Item	Number of Items	Cost per Item	Total Cost
Nuts	N	5	$5N$
Bolts	B	10	$10B$

We can write the first equation as

"The total value of an order of nuts and bolts is 140"

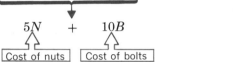

$5N + 10B = 140$

Cost of nuts   Cost of bolts

The second equation would be

". . . the number of bolts is four more than twice the number of nuts. . . ."

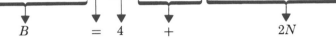

$B = 4 + 2N$

The system of equations to be solved is:

$$5N + 10B = 140$$
$$B = 4 + 2N\ .$$

Use substitution. Since B is equal to $4 + 2N$, replace B in the first equation with $4 + 2N$.

$$5N + 10(4 + 2N) = 140$$
$$5N + 40 + 20N = 140$$
$$25N + 40 = 140 \qquad \text{collecting like terms.}$$
$$25N = 100$$
$$N = 4$$

When we replace N with 4 in the second equation,

$$B = 4 + 2\,(4)$$
$$B = 4 + 8$$
$$B = 12$$

There are 4 nuts and 12 bolts.

Check:

$$5N + 10B = 140 \qquad B = 4 + 2N$$
$$5(4) + 10(12) = 140 \qquad 12 = 4 + 2(4)$$
$$20 + 120 = 140 \qquad 12 = 4 + 8$$
$$140 = 140 \qquad 12 = 12$$

(b) Recalling the formula for the perimeter of a rectangle, we have:

"The perimeter of a rectangular lot is 350 ft."

$2L + 2W = 350$

The second sentence gives us:

"The length of the lot is 10 ft more than twice the width."

$L = 10 + 2W$

The system of equations to be solved is:

$$2L + 2W = 350$$
$$L = 10 + 2W$$

Using substitution, we replace L with $10 + 2W$ in the first equation.

$$2(10 + 2W) + 2W = 350$$
$$20 + 4W + 2W = 350$$
$$20 + 6W = 350$$
$$6W = 330$$
$$W = 55'$$

Now substitute 55 for W in the second equation:

$$L = 10 + 2(55)$$
$$L = 10 + 110$$
$$L = 120'$$

The lot is 120 ft long and 55 ft wide.

Check:

$$2L + 2W = 350 \qquad L = 10 + 2W$$
$$2(120) + 2(55) = 350 \qquad 120 = 10 + 2(55)$$
$$240 + 110 = 350 \qquad 120 = 10 + 110$$
$$350 = 350 \qquad 120 = 120$$

(c) First set up the following table:

Item	Amount (cu ft)	Cost per cu ft	Total Cost
Cheaper rock	x	\$1.40	$1.40x$
More expensive rock	y	\$2.60	$2.60y$

The first equation would come from the statement:

". . . a 900 cubic foot mixture of two different types of rock."

$$900 = x + y$$

Consulting the table, we would write the second equation as follows:

". . . the cost of the mixture is to run \$1530. . . ."

$$1.40x + 2.60y = 1530$$

Multiply this last equation by 10 to get the system of equations

$$x + y = 900$$

$$14x + 26y = 15{,}300$$

Multiply the first equation by -14 and add it to the second equation:

$$-14x + (-14y) = -12600$$
$$14x + 26y = 15300$$
$$12y = 2700$$
$$y = 225 \text{ cu ft}$$

Replacing y with 225 in the first equation, we have:

$$x + (225) = 900$$
$$x = 675 \text{ cu ft}$$

There should be 225 cu ft of the \$2.60 per cu ft mixture, and 675 cu ft of the \$1.40 per cu ft mixture.

Check:

$$\begin{aligned} x + y &= 900 \\ 675 + 225 &= 900 \\ 900 &= 900 \end{aligned} \qquad \begin{aligned} 1.40x + 2.60y &= 1530 \\ 1.40(675) + 2.60(225) &= 1530 \\ 945 + 585 &= 1530 \\ 1530 &= 1530 \end{aligned}$$

(d)

Solution	Amount (Liters)	Salt Fraction	Total Salt
5%	A	0.05	$0.05A$
15%	B	0.15	$0.15B$

Since the final solution is to contain 4 liters, we have:

$A + B = 4$

The second equation represents the total amount of salt:

$0.05A + 0.15B = (0.12)4$

Multiplying this last equation by 100 to eliminate the decimals, we have the system:

$$\begin{aligned} A + B &= 4 \\ 5A + 15B &= 48 \end{aligned}$$

To solve this system, multiply each term in the first equation by -5 and add to get

$$\begin{aligned} -5A + (-5B) &= -20 \\ 5A + 15B &= 48 \\ \hline 10B &= 28 \\ B &= 2.8 \text{ liters} \end{aligned}$$

Substituting back into the first equation, we have:

$$\begin{aligned} A + 2.8 &= 4 \\ A &= 1.2 \text{ liters} \end{aligned}$$

Check:

$$\begin{aligned} A + B &= 4 \\ 1.2 + 2.8 &= 4 \\ 4 &= 4 \end{aligned} \qquad \begin{aligned} 5A + 15B &= 48 \\ 5(1.2) + 15(2.8) &= 48 \\ 6 + 42 &= 48 \\ 48 &= 48 \end{aligned}$$

Now turn to **19** for a set of exercises on systems of equations.

Exercises 10-1 Systems of Equations

19 A. Solve each of the following systems of equations using the method of substitution. If the system is inconsistent or dependent, say so.

1. $y = 10 - x$, $2x - y = -4$
2. $3x - y = 5$, $2x + y = 15$
3. $2x - y = 3$, $x - 2y = -6$
4. $2y - 4x = -3$, $y = 2x + 4$
5. $3x + 5y = 26$, $x + 2y = 10$
6. $x = 10y + 1$, $y = 10x + 1$

B. Solve each of the following systems of equations. If the system is inconsistent or dependent, say so.

1. $x + y = 5$, $x - y = 13$
2. $2x + 2y = 10$, $3x - 2y = 10$
3. $2y = 3x + 5$, $2y = 3x - 7$

4. $2y = 2x + 2$
$4y = 5 + 4x$

5. $x = 3y + 7$
$x + y = -5$

6. $3x - 2y = -11$
$x + y = -2$

7. $5x - 4y = 1$
$3x - 6y = 6$

8. $y = 3x - 5$
$6x - 3y = 3$

9. $y - 2x = -8$
$x - \frac{1}{2}y = 4$

10. $x + y = a$
$x - y = b$

C. Practical Problems

Translate each problem statement into a system of equations and solve.

1. The sum of two numbers is 39 and their difference is 7. What are the numbers?

2. The sum of two numbers is 14. The larger is two more than three times the smaller. What are the numbers?

3. Separate a collection of twenty objects into two parts so that twice the larger amount equals three times the smaller amount.

4. The average of two numbers is 25 and their difference is 8. What are the numbers?

5. Four bleebs and three freems cost \$11. Three bleebs and four freems cost \$10. What does a bleeb cost?

6. The perimeter of a rectangular window is 14 ft, and its length is 2 ft less than twice the width. What are the dimensions of the window?

7. Harold exchanged a \$1 bill for change and received his change in nickels and dimes, with seven more dimes than nickels. How many of each coin did he receive?

8. If four times the larger of two numbers is added to three times the smaller, the result is 26. If three times the larger number is decreased by twice the smaller, the result is 11. Find the numbers.

9. Mr. Brown bought five cans of peas and four cans of corn, but he forgot what each cost. He knows that the total cost was \$2.90 and he recalls that a can of peas cost 5¢ less than a can of corn. How much did each can cost?

10. The length of a piece of sheet metal is twice the width. The difference in length and width is 20 inches. What are the dimensions?

11. A 30 inch piece of wire is to be cut into 2 parts, one part being 4 times the length of the other. Find the length of each.

12. A painter wishes to mix paint worth \$6 per gallon with paint worth \$10 per gallon to make a 12 gallon mixture worth \$7 per gallon. How many gallons of each should he mix?

13. A chemist wishes to mix a 10% salt solution with a 2% salt solution to obtain 6 liters of a 4% salt solution. How many liters of each should be added?

14. The perimeter of a rectangular field is 520 ft. The length of the field is 20 ft more than three times the width. Find the dimensions.

15. A 24 ton mixture of crushed rock is needed in a construction job; it will cost \$360. If the mixture is composed of rock costing \$13 per ton and \$19 per ton, how many tons of each should be added?

When you have completed these problems check your answers on page 572. Then turn to **20** to learn about quadratic equations.

10-2 QUADRATIC EQUATIONS

20 Thus far in your study of algebra you have worked only with linear equations. In a linear equation the variable appears only to the first power. For example, $3x + 5 = 2$ is a linear equation. The variable appears as x or x^1. No powers of x such as x^2, x^3, or x^4 appear in the equations.

An equation in which the variable appears to the second power is called a *quadratic equation.*

Which of the following are quadratic equations?

(a) $x^2 = 49$
(b) $2x - 1 = 4$
(c) $3x - 2y = 19$
(d) $5x^2 - 8x + 3 = 0$
(e) $x^3 + 3x^2 + 3x + 1 = 0$

Check your answer in **21**.

---

21 Equations (a) and (d) are quadratic equations. Equation (b) is a linear equation in one variable. Equation (c) is a linear equation in two variables, x and y. Equation (e) is a cubic or third-order equation. Because (e) contains an x^3 term it is not a quadratic.

Every quadratic equation can be put into a *standard quadratic form.*

$ax^2 + bx + c = 0$ where a cannot equal zero.

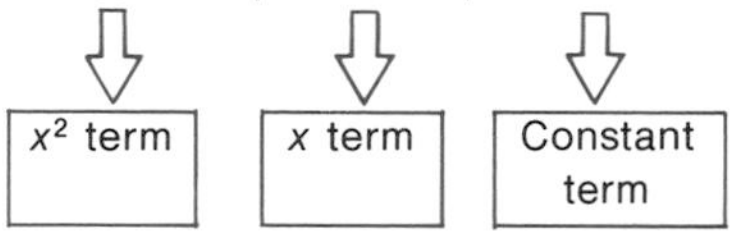

Every quadratic equation must have an x^2 term, but the x term and the constant term may be missing. For example,

$2x^2 + x - 5 = 0$ is a quadratic equation in standard form: the x^2 term is first, the x term second, and the constant term last on the left.

$x^2 + 4 = 0$ is also a quadratic equation in standard form. The x term is missing, but the other terms are in the proper order. We could rewrite this equation as

$x^2 + 0 \cdot x + 4 = 0$

Which of the following quadratic equations are written in standard form?

(a) $7x^2 - 3x + 6 = 0$ (b) $8x - 3x^2 - 2 = 0$
(c) $2x^2 - 5x = 0$ (d) $x^2 = 25$
(e) $x^2 - 5 = 0$ (f) $4x^2 - 5x = 6$

Check your answer in **22**.

---

22 Equations (a), (c), and (e) are in standard form.

In order to solve a quadratic equation, the first step is usually to rewrite it in standard form. For example, the equation

$x^2 = 25x$	becomes	$x^2 - 25x = 0$	in standard form.
$8x - 3x^2 - 2$	becomes	$-3x^2 + 8x - 2 = 0$	in standard form.
	or	$3x^2 - 8x + 2 = 0$	if we multiply all terms by -1.
$4x^2 - 5x = 6$	becomes	$4x^2 - 5x - 6 = 0$	in standard form.

In each case we add or subtract a term on both sides of the equation until all terms are on the left, then rearrange terms until the x^2 is first on the left, the x term next, and the constant term third.

Try it. Rearrange the following quadratic equations in standard form.

(a) $5x - 19 + 3x^2 = 0$	(b) $7x^2 = 12 - 6x$
(c) $9 = 3x - x^2$	(d) $2x - x^2 = 0$
(e) $5x = 7x^2 - 12$	(f) $x^2 - 6x + 9 = 49$
(g) $3x + 1 = x^2 - 5$	(h) $1 - x^2 + x = 3x + 4$

Check your answers in **23**.

23

(a) $3x^2 + 5x - 19 = 0$	(b) $7x^2 + 6x - 12 = 0$
(c) $x^2 - 3x + 9 = 0$	(d) $x^2 - 2x = 0$
(e) $7x^2 - 5x - 12 = 0$	(f) $x^2 - 6x - 40 = 0$
(g) $x^2 - 3x - 6 = 0$	(h) $x^2 + 2x + 3 = 0$

The solution to a linear equation is a single number. The solution to a quadratic equation is usually a *pair* of numbers, each of which satisfies the equation. For example, the quadratic equation

$$x^2 - 5x + 6 = 0$$

has the solutions

$$x = 2 \qquad \text{or} \qquad x = 3$$

To see that either 2 or 3 are solutions, substitute them into the equation.

For $x = 2$

$$(2)^2 - 5(2) + 6 = 0$$
$$4 - 10 + 6 = 0$$
$$10 - 10 = 0$$

For $x = 3$

$$(3)^2 - 5(3) + 6 = 0$$
$$9 - 15 + 6 = 0$$
$$15 - 15 = 0$$

Show that $x = 5$ or $x = 3$ give solutions of the quadratic equation

$$x^2 - 8x + 15 = 0$$

Check your work in **24**.

---

24 $x^2 - 8x + 15 = 0$

For $x = 5$

$$(5)^2 - 8(5) + 15 = 0$$
$$25 - 40 + 15 = 0$$
$$40 - 40 = 0$$

For $x = 3$

$$(3)^2 - 8(3) + 15 = 0$$
$$9 - 24 + 15 = 0$$
$$24 - 24 = 0$$

The easiest kind of quadratic equation to solve is one in which the linear term is missing. For example, to solve the quadratic equation

$$x^2 - 25 = 0$$

simply rewrite it as

$$x^2 = 25$$

and take the square root of both sides of the equation.

$$\sqrt{x^2} = \sqrt{25}$$

or $x = \pm\sqrt{25}$

or $x = 5 \qquad$ or $\qquad x = -5$

Both 5 and -5 satisfy the original equation.

$$x^2 - 25 = 0 \quad \text{and} \quad x^2 - 25 = 0$$
$$(5)^2 - 25 = 0 \qquad (-5)^2 - 25 = 0$$
$$25 - 25 = 0 \qquad 25 - 25 = 0$$

Every positive number has two square roots, one positive and the other negative. Both of them may be important in solving a quadratic equation.

Solve each of the following quadratic equations and check *both* solutions. (Round to two decimal places if necessary.)

(a) $x^2 - 36 = 0$ (b) $x^2 = 8$ (c) $x^2 - 64 = 0$

(d) $4x^2 = 81$ (e) $3x^2 = 27$ (f) $x^2 - 1 = 2$

Our solutions are in **25**.

---

25 (a) $x^2 = 36$
$x = \pm\sqrt{36}$
$x = 6$ or $x = -6$

(b) $x^2 = 8$
$x = \pm\sqrt{8}$
$x = 2.83$ or $x = -2.83$, rounded

(c) $x^2 = 64$
$x = \pm\sqrt{64}$
$x = 8$ or $x = -8$

(d) $4x^2 = 81$
$x^2 = \dfrac{81}{4}$
$x = \pm\sqrt{\dfrac{81}{4}}$
$x = \dfrac{9}{2}$ or $x = -\dfrac{9}{2}$

(e) $3x^2 = 27$
$x^2 = 9$
$x = \pm\sqrt{9}$
$x = 3$ or $x = -3$

(f) $x^2 = 3$
$x = \pm\sqrt{3}$
$x = 1.73$ or $x = -1.73$, rounded

Notice in each case that first we rewrite the equation so that x^2 appears alone on the left and a number appears alone on the right. Second, take the square root of both sides. The equation will have two solutions.

You should also notice that an equation such as

$x^2 = -4$

has no solution. There is no number x whose square is a negative number.

Solve the following quadratic equations. (Round to two decimal places if necessary.)

(a) $x^2 - 3.5 = 0$ (b) $x^2 = 18$ (c) $6 - x^2 = 0$

(d) $9x^2 = 49$ (e) $7x^2 = 80$ (f) $\dfrac{3x^2}{5} = 33.3$

Check your answers in **26**.

---

26 (a) $x^2 = 3.5$
$x = \pm\sqrt{3.5}$
$x = 1.87$ or $x = -1.87$

(b) $x^2 = 18$
$x = \pm\sqrt{18}$
$x = 4.24$ or $x = -4.24$

(c) $6 - x^2 = 0$
$x^2 = 6$
$x = \pm\sqrt{6}$
$x = 2.45$ or $x = -2.45$

(d) $9x^2 = 49$
$x^2 = \dfrac{49}{9}$
$x = \pm\sqrt{\dfrac{49}{9}}$
$x = \dfrac{7}{3}$ or $x = -\dfrac{7}{3}$

(e) $7x^2 = 80$

$$x^2 = \frac{80}{7}$$

$$x = \pm\sqrt{\frac{80}{7}}$$

$$x = \pm\sqrt{11.4286}$$

$$x = 3.38 \text{ or } x = -3.38$$

(f) $\dfrac{3x^2}{5} = 33.3$

$$x^2 = \frac{(33.3)(5)}{3} = 55.5$$

$$x = \pm\sqrt{55.5}$$

$$x = 7.45 \text{ or } x = -7.45$$

In general, a quadratic equation will contain all three terms, an x^2 term, an x term, and a constant term. The solution of any quadratic equation

$$ax^2 + bx + c = 0$$

is

$$x = \frac{-b + \sqrt{b^2 - 4ac}}{2a} \quad \text{or} \quad x = \frac{-b - \sqrt{b^2 - 4ac}}{2a}$$

or

$$x = \frac{-b \pm \sqrt{b^2 - 4ac}}{2a} \qquad \text{The Quadratic Formula}$$

For example, to solve the quadratic equation

$$2x^2 + 5x - 3 = 0$$

follow these steps.

Step 1 *Identify* the coefficients a, b, and c for the quadratic equation.

$$2x^2 + 5x - 3 = 0$$

$a = 2$ $b = 5$ $c = -3$

Step 2 *Substitute* these values of a, b, and c into the quadratic formula.

$$x = \frac{-(5) \pm \sqrt{(5)^2 - 4(2)(-3)}}{2(2)}$$

Step 3 *Simplify* these equations for x.

$$x = \frac{-5 \pm \sqrt{25 + 24}}{4}$$

$$x = \frac{-5 \pm \sqrt{49}}{4}$$

$$x = \frac{-5 \pm 7}{4}$$

$$x = \frac{-5 + 7}{4} \quad \text{or} \quad x = \frac{-5 - 7}{4}$$

$$x = \frac{2}{4} \quad \text{or} \quad x = \frac{-12}{4}$$

$$x = \frac{1}{2} \quad \text{or} \quad x = -3$$

The solution is $x = \frac{1}{2}$ or $x = -3$.

Step 4 *Check* the solution numbers by substituting them into the original equation.

Check: $2x^2 + 5x - 3 = 0$

For $x = \frac{1}{2}$

$$2\left(\frac{1}{2}\right)^2 + 5\left(\frac{1}{2}\right) - 3 = 0$$
$$2\left(\frac{1}{4}\right) + 5\left(\frac{1}{2}\right) - 3 = 0$$
$$\frac{1}{2} + 2\frac{1}{2} - 3 = 0$$
$$3 - 3 = 0$$

For $x = -3$

$$2(-3)^2 + 5(-3) - 3 = 0$$
$$2(9) + 5(-3) - 3 = 0$$
$$18 - 15 - 3 = 0$$
$$18 - 18 = 0$$

Your turn. Use the quadratic formula to solve $x^2 + 4x - 5 = 0$.

Check your work in **27**.

---

27 **Step 1** $x^2 \quad + \quad 4x \quad - \quad 5 \quad = 0.$

$a = 1$ $\quad b = 4$ $\quad c = -5$

Step 2 $x = \dfrac{-(4) \pm \sqrt{(4)^2 - 4(1)(-5)}}{2(1)}$

Step 3 Simplify

$$x = \frac{-4 \pm \sqrt{36}}{2}$$
$$x = \frac{-4 \pm 6}{2}$$
$$x = \frac{-4 + 6}{2} \quad \text{or} \quad x = \frac{-4 - 6}{2}$$

or $x = 1$ or $x = -5$ The solution is $x = 1$ or $x = -5$.

Step 4

Check: $x^2 + 4x - 5 = 0$

For $x = 1$

$$(1)^2 + 4(1) - 5 = 0$$
$$1 + 4 - 5 = 0$$
$$5 - 5 = 0$$

For $x = -5$

$$(-5)^2 + 4(-5) - 5 = 0$$
$$25 - 20 - 5 = 0$$
$$25 - 25 = 0$$

Here is one that is a bit tougher. Solve $3x^2 - 7x = 5$

Check your work in **28**.

---

28 **Step 1** Rewrite the equation in standard form.

$$3x^2 - 7x - 5 = 0$$

$a = 3$ | $b = -7$ | $c = -5$

Step 2 $x = \dfrac{-(-7) \pm \sqrt{(-7)^2 - 4(3)(-5)}}{2(3)}$

Step 3 Simplify

$$x = \frac{7 \pm \sqrt{109}}{6}$$

or $x = \dfrac{7 + \sqrt{109}}{6}$ or $x = \dfrac{7 - \sqrt{109}}{6}$

$x = \dfrac{7 + 10.44}{6}$ or $x = \dfrac{7 - 10.44}{6}$ rounded to two decimal places

$x = \dfrac{17.44}{6}$ or $x = \dfrac{-3.44}{6}$

$x = 2.91$ or $x = -0.57$ The solution is approximately $x = 2.91$ or $x = -0.57$.

Step 4 Check both answers by substituting them back into the original quadratic equation.

Use the quadratic formula to solve each of the following equations. (Round to two decimal places if necessary.)

(a) $6x^2 - 13x + 2 = 0$ (b) $3x^2 - 13x = 0$
(c) $2x^2 - 5x + 17 = 0$ (d) $8x^2 = 19 - 5x$

Check your work in **29**.

---

29 (a) $6x^2 - 13x + 2 = 0$

$a = 6$ | $b = -13$ | $c = 2$

$$x = \frac{-(-13) \pm \sqrt{(-13)^2 - 4(6)(2)}}{2(6)}$$

$$x = \frac{13 \pm \sqrt{121}}{12}$$

$$x = \frac{13 \pm 11}{12}$$

The solution is $x = \dfrac{13 + 11}{12}$ or $x = \dfrac{13 - 11}{12}$

or $x = 2$ or $x = \dfrac{1}{6}$ The solution is $x = 2$ or $x = \frac{1}{6}$.

Check it.

(b) $3x^2 - 13x = 0$

$3x^2 \quad - \quad 13x \quad + \quad 0 \quad = 0$

⇩ ⇩ ⇩

$a = 3$ | $b = -13$ | $c = 0$

$$x = \frac{-(-13) \pm \sqrt{(-13)^2 - 4(3)(0)}}{2(3)}$$

$$x = \frac{13 \pm \sqrt{169}}{6}$$

The solution is $x = \dfrac{13 + 13}{6}$ or $x = \dfrac{13 - 13}{6}$

or $x = \dfrac{13}{3}$ or $x = 0$ The solution is

$x = 0$ or $x = \frac{13}{3}$.

(c) $2x^2 \quad - \quad 5x \quad + \quad 17 \quad = 0$

⇩ ⇩ ⇩

$a = 2$ | $b = -5$ | $c = 17$

$$x = \frac{-(-5) \pm \sqrt{(-5)^2 - 4(2)(17)}}{2(2)}$$

$$x = \frac{5 \pm \sqrt{-111}}{4}$$

But the square root of a negative number is impossible to find if our answer must be a real number. This quadratic equation has no numerical solution.

(d) $8x^2 = 19 - 5x$

or $8x^2 \quad + \quad 5x \quad - \quad 19 \quad = 0$

⇩ ⇩ ⇩

$a = 8$ | $b = 5$ | $c = -19$

$$x = \frac{-(5) \pm \sqrt{(5)^2 - 4(8)(-19)}}{2(8)}$$

$$x = \frac{-5 \pm \sqrt{633}}{16}$$

$$x = \frac{-5 \pm 25.16}{16}$$ rounded to two decimal places.

The solution is $x = -1.89$ or $x = 1.26$, rounded.

When you check a solution that includes a rounded value, the check may not give an exact fit. The differences should be very small if you have the correct solution.

Consider the following problem.

In a DC circuit the power P dissipated in the circuit is given by the equation $P = RI^2$ *where R is the circuit resistance in ohms and I is the current in amperes. What current will produce 1440 watts of power in a 10-ohm resistor?*

Substituting into the equation,

$$1440 = 10I^2$$

or $$I^2 = 144$$
$$I = \pm\sqrt{144}$$

The solution is $I = 12$ amp or $I = -12$ amp.

The negative current is not a valid solution; negative currents have no meaning in this case. The answer is $I = 12$ amp.

Another example:

One side of a rectangular opening for a heating pipe is 3 in. longer than the other side. If the total cross-sectional area is 70 sq in., find the dimensions of the cross section.

Let L = length, W = width.

Then $L = 3 + W$

and the area is

$$\text{Area} = LW$$

or $$70 = LW$$

Substituting $L = 3 + W$ into the area equation,

$$70 = (3 + W)W$$

or $$70 = 3W + W^2$$

or $$W^2 + 3W - 70 = 0$$ in standard quadratic form

$a = 1$ $\quad$ $b = 3$ $\quad$ $c = -70$

To find the solution, substitute a, b, and c into the quadratic formula.

$$W = \frac{-(3) \pm \sqrt{(3)^2 - 4(1)(-70)}}{2(1)}$$

or $$W = \frac{-3 \pm \sqrt{289}}{2}$$

$$W = \frac{-3 \pm 17}{2}$$

The solution is

$$W = -10 \qquad \text{or} \qquad W = 7$$

Only the positive value makes sense. The answer is $W = 7$ in.

Substituting back into the equation $L = 3 + W$, we find $L = 10$ in.

Check to see that the area is indeed 70 sq in.

Solve the following problems. (Round each answer to two decimal places if necessary.)

(a) Find the side length of a square whose area is 200 sq m.

(b) The cross-sectional area of a rectangular duct must be 144 sq in. If one side must be twice as long as the other, find the length of each side.

(c) Find the diameter of a circular pipe whose cross-sectional area is 3.00 sq in.

(d) The SAE horsepower rating of an engine is given by $H.P. = \dfrac{D^2N}{2.5}$, where D is the bore of the cylinder in inches, and N is the number of cylinders. What must the bore be for an 8-cylinder engine to have a horsepower rating of 300?

(e) One side of a rectangular plate is six inches longer than the other. The total area is 216 sq inches. How long is each side?

(f) A man has a long strip of sheet steel 12 ft wide. He wishes to make an open-topped water channel with a rectangular cross section. If the cross-sectional area must be 16 sq ft, what should the dimensions of the channel be? (*Hint:* $H + H + W = 12''$.)

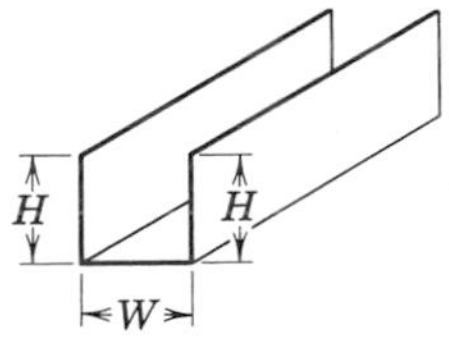

Our worked solutions are in **30**.

---

30 (a) $A = s^2$

$200 = s^2$

$s = \pm\sqrt{200}$

$s = 14.14$ m, rounded The negative solution is not possible.

(b) $L \cdot W = A$

and $L = 2W$

Therefore $(2W)W = A$

or $2W^2 = A$

$2W^2 = 144$

$W^2 = 72$

$W = +\sqrt{72}$

$W = 8.49$ in., rounded for the positive root.

Then $L = 2W$

or $L = 16.98$ in., rounded

(c) $A = \dfrac{\pi D^2}{4}$

$3.00 = \dfrac{\pi D^2}{4}$

or $D^2 = \dfrac{4(3.00)}{3.1416}$

$D^2 = 3.8197$

$D = +\sqrt{3.8197}$

$D = 1.95$ in., rounded

(d) $HP = \dfrac{D^2N}{2.5}$

or $D^2 = \dfrac{(300)(2.5)}{8}$

$D^2 = 93.75$

$D = +\sqrt{93.75}$

or $D = 9.68$ in., rounded

(e) "One side . . . must be six inches longer than the other."

$$L = 6 + W$$

For a rectangle

Area = LW

Then

$$216 = LW$$

or $216 = (6 + W)W$
$216 = 6W + W^2$

$W^2 + 6W - 216 = 0$ in standard quadratic form

$a = 1$ $b = 6$ $c = -216$

When we substitute into the quadratic formula,

$$W = \frac{-(6) \pm \sqrt{(6)^2 - 4(1)(-216)}}{2(1)}$$

$$W = \frac{-6 \pm \sqrt{900}}{2}$$

$$W = \frac{-6 \pm 30}{2}$$

The solution is

$W = -18$ or $W = 12$ Only the positive value is a reasonable solution. The answer is $W = 12$ in.

Substituting into the first equation,

$L = 6 + W$
$L = 6 + (12) = 18$ in.

(f) $H + H + W = 12$ or $2H + W = 12$
or $W = 12 - 2H$

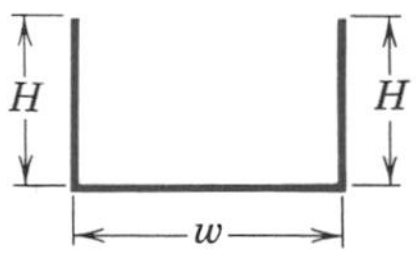

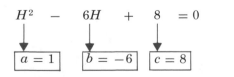

Area = HW
or $16 = HW$ Substitute $W = 12 - 2H$ into the area equation.

$16 = H(12 - 2H)$

$16 = 12H - 2H^2$ or $8 = 6H - H^2$ Divide each term by 2 to get a simpler equation.

$H^2 - 6H + 8 = 0$ in standard quadratic form.

$a = 1$ $b = -6$ $c = 8$

Substituting a, b, and c into the quadratic formula,

$$H = \frac{-(-6) \pm \sqrt{(-6)^2 - 4(1)(8)}}{2(1)}$$

$$H = \frac{6 \pm \sqrt{4}}{2}$$

$$H = \frac{6 \pm 2}{2}$$

The solution is $H = 4$ ft or $H = 2$ ft

For $H = 4$ ft, $W = 4$ ft. Since $16 = HW$.
For $H = 2$ ft, $W = 8$ ft.

The channel can be either 4 ft by 4 ft or 2 ft by 8 ft. Both dimensions give a cross-sectional area of 16 sq ft.

Now turn to **31** for a set of exercises on solving quadratic equations.

---

31 **Exercises 10-2 Quadratic Equations**

A. Which of the following are quadratic equations?

1. $5x - 13 = 23$
2. $2x + 5 = 3x^2$
3. $2x^3 - 6x^2 - 5x + 3 = 0$
4. $x^2 = 0$
5. $8x^2 - 9x = 0$

Which of the following quadratic equations are in standard form? For those that are not, rearrange them into standard form.

6. $7x^2 - 5 + 3x = 0$
7. $14 = 7x - 3x^2$
8. $13x^2 - 3x + 5 = 0$
9. $23x - x^2 = 5x$
10. $4x^2 - 7x + 3 = 0$

B. Solve each of these quadratic equations. (Round to two decimal places if necessary.)

1. $x^2 = 25$
2. $3x^2 - 27 = 0$
3. $5x^2 = 22x$
4. $2x^2 - 7x + 3 = 0$
5. $4x^2 = 81$
6. $6x^2 - 13x - 63 = 0$
7. $15x = 12 - x^2$
8. $4x^2 - 39x = -27$
9. $0.4x^2 + 0.6x - 0.8 = 0$
10. $0.001 = x^2 + 0.03x$

C. Practical Problems

1. The area of a square is 625 sq mm. Find its side length.
2. The capacity in gallons of a cylindrical tank can be found using the formula

$$C = \frac{0.7854D^2L}{231}$$

where C = capacity in gallons
D = diameter of tank in inches
L = length of tank in inches

(a) Find the diameter of a 42 in. long tank that has a capacity of 30 gallons.
(b) Find the diameter of a 60 in. long tank that has a capacity of 50 gallons.

3. The length of a rectangular pipe is three times longer than its width. Find the dimensions that give a cross-sectional area of 75 sq in.
4. Find the radius of a circular vent that has a cross-sectional area of 250 sq cm.
5. The length of a rectangular piece of sheet steel must be 5 in. longer than the width. Find the exact dimensions that will provide an area of 374 sq in.

6. If the horsepower rating of an engine is given by

$$H.P. = \frac{D^2N}{2.5}$$

Find D (the bore of the cylinder in inches) if

(a) N (the number of cylinders) is 4 and the horsepower rating is 80.
(b) N is 6 and $H.P. = 200$.

7. Find the diameter of a circular pipe whose cross-sectional area is 40 sq in.

8. If $P = RI^2$ for a direct current circuit, find I (current in amperes) if:

(a) The power (P) is 405 watts and the resistance (R) is 5 ohms.
(b) The power is 800 watts and the resistance is 15 ohms.

9. Total piston displacement is given by the following formula:

$$P.D. = 0.7854D^2LN$$

where

$P.D.$ = piston displacement in cubic inches
D = diameter of bore of cylinder
L = length of stroke in inches
N = number of cylinders

Find the diameter if

(a) $P.D. = 400$ cu in.
$L = 4.5$ in.
$N = 8$

(b) $P.D. = 392.7$ cu in.
$L = 4$ in.
$N = 6$

10. An open-topped channel must be made out of a 20 ft wide piece of sheet steel. What dimensions will result in a cross-sectional area of 48 sq ft?

Check your answers on page 573. Then turn to **32** for a problem set covering both systems of equations and quadratic equations.

10 Advanced Algebra

PROBLEM SET 10

Advanced Algebra

Answers are on page 573.

32

A. Solve and check each of the following systems of equations. If the system is inconsistent or dependent, say so.

1. $x + 4y = 27$
   $x + 2y = 21$
2. $3x + 2y = 17$
   $x = 5 - 2y$
3. $5x + 2y = 20$
   $3x - 2y = 4$
4. $x = 10 - y$
   $2x + 3y = 23$
5. $3x + 4y = 45$
   $x - \frac{1}{3}y = 5$
6. $2x - 3y = 11$
   $4x - 6y = 22$
7. $2x + 3y = 5$
   $3x + 2y = 5$
8. $5x = 1 - 3y$
   $4x + 2y = -8$
9. $3x - 2y = 10$
   $4x + 5y = 12$

B. Solve and check each of the following quadratic equations.

1. $x^2 = 9$
2. $x^2 - 3x - 28 = 0$
3. $3x^2 + 5x + 1 = 0$
4. $4x^2 = 3x + 2$
5. $x^2 = 6x$
6. $2x^2 - 7x - 15 = 0$
7. $x^2 - 4x + 4 = 9$
8. $x^2 - x - 30 = 0$
9. $\dfrac{5x^2}{3} = 60$
10. $7x + 8 = 5x^2$

C. Practical Problems

For each of the following, set up either a system of equations in two variables or a quadratic equation and solve. (Round to two decimal places if necessary.)

1. The sum of two numbers is 38. Their difference is 14. Find them.
2. The area of a square is 196 sq in. Find its side length.
3. The difference of two numbers is 21. If twice the larger is subtracted from five times the smaller, the result is 33. Find the numbers.
4. The perimeter of a rectangular door is 22 ft. Its length is 2 ft more than twice its width. Find the dimensions of the door.
5. One side of a rectangular heating pipe is 4 times as long as the other. The cross-sectional area is 125 sq cm. Find the dimensions of the pipe.
6. A mixture of 650 nails costs \$23.50. If some of the nails cost 5¢ apiece, and the rest cost 3¢ apiece, how many of each are there?
7. In the formula $P = RI^2$ find the current I in amperes if:
   (a) The power P is 1352 watts and the resistance R is 8 ohms.
   (b) The power P is 1500 watts and the resistance is 10 ohms.
8. The length of a rectangular piece of sheet steel is 2 in. longer than the width. Find the exact dimensions if the area of the sheet is 168 sq in.
9. A 42 in. piece of wire is to be cut into 2 parts. If one part is 2 in. less than three times the length of the other, find the length of each piece.
10. The perimeter of a rectangular field is 750 ft. If the length is four times the width, find the dimensions of the field.
11. Find the diameter of a circular vent with a cross-sectional area of 200 sq in.
12. A painter mixes paint worth \$4 per gallon with paint worth \$8.50 per gallon. He wishes to make 15 gallons of a mixture worth \$7 per gallon. How many gallons of each kind of paint must be included in the mixture?

Name

Date

Course/Section

13. In the formula $H.P. = \dfrac{D^2N}{2.5}$ find the diameter of the cylinder bore D if:

(a) The number of cylinders N is 6 and the horsepower rating $H.P.$ is 60.
(b) The number of cylinders is 8 and the horsepower rating is 125.

14. The length of a rectangular pipe is 3 in. less than twice the width. Find the dimensions if the cross-sectional area is 20 sq. in.

15. Fifty tons of a mixture of decorative rock cost $1750. If the mixture consists of rock costing $30 per ton and a more expensive rock costing $55 per ton, how many tons of each are used to make the mixture?

APPENDIX A

Using an Electronic Calculator

The hand-held electronic calculator is not a fad. In a very short time it has become an essential tool for people in every occupation, from engineers to shoppers, from clerks to bank presidents. Just as large-scale computers have revolutionized society, the inexpensive pocket calculator has changed personal calculating. For millions of people arithmetic will never be the same again.

It is reasonable to ask, if calculators are available why study arithmetic at all? The answer is that while a calculator can be a very useful tool, it must not become a crutch. A calculator will help you to make mathematical computations more quickly, but it will not tell you *what* to do or *how* to do it. Because the calculator allows you to make very long and difficult calculations quickly, the basic mathematical skills are more important than ever. The gadget itself can't make a mistake, but you can. If you know the basic mathematics skills, you will recognize when an answer must be wrong.

To use a calculator intelligently and effectively in your work, you must be able to

(a) Multiply and add one-digit numbers quickly and correctly, almost like an instant reflex.
(b) Read and write any number, very large or very small, decimal, whole number, or fraction.
(c) Work with fractions and percents.
(d) Estimate answers and check your calculations.

We have included a very careful review of these basic skills in Chapters 1 through 4 in this textbook. If you understand these basic skills the calculator will be a useful tool. If you do not understand these basic skills, you will find that a calculator is simply a lightning-fast way of arriving at a wrong answer.

This appendix is designed to help you to understand the calculator and to use it correctly and effectively.

There are dozens of electronic calculators available. They differ in size, shape, color, cost and, most important, in how they work and what they can do. In this brief introduction we assume that the calculator you are using has the following characteristics.

★ It has at least an 8-digit display.
★ It performs at least the four basic arithmetic functions: Addition (+), subtraction(−), multiplication(×), and division(÷).
★ It uses floating-point arithmetic, so that the position of the decimal point is given automatically in any answer.
★ It uses algebraic logic. (This is by far the most popular type of calculator in use today. If your calculator has a key marked [+=] it is probably an "arithmetic logic" device, and some of this appendix will not be very helpful to you.)

Your calculator may have a great many additional features of course, including square root [√] or square [x^2], trigonometric or more advanced functions, one or more memories, and perhaps a programming capability. We'll consider only the simple calculations here. In the following examples, the displays shown for our "average" calculator may differ slightly from what you find with your calculator, but the differences will not be confusing.

THE CALCULATOR

First, check out the machine. You'll find an on-off switch somewhere, a display, and a keyboard with both *numerical* (0, 1, 2, 3, . . . , 9) and basic *function* (+, −, ×, ÷, =, .) keys usually arranged like this:

7	8	9	÷
4	5	6	×
1	2	3	−
0	.	=	+

In addition, you will find a *clear* [C] key somewhere on the keyboard, and perhaps other keys will appear such as [%] [M+] [MC] [√] and so on.

In order to understand the operation of the calculator completely, it helps if you know a little about what goes on inside the case. As the following diagram shows, all information, numbers, or arithmetic instructions that are entered on the keyboard go immediately to an *operation register* or Op. Reg., where they are stored until they are used. The number in the operation register is also shown in the display.

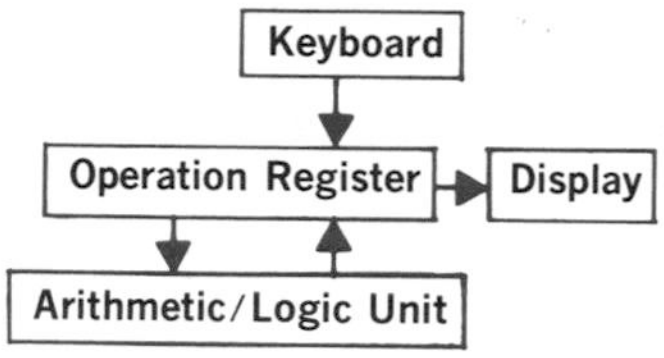

All arithmetic calculations or other operations are performed in the *Arithmetic/Logic Unit* or ALU, the heart of the calculator.

ENTERING NUMBERS

The *clear* key [C] "clears" all entries from the Op. Reg. and ALU, and causes the number zero to appear in the display. Pressing the clear key means that you want to start the entire calculation over or begin a new calculation.

Keyboard	Display
[C]	0.

The *clear entry* key [CE] "clears" the display only, leaves the Op. Reg. and ALU unchanged, and causes a zero to appear in the display. Pressing the clear entry key [CE] means that you wish to remove the last entry only.

Keyboard	Display
[CE]	0.

Every number is entered into the calculator digit by digit, left to right. For example, to enter the number 37.4 the sequence looks like this:

Keyboard	Display
[3]	3
[7]	37
[.]	37.
[4]	37.4

In some calculators the decimal point is always displayed on the right.

Don't worry about leading zeros to the left of the decimal point (as in 0.5, 0.664, or 0.15627) or final zeros after the decimal point (as in 4.70, 32.500, or 8.25000). You need not enter these zeros; the calculator will automatically interpret the number correctly and display the zeros if they are needed. Of course, no harm will be done if you do enter them.

	Keyboard	Display
	[.] [5]	0.5
or	[0] [.] [5]	0.5
or	[.] [5] [0]	0.50

If you make an error in entering the number, press either the clear [C] or clear entry [CE] key. For example, if you wanted to enter the number 46 and made an error, the following sequence might result:

Keyboard	Display	
[4]	4	
[7]	47	⇐ . . . an error
[CE]	0	⇐ . . . Pressing the [CE] key clears the entire number. Start over.
[4]	4	
[6]	46	⇐ . . . the correct entry

Fractions must be translated to decimal form before they can be entered. The fraction $4\frac{1}{4}$ is entered as 4.25 and $5\frac{1}{9}$ as 5.11111. . . .

Keyboard	Display
[4] [.] [2] [5]	4.25
[5] [.] [1] [1] [1] [1]	5.1111

Converting numbers from fraction to decimal form is covered in pages 101 to 106.

Exercises A-1 Entering Numbers

Enter each of the following numbers into your calculator. Convert to decimal form if necessary.

1. 327
2. 46,002
3. 0.0120
4. 137,620.4
5. Seven tenths
6. $6\frac{1}{2}$ lb
7. 3 dollars and 37¢ (in dollars)
8. Four and one-quarter inches
9. $14.75
10. 16¢ (in dollars)

Check your answers on page 538.

When you solve a problem using a calculator, your result will be a number that appears in the display. To find the actual answer from this display may require a bit of interpretation on your part. For example, if you solve a business problem where the answer is in dollars, the display

37.6	means	$37.60	
0.07	means	$0.07	or 7¢
14250.1	means	$14,250.10	. . . and so on

The calculator is capable of displaying an answer with eight numerical digits, so that in general you must round the displayed number to find the correct answer to your problem. For example, if the answer is in dollars, the display

6.7912	means	$6.79	rounding to the nearest cent,

and

145.3086 means $145.31 rounding to the nearest cent.

Rounding is discussed in detail on page 95.

Exercises A-2 Interpreting Numbers

Interpret each of the following calculator displays as an amount of money.

1. 4.2	2. 0.4	3. 0.02	4. 11.7426
5. 431.751	6. 3786506.1	7. 0.3881	8. 6.007143

Round each of the following displays to the nearest tenth.

9. 4.107	10. 189.01	11. 3278.5674	12. 4.25

Round each of the following displays to the nearest hundredth.

13. 137.515	14. 6.845	15. 4.0076

Notice that there are no commas in large numbers displayed on the calculator. The display in Problem 6 above 3786506.1 should be written as $3,786,506.10. You need to count over from the decimal point and insert the commas.

OPERATIONS

Every numerical entry is built up digit by digit. The calculator does not know if the number is 3 or 37 or 37.4 or 37.492116 until you stop entering numerical digits and press a function key: [+], [−], [×], [÷] or [=]. For example, to add 48 + 37 = ? use this sequence:

Keyboard	Display	
[4][8]	48	
[+]	48.	← Pressing the [+] key completes the entry of the first number, and tells the ALU to add the next entry to the number in the Op. Reg.
[3][7]	37	
[=]	85.	← Pressing the [=] key completes the entry of the second number, the two numbers are added, and the sum is placed in the Op. Reg. The display shows what is in the Op. Reg.

Another example: 7.42 × 3.5 = ?

Keyboard	Display	
[7][.][4][2]	7.42	
[×]	7.42	← Pressing the [×] key completes the entry of the first number into the Op. Reg.
[3][.][5]	3.5	
[=]	25.97	← Pressing the [=] key completes the entry of the second number. The two numbers are multiplied and the product is placed in the Op. Reg. The display shows this product.

Suppose that you made an error in entering one of the numbers to be multiplied. The sequence of actions might look like this:

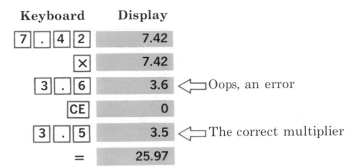

Keyboard	Display	
7 . 4 2	7.42	
×	7.42	
3 . 6	3.6	⇦ Oops, an error
CE	0	
3 . 5	3.5	⇦ The correct multiplier
=	25.97	

Notice that pressing the CE key clears only the second entry (3.6) and does not disturb the earlier part of the calculation. Pressing the C key would clear the entire calculation from the calculator. The CE key clears only the display and does not change any previous entries in the Op. Reg.

ESTIMATING

In order to use a calculator most effectively and accurately you need to develop the skill of *estimating*. Before beginning any calculation you should first estimate the answer by doing simple arithmetic. This estimate gives you a quick and easy check on the calculator answer.

For example, before calculating the addition

$4820 + 1241 = ?$

mentally round each number to the nearest thousand and add:

$5000 + 1000 = 6000$

Your final calculator answer should be roughly 6000. The actual sum is 6061 in this case.

Estimating the answer in this way greatly reduces the possibility that you will make an error. The first law of effective calculating is

> *Never use the calculator until you have a good estimate of the answer.*

Here are a few examples of the process of estimating:

Problem	*Simplified Problem*	*Estimate*	*Actual Answer*
82.2×47.1	80×50	4000	3871.62
0.0912×0.615	0.09×0.6	0.054	0.056088
\$117.92 + \$6.37	\$120 + \$6	\$126	\$124.29
$217.64 \div 41.62$	$200 \div 40$	5	5.2292167

In each problem round the numbers to form a simplified problem that can be solved mentally—"in your head." Use this mental arithmetic estimate to check the answer obtained on the calculator.

Try the following simple problems.

Exercises A-3 Simple Operations

Perform the following calculations

1. $3569 + 407$
2. 7.6×12.8

3. 94.7 − 65.9
4. 4.17 ÷ 3.05
5. 3675.42 − 1996.87
6. 14 ÷ 384.17
7. \$42.76 + \$5.48
8. \$377 ÷ 23
9. Add "two and three-quarters" to "five and one-half."
10. Find the total cost of 6 shirts costing \$9.75 each.
11. If hiking socks are on sale at 3 pairs for \$6.59, what will one pair cost? (Divide by 3 and round to the nearest cent.)
12. What is the cost of 135 sq ft of flooring at \$8.75 per sq ft?

COMBINED OPERATIONS

A calculator can only operate on two numbers at a time, but very often you need to use the result of one calculation in a second calculation or in a long string of calculations. Because the answer to any calculation remains in the Op. Reg., you can carry out a very complex chain of calculations without ever stopping the calculator to write down an intermediate step. For example, to add a list of numbers simply continue the addition step.

4.1 + 0.72 + 12.68 + 3.2 = ? **Estimate:** 4 + 1 + 12 + 3 = 20.

Keyboard	Display	
[4] [.] [1]	4.1	
[+]	4.1	
[.] [7] [2]	0.72	
[+]	4.82	⇐ The sum of the first two numbers
[1] [2] [.] [6] [8]	12.68	
[+]	17.5	⇐ The sum of the first three numbers
[3] [.] [2]	3.2	
[=]	20.7	⇐ The total sum of all four numbers

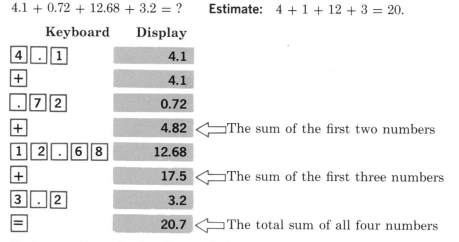

Each sum sits in the Op. Reg, and the next number entered is added to the Op. Reg.

Of course, we can combine addition and subtraction in the same sequence of operations. For example, 462 − 294 + 31 = ? **Estimate:** 460 − 300 + 30 = 190.

Keyboard	Display	
[4] [6] [2]	462	
[−]	462.	
[2] [9] [4]	294	
[+]	168.	⇐ The difference of the first two numbers
[3] [1]	31	
[=]	199.	⇐ The answer

Multiplication of a list of factors is similar to addition and quite simple. For example, $3.2 \times 4.1 \times 0.53 \times 2.6 = ?$ **Estimate:** $3 \times 4 \times 0.5 \times 3 = 18$.

Keyboard	Display	
[3] [.] [2]	3.2	
[×]	3.2	
[4] [.] [1]	4.1	
[×] [ ]	13.12	⇦ Product of the first two numbers
[.] [5] [3]	0.53	
[×]	6.9536	⇦ Product of the first three factors
[2] [.] [6]	2.6	
[=]	18.07936	⇦ Final product of all four factors

Each product sits in the Op. Reg. and the next number multiplies the number in the Op. Reg.

Another example: $2.1^3 = ?$ **Estimate:** $2^3 = 8$

Keyboard	Display
[2] [.] [1]	2.1
[×]	2.1
[2] [.] [1]	2.1
[×]	4.41
[2] [.] [1]	2.1
[=]	9.261

We can easily combine multiplication and division operations. For example, $2 \times 4 \div 5 = ?$

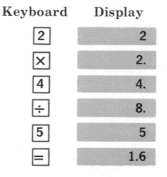

Keyboard	Display
[2]	2
[×]	2.
[4]	4.
[÷]	8.
[5]	5
[=]	1.6

The operations of multiplication and division can be combined with the operations of addition and subtraction using a calculator, but it may be necessary to write down the intermediate results. If your calculator has a memory, these intermediate results may be stored in the calculator and used in the rest of the calculation.

In general you must be very careful when combining unlike arithmetic operations. For example, punch this problem into your calculator:

[2] [+] [3] [×] [4] [=] ?

If you press the keys in the order given and your calculator displays the answer as 20, then your calculator is doing the operations sequentially, left to right. It has interpreted these instructions as

$(2 + 3) \times 4$ or 5×4

A different model of calculator may display the answer 14. This second kind of calculator interprets the instructions as

$2 + (3 \times 4)$ or $2 + 12$

The second calculator follows an order of operations where multiplications or divisions are done first, then additions and subtractions are done.

To be certain there is no misinterpretation, parentheses are used to show which calculation should be done first. For example,

$(3.1 \times 4.2) + 1.5 = ?$ **Estimate:** $(3 \times 4) + 1 = 12 + 1 = 13$

Keyboard	Display	
[3] [.] [1]	3.1	
[×]	3.1	
[4] [.] [2]	4.2	
[=]	13.02	⇦ Pressing the [=] key gives the product of the multiplication in the parentheses.
[+]	13.02	
[1] [.] [5]	1.5	
[=]	14.52	⇦ The final answer

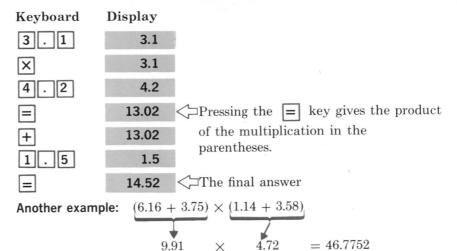

Another example: $\underbrace{(6.16 + 3.75)}_{9.91} \times \underbrace{(1.14 + 3.58)}_{4.72} = 46.7752$

In this problem we must write down or store in memory the first sum (9.91) while the second sum is being performed. Once both sums are available, multiply them to find the answer.

Another example: $\underbrace{(37.01 - 14.65)}_{22.36} \div \underbrace{(8.4 - 4.791)}_{3.609} = 6.195622$

A fraction bar is also used to group numbers in a complex calculation. For example, in the following calculation the fraction bar tells you that the upper number $(2 + 4)$ is to be divided by the lower number $(3 + 7)$.

$\frac{2 + 4}{3 + 7} = \frac{6}{10}$ Divide 6 by 10 to get $6 \div 10 = 0.6$

Another example: $\frac{3.15 + 1.77}{9.42 - 3.67} = \frac{4.92}{5.75}$ Divide: $4.92 \div 5.75 = 0.8556521$

Calculate and write down or store in memory the denominator; then calculate the numerator and divide it by the denominator.

Another example: $\frac{(3.18 \times 1.4) + 0.66}{14.4 - 12.007} = \frac{5.112}{2.393}$

Dividing $5.112 \div 2.393$ gives 2.1362306

Exercises A-4 Combined Operations

Perform the following calculations.

1. $16.4 + 97.1 + 112 + 4.67$
2. $72 - 117 - 6.5 + 68$
3. $3468 - 3061 + 1109 - 702$
4. $1.06 - 0.005 + 0.91 - 0.07$
5. 37¢ + \$1.14 − 79¢ − 43¢
6. $6.25 \times 7.04 \times 13.67 \times 2$
7. 3.14×1.17^2
8. 8.04^2
9. $13.19 \div 175.15 \div 2$
10. $\dfrac{81.06}{39.9} \times 1.05$
11. $\dfrac{23.82 \times 31.4}{8.6}$
12. $\dfrac{3.14 \times 1.25^2}{4}$
13. $\dfrac{(4.1 \times 3.2) - 1.8}{1.20}$
14. $\dfrac{(5.92 + 3.08) \times 3.1}{19.1}$
15. $\dfrac{3.5 \times 7.5}{6} + 8.25$
16. $14.9 - \dfrac{3.14 \times 2.8^2}{4}$
17. An employee in the Acme Store is paid \$3.91 per hour and works 32 hours per week. How much does she earn in $5\frac{1}{2}$ weeks?
18. Use the formula $C = \dfrac{5 \times (F - 32)}{9}$ to find C when F is equal to 65. Round to the nearest tenth.

DIVISION WITH A REMAINDER

Division is easy with a calculator. The answer, or quotient, is given as a decimal number. The answer will either be exact or will be given to the limit of the 8-digit display. For example,

$17 \div 7 = 2.4285714$ to eight digits.

In many practical situations it is useful to write the answer to a division as a whole number quotient plus a remainder. For example,

$17 \div 7 = 2$ with remainder 3 In longhand division:

```
      2  <= Quotient
    7)17
      14
      --
       3 <= Remainder
```

To do division with a remainder on your calculator follow this sequence of operations:

Keyboard	Display	
[1] [7]	17	
[÷]	17.	
[7]	7	
[=]	2.4285714	⇐ The answer to 8 digits
[−]	2.4285714	
[2]	2	⇐ The quotient is 2.
[=]	0.4285714	
[×]	0.4285714	
[7]	7	
[=]	2.9999999	⇐ The remainder. Round it to 3.

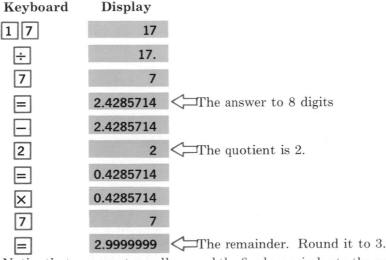

Notice that you must usually round the final remainder to the nearest whole number. In some calculators the rounding is performed automatically and the answer displayed is exact.

Division with a remainder is useful in solving many problems involving units. For example, write the length measurement 191 inches in feet and inches.

Keyboard	Display	
[1] [9] [1]	191	
[÷]	191.	
[1] [2]	12	
[=]	15.916666	
[−]	15.916666	
[1] [5]	15	⇦ The quotient is 15 ft.
[=]	0.9166666	
[×]	0.9166666	
[1] [2]	12	
[=]	10.999999	⇦ The remainder is 11 inches.

191 inches = 15 ft 11 in.

The following set of problems will provide you with a bit of practice in this kind of division.

Exercises A-5 Division with a Remainder

Find the quotient and remainder for each division.

1. $41 \div 9$
2. $318 \div 13$
3. $999 \div 123$
4. $1001 \div 333$
5. $5046 \div 1070$
6. $23781 \div 5107$
7. $1765 \div 60$
8. Convert 112 hours to days and hours.
9. Convert 370 inches to feet and inches.
10. Convert 212 minutes to hours and minutes.

SQUARE ROOTS

The square of a number is easy to calculate. For example, $2^2 = 2 \times 2$ or 4. The square root of a number is usually much more difficult to find. For example,

$\sqrt{2} = 1.4142135$ to eight digits.

(Check it: $(1.4142135)^2 = 1.9999998$ or approximately 2.)

The easiest way to find a square root is to look it up in a printed table of square roots or to use the square root function key [√] if one appears on your calculator keyboard. For example, to find $\sqrt{3}$

Keyboard	Display
[3]	3
[√]	1.7320508

If your calculator does not have an automatic square root function key, the square root of any number can be found through an approximation or "trial and error" method. Follow these steps:

First, find an approximate value for the square root.

Example: $\sqrt{20} \cong 4$ 4 is a reasonable first guess to $\sqrt{20}$ since 4×4 is nearly equal to 20.

Second, find a closer approximation by calculating the number

$$\frac{4^2 + 20}{2 \times 4}$$

Example: For $\sqrt{20}$ the first approximation is 4. The second approximation is found from the following sequence of steps:

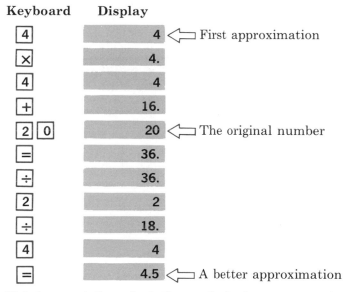

Keyboard	Display	
4	4	⇦ First approximation
×	4.	
4	4	
+	16.	
2 0	20	⇦ The original number
=	36.	
÷	36.	
2	2	
÷	18.	
4	4	
=	4.5	⇦ A better approximation

Third, repeat the calculation to find a better approximation.

Keyboard	Display	
4 . 5	4.5	⇦ The second approximation
×	4.5	
4 . 5	4.5	
+	20.25	
2 0	20	⇦ The original number
=	40.25	
÷	40.25	
2	2	
÷	20.125	
4 . 5	4.5	
=	4.4722222	⇦ An even better approximation to $\sqrt{20}$

This process can be repeated as many times as you wish, with each calculation giving a better approximation to $\sqrt{20}$. The actual value of $\sqrt{20}$ to 8 digits is 4.4721359.

In theoretical terms, if you want to find $\sqrt{N}$ and if the number A is a first approximation to $\sqrt{N}$, then an even better approximation to $\sqrt{N}$ can be found by calculating the quantity

$$\frac{A^2 + N}{2A}$$

The same sort of approximation method can be used to find cube roots. If you wish to find $\sqrt[3]{N}$ and if the number A is a first approximation to $\sqrt[3]{N}$, then an even better approximation is

$$\frac{2A^3 + N}{3A^2}$$

Exercises A-6 Square Roots

Use the approximation method or the $\boxed{\sqrt{\ }}$ key to find the following square roots. Round to two decimal places.

1. $\sqrt{5}$	2. $\sqrt{10}$	3. $\sqrt{51}$
4. $\sqrt{161}$	5. $\sqrt{0.5}$	6. $\sqrt{1.1}$

7. If a square lot has an area of 12,000 sq ft, what is the length of one side? (Find $\sqrt{12000}$. **Hint:** Try $A = 100$ as a first guess.)

8. One acre is an area of 43,560 sq ft. What is the length of side of a square lot whose area is 1 acre? (Find $\sqrt{43560}$. **Hint:** Try $A = 200$ as a first guess.)

9. A square table top has an area of 460 sq inches. What are its dimensions? (Find $\sqrt{460}$. **Hint:** Try $A = 20$ as a first approximation.)

10. Suppose you have a square lct of area 2.8 acres fronting on a highway and you receive the following offers to purchase it:

    Offer A. $60 per frontage foot
    Offer B: $7500 per acre
    Offer C: $20,000 for the lot

    Which is the best offer? (**Hint:** Use 1 acre = 43,560 sq ft.)

ANSWERS TO EXERCISES

A-1	1.	327.	2.	46002.	3.	0.0120		
	4.	137620.4	5.	0.7	6.	6.5		
	7.	3.37	8.	4.25	9.	14.75		
	10.	0.16						
A-2	1.	$4.20	2.	$.40 or 40¢	3.	$.02 or 2¢	4.	$11.74
	5.	$431.75	6.	$3,786,506.10	7.	$0.39 or 39¢	8.	$6.01
	9.	4.1	10.	189.0	11.	3278.6	12.	4.3
	13.	137.52	14.	6.85	15.	4.01		
A-3	1.	3976	2.	97.28	3.	28.8	4.	1.3672131
	5.	1678.55	6.	0.0364422	7.	$48.24	8.	$16.391304
	9.	8.25	10.	$58.50	11.	$2.20	12.	$1181.25
A-4	1.	230.17	2.	16.5	3.	814	4.	1.895
	5.	$0.29	6.	1202.96	7.	4.298346	8.	64.6416
	9.	0.0376534	10.	2.1331578	11.	86.970697	12.	1.2265625
	13.	9.4333333	14.	1.4607329	15.	12.625	16.	8.7456
	17.	$688.16	18.	18.333333				
A-5	1.	4 rem. 5	2.	24 rem. 6	3.	8 rem. 15	4.	3 rem. 2
	5.	4 rem. 766	6.	4 rem. 3353	7.	29 rem. 25	8.	4 days 16 hr
	9.	30 ft 10 in.	10.	3 hr 32 min				
A-6	1.	2.24	2.	3.16	3.	7.14	4.	12.69
	5.	0.71	6.	1.05	7.	109.54 ft	8.	208.71 ft
	9.	21.45 in.						
	10.	Offer A is $20,954, Offer B is $21,000, Offer C is $20,000. Offer B is best.						

APPENDIX B
Table of Square Roots

Number	Square root	Number	Square root	Number	Square root	Number	Square root
1	1.0000	51	7.1414	101	10.0499	151	12.2882
2	1.4142	52	7.2111	102	10.0995	152	12.3288
3	1.7321	53	7.2801	103	10.1489	153	12.3693
4	2.0000	54	7.3485	104	10.1980	154	12.4097
5	2.2361	55	7.4162	105	10.2470	155	12.4499
6	2.4495	56	7.4833	106	10.2956	156	12.4900
7	2.6458	57	7.5498	107	10.3441	157	12.5300
8	2.8284	58	7.6158	108	10.3923	159	12.5698
9	3.0000	59	7.6811	109	10.4403	159	12.6095
10	3.1623	60	7.7460	110	10.4881	160	12.6491
11	3.3166	61	7.8102	111	10.5357	161	12.6886
12	3.4641	62	7.8740	112	10.5830	162	12.7279
13	3.6056	63	7.9373	113	10.6301	163	12.7671
14	3.7417	64	8.0000	114	10.6771	164	12.8062
15	3.8730	65	8.0623	115	10.7238	165	12.8452
16	4.0000	66	8.1240	116	10.7703	166	12.8841
17	4.1231	67	8.1854	117	10.8167	167	12.9228
18	4.2426	68	8.2462	118	10.8628	168	12.9615
19	4.3589	69	8.3066	119	10.9087	169	13.0000
20	4.4721	70	8.3666	120	10.9545	170	13.0384
21	4.5826	71	8.4261	121	11.0000	171	13.0767
22	4.6904	72	8.4853	122	11.0454	172	13.1149
23	4.7958	73	8.5440	123	11.0905	173	13.1529
24	4.8990	74	8.6023	124	11.1355	174	13.1909
25	5.0000	75	8.6603	125	11.1803	175	13.2288
26	5.0990	76	8.7178	126	11.2250	176	13.2665
27	5.1962	77	8.7750	127	11.2694	177	13.3041
28	5.2915	78	8.8318	128	11.3137	178	13.3417
29	5.3852	79	8.8882	129	11.3578	179	13.3791
30	5.4772	80	8.9443	130	11.4018	180	13.4164
31	5.5678	81	9.0000	131	11.4455	181	13.4536
32	5.6569	82	9.0554	132	11.4891	182	13.4907
33	5.7446	83	9.1104	133	11.5326	183	13.5277
34	5.8310	84	9.1652	134	11.5758	184	13.5647
35	5.9161	85	9.2195	135	11.6190	185	13.6015
36	6.0000	86	9.2736	136	11.6619	186	13.6382
37	6.0828	87	9.3274	137	11.7047	187	13.6748
38	6.1644	88	9.3808	138	11.7473	188	13.7113
39	6.2450	89	9.4340	139	11.7898	189	13.7477
40	6.3246	90	9.4868	140	11.8322	190	13.7840
41	6.4031	91	9.5394	141	11.8743	191	13.8203
42	6.4807	92	9.5917	142	11.9164	192	13.8564
43	6.5574	93	9.6437	143	11.9583	193	13.8924
44	6.6332	94	9.6954	144	12.0000	194	13.9284
45	6.7082	95	9.7468	145	12.0416	195	13.9642
46	6.7823	96	9.7980	146	12.0830	196	14.0000
47	6.8557	97	9.8489	147	12.1244	197	14.0357
48	6.9282	98	9.8995	148	12.1655	198	14.0712
49	7.0000	99	9.9499	149	12.2066	199	14.1067
50	7.0711	100	10.0000	150	12.2474	200	14.1421

APPENDIX C

Table of Trigonometric Functions

Angle	Sine	Cosine	Tangent	Angle	Sine	Cosine	Tangent
0°	0.000	1.000	0.000	46°	.719	.695	1.036
1°	.018	1.000	.018	47°	.731	.682	1.072
2°	.035	.999	.035	48°	.743	.669	1.111
3°	.052	.999	.052	49°	.755	.656	1.150
4°	.070	.998	.070	50°	.766	.643	1.192
5°	.087	.996	.088	51°	.777	.629	1.235
6°	.105	.995	.105	52°	.788	.616	1.280
7°	.122	.993	.123	53°	.799	.602	1.327
8°	.139	.990	.141	54°	.809	.588	1.376
9°	.156	.988	.158	55°	.819	.574	1.428
10°	.174	.985	.176	56°	.829	.559	1.483
11°	.191	.982	.194	57°	.839	.545	1.540
12°	.208	.978	.213	58°	.848	.530	1.600
13°	.225	.974	.231	59°	.857	.515	1.664
14°	.242	.970	.249	60°	.866	.500	1.732
15°	.259	.966	.268				
16°	.276	.961	.287	61°	.875	.485	1.804
17°	.292	.956	.306	62°	.883	.470	1.881
18°	.309	.951	.325	63°	.891	.454	1.963
19°	.326	.946	.344	64°	.899	.438	2.050
20°	.342	.940	.364	65°	.906	.423	2.145
21°	.358	.934	.384	66°	.914	.407	2.246
22°	.375	.927	.404	67°	.921	.391	2.356
23°	.391	.921	.425	68°	.927	.375	2.475
24°	.407	.914	.445	69°	.934	.358	2.605
25°	.423	.906	.466	70°	.940	.342	2.747
26°	.438	.899	.488	71°	.946	.326	2.904
27°	.454	.891	.510	72°	.951	.309	3.078
28°	.469	.883	.532	73°	.956	.292	3.271
29°	.485	.875	.554	74°	.961	.276	3.487
30°	.500	.866	.577	75°	.966	.259	3.732
31°	.515	.857	.601	76°	.970	.242	4.011
32°	.530	.848	.625	77°	.974	.225	4.331
33°	.545	.839	.649	78°	.978	.208	4.705
34°	.559	.829	.675	79°	.982	.191	5.145
35°	.574	.819	.700	80°	.985	.174	5.671
36°	.588	.809	.727	81°	.988	.156	6.314
37°	.602	.799	.754	82°	.990	.139	7.115
38°	.616	.788	.731	83°	.993	.122	8.144
39°	.629	.777	.810	84°	.995	.105	9.514
40°	.643	.766	.839	85°	.996	.087	11.43
41°	.656	.755	.869	86°	.998	.070	14.30
42°	.669	.743	.900	87°	.999	.052	19.08
43°	.682	.731	.933	88°	.999	.035	28.64
44°	.695	.719	.966	89°	1.000	.018	57.29
45°	.707	.707	1.000	90°	1.000	.000	∞

Answers

Chapter 1

Exercises 1-1, page 6

A.

10	11	11	12	16	7	9	15	12	13
10	13	15	8	14	9	13	18	9	11
16	8	10	17	9	14	12	7	10	11
13	13	12	16	9	11	10	10	14	17
11	14	11	10	11	14	13	12	13	15

B.

11	12	14	17	18	21	11	20	18	15
15	14	15	18	22	17	14	14	18	16
12	19	8	10	22	12	19	15	14	16

Exercises 1-2, page 11

A.

1. 70	2. 65	3. 80	4. 103	5. 123
6. 124	7. 132	8. 136	9. 393	10. 1003
11. 1390	12. 831	13. 1009	14. 806	15. 861
16. 5525	17. 9461	18. 9302	19. 11,428	20. 15,715
21. 25,717	22. 47,111	23. 11,071	24. 14,711	25. 175,728

B.

1. 1042	2. 5211	3. 2442	4. 6441
5. 7083	6. 16,275	7. 6352	8. 7655
9. 6514	10. 9851	11. 64	12. 141
13. 55	14. 148	15. 357	

C. 1. 4861 ft 2. 11,365 fbm 3. 1636 screws 4. $1148
5. 1129 minutes 6. 6670 shingles 7. (a) 3607 watts (b) 1997 watts (c) 850 watts

D. 1. (a) *Daily Totals* (b) *Machine Totals* (c) Yes

(a) Daily Totals	(b) Machine Totals
1. 1919	A. 4074
2. 2125	B. 4462
3. 1958	C. 2185
4. 1594	D. 5665
5. 2192	E. 2883
6. 1953	
7. 1974	
8. 1988	
9. 1667	
10. 1899	

2. (a) $307,225 (b) $732,813 (c) $2,298,502 (d) $7156
3. (a) $7788 (b) $7735 (c) SELL
4. (a) #12 BHD: 11,453; #Tx: 258; 410 AAC: 12,715; 110 ACSR: 8792; 6 B: 7425
   (b) A3: 3530; A4: 8412; B1: 4294; B5: 5482; B6: 5073; C4: 6073; C5: 7779

Exercises 1-3, page 18

A.

1. 6	2. 7	3. 2	4. 8	5. 4
6. 9	7. 3	8. 0	9. 3	10. 3
11. 8	12. 8	13. 9	14. 9	15. 9
16. 9	17. 3	18. 6	19. 8	20. 5
21. 7	22. 18	23. 7	24. 8	25. 0

B. 1. 13 2. 29 3. 12 4. 19 5. 15
6. 36 7. 22 8. 25 9. 38 10. 189
11. 85 12. 281 13. 154 14. 273 15. 715
16. 574 17. 29 18. 2809 19. 5698 20. 12,518
21. 56,042 22. 4741 23. 9614 24. 47,593 25. 22,422

C. 1. $247 2. 2353 sq ft 3. 1758 ft 4. 29 cm 5. $330,535
6. $3144 7. 3 drums, by 44 liters 8. (a) 9″ (b) 4″
(c) 4″ (d) 5″

D. 1.

Truck No.	1	2	3	4	5	6	7	8	9	10
Mileage	1675	1167	1737	1316	1360	299	1099	135	1461	2081

Total Mileage 12,330

2. First sum: 1,083,676,269
Second sum: 1,083,676,269
They are the same.
3. $7286
4. $16,805
5. (a) $2065
(b)

Deposits	Withdrawals	Balance
		$6375
	$ 379	5996
$1683		7679
474		8153
487		8640
	2373	6267
	1990	4277
	308	3969
	1090	2879
	814	2065

Exercises 1-4, page 23

A.

12	32	63	36	12	18	0	24	14	8
48	16	45	30	10	9	72	35	18	4
28	15	36	49	8	40	42	54	64	24
20	0	25	27	81	6	1	48	16	63

B.

16	30	9	35	18	20	28	48	12	63
32	0	18	24	9	25	24	45	10	72
15	49	40	54	36	8	42	64	0	4
25	27	7	56	36	12	81	0	2	56

Exercises 1-5, page 28

A. 1. 42 2. 56 3. 48 4. 72 5. 63
6. 87 7. 576 8. 423 9. 320 10. 156
11. 290 12. 564 13. 153 14. 282 15. 308
16. 720 17. 1728 18. 5040 19. 7138 20. 1938
21. 1650 22. 4484 23. 928 24. 3822 25. 8930

B. 1. 37,515 2. 375,750 3. 297,591 4. 38,023

5. 378,012 6. 41,064 7. 30,780 8. 1,368,810
9. 397,584 10. 60,241 11. 7281 12. 4263
13. 25,000 14. 325,200 15. 3,532,536

C. 1. $760 2. 2100 ft 3. 1300 ft 4. $14,184
5. 8000 envelopes 6. 225″ 7. 2430 8. 132″

D. 1. $183.45 left 2. (a) nets $36,500 (b) nets $36,400
(c) nets $31,200 (d) nets $671,088.63
Therefore (d) gives you the most money.

3. (a) 111,111,111; 222,222,222; 333,333,333
(b) 111,111; 222,222; 333,333
(c) 1; 121; 12,321; 1,234,321; 123,454,321
(d) 42; 4422; 444,222; 44,442,222; 4,444,422,222

4. 34,600 lb

5. Alpha Beta Gamma Delta Tau
$3510 $5695 $4640 $9065 $16,020

Exercises 1-6, page 35

A. 1. 9 2. 11 rem. 4 3. No solution 4. 7 rem. 2
5. 10 rem. 1 6. 1 7. 8 8. 4
9. 6 10. 35 11. 23 rem. 6 12. 57
13. 51 rem. 4 14. 1103 rem. 1 15. 21 16. 52
17. 37 18. 50 rem. 1 19. 23 20. 20 rem. 2
21. 39 22. 25 23. 9 rem. 1 24. 53
25. 22

B. 1. 120 2. 9 rem. 6 3. 56 rem. 8 4. 95 rem. 6
5. 96 6. 142 rem. 6 7. 222 rem. 2 8. 32
9. 305 rem. 5 10. 84 rem. 41 11. 119 12. 3001
13. 501 14. 8001 rem. 3 15. 604 16. 20,720
17. 200 18. 50 rem. 4 19. 108 rem. 4 20. 2009 rem. 2
21. 600 22. 61 23. 102 rem. 98 24. 81
25. 100 rem. 11

C. 1. 27″ 2. 6 3. 13 hr 4. $W = 12''$, $H = 8''$
5. 28 6. 50 7. 7″ 8. 373 HP
9. $11 10. 11 hr

D. 1. (a) 21,021,731 (b) 449 (c) 93 (d) 27,270

2. (a)

Truck No.	1	2	3	4	5	6	7	8	9	10
Miles per Day	76	53	79	60	62	14	50	6	66	95

(b) $935

3. 19.148255 or 20 rivets to be sure
4. 123,091 lb 5. 51 hours 40 minutes
6. (a) 16,094 (b) 201,023 (c) 2283 (d) 357

Problem Set 1, page 39

A. 1. 93 2. 83 3. 528 4. 860
5. 934 6. 2980 7. 15 8. 26
9. 649 10. 196 11. 195 12. 2615
13. 1407 14. 3690 15. 13,041 16. 290,764
17. 230,384 18. 1,575,056 19. 57 20. 62
21. 8 rem. 13 22. 43 23. 18 24. 69
25. 3 rem. 508

B. 1. 43 ft 2. 64 rods 3. 1892 sq ft 4. 445 lb
5. 6 6. 24 hr 7. 207 lb 8. $338
9. $205 10. 252 ft 11. 563,988 mi 12. $10,535
13. 650 gpm 14. (a) 513 (b) 5068 ft
15. 2839 lb 16. 263 17. 24 hr 18. $10.16
19. 87,780 cu in. 20. 646 21. 193 rpm

Chapter 2

Exercises 2-1, page 52

A. 1. $\frac{7}{3}$ 2. $\frac{15}{2}$ 3. $\frac{67}{8}$ 4. $\frac{17}{16}$ 5. $\frac{23}{8}$
6. $\frac{2}{1}$ 7. $\frac{8}{3}$ 8. $\frac{259}{64}$ 9. $\frac{29}{6}$ 10. $\frac{29}{16}$

B. 1. $8\frac{1}{2}$ 2. $1\frac{3}{5}$ 3. $1\frac{3}{8}$ 4. $2\frac{8}{16}$ or $2\frac{1}{2}$ 5. $1\frac{1}{2}$
6. $3\frac{2}{3}$ 7. $16\frac{4}{6}$ or $16\frac{2}{3}$ 8. $1\frac{1}{3}$ 9. $2\frac{16}{32}$ or $2\frac{1}{2}$ 10. $2\frac{1}{2}$

C. 1. $\frac{3}{4}$ 2. $\frac{2}{3}$ 3. $\frac{3}{8}$ 4. $\frac{9}{2}$ 5. $\frac{2}{5}$
6. $\frac{7}{6}$ 7. $\frac{4}{5}$ 8. $\frac{5}{2}$ 9. $4\frac{1}{4}$ 10. $\frac{17}{16}$

D. 1. 14 2. 12 3. 8 4. 24 5. 20
6. 92 7. 36 8. 34 9. 5 10. 24

E. 1. $\frac{3}{5}$ 2. $\frac{13}{8}$ 3. $1\frac{1}{2}$ 4. $\frac{13}{16}$ 5. $\frac{7}{8}$
6. $2\frac{1}{2}$ 7. $\frac{6}{4}$ 8. $\frac{25}{60}$ 9. $\frac{13}{5}$ 10. $2\frac{7}{4}$

F. 1. $15\frac{3}{4}''$ 2. $\frac{3}{4}$ 3. $\frac{19}{6}$; $\frac{25}{8}$ 4. $\frac{13}{64}''$ fastener 5. No.
6. $2\frac{1}{2}''$ 7. $\frac{3}{5}$ 8. $1\frac{1}{2}$

Exercises 2-2, page 57

A. 1. $\frac{1}{8}$ 2. $\frac{4}{15}$ 3. $\frac{2}{15}$ 4. 3 5. $2\frac{2}{3}$
6. $\frac{11}{45}$ 7. $1\frac{1}{9}$ 8. $\frac{13}{16}$ 9. $2\frac{1}{2}$ 10. 14
11. 3 12. 8 13. $3\frac{1}{4}$ 14. $1\frac{1}{21}$ 15. 69
16. $35\frac{3}{4}$ 17. 74 18. $9\frac{7}{8}$ 19. $10\frac{3}{8}$ 20. $21\frac{1}{3}$
21. $\frac{1}{8}$ 22. $\frac{3}{10}$ 23. $\frac{1}{15}$ 24. $1\frac{1}{3}$ 25. 2

B. 1. $\frac{1}{6}$ 2. $\frac{3}{32}$ 3. $\frac{1}{2}$ 4. $\frac{7}{16}$ 5. $\frac{3}{4}$
6. $\frac{15}{16}$ 7. $1\frac{5}{16}$ 8. $2\frac{7}{9}$ 9. 1 10. $\frac{1}{2}$

C. 1. $137\frac{3}{4}''$ 2. $99\frac{3}{4}''$ 3. $11\frac{2}{3}'$
4. $4\frac{5}{16}''$ 5. $14'1\frac{1}{2}''$ 6. $110\frac{1}{4}''$
7. $319\frac{1}{5}$ mi 8. $36\frac{3}{4}''$ 9. $356\frac{1}{2}$ lb
10. 118′ 11. $\frac{3}{4}''$ 12. 210″ or 17′6″
13. $9\frac{3}{4}''$ 14. $431\frac{41}{64}$ cu in. 15. $348\frac{3}{4}$ min
16. 1001 cu in. 17. $10\frac{2}{3}$ hr 18. 126 in.
19. $5\frac{2}{5}''$

Exercises 2-3, page 63

A. 1. $1\frac{2}{3}$ 2. 9 3. $\frac{5}{16}$ 4. 32 5. $\frac{1}{2}$
6. 1 7. $\frac{1}{4}$ 8. $2\frac{2}{5}$ 9. 9 10. 4
11. $1\frac{1}{3}$ 12. $1\frac{3}{4}$ 13. $1\frac{1}{5}$ 14. $8\frac{1}{3}$ 15. 16
16. $\frac{1}{9}$ 17. 18 18. $\frac{6}{7}$ 19. $7\frac{1}{2}$ 20. $\frac{3}{5}$

B. 1. 8 ft 2. $4\frac{1}{2}''$ 3. 48 4. 12 5. 84
6. $40\frac{1}{2}$ ft 7. 18 8. 8 9. 210 10. 29′ by 34′

Exercises 2-4, page 72

A. 1. $\frac{1}{4}$ 2. $1\frac{1}{3}$ 3. $\frac{3}{4}$ 4. $\frac{5}{6}$ 5. $\frac{1}{2}$

6. $\frac{5}{8}$ 7. $\frac{2}{5}$ 8. $\frac{1}{4}$ 9. $\frac{15}{16}$ 10. $1\frac{3}{8}$

11. $\frac{3}{4}$ 12. $\frac{13}{16}$ 13. $\frac{11}{24}$ 14. $\frac{29}{48}$ 15. $\frac{1}{8}$

16. $\frac{9}{16}$ 17. $\frac{7}{16}$ 18. $\frac{13}{32}$ 19. $\frac{5}{8}$ 20. $\frac{11}{48}$

21. $1\frac{3}{4}$ 22. $2\frac{11}{16}$ 23. $4\frac{1}{8}$ 24. $3\frac{5}{16}$ 25. $\frac{3}{4}$

B. 1. $5\frac{1}{8}$ 2. $1\frac{13}{16}$ 3. $2\frac{13}{16}$ 4. $1\frac{17}{60}$

5. $\frac{7}{8}$ 6. $20\frac{3}{8}$ 7. $2\frac{1}{8}$ 8. $3\frac{3}{8}$

9. $2\frac{11}{16}$ 10. $5\frac{5}{12}$ 11. $3\frac{9}{10}$ 12. $\frac{15}{56}$

C. 1. $10\frac{13}{20}$ min 2. $\frac{13}{16}''$ 3. $7\frac{31}{32}''$ 4. $23\frac{5}{16}''$

5. $1\frac{9}{16}''$ 6. $\frac{15}{16}''$ 7. $\frac{3}{8}''$ 8. $2\frac{1}{6}$ ft

9. $1\frac{1}{8}''$ 10. $\frac{843}{1000}$ 11. $\frac{3}{32}''$ 12. $1\frac{5}{8}''$

13. $2\frac{13}{16}''$ 14. No. 15. $25\frac{1}{4}$ c.i. 16. $23\frac{3}{4}''$

Problem Set 2, page 75

A. 1. $\frac{9}{8}$ 2. $\frac{21}{5}$ 3. $\frac{5}{3}$ 4. $\frac{35}{16}$ 5. $\frac{99}{32}$

6. $\frac{33}{16}$ 7. $\frac{13}{8}$ 8. $\frac{55}{16}$ 9. $2\frac{8}{16}$ or $2\frac{1}{2}$ 10. $9\frac{1}{2}$

11. $8\frac{1}{3}$ 12. $1\frac{1}{8}$ 13. $1\frac{18}{32}$ or $1\frac{9}{16}$ 14. $1\frac{5}{16}$ 15. $12\frac{4}{8}$ or $12\frac{1}{2}$

16. $2\frac{5}{15}$ or $2\frac{1}{3}$ 17. $\frac{3}{16}$ 18. $\frac{1}{4}$ 19. $\frac{3}{8}$ 20. $\frac{3}{4}$

21. $\frac{1}{6}$ 22. $1\frac{4}{7}$ 23. $1\frac{4}{5}$ 24. $3\frac{2}{5}$ 25. 9

26. 28 27. 44 28. 44 29. 68 30. 18

31. 15 32. 26 33. $\frac{7}{16}$ 34. $\frac{2}{3}$ 35. $\frac{7}{8}$

36. $1\frac{1}{4}$ 37. $\frac{3}{5}$ 38. $\frac{2}{10}$ 39. $\frac{7}{4}$ 40. $\frac{1}{9}$

B. 1. $\frac{3}{32}$ 2. $\frac{1}{2}$ 3. $\frac{7}{12}$ 4. $\frac{1}{256}$ 5. $1\frac{1}{4}$

6. $\frac{49}{80}$ 7. $\frac{5}{64}$ 8. $1\frac{1}{2}$ 9. $7\frac{1}{2}$ 10. $\frac{2}{3}$

11. 27 12. 34 13. 2 14. $\frac{4}{5}$ 15. 32

16. $10\frac{2}{3}$ 17. $\frac{1}{6}$ 18. $\frac{3}{4}$ 19. $\frac{7}{10}$ 20. $\frac{5}{6}$

C. 1. $1\frac{1}{4}$ 2. $1\frac{1}{4}$ 3. $\frac{7}{32}$ 4. $1\frac{5}{8}$ 5. $\frac{3}{8}$

6. $\frac{3}{8}$ 7. $\frac{7}{16}$ 8. $1\frac{13}{32}$ 9. $3\frac{3}{8}$ 10. $2\frac{7}{16}$

11. $4\frac{1}{2}$ 12. $1\frac{1}{8}$ 13. $1\frac{19}{24}$ 14. $1\frac{5}{12}$ 15. $1\frac{1}{30}$

16. $3\frac{19}{20}$ 17. $1\frac{1}{6}$ 18. $1\frac{1}{3}$ 19. $\frac{2}{5}$ 20. $3\frac{1}{3}$

D. 1. $37\frac{1}{8}''$ 2. 22 3. $25\frac{5}{8}''$; $23\frac{1}{8}''$ 4. $7\frac{9}{16}''$

5. $\frac{25}{32}''$ 6. Yes. 7. $4\frac{3}{5}$ cu ft 8. $15\frac{15}{16}''$

9. $92\frac{13}{16}''$ 10. $13'$ 11. $\frac{15}{16}''$

12. *A:* $2\frac{11}{16}''$ *B:* $2\frac{5}{32}''$ *C:* $6\frac{9}{32}''$ *D:* $8\frac{7}{16}''$

13. $4\frac{3}{32}''$ 14. $\mathbf{21\frac{3}{8}}$ 15. $244\frac{1}{2}''$ 16. $859\frac{3}{8}''$ or $71'7\frac{3}{8}''$

Chapter 3 *Exercises 3-1, page 87*

A. 1. 21.01 2. 78.17 3. \$15.02 4. \$151.11

5. 1.617 6. 5.916 7. 828.6 8. 238.16

9. 63.7305 10. 462.04 11. 6.97 12. 1.04

13. \$15.36 14. \$6.52 15. 42.33 16. 36.18

17. \$22.02 18. \$24.39 19. 113.96 20. 13.22

21. 45.195 22. 245.11 23. \$27.51 24. 151.402

25. 95.888

B. 1. 0.473 in. 2. 11.85 lb 3. (a) 0.013″ (b) smaller; 0.021″ (c) #14
4. $1426.75 5. *A*: 2.246″ *B*: 0.455″ *C*: 4.21″ 6. 2.7′; 8.275″
7. 2.267″

C. 1. $308.24 2. $5715.85
3. (a) 0.7399 (b) 4240.775 (c) 510.436 (d) 7.4262

Exercises 3-2, page 99

A.

1.	0.00001	2.	21.5	3.	4	4.	0.09
5.	0.84	6.	0.00006	7.	0.00003	8.	2.18225
9.	0.07	10.	0.03	11.	2.16	12.	3.6225
13.	6.03	14.	120	15.	20	16.	130
17.	0.045	18.	126	19.	60	20.	10,000
21.	400	22.	0.037	23.	6.6	24.	3256.25
25.	605						

B.

1.	3.33	2.	0.83	3.	10.53	4.	37.04
5.	0.12	6.	2.62	7.	33.86	8.	4.96
9.	33.3	10.	0.2	11.	0.3	12.	11.1
13.	0.2	14.	0.3	15.	0.143	16.	0.224
17.	65	18.	13.268	19.	2.999	20.	1109.001

C. 1. $126 2. $26.35 3. 18.8 lb 4. 7.8 hr

5.

	W	*C*
A	16.65 lb	$16.32
B	23.46	20.88
C	8.55	8.98
D	2.775	5.97

$T = \$52.15$

6. 40.035 lb 7. 48 lb 8. 8.36 lb

D. 1. 8.0000008 2. (a) 0.01234567 (b) 0.00112233 (c) 0.000111222
3. 4.2435″ 4. 0.0659 mm 5. $580

Exercises 3-3, page 106

A.

1.	0.25	2.	0.67	3.	0.75	4.	0.4	5.	0.8
6.	0.83	7.	0.29	8.	0.57	9.	0.86	10.	0.38
11.	0.75	12.	0.1	13.	0.3	14.	0.17	15.	0.42
16.	0.19	17.	0.38	18.	0.56	19.	0.81	20.	0.15

B.

1.	4.385	2.	1.77	3.	1.5	4.	0.7681
5.	7.88	6.	2.33	7.	1.43	8.	2.98

C. 1. 1.375 g: 5.2
2. (a) 420 sq ft (b) 3712.5 sq ft of 4″ and 4851 sq ft of 6″
(c) $1789.02
3. 2.3 squares 4. 2″
5. (a) $4.57 (b) $26.60 (c) $23.24 (d) $22.84, Total: $77.25
6. 523 full lots 7. 0.396″

D. 1.

Gauge No.	*Thickness (in.)*	*Gauge No.*	*Thickness (in.)*
7–0	0.5	14	0.078
6–0	0.469	15	0.070
5–0	0.438	16	0.063
4–0	0.406	17	0.056
3–0	0.375	18	0.05
2–0	0.344	19	0.044
0	0.313	20	0.038

1	0.281	21	0.034
2	0.266	22	0.031
3	0.25	23	0.028
4	0.234	24	0.025
5	0.219	25	0.022
6	0.203	26	0.019
7	0.188	27	0.017
8	0.172	28	0.016
9	0.156	29	0.014
10	0.141	30	0.013
11	0.125	31	0.011
12	0.109	32	0.010
13	0.094		

2. \$768.30; \$1871.54; \$2731.68; \$1866.24; Total: \$7237.76 3. \$295.59

Box page 115

1. 6.43 2. 1.20 3. 10.66 4. 0.27 5. 20.62
6. 4.48 7. 2.84 8. 0.08 9. 25.51 10. 14.49

Exercises 3-4, page *115*

A. 1. 16 2. 9 3. 64 4. 125 5. 1000
6. 49 7. 256 8. 36 9. 512 10. 81
11. 625 12. 100,000 13. 8 14. 243 15. 729
16. 1 17. 5 18. 1 19. 32 20. 64
21. 196 22. 441 23. 3375 24. 256 25. 108
26. 576 27. 1125 28. 2744 29. 2700 30. 9216

B. 1. 9 2. 12 3. 6 4. 4 5. 5
6. 3 7. 16 8. 20 9. 15 10. 7
11. 18 12. 11 13. 2.12 14. 22.36 15. 3.52
16. 26.46 17. 14.49 18. 17.92 19. 28.46 20. 9.62
21. 31.62 22. 44.72 23. 158.11 24. 50 25. 12.25
26. 17.32 27. 54.77 28. 173.21 29. 1.12 30. 1.00

C. 1. 13.6′ 2. 35.6′ 3. 13′ × 13′ 4. 15.9 ft
5. 111.2″ 6. 9′ 7. 2.5″ 8. 13′
9. 294 yd

Exercises 3-5, page 122

A. 1. −2 2. −4 3. −19 4. −20 5. −16
6. −2.1 7. 2.6 8. 2.1 9. 0.19 10. −1.5
11. −21 12. 45 13. −6 14. −56 15. −1
16. −1.55 17. −8.75 18. 3.6 19. 2.25 20. −6
21. −4 22. 1 23. −7 24. 90 25. 120

B. 1. 14,698 ft 2. 263° 3. 6 yards; 0.9 yard per carry
4. 39° 5. +\$22 6. 4423 ft 7. \$3113

Problem Set 3, page 123

A. 1. 23.19 2. 174.96 3. \$19.29 4. 26.06 5. 1.94
6. \$3.95 7. 88.26 8. 12.55 9. 277.104 10. 5.311
11. 239.01 12. 253.01 13. 83.88 14. 3.985 15. 33.672
16. 4.71 17. 4.28 18. 0.34 19. 1.92 20. 6.77

B. 1. 0.00008 2. 0.003 3. 0.84 4. 37.68 5. 0.108
6. 2.12 7. 19.866 8. 1.24 9. 61.7 10. 0.26
11. 3.8556 12. 0.23 13. 1.50 14. 2.78 15. 0.22
16. 0.526 17. 0.503 18. 214.634 19. 1.5 20. 23.5

C. 1. 0.0625 2. 0.875 3. 0.15625 4. 0.4375

5. 1.375 6. 1.1875 7. 2.34375 8. $1.1\overline{6}$

9. $2.\overline{6}$ 10. 1.03125 11. 2.3125 12. $\frac{3}{50}$

13. $\frac{2}{25}$ 14. $\frac{7}{20}$ 15. $\frac{16}{25}$ 16. $1\frac{1}{4}$

17. $2\frac{3}{4}$ 18. $3\frac{1}{16}$ 19. $2\frac{1}{100}$ 20. $14\frac{11}{25}$

21. 19.28 22. 30.67 23. 1.92 24. 78.57

25. 0.96 26. 0.13

D. 1. 64 2. 243 3. 289 4. 12,167 5. 0.25

6. 1.728 7. 0.0004 8. 0.000027 9. 0.000000001 10. 4.4041

11. 16.1604 12. 37 13. 28 14. 2.1 15. 0.4

16. 2.3 17. 8.94 18. 10.30 19. 17.61 20. 1.34

21. 2.05 22. 1.74 23. 1.04

E. 1. −13 2. −21 3. −35 4. −36 5. 9.5

6. 10.3 7. −12.5 8. 9.31 9. 0.48 10. −0.279

11. −0.27 12. −0.36 13. −8.2 14. 114.5 15. −214

16. 0.084

F. 1. 291 2. 2.80″ 3. 9.772″ 4. $322.13

5. 0.00857″ 6. 1.132 ohms 7. 489.4 lb 8. 0.02374″

9. (a) 0.1875″ (b) 0.15625″ (c) 0.375″ (d) 0.203125″

10. (a) $\frac{9}{64}$″ (b) $\frac{6}{64}$″ (c) $\frac{19}{64}$″ (d) $\frac{12}{64}$″

11. (a) 0.0089″ (b) 0.0027″ (c) 0.55215″ (d) 0.306175″

12. $1408.74 13. 17.5 ft 14. 0.0423 in.

15. Max: 2.53125″ Min: 2.46875″ 16. $223.13 17. $112.53

18. $7.30 19. 56.241 cu ft 20. 27″

21. 19 22. (a) $342.13 (b) $26.25 (c) $146.88
(d) $78.84 (e) $90.16 23. −38° C 24. −$848

CHAPTER 4

Exercises 4-1, page 136

A. 1. 32% 2. 100% 3. 50% 4. 210%

5. 25% 6. 375% 7. 4000% 8. 67.5%

9. 200% 10. 7.5% 11. 50% 12. $16\frac{2}{3}$%

13. 33.5% 14. 0.1% 15. 0.5% 16. 30%

17. 150% 18. 7.5% 19. 330% 20. 20%

B. 1. 0.06 2. 0.45 3. 0.01 4. 0.33

5. 0.71 6. 4.56 7. 0.005 8. 0.0005

9. 0.0625 10. 0.0875 11. 0.3 12. 0.021

13. 8 14. 0.08 15. 0.0025 16. $0.16\overline{3}$

Exercises 4-2, page 146

A. 1. 80% 2. 64 3. 15 4. 100%

5. 54 6. 12.5% 7. 150 8. 80

9. $66\frac{2}{3}$% 10. 500% 11. 100 12. 52%

13. 21.25 14. 600% 15. 1.5 16. $23\frac{1}{3}$

17. 43.75% 18. $133\frac{1}{3}$%

B. 1. 225 2. 25% 3. 20¢ 4. 180%

5. 160 6. $2.72 7. 150% 8. 2%

9. 17.5 10. 400% 11. 427 12. 22.1

13. 2% 14. 3.625 15. 460 16. 1400

17. 50 18. 2.475

C. 1. 88% 2. $48,200 3. 5.5% 4. $6900

5. $805.60 6. 47¢ 7. $5.32 8. $35.15

Exercises 4-3, page 158

1. 64% 2. 3.78 kw 3. 63.6%
4. 5035 to 5565, 2695 to 2805, 6120 to 7480, 4536 to 6804
5. 67.5% 6. 550 ohms 7. 5%
8. $504 9. 12.5% 10. $16\frac{2}{3}$% ($\frac{2}{3}$ lb) tin; 2% (0.08 lb) zinc; $81\frac{1}{3}$% (3.25 lb) copper
11. 0.9% 12. 1660 lb 13. 81.25 hp 14. Yes (1910)
15. 0.13% 16. $31.96
17. (a) 0.029% (b) 0.438% (c) 0.007″
18. $9.69 19. 37.5% 20. $82.12

Problem Set 4, page 161

A. 1. 72% 2. 86% 3. 60% 4. 35% 5. 130%
6. 303% 7. 10% 8. 70% 9. $16\frac{2}{3}$% 10. 260%

B. 1. 0.04 2. 0.37 3. 0.11 4. 0.94 5. 0.0125
6. 0.0009 7. 0.002 8. 0.017 9. 0.03875 10. 0.0802
11. 1.15 12. 2.1

C. 1. 60% 2. $6 3. 5.6 4. 200% 5. 42
6. 15¢ 7. $28.97 8. 125 9. 8000 10. 0.099

D. 1. 38.636″ 2. 0.9 lb of carbon in cast iron, 0.016 lb in wrought iron
3. (a) $3815 (b) $79.50 (c) $429
4. $107.58 5. (a) $2.10 (b) $3.29 (c) 73¢
(d) $14.42 (e) $260.71
6. $200 7. 873.6 bd ft 8. 4%
7. 873.6 bd ft 8. 4%
9. (a) 0.056% (b) 2.8% (c) 0.009″ (d) 0.032% (e) 0.62 mm
10. 22% 11. $1466.25 12. 14.71 cm
13. ±135 ohms, 4365 to 4635 ohms
14. 68.6% 15. 0.92% 16. $5.06

CHAPTER 5

Exercises 5-1, page 175

A. 1. 11″ 2. 50 lb 3. 9 in. 4. 38.6 psi
5. 0.334 in. 6. 0.74 ft 7. 47.5 mph 8. 2.8 gal
9. 0.08 in. 10. 25.4 psi

B. 1. 28 sq ft 2. 5.3 sq in. 3. 7.2 sq ft 4. 25.2 miles
5. 1.9 sq ft 6. 384 sq in. 7. 12.1 cu ft 8. 8.6 ft
9. 20 mph 10. $0.43/lb 11. 1.6 hr 12. 1.9 ft

C. 1. (a) 1.88 in. (b) 4.05″ (c) 3.13 sec (d) 0.06 in.
(e) 0.09 in. (f) 1.23 lb (g) 2.27 in. (h) 0.59 in.
(i) 1.14 lb
2. (a) $\frac{15}{16}$″ (b) $2\frac{9}{16}$″ (c) $1\frac{13}{16}$″ (d) $3\frac{11}{16}$″
(e) $\frac{13}{16}$″ (f) $\frac{5}{16}$″ (g) $1\frac{15}{16}$″ (h) $1\frac{9}{16}$″
(i) $\frac{13}{16}$″
3. (a) $1\frac{29}{32}$″ (b) $\frac{27}{32}$″ (c) $2\frac{11}{32}$″ (d) $\frac{21}{32}$″
(e) $2\frac{3}{32}$″ (f) $\frac{9}{32}$″ (g) $\frac{19}{32}$″ (h) $\frac{22}{32}$″
(i) $1\frac{17}{32}$″
4. (a) $\frac{15}{64}$; error of 0.0006 (b) $\frac{33}{64}$; error of 0.0006
(c) $1\frac{51}{64}$; error of 0.0031 (d) $2\frac{27}{64}$; error of 0.0019
(e) $3\frac{11}{64}$; error of 0.0031 (f) $2\frac{55}{64}$; error of 0.0006
(g) $1\frac{60}{64}$; error of 0.0025 (h) $\frac{41}{64}$; error of 0.0044
(i) $\frac{31}{64}$; error of 0.0044

D. 1. Yes 2. $\frac{13}{32}$; error of 0.0033 3. $\frac{30}{64}$ in.
4. $\frac{24}{32}$ in. 5. 4.28″ 6. 0.02325″

Box page 183

1. (a) 2 bf (b) 16 bf (c) 8 bf (d) $10\frac{2}{3}$ bf (e) 7 bf (f) 3 bf
2. 1280 bf 3. 4410 bf

Exercises 5-2, page 185

A. 1. 51 in. 2. 99 ft 3. 17,952 ft 4. 272
5. 4 bbl 6. 14,960 yd 7. 3.1 atm 8. 3.75 ton
9. 582.8 gal 10. $2\frac{2}{3}$ ft 11. $\frac{1}{2}$ mi 12. 104.5 rods

B. 1. 864 2. 108,900 3. 72.6 4. 42.8
5. 0.563 6. 5.42 7. 0.03 8. 7.1
9. $0.\overline{3}$ or $\frac{1}{3}$ 10. 500 11. 12,800 12. 95

C. 1. (a) 43,560 (b) 325,851 (c) 10,344 (d) 1613
2. (a) $\frac{1}{12}$ (b) 144 3. 4
4. (a) $12\frac{1}{2}$ gal (b) \$13.57 (c) 2000 sq ft (d) 2.3 gal (e) \$41.18
5. 32.6 ± 0.6 in. 6. (a) 5400″ × 900″ × 540″ or 450′ × 75′ × 45′
(b) 117″ or 9 ft 9 in.
7. 141 sq yd 8. 0.1 lb/cu in. 9. 10.2 rps

Exercises 5-3, page 203

A. 1. (c) 2. (b) 3. (a) 4. (a) 5. (b) 6. (c) 7. (a)
8. (c) 9. (a) 10. (b) 11. (c) 12. (c) 13. (a) 14. (c)
15. (a) 16. (a) 17. (a) 18. (b) 19. (a) 20. (b)

B. 1. (b) 2. (c) 3. (a) 4. (a) 5. (b) 6. (c)
7. (b) 8. (b) 9. (b) 10. (c) 11. (a) 12. (a)

C. 1. 7.6 2. 121.9 3. 3.2 4. 6.2 5. 23.5
6. 66 7. 69.1 8. 1.5 9. 18.7 10. 294.8
11. 11.4 12. 2.6 13. 6.1 14. 99.8 15. 5.0
16. 97.8 17. 50 18. 88.9 19. 19.5 20. 6.6

D. 1.

in.	mm
0.030	0.762
0.035	0.889
0.040	1.016
0.045	1.143

in.	mm
1/16	1.588
5/64	1.984
3/32	2.381
1/8	3.175

in.	mm
5/32	3.969
3/16	4.763
3/8	9.525
11/64	4.366

2. (a) 2.2 lb (b) 28.3 kg (c) 1 kg
3. 11.9 mph; 19.1 kmh
4. (a) 103,000 (b) 689,000 (c) 2000 (d) 120,000
5. (a) 0.6093 (b) 5.11 (c) 564.6 (d) 0.891 (e) 0.057
(f) 1.023 (g) 204.6
6. 57.15 mm by 69.85 mm or 57 by 70, rounded
7. 12 km/liter 8. \$1.39 9. 0.93 sq m
10. (a) 2.4 m × 1.2 m × 1.2 m (b) 3.5 cu m
11. (a) 1.6093 (b) 60.96 (c) 21.59 cm × 27.94 cm (d) 2.54
(e) 804.65 (f) 0.0648 (g) 28.35; 453.6 (h) 0.0648
(i) 7.57 (j) 37.85 (k) 8810 12. 6.6 l

Exercises 5-4, page 226

A. 1. (a) 1782 kwh (b) 7173 kwh (c) 5407 kwh (d) 3580 kwh
(e) 5214 kwh (f) 681 kwh
2. (a) 737,700 cu ft (b) 208,800 cu ft (c) 953,900 cu ft (d) 810,700 cu ft

B. 1. (a) $\frac{5}{8}''$ (b) $1\frac{7}{8}''$ (c) $2\frac{1}{2}''$ (d) $3''$
(e) $\frac{1}{4}''$ (f) $1\frac{1}{16}''$ (g) $2\frac{5}{8}''$ (h) $3\frac{1}{2}''$
2. (a) $\frac{5}{16}''$ (b) $\frac{3}{4}''$ (c) $1\frac{1}{32}''$ (d) $1\frac{9}{16}''$
(e) $\frac{1}{8}''$ (f) $\frac{17}{32}''$ (g) $\frac{53}{64}''$ (h) $1\frac{7}{16}''$
3. (a) $\frac{2}{10}'' = 0.2''$ (b) $\frac{5}{10}'' = 0.5$ (c) $1\frac{3}{10}'' = 1.3''$ (d) $1\frac{6}{10}'' = 1.6''$
(e) $\frac{25}{100}'' = 0.25''$ (f) $\frac{72}{100}'' = 0.72''$ (g) $1\frac{49}{100}'' = 1.49''$
(h) $1\frac{75}{100}'' = 1.75''$

C. 1. 0.65″ 2. 0.705″ 3. 0.287″ 4. 0.061″
5. 0.850″ 6. 0.441″ 7. 0.4068″ 8. 0.5997″
9. 0.2581″ 10. 0.7796″ 11. 0.0888″ 12. 0.8451″
13. 11.57 mm 14. 3.41 mm 15. 22.58 mm 16. 18.21 mm
17. 18.94 mm 18. 13.27 mm

D. 1. 3.255″ 2. 0.836″ 3. 2.078″ 4. 0.609″
5. 2.905″ 6. 1.682″ 7. 2.040″ 8. 0.984″
9. 0.826″ 10. 2.655″

E. 1. 62°21′ 2. 20°38′ 3. 35°56′ 4. 65°10′
5. 35°34′ 6. 64°48′ 7. 20°26′ 8. 44°15′

F. 1. 2.000″ + 0.200″ + 0.050″ + 0.057″ + 0.0503″
2. 1.000″ + 0.500″ + 0.200″ + 0.140″ + 0.056″ + 0.0507″
3. 0.200″ + 0.120″ + 0.053″ + 0.0502″
4. 2.000″ + 1.000″ + 0.500″ + 0.200″ + 0.050″ + 0.058″ + 0.0507″
5. 1.000″ + 0.500″ + 0.070″ + 0.058″ + 0.0509″
6. 2.000″ + 0.500″ + 0.150″ + 0.058″ + 0.0508″
7. 0.050″ + 0.053″ + 0.0509″
8. 2.000″ + 0.500″ + 0.300″ + 0.120″ + 0.057″ + 0.0506″
9. 2.000″ + 0.140″ + 0.0505″
10. 1.000″ + 0.400″ + 0.056″ + 0.0506″

Problem Set 5, page 231

A. 1. 3.37 sec 2. 48 sq ft 3. 0.75 in 4. 10.24 sq cm
5. $16\frac{7}{8}''$ 6. 3 lb 14 oz 7. 21.25 mi/gal 8. 9.6 cm
9. 7.65 psi 10. 2.95 ft 11. 3.869″ 12. 403.7 lb
13. 21′1″ 14. $12\frac{25}{64}''$

B. 1. 15.24 2. 73.33 3. 71.6 4. 2.75
5. 45,792 6. −25 7. 31.53 8. 8.41
9. 23.33 10. 5735.47 11. 60.22 12. 14.86
13. 11.48 14. 1.82 15. 979.2 16. 11.59

C. 1. (b) 2. (b) 3. (a) 4. (a) 5. (b)
6. (c)

D. 1. (a) $\frac{5}{10}'' = 0.5''$ (b) $1\frac{3}{10}'' = 1.3''$ (c) $\frac{17}{100}'' = 0.17''$
(d) $1\frac{35}{100}'' = 1.35''$ (e) $\frac{3}{4}''$ (f) $3\frac{3}{4}''$ (g) $\frac{5}{16}''$
(h) $1\frac{19}{32}''$ (i) $\frac{3}{8}''$ (j) $1\frac{27}{64}''$
2. (a) 1790 (b) 3135 (c) 893 (d) 7501 (e) 5204 (f) 4444
3. (a) 179,200 (b) 850,400 (c) 520,000 (d) 622,700 (e) 799,700
(f) 614,600
4. (a) 0.807 (b) 0.550 (c) 12.13 mm (d) 19.07 (e) 0.5469″
(f) 0.8179″
5. (a) 2.157″ (b) 3.030″ (c) 0.612″ (d) 1.925″ (e) 1.706″
(f) 1.038″

E. 1. $22\frac{3}{64}''$ 2. 53 mph 3. 6.1 mph 4. 0.0505″, 0.057″, 0.130″ 0.200″, 2.000″
5. $\frac{15}{32}$ 6. 2741° F 7. 936 sq in. 8. 18.9 gallons

9. 692 CCF 10. 899° C 11. 97.6 cu in. 12. 7.623″
13. (a) 3.85 cu m/hr (b) 7.62 m (c) 6.35 cm 14. 0.28 lb/cu in.
15. 28.3 liter 16. 5.08 cm × 10.16 cm or 5 cm by 10 cm 17. 1.5 m

---

CHAPTER 6

Exercises 6-1, page 251

A. 1. $9y$ 2. $9x^2y$ 3. $6E$ 4. $-4ax$
5. $7B$ 6. 0 7. $-2x^2$ 8. $10x + 16y$
9. $5R + 2R^2$ 10. $1.45A - 0.8A^2$ 11. $\frac{1}{8}x$ 12. $-2x$
13. $1\frac{1}{2} - 3.1W$ 14. $q - 2\frac{1}{2}p$ or $q - \dfrac{5p}{2}$

B. 1. $20x$ 2. a^5 3. $6R^2$ 4. $9x^2$
5. $-8x^3y^4$ 6. $8x^3$ 7. $0.6a^2$ 8. $2x^4A^3$
9. $\frac{1}{16}Q^4$ 10. $1.2p^4q^4$ 11. $24M^6$ 12. $6x^2 - 2x$
13. $-2 + 4y$ 14. $a^3b - ab^3$

C. 1. $A = 6$ 2. D = 7 3. $T = 13$ 4. $H = 9$
5. $K = 8$ 6. $Q = 38$ 7. $F = 19$ 8. $W = 6$
9. $L = 1$ 10. $A = 8$ 11. $B = 13$ 12. $F = 9$

D. 1. $A = 11$ 2. $V = 106.14$ 3. $I = 15$ 4. $H = 10$ 5. $T = 22.902$
6. $V = 220.78$ 7. $P = 40$ 8. $W = 2835.2$ 9. $V = 120.9$ 10. $V = 2.73$

E. 1. $P = 39$ in. 2. $i = 12$ amperes 3. $P = 96$ watts 4. $A = 1256$ sq cm
5. $F = 104°$ 6. $V = 141.3$ cu in. 7. $A = \$1200$ 8. $R = 120$ ohms
9. $T = 19$ sq cm 10. (a) 0.3125″ (b) 0.1185″ (c) 0.2625 cm (d) $\frac{5}{32}$″
11. $L = 18\frac{1}{4}$″ 12. $L = 16\frac{1}{2}$″ 13. $L = 5670$ in. 14. $L = 116.15$″
15. $x = 1000$ ft 16. $C = 178.7$ ft/min 17. $P = 26.4$ kw 18. $V = 600$ mph
19. $L = 920$ cm 20. $F = 1250$ lb

F. 1. $V = 4.2850$ cu in. 2. $T = 39.3$ minutes
3. (a) 0.7860 sq in. (b) 0.6013 sq in. (c) 13.3803 sq in. (d) 3.3345 sq cm
4. 79.84 ft

Box, page 266

1. $x = 8$ 2. $x = 17$ 3. $a = \frac{7}{2}$ 4. $R = 5$ 5. $x = -15$
6. $x = 2\frac{1}{2}$ 7. $T = -20$ 8. $y = \frac{16}{3}$ 9. $x = -\frac{1}{4}$ 10. $x = -\frac{3}{2}$

Exercises 6-2, page 276

A. 1. $x = 9$ 2. $A = 17$ 3. $x = -24$ 4. $z = 47$
5. $a = 8\frac{1}{2}$ 6. $x = -62$ 7. $y = -24.66$ 8. $R = 7.9$
9. $x = -13$ 10. $y = -13$ 11. $a = 0.006$ 12. $x = 21$
13. $z = 0.65$ 14. $N = \frac{3}{4}$ 15. $x = \frac{1}{4}$ 16. $x = 10$
17. $a = 7$ 18. $x = 14$ 19. $z = -5\frac{1}{2}$ 20. $x = -3$
21. $x = \frac{2}{5}$ 22. $A = -\frac{3}{5}$ 23. $P = 3$ 24. $x = -6$

B. 1. $L = \dfrac{S}{W}$ 2. $B = \dfrac{2A}{H}$ 3. $I = \dfrac{V}{R}$
4. $D = 2H + R$
5. $T = \dfrac{2S - WA}{W}$ 6. $H = \dfrac{V + AB}{\pi R^2}$ 7. $B = \dfrac{P - 2A}{2}$
8. $R = 2H - 0.1$ 9. $R = \dfrac{2T}{T - P}$ 10. $V = IR - E$

C. 1. $H = 1.4\,W$ 2. $W_1 + W_2 = 167$ 3. $V = \frac{1}{4}h\pi d^2$
4. $K = 0.454\,P$ 5. $V = AL$ 6. $V = \frac{1}{3}\pi hr^2$
7. $D = N/P$ 8. $T = \dfrac{L}{FR}$ 9. $W = 0.785\,hDd^2$ 10. $V = iR$

D. 1. (a) $a = \dfrac{360L}{2\pi R}$ (b) $R = \dfrac{360L}{2\pi a}$ (c) $L = 5.23$ in.

2. (a) $a = \frac{360A}{\pi R^2}$ (b) $A = 56.52$ sq in.

3. (a) $R_1 = \frac{V - R_2 i}{i}$ (b) $R_2 = 50$ ohms

4. (a) $N = \frac{3.78P}{t^2 d}$ (b) $P = 0.05$ hp

5. (a) $V_1 = \frac{V_2 P_2}{P_1}$ (b) $V_2 = \frac{V_1 P_1}{P_2}$ (c) $P_1 = \frac{V_2 P_2}{V_1}$
   (d) $P_2 = \frac{V_1 P_1}{V_2}$ (e) $P_1 = 300$ psi

6. (a) $L = \frac{6V}{\pi T^2}$ (b) $T^2 = \frac{6V}{\pi L}$

7. (a) $D = \frac{CA + 12C}{A}$ (b) 0.14 grams

8. $n = 1 + \frac{rR}{F(r + R)}$

9. (a) $i_L = \frac{i_s T_p}{T_s}$ (b) $i_s = \frac{i_L T_s}{T_p}$ (c) $i_L = 22.5$ amps

10. (a) $P = IV$ (b) $I = \frac{P}{V}$ (c) $V = 375$ volts

11. $L = 4138$ in. or 4100 in. rounded

12. (a) 1.11 in. (b) 0.989 in. (c) 0.741

Exercises 6-3, page 294

A. (In each case only the missing answer is given.)

1. (a) 7:1 (b) 12:7 (c) 6 (d) 6
   (e) 45 (f) 9 (g) 16 (h) 50
   (i) 3:2 (j) 2:5
2. (a) 8:3 (b) 5:4 (c) 16″ (d) 6 cm
   (e) 40 cm (f) $2\frac{1}{2}$ (g) 1:2.1 (h) 6.5 cm
   (i) 5.4 cm (j) 17.8 cm
3. (a) 2:3 (b) 8′ (c) 28″ (d) 4:7
   (e) 0.54 (f) 4′ (g) 20′ (h) 5′1″

B.
1. $x = 12$ 2. $R = 86.4$ 3. $y = 100$ 4. $H = 60$
5. $P = \frac{7}{6}$ 6. $x = 102$ 7. $x = 3$ 8. $A = 1.3$
9. $x = 5.04$ 10. $R = 6\frac{1}{2}$ 11. $L = 5'$ 12. $x = 58.9$ cm
13. $x = 4$ cm 14. $W = \frac{1}{8}''$ or 0.125″

C.
1. (a) $\frac{5}{8}''$ (b) 31.2 cm
2. (a) $2\frac{4}{5}''$ (b) $12\frac{1}{2}$ ft
3. (a) 100 (b) 9 (c) 25 (d) 1500
4. (a) 150 (b) 15 (c) 8 (d) 700

D.
1. 24″ 2. 15.4 to 1 3. 139 hp 4. 120 min
5. 980 rpm 6. 23 lb 7. 2500 rpm 8. 500 rpm
9. 21.6 watts 10. 33 turns

E.
1. 4.766 ohms 2. $219.40
3. $A = 5.95$ cm, $B = 5.1$ cm, $C = 4.25$ cm,
   $D = 3.4$ cm, $E = 2.55$ cm, $F = 1.7$ cm, $G = 0.85$ cm

Problem Set 6, page 299

A.
1. $10x$ 2. $12y$ 3. $12xy$ 4. $15xy^3$
5. $1\frac{11}{24}x$ 6. 0 7. $0.4G$ 8. $16y - 8y^2$
9. $21x^2$ 10. $8m^3$ 11. $-10x^2y^2z$ 12. $27x^3$
13. $12x - 21$ 14. $6a^3b - 10ab^3$

B. 1. $L = 13$ 2. $M = 8$ 3. $I = 144$ 4. $I = 183{,}333\frac{1}{3}$
5. $V = 42\frac{21}{64}$ 6. $N = -95$ 7. $L = 115.2$ 8. $f = \frac{1}{48}$
9. $t = 4$ 10. $V = 270.64$

C. 1. $x = 12$ 2. $m = \frac{1}{8}$ 3. $e = 44$ 4. $a = -\frac{1}{6}$
5. $x = -5$ 6. $x = 10\frac{1}{2}$ 7. $x = 4$ 8. $m = -6$
9. $y = 32$ 10. $M = 60$ 11. $B = 3.5$ 12. $x = 6.23$
13. $y = 40$ 14. $E = 48$ 15. $F = 0.72$ 16. $x = 3$
17. $x = -6$ 18. $f = -1.15$ 19. $x = -4$ 20. $y = 2$
21. $g = 10$ 22. $m = 1.5$ 23. $x = 16$ 24. $x = 0$

25. $b = \dfrac{A}{H}$ 26. $P = R - S$ 27. $L = \dfrac{P - 2W}{2}$ 28. $F = \dfrac{w}{P}$

29. $g = \dfrac{2S + 8}{t}$ 30. $A = \dfrac{\pi R^2 H - V}{B}$

D. 1. (a) 31.25% (b) $C = \dfrac{100(H - X)}{CG}$; $C = 166.7''$

(c) $H = \dfrac{C \cdot CG + 100x}{100}$; $H = 140''$ (d) $x = \dfrac{100H - C \cdot CG}{100}$; $x = 156''$

2. (a) 12:5 or 2.4 to 1 (b) 3600 rpm

3. (a) $L = \dfrac{FDU}{A}$ (b) 42 foot candles (c) $A = \dfrac{FDU}{L}$ (d) 66.7 sq ft

4. $3\frac{3}{4}$ hr 5. $8W + 2W = 80$; $W = 8$ cm, $L = 32$ cm
6. (a) 9′ (b) 30′ (c) $\frac{1}{8}$ ft (d) $7\frac{1}{2}$ ft
7. 6 sacks 8. $L = 225$ tons
9. 1:4 10. $4\frac{1}{16}$ in. 11. 360 turns
12. $2x + x = 0.048$; $x = 0.016''$, $2x = 0.032''$ 13. 10.5 to 1

E. 1. \$378.56 2. 546 rpm 3. 999600 BTU
4. 7441 5. 1.984″ 6. 83.96 psi 7. 0.6168″
8. 0.1139″ 9. 1,766,667

Chapter 7 *Exercises 7-1, page 312*

A. 1. (a) $\angle B$, $\angle x$, $\angle ABC$ (b) $\angle O$, $\angle POQ$ (c) $\angle a$
(d) $\angle T$ (e) $\angle 2$ (f) $\angle s$, $\angle RST$
2. (a) $\angle AOB$, $\angle AOC$ (b) $\angle HOG$, $\angle HOF$ (c) $\angle AOF$, $\angle AOG$
(d) $\angle AOE$, $\angle HOE$
(e) $\angle AOB = 25°$, $\angle GOF = 39°$, $\angle HOC = 128°$,
$\angle FOH = 75°$, $\angle BOF = 80°$, $\angle COG = 97°$
4. (a) 145° (b) 47° (c) 30° (d) 70° (e) 75°
(f) 90° (g) 45° (h) 45° (i) 150°

B. 1. (a) $a = 55°$, $b = 125°$, $c = 55°$ (b) $d = 164°$, $e = 16°$, $f = 164°$
(c) $p = 110°$ (d) $x = 75°$, $t = 105°$, $k = 115°$
(e) $s = 60°$, $t = 120°$ (f) $p = 30°$, $q = 30°$, $w = 90°$
2. (a) $A = 68°$ (b) $E = 137°$
(c) $G = 42°$ (d) 133°
(e) $a = 70°$ (f) 53°
(g) 63° (h) 45°5′ (i) 60°0′30″
3. (a) $w = 80°$, $x = 100°$, $y = 100°$, $z = 100°$
(b) $a = 60°$, $b = 60°$, $c = 120°$, $d = 120°$, $e = 120°$, $f = 120°$, $g = 60°$
(c) $n = 100°$, $q = 80°$, $m = 100°$, $p = 100°$, $h = 80°$, $k = 100°$
(d) $z = 140°$, $s = 40°$, $y = 140°$, $t = 140°$, $u = 40°$, $w = 140°$, $x = 40°$

C. 1. $\angle ACB = 67°$ 2. 45° 3. $B = 125°$, $A = 27\frac{1}{2}°$, $C = 27\frac{1}{2}°$
4. $a = 35°$ 5. 2″ 6. 45° 7. 29°5′36″

8. 60°   9. $a = 112°$, $b = 68°$ 10. $22\frac{1}{2}°$
11. $\angle 2 = 106°$ 12. $\angle y = 48°$

Box, page 353

1. 4 in.   2. 8.5 in.   3. 5″   4. 10.4 in.   5. 5.1 in.

Exercises 7-2, page 355

A.

	Polygon	Sides	Vertices	Diagonals
1.	Triangle (Isosceles)	AB, AC, BC	A, B, C	None
2.	Square	LN, LM, NP, MP	L, M, N, P	LP, MN
3.	Parallelogram	RS, SP, PQ, RQ	R, S, P, Q	PR, SQ
4.	Triangle (isosceles)	AB, AC, BC	A, B, C	None
5.	Hexagon	DE, EF, FG, GH, HI, ID	D, E, F, G, H, I	DF, DG, DH, EG, EH, EI, FH, FI, GI
6.	Trapezoid	JK, KL, LM, MJ	J, K, L, M	JL, KM
7.	Rectangle	NO, OP, PQ, QN	N, O, P, Q	NP, QO
8.	Pentagon	RS, ST, TU, UV, VR	R, S, T, U, V	RT, RU, SU. SV, TV
9.	Triangle	XY, YZ, ZX	X, Y, Z	None
10.	Quadrilateral	AB, BC, CD, DA	A, B, C, D	AC, BD
11.	Triangle (equilateral)	EF, FG, GE	E, F, G	None
12.	Trapezoid	MN, NO, OP, MP	M, N, O, P	MO, NP

B.

	Perimeter	Area		Perimeter	Area
1.	20″	25 sq in.	2.	69.1″	380.1 sq in.
3.	60 mm	173.2 sq mm	4.	$22\frac{2}{3}$ yd	$29\frac{5}{9}$ sq yd
5.	70.6″	285 sq in.	6.	127.7′	520 sq ft
7.	78 cm	439.1 sq cm	8.	98′	420 sq ft
9.	66.0″	346.4 sq in.	10.	96′	336 sq ft
11.	42′	53.2 sq in.	12.	56 m	160 sq m
13.	60″	259.8 sq in.	14.	17.3 yd	18.8 sq yd
15.	45″	74.2 sq in.	16.	124 m	468 sq m
17.	42 cm	84.9 sq cm	18.	42.5 ft	109.4 sq ft
19.	30.8 yd	36 sq yd	20.	61.5″	162 sq in.

C.

	Perimeter	Area		Perimeter	Area
1.	116′	432 sq ft	2.	75″	234 sq in.
3.	77.2′	364.8 sq ft	4.	125.7 cm	238.8 sq cm
5.	150 ft	354 sq ft	6.	4.5″	0.7 sq in.
7.	45.7 cm	135.3 cm^2	8.	268 m	2310 sq m
9.	108 in.	120 sq in.	10.	25.1 in.	17.1 sq in.
11.	67.4 in.	246.1 sq in.	12.	11.4 cm	5.1 sq cm
13.	$45\frac{1}{2}$″	46.5 sq in.	14.	466.5′	8037.5 sq ft

D. 1. 10.63″   2. 13.42′   3. 20 cm   4. 28.62 yd
5. 46.76 mm   6. 39″   7. 27.71″   8. 54″
9. 28′   10. 16.5 mm   11. 21.17″   12. 28″
13. 3″   14. 10.49 cm

E. 1. 30.13″   2. 7 gallons   3. 0.6″   4. 0.05 sq in.
5. 4.27 sq ft or 615.4 sq in.   6. 7.071″   7. about 960 bricks
8. 2.418 sq in.   9. $1\frac{3}{4}$″   10. 314   11. 4.24 cm
12. 35.875 sq ft   13. 268 sq in.   14. $6.17   15. 99.52 meters
16. 174 in.   17. 10′4.5″   18. 2.4′   19. 67.0 sq yds
20. 5 squares   21. 74 ft   22. $3633.06   23. 146.91 cm
24. 2.19625″ or about a $2\frac{1}{4}$″ wrench   25. 11.3″   26. 174.5 sq ft
27. 2.6 cm   28. 15.2 ft

F. 1.

	Area	Perimeter
(a)	33.13 sq cm	27.91 cm
(b)	16.81 sq cm	16.4 cm
(c)	39,496.69 sq ft	847.5′
(d)	722.2 sq m	114.9 m
(e)	333,814 sq ft	2618 ft
(f)	491.2 sq ft	82.5 ft

2. (a) 9.369 cm (b) 8.288″ (c) 0.333″ (d) 21.35″
   (e) 117.776′ (f) 43.91 m
3. 2.771 sq cm
4. 3.23′
5. 543 ft
6. $x = 64.91'$
   $y = 36.44'$

Exercises 7-3, page 384

A.

	Lateral Surface Area	Volume		Lateral Surface Area	Volume
1.	280 sq ft	480 cu ft	2.	453.6 sq cm	519.6 cu ft
3.	244.9 sq in.	367.4 cu in	4.	608 sq ft	1120 cu ft
5.	10.5 sq in.	2.3 cu in	6.	646.8 sq ft	1884.0 cu ft
7.	784 sq mm	2744 cu mm	8.	360 sq in.	480 cu in.
9.	384 sq in.	616.7 cu in.	10.	55.2 sq in.	31.2 cu m
11.	301.4 sq yd	602.9 cu yd	12.	427.0 sq in.	1004.8 cu in.

B.

	Total Surface Area	Volume		Total Surface Area	Volume
1.	150 sq yd	125 cu yd	2.	4071.5 sq ft	24,429.1 cu ft
3.	1949.4 sq mm	6355.5 cu mm	4.	162.0 sq in.	105.2 cu in.
5.	2285.5 sq ft	7068.6 cu ft	6.	5290 sq cm	23,940 cu cm
7.	7.1 sq in.	1.8 cu in.	8.	311.3 sq m	357.9 cu m
9.	172.8 sq ft	144 cu ft	10.	26.7 sq in.	6.2 cu in.
11.	734.2 sq yd	849.4 cu yd	12.	112 sq ft	69.3 cu ft
13.	238.3 sq m	245.0 cu m			
14.	601.9 sq in.	1055.0 cu in.			

C.
1. 333.3 cu ft
2. 20.4 lb
3. 47,124 gal
4. 104.7 cu m
5. 8 gallons
6. 12.3 lb
7. 450 trips
8. 0.24 ft
9. 746.1 sq cm
10. 6283.2 liters
11. 2.6 lb
12. 322.4 bushels
13. 2011.7 lb
14. 71.3 cu yd
15. 57.4 in.
16. 384 sq in.
17. No. (4512 gal)
18. 4071.5 sq in.
19. $\frac{1}{27} \cdot \frac{1}{3} \cdot 2(729 + 1089 + \sqrt{793881})$ or 66.9 cu yd
20. 19.8 lb

D.
1. 8.1 gal
2. 32.3 lb
3. 1073.6 sq cm
4. 0.4 ft
5. 34.6 lb
6. 581.7 gal
7. 2.0 lb
8. 5671.3 lb
9. 8.9 cu yd
10. 0.4 lb

Problem Set 7, page 405

A.
1. $\measuredangle 1$, acute
2. $\measuredangle ABC$ or $\measuredangle B$, obtuse
3. $\measuredangle m$, acute
4. $\measuredangle E$, right
5. $\measuredangle PQR$ or $\measuredangle Q$, obtuse
6. $\measuredangle G$, right
7. 65°
8. 50°
9. 68°
10. 90°
11. 180°
12. 153°
13. 55°
14. 70°
15. 102°
17. (a) 55° (b) $a = 95°,\ b = 85°,\ c = 95°$
    (c) $a = 130°,\ b = 50°,\ c = 130°,\ d = 50°,\ e = 130°,\ f = 50°,\ g = 130°$
    (d) 131° (e) 100° (f) all equal 60° (g) 58°

B.

	Perimeter	Area		Perimeter	Area
1.	12″	6 sq in.	2.	30′	48 sq ft
3.	31.4 cm	78.5 sq cm	4.	3″	$\frac{9}{16}$ sq in.
5.	12′	10.4 sq ft	6.	74.4″	289.8 sq in.

7.	17 yd	8 sq yd	8.	100 mm	561 sq mm
9.	36″	53.7 sq in.	10.	42′	102 sq ft
11.	12 in.	6.9 sq in.	12.	94.2 cm	706.9 sq cm
13.	17.2 ft	12.0 sq ft	14.	74 in.	145 sq in.
15.	51.7 ft	169.3 sq ft	16.	70 m	284 sq m
17.	28.3″	51.1 sq in.	18.	7.5″	2.3 sq in.
19.	5.77 mm		20.	15.59″	
21.	3.46″		22.	23.56 ft	
23.	$\frac{3}{4}''$		24.	11.70″	

C.

	Lateral Surface Area	*Volume*		*Lateral Surface Area*	*Volume*
1.	226.2 sq in.	452.4 cu in.	2.	240 sq in.	504 cu in.
3.	43.2 sq ft	$26\frac{2}{3}$ cu ft	4.	816.9 sq ft	1638 cu ft
5.	259.2 sq mm	443.5 cu mm	6.	630 sq in.	1909.5 cu in.
7.	376.8 sq in.	979.7 cu in.	8.	300 sq ft	912 cu ft

	Total Surface Area	*Volume*		*Total Surface Area*	*Volume*
9.	384 sq ft	512 cu ft	10.	107.8 sq in.	56.6 cu in.
11.	1767.2 sq cm	5301.5 cu cm	12.	256.2 sq in.	259.8 cu in.
13.	450 sq ft	396 cu ft	14.	879.6 sq in.	1570.8 cu in.
15.	1256.6 sq cm	4188.8 cu cm	16.	648 sq in.	1008 cu in.
17.	530.9 sq in.	1150.3 cu in.	18.	1381.6 sq in.	3479.1 cu in.
19.	1263.6 sq in.	2520 cu in.	20.	35.3 sq in.	14.7 cu in.
21.	329.2 sq in.	312 cu in.			

E.

1.	$51,425	2.	1125	3.	7.07″	4.	45°
5.	1204 cu yd	6.	640 sq ft	7.	251 ft	8.	108°
9.	$693.24	10.	423.0 gal	11.	32.5 in.	12.	18″
13.	36′5″	14.	2652 lb	15.	36	16.	40 qt
17.	84 trips	18.	8 ft	19.	34.7″	20.	108°
21.	15 ounces	22.	1.5 cu ft	23.	94 ft	24.	0.707″
25.	294 lb	26.	22,000 sq in.	27.	$23.40	28.	14 cu yd
29.	738 sq in.	30.	149 sq ft	31.	69.1 cm	32.	68.0″
33.	$552.64	34.	2.8 gal	35.	18.5 lb	36.	118°

Exercises 8-1, page 431

Chapter 8

A.
1. (a) 20 (b) 1.6% (c) Yes. In 1969. (d) 40,000 (e) 2,000,000
2. (a) 100 billion (b) 750 billion (c) 350 billion
3. (a) 15% (b) 3% (c) $25,200; $63,000
4. (a) 3% (b) $36 million (c) $36 million (d) $62.5 million

B. 1. (a) Absorption Capacities of Five Typical Soils

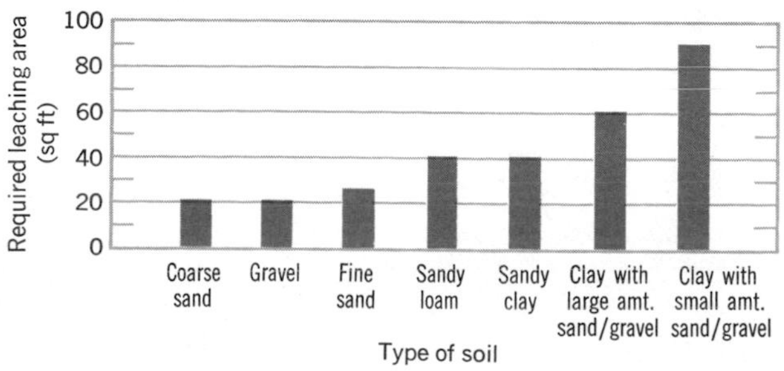

(b) Average Modulus of Elasticity of Various Materials

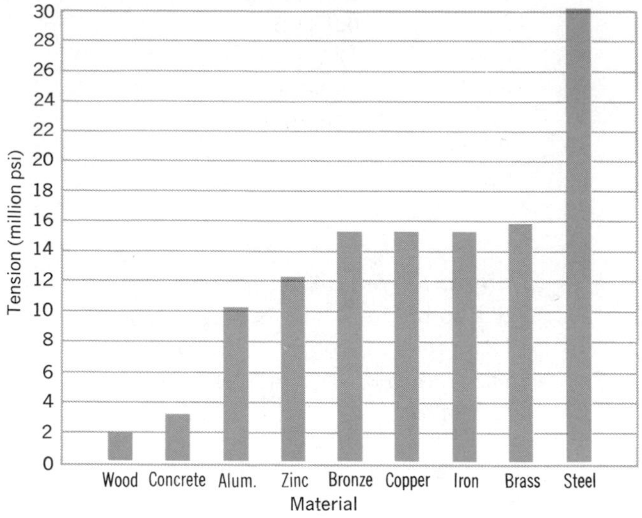

(c) Distribution of Passenger Car Trips

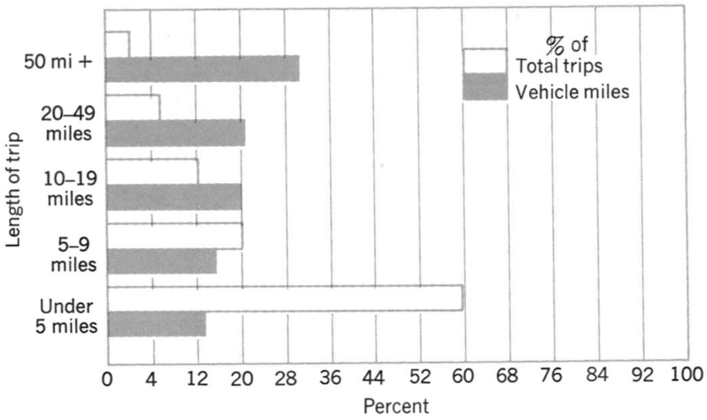

(d) Amount of Steel Used (Williams Manufacturing Co.)

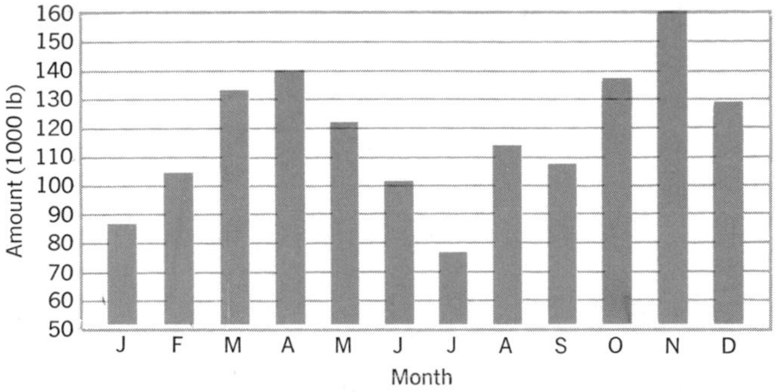

2. (a) Auto Factory Sales in the U.S.

(b) High and Low Temperatures During a Week in Helena, Montana

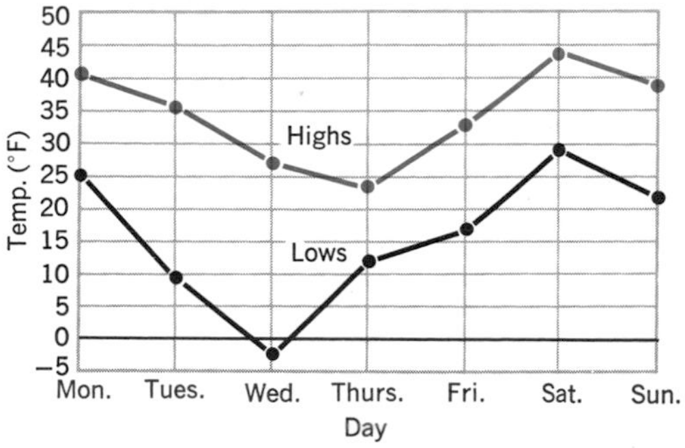

(c) New Business Incorporations: 1976

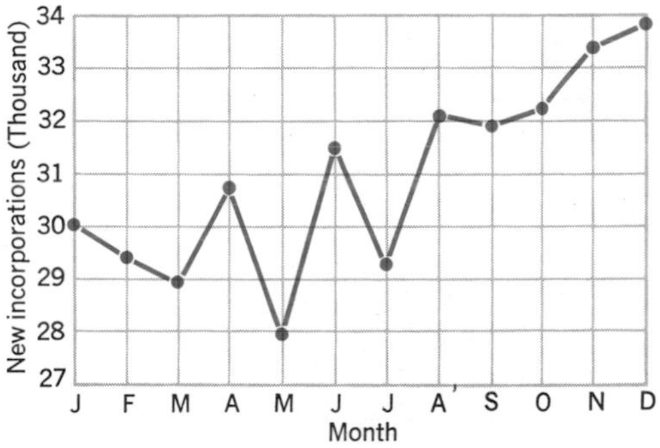

(d) Sales of Construction Material: Smyth Building Supplies

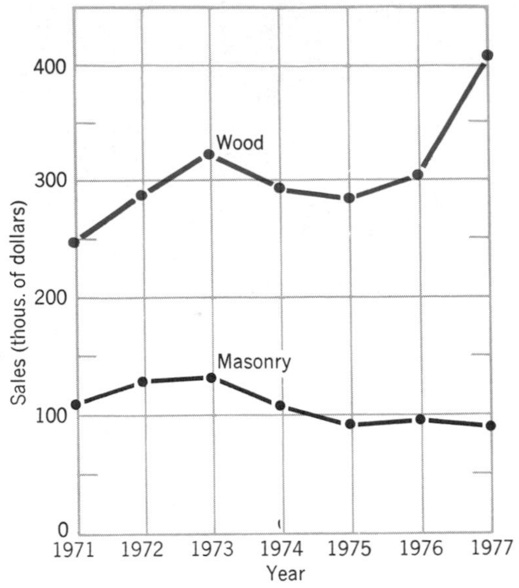

3. (a) California's Present Sources of Electricity

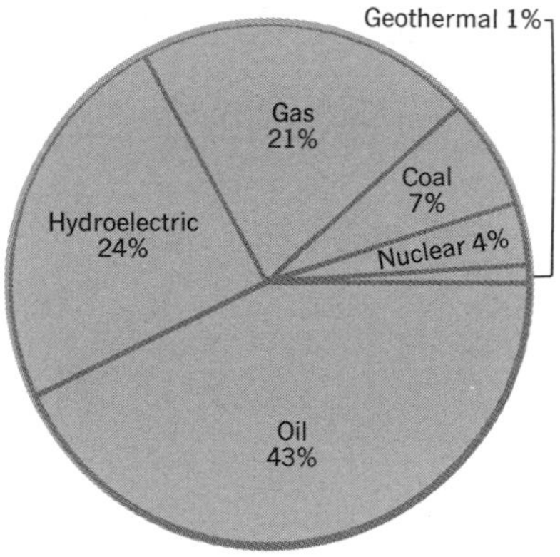

(b) Family Budget: Robbins Family

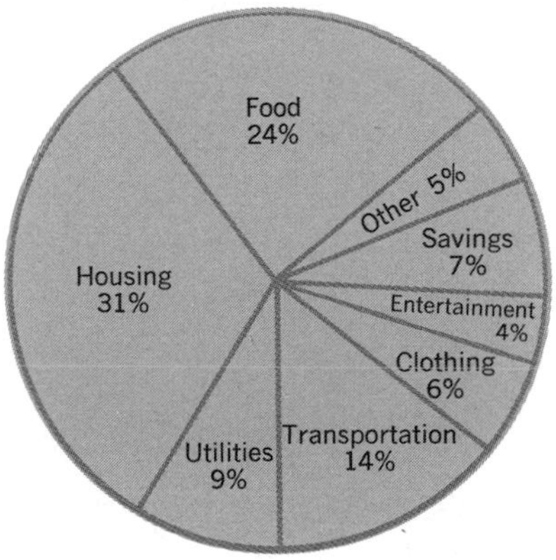

(c) Housing Situation: Pine Valley

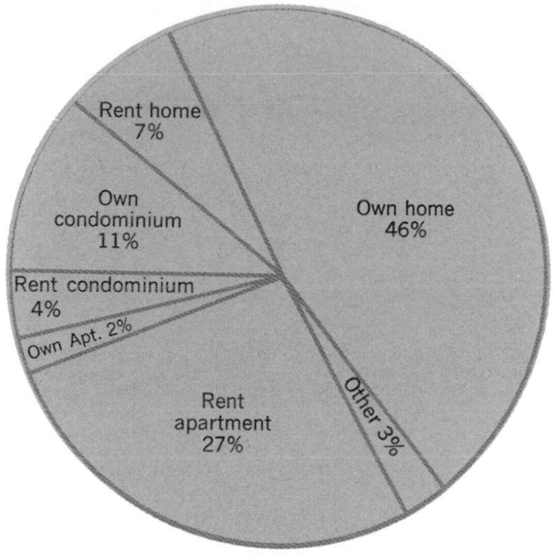

(d) Government Budget: Center City

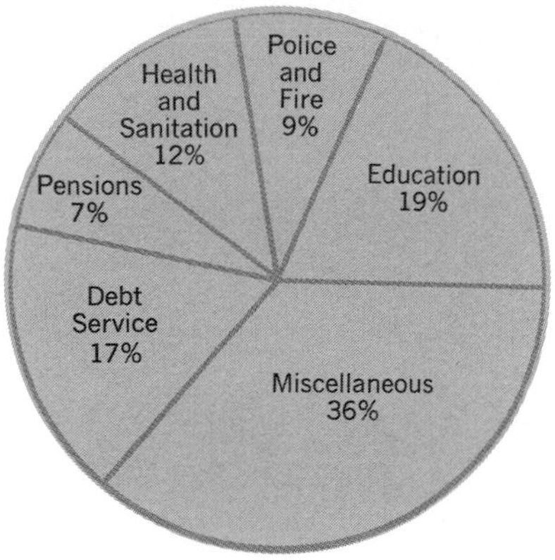

Exercises 8-2, page 444

A. 1. (a) Increases (b) 140° F (c) 155 sq in. (d) $6\frac{1}{2}$
(e) 10° C (f) 8″ (g) 2000 watts (h) 380 sq in.
(i) 80° F (j) $9\frac{1}{2}$″

2. (a) More (b) 35 gpm (c) 63 gpm (d) 37 (e) 40

B. 1. Stretch of a Copper Wire

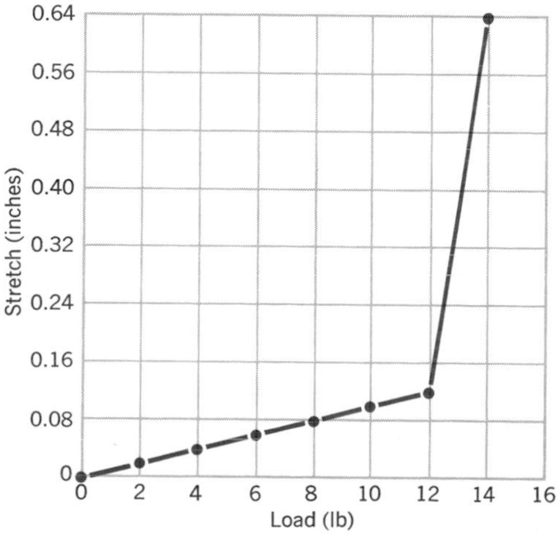

2. Growth of a Carrot Plant

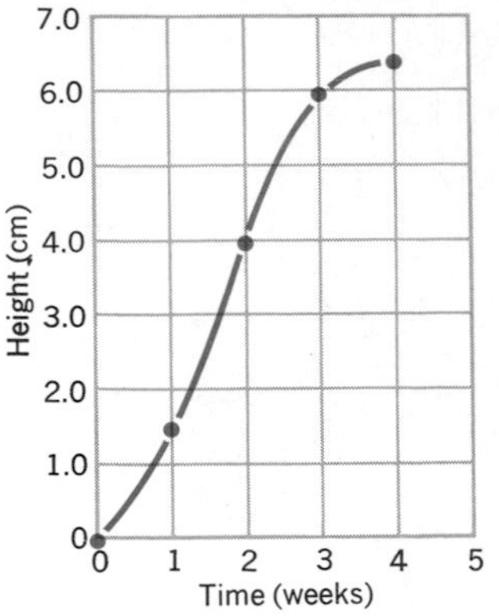

3. Sliding Object

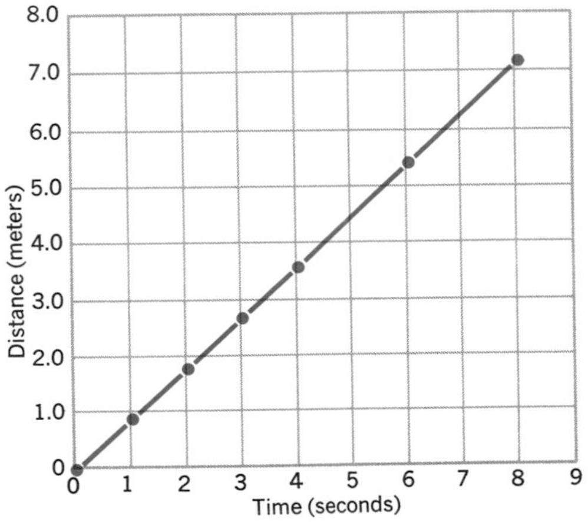

4. Horsepower and Efficiency

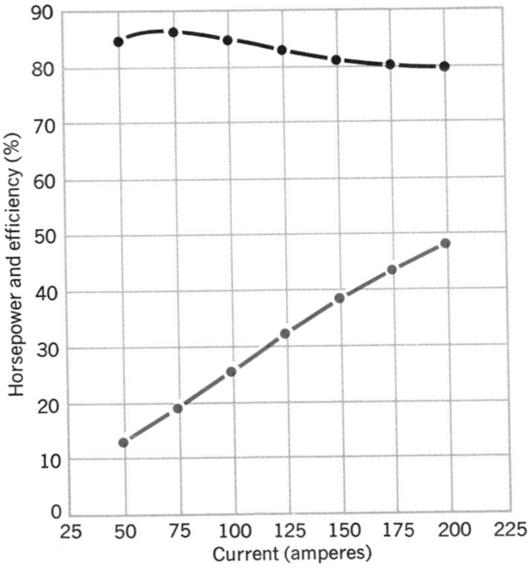

5. Voltage and Specific Gravity

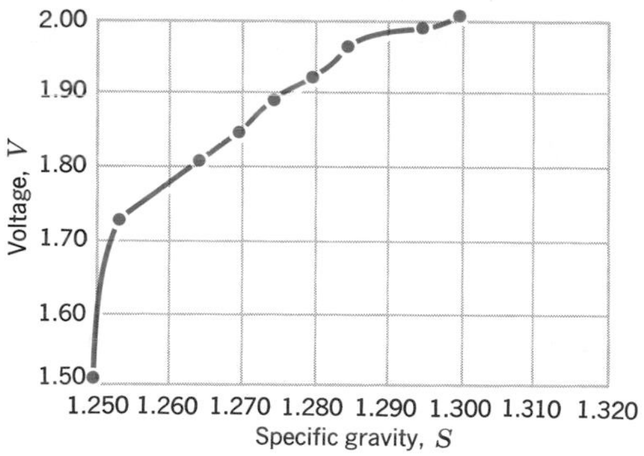

6. Air Temperature at Varying Altitudes

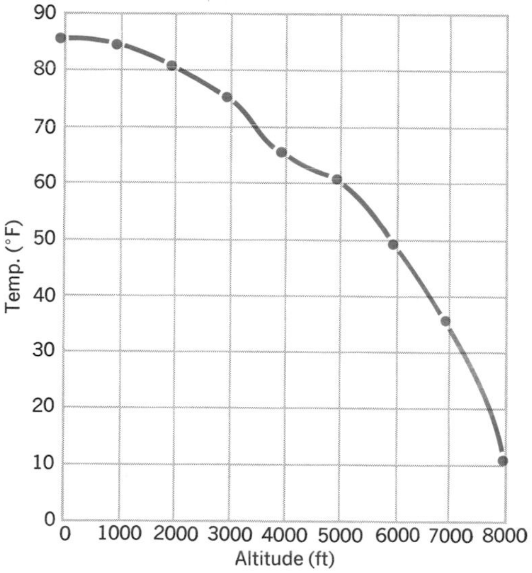

C. 1. Resistance of a Copper Wire

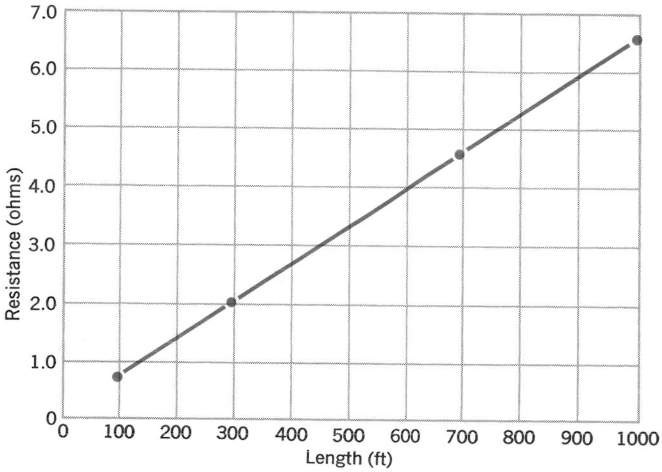

2. Bricks Needed for Different Wall Volumes

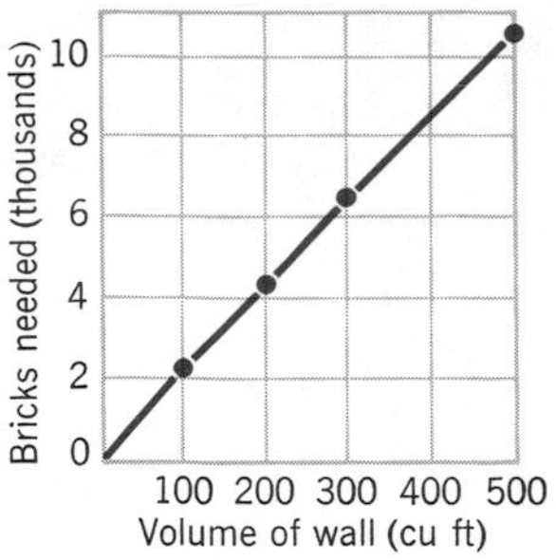

3. Power Dissipation in DC Circuit

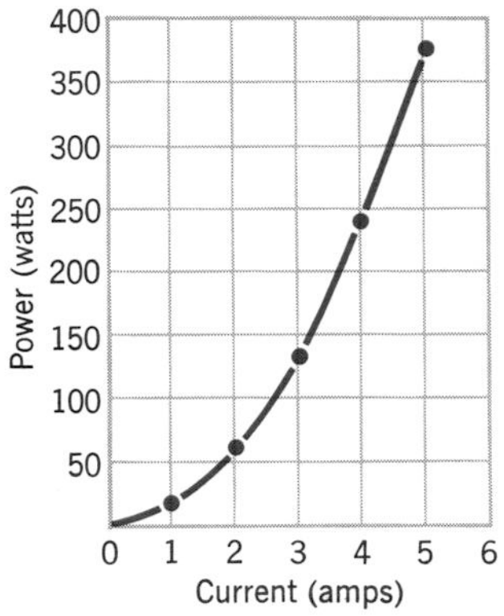

4. Power vs Current (2 phase, 3 wire, 120 volt circuit)

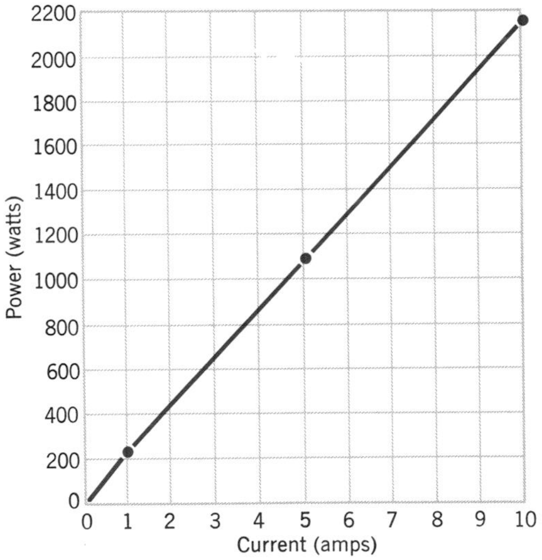

5. Lift of Airplane Wing

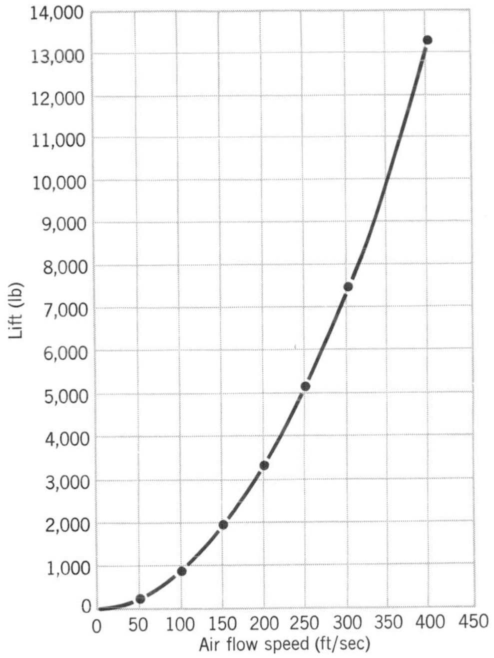

6. (a)

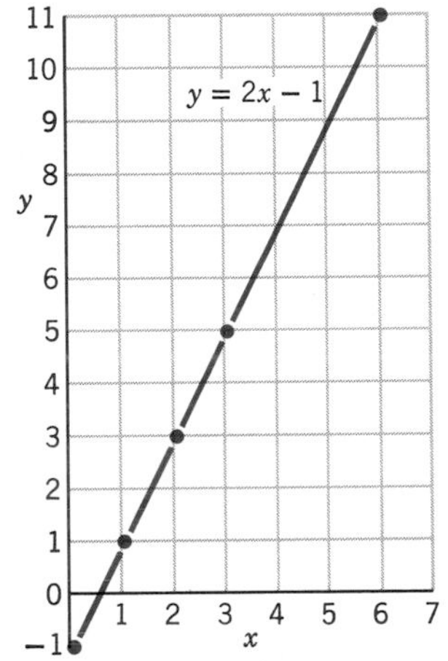

(c)

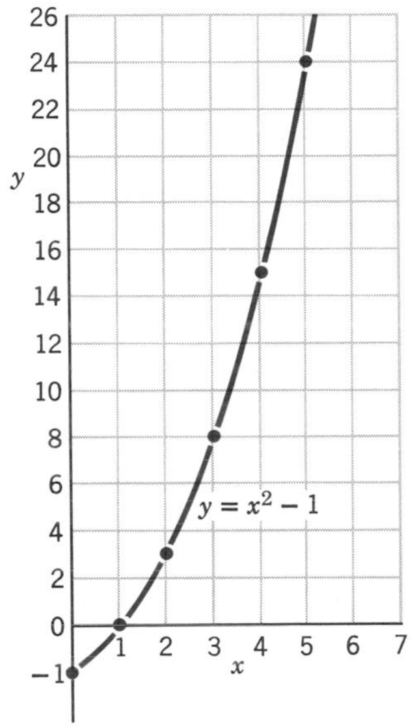

(b)

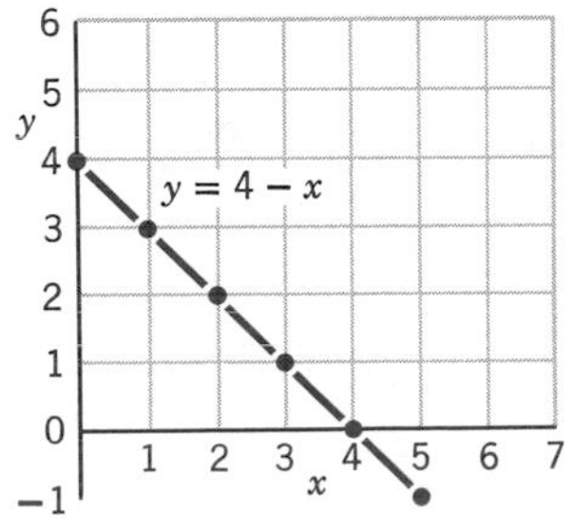

(d)

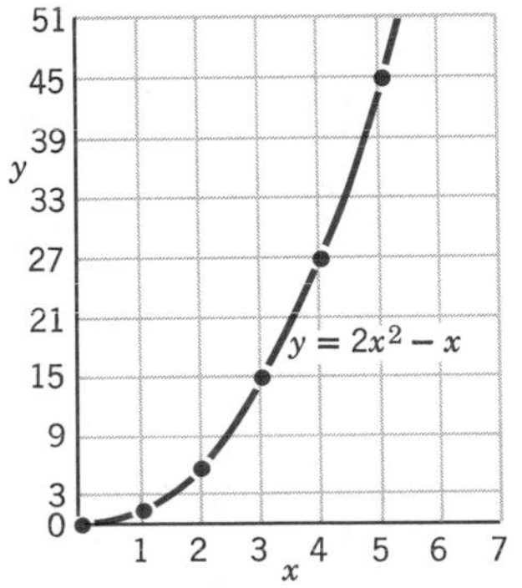

Problem Set 8, page 447

A.
1. Decreases
2. 10
3. 14
4. 40 or 50 amps
5. All circuits in the graph
6. 40,000 approx.
7. 75,000 approx.
8. A.G.
9. B.C.
10. A.G.
11. B.C.
12. B.C.
13. H.V., P.T., M.M., A.G.
14. Approx. 298,000 (Your answer may vary slightly.)
15. Approx. 240,000 (Your answer may vary slightly.)
16. 1977; about \$3.50
17. 1970; \$0.50
18. in 1972; \$1.50
19. 1974; \$1.50
20. \$1.25
21. \$26,000,000
22. \$3,250,000
23. \$30,000
24. \$35,000
25. December
26. July
27. January
28. January
29. January, February, March, June, July, August, September, October, November
30. November; \$20,000
31. 16%
32. 55%
33. 119°
34. 40 oz
35. 0.73 g
36. 250 hp
37. 200 hp
38. 40 mph
39. 50 mph
40. 150 hp
41. 175 hp
42. Decreases
43. 30 ohms
44. 3 ohms
45. 22 mils
46. 12 mils

B. 1. Fire Resistance Ratings of Various Plywoods

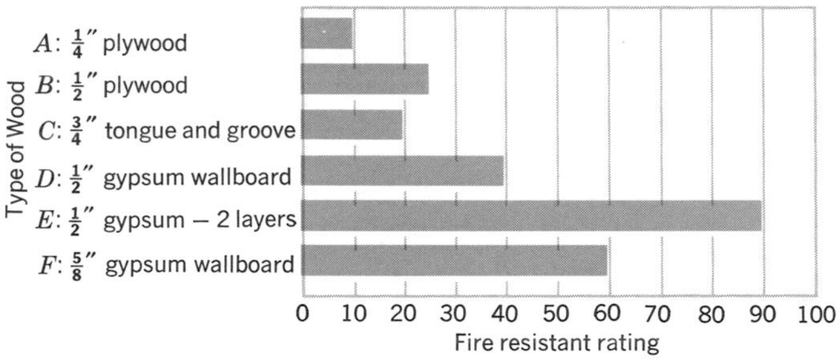

2. Output of Bricklayers: ABC Construction

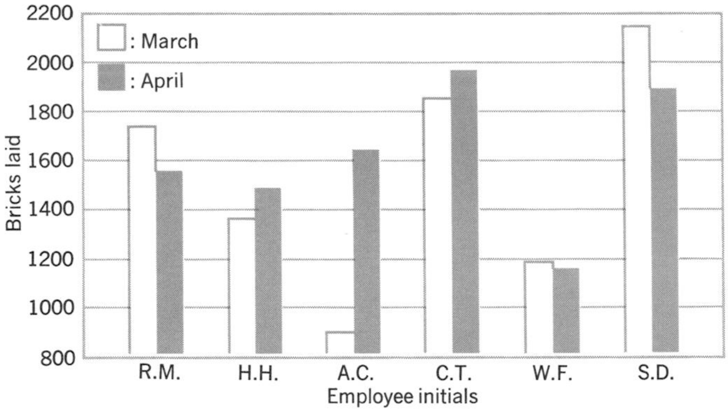

3. Temperature Range for the Week

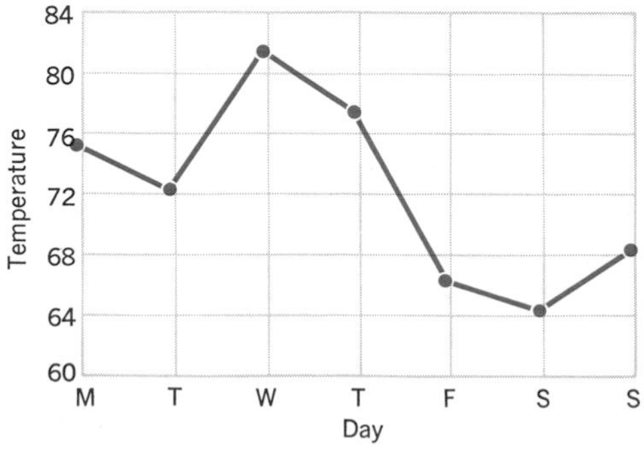

4. Earnings vs. Dividends

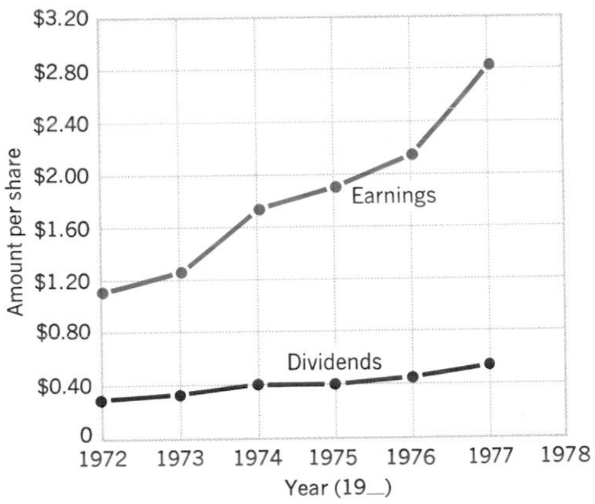

5. Alloy Composition

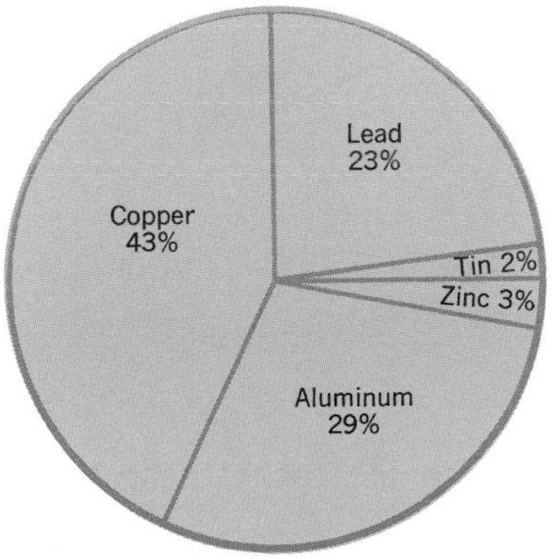

6. Worker Experience: Evans Tool and Dye

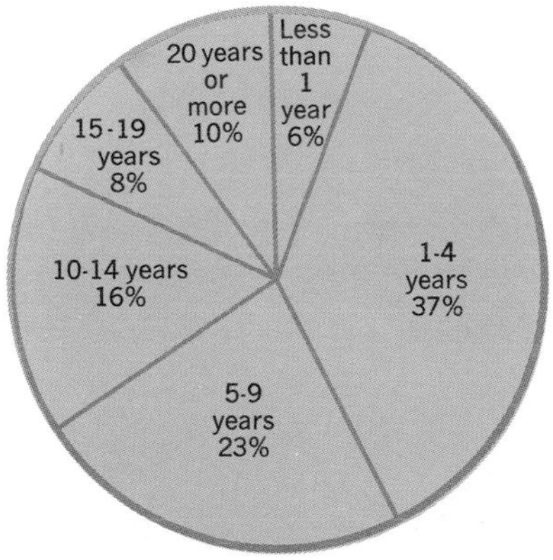

7. Amount of Gas Supplied by Different Lengths of $\frac{3}{4}''$ Pipe

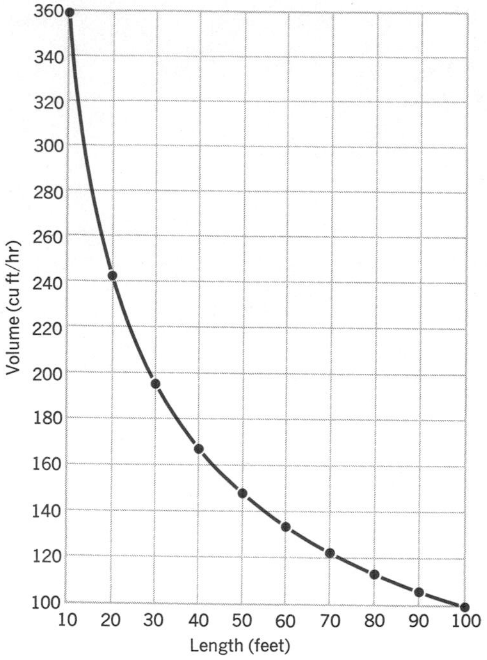

8. Cutting Depth vs. Nozzle Pressure PVC and Acrylic Pipe

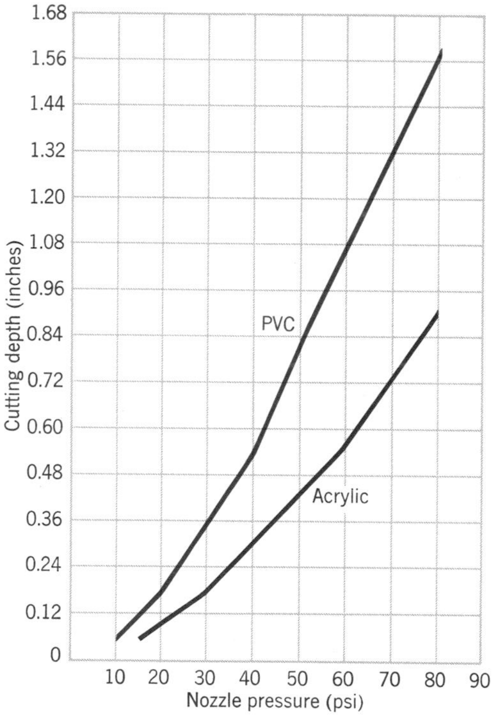

C. 1. Power Dissipation in an Electronic Circuit

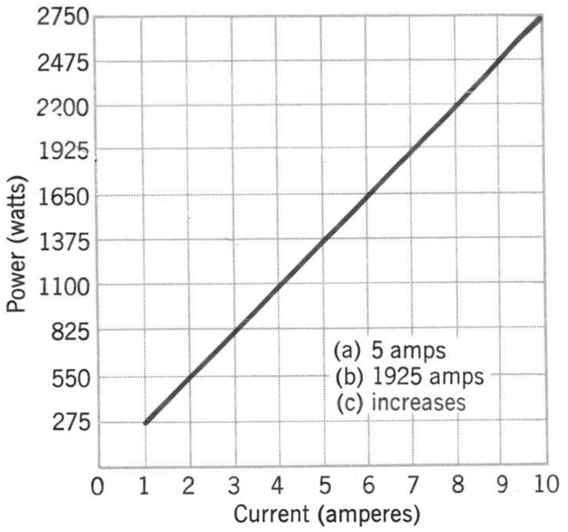

2. Output of a Machine Shop Worker

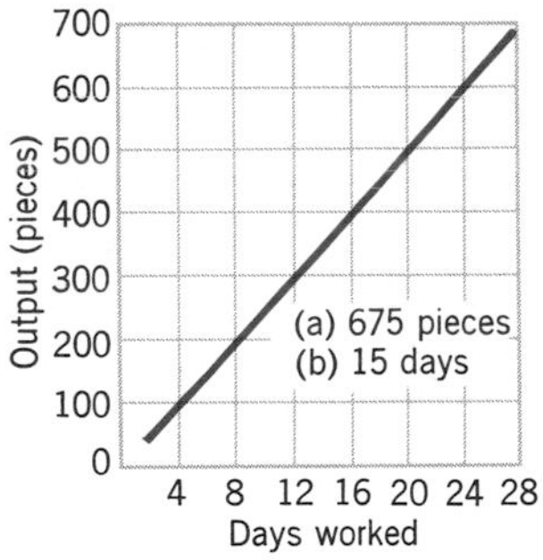

3. Simple Interest on $1000 at 6%

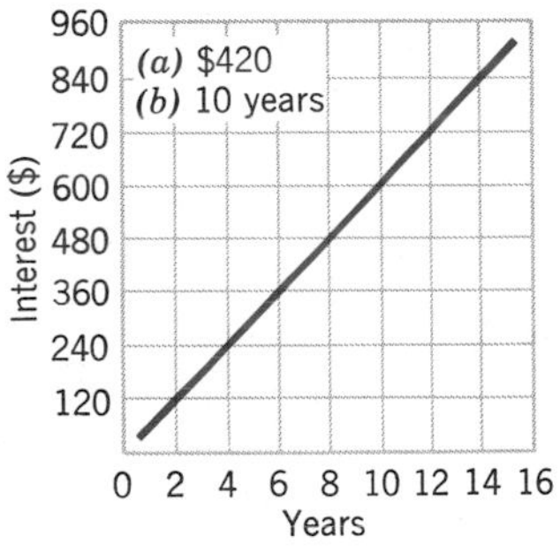

4. Area of a Square

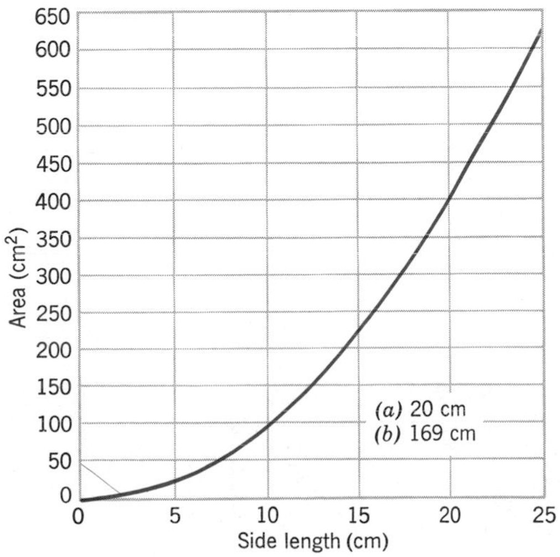

5. Horsepower Produced by 8-Cylinder Engine

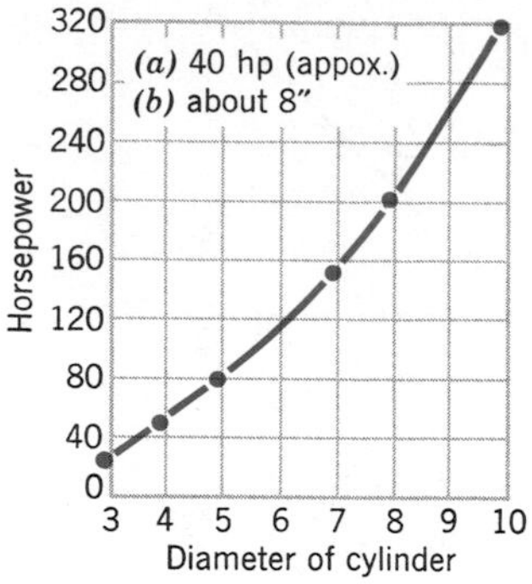

6. Power Dissipation in a Circuit

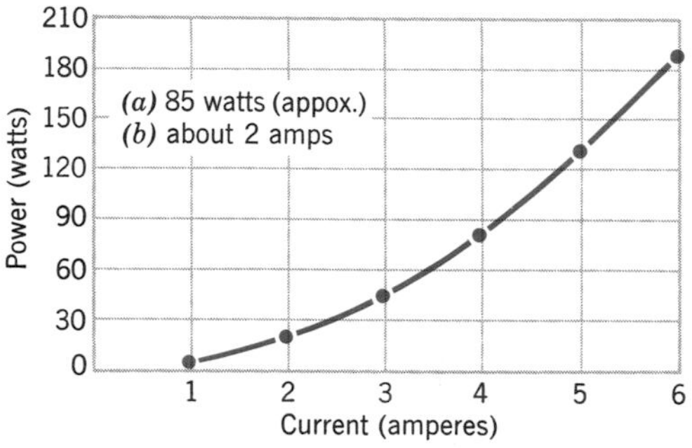

CHAPTER 9

Box, page 459

1.	(a) 0.18	(b) 0.61	(c) 1.58	(d) 2.1
2.	(a) 17°11′	(b) 85°57′	(c) 45°50′	(d) 2°52′

Exercises 9-1, page 464

A.

1. Acute	2. Right	3. Obtuse	4. Right
5. Straight	6. Obtuse	7. Acute	8. Acute
9. Right	10. Acute		

B.

1. 36°15′	2. 73°36′	3. 65°27′	4. 17°30′
5. 47°23′	6. 16°7′	7. 24°45′	8. 185°48′
9. 1°3′	10. 216°26′	11. 9°39′	12. 80°36′

C.

1.
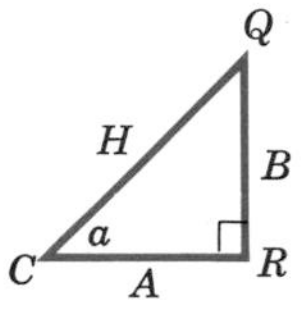

2.
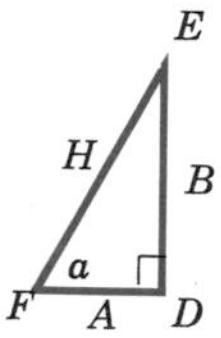

3.
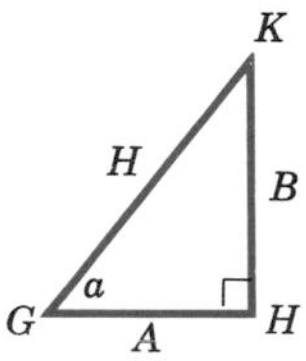

4.
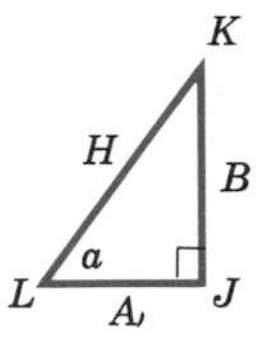

5.
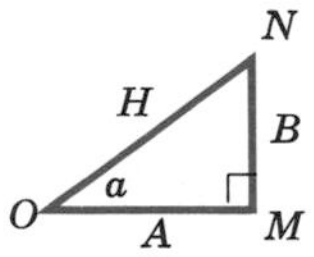

6.
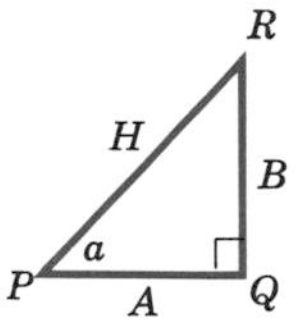

7.
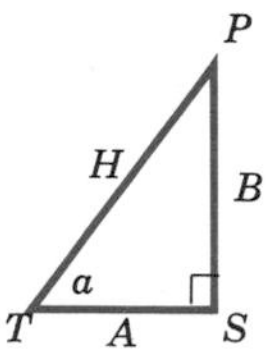

8.
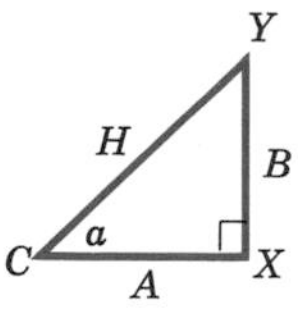

9.
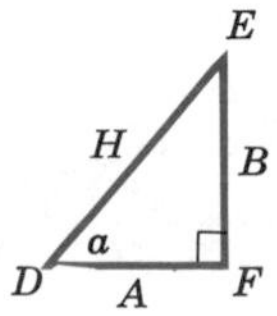

10.
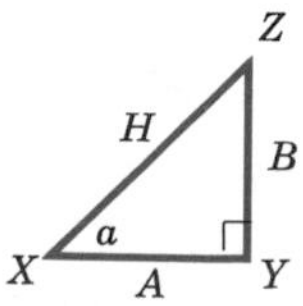

11. 26 12. 9.7 13. 24 14. 18.6
15. 12′ 16. 3.6″ 17. 1.5 18. 3.3 cm.
19. 8.0′ 20. 1.7″ 21. (a) 45° (b) 4.2′
22. (a) 6.9″ (b) 8″ 23. (a) 4 (b) 53° (c) 37°
24. (a) 3 cm (b) 5.2 cm (c) 60° 25. (a) 7 (b) 45°
26. (a) 5″ (b) 53° 27. (a) 8.7′ (b) 10′ (c) 30°
28. (a) 0.4″ (b) 0.3″ (c) 90° 29. (a) 10″ (b) 45°
30. (a) 3.48 cm (b) 60° (c) 30°

Box, page 475

(a) −0.342 (b) 0.292 (c) −1.428
(d) 0.743 (e) −0.158 (f) −0.809

Exercises 9-2, page 477

A. 1. 0.54 2. 0.74 3. 0.42 4. 0.53
5. 0.44 6. 0.69

B. 1. 0.454 2. 0.788 3. 0.213 4. 0.998
5. 0.191 6. 0.105 7. 0.052 8. 0.994
9. 1.206 10. 0.969 11. 0.752 12. 0.149

C. 1. 69° 2. 35° 3. 64° 4. 5°
5. 76° 6. 22° 7. 9°40′ 8. 69°45′

Exercises 9-3 page 482

A. 1. $X = 11.03''$, $Y = 8.62''$ 2. $X = 4.86'$, $Y = 7.72'$
3. $X = 29.7$ cm, $Y = 32$ cm 4. $a = 32°28'$, $X = 26.08'$
5. $a = 60°$, $X = 1.30''$ 6. $X = 1.84$ mm, $Y = 3.33$ mm
7. $X = 44.40''$, $Y = 14.46''$ 8. $a = 50°38'$, $X = 4.65$ yd
9. $a = 35°54'$, $X = 1.54''$ 10. $X = 33.52$ cm, $Y = 22.85$ cm

B. 1. 31.07′ 2. 43°46′ 3. (a) 72.72′ (b) No. 4. 4478′
5. 51°42′ 6. 4°17′ 7. 9.58′ 8. $A = 3.08''$, $B = 4.66''$
9. 3.74″ 10. Yes

C. 1. (a) $x = 1.160''$, $Y = 1.473''$
(b) $a = 61°21'$, $X = 1.47''$
(c) $m = 41°34'$, $X = 27.7$ cm
(d) $X = 28.39''$, $Y = 15.75''$
(e) $X = 1.27''$, $Y = 1.42''$
(f) $m = 16°27'$, $X = 17.973$ cm
2. (a) 68°45′ (b) 1.11″ (c) 274′
2. (d) 0.5456″ (e) 33.39 cm (f) 6°37′

Problem Set 9, page 487

A. 1. Yes 2. Yes 3. No 4. No 5. Yes
6. Yes 7. No 8. No 9. Yes 10. No

B. 1. 0.391 2. 6.314 3. 0.070 4. 0.451
5. 0.720 6. 0.488 7. 14° 8. 57°
9. 68° 10. 72° 11. 11° 12. 56°43′
13. 39°27′ 14. 42°14′

C. 1. $X = 37.1$, $Y = 32.5$ 2. $X = 20.2$, $Y = 16.4$
3. $a = 46°$, $X = 4.9$ 4. $X = 5.3$, $Y = 8.4$
5. $a = 54°$, $X = 5.6$ 6. $X = 70.3'$, $Y = 77.1'$
7. $t = 39°$, $X = 9.8$

D. 1. 1.3″ 2. $a = 9°37'$, $V = 118.3$ mph 3. 2″
4. 6°7′ 5. (a) $X = 28.9'$ (b) 57.8′
6. (a) 80.9′ (b) 8°32′ 7. (a) 1°7′ (b) 0.3″
8. $X = 16.6'$, $Y = 25.7'$ 9. 0.262″ 10. 67°34′

CHAPTER 10

Exercises 10-1, page 511

A. 1. $x = 2, y = 8$ 2. $x = 4, y = 7$ 3. $x = 4, y = 5$
4. Inconsistent 5. $x = 2, y = 4$ 6. $x = -\frac{1}{9}, y = -\frac{1}{9}$

B. 1. $x = 9, y = -4$ 2. $x = 4, y = 1$ 3. Inconsistent
4. Inconsistent 5. $x = -2, y = -3$ 6. $x = -3, y = 1$
7. $x = -1, y = -1\frac{1}{2}$ 8. $x = 4, y = 7$ 9. Dependent
10. $x = \frac{a+b}{2}, y = \frac{a-b}{2}$

C. 1. $x + y = 39$, $x - y = 7$ Solution: $x = 23$, $y = 16$
2. $x + y = 14$, $x = 2 + 3y$ Solution: $x = 11$, $y = 3$
3. $x + y = 20$, $2x = 3y$ Solution: $x = 12$, $y = 8$
4. $\frac{1}{2}(x + y) = 25$, $x - y = 8$ Solution: $x = 29$, $y = 21$
5. $4b + 3f = 11$, $3b + 4f = 10$ Solution: $b = \$2$, $f = \$1$
6. $2L + 2W = 14$, $L = 2W - 2$ Solution: $L = 4'$, $W = 3'$
7. $10d + 5n = 100$, $d = n + 7$ Solution: $n = 2$, $d = 9$
8. $4L + 3S = 26$, $3L - 2S = 11$ Solution: $L = 5$, $S = 2$
9. $5p + 4c = 290$, $p = c - 5$ Solution: $c = 35¢$, $p = 30¢$
10. $L = 2W$, $L - W = 20$ Solution: $L = 40$ in., $W = 20$ in.
11. $x + y = 30$, $x = 4y$ Solution: $x = 24''$, $y = 6''$
12. $x + y = 12$, $6x + 10y = 84$ Solution: $x = 9$ gal (of \$6), $y = 3$ gal (of \$10)
13. $x + y = 6$, $0.1x + 0.02y = 0.24$ Solution: $x = 1.5$ liter (10%), $y = 4.5$ liter (2%)
14. $2L + 2W = 520$, $L = 20 + 3W$ Solution: $L = 200'$, $W = 60'$
15. $x + y = 24$, $13x + 19y = 360$ Solution: $x = 16$ ton (\$13 rock), $y = 8$ ton (\$19 rock)

Exercises 10-2, page 523

A. 1. No 2. Yes 3. No
4. Yes 5. Yes 6. $7x^2 + 3x - 5 = 0$
7. $3x^2 - 7x + 14 = 0$ 8. OK 9. $x^2 - 18x = 0$
10. OK

B. 1. $x = 5$ or $x = -5$ 2. $x = 3$ or $x = -3$
3. $x = 0$ or $x = 4\frac{2}{5}$ 4. $x = \frac{1}{2}$ or $x = 3$
5. $x = 4\frac{1}{2}$ or $x = -4\frac{1}{2}$ 6. $x = 4\frac{1}{2}$ or $x = -2\frac{1}{3}$
7. $x = 0.76$ or $x = -15.76$ 8. $x = 9$ or $x = \frac{3}{4}$
9. $x = -2.35$ or $x = 0.85$ 10. $x = 0.02$ or $x = -0.05$

C. 1. 25 mm 2. (a) 14.49 in. (b) 15.66 in.
3. 5″ by 15″ 4. 8.92 cm 5. 17″ by 22″
6. (a) 7.07″ (b) 9.13″ 7. 7.14″ 8. (a) 9 amps (b) 7.30 amps
9. (a) 3.76″ (b) 4.56″ 10. 6′ by 8′ or 4′ by 12′

Problem Set 10, page 525

A. 1. $x = 15, y = 3$ 2. $x = 6, y = -\frac{1}{2}$ 3. $x = 3, y = 2\frac{1}{2}$
4. $x = 7, y = 3$ 5. $x = 7, y = 6$ 6. Dependent
7. $x = 1, y = 1$ 8. $x = -13, y = 22$ 9. $x = \frac{74}{23}, y = -\frac{4}{23}$

B. 1. $x = 3$ or $x = -3$ 2. $x = 7$ or $x = -4$
3. $x = -1.43$ or $x = -0.23$ 4. $x = 1.18$ or $x = -0.43$
5. $x = 0$ or $x = 6$ 6. $x = 5$ or $x = -1\frac{1}{2}$
7. $x = 5$ or $x = -1$ 8. $x = 6$ or $x = -5$
9. $x = 6$ or $x = -6$ 10. $x = 2.15$ or $x = -0.75$

C. 1. $x + y = 38$, $x - y = 14$, $x = 26$, $y = 12$
2. $s^2 = 196$, $s = 14$ in.
3. $x - y = 21$, $5y - 2x = 33$, $x = 46$, $y = 25$
4. $2L + 2W = 22$, $L = 2 + 2W$, $W = 3'$, $L = 8'$
5. $L = 4W$, $4W^2 = 125$, $W = 5.59$ cm, $L = 22.36$ cm
6. $x + y = 650$, $5x + 3y = 2350$, $x = 200$ (5¢), $y = 450$ (3¢)
7. (a) $I = 13$ amps (b) $I = 12.25$ amps
8. $L = 2 + W$, $W(2 + W) = 168$, $W = 12''$, $L = 14''$
9. $x + y = 42$, $x = 3y - 2$, $y = 11''$, $x = 31''$
10. $2L + 2W = 750$, $L = 4W$, $W = 75'$, $L = 300'$
11. $d = 15.96$ in.
12. $x + y = 15$, $4x + 8.5y = 105$, $x = 5$ gal, $y = 10$ gal
13. (a) $D = 5''$ (b) $D = 6.25''$
14. $L = 2W - 3$, $W(2W - 3) = 20$, $W = 4''$, $L = 5''$
15. $x + y = 50$, $30x + 55y = 1750$, $x = 40$ tons ($30), $y = 10$ tons ($55)

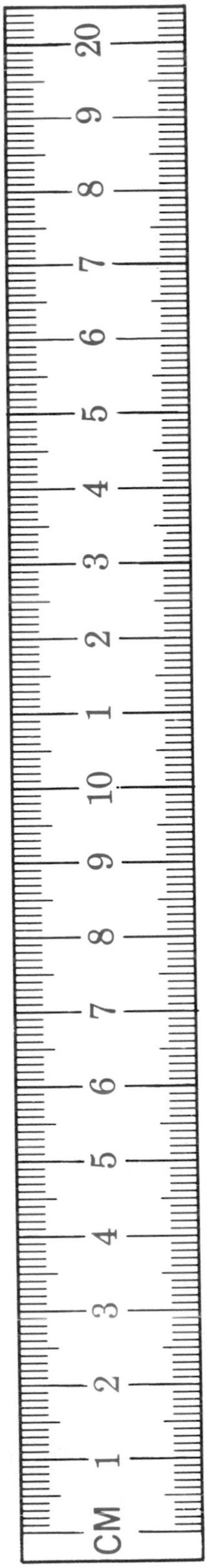

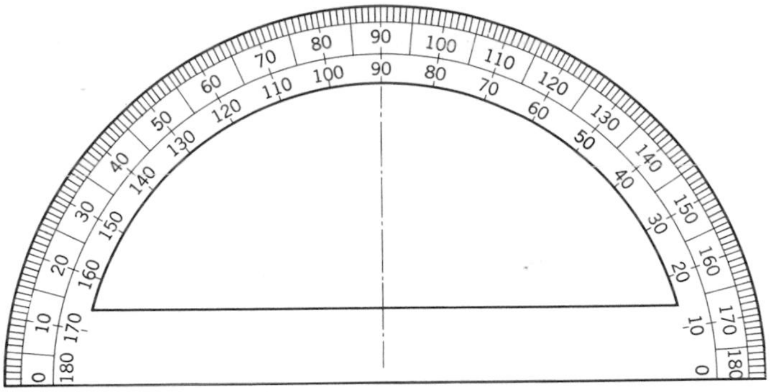

×	0	1	2	3	4	5	6	7	8	9	10
0	0	0	0	0	0	0	0	0	0	0	0
1	0	1	2	3	4	5	6	7	8	9	10
2	0	2	4	6	8	10	12	14	16	18	20
3	0	3	6	9	12	15	18	21	24	27	30
4	0	4	8	12	16	20	24	28	32	36	40
5	0	5	10	15	20	25	30	35	40	45	50
6	0	6	12	18	24	30	36	42	48	54	60
7	0	7	14	21	28	35	42	49	56	63	70
8	0	8	16	24	32	40	48	56	64	72	80
9	0	9	18	27	36	45	54	63	72	81	90
10	0	10	20	30	40	50	60	70	80	90	100

Index

'OM
nd the HE
a Go Go
PETTY
THE
SOLD
OUT
REAKERS

T
ar

RUNNIN' DOWN A DREAM

PETTY

ARTBREAKERS

FOREWORD AND INTERVIEWS BY PETER BOGDANOVICH

EDITED AND ADDITIONAL INTERVIEWS BY WARREN ZANES

CHRONICLE BOOKS
SAN FRANCISCO

Editor's Note

In 2006, Tom Petty was approached with the idea of filming a documentary on the subject of his life and career. The plan was to put director Peter Bogdanovich together with Petty, a meeting of two American mavericks. The project quickly grew in its proportions. Bogdanovich conducted extensive interviews with Petty, the Heartbreakers, and the artist's inner circle of friends, associates, and fellow performers. The wealth of materials that accumulated sparked the idea that a book project would be a natural outgrowth of all the film-related activities. What emerged is *Runnin' Down a Dream*, a volume that puts selected Bogdanovich interview material side-by-side with artifacts and images drawn from the Heartbreakers' archives. Personal, certainly intimate, *Runnin' Down a Dream* takes you straight into the world of one of American popular music's treasures, Tom Petty and the Heartbreakers.

Library of Congress Cataloging-in-Publication Data is available.

ISBN-10: 0-8118-6201-1
ISBN-13: 978-0-8118-6201-1

Manufactured in China

Designed by Jeri Heiden, SMOG Design, Inc.
Front cover photograph by Dennis Callahan

10 9 8 7 6 5 4 3 2 1

Chronicle Books LLC
680 Second Street
San Francisco, California 94107

www.chroniclebooks.com